Study Edition

Dr. Tim Higgins
Chemistry Department
National University of
Ireland
Galway, Ireland.
Tel: 353-91-524411

Cover picture: stereo diagram of a hydrated oligonucleotide (Chapter 24, reference [865]).

How to see stereo without stereoglasses: Adapt your eyes to "infinity". Slide the stereo diagram into field of vision and keep eyes adapted to "infinity". Now you should see *three* pictures. Focus your eyes on the *center* picture and you will see a right-handed DNA double helix, studded with water molecules drawn black. Hydrogen bonds to water molecules drawn with thin lines.

G. A. Jeffrey W. Saenger

Hydrogen Bonding in Biological Structures

With 315 Figures in 412 Separate Illustrations
and 111 Tables

Springer-Verlag
Berlin Heidelberg New York
London Paris Tokyo
Hong Kong Barcelona
Budapest

Professor Dr. George A. Jeffrey
Department of Crystallography
The University of Pittsburgh
Pittsburgh, PA 15260, USA

Professor Dr. Wolfram Saenger
Institut für Kristallographie
Freie Universität Berlin
Takustraße 6, D-14195 Berlin, Germany

Second Printing 1994

ISBN 3-540-57903-6 Springer-Verlag Berlin Heidelberg New York
ISBN 0-387-57903-6 Springer-Verlag New York Berlin Heidelberg

CIP data applied for.

This work is subject to copyright. All rights are reserved, whether the whole or part of the material is concerned, specifically the rights of translation, reprinting, reuse of illustrations, recitation, broadcasting, reproduction on microfilm or in any other way, and storage in data banks. Duplication of this publication or parts thereof is permitted only under the provisions of the German Copyright Law of September 9, 1965, in its current version, and permission for use must always be obtained from Springer-Verlag. Violations are liable for prosecution under the German Copyright Law.

© Springer-Verlag Berlin Heidelberg 1991, 1994
Printed in Germany

The use of general descriptive names, registered names, trademarks, etc. in this publication does not imply, even in the absence of a specific statement, that such names are exempt from the relevant protective laws and regulations and therefore free for general use.

Product liability: The publishers cannot guarantee the accuracy of any information about dosage and application contained in this book. In every individual case the user must check such information by consulting the relevant literature.

Typesetting: K+V Fotosatz GmbH, Beerfelden; Printing: Beltz, Hemsbach; Binding: Schäffer, Grünstadt
SPIN: 10465561   39/3130-5 4 3 2 1 0 – Printed on acid-free paper

Preface

When writing a book on as vast a subject as Hydrogen Bonding, a major problem is to circumscribe the subject matter. In this respect, we have adopted the crystallographer's point of view. We have focused our attention on the structure of the hydrogen-bonding patterns, as revealed by crystal structure analyses and described by the relative positions of atoms in three dimensions on the nanometer scale in the crystalline state. The molecules which are important in biological reactions have dimensions ranging from approximately 10 to 1000 Å.[1] The cohesive forces between them are such that the nearest neighbor interatomic distances between these molecules fall in the relative narrow range of 1.5 Å for strong cohesive forces between atoms, to 3.5 Å for the weak forces. The topology of the assemblages of biological molecules which take part in biological processes must therefore be studied on this scale.

To further circumscribe our subject matter, we have limited our attention to those aspects of hydrogen-bond research which we perceive to have the most bearing on the objective of more complete understanding of the role that hydrogen bonds play in organizing the three-dimensional structure and interactions of the biological macromolecules on the atomic scale. We have placed the most reliance on the results of neutron diffraction single crystal structure analyses, since this method is unique in its ability to determine the positions of the hydrogen atoms with a precision comparable to those of the carbon, nitrogen, and oxygen atoms in biologically interesting molecules. Unfortunately, neutron diffraction analyses are relatively few, and much of the structural data reported and discussed has to rely on carefully selected X-ray crystal structure analyses. The availability of the results of over 60000 crystal structure analyses in computer-readable format through the Crystallographic Data Files has made this possible.

Of all the physical sciences, Crystallography has benefited the most from the spectacular advances in computer technology in the past decade. The Cambridge Crytallographic Crystal Structure Data Base contains the results of the published crystal structure analyses of organic and organo-metallic compounds. The Protein

---

[1] For historical reasons, the Ångstrom unit, i.e., 0.1 nanometer, although not the "officially approved" unit, is the more familiar to crystallographers and molecular biologists, and is used throughout this text.

Crystal Structure Data Base has more than 400 entries, including nucleic acids and the viruses. Incorporated in this structural information is that relating to the molecular cohesive forces, the most biologically important of which is the hydrogen bond. The crystallographic data bases were designed by crystallographers, originally for use by crystallographers. Although significant efforts are made to design software which makes these data bases user-friendly to chemists and biological scientists, learning the crystallographic regimen of lattice and space-group symmetry can be an obstacle. In this monograph we have attempted to extract and rationalize the relevant structure information on hydrogen bonding that is currently available, setting the stage for later studies when it will be possible to survey the hydrogen-bonding patterns from much larger samples of small and large biological molecules, and place some of our present conclusions on a firmer statistical basis.

We have not discussed quantitative hydrogen bond energies and how these are measured and calculated. Neither have we discussed in any detail the interpretation of the many facets of molecular and solution spectroscopy. While these are important for analysis of the dynamical properties of biological molecules, they are not methodologies that provide the quantitative information relating to the static three-dimensional structure that is the central focus of this monograph.

It is interesting that the sixteenth century argument whether mineralogy was a separate science from botany has turned full circle. The idea that biologically active enzymes and viruses can be crystalline comes as a surprise to non-scientists, or that crystallography is a vigorous component of the life-science research weaponry.

Why is the hydrogen bond so important in biological structures? It can be argued that the most important biological molecule is water, which is the hydrogen bonding molecule *par excellence*. Several volumes have been written on the research relating to $(H_2O)_n$ alone. We have selected those aspects of the Ices, the related Ice-like clathrate structures, and the hydrates of small biological molecules which we believe have some relevance presently and in the future to the interpretation of water structure in heavily hydrated macromolecules, such as the proteins and nucleic acids.

A glance through an elementary text-book on Biochemistry reveals the following chapter headings: amino acids, peptides, proteins; mono-, oligo- and polysaccharides; pyrimidines, purines, nucleosides, nucleotides and nucleic acids. All these molecules have one common feature. They contain many hydrogen-bonding functional groups. This is surely connected with the evolution of life from an aqueous medium. Only with the lipids do molecules appear with no hydrogen-bonding functional groups, but even these molecules generally function biologically with a hydrogen-bonding head group attached to them. The special properties and function of the glycolipids, for example, are associated with molecular self-organization, midway between the crystalline and liquid states, in which hydrogen bonding plays a role.

It may sound paradoxical, but the relative unspecificity and the weakness of the hydrogen bond of only ~3 kcal/mol are of prime importance for life, so much that *life without hydrogen bond were impossible*. The reason is that all biological processes involve intermolecular recognition which has to be rapid, close to the time scale of $\sim 10^9 \, \text{s}^{-1}$ where diffusion controlled events take place. It consequently requires a weak interaction like the hydrogen bond which permits very fast association and dissociation so that in a short time many possible combinations can be checked before the correct partners associate. The specificity which is so typical for biological processes is not achieved by a very specific, single interaction. In contrast, the unspecific hydrogen bond is employed and specificity is due to the simultaneous formation of several hydrogen bonds, between sterically complementary organized donors and acceptors to form a pattern just as, in Fischer's old concept, a key fits into a lock.

This book was initiated during the tenure of an Alexander von Humboldt U.S. Senior Scientist Award. We also wish to acknowledge support of our research from the U.S. National Institutes of Health (G. A. J.) and the Deutsche Forschungsgemeinschaft, Fonds der Chemischen Industrie, and Bundesministerium für Forschung und Technologie (W. S.). Of the many colleagues who worked with us in topics relating to hydrogen bonding, we wish to thank particularly Hanna Maluszynska, Jayati Mitra, Bogdan Lesyng, Christian Betzel, Volker Zabel and Thomas Steiner. Nothing would have been possible without the skills of our secretaries, Joan Klinger and Ilse Kriegl, and the patience and understanding of our wives, Maureen and Barbara.

The Cambridge Crystallographic Data Base was an indispensible source of data and we are grateful for permission to use the Refcodes for indexing.

Pittsburgh and Berlin, March 1991   G. A. Jeffrey
W. Saenger

Contents

Part IA Basic Concepts

1 The Importance of Hydrogen Bonds 3

1.1 Historical Perspective 3
1.2 The Importance of Hydrogen Bonds
     in Biological Structure and Function 8
1.3 The Role of the Water Molecules 10
1.4 Significance of Small Molecule Crystal Structural Studies 11
1.5 The Structural Approach 14

2 Definitions and Concepts 15

2.1 Definition of the Hydrogen Bond – Strong and Weak Bonds 15
2.2 Hydrogen-Bond Configurations: Two- and Three-Center
     Hydrogen Bonds; Bifurcated and Tandem Bonds 20
2.3 Hydrogen Bonds Are Very Different from Covalent Bonds 24
2.4 The van der Waals Radii Cut-Off Criterion Is Not Useful 29
2.5 The Concept of the Hydrogen-Bond Structure 33
2.6 The Importance of σ and π Cooperativity 35
2.7 Homo-, Anti- and Heterodromic Patterns 38
2.8 Hydrogen Bond Flip-Flop Disorder:
     Conformational and Configurational 40
2.9 Proton-Deficient Hydrogen Bonds 42
2.10 The Excluded Region 43
2.11 The Hydrophobic Effect 44

3 Experimental Studies of Hydrogen Bonding 49

3.1 Infrared Spectroscopy and Gas Electron Diffraction 50
3.2 X-Ray and Neutron Crystal Structure Analysis 52
3.3 Treatment of Hydrogen Atoms in Neutron Diffraction Studies ... 58
3.4 Charge Density and Hydrogen-Bond Energies 63

3.5	Neutron Powder Diffraction	67
3.6	Solid State NMR Spectroscopy	69
4	**Theoretical Calculations of Hydrogen-Bond Geometries**	**71**
4.1	Calculating Hydrogen-Bond Geometries	71
4.2	Ab-Initio Molecular Orbital Methods	74
4.3	Application to Hydrogen-Bonded Complexes	77
4.4	Semi-Empirical Molecular Orbital Methods	84
4.5	Empirical Force Field or Molecular Mechanics Methods	85
5	**Effect of Hydrogen Bonding on Molecular Structure**	**94**

Part IB Hydrogen-Bond Geometry

6	**The Importance of Small Molecule Structural Studies**	**103**
6.1	Problems Associated with the Hydrogen-Bond Geometry	103
6.2	The Hydrogen Bond Can Be Described Statistically	104
6.3	The Problems of Measuring Hydrogen-Bond Lengths and Angles in Small Molecule Crystal Structures	107
7	**Metrical Aspects of Two-Center Hydrogen Bonds**	**111**
7.1	The Metrical Properties of O—H···O Hydrogen Bonds	111
7.1.1	Very Strong and Strong OH···O Hydrogen Bonds Occur with Oxyanions, Acid Salts, Acid Hydrates, and Carboxylic Acids	111
7.1.2	OH···O Hydrogen Bonds in the Ices and High Hydrates	116
7.1.3	Carbohydrates Provide the Best Data for OH···O Hydrogen Bonds: Evidence for the Cooperative Effect	121
7.2	N—H···O Hydrogen Bonds	128
7.3	N—H···N Hydrogen Bonds	132
7.4	O—H···N Hydrogen Bonds	133
7.5	Sequences in Lengths of Two-Center Hydrogen Bonds	133
7.6	H/D Isotope Effect	134
8	**Metrical Aspects of Three- and Four-Center Hydrogen Bonds**	**136**
8.1	Three-Center Hydrogen Bonds	136
8.2	Four-Center Hydrogen Bonds	145
9	**Intramolecular Hydrogen Bonds**	**147**

10	Weak Hydrogen-Bonding Interactions Formed by C–H Groups as Donors and Aromatic Rings as Acceptors	156
11	Halides and Halogen Atoms as Hydrogen-Bond Acceptors	161
12	Hydrogen-Bond Acceptor Geometries	164

Part II Hydrogen Bonding in Small Biological Molecules

13	Hydrogen Bonding in Carbohydrates	169
13.1	Sugar Alcohols (Alditols) as Model Cooperative Hydrogen-Bonded Structures	172
13.2	Influence of Hydrogen Bonding on Configuration and Conformation in Cyclic Monosaccharides	178
13.3	Rules to Describe Hydrogen-Bonding Patterns in Monosaccharides	187
13.4	The Water Molecules Link Hydrogen-Bond Chains into Nets in the Hydrated Monosaccharide Crystal Structures	192
13.5	The Disaccharide Crystal Structures Provide an Important Source of Data About Hydrogen-Bonding Patterns in Polysaccharides	195
13.6	Hydrogen Bonding in the Tri- and Tetrasaccharides Is More Complex and Less Well Defined	210
13.7	The Hydrogen Bonding in Polysaccharide Fiber Structures Is Poorly Defined	214
14	Hydrogen Bonding in Amino Acids and Peptides: Predominance of Zwitterions	220
15	Purines and Pyrimidines	232
15.1	Bases Are Planar and Each Contains Several Different Hydrogen-Bonding Donor and Acceptor Groups	232
15.2	Many Tautomeric Forms Are Feasible But Not Observed	235
15.3	π-Bond Cooperativity Enhances Hydrogen-Bonding Forces	235
15.4	General, Non-Base-Pairing Hydrogen Bonds	237
16	Base Pairing in the Purine and Pyrimidine Crystal Structures	247
16.1	Base-Pair Configurations with Purine and Pyrimidine Homo-Association	248
16.2	Base-Pair Configurations with Purine-Pyrimidine Hetero-Association: the Watson-Crick Base-Pairs	259
16.3	Base Pairs Can Combine to Form Triplets and Quadruplets	266

17	Hydrogen Bonding in the Crystal Structures of the Nucleosides and Nucleotides	269
17.1	Conformational and Hydrogen-Bonding Characteristics of the Nucleosides and Nucleotides	270
17.2	A Selection of Cyclic Hydrogen-Bonding Patterns Formed in Nucleoside and Nucleotide Crystal Structures	276
17.3	General Hydrogen-Bonding Patterns in Nucleoside and Nucleotide Crystal Structures	281

Part III Hydrogen Bonding in Biological Macromolecules

18	$O-H \cdots O$ Hydrogen Bonding in Crystal Structures of Cyclic and Linear Oligoamyloses: Cyclodextrins, Maltotriose, and Maltohexaose	309
18.1	The Cyclodextrins and Their Inclusion Complexes	309
18.2	Crystal Packing Patterns of Cyclodextrins Are Determined by Hydrogen Bonding	313
18.3	Cyclodextrins as Model Compounds to Study Hydrogen-Bonding Networks	315
18.4	Cooperative, Homodromic, and Antidromic Hydrogen-Bonding Patterns in the α-Cyclodextrin Hydrates	320
18.5	Homodromic and Antidromic $O-H \cdots O$ Hydrogen-Bonding Systems Analyzed Theoretically	330
18.6	Intramolecular Hydrogen Bonds in the α-Cyclodextrin Molecule are Variable – the Induced-Fit Hypothesis	332
18.7	Flip-Flop Hydrogen Bonds in β-Cyclodextrin $\cdot 11\,H_2O$	333
18.8	From Flip-Flop Disorder to Ordered Homodromic Arrangements at Low Temperature: The Importance of the Cooperative Effect	340
18.9	Maltohexaose Polyiodide and Maltotriose – Double and Single Left-Handed Helices With and Without Intramolecular $O(2) \cdots O(3')$ Hydrogen Bonds	344

19	Hydrogen Bonding in Proteins	351
19.1	Geometry of Secondary-Structure Elements: Helix, Pleated Sheet, and Turn	352
19.2	Hydrogen-Bond Analysis in Protein Crystal Structures	359
19.3	Hydrogen-Bonding Patterns in the Secondary Structure Elements	362
19.4	Hydrogen-Bonding Patterns Involving Side-Chains	365
19.5	Internal Water Molecules as Integral Part of Protein Structures	372
19.6	Metrical Analysis of Hydrogen Bonds in Proteins	374

19.7	Nonsecondary-Structure Hydrogen-Bond Geometry Between Main-Chains, Side-Chains and Water Molecules	379
19.8	Three-Center (Bifurcated) Bonds in Proteins	382
19.9	Neutron Diffraction Studies on Proteins Give Insight into Local Hydrogen-Bonding Flexibility	383
19.10	Site-Directed Mutagenesis Gives New Insight into Protein Thermal Stability and Strength of Hydrogen Bonds	388
20	**The Role of Hydrogen Bonding in the Structure and Function of the Nucleic Acids**	**394**
20.1	Hydrogen Bonding in Nucleic Acids is Essential for Life	394
20.2	The Structure of DNA and RNA Double Helices is Determined by Watson-Crick Base-Pair Geometry	397
20.3	Systematic and Accidental Base-Pair Mismatches: "Wobbling" and Mutations	403
20.4	Noncomplementary Base Pairs Have a Structural Role in tRNA	406
20.5	Homopolynucleotide Complexes Are Stabilized by a Variety of Base-Base Hydrogen Bonds – Three-Center (Bifurcated) Hydrogen Bonds in A-Tracts	408
20.6	Specific Protein-Nucleic Acid Recognition Involves Hydrogen Bonding	411

Part IV Hydrogen Bonding by the Water Molecule

21	**Hydrogen-Bonding Patterns in Water, Ices, the Hydrate Inclusion Compounds, and the Hydrate Layer Structures**	**425**
21.1	Liquid Water and the Ices	426
21.2	The Hydrate Inclusion Compounds	432
21.3	Hydrate Layer Structures	449
22	**Hydrates of Small Biological Molecules: Carbohydrates, Amino Acids, Peptides, Purines, Pyrimidines, Nucleosides and Nucleotides**	**452**
23	**Hydration of Proteins**	**459**
23.1	Characterization of "Bound Water" at Protein Surfaces – the First Hydration Shell	461
23.2	Sites of Hydration in Proteins	466
23.3	Metrics of Water Hydrogen Bonding to Proteins	470
23.4	Ordered Water Molecules at Protein Surfaces – Clusters and Pentagons	479

24	**Hydration of Nucleic Acids**	487
24.1	Two Water Layers Around the DNA Double Helix	487
24.2	Crystallographically Determined Hydration Sites in A-, B-, Z-DNA. A Statistical Analysis	490
24.3	Hydration Motifs in Double Helical Nucleic Acids	494
24.3.1	Sequence-Independent Motifs	494
24.3.2	Sequence-Dependent Motifs	499
24.4	DNA Hydration and Structural Transitions Are Correlated: Some Hypotheses	500
25	**The Role of Three-Center Hydrogen Bonds in the Dynamics of Hydration and of Structure Transition**	505

References 511

Refcodes 547

Subject Index 563

Part IA
Basic Concepts

Chapter 1
The Importance of Hydrogen Bonds

> "The discovery of the Hydrogen Bond could have won someone the Nobel Prize, but it didn't."

1.1 Historical Perspective

Until a concept is given a *descriptor*, it is difficult to retrieve references to it in the literature. Hydrogen bonding is a typical example. Prior to modern methods of abstracting scientific papers and the use of key words, it is difficult to trace the origin of concepts. We know that before 1932, a number of chemical phenomena were observed that led investigators to postulate the existence of weak or secondary bonds involving the interactions between functional groups, in one of which a hydrogen atom is covalently bonded to oxygen or nitrogen atoms.

Early concepts of hydrogen bonding. Some of the very early examples of hydrogen bonding are given in Table 1.1, with the formulations used in the original literature. There is no doubt that the German chemists such as Hantzsch and Werner were aware of hydrogen bonding, and the term "Nebenvalenz" (minor valence) is, in fact, a good description of the hydrogen bond phenomenon. The suggestions by Huggins [7, 8] and by Latimer and Rodebush [9] that hydrogen could sometimes have a valence of two was heresy in the golden age of paired electrons and Lewis's octet rule, and the impact of this suggestion at that time was minimal.

There is a literature gap between 1922 and 1935 when the concept of the hydrogen bond appears to have been forgotten. This was the period when both theory and experiment were focusing on the strong interatomic interactions in chemistry as represented by the covalent, ionic, and metallic bonds. The Braggs and collaborators were determining the crystal structures of the simple inorganics to form the basis of a new science, Crystal Chemistry [11].

In none of the very early crystal structural studies of urea, from 1923 to 1934 [12 to 14], nor in the classical paper on the tetrahedral coordination of water molecules by Bernal and Fowler in 1933 [15], which forms the foundation of modern thought concerning the structure of water, were there any references to the hydrogen bond per se. Similarly, in Astbury's pioneering work on the X-ray diffraction of the polypeptide fibers of hair, silk, and wool, the concept of the hydrogen bond is not used in his explanation of the structural changes induced by folding and unfolding of the polypeptide chain [16, 17]. Had Astbury been aware of this concept, he would clearly have substituted "hydrogen bridge bonds" for "bridge atoms" in his statement: "The basis of the unstretched fibrous keratins is a series of hexagonal ring systems linked along the fibre axis by *hydrogen bridge atoms*."

Table 1.1. Some early formulations of hydrogen bonds

Year	Formulation	Description
1902	$X(H \cdots NH_3)^a$ $H_3N \cdots HCl$	Nebenvalenz [1, 2]
1910	(acetylacetone methyl ether structure)	Nebenvalenz [3]
1912	$(H_3C)_3N \cdots H-OH$; $((H_3C)_3N-H)\cdots OH$	Weak unions [4]
1913	(anthraquinone–OH intramolecular structure)	Innere Komplexsalzbildung [5]
1914	$\underset{HO}{\overset{R}{>}}C=O \cdots HOC\underset{O}{\overset{R'}{<}}$	Nebenvalenz [6]
1919	$A + :B \rightarrow A:B$	Electron pair bonds [7]
1920 / 1922	$H:\underset{H}{\overset{H}{N}}:H:\overset{..}{\underset{..}{O}}:H^a$	Hydrogen nucleus between two octets constitutes a weak bond [8, 9]

[a] The complex of NH_3 and H_2O in the gas phase was studied by microwave rotation spectroscopy in 1986, and shown to have a $HOH-NH_3$ hydrogen bond [10].

The illustrations from Astbury's 1933 paper shown in Fig. 1.1 imply some kind of bonding between NH and C=O groups, but the term "hydrogen bridge" or "bond" is never used in his model of polypeptide folding.

Surprisingly, it was not until 10 years later that Huggins published a series of definitive papers on hydrogen bonding in organic molecules [18], in ice and liquid water [19], and Bernal and Megaw in hydroxide minerals [20]. It appears to have been these three authors who supplied the first *descriptors*. Huggins used the term

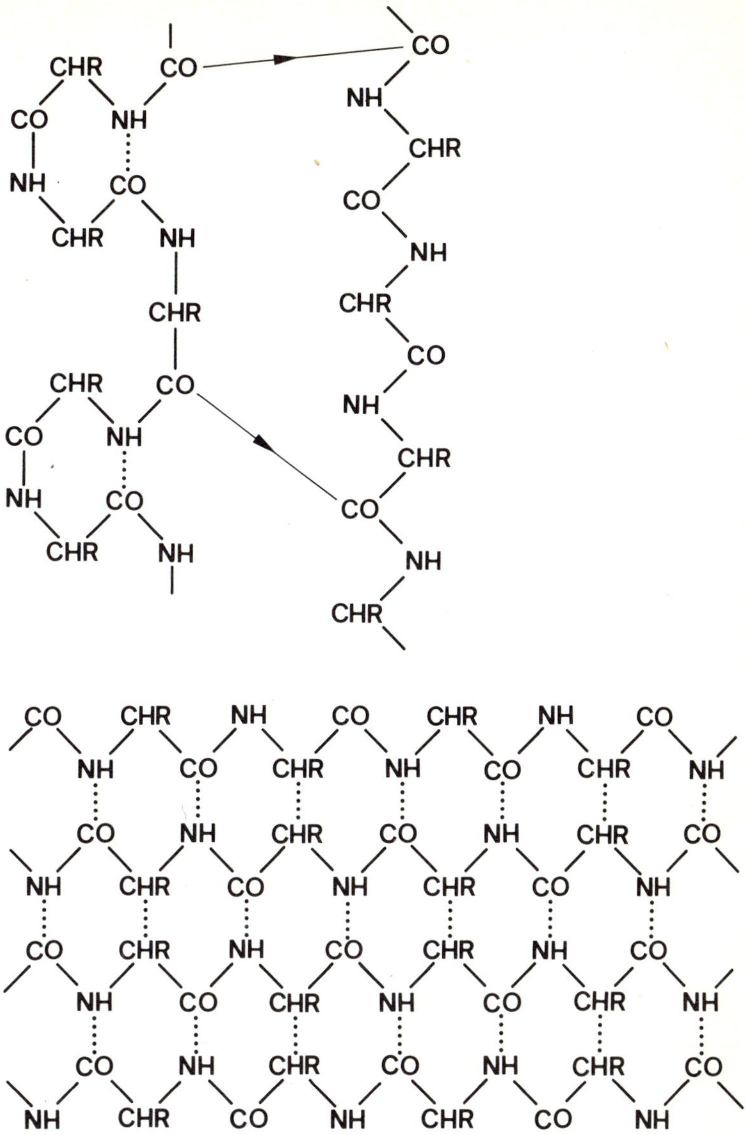

Fig. 1.1. Asbury's polypeptide folded chains and sheets (1933). [16, 17]

"hydrogen bridge" which is still used in the German "Wasserstoffbrücken-Bindung" and proposed the sheet and helical structures for polypeptide chains stabilized by intra-residue NH $\cdots$ O–C hydrogen bonds [21], shown in Fig. 1.2.

When the concept of hydrogen bonding was accepted, it seemed necessary to explain the difference between the strong –OH $\cdots$ O=C hydrogen bonds of the carboxylic acids and the weaker –OH $\cdots$ OH hydrogen bonds of the polyhydrox-

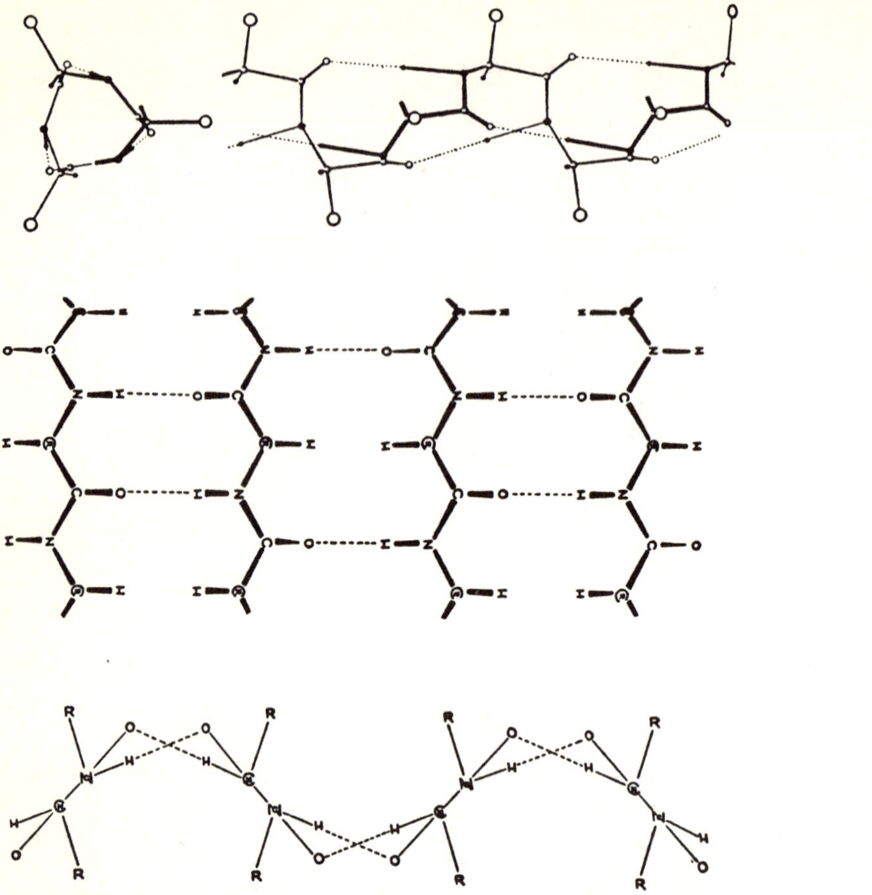

Fig. 1.2. Huggin's proposals for hydrogen-bonded helical and sheet structures for polypeptide folding (1943). [21] (Note: the helical structure (top) is now known as the 3.10 helix)

yl compounds. Bernal therefore introduced the distinction between "hydrogen bonds" and "hydroxyl bonds" [20]. The dual nomenclature, however, did not persist for more than a few years, although there are clear distinctions between these two types of interactions. In 1935 and 1937, the OH $\cdots$ O hydrogen bonds in the crystal structures of α-resorcinol [22] and pentaerythritol [23] were recognized and identified as such from single crystal X-ray analyses (see Fig. 1.3).

The hydrogen bond was first introduced as an important principle in structural chemistry to a wider audience by the chapter on Hydrogen Bonding in Pauling's first edition of *The Nature of the Chemical Bond* in 1939 [24]. Subsequently, hydrogen bonding was accepted as a major cohesive force between molecules containing hydroxyl and amino groups, with carbonyl and hydroxyl groups as the major hydrogen bond acceptors. Evidence for other groups being involved in hydrogen bonding was considered debatable, as illustrated by the interesting Faraday

Historical Perspective

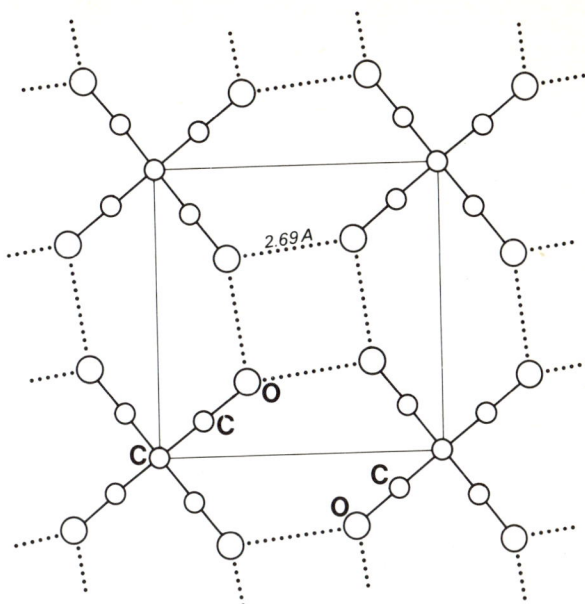

Fig. 1.3. The cyclic hydrogen bond structure in pentaerythritol, 1937. [23]

discussion of 1937, where Glasstone argued for a hydrogen-bonded complex between chloroform and acetone [25].

The early crystal structure analyses of the amino acids, carbohydrates, pyrimidines, and purines, carried out in the period from 1939–1953, all showed strong evidence that the principal cohesive forces between these subunits of biological structure were hydrogen bonds. These were recognized as stronger than van der Waals forces and more directional, leading to compounds with high melting points and generally harder crystals.

Because the electron density describing the hydrogen atom positions could not be observed in those early X-ray crystal structure analyses for technical reasons, the myth persists that hydrogen atoms cannot be "seen" by X-ray crystal diffraction. This led to the unfortunate association of the descriptor "hydrogen bond length" with the separation between the two hydrogen-bonded nonhydrogen atoms.

It was these earlier studies that set the stage for two major discoveries pointing to the significance of hydrogen bonding in biological structure, the α-helix and β-pleated sheet structures of protein architecture in 1951 [26] and the Watson-Crick base-pairing in the DNA double helix in 1953 [27]. Since then, investigators in the field of molecular biology have been compelled to realize that the hydrogen bond, with its special relationship to the water molecules, is the most important intra- and intermolecular cohesive force, determining geometry and mode of recognition and association of biological molecules.

The literature on hydrogen bonding is steadily increasing. Once the descriptor "the hydrogen bond" was accepted, it was possible to monitor the rapidly develop-

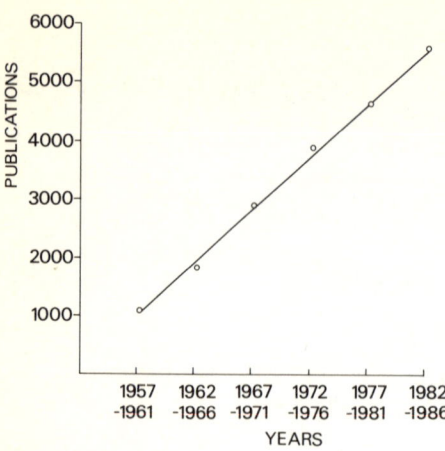

Fig. 1.4. Number of papers abstracted by *Chemical Abstracts* under the subject index "Hydrogen Bonding"

ing interest in the subject. The first major text devoted exclusively to this subject, *The Hydrogen Bond*, in 1960, gave 1500 references [28]. As shown in Fig. 1.4, the number of references in *Chemical Abstracts* under that heading has increased progressively since then. A current estimate of 20 000 publications on the subject of hydrogen bonding to date is probably conservative. It is reasonable to predict that a paper on hydrogen bonding is now being published somewhere in the world every half hour of the working day.

1.2 The Importance of Hydrogen Bonds in Biological Structure and Function

"Hydrogen Bond. A weak electrostatic chemical bond which forms between covalently bonded hydrogen atoms and a strongly electronegative atom with a lone pair of electrons. The hydrogen bond is of enormous importance in biochemical processes, especially the N–H $\cdots$ N (and N–H $\cdots$ O) bond which enables complex proteins to be built up. Life would be impossible without this type of bond."

The Penguin Dictionary of Science [29]
(with our addition in parentheses)

Realization of the importance of hydrogen bonds for the structural organization of biological macromolecules came long before the determination at atomic resolution of protein and nucleic acid structures was made possible by the methods of X-ray crystallography. Based on the combination of X-ray fiber diffraction and the results of small molecule crystal structure analyses, Watson, Crick, and Franklin derived the topology of the DNA double helix [27] in much the same way that Huggins [21], then Pauling, Corey, and Branson [26] had modeled the spatial folding of polypeptides from the results of crystal structure analyses of the amino acids. The Watson-Crick base-pair hydrogen bonding has subsequently been shown to be a feature of all known double helical structures of naturally occurring nucleic acids and the basis for genetic coding in all living organisms. Similarly, two of the

Fig. 1.5. Gulliver: a giant, constrained by a multitude of weak bonds. Illustration by Ulrik Schramm in J. Swift, Gullivers Reisen, reprinted with permission from Ueberreuter Verlag, Vienna

proposed polypeptide hydrogen-bonding schemes, the α-helix and the β-pleated sheet, have been found to be the most recurrent features which stabilize the secondary structures of the proteins. All subsequent structural research on proteins and nucleic acids has reinforced the concept that although hydrogen bonds are weak interactions [30], they are the single most important atomic interactions determining the three-dimensional folding of these biological macromolecules.

Hydrogen bonds have functional properties that are essential for life processes. They are weak interactions relative to covalent or ionic bonds, and can therefore be switched on or off with energies that are within the range of thermal fluctuations at life temperatures. This means that processes which require fast intermolecular recognition and reaction can easily occur. Stronger interactions, with bonding energies well in excess of those attained by hydrogen bonding, would seriously impede the flow of biological information and events.

On the other hand, the weakness of the individual bond is such that it is often not sufficient to provide the strength and specificity necessary for biological processes. This can be overcome because hydrogen bonds have vectorial properties and are sensitive to stereochemistry. If hydrogen-bond donors and acceptors are arranged in particular geometries, the hydrogen-bonding interactions become very specific, with additive and often cooperative strengths. For the description of these properties the concepts of the hydrogen-bond structure are important.

How these weak forces, little greater than van der Waals forces, contrive to determine the shapes of these great, complicated biological molecules is best illustrated with Fig. 1.5, taken from Swift's *Gullivers Travels*; by *cooperative action*, the Lillyputians force Gulliver into a particular spatial situation. There is a difference, however: Gulliver is fixed by the bonds so as to be in an inactive state until he wakes up. In biological molecules, the hydrogen bonds are essential for maintaining the molecules in the active state for biological reactions to take place.

1.3 The Role of the Water Molecules

Life as we know it does not exist without water molecules, which in the liquid and solid state can be described as oxygen atoms bonded to themselves and to other atoms entirely by means of hydrogen bonds.

In addition to their thermodynamic role of temperature regulation in living organisms, where they perform the same function for man as they do for the inorganic surface of the earth, water molecules are important for structural and mechanistic reasons.

Water is the most abundant biological molecule, constituting 65% by weight of the human body and higher proportions in brain, lung, and muscle tissue. There is little doubt that water played *the* essential role in the evolution of life processes on the earth. Nevertheless, the functions of this ubiquitous molecule in biological systems are relatively poorly understood, despite a library of literature, from *The Properties of Ordinary Water Substance* 1940 [31], to *Water. A Comprehensive Treatise* Vols. 1–7, 1972–1982 [32]. One role is certainly that of a mobile high-dielectric insulator. It prevents the oppositely charged moieties of biological macromolecules from aggregating and thereby collapsing a three-dimensional topology that is necessary for a particular biological function. Water molecules also have a role in the entropic formation of the hydrophobic interactions which are of importance in the globular folding of proteins (the so-called oildrop model); and in the assembly of individual subunits of more complex proteins with defined quaternary structure. Water forms a hydration shell around solute molecules and if a substrate enters the active site of an enzyme or if molecules aggregate, the water molecules have to move out in a coordinated and cooperative manner with the least expenditure of energy. Hydrogen bonding must be important in this substitution process and one can speculate that the dynamical properties of both two- and three-center hydrogen bonds are involved. Through the cooperative effect, inherent in water hydrogen bonding, the influence that these weak bonds can have on modulating and controlling the shapes and dynamics of the large complex biological molecules is considerably intensified.

Because water is regarded as a centrally important biological molecule, the hydrogen-bonding patterns of assemblages of water molecules alone, and in combination with biological molecules, are a special theme in this monograph. Nevertheless, the characterization of the hydrogen-bonding properties of water structure in biological systems has proved to be particularly difficult. This, we believe, arises from the versatility of the hydrogen-bonding stereochemistry of assemblages of water molecules, giving rise to a multiplicity of configurations with comparable energy and therefore similar probabilities, all of which result in superimposed structures in any long-time experimental study. Investigating this stereochemistry in complex biological structures by crystal X-ray diffraction methods is made even more difficult by the problem of locating the position of the hydrogen atoms in low-resolution electron density maps. It is therefore necessary to understand the hydrogen-bonding patterns of water molecules in small molecule hydrates and to extrapolate to the more complicated situation in the macromolecules.

The hydrogen-bonding patterns in the cyclodextrin hydrates, discussed in Part III, Chap. 13, are particularly informative in this regard. These molecules approach the complexity of the biological macromolecules, but because of their excellent crystallinity, they provide direct structural information relating to the positions of the hydrogen atoms in the hydrogen-bond structures. These structures form an important bridge between the information derived from the many structures of small molecule hydrates and that deduced by indirect means concerning hydrogen bonding of water molecules in the proteins and nucleic acids.

1.4 Significance of Small Molecule Crystal Structural Studies

Until 1970, the main source of information about hydrogen bonding was from solution studies using spectroscopic and thermodynamic methods. This information was diagnostic, indicating the formation of hydrogen bonds, and is quantitative only with regard to the thermodynamics of hydrogen bonding. This aspect of the subject has been reviewed in several textbooks and monographs since 1970 [33–35], and is not repeated in this monograph, except as it relates to a particular structural feature of hydrogen bonding.

As a result of the dramatic advances in computer technology in the past decade, the number-crunching computations of a crystal structure analysis ceased to be a major obstacle to the research. In addition, computer-controlled diffractometers have replaced the otherwise very tedious photographic methods of measuring the intensities of the diffraction spectra. With the development of the so-called Direct Method for solving the phase problem and the dissemination of computer programs to implement the method, the crystal structure analysis of small organic or inorganic molecules has become a routine procedure which can often be accomplished in a matter of hours. The structure analysis of macromolecules with molecular weights of several hundred thousand can now be tackled, with a reasonable expectation of success in an acceptable period of time, once procedures have been established to grow crystals which promote adequate X-ray diffraction data. Even the necessity for a number of different heavy atom derivatives is receding in favorable cases [36, 37]. The problem of collecting a large number of diffraction data from as few crystals as possible is being overcome by the development of area detectors and the use of high-powered synchrotron X-ray sources.

These advances have made X-ray and neutron single crystal analysis the principal methods for exploring and mapping the nanometer world of interatomic interactions for the past decade. It is a world where the nanometer, 10 Å, has the same place as the meter in the macroscopic world of everyday life. It is a striking observation that the level of understanding of biological structure and function has, over the past three decades, been directly proportional to the level of crystallinity of the molecules involved. Most progress is in the protein field, with nearly 400 single crystal structure analyses, least progress is in the polysaccharide-protein-lipid area where homogeneity and crystallinity are most difficult to achieve. The third major component of biological structure, the nucleic acids and their components, lies midway between these two in the depth of structure-functional knowledge [38].

Hydrogen bonding between the functional groups which occur in biological molecules displays great variety both in the energetics of the bonds and their structural features. The development of the concepts of hydrogen bonding of the base pairs in the nucleic acids and of the peptide links in the proteins all involve molecular configurations where the position of the hydrogen atom can be inferred, with reasonable certainty, from that of the atom to which it is covalently bonded, as in $>$N–H $\cdots$ O=C$<$ and $>$N–H $\cdots$ N$<$. But with many of the other functional groups which are important in biological molecules, such as P–OH, C–OH, C–NH$_2$, C–N$\overset{+}{\text{H}}_3$, the hydrogen atom positions can only be placed, in the absence of direct experimental observation, on the line of centers between the (nonhydrogen) donor and acceptor atoms. This is an assumption which is sometimes not even a good approximation. It can greatly oversimplify the real nature of the hydrogen-bond interaction and it is therefore important to have a detailed knowledge of hydrogen-bonding stereochemistry. This can be provided by the large data base which has been built up steadily by the increasing number of accurate X-ray and, more recently, neutron diffraction studies of small molecules with hydrogen-bonding functional groups as found in the biological macromolecules.

Interpreting the hydrogen bonding in the structure of biological molecules is therefore analogous to flying over an unknown land in a plane. At first, the details of the topography are well defined and distinct, as in the structural studies of small molecules. At higher altitudes, the definition is lost and only the grosser features are apparent, as in the diffraction studies of biological macromolecules. However, we know that the trees, i.e., the hydrogen bonds, are still present, because we saw them at the lower altitude. It is in this spirit that the material of this book is organized. Extrapolation from observations on small molecules to the hydrogen-bonding patterns of biological macromolecules is more complex than the macroscopic aerobatic analogy, however, since it requires topological cognition in three dimensions, while for the observer in a plane at high altitude, the earth is flat, as a first approximation.

The crystallographic data explosion. A consequence of the development of the direct methods of crystal structure analysis combined with automation of the experimental measurements has been the dramatic increase in the number of crystal structure analyses reported, as illustrated in Fig. 1.6. Fortunately, this explosion in crystallographic data has been accompanied by the development of data bases where this vast amount of structural information at atomic resolution is stored. Equally important has been the computer software development, whereby this information can be retrieved in forms appropriate to the interests of the interrogators of the data bases [39, 40].

As shown in Fig. 1.7, with the protein structures excepted, there has not been a correspondingly large increase in the complexity of the structures analyzed. This is because of the increased difficulty in growing single crystals of adequate size as

Significance of Small Molecule Crystal Structural Studies 13

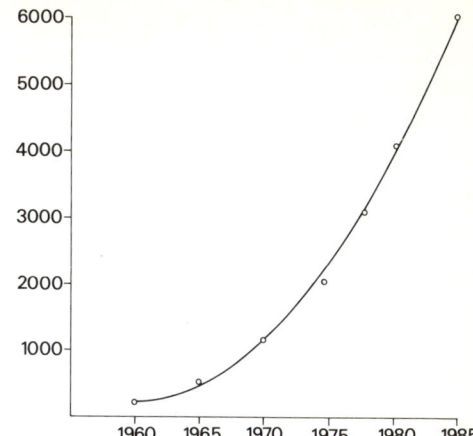

Fig. 1.6. Increase in the number of organic and organometallic crystal structure analyses reported each year since 1960, as reported in the Cambridge Crystallographic Data File

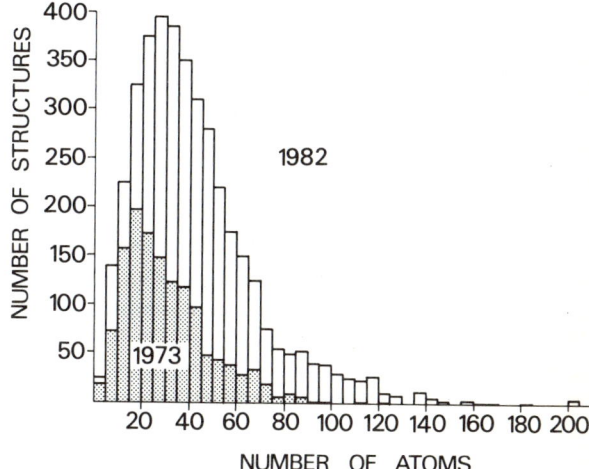

Fig. 1.7. Distribution of the number of atoms in the small molecule organic and organometallic structures included in the Cambridge Crystallographic Data File in the years 1973 and 1982

the molecules become larger. There is an obvious reluctance of molecules in the molecular weight range 1000 to 5000 to crystallize in a form suitable for X-ray diffraction studies, if at all. Unfortunately, the techniques of growing organic and small biological crystals from solution has not experienced the computer technology "revolution" and in consequence has not kept pace with the techniques of solving crystal structure analyses.[1]

Since this structural data consists of atomic parameters which describe the interatomic vectors in three dimensions, the simultaneous evolution of computer graphics has played an important role in the way the data can be used. The data base which is the particular source for the hydrogen bond data analyzed in this monograph is the Cambridge Crystallographic Structure Data Base [39, 40]. There is also a vast amount of structural information in the protein and nucleic acid data

---

[1] Crystal-growing robots are now being developed.

stored in the Brookhaven Protein Data Bank [41]. However, these rarely provide information relating to hydrogen atom positions, and they have had only a minor role in the preparation of this monograph, except for the few protein crystal structure analyses determined on the basis of high-resolution X-ray data.

1.5 The Structural Approach

Our approach to understanding the role of the hydrogen bond in determining the *three-dimensional structure* of molecular shapes and interactions in biological systems is analogous to the modern meaning of epidemiology. That is, the prediction of the most probable behavior by means of surveys of the behavior of similar species, or the same species in different habitats.

With this concept in mind, we have analyzed the metrical properties and pattern characteristics of the hydrogen bonds in terms of the types of molecules involved, rather than survey a type of bond in a variety of different molecules.

Hydrogen-bonding metrics and patterns are characteristic properties of molecules. It is by no means certain that a $>$NH $\cdots$ O=C bond will have exactly the same metrical properties when it occurs in a peptide, a barbiturate, and a nucleoside. In fact, as will be shown later, there are small, but significant, differences in these metrical properties. There are also differences in the pattern characteristics. The patterns that are observed in the crystal structures of these various types of small biological molecules provide a portfolio of observed hydrogen bond structures. These can be used as models for interpretations of structure and for hypotheses concerning intermolecular interactions in those situations where direct observation of the position of the hydrogen atom is experimentally impossible.

As will become apparent in the later chapters, the statistical nature of the conclusions is such that the present amount of data in the crystallographic data base is inadequate. We find ourselves presenting results that show well-defined trends, but in doing so, have come to realize that we need a tenfold increase in the information from X-ray and neutron crystallography before the conclusions drawn can become really definitive. This tenfold increase in structural data will surely come within the next decade, and will be useful only if the publication, data storage, and retrieval mechanisms keep pace with the accelerating rate of data acquisition.

We are cognizant that the structural data on hydrogen bonds derived from crystal structure analyses refer particularly to the hydrogen bond in the solid state. These data are subject to crystal field effects caused by other intermolecular forces. Just as with any discussion of covalent or ionic bonds observed in crystals, these crystal field effects have to be taken into consideration when extrapolating from the precise data available from the crystalline state to the imprecise data that applies to the liquid state in which most chemical and biochemical reactions take place.

The problem of transference from the hydrogen bonds in the crystal to those in a biological process is not different, in principle, from the transference of molecular structural information determined by crystal structure analysis to the interpretation of the mechanism of a chemical reaction. In Chapter 4, we discuss the differences between the geometry of hydrogen bonds in crystals and in the free molecule models that are necessarily used by the theoretical methods.

Chapter 2
Definitions and Concepts

2.1 Definition of the Hydrogen Bond – Strong and Weak Bonds

The modern concept of the hydrogen bond has its basis in the principle of the relative atomic electronegativities as put forward by Pauling in his *Nature of the Chemical Bond* [24]. As one proceeds from left to right of the Periodic Table, the electronegativity increases the electron density over that necessary to balance the nuclear charge and the electric potential is negative; the additional charge on the nuclei is increasingly less screened by the additional electron, even though the overall charge is neutral. This means that in the direction of an X–H bond, the proton is increasingly descreened as X proceeds from carbon to fluorine and its electric potential is increasingly positive. This concept is supported by the observation that the strongest bonds involve hydrogens attached to fluorine atoms and the weakest hydrogen bonds involve those attached to carbon atoms. It has long been known that with very electronegative donor and acceptor atoms, the hydrogen bond resembles a covalent bond, whereas, with weakly electronegative atoms, it is primarily electrostatic [42].

A hydrogen bond is therefore the attractive force that arises between the *donor* covalent pair X–H in which a hydrogen atom H is bound to a more electronegative atom X, and other noncovalently bound nearest neighbor electronegative *acceptor* atoms A, A'. The electron formally associated with the hydrogen atom is involved in the covalent X–H bond. Its center of mass is displaced relative to the hydrogen atom position in the direction of the center of the bond and the hydrogen atom becomes descreened. This gives rise to a dipole with positive charge at the hydrogen end of the X–H bond, irrespective of whether X carries a net charge. It is the Coulombic interaction of the dipole with the excess electron density at the acceptor atoms that forms the hydrogen bond interaction:

$$\overset{\delta^{\ominus}}{X} \text{———} \overset{\delta^{\oplus}}{H} \cdots\cdots \overset{\delta^{\ominus}}{A} \text{———} \overset{\delta^{\oplus}}{Y}$$

center of mass of electron density

This general view of hydrogen bonding is supported by experimental observations which determine directly the electron distribution around the hydrogen atom, such as X-ray and neutron charge density studies (see Chap. 3.4), the large changes

Table 2.1. Hydrogen bond functional groups in biological molecules

Donors	Acceptors
O–H	Ō=P
N–H	Ō=C, O=C⟨, O⟨^C_H, O⟨^C_C
$\overset{+}{N}$–H	Ō=S
S–H	N⟨, N⟵
C–H	S=C⟨

Most common hydrogen bonds in biological molecules

$\overset{+}{N}$–H···Ō=C	Amino acid zwitterions, proteins
O–H···Ō=C	Carboxylic acids, acid hydrates
O–H···O=C	Peptides, proteins, nucleosides,
N–H···O=C	nucleotides, nucleic acids
O–H···O–H	Hydrates
O–H···O⟨^H_C	⎫
	⎬ Carbohydrates
O–H···O⟨^C_C	⎭
O–H···N⟨	⎫
N–H···N⟨	⎪ Proteins, nucleosides,
O–H···Ō=P	⎬ nucleotides, nucleic acids
N–H···Ō=P	⎭

in proton NMR shielding anisotropy on hydrogen bonding [43–45], and the IR spectral manifestations of hydrogen-bond formation [46, 47]. As has been frequently emphasized, there is some directional character to the hydrogen bond which, however, is much less pronounced than in covalent bonding, especially for the moderate to weak hydrogen bonds, where the interaction is primarily Coulombic. This is supported by theoretical discussions of the geometrical properties of the hydrogen bond and is consistent with crystallographic studies.

The hydrogen-bonding donor and acceptor groups which are common in biological structures are shown in Table 2.1. They range from moderately strong bonds such as P–OH···Ō=P, to very weak bonds such as C–H···O=C. The hydrogen bond H···A distances associated with these groups range from about 1.5–3.0 Å. Since electrostatic attractive forces attenuate as r^{-1}, compared with the r^{-6} attenuation of dispersion forces, they must predominate at the longer distances. There is therefore *no* justification for the pragmatic definition sometimes used that "a hydrogen bond is said to exist when the distance between the hydrogen and the acceptor atom is shorter by at least 0.20 Å than the sum of the van der Waals contact radii" [48].

An ambiguity in the definition of the hydrogen bond. Huggins, by using the term "hydrogen bridge", implicated the electronic structure of both the donor and acceptor atoms, X and A, in his definition of a hydrogen bond. Similarly, Pauling in 1939 described the hydrogen bond as "largely ionic in character and *only* *formed between electronegative atoms*" [24]. This definition was broadened by Pimentel and McClellan [28]: "a hydrogen bond is said to exist when (1) there is evidence of a bond, and (2) there is evidence that this bond sterically involves a hydrogen atom already bonded to another atom." In more modern terms, this would be evidence of an attraction between the electric potentials at the position of a covalently bonded hydrogen atom.

The nature of the donor and acceptor atoms does not enter into these definitions, which has given rise to some controversy concerning the hydrogen-bonding potential of C−H, P−H, As−H, Si−H and Ge−H groups. In these, the proton will always be deshielded in the forward direction with respect to the covalent bond, because the center of charge of the hydrogen electron will be displaced toward the heavy atom. Are these bonds therefore always potential hydrogen-bond donors, even though there is no accumulation of extra electron density on the nonhydrogen atoms? In the biological molecules, the controversy is concerned only with C−H···O=C hydrogen bonds in the crystal structures, where it is maintained that there is strong evidence of bond formation, irrespective of any evidence that the carbon atom carries a net negative charge.

The question is, therefore, should the C−H···O=C interaction be referred to as a hydrogen bond, even though there is every reason to suspect that the carbon atom is not electronegative and may even carry a net positive charge? By Pauling's definition, the answer is no. By Pimental and McClellan's definition, the answer is yes.

From the observation of crystal structures, there is a clear experimental distinction between C−H···A hydrogen bonds and those where the O−H and N−H bonds are the hydrogen-bond donors. It is very rarely that the hydrogen atom in a O−H or N−H bond does not have one or more electronegative atoms as first neighbors in a direction which is forward with respect to the O−H or N−H bond. That is, it is rare to find no structural evidence of hydrogen-bond formation for O−H or N−H bonds. The reverse is true of C−H bonds. Unambiguous structural evidence of C−H···A hydrogen-bond formation is relatively infrequent. This evidence is discussed in Part IB, Chapter 10.

Strong and weak hydrogen bonds have very different properties. A case can be made for the view that *strong* hydrogen bonds are quantitatively different in most of their properties from moderate or weak bonds [49].

The concept of the hydrogen bond given in the preceding section has to be modified in the special cases when X is an exceedingly electronegative atom and A has an exceptionally large excess of electron charge. Under these circumstances, strong hydrogen bonds are formed. With very strong hydrogen bonds, such as F−H···$\bar{\text{F}}$, the structural distinction between the covalent F−H bond and the H···F hydrogen bond disappears, since the hydrogen atom lies at or close to the mid-point of the F···F line of centers, F···H···$\bar{\text{F}}$. Strong, almost symmetrical, hydrogen bonds are also observed when the donor group is a cation or the acceptor group is an anion, as in $\overset{+}{\text{O}}$−H···O, O−H···$\bar{\text{O}}$, see Table 2.2. The range of bond lengths observed in crystal structures for a particular type of strong hydrogen bond is very narrowly defined and characteristic for that type, since crystal field forces are relatively weak and of almost no influence.

Table 2.2. Properties of very strong and "normal" or weak hydrogen bonds

Property	Very strong bonds	Normal or weak bonds
Types of bonds	F−H···F⁻ O−H···O⁻ O⁺−H···O Only two-center bonds	X−H···A, where A is an electronegative atom Two-, three- and four-center bonds[a]
Bond lengths	Narrow range H···A 1.2 to 1.5 Å H···A ≈ X−H	Broad range H···A 1.5 to 3.0 Å H···A > X−H
Bond angles	Strongly directional X−H···A ≈ 180°	Weakly directional X−H···A ≈ 160±20°
Bond energy	>40 kJ mol⁻¹	<20 kJ mol⁻¹
IR vibration frequency	<1600 cm⁻¹	2000−3000 cm⁻¹
H¹ Chemical shift	>17 ppm>	

[a] for definition, see Chapter 2.2.

Because the formation of strong hydrogen bonds results in major perturbation of the spectral and structural properties of the original covalent X−H bonds (Table 2.2) they have received much of the attention directed to the subject, especially in the earlier history, before the preoccupation with the role of hydrogen bonding in biological structure and function began to take the center stage. In consequence, some of the special properties associated with the strong hydrogen bonds, such as their X−H···A linearity and short H···A distances (much shorter than van der Waals distances), have been incorrectly associated with all types of hydrogen bonds.

The distinction between strong and moderate hydrogen bonds is blurred. There is a continuum with respect to most properties between the strongest of the hydrogen bonds and those represented by the donor and acceptor functional groups shown in Table 2.1. With exception of the P−OH···Ō=P bonds and hydrogen bonds involving "counter-cations" in solvated macromolecules which might be described as strong − *strong hydrogen bonds are of minor importance in biological structures.*

For the hydrogen bonds discussed in this monograph, the hydrogen atom is always unsymmetrically located so that one bond can be clearly associated with the covalent bond, while the other interactions can be identified as hydrogen bonds, the X−Ĥ···A angles peak in the range 160±20°, and the interaction is primarily electrostatic in nature. In contrast to the very strong bonds, the weak hydrogen bonds are compressed and expanded by crystal field packing forces and the characteristic equilibrium bond length is a statistical quantity.

Since weak hydrogen bonds are primarily electrostatic[1], the weaker the hydrogen bond, the less are the associated directional characteristics. This is demonstrated by the metrical properties of the hydrogen bonds that are discussed in the following chapters. The strong P−OH···Ō hydrogen bonds tend to have OH···Ō angles closer to 180° than do the weaker N(H)H···O bonds. Similarly, the hydro-

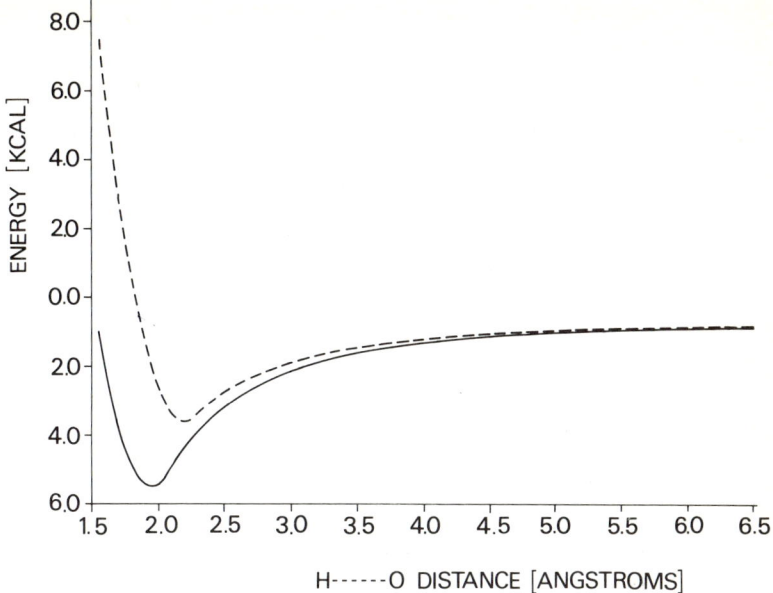

Fig. 2.1. Potential energy function along O– – –O direction for the water dimer, calculated by ab-initio MO method at HF/6-31G level [50]. (———) Total energy. (– – –) electrostatic+exchange+mixing energy only

gen atom of a P–OH group rarely has more than one acceptor atom, while a significant proportion, 25%, of the weaker >NH···O=C<, O–H···O, O–H···N< bonds are of the three-center type, defined in Chapter 2.2.

Weak hydrogen bonds are long-range interactions[1] **falling off with r^{-1}.** Support for the electrostatic character of the hydrogen bond at longer distances comes from the study of the potential surface of the water dimer using ab-initio molecular orbital methods. Figure 2.1 shows the potential energy curve along the O···O line of centers calculated at the Hartree-Fock level with a 6-31G basis set [50]. When the curve is separated into its various components according to the regimen of Moromoto [51], the electrostatic + exchange + mixing energy contributions are as indicated by the dashed line.

The significant result is that at an H···O separation of 3.5 Å, the electrostatic component of the bond energy is reduced by only 50% compared to the value at the equilibrium separation. This is consistent with the electrostatic attenuation as r^{-1}. In comparison, the exchange and dispersion components which fall off with r^{-6} have become negligible. These results indicate that first neighbor hydrogen-bond interactions will still be significant at distances as far away as 3.5 Å from the hydrogen atom. In fact, it is observed in the analyses of hydrogen-bond geome-

---

[1] Work = (Coulombic) force × distance = $\dfrac{e_i \cdot e_j}{\varepsilon \cdot r_{ij}^2} \times r_{ij} = \dfrac{e_i \cdot e_j}{\varepsilon \cdot r_{ij}}$.

2.2 Hydrogen-Bond Configurations: Two- and Three-Center Hydrogen Bonds; Bifurcated and Tandem Bonds

The hydrogen bonds observed in crystal structures are rarely linear. Before high-precision X-ray or neutron diffraction crystal structure analyses permitted the direct location of the hydrogen atom, it was already recognized that the $X-H\cdots A$ configuration might not be linear [52–54]. Even when there is only one first neighbor electronegative acceptor atom present, the probability of observing the hydrogen atom lying exactly on the $X\cdots A$ line of centers is very low according to a $\sin\theta$ geometrical factor [53, 54]. Because of the "softness" of the hydrogen-bond bending force constants, there is a balance between the equilibrium 180° and the greater probability of bonds making a smaller angle. The most probable value of θ observed for various samples of hydrogen bonds is around 165°, a value that is consistent with theoretical calculations of $O-H\cdots O$ bonds using ab-initio quantum mechanics on model systems [55]. This does not imply, however, that linearity is not the most stable configuration for a hydrogen bond between isolated donor and acceptor molecules or between hydrogen-bonded dimers in the gas phase.

There is, however, another reason why a hydrogen bond may not be linear. This is due to an attractive force to a second acceptor, as in the configuration shown in **1**.

Although there was strong evidence to the contrary as early as 1939, there has been a curious hesitation to consider the possibility that there can be more than one simultaneous hydrogen-bond acceptor group. Chemists rather reluctantly permitted hydrogen to have a valence of two 50 years ago, and valences of three or four were rarely considered until very recently.

As neutron diffraction and more accurate X-ray crystal structure analyses have become available, it is clear that this configuration **1** is quite common in the crystal structures of many biological small molecules. Surveys of the peptide $NH\cdots O=C$ hydrogen bond [56] and hydrogen bonds in carbohydrates [57–59], amino acids [60], purines and pyrimidines [61] and nucleosides and nucleotides [62] indicate that between 25% and 40% of the bonds in a sample of structures may be of this type.

This configuration, **1**, was first proposed by Albrecht and Corey, who applied the descriptor *bifurcated bond* in 1939 to account for the nonhydrogen bond geometry of the $\overset{+}{N}H$ and $\overset{-}{C}=O$ groups in the crystal structure of α-glycine [63]. This suggestion was confirmed both by a very careful X-ray study [64] and by neutron diffraction in 1972 [65], with the dimensions shown in Fig. 2.2.

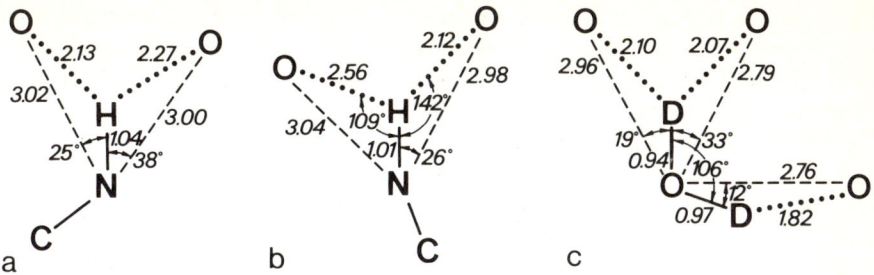

Fig. 2.2a–c. Early examples of three-center bonds from neutron diffraction studies. **a** In α-glycine [65]; **b** in glycine hemihydrochloride [66]; **c** in perdeuterated violuric acid [67]

Unfortunately the descriptor "bifurcated hydrogen bond" was also used for the quite different configuration 2 in the very influential 1960 edition of *The Hydrogen Bond* [28]. Because of its possible relevance to the structure of liquid water, this configuration has received considerable theoretical attention [50, 51], although, in fact, it is rarely observed in the crystalline state, and is not found in any of the ice structures.

There are, however, some examples of this bifurcated bond configuration observed in the nucleosides, but only in conjunction with three-center bonds, and therefore referred to as three-center bifurcated bonds, as in **3** and **4**:

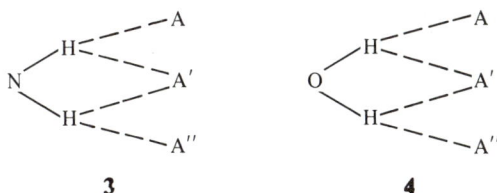

More recently the term "bifurcated" has also been applied to the configuration **5**, proposed for interactions between water and dimethyl phosphate or formate ions [68]. An unusual bifurcated configuration **6** is reported in 1,3-dimethyl uracil [69].

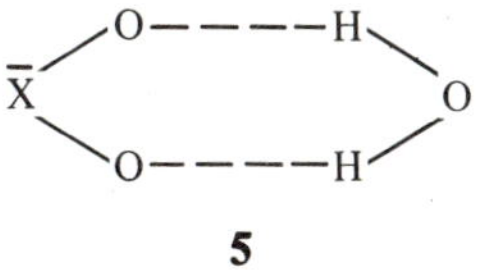

To avoid confusion, the term *three-centered* has been applied to the configuration where the hydrogen atom is located between three electronegative atoms, being covalently bound to one and hydrogen bonded to the other two[2]. This configuration, **1**, is defined geometrically by the condition that the hydrogen atom lies in or close to the plane of the three-bonded atoms, as required by an attractive

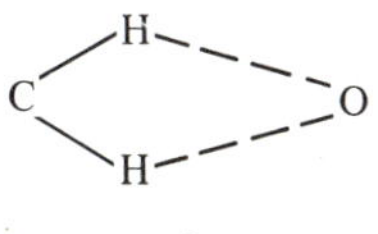

---

[2] An alternative descriptor is bifurcated-donor, to distinguish it from **2**, the bifurcated-acceptor configuration.

force to the two acceptors. A simple test, first applied by Parthasarathy in 1969 to 15 possible examples in the X-ray crystal structure analysis of glycylglycine hydrochloride [70], was that in **1**, $\theta_1 + \theta_2 + \alpha \approx 360°$. An alternative, more sensitive, criterion that has been used is that the hydrogen atom is within 0.2 Å of the plane defined by X, A, and A' [56].

Further evidence for three-center bonds came from the X-ray analyses of the β and γ forms of glycine [71, 72] and glycine hemihydrochloride [66], but the first definitive neutron diffraction evidence for the existence of the three-center bond, as a relative rarity, was from the 1964 crystal structure analysis of perdeuterated violuric acid [67], the hydrogen-bond geometry of which is shown in Fig. 2.2c. In 1968, Donohue reviewed examples of crystal structures where there was good evidence for the existence of three-center bonds [73].

Three-center hydrogen bonds are most common in crystal structures which are proton-deficient. Evidence for the much more common occurrence of the three-center bond in the crystal structures of the simple carbohydrates and amino acids came from the analysis of a series of neutron diffraction studies [57–60, 74]. In the pyranose and pyranoside sugars, the number of acceptor oxygens from the hydroxyl groups and the ring and glycosidic oxygens exceeds the number of donors by about 25%. This corresponds approximately to the proportion of three-center bonds [59]. The same situation occurs in the amino acids, where the primary donors, the NH_3^+, have three protons, but the primary acceptors, CO_2^-, commonly accept four bonds, two to each oxygen atom. An analysis of the neutron data for the amino acids [60, 74] showed that the majority, 46 out of 64, of the hydrogen bonds between the zwitterion groups are three-centered. Recent analyses of the hydrogen bonding in barbiturates, purines and pyrimidines, nucleosides and nucleotides show similar proportions of three-center bonds as observed in the carbohydrates, as shown in Table 2.3.

Three-center bonding in the carbohydrates and the amino acids is therefore a result of excess acceptors over donors, or of *proton deficiency.* Hence the analogy with the three-center electron bond from electron deficiency in the boron hydrides [76].

In crystal structure analyses where the hydrogen atoms are not directly observed, as in macromolecules and structures with disordered water molecules, *the signature of the three-center hydrogen bond is a triangle* of potential donor or acceptor nonhydrogen atoms at distances in the range observed for weak hydrogen-bonding interactions, i.e., 2.8 to 3.5 Å.

Three-center bonds can be symmetrical with $r_1 \approx r_2$, $\theta_1 \approx \theta_2$ and unsymmetrical, in which r_1 and r_2 differ by as much as 1.0 Å, θ_1 is close to 180° and θ_2 is close to 90°. Examples of these have been well established by the neutron diffraction studies of the amino acids [60, 74] and the pyranose sugars [58]. A case where the decision between two- and three-centered bonding is difficult is illustrated from the neutron analysis of erythritol [77], shown in Fig. 2.3. Although one bond is much longer than the other, both are primarily electrostatic. It is difficult to refer to one as a hydrogen bond and the other as an electrostatic attraction. In this monograph, we call these the *major and minor component* of a three-center bond.

Hydrogen-Bond Configurations

Table 2.3. Percentage of three-center bonds in the crystal structures of biological small molecules

Donor	Acceptor	Total No. of bonds	Percent three-centered	References
All types of molecules				
$\overset{+}{\text{NH}}$	$O=\bar{C}$ }	1509	20	[75]
NH	$O=C$			
Amino acids (neutron data only)				
$\overset{+}{\text{NH}}$	$O=\bar{C}$	64	72	[60]
Carbohydrates (neutron data only)				
OH	O<	100	25	[58]
Nucleosides and nucleotides (X-ray data)				
All types of H-bonds		529	24	[62]
Barbiturates, purines and pyrimidines (X-ray data)				
All types of H-bonds		832	27	[61]

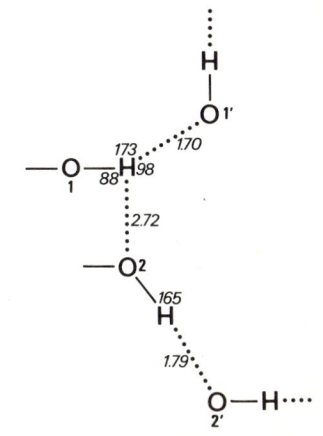

Fig. 2.3. A questionable three-center bond from the neutron diffraction crystal structure analysis of erythritol [77]

When both acceptor atoms are covalently bonded to the same atom, as in $X-H \overset{A}{\underset{A}{\cdots}} Y$, the acceptor is said to be a *chelate*, and these are referred to as *chelated three-center bonds*. They occur with carboxylate, nitrate, phosphate, phosphoric acid, and the amidine groups in nucleic acid components.

Four-center hydrogen bond configurations, as in **7**, are also observed, but much less frequently. In the four-center hydrogen bonds, the proton makes three first neighbor contacts to potential hydrogen-bond acceptor atoms in the forward direction with respect to the donor X–H bond. The geometrical definition is less rigorous than that for three-center bonds, since it requires only that the X–H···A, A′, A″ angles are greater than 90°.

$$X-H\begin{matrix}---A\\---A'\\---A''\end{matrix}$$

7

In the study of 1352 hydrogen bonds of type NH···O=C [56], only 6 were regarded as four-centered, but this analysis used an arbitrary van der Waals cut-off criterion which may have biased the conclusions. In the carbohydrates, purines, pyrimidines, nucleosides, and nucleotides, the proportion was about or less than 1%.

Tandem hydrogen bonds. The configuration **8** has also been considered, mainly in connection with water structure, and the descriptor *planar ring* was applied [28]. In some theoretical studies [50], this configuration was also referred to as *cyclic* hydrogen bonding. Since we wish to use that term cyclic for the configurations of four or more hydrogen bonds which have been observed experimentally in a number of crystal structures, we prefer the term *tandem* to distinguish it from the cyclic arrangements which do not contain the obviously short nonbonded H···H interaction. This makes **8** an unlikely configuration unless the hydrogen atom positions are disordered, so that both are not simultaneously occupied.

8

This configuration **8** has, however, been postulated to occur in a few crystal structures, including that of α-cyclodextrin 7.57 H₂O [78], which is discussed in Part III, Chap. 18. All these analyses are room-temperature X-ray studies of high to moderate accuracy and the evidence for hydrogen positions is not strong. A tandem hydrogen bond with a H···H separation of 1.75 Å reported from an X-ray structure analysis of the A-form of potassium gluconate [79] was later shown to be incorrect when the analysis was repeated using neutron diffraction [80].

2.3 Hydrogen Bonds Are Very Different from Covalent Bonds

Perhaps we should have retained the Huggin's descriptor, hydrogen bridge, since the moderate to weak hydrogen bonds are very different from covalent bonds in two important respects. One is that they are "soft" bonds which are easily deformed by the other intermolecular interactions, which may be other hydrogen bonds, and by van der Waals forces; the other is that they display group properties.

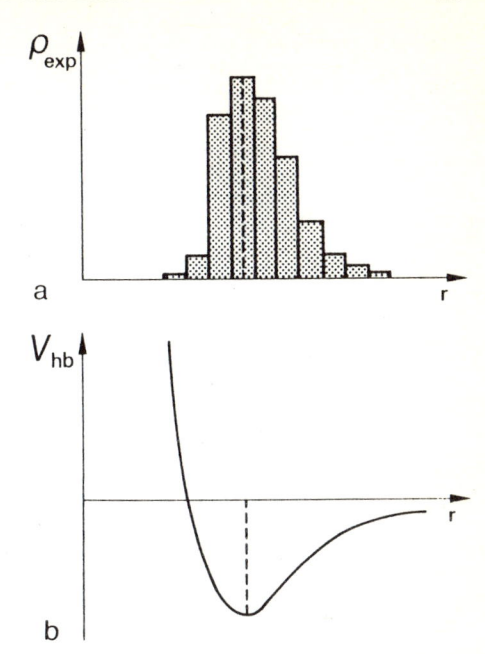

Fig. 2.4. a Distribution of r(H···O) bond lengths for 1509 N−H···O=C bonds; **b** H-bond potential energy curve [75]

The stretching and bending force constants of hydrogen bonds are about 15 times smaller than for covalent bonds. Therefore, from a structural point of view, the hydrogen-bond lengths or hydrogen-bond angles observed in any particular molecular structure are dependent on the environment in which they are observed. In any one crystal structure, the hydrogen-bond X−H···A geometries will be compressed or expanded by up to 20% of their equilibrium distances, that is, between 1.4 and 2.1 Å for an equilibrium bond length of 1.8 Å.

In consequence, a hydrogen-bond length and even more so a hydrogen-bond angle observed in one structure is not transferable to the same type of bond when it occurs in another structure. What can be characteristic of a particular type of hydrogen bond is the *most probable* hydrogen-bond length, obtained by statistical surveys of a large number of structures in which they occur. The assumption is made that the distortions from the equilibrium values will cancel out over a large enough number of crystal structures. This assumption is examined critically in Chapter 4. An illustration of such a survey in which extensive use is made of the Crystallographic Data Base is given in Fig. 2.4. The N−H···O=C hydrogen bonds investigated [75] have an H···O bond length distribution which reflects that expected from an asymmetrical r(H···O) potential energy curve, while the maximum of this curve corresponds approximately to the "equilibrium" hydrogen bond length in the *effective* potential energy curve.

The second important difference from covalent bonds is that hydrogen bonds are not atom-pair but group-pair properties. Since covalent and ionic bonds are atom-pair properties as a first approximation, they can be decomposed into atomic properties such as covalent and ionic radii, which are additive. They are almost un-

Table 2.4. Hydrogen bond geometry in the α- and β-deutero oxalic acid dideuterates from neutron diffraction analysis [83]

	$-C\begin{smallmatrix}O\\OD\cdots O_WD_2\end{smallmatrix}$			$\begin{smallmatrix}HO\\DO_WD\cdots O\end{smallmatrix}C-$		
	OD⋯O$_W$	O⋯O$_W$	O–D̂⋯O$_W$	O$_W$D⋯O	O⋯O$_W$	O$_W$–D̂⋯O
α-form	1.493(2)	2.524(2)	177.4°(1)	1.939(2)	2.879(2)	167.8°(2)
				2.008(2)	2.906(2)	156.0(2)
β-form	1.520(2)	2.538(2)	174.2(2)	1.960(2)	2.855(2)	157.2(2)
				1.895(2)	2.834(2)	170.3(1)

affected by the environment of the molecules considered because compression/expansion effects in bond distances seldom exceed 2%, and valence angles vary over only a few degrees. Based on this observation, Pauling derived effective covalent and ionic bond radii as atomic properties which can be used to predict successfully the geometry of molecules for structural, molecular mechanical, and dynamical calculations.

This is not true for hydrogen bonds. The properties of a hydrogen bond are group properties, depending not only upon the first neighbor atoms of both X and A, but also upon the sequential nature of the *total pattern of bonding*. In crystals, there is evidence that the shape of the molecules and the other forces that determine molecular packing will also influence the hydrogen-bonding geometry. Because of this, an attempt to rationalize hydrogen-bond lengths in terms of hydrogen-bond atom radii [81] was doomed to fail. Attempts to model hydrogen bond interactions as if they were atom-pair functions has been equally unsuccessful, being replaced in recent years by modeling in terms of donor and acceptor group potentials [82].

The group property of the hydrogen bonds is well illustrated by the neutron diffraction data on the α and β deutero oxalic acid hydrates given in Table 2.4 [83]. While the O–H⋯O hydrogen bonds from acid hydroxyls to water oxygens are short strong bonds, those from water hydroxyls to acid carboxylic oxygens are long weak bonds. This difference, which is not surprising in terms of the concept of acid-base interactions, is not adequately represented if both bonds are simply referred to as O–H⋯O bonds.

Similarly, the $\overset{+}{N}$–H⋯O=C bond lengths show a most probable value of 1.72 Å for the $R_3NH^+\cdots\,^-O_2C-$ hydrogen bond which is shorter than the total average of 1.86 Å, whereas that of $\ggN-H\cdots O=C(OH)-$ is longer, 1.98 Å [75]. The general applicability and usefulness of this concept was not generally realized until a significant number of neutron diffraction studies were available, which in turn prompted more detailed analyses of the available X-ray data.

Hydrogen-bond lengths are statistical properties. As a first approximation, covalent bond lengths are independent of the state of matter for which they are measured or the type of molecule in which they occur. The C–C bond length differs

Table 2.5a. Selected values of hydrogen-bond energies in kJ mol^{-1} (from ref. 28, except where noted)

C−OH···O=C		$O_WH···O_W$	
Formic acid (gas)	14.5 – 15.2	Water (gas)	22.6 [86]
Acetic acid (gas)	14.1 – 15.7	(liquid)	14.2
(in CCl$_4$)	11.3	(ice)	12.5 – 32.2
(in C$_6$H$_6$ + H$_2$O)	8.3 – 9.2	(theory-dimer)	14.5 – 25.0 [87]
(solid)	32.2 [88]		
C−OH···OH		**$O_WH···O=CH_2$**	
Methanol (gas)	13.4 – 18.8	(theory)	14.3 [89]
(in CCl$_4$)	19.2 – 21.7		
(theory-dimer)	23.4 [90], 28.9 [91]		
Ethanol (gas)	14.2 – 16.7		
(in CCl$_4$)	7.5		
>NH···O=C			
Formamide			
(theory-dimer)	33.0 [92]		
N-methyl formamide			
(in C$_6$H$_6$)	14.6		
N-methyl acetamide			
(in C$_6$H$_6$)	15.1		
N(R$_2$)H···NH(R)$_2$			
Methylamine (gas)	14.2		
Dimethylamine (gas)	13.8		
Diethylamine (gas)	13.8		

only by a few percent between diamond and gaseous ethane. Therefore values for covalent bond lengths can be obtained which are transferable between different molecules in different states. The same is true for covalent bond energies. For this reason, useful, albeit empirical, quantitative relationships between bond length and energy have been derived [24].

The same opportunity does not exist with hydrogen bonds. The $O_WH···O_W$ hydrogen bond in ice is 1.72 Å, whereas in the gas-phase water dimer, it is 17 percent longer, 2.02 Å. In water, the best accepted value is intermediate, 1.85 Å. Corresponding $O_WH···O_W$ bond energies have been reported which range from 12 to 32 kJ mole^{-1} depending upon the state of matter and the type of measurement [28].

The hydrogen-bond lengths observed for a particular donor/acceptor combination in the crystal structures of the same class of small biological molecules range between ±10 percent of the mean value, depending upon the crystal structure in which they are observed [84]. It is questionable whether this statistical data can be quantitatively correlated with bond energies [85].

To obtain a comparative set of bond energy values, it is necessary to be consistent about the state of matter or type of experiment or calculation used.

Table 2.5b. Hydrogen-bond energy in competition with solvent water. Free energy of transfer (ΔG_{tr}) for polar and charged groups in amino acids, separated into contributions of the non-polar environment and hydrogen bonding. (Courtesy Prof. N. Pace, work to be published)

Group		$\Delta G_{tr}{}^a$ = Nonpolar[b] + Hydrogen bonding		Number of hydrogen bonds formed[c]	Energy per hydrogen bonding kcal/mol
Peptide	$-\text{N}-\text{C}-$ $\;\;\;\|\;\;\;\;\;\|\|$ $\;\;\text{H}\;\;\;\text{O}$	−5.3	−1.1	−4.2	3	−1.4
Tyr:	−OH	−3.1	−1.1	−2.0	2	−1.0
Ser.+Thr:	−OH	−4.3	−1.0	−3.3	3	−1.1
Asn+Gln:	$-\text{C}-\text{NH}_2$ $\;\;\;\|\|$ $\;\;\;\text{O}$	−7.4	−1.7	−5.7	4	−1.4
His:	⟩N and NH	−7.2	−2.4	−4.8	2	−2.4
Asp+Glu:	$-\text{COO}^{(-)}$	−9.0	−2.1	−6.9	4	−1.7
Lys:	$-\text{NH}_3^{(+)}$	−9.0	−4.4	−4.6	3	−1.5
Arg:	$-\text{NH}-\text{C}^{(+)}-\text{NH}_2$ $\;\;\;\;\;\;\;\;\;\;\;\;\|\|$ $\;\;\;\;\;\;\;\;\;\;\;\text{NH}_2$	−17.5	−3.6	−13.9	5	−2.8

[a] (ΔG_{tr}) cyclohexane to water [93a].
[b] (ΔG_{tr}) octanol to water [93b].
[c] See Ref. [596].

As shown in Table 2.5a, measured and calculated hydrogen-bond energies generally lie in the range between 8 and 22 kJ mole^{-1}. There are no trends which correspond to the sequences of mean bond lengths that are described in Part IB, Chapt. 7. As pointed out elsewhere [28] more accurate and systematic experimental measurements are required. It now seems likely that the most consistent set of bond energy values will come from dimer calculations at high levels of ab-initio theory on a wider range of molecules than is presently possible.

Very high precision X-ray charge density analyses, based on low temperature X-ray and neutron Bragg diffraction measurements, do offer the opportunity to measure hydrogen bond geometry and the corresponding interaction energies in the same crystal, as discussed in Part IA, Chap. 3. This research, however, is at the frontier of X-ray crystallography. It requires high diffraction quality crystals and very precise measurements. There have been too few examples to assess the general usefulness of the method [93].

Competition of solvent water reduces hydrogen-bonding strength. Biological molecules are usually immersed in water. Consequently, intramolecular hydrogen bonds stabilising their three dimensional structure and intermolecular hydrogen bonds between associated molecules have to compete with hydrogen bonds formed between donor/acceptor groups and water molecules. However, there must be a net hydrogen-bonding energy, which gives rise to the intra- or intermolecular cohesive force. It can be estimated from partitioning experiments where, in case of proteins, amino acid derivatives are transferred from water to 'wet' hexane containing only ~2.5 mM water (where hydrogen bonding makes a large contribution to the transfer free energy ΔG_{tr}), and to 'wet' octanol with 2.3 M water (where hydrogen bonding makes a small contribution to ΔG_{tr}). The difference between these values for each amino acid derivative gives the hydrogen-bonding energy for the polar group and, if their energy is divided by the number of hydrogen bonds formed by each group, the energy contribution per hydrogen bond is obtained, see Table 2.5b. For polar, uncharged groups it is in the range 1.0–1.4 Kcal/mol and for charged groups 1.5–2.8 Kcal/mol.

2.4 The van der Waals Radii Cut-Off Criterion Is Not Useful

In the absence of direct experimental evidence concerning the position of the hydrogen atom, the term hydrogen-bond length was applied to the separation of the nonhydrogen atoms $X \cdots A$, as illustrated by Table 2.6 taken from a popular 1968 crystallography textbook [94]. Unfortunately, this definition persists, particularly in discussions of hydrogen bonding in biological structures.

The use of $X \cdots A$ cut-off criteria should be discouraged because the $X \cdots A$ separation is a function of the covalent bond length X–H, the hydrogen-bond length, $H \cdots A$, and the angle $X-\hat{H} \cdots A$, which, as we know, is rarely 180° in the solid state. Since, until 1965, hydrogen bonds were identified structurally by distances between the nonhydrogen atoms, there was a need to distinguish them from van der Waals contacts. A hydrogen bond was said to exist "if the $X \cdots A$ separation was closer than a van der Waals contact" [48].

Alternately, a hydrogen bond permitted the insertion of a hydrogen atom between two nonhydrogen atoms without extending their separation. The implication

Table 2.6. $N \cdots X$ and $O \cdots X$ hydrogen-bond lengths and ranges [94]

Bond	Mean, Å	Range, Å
$N-H \cdots N$	3.10	2.88 – 3.38
$N-H \cdots O$		
Ammonia	2.88	2.68 – 3.24
Amides	2.93	2.55 – 3.04
Amines	3.04	2.57 – 3.22
$N-H \cdots F$	2.78	2.62 – 3.01
$N-H \cdots Cl$	3.21	2.91 – 3.52
$N-H \cdots Br$	3.37	3.28 – 3.44
$O-H \cdots N$	2.80	2.62 – 2.93
$O-H \cdots O$		
Oximes, inorganic acids	2.58	2.44 – 2.84
Carboxylic acids	2.63	2.45 – 2.75
H_2O in org.-inorg.	2.71	2.49 – 3.07
Alcohols	2.74	2.55 – 2.96
H_2O in inorg.	2.75	2.49 – 3.15
H_2O in org.	2.80	2.65 – 2.93
Hydroxides	2.82	2.36 – 3.36
$O-H \cdots Cl$	3.07	2.86 – 3.21
$O-H \cdots Br$	3.30	3.17 – 3.38

Table 2.7. Experimentally observed XH···A hydrogen-bond distances versus van der Waals radii criteria

	van der Waals radii (Å)		
	Pauling[a]	Bondi[b]	Allinger[c]
H	1.2	1.2	1.50
C	1.2	1.70	1.75
N	1.5	1.55	1.70
O	1.4	1.52	1.65
Cl⁻	1.8	1.75	1.95

Type of bond XH···A	Observed range of H···A distances (Å)	Covalent R_{X-H} (neutron value)	Criteria for H-bonds based on sums of van der Waals radii Top line: $< W_X + W_A - R_{X-H}$ Bottom line: $< W_H + W_A - 0.3$		
			Pauling	Bondi	Allinger
OH···O	1.44–2.10	0.97	1.83	2.07	2.33
			2.30	2.42	2.95
OH···Cl⁻	1.95–2.33	0.97	2.23	2.30	2.63
			2.70	2.65	3.25
NH···O=C	1.58–2.05	} 1.03	1.87	2.04	2.32
NH···O	1.60–2.40		2.30	2.42	2.95
NH···Cl⁻	2.10–2.20	1.03	2.27	2.27	2.62
			2.70	2.65	3.25
CH···O	2.04–2.65	1.10	1.83	2.22	2.30
			2.30	2.42	2.95
CH···N	2.50–2.75	1.10	1.93	2.15	2.35
			2.40	2.45	3.00
CH···Cl⁻	2.60–2.95	1.10	2.23	2.35	2.60
			2.70	2.65	3.25

[a] Pauling van der Waals radii [24]. [b] Bondi van der Waals radii [97]. [c] Allinger van der Waals constants radii [98].

was that the insertion of a hydrogen atom between X and A leads to a contraction relative to the sums of the van der Waals radii of X, or H, and A, when hydrogen bonding takes place. The degree of this contraction was arbitrary and "cut-off" criteria using values of 0.2, 0.4, and 0.6 Å have been applied [48, 95, 96].

There are two objections to this procedure. One is fundamental, because the cut-off is inconsistent with theory. The other objection to the use of van der Waals criterium is that it can sometimes give rise to incorrect conclusions.

In Table 2.7, the range of observed hydrogen-bond lengths for different donor-acceptor atom combinations is compared with criteria based on the three most commonly used van der Waals radii. For all bonds, except the NH···Cl⁻, the criteria based on the Pauling W_X and W_A van der Waals radii is too restrictive. Only by using the Allinger van der Waals constants radii given in Table 2.7, it is possible to use the criterion that the distances between neighboring atoms of the donor and

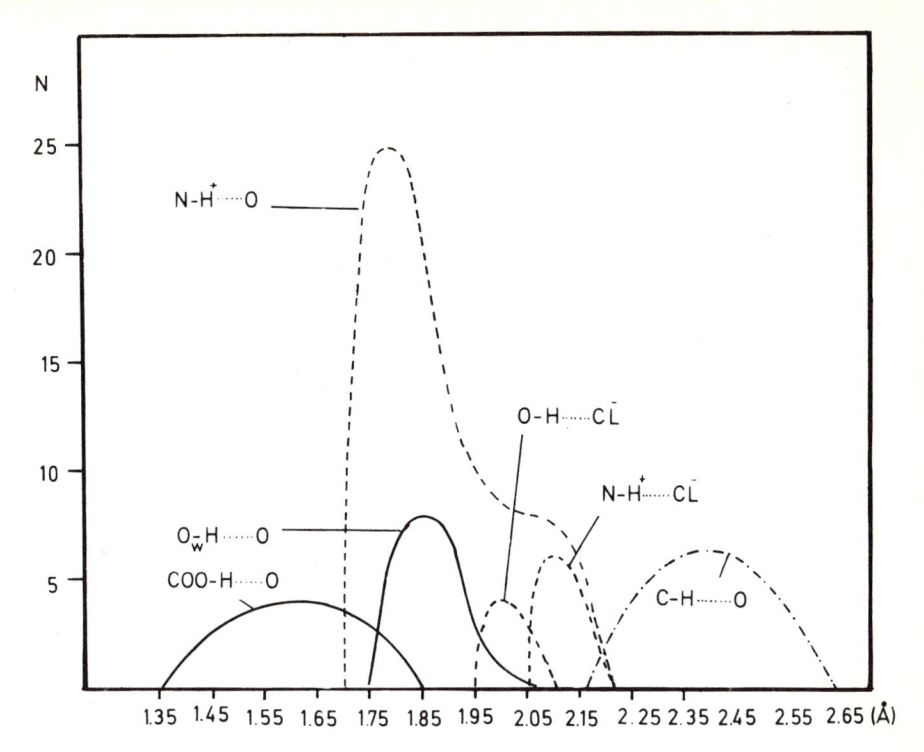

Fig. 2.5. Distribution of hydrogen-bond distances observed by neutron diffraction crystal structure analyses of the amino acids [60]

acceptor groups must be substantially less than the sum of the van der Waals radii [98]. These van der Waals constant radii are based on the minima in potential energy curves and require compression terms to be applicable to van der Waals distances in the solid state.

The use of a van der Waals distance "cut-off" criterion carries the wrong implication that hydrogen bonds become van der Waals interactions at longer distances. This becomes clear from the concept that the van der Waals attractive forces will be negligible at distances where the electrostatic hydrogen-bonding attractions are still large (see Fig. 2.1). For molecules with hydrogen-bonding functional groups, strong electrostatic attractive and repulsive forces will occur at intermolecular distances greater than the sum of the respective Pauling van der Waals radii.

A survey of the hydrogen-bond lengths observed by neutron diffraction in the amino acids [60], for example, shows that they have distributions with minimum and maximum values which correspond to those expected from a typical bonding potential energy curve, as shown in Fig. 2.5. There is no evidence of a cut-off distance beyond which another type of cohesive force takes over. Instead, the number of H···A distances steadily decreases as they exceed the most probable H···A bond length. In practice, it is observed that in crystal structures the first neighbor

Table 2.8. Selected hydrogen-bond geometries determined by neutron diffraction

Structure	Hydrogen bond	N···O (Å)	H···O (Å)	Refs.
L-Lysine monohydrochloride dihydrate	N(2)–H(4)···O(2)	2.887(3)	2.083(6)	[54]
Hydrazinium hydrogen oxalate	N(2)–H(2)···O(1)	2.884(1)	1.935(2)	[416]
9-Methyladenine-1-methylthymine	N(10′)–H(4)···O(8)	2.872(3)	1.932(5)	[537]
L-Glutamic acid (α-form)	N(1)–H(2)···O(11)	2.895(1)	1.926(2)	[100]
L-Tyrosine	N(1)–H(3)···O(2)	2.884(2)	1.853(4)	[54]
DL-Serine	N(1)–H(2)···O(2)	2.876(1)	1.844(2)	[54]

distances between hydrogen atoms and electronegative acceptor atoms very rarely exceed 3.0 Å for the biological molecules under consideration. The next level of intermolecular separations are the weaker H···H van der Waals interactions, which have similar probability distributions centering about 3.5 Å, and extending to about 4 Å.

A further difficulty arises because the van der Waals radii of hydrogen and oxygen are particularly ill-defined concepts. For other than monoatomic ions, the van der Waals surface is far from spherical. That of hydrogen is ill-defined because of the diffuse and easily polarizable nature of the hydrogen atom electron density. This is illustrated by the wide variation in the potential energy curves for H···H interactions, that have been proposed at various times [98]. In the environment of hydrogen-bond formation, the uncertainty about the van der Waals "surface" of a hydrogen atom is even greater, as reflected in the examples given in Table 2.7. For atoms with lone electron pairs such as oxygen and nitrogen, the fact that the van der Waals surface is not spherical [99, 99a], is usually neglected.

The use of the nonhydrogen distances as a criterion of hydrogen bonding can also obscure interesting and relevant aspects of hydrogen bond character, as illustrated in Table 2.8, which shows data from some neutron diffraction analyses.

The N···O distances imply that the hydrogen bonds are very similar, ranging in length from 2.872 to 2.895 Å, but the H···O hydrogen-bond distances show that they differ by 0.24 Å from 1.844 to 2.083 Å. As will be demonstrated later, by making appropriate assumptions and corrections, reliable hydrogen-bond lengths can be derived even when the hydrogen atomic positions cannot be accurately obtained from the experimental data.

The use of the van der Waals criterion delayed the recognition of the importance of the three-center bond. The use of criteria based on comparison with van der Waals distances is particularly misleading when the three-center hydrogen bonds are involved. The geometry of the three-center bond in α-glycine, shown in Fig. 2.2a, illustrates this problem. *In this configuration, the shorter H···O bond length is associated with the longer N···O separation.*

Using the criterion N···O $< W_N + W_O$ with Pauling's van der Waals radii (=1.4+1.5 Å), neither of these is a hydrogen bond; using H···O $< W_O +$

$W_H - 0.2$ (= 1.4 + 1.2 − 0.2), both are hydrogen bonds; using $H \cdots O < W_O + W_H - 0.4$ (= 1.4 + 1.2 − 0.4), one is a hydrogen bond and one is not.

2.5 The Concept of the Hydrogen-Bond Structure

In the majority of biological structures, hydrogen bonds form one, two, or three-dimensional arrays, where the donor X−H groups are simultaneously acceptors X. These are either connected directly, as in the donor-acceptor sequences of $\cdots O - H \cdots O - H \cdots O - H \cdots$ in the ices, high hydrates, carbohydrates and cyclodextrins, or they may be connected indirectly through polarizable covalent bonds, as in the carboxylic acids, purines, pyrimidine, nucleosides, nucleotides and proteins. Some of these examples are shown in Table 2.9.

Table 2.9. Cooperative hydrogen-bond configurations observed in crystal structures
Finite or infinite chains in carbohydrates and some hydrated nucleosides:

$O-H \rightarrow O-H \rightarrow O-H \rightarrow O\langle$

$\rightarrow O-H \rightarrow O-H \rightarrow O-H \rightarrow O\langle$

$\rightarrow O-H \rightarrow O-H \begin{smallmatrix} \nearrow O\langle \\ \searrow OH \rightarrow OH \rightarrow \end{smallmatrix}$

Intersecting chains form four-connected nets, observed in carbohydrate hydrates, cyclodextrin hydrates, some hydrated nucleosides, also in some hydrochlorides:

```
        ↑                           ↑
        O                           O
        H                           H
        ↓                           ↓
        O                           O
        H                           H
        ↓                           ↓
→ OH → OH → OH → OH → OH →     → OH → OH → Cl⁻ ← HO ← HO ←
              H                             ↑
              ↓                             H
              O                             O
              H                             ↓
              ↓                             H
                                            O
                                            ↓
```

Cycles observed in ices, clathrate hydrates, layer hydrates, and cyclodextrin hydrates:

```
    ⟋H → O−H⟍
   O           O
   ↑           ↓
   H⟍       ⟋H
      O ← H−O
```

Table 2.9 (continued)

Dimers, trimers and chains in the carboxylic acids, purines, pyrimidines, nucleosides, peptides and proteins:

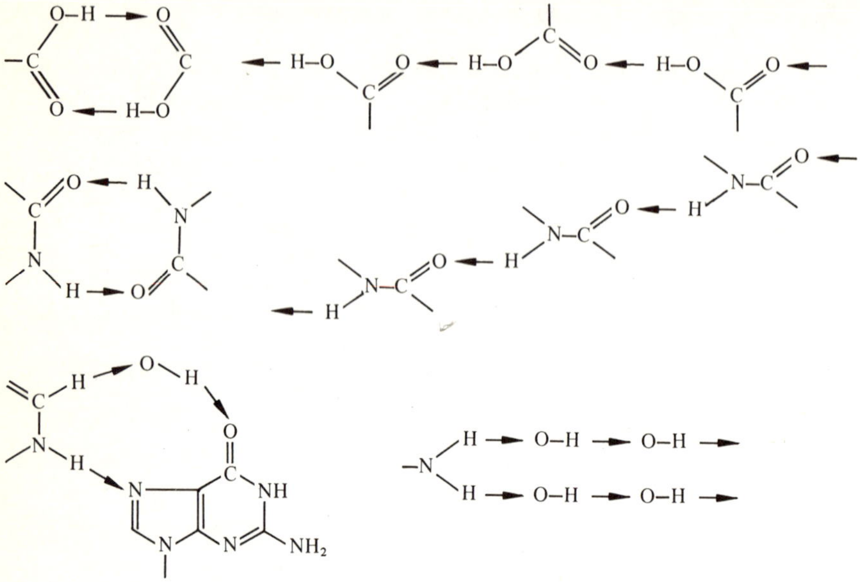

We refer to this as the *hydrogen-bond structure* and describe it in the same progressive way as is used for covalently bonded molecular structures. The *constitution* of the hydrogen bonding describes the types of donor and acceptor functional groups involved. The *configuration* describes the connectivity of these groups and the *conformation* describes their complete stereochemistry in terms of hydrogen-bond lengths, bond angles, and torsion angles.

While it has long been recognized as essential to structural chemistry to describe molecules in terms of their configuration and conformation, it is only recently that the same concept has been applied to the repeating hydrogen-bond patterns that occur in crystals. This may have been because, prior to the development of computer graphics, these patterns were difficult to visualize. With computer graphics the hydrogen-bonding functional groups can be displayed alone with their observed stereochemistry, which is maintained as the structure is rotated for scrutiny in three dimensions. The nonfunctional atoms, such as carbon, which confuse the picture, can be eliminated and the model does not collapse, as it would with the old ball and stick model.

Figure 2.6 shows the configuration, i.e., connectivity, of the hydrogen-bond structure of a hydrated disaccharide, maltose monohydrate. A more detailed picture with hydrogen-bond distances and angles is given in Fig. 13.38. Without diagrams of this type, derived using computer graphics, it is impossible to begin to rationalize the hydrogen-bonding patterns of complex biological molecules.

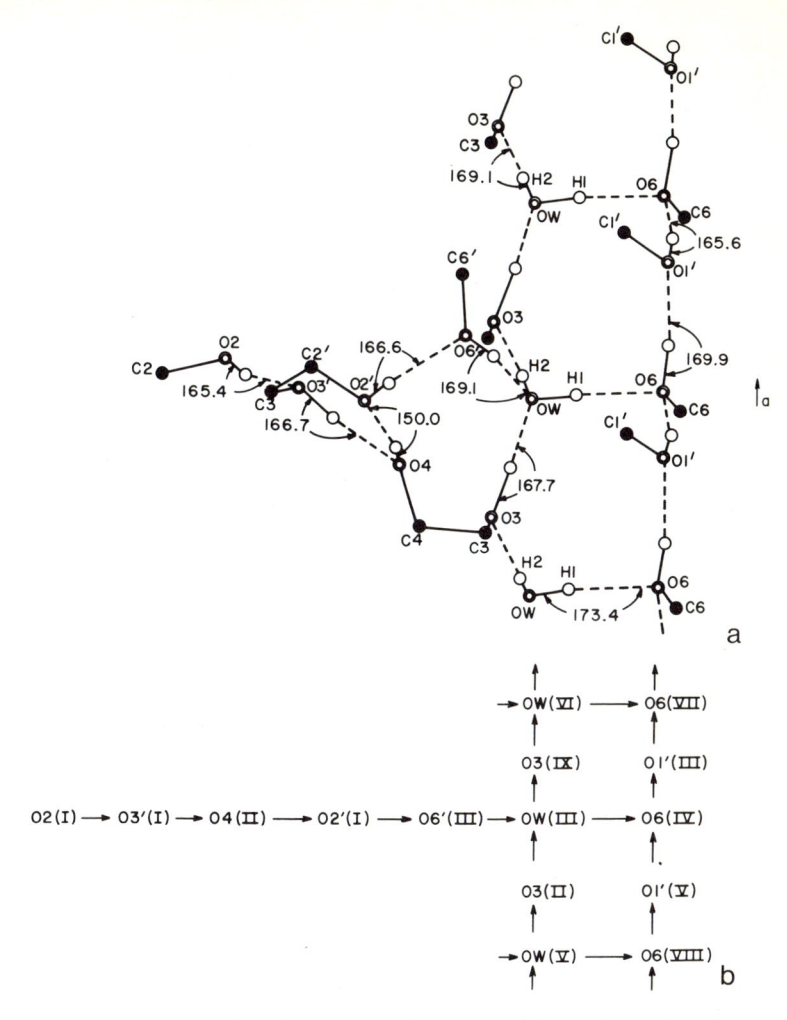

Fig. 2.6a, b. Hydrogen-bond connectivity diagram for maltose monohydrate. **a** Projection of a section of the crystal structure; **b** schematic connectivity diagram

2.6 The Importance of σ and π Cooperativity

Hydrogen bonding in all classes of biological molecules displays the property of cooperativity. As displayed in Table 2.9, hydrogen bonds can form continuous structures, which in the case of crystals, extend and link together molecules into periodic two- or three-dimensional arrays. Because of the polarizability or charge-transfer nature of these continuous patterns, the binding energy of a hydrogen bond structure is greater than that of the sum of the individual bonds,

$$E(H\cdots A)_n > nE(H\cdots A) \ .$$

This concept of cooperativity, or nonadditivity, of hydrogen bonds was inferred conceptually from the early theories of liquid water and aqueous solutions [101, 102]. It was clearly demonstrated and defined by the early ab-initio quantum mechanical calculations on water polymers [103 to 105]. The first systematic studies of hydrogen-bonding patterns in carbohydrate crystal structures revealed these cooperative patterns [106] and the effect they had on the mean hydrogen-bond lengths [107]. Later theoretical calculations estimated the bond shortening and energy gain due to the cooperative effect in hydrogen-bonded methanol trimers to be 0.08 Å, with an energy gain per hyrogen bond of about 12% [108].

Two types of hydrogen-bond cooperativity can be distinguished. In the crystal structures of the ices, hydrates, carbohydrates, and cyclodextrins, the energetic advantage of forming finite or infinite chains is clearly reflected structurally in the patterns observed and in a trend toward shorter O−H···O hydrogen bonds. This is possible because the hydroxyl group has both donor and acceptor properties. On the other hand, the hydrogen bonding in the carboxylic acids, peptides, and purines and pyrimidines involves functional groups which do not have simultaneously donor and acceptor properties. These are the functional groups such as >NH, NH$_2$ which, with very few exceptions, are donors only, and C=O, N<, which can only be acceptors. In these structures, chains or cyclic patterns are also formed, because the cooperative effect can take place through the polarization of the π-bonding in the covalent bonds adjacent to the donor and acceptor atoms.

σ- and π-bond cooperativity are both consequences of polarization. When describing the hydrogen-bonding patterns in crystal structures, we found it necessary to distinguish between *σ-bond cooperativity* and *π-bond cooperativity*.

σ-Bond cooperativity occurs with functional groups having simultaneously donor and acceptor properties. In practice, this applies only to the O−H groups, since the >NH is never, and the −NH$_2$ (in nucleic acid bases) is rarely a hydrogen-bond acceptor. Polarization occurs through the O−H σ bonds.

$$\overset{\delta+}{O}-\overset{\delta-}{H}\cdots\overset{\delta+}{O}-\overset{\delta-}{H}\cdots\overset{\delta+}{O}-\overset{\delta-}{H}\cdots\overset{\delta+}{O}-H \ .$$

The effect on the covalent O−H bond length is so small that it has not been observed experimentally.

π-Bond cooperativity requires that adjacent hydrogen-bonding functional groups are linked by bonds with π electron character. It involves the polarization of the electrons in the multiple covalent bonds, as in the carboxylic acid dimers and polymers or the formamide dimers and polymers.

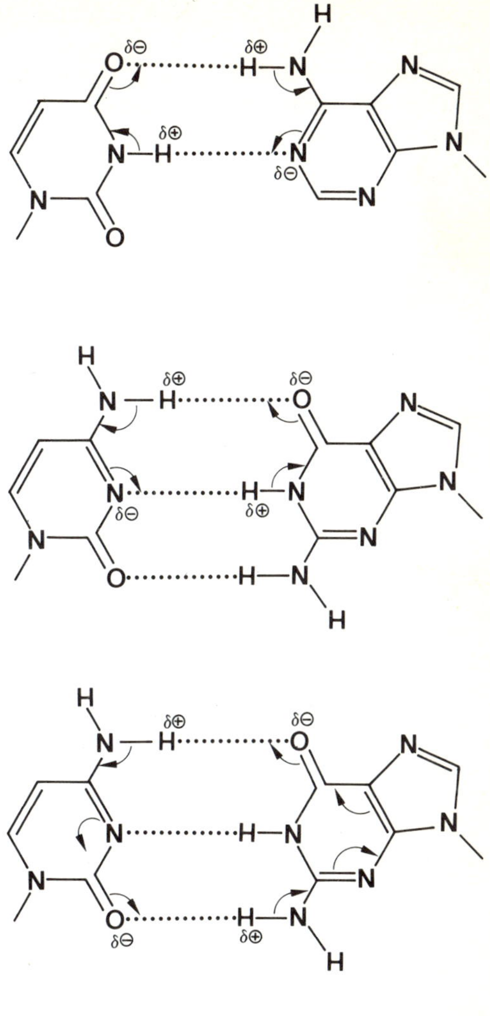

Fig. 2.7. Hydrogen-bond π-cooperativity in purine-pyrimidine base pairing

In this case, the lengthening of the N–H and C=O bonds, and the shortening of the C–N bonds as a consequence of hydrogen-bond formation can be measured by careful X-ray or neutron diffraction analyses, as described in Chapter 5. This type of π-cooperativity is especially important in the main-chain hydrogen bonding of proteins, where it increases the double-bond character and hence the torsional rigidity of the peptide C–N bond.

π-Bond cooperativity also contributes to the stabilization of the Watson-Crick base pairs in the nucleic acids by electron delocalizations shown in Fig. 2.7. In the enol form of some β-diketone moieties, it is referred to as *resonance-assisted hydrogen bonding* [108a].

2.7 Homo-, Anti- and Heterodromic Patterns

Early quantum mechanical calculations [105] on the water trimer showed clearly that the sequential hydrogen bonding, **1**, was energetically favored over the alternative double donor, **2**, or double acceptor, **3**, hydrogen bonding.

This result, and others [55, 103, 104] from model calculations, gave an early theoretical basis for understanding the predominance of finite and infinite chains of hydrogen bonds in the carbohydrate and cyclodextrin crystal structures in which there is a uniform donor-acceptor direction, as in **4**,

$$\rightarrow OH \rightarrow OH \rightarrow OH \rightarrow OH \rightarrow OH \rightarrow \quad . \qquad 4$$

In the cyclic systems, more commonly observed in the cyclodextrin hydrates, all three alternatives, **1**, **2**, **3**, are expanded to correspond to the configurations **5**, **6**, and **7**, for which the descriptors homodromic, **5**, antidromic, **6**, and heterodromic, **7**, have been applied[3] [109], which can also be used for the description of linear chains with homo-, anti- or heterodromic orientation of the O–H···O hydrogen bonds. These systems have hitherto been too large for useful ab-initio calculations, but semi-empirical quantum mechanical calculations suggest that, as with the water trimer, the sequential bonding of **5** is energetically favored over **6** and **7** [110,

---

[3] Greek: homo = the same, anti = contrary, hetero = mixed; dromon (δρομον) = to run: i.e., homodromic = running in the same direction. Compare with "Velodrom", where bicycles are all running in the same direction on the same course.

111]. The same reversal of the cooperative hydrogen bonding as in **6** and **7** will occur when an anion is incorporated in the hydrogen-bonding sequence, as in **8**, or a cation, as in **9**.

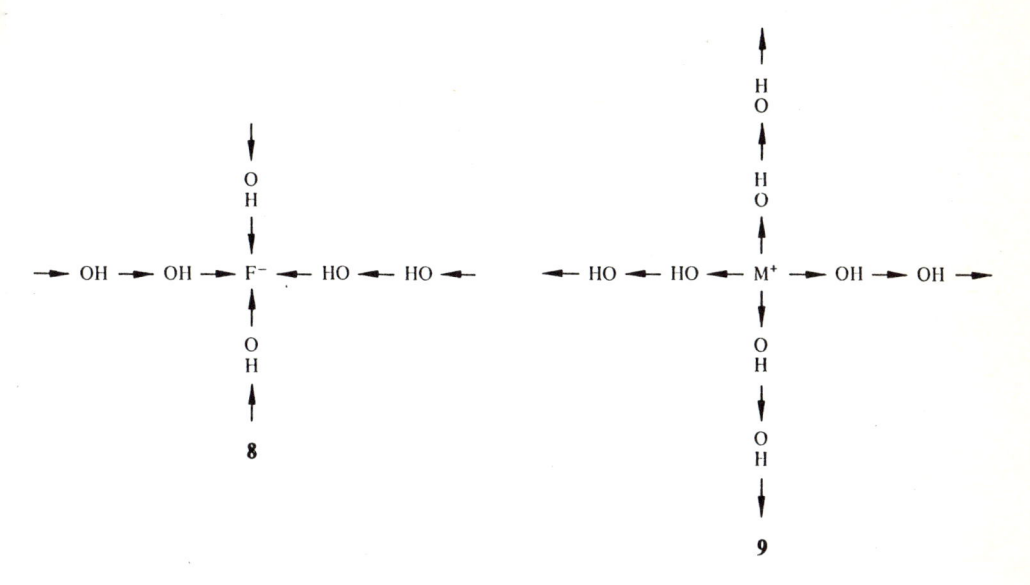

It is important to note that water molecules in hydrates do not necessarily interrupt cooperativity, as in the antidromic cyclic system **6**, if they are at the intersection of chains, as in **10**,

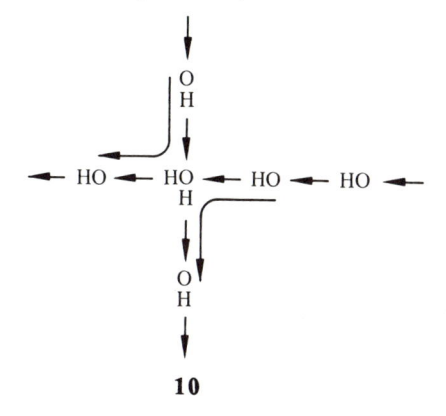

Cyclic hydrogen-bond patterns involving 4, 5, 6 and 7 bonds are commonly observed in the ices and the clathrate hydrates, described in Part IV. Similar cyclic systems are also observed in the hydrates of strong acids and salts which contain the so-called hydrated proton, for example, $(H_2O)_n H^+$ in $HBr \cdot 4\ H_2O$ [112, 113].

2.8 Hydrogen Bond Flip-Flop Disorder: Conformational and Configurational

In crystals, hydrogen-bond structures are sometimes found where *two half-hydrogen* atoms are located approximately along the lines of centers, $(X, A) - \frac{1}{2}H \cdots \frac{1}{2}H - (A, X)$, with $\frac{1}{2}H \cdots \frac{1}{2}H$ separations of about 1.0 Å. This distance is considerably shorter than the 2.4 Å expected for a van der Waals $H \cdots H$ separation, and requires the two H atoms to be in mutually incompatible positions, corresponding to a time- or space-averaged equilibrium, and that X and A be the same atom type

$$X-H \cdots A \rightleftharpoons X \cdots H-A .$$

In all instances where this disorder was actually observed by X-ray or neutron diffraction methods, X and A are hydroxyl oxygen atoms, as in the ices Ih and Ic, in certain of the high pressure ices, and in the cyclodextrin hydrates, see Parts III and IV. In ice Ih the disorder gives rise to the well-known residual entropy of 0.82 ± 0.05 cal deg^{-1} [114 to 117].

The tandem hydrogen-bond configuration may also be an example of hydrogen bond disorder, since the bonds involved are also hydroxyls and the hydrogen to hydrogen separation is less than 2.4 Å.

Two mechanisms for hydrogen-bond disorder have to be considered. One is the *conformational* mechanism which involves the reorientation of hydrogen bonds by rotation of X–H groups. This disorder does not require breaking of covalent X–H bonds, and propagates as a concerted process,

When this type of disorder occurs in long cooperative chains or in cyclic systems of hydrogen bonds, it has been described as a *flip-flop* mechanism [118], because it must proceed from bond to bond like a domino effect and involves only two well-defined states.

The other is a *configurational* mechanism, in which the covalent X–H bonds are broken as the hydrogen atoms shift across the line of centers to form the new hydrogen bond. This must also be a concerted process, which is only possible due to quantum mechanical tunneling:

Because in this mechanism covalent X–H bonds have to be broken, it appears more probable with very strong hydrogen bonds where the bonding is primarily covalent and the hydrogen atom is close to being symmetrically located, see Fig. 2.8.

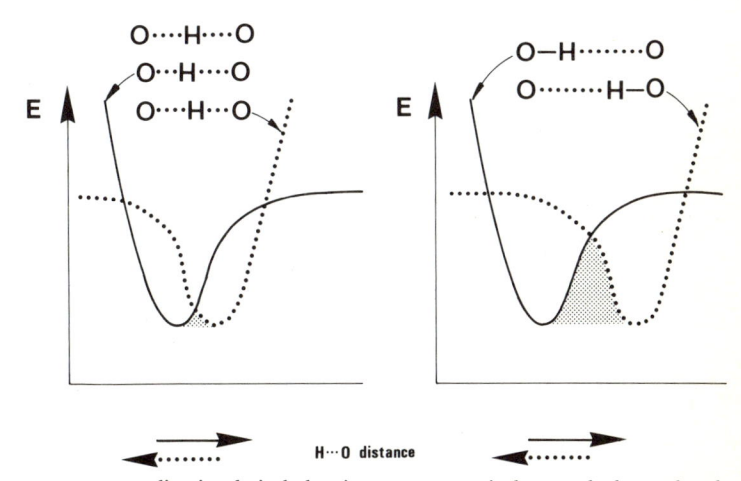

Fig. 2.8. Barrier to proton tunneling is relatively low in near-symmetrical strong hydrogen bonds (*left*), and high in the unsymmetrical, moderate to weak hydrogen bonds (*right*), observed in biological structures. Energy barriers are indicated by *shading*

It becomes less probable with unsymmetrical hydrogen bonds, and may be negligible in the moderate to weak hydrogen bonds.

In biological structures, the conformational flip-flop disorder will occur in the carbohydrates, the cyclodextrins, in water, or generally, in all systems where hydroxyl groups are involved.

In the proteins and nucleic acids the configurational mechanism has to be invoked to explain any hydrogen-bond disorder involving >N−H or >N̈H groups where there is no orientational flexibility. This is shown in the two examples, below (the second describing amino/imino and keto/enol tautomeric states):

amino   keto   imino   enol

Although this tautomerism has been vividly discussed in the literature [119 to 123], there is no evidence for its occurrence to significant proportions in biological structures, and there is no X-ray or neutron diffraction study where such configurational hydrogen atom disorder was observed. In nucleic acids, there is some support from spectroscopic data that keto/enol tautomerism could exist in solution in equilibrium to less than 0.01%, but these numbers are subject to conditions and experimental error. We can confidently state that *configurational hydrogen-bonding disorder do not take place in the nucleic acid bases, where it would upset the genetic code* (see Part IIB Chap. 15.2 and Part III Chap. 20.3), *and it probably does not play a role in NH···O=C hydrogen bonds stabilizing protein secondary structure*.

2.9 Proton-Deficient Hydrogen Bonds

Three-, and to a lesser extent, four-centered hydrogen bonds are favored in crystal structures when there is a proton deficiency, such as the amino acids discussed in Part IIB, Chapter 14. The energetic advantage comes not only from the additional bonds, but also from the extra cooperativity of the extended chains or nets of hydrogen bonds which are formed versus single links or short chains. A well-defined, but different, example of proton deficiency is provided by the high-temperature polymorphs of tetramethyl ammonium hydroxide pentahydrate [124]. In the low-temperature form, the $(CH_3)_4N^+$ ions are enclosed in a *broken* truncated octahedron, shown in Fig. 2.9 (left), by $[20\ H_2O + 4\ OH]^-$ [125]. This open cage is constructed from 33 hydrogen bonds. In the high-temperature phase (between 42°C and 68°C), the cations are enclosed in a *closed* truncated octahedron $[4^6 \cdot 6^8]$ shown in Fig. 2.9 (right), which has 14 faces, 24 vertices, and 36 edges (F+V = E+2), i.e., 36 hydrogen bonds. The three nonbonded edges of the "bro-

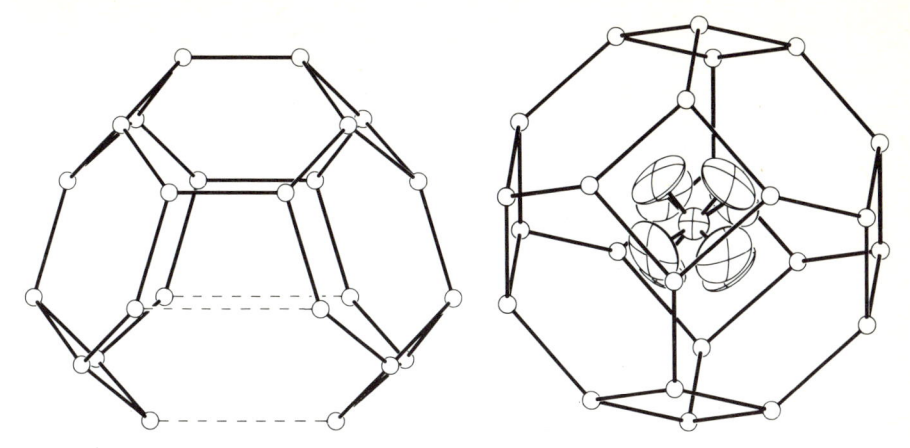

Fig. 2.9. The "broken" (*left*) and closed (*right*) truncated octahedron of hydrogen bonds in the low and high temperature structure of tetramethyl ammonium hydroxide tetrahydrate [124]

ken" octahedron with an O···O separation of 4.46 Å become hydrogen bonded with O···O = 2.88 Å. Since no more protons are added to the structure, this implies a *proton-deficient hydrogen-bond framework*, with 32 n protons[4] forming 36 n hydrogen bonds. Each hydrogen bond contains 8/9 of a proton and these are presumably disordered into two 4/9 occupancy sites across each bond. As in the case of the three-center bonds, the energetic advantage for this must arise from the additional bonds plus the additional cooperative energy of the complete four-connected lattice over a related incomplete lattice. If this is other than an isolated example, which seems unlikely, the possibility of forming more two-center hydrogen bonds than there are protons available in order to achieve better cooperativity in water networks will add further complexity to the problem of describing the structure of water in heavily hydrated biological systems.

2.10 The Excluded Region

The concept of an *Excluded Region* can be very useful for assigning hydrogen bonds in macromolecular structures [126, 127]. In contrast to the attractive components of hydrogen bonding, the repulsive forces between the atoms involved are relatively "hard", attenuating approximately as r^{-6}. They can therefore be used to define a region in conformational space in which hydrogen atoms will not be found. When dealing with groups such as $-OH$, $-NH_2$, $-\overset{+}{N}H_3$ and the water molecule, H_2O, this places a useful limitation on the hydrogen-bond configurations that need to be considered when hydrogen atoms cannot be directly observed.

---

[4] Since the water framework is formed by a space-filling arrangement of face-sharing polyhedra, 12 (i.e., 24/2) of the 44 protons in the $[20H_2O + 4OH]^+$ form disordered bonds with the adjacent polyhedra.

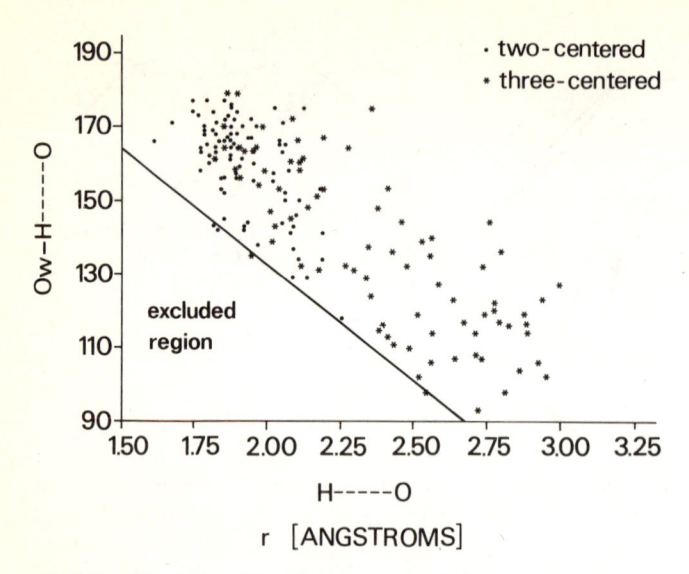

Fig. 2.10. Plot of $O_W-H\cdots O$ angles versus $H\cdots O$ distances from two- and three-center bonds from the hydrates of small biological molecules discussed in Chap. 22

Figure 2.10 shows a plot of the excluded region for the $O_w-H\cdots O$ bonds discussed in Part IV, Chap. 22.

This concept promises to be as useful for analyzing water structure in macromolecular crystals as were the Ramachrandran plots for limiting the conformation possibilities in polypeptides.

2.11 The Hydrophobic Effect

The hydrophobic effect arises from the age-old adage that oil and water do not mix. On a molecular basis, the solvation of small molecules in water may be considered as occurring in two steps. One is the formation of a cavity in water for the solute molecule which requires free energy because strongly interacting, hydrogen-bonded water molecules are to be separated. The second step is the insertion of the solute molecule into this cavity. It is energetically favorable and compensates for the cavity-forming energy if the solute is polar and forms hydrogen bonds with water molecules.

If the solute is nonpolar, there is only weak van der Waals attraction with water, and water molecules arrange around the nonpolar solute such that they form the most extensive number of hydrogen bonds, with the ice clathrates (Part IV, Chap. 21) the extreme case. The ordering of water molecules is entropically unfavorable, since they lose orientational and translational freedom. This can be compensated for if the solvated solute molecules aggregate and the ordered water molecules are released from their surface into bulk water, a process which is entropically favorable and the main driving force for the "hydrophobic effect" [128 to 134].

The Hydrophobic Effect

Table 2.10. Thermodynamics of the transfer cycle of methane in the gas phase, in inert solvent (CCl$_4$), and in water [135]

Transfer	ΔG kcal mol^{-1}	ΔH kcal mol^{-1}	ΔS kcal mol^{-1} K^{-1}
CH$_4$ in CCl$_4$ → gas	−3.5	+0.5	+14
CH$_4$ gas → water	+6.3	−3.2	−32
CH$_4$ in CCl$_4$ → water	+2.8	−2.7	−18

The hydrophobic effect can be measured in terms of transfer of a molecule from gaseous phase or dissolved in nonpolar solvent to water. Since the change in Gibbs free energy ΔG is associated with changes in ethalpy ΔH, and entropy ΔS according to

$$\Delta G = \Delta H - T\Delta S \ ,$$

negative changes in ΔG and ΔH and positive changes in ΔS are energetically favorable, as are elevated temperatures.

The transfer cycle for methane ([135], Table 2.10) shows that transfer from the inert solvent to the gas phase is favorable due to the large increase in entropy ($\Delta S = +14$ kcal/mol^{-1} K^{-1}) which compensates for the loss in van der Waals contacts indicated by a positive enthalpy ($\Delta H = +0.5$ kcal mol^{-1}). The transfer from the inert solvent and from the gas phase to water is unfavorable due to the large decrease in entropy (the water molecules arrange around methane in "clathrate" form), which outweighs the favorable change in enthalpy ($\Delta H = -2.7$ and -3.2 kcal mol^{-1}) arising from the increase in van der Waals contacts with water.

A hydrophobicity scale for amino acids. The free energy of transfer from organic solvent to water depends linearly on the accessible surface area of a solute molecule [136 to 138], as illustrated for a number of hydrocarbons and amino acids in Fig. 2.11. For the hydrocarbons and the amino acids with nonpolar side chains Ala, Val, Leu, Phe, the lines with a slope of 25 cal Å^{-2} and 22 cal Å^{-2} respectively pass through the origin. The line for the amino acids with polar side chains Ser, Thr, His, Met, Tyr, Trp also has a slope of about 25 cal Å^{-2}, but the free energy of transfer is systematically lower than expected from their surface areas.

We can associate these data directly with the concept of ordered water molecules around a nonpolar solute molecule, viz. the larger the surface area, the more water molecules are arranged around the solute molecule, with a loss in energy of 22 to 25 cal Å^{-2}. Based on arguments similar to those given above, the hydrophobicity scale was extended to all amino acids and is compared in Table 2.11 with a scale derived on the basis of the actual distribution of amino acids in the interior or at the surface of a sample of 46 monomeric proteins.

A consideration of thermodynamic properties of the aqueous solution of rare gases and hydrocarbons led to the "iceberg" model for water structure around nonpolar molecules [139], which later had to be abandoned (see Part IV, Chap. 23.4). The gas hydrate clathrate structures described in Part IV, Chap. 21 provided

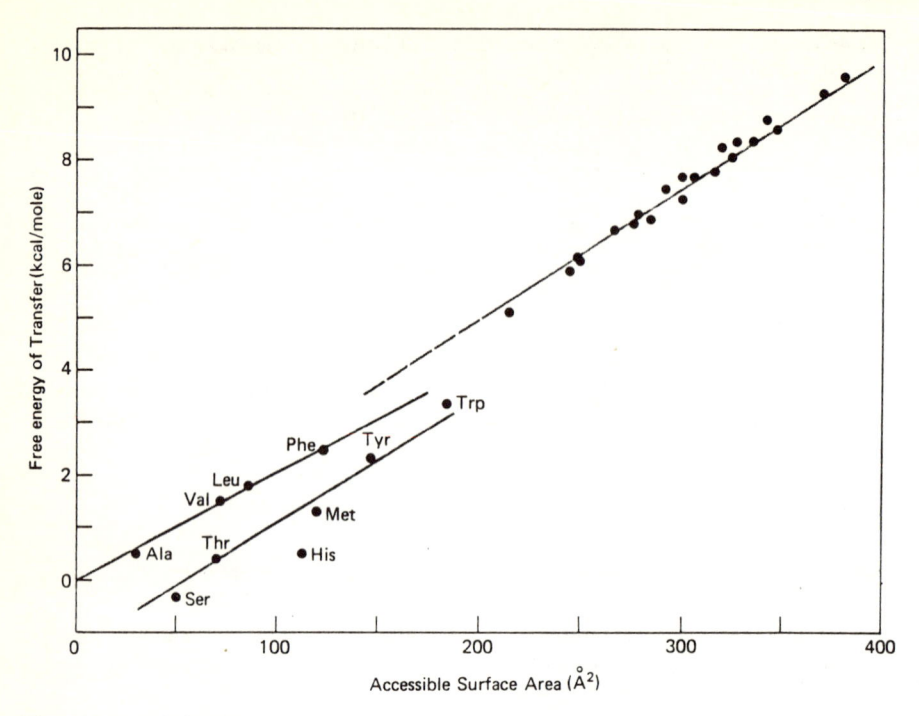

Fig. 2.11. Correlation between accessible surface area and hydrophobicity expressed as free energy of transfer between organic solvent and water for various hydrocarbons (*unlabeled dots*), and for amino acids. The accessible surface area is obtained by rolling a water molecule (sphere 1.4 Å) around the solute molecule and calculating the contact surface. The slopes of the lines are 25 cal Å^{-2} for hydrocarbons and polar amino acids and 22 cal Å^{-2} for nonpolar amino acids [137]

a factual model for this local ice-like structure, and in consequence, earlier concepts envisioned four-coordinated water molecules forming connected nets with a preponderance of edge-sharing pentagons [140, 141].

However, recent theoretical work using Monte Carlo simulations do not support this idea. The water networks resulting from these computations involve 3, 4, and 5 coordinated water molecules forming irregular patterns in which the pentagon is not especially conspicuous [142–145].

Three, four, and five coordinated water molecules are a feature of the hydrates of the small biological molecules described in Part IV, Chap. 22. At present, there seems to be little support for the view that the structure of water at a hydrophobic/hydrophilic interface is significantly different from that at the surface of pure water.

The hydrophobic effect stabilizes the three-dimensional structures of macromolecules. In the nucleic acid double helical structures, the hydrophobic bases are stacked along the helix axis and shielded from solvent by the hydrophilic sugar-phosphate backbone, which is heavily hydrated. A comparable scheme is found in many crystal structures of nucleosides and nucleotides, where the bases are stacked

Table 2.11. Hydrophobicity scale for amino acids based on partition of residues between protein surface and interior. Data in kcal mol^{-1} give transfer free energies surface/interior, $\Delta G_{tr,s/i}$ [138]

Residue	$\Delta G_{tr,s/i}$	S.D.
Ala	0.20	0.06
Arg	−1.34	0.25
Asn	−0.69	0.12
Asp	−0.72	0.11
Cys	0.67	0.10
Gln	−0.74	0.15
Glu	−1.09	0.17
Gly	0.06	0.06
His	0.04	0.12
Ile	0.74	0.08
Leu	0.65	0.07
Lys	−2.00	0.30
Met	0.71	0.14
Phe	0.67	0.09
Pro	−0.44	0.12
Ser	−0.34	0.08
Thr	−0.26	0.09
Trp	0.45	0.13
Tyr	−0.22	0.10
Val	0.61	0.07
N terminus	−1.25	0.32
C terminus	(−2.0)	

to form a hydrophobic section clearly separated from the hydrophilic section formed by sugar/phosphate moieties and by cations and/or water of hydration molecules.

In globular proteins, the folding of the polypeptide chain is such that the amino acids with nonpolar side chains are assembled in the interior to form a hydrophobic core, whereas the amino acids with polar and charged side chains tend to be at the surface to interact with the (aqueous) solvent. This oil-drop-like distribution of hydrophilic and hydrophobic amino acids is of importance for the functionallity and stability of a protein because pK values of acidic and basic side chains can be shifted in nonpolar environment by several units, and internal hydrogen bonds are strengthened because the donors and acceptors do not have to compete with water molecules [133, 134].

Formation of membrane structures is also due to the hydrophobic effect. When a molecule is composed of both polar and nonpolar moieties, it is said to be *amphiphilic*. Since the hydrophobic effect is concerned with the tendency for nonpolar molecules or nonpolar components of molecules to associate spontaneously in aqueous solutions, there is a wide range of small molecules such as fatty acids and phospho- and glycolipids which, when dissolved in water, segregate to form membranes or micelles [146]. The same type of segregation occurs in the crystalline state, as in the structures of the long chain alkyl glycosides [147], cerebrosides, and the glycosphingolipids and phospholipids [148–150], in which the hydrophilic moieties are hydrogen-bonded in layers, while the hydrocarbon chains associate by van der Waals forces. When the crystals melt or are dissolved in polar or nonpolar solvents, this segregation persists, forming the molecular clusters that cause the appearance of thermotropic or lyotropic liquid crystal behavior, and the aggregation in the form of membranes or micelles [151, 152].

Chapter 3
Experimental Studies of Hydrogen Bonding

Hydrogen bonding is a phenomenon that can be recognized and studied by almost all the methods of physical chemistry, spectroscopy, and diffraction. The anomalies in the physical properties of gases, liquids, solids, and solutions having hydrogen-bonding functional groups played the significant role in the recognition of hydrogen bonding as an intra- or intermolecular cohesive force which is distinctively different from a van der Waals interaction [28, 42, 46, 153 – 155] (Box 3.1).

In liquids and in solutions, hydrogen bonding gives rise to *self-association* between like molecules [34] and *mixed association* between unlike molecules. If one molecular species is a solvent, this is referred to as solvation [33]. These interactions are dynamic in nature so that any one-time snapshot of structure is different from another. In contrast, the hydrogen bonding in solids results in long-range patterns which are persistent and reproducible. In crystals, these patterns are periodic in three dimensions. The individual molecules are associated into infinite macromolecules through the structure of the hydrogen bonds.

These long-range association patterns are also controlling factors in the interaction between biological macromolecules and small substrate molecules, i.e., drugs which inhibit or modify biological activity. Only the single crystal X-ray or neutron diffraction methods can provide the complete metrical description of these architectures, which is as necessary a base for understanding biological activity as it is for chemistry. For this reason, the main emphasis later in this chapter

---

Box 3.1. Phenomenological Manifestations of Hydrogen Bonding

Deviations from Raoult's law
Abnormal melting and boiling points
Lowering of solubilities
Complex formation in freezing point diagrams
Abnormal enthalpies of mixing
Abnormal dipole moments
Abnormal ionization constants for acids
Excess viscosity
Vibration spectroscopy: frequency and intensity changes in infrared and Raman spectra
Nuclear magnetic resonance spectroscopy: large ^{1}H chemical shifts, small ^{13}C chemical shifts
Internuclear H···A distances: less than van der Waals, but greater than covalent or ionic

will be on the determination of the positions of hydrogen atoms by diffraction methods.

3.1 Infrared Spectroscopy and Gas Electron Diffraction

Prior to 1965, the principal methods for studying hydrogen bonding were spectroscopic, particularly infrared, Raman, and NMR spectroscopy. The shift to lower frequencies and the broadening of the infrared X–H stretch-absorption bands resulting from a weakening of the force constant due to hydrogen-bond formation was a major diagnostic tool. The reduction of the X–H bending motion on hydrogen bonding also causes changes in frequencies, but these are smaller and not necessarily accompanied by changes in band intensities.

Infrared spectroscopy has the advantage that it can be applied to the same compound in all three states of matter. However, there still appears to be insufficient data from gas-phase studies of hydrogen-bonded complexes to permit a correlation between X–H stretching frequencies and bond lengths, so as to improve on the long-standing empirical Badger's Rule [158].

The O–H bending frequencies in crystalline hydrates have a distribution curve about $1620\,\text{cm}^{-1}$ that is similar in form to that of the H···O bond-length distributions. Extensive studies have been made of the effect of hydrogen bonding on the vibrational spectra of water molecules in crystalline hydrates [95, 157]. By deuterium substitution, the fundamental frequencies of all nine degrees of freedom of the water molecules could be assigned in favorable cases. Most of this work was directed toward inorganic salt hydrates with very little, or no attention, given to the many hydrates formed by the small biological molecules discussed later.

Similarly, the down-field chemical shift of the proton resonance in H-NMR spectroscopy, which is a diagnostic tool for hydrogen bonding in solution, might be related to crystal structural data through the recent development of solid-state NMR spectroscopy.

Spectroscopic data provide only qualitative information. Unfortunately, the hydrogen bonding in most biological molecules, except water and the simpler monosaccharides and amino acids, is too complex to permit other than qualitative results from spectroscopic data. For any particular class of hydrogen-bonded complexes, it is possible to use infrared spectroscopy measurements to order the hydrogen-bonding strengths of donor groups qualitatively. For example, the studies of the alcohol-amine complexes in the gas phase show that for any given alcohol, the hydrogen-bonding strengths for a R_1R_2 group increase in the sequence $NH_3 < CH_3NH_2 < C_2H_5NH_2 < (CH_3)_2NH < (C_2H_5)_2NH$ [158]. These conclusions correlate well with the bond-length sequences reported for R_1R_2NH ···O=C bonds [75]. Other hydrogen-bonded complexes have been studied in the gas phase by this and related methods [158–163].

Empirical correlations have long been sought between changes in stretching frequencies due to hydrogen bonding and hydrogen-bond lengths. Due to lack of neutron diffraction or high-precision low-temperature X-ray analyses, these correlations have been with the nonhydrogen atom separations rather than the actual hydrogen-bond lengths [164–169]. Recent experiments [170], possibly using the derivatives of the infrared spectra [171–174], suggest that coordinated experiments involving both the spectroscopy and high-precision X-ray or neutron analysis on carefully selected crystals could provide better hydrogen-bond length/spectral shift relationships for the different types of hydrogen bonds. Such experimental data would be very useful for correlation with theory.

An interesting example where infrared O–H frequencies were used to correlate structures is for choline chloride dihydrate, which is postulated to have a semiclathrate hydrate structure by analogy with the known crystal structure of tetraethyl ammonium fluoride pentahydrate [162].

Solid-state infrared spectroscopy has also been used to identify nonbonded or weakly bonded hydroxyl groups in carbohydrate crystal structures. In general, it is not possible to resolve the stretching frequencies of O–H bonds in polyhydroxy compounds, since they coalesce into a broad band in the region of 3000–3600 cm^{-1}. However, in those crystal structures where a hydroxyl does not form a hydrogen bond (as in the crystal structure of the disaccharide turanose), or where it forms a very weak bond (as in β-D-fructopyranose and β-L-arabinose) (see Part II, Chap. 13), the shift in frequency to shorter wave lengths is clearly distinguishable. Again, this is a useful diagnostic tool, but theoretical correlation with bond geometry or energy is not straightforward [162a].

There is no direct experimental relationship between hydrogen-bond lengths and hydrogen-bond strengths (see Part I, Chap. 2.3). It seems axiomatic that stronger hydrogen bonds will have shorter bond lengths than weaker bonds. It is certainly true that the very strong bonds as a class have shorter bond lengths than the moderate or weak bonds formed between biological molecules. However, for a particular type of donor or acceptor group, the correlation is not straightforward. Bond energies are derived from thermochemical data such as virial coefficients, heats of fusion, and transport properties such as viscosity, thermal conductivity, and spectroscopic data. Hydrogen-bond lengths come solely from diffraction experiments, with the exception of a few complexes of simple molecules which are stable in the gas phase and can be studied by microwave spectroscopy, or high-resolution infrared spectroscopy [163].

The connection between energies and geometry must be empirical or indirect through theoretical hypotheses and calculations. This connection is further complicated by the other intermolecular interactions in crystals. The hydrogen-bond lengths observed in crystals, which provide the vast majority of the data available, are modulated by the crystal field effects and any one observation is therefore not necessarily representative of the potential energy minima of the isolated hydrogen-bonded adducts.

Some values for the enthalpy of hydrogen-bond formation are given in Table 2.5. The sequence of energies is in qualitative agreement with the bond lengths observed in crystals, $OH \cdots O=C < O_W H \cdots O_W < NH \cdots O=C < N(H)H \cdots N$, but no generally accepted quantitative relationships could be proposed.

Gas-phase studies of hydrogen-bonded complexes give data on hydrogen-bond geometry which are unperturbed by crystal field or solvation effects. The experimental methods available for studying hydrogen-bonded complexes in the gas phase are electron diffraction [175], pulsed-nozzle Fourier transform microwave spectroscopy [176, 177] and high-resolution infrared spectroscopy [178]. Gas-phase electron diffraction has the disadvantage that it is rarely possible to use all the structural features as variable parameters and the corrections for thermal motion reduce the precision. The other methods can provide high precision data, but very few of the complexes studied involve the moderate or weak hydrogen bonds that are so important in biological structure. These methods are most effective for the complexes formed by the strong or very strong hydrogen bonds [179].

The hydrogen-bonded dimers with biologically interesting functional groups that have been studied by these methods are the $O_W H \cdots O_W$ bond in the water dimer and the $OH \cdots O=C$ bond in the formic and acetic acids dimers. They are important because they refer more directly to the intrinsic properties of a particular bond and provide experimental data for comparison with the ab-initio molecular orbital calculations on simple systems described in Chapter 4.

3.2 X-Ray and Neutron Crystal Structure Analysis

Crystal structure analyses by X-ray and to a lesser extent, neutron diffraction, have become increasingly important to the biological sciences over the past decade. We discuss this experimental method in more detail since a principal objective of this monograph is to analyze the hydrogen-bonding geometries and patterns observed in the crystal structure analyses of the small monomer components of these macromolecules, where the hydrogen atom positions can be directly observed, and extrapolate these findings to the macromolecules.

For many years it was believed that hydrogen atoms "could not be seen" in the electron density maps produced by X-ray diffraction. The reason for this is that the atomic X-ray scattering power is proportional to the square of the atomic number. This statement was generally, but not invariably, true until the demise of film-recording methods in the mid-1960's and the advent of the computer-controlled X-ray diffractometers which could provide very accurate X-ray diffraction intensities.

Unfortunately, old habits do not necessarily change with new instrumentation. The crystallographer's attitude toward hydrogen atoms can only be described as "cavalier". The reason for this is that the purpose of the X-ray crystal structure analysis was, and still is, almost invariably to determine either molecular configuration or conformation. While a reasonable level of standardization has been developed to define a "satisfactory" determination of a crystal structure with respect to the nonhydrogen atoms [94, 180–182], there is no such uniformity in the way in which hydrogen atomic coordinates are derived or reported. The hydrogen atomic positions are determined in the final stages of an X-ray crystal structure analysis from inspection of "difference maps". These difference Fourier syntheses use $|F_{obs}| - |F_{calc}|$ as the structure amplitudes, where $|F_{calc}|$ excludes the hydrogen atoms. The calculated phases also exclude the hydrogen atoms and so are biased in the general non-centrosymmetrical cases. These include all biological compounds and derivatives thereof, except in the rare cases of D,L mixtures or isomers with meso configurations. These difference syntheses are calculated, generally, with isotropic temperature factors on the nonhydrogen atoms, since anisotropic factors would adjust to fit the hydrogen electron density. Hydrogen atoms are fitted to the peaks of the difference synthesis when the positions are consistent with the nonhydrogen stereochemistry. In room-temperature analyses of reasonable quality, hydrogen atoms attached to carbons can generally be observed, whereas those attached to oxygen or nitrogen atoms may not. At liquid nitrogen temperatures, the thermal motion of the hydrogen atoms is reduced and the atoms are usually located, except when there is orientational disorder. The precision with which hydrogen atoms are located from X-ray difference syntheses is low, ~ 0.1 to 0.3 Å, compared with ~ 0.005 Å for nonhydrogen atoms. This is improved if the diffraction measurements are made at liquid nitrogen temperature, ~ 120 K. But, in general, when the coordinates can be derived from the nonhydrogen positions, they are more reliable than those derived directly from the difference synthesis.

A variety of levels of reporting hydrogen atom positions can be found in the literature. These are (1) no hydrogen atomic coordinates reported (in some journals, deposition is required); (2) a selection of hydrogen atomic coordinates are reported. These could be only those which could be calculated from the nonhydrogen positions, or only those observed on the difference maps, or a combination of both; (3) all hydrogen coordinates reported (especially in more recent publications). Here again, practices differ. Some authors prefer to report calculated positions when they can be calculated, and observed positions only when they cannot be calculated.

In the surveys of the hydrogen bonding in the crystal structures of the purines and pyrimidines [61] and the nucleosides and nucleotides [62], less than half and one quarter respectively of the crystal structures surveyed were judged to provide reliable positions for all the hydrogen atoms involved in hydrogen-bonding functionality. The larger molecules generally provide poorer crystals and, consequently, poorer data.

As the use of low-temperature techniques becomes more routine, the quality of the analyses in this regard should improve, but this will be affected by a trend toward the study of more complex molecules where the diffraction quality of the crystals and the observation-to-parameter ratio becomes less favorable for locating the hydrogen atoms.

The refinement procedures for hydrogen atoms also vary. Isotropic temperature factors are assigned to the hydrogen atoms, but are generally not refined. They are fixed at the isotropic equivalent values of the anisotropic factors for the atom to which the hydrogen is covalently bonded, or, more appropriately, twice those values.

Sometimes the hydrogen positional coordinates are refined with those of the nonhydrogen atoms in the final stages of the structure analysis and sometimes they are not. It is not unusual to omit from the publication the description of which procedure was used. There are even a few cases where the coordinates of the hydrogen atoms refer to a different symmetry-related molecule to those of the nonhydrogen atoms. Mathematically, this is of no consequence, but can mystify someone not familiar with space group theory.

In the report of a crystal structure analysis the same results are reported twice: once as fractional atomic coordinates and unit cell dimensions, and a second time

Box 3.2. Comparison of X-ray and Neutron Diffraction Single Crystal Analysis

X-ray diffraction	Neutron diffraction
X-rays available on demand from laboratory instruments	Neutrons available from national or international centers
Data collection time, a few days	Data collection time, a few weeks
Temperatures not generally available below 120 K	Temperatures down to 10 K conveniently available
Hydrogen atoms poorly located, especially O–H. Accuracy ~ 0.1 Å	Hydrogen atomic parameters comparable in accuracy to C, N and O, ~ 0.001 Å
Cannot analyze anisotropic thermal motion or disorder of H atoms	Analysis comparable to that of C, N, O
Small crystals can be used, ≈ 0.01 mm³, ≈ 0.01 mg	Large crystals required, 4 mm³, ≈ 5 mg
Number of variable parameters 9N+4H+1, where N is number of nonhydrogen atoms	For comparable observation/parameter ratio, need more observations. Number of variables 9N′+10, where N′ is number of all atoms and nine anisotropic extinction parameters are used
Not necessary to make deuterium substitution	If structure contains many hydrogens, deuterium substitution may be essential to reduce incoherent background scattering
Careful absorption corrections necessary	Absorption negligible, except for crystals containing B, Cd, Sm, Li (corrections for H advisable. $\mu/\sigma = 0.11$ cm² g⁻¹, for molecules with large H content)
Extinction generally not serious for organic compounds	Extinction serious and pervasive. Careful corrections necessary
Radiation damage can occur and must be monitored by repeating selected measurements	No radiation damage

as bond lengths, valence angles, torsional angles, and nearest nonbonded atom-pair distances. Since typographical errors can occur, a computer check for consistency between these two sets of data is always desirable.

Single crystal neutron diffraction is the optimum method for studying hydrogen bond geometry in crystals. However, this method has some operational disadvantages relative to single crystal X-ray diffraction. The pros and cons of these two methods are shown in Box 3.2. Modern X-ray crystal structure analysis is so rapid that the neutron analyses serve to complement the deficiencies of the X-ray method. Since the neutron scattering cross-sections (Table 3.1) vary over a narrow range compared with the X-ray scattering factors, which increase with atomic number, neutron diffraction is most effective for determining the positions of light atoms (i.e., H, D) in the presence of heavy atoms (e.g., C, N, O, P, S, Cl, etc.). It is also useful for distinguishing between atoms which are adjacent in the Periodic Table where there is only a small proportional difference in the X-ray scattering factors. Because the nucleus is very small compared with the neutron wave lengths (~ 1.0 to 1.5 Å), the scattering power does not fall off with scattering angles as in X-ray diffraction. This makes single crystal neutron analysis particularly effective for

Table 3.1. Neutron coherent scattering cross-sections b and neutron absorption coefficients μ/σ ($\lambda = 1.08$ Å) for elements found in biological molecules. [183]

	b (10^{-13} cm)	μ/σ (cm^2 g^{-1})		b (10^{-13} cm)	μ/σ (cm^2 g^{-1})
H	-3.741	0.11	Li	-2.03	3.5
D	6.67	<0.01	Na	3.63	<0.01
C	6.649	<0.01	K	3.71	0.02
N	9.38	0.05	Mg	5.38	<0.01
O	5.81	<0.01	Ca	4.90	<0.01
P	5.13	<0.01	Fe	9.54	0.02
S	2.85	<0.01	Cu	7.69	0.02
Cl	9.58	0.33	Zn	5.7	<0.01

high-precision studies, since a good observation-to-parameter ratio can usually be maintained.

The fact that both hydrogen and deuterium have scattering power comparable to that of the other elements which occur in biological molecules makes this method important in the study of hydrogen bonding in these materials. The negative phase of hydrogen neutron scattering provides a valuable tool, since by adjusting the H/D ratio by substitution experiments, the scattering cross-section can be manipulated between the two extreme values.

The three major disadvantages of neutron diffraction for biological molecules are size of crystals, hydrogen incoherent scattering, and extinction. The more complex the organic molecule, the greater the difficulty of growing large single crystals. This accounts for the paucity of neutron diffraction data on the oligopeptides, oligosaccharides, nucleosides, and nucleotides, whereas the amino acids [60, 100] and the pyranoses and methylpyranosides [57, 58] have been studied more extensively. This problem could be overcome by increasing the flux from the nuclear reactors, but this presents severe technical and sociological obstacles.

Since the number of hydrogen atoms in most biological molecules equals or exceeds the number of nonhydrogen atoms, the large incoherent neutron-scattering cross-section for hydrogen is a disadvantage, since it adversely affects the signal-to-noise ratio. For water molecules and hydrogen-bonding functional hydrogens, this problem can be overcome by deuterium substitution, but this method has problems which are discussed later.

Extinction, which is the failure of the kinematic scattering theory ($I_{hkl} \propto F_{hkl}^2$), is only a minor problem in X-ray diffraction. In neutron diffraction, extinction is serious and pervasive throughout the whole data, as shown by the examples in Table 3.2. The best methods available for extinction correction require careful measurement of crystal dimensions. Although somewhat empirical, it has proved to be very effective [184, 185]. At least one, and sometimes six, additional extinction parameters, g_{iso} or g_{ij}, have to be added to the variable parameters. Uncertainty in the validity of these extinction parameters appears to have very little effect on atomic positional coordinates, but may influence the absolute values of the atomic temperature factors. This is important in charge density or electrostatic potential

Table 3.2. Extinction and refinement data for some neutron diffraction precision structure analyses[a]

Compound	Crystal dimensions (mm^3)	Extinction Max F_o/F_c[b]	$F_o/F_c \leqslant 0.95$[c]	g_{iso}[d]	g_{ii}/g_{jj}[e]	Obs/par ratio[f]	wR(F^2)	ESD (Å) Bond length[g] C–N	N–H
α-Cyano-acetohydrazide C$_3$H$_5$N$_3$O	6	0.55	30%		3	15.8	0.042	0.0006	0.0010
Carbonohydrazide (NH$_2$NH)$_2$CO	7.1	0.35	33%		7	16.4	0.046	0.0008	0.0010
Glyoxime C$_2$H$_4$N$_2$O$_2$	3.4	0.45	13%		54	14.8	0.044	0.0008	0.0012 (C–H)
1,2,4-Triazole C$_2$H$_3$N$_3$	3.5	0.65	2%	0.2		16.8	0.050	0.0007	0.0014
Formamide oxime CH$_4$N$_2$O	7.7	0.58	5%		2	15.6	0.023	0.0004	0.0008
γ-Aminobutyric acid C$_4$H$_9$NO$_2$ (122 K)	13.6	0.75			6	11.1	0.044	0.0010	0.0020
Parabanic acid C$_3$H$_2$N$_2$O$_3$ (123 K)	1.8	0.45				13.5	0.043	0.0010	0.0020
Monofluoroacetamide CH$_2$FCONH$_2$	4.8	0.63	22%		1	15.6	0.025	0.0004	0.0008
Thioacetamide CH$_3$CSNH$_2$	2.0	0.56	14%		43	11.9	0.047	0.0008	0.0014
N,N'-Diformohydrazide	3.5	0.85	5%	0.05		15.1	0.024	0.0004	0.0007

[a] For individual references see [186, 187, 222].
[b] Max F_o/F_c is the largest ratio observed between observed and calculated structure factors, i.e. the maximum extinction effect.
[c] $F_o/F_c \leq 0.95$ is the percentage of observed structure factors for which the F_o/F_c ratio is equal to or greater than 0.95.
[d] g_{iso} is the isotropic extinction parameter, in 10^4 rad^{-1}.
[e] g_{ii}/g_{jj} is the ratio between the largest and smallest values of the extinction parameter tensors and is a measure of the extinction anisotropy.
[f] Ratio of observed data to refined parameters.
[g] ESD is the estimated standard deviation obtained from the least squares refinement correlation matrix.

analyses using the X−N method, where precise correspondence between the X-ray and neutron temperature factors for the same compound at the same temperature is desirable. Some examples of crystal size, extinction corrections, and estimated standard deviations of the C−N and N−H bond lengths are given in Table 3.2.

Use of the Cambridge Crystallographic Data Base for studying hydrogen bonding. All the structural information relating to hydrogen bonding in the crystal structures of small biological molecules used in this monograph was obtained from the Cambridge Crystallographic Data Base [39]. These computer tapes contain the compound name, formula, bibliographical reference, REFCODE[1], and crystallographic data, including unit cell dimensions and atomic coordinates for all reported organic and organometallic crystal structures. Thermal motion parameters are not reported. These data are more reliable than the original papers because they are checked for inconsistencies with the reported molecular bond lengths, valency angles, and torsion angles. Where discrepancies are found, they are noted and, where possible, appropriate corrections are made. Where this is not possible, the original authors are contacted and those corrections are recorded as a comment. Neutron diffraction analyses are identified as such. In the organic section, the data are separated into classes. When structures in particular classes are being interrogated, it is generally useful to separate the classes onto discs rather than manipulate the complete tapes. The classes used primarily in this monograph are:

Class 44:   Purines and pyrimidines

Class 45:   Carbohydrates

Class 47:   Nucleosides and nucleotides

Class 48:   Amino acids and peptides.

Hydrogen-bond geometry from data reported in publications. For the purpose of displaying the connectivity of hydrogen-bond structure in crystals, the reference molecule at the particular position x, y, z in the unit cell is transformed by the appropriate symmetry operations given by the crystal space group to fill the entire unit cell. In addition, molecules are generated in adjacent unit cells so that the hydrogen-bonding distances and angles, together with the hydrogen-bonding connectivity or configuration, can be described using the usual distance and angle programs. This is illustrated in Table 3.3 for the crystal structure of methyl-α-D-xylopyranoside [MXLPYR][1]. It contains two molecules, A and A′, in the asymmetric unit each at x, y, z and at the symmetry equivalent position (1−x, $\frac{1}{2}$+y, 1−z). Since the data are based on an X-ray analysis, the X−H hydrogen bond lengths are normalized to the standard value 0.97 Å to provide for more realistic hydrogen-bonding geometry (Part IB, Chap. 6.3). The hydrogen-bonding pattern is converted to the connectivity diagram displayed graphically in Fig. 3.1. In this diagram, seven molecules are shown with a number of carbon and hydrogen atoms not involved in the hydrogen-bonding scheme omitted for clarity. As with

---

[1] REFCODES used in the text are listed with biographical information at the end of this book.

Table 3.3. Displays of hydrogen-bonding patterns in methyl α-D-xylopyranoside

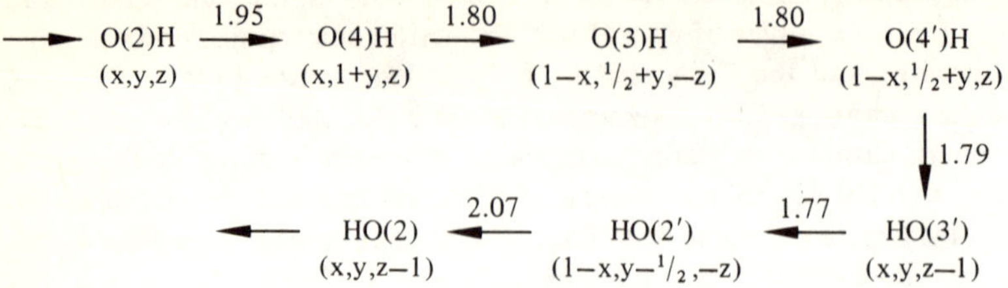

$$\text{O(2)H} \xrightarrow{1.95} \text{O(4)H} \xrightarrow{1.80} \text{O(3)H} \xrightarrow{1.80} \text{O(4')H}$$
$$(x,y,z) \qquad (x,1+y,z) \qquad (1-x,{}^1/_2+y,-z) \qquad (1-x,{}^1/_2+y,z)$$

$$\downarrow 1.79$$

$$\text{HO(2)} \xleftarrow{2.07} \text{HO(2')} \xleftarrow{1.77} \text{HO(3')}$$
$$(x,y,z-1) \qquad (1-x,y-{}^1/_2,-z) \qquad (x,y,z-1)$$

Molecule A

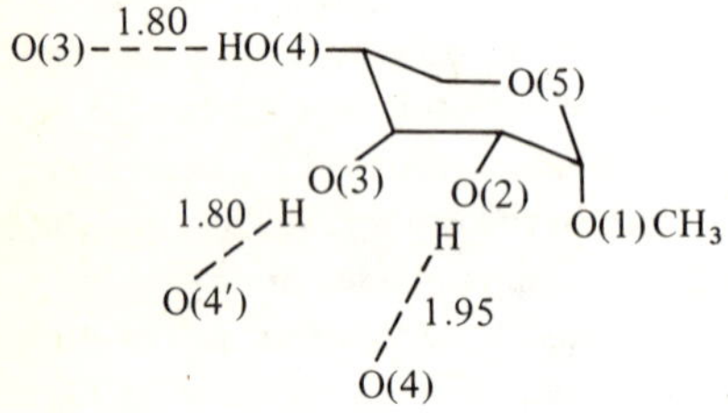

$$\text{O(3)} \cdots \underset{1.80}{\cdots} \text{HO(4)}$$

with O(5), O(3), O(2), O(1)CH₃, O(4'), H (1.80), O(4) (1.95)

Molecule A'

$$\text{O(3')} \cdots \underset{1.79}{\cdots} \text{HO(4')}$$

with O, O(3')H, O(2')H, O(1')CH₃, O(2') (1.77), O(2) (2.07)

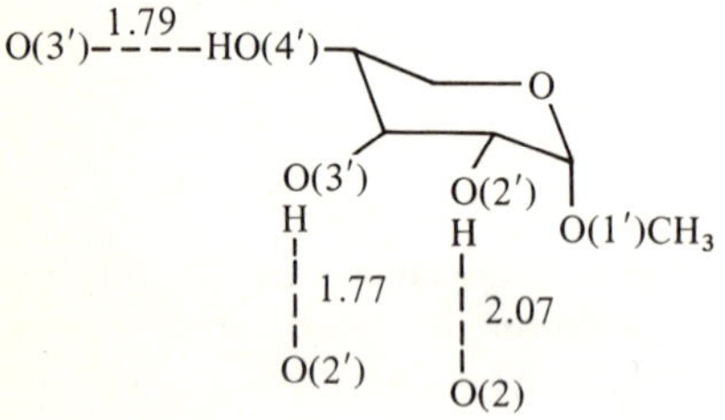

the arrangement of the molecules in the crystal, the hydrogen bonding has a periodic structure, which forms an infinite winding chain in the direction of the crystallographic c-axis. It involves only the hydroxyl oxygen atoms as acceptors with the ring O(5), O(5') and glycosidic O(1), O(1') oxygens excluded.

3.3 Treatment of Hydrogen Atoms in Neutron Diffraction Studies

Hydrogen atoms have special properties in neutron diffraction crystal structure analyses. These properties are both advantageous and disadvantageous. As shown in Table 3.2, the hydrogen nucleus has a negative scattering cross-section. In the low resolution scattering density maps frequently obtained with biological macro-

--→ O(2)H --→ O(4)H --→ O(3)H --→ O(4')H --→ O(3')H --→ O(2')H --→ O(2)H --→

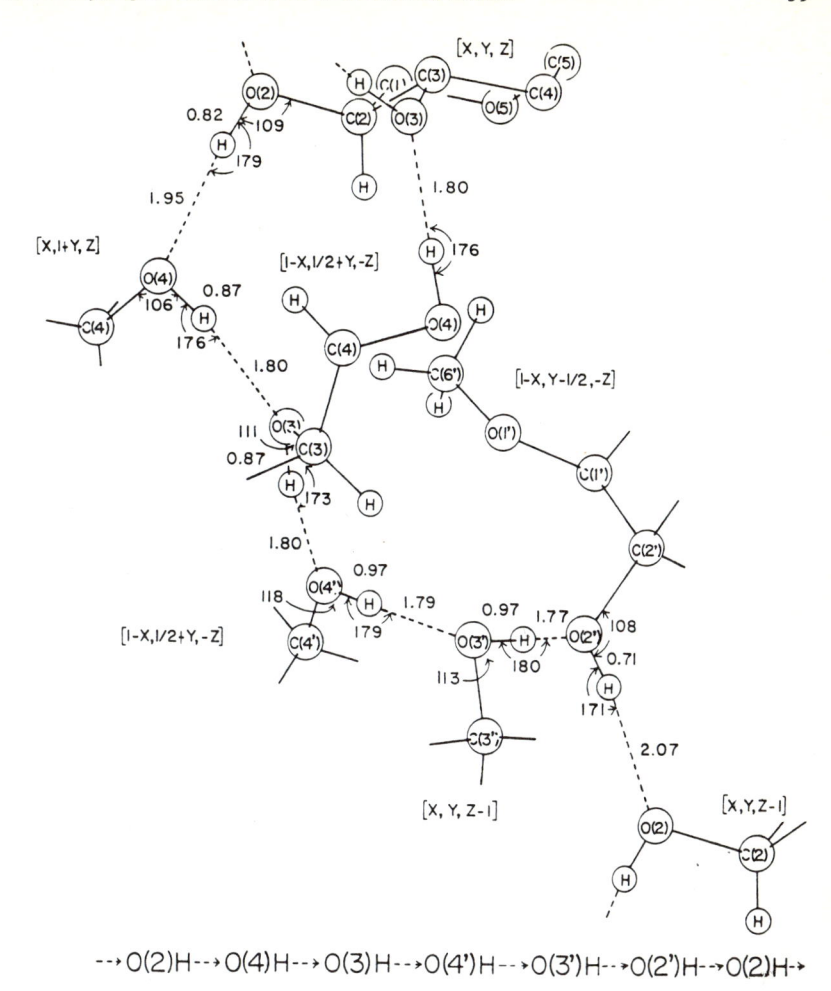

Fig. 3.1. The hydrogen-bond structure in the crystals of methyl α-xylopyranoside, illustrating the infinite chain of hydrogen bonds described in Table 3.3

molecules, this means that certain groups containing hydrogen atoms can almost disappear (e.g., $\rangle CH_2$, NH_2 and water, as shown in Table 3.4 [188]). For this reason, the neutron scattering radial distribution curve for liquid water is almost featureless and the water molecules "disappear" in neutron diffraction analyses of proteins. At very low resolution, the scattering of the five carbon atoms and eight hydrogen atoms in a deoxyribosyl moiety almost cancel.

Deuterium atoms are indistinguishable from carbon atoms and the difference from oxygen is not large. This is seldom a problem, however, since neutron diffraction crystal structural studies are always preceded by X-ray studies, and the configuration of the molecules is sufficiently well known to avoid any ambiguity.

Table 3.4. Coherent scattering cross-sections b (10^{-13} cm) scattering by atomic groups occurring in biological macromolecules, based on Table 3.1

H_2O	-1.67	$-OD$	$+12.48$
$-\overset{+}{N}H_3$	-1.84	$>ND$	$+16.05$
$>CH_2$	-0.83	D_2O	$+19.14$
$-CH_3$	-4.57	$-ND_2$	$+23.72$
$-NH_2$	$+1.90$	$-CD_3$	$+26.66$
$-OH$	$+2.07$	$-\overset{+}{N}D_3$	$+29.34$
$>NH$	$+5.64$		

To deuterate or not to deuterate? That is the question! A tempting way of overcoming the hydrogen incoherent background scattering problem mentioned earlier is to replace the hydrogen atoms by deuterium which has a low incoherent scattering cross-section. This requires special synthesis, except for the hydrogens in hydrogen-bonded functional groups and water molecules. Biological molecules are generally sufficiently soluble in water for deuterium substitution to be obtained by recrystallization from D_2O. However, 100% deuterium exchange by recrystallization is a slow and uncertain procedure requiring long equilibrium times and repeated recrystallizations. If the replacement is not complete, the crystal structure analysis is complicated by the necessity to determine individual atomic H/D occupation factors. This poses additional problems if high accuracy is sought, because the small differences in the X–H and X–D distances and stretching and bending frequencies causes difficulties in deconvoluting the combined Fermi density.

In protein neutron crystal structure analyses, complete H/D exchange is rarely obtained and the partial deuterium substitution has become a standard method of exploring the relative accessibility of hydrogen-bonding functional groups to solvent exposure.

Neutron diffraction determines nuclear atomic coordinates whereas X-ray diffraction refers to the maxima of the electron density distribution. Except for hydrogen or deuterium atoms, the difference between the nuclear positions and the electron density peaks is observable only in very high-precision structure analyses. Even then it is only apparent for atoms such as oxygen, where the lone-pair electron density is not symmetrical about the nucleus, or for heteroatom multiple bonds. A systematic study of the differences in bond lengths from X-ray and neutron diffraction analyses of the same crystal structures gave very small values for $d_X - d_N$ for C–OH bonds, $+0.054(3)$ Å, whereas for C–H and O–H bonds, the differences were much larger and significant, i.e., $-0.096(7)$ and $-0.155(10)$ Å [189]. It is for this reason that the X-ray X–H bond lengths were normalized to standard (neutron diffraction) values for the hydrogen-bond length correlations (see subsequent chapters).

In neutron diffraction structure analyses, the correction for thermal motion is much more complex for hydrogen atoms than for nonhydrogen atoms. It has long been recognized that vibrational thermal motion causes an apparent reduction in interatomic distances relative to those for the atoms at rest, and the methods for thermal motion analysis and geometry correction are well developed [190–192].

Fig. 3.2. Thermal ellipsoids (at 99% probability) for 1,2,4-triazole by neutron diffraction at 15 K illustrating the relative thermal motion of hydrogen and nonhydrogen atoms. That of the hydrogen bonded H(1) is only slightly less than that of H(3) and H(5), and the corrections of the X−H bond lengths are +0.005 Å for N−H versus +0.006 for the C−H bonds at 15 K [199]

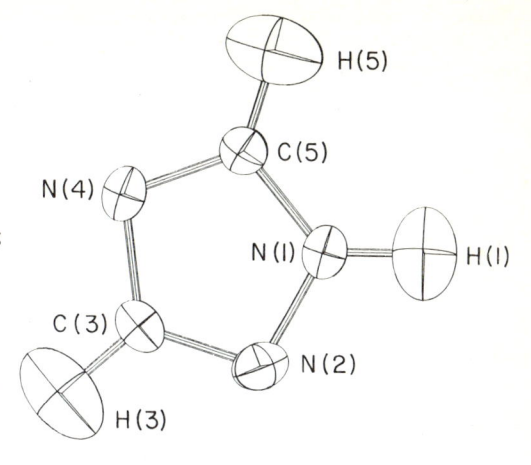

In very precise crystal structure analyses, where the objective is to compare the geometry of the molecule in the crystal with that calculated by ab-initio quantum theory, thermal motion corrections are applied to give "at rest" values. The vibrations of rigid molecules in crystals consist of translational and librational modes [193]. The librational motion can cause one atom to move in an arc relative to another. This is particularly so for hydrogen atoms and affects both the X−H and H···A bond lengths. Corrections are made by assuming that the motion is harmonic and that the molecules librate as a rigid body about their center of mass [194], or as rigid segmented bodies joined by flexible links [195]. For the bond lengths between nonhydrogen atoms, the thermal motion corrections are very small at low temperatures for hydrogen-bonded crystals, i.e., < 0.005 Å. At room temperatures, they can be as much as 0.1 Å, depending on the melting point of the crystals [196, 197].

For hydrogen atoms, the thermal motion is significantly larger because the X−H bending and stretching force constants are an order of magnitude smaller than for covalent bonds between nonhydrogen atoms [198]. As illustrated in Fig. 3.2, the hydrogen atoms "ride" on the nonhydrogen skeleton. For this reason, it is common practice in X-ray analyses to assign isotropic temperature factors to the hydrogen atoms which are twice the isotropic equivalent factors for the nonhydrogen atoms to which they are covalently bonded.

In neutron diffraction analyses, the anisotropic temperature factors of the hydrogen atoms are determined from the experimental data by least squares refinement in the same way as those for the nonhydrogen atoms. For nonhydrogen atoms, it is a reasonable assumption that the internal bending and stretching thermal motion is negligible compared with the overall molecular motion, hence the "rigid-body" concept. For precision analyses that are usually carried out at low temperatures, at liquid helium (∼ 10 K) or liquid nitrogen (∼ 120 K) temperatures, the corrections for thermal motion of the nonhydrogen atoms are small enough

Table 3.5. Riding motion and anharmonicity corrections for C−H and N−H bond lengths in the neutron diffraction analyses at 15 K of deuterated benzene [203], thioacetamide [204], and 1,2,4-triazole [199]

	Observed X−H	Wagging motion correction	Anharmonicity correction	Corrected X−H	Theoretical[a] X−H
Deuterated benzene					
C(1)−D(1)	1.0879(9)	+0.0130	−0.0098	1.0911	
C(2)−D(2)	1.0869(4)	+0.0130	−0.0119	1.0880	
C(3)−D(3)	1.0843(8)	+0.0127	−0.0113	1.0857	
			Mean	1.0878	1.087
Thioacetamide					
N(1)−H(1)	1.0256(14)[b]	+0.013	−0.020	1.019	0.999
	1.0263(14)	+0.016	−0.017	1.024	
N(1)−H(2)	1.0246(15)	+0.015	−0.019	1.021	1.000
	1.0231(14)	+0.016	−0.017	1.022	
C(2)−H(3)	1.0903(17)	+0.024	−0.017	1.097	1.083
	1.0880(17)	+0.030	−0.018	1.100	
C(2)−H(4)	1.0903(16)	+0.026	−0.015	1.101	1.082
	1.0855(16)	+0.031	−0.020	1.097	
C(2)−H(5)	1.0887(16)	+0.025	−0.017	1.097	1.082
	1.0832(16)	+0.031	−0.016	1.098	
1,2,4-Triazole					
N(1)−H(1)	1.0478(14)	+0.003	−0.017	1.033	0.993
C(3)−H(3)	1.0855(16)	+0.002	−0.016	1.072	1.061
C(5)−H(5)	1.0858(16)	+0.002	−0.015	1.073	1.063

[a] MP2/6−31G* for deuterated benzene; HF/3−21G for thioacetamide and 1,2,4-triazole.
[b] First line, molecule A; second line, molecule B.

that including internal motion would result in further corrections of an order of magnitude less than the experimental standard deviations. This is not so for bonds involving hydrogen atoms [200−202].

In this case, the riding motion effect due to the soft bending force constants and the anharmonicity effect due to the soft stretching force constants result in corrections which are significant in terms of the experimental standard deviations. These corrections are (fortunately) in opposite directions. At very low temperatures, ~15 K, they tend to cancel, so that the difference is within the experimental errors, as shown in Table 3.5. At room temperatures, the riding motion correction is predominant.

Although these effects are small, they affect not only the covalent X−H bond lengths, but also the H···A hydrogen-bond lengths. For very high-precision low-temperature neutron diffraction analyses [186, 187], they have to be taken into account. An adequate way to do this, without the complexity of a complete anharmonic thermal motion analysis, is to use the experimentally determined anisotropic temperature factors. As shown by comparisons between experimental and theoretical X−H values from low temperature neutron analyses and theoretical ab-initio molecular orbital calculations, one can assume that the motion of the two atoms is uncorrelated. In this case, the riding motion can be corrected from the difference of the mean square amplitudes of the atoms X and H *perpendicular* to the direction of the X–H bond, i.e., $\perp U_X^2$ and $\perp U_H^2$, using the formula

$$(X-H)_{corr} = (X-H_{obs}) + \frac{(\perp U_X^2 + \perp U_H^2)}{2(X-H)_{obs}} .$$

The correction for stretching anharmonicity can be made in a similar way using the differences in the mean square amplitudes *parallel* to the direction of the X–H bond using the leading term in the correction for a Morse-curve oscillation [205–207]

$$(X-H)_{corr} = (X-H)_{obs} - \tfrac{3}{2} K (\| U_H^2 - \| U_X^2) ,$$

where K is an arbitrary constant for each type of X–H bond. In practice, a value of 1.98 has been used, which is derived from the experience of gas-phase diffraction, where the effect is much more serious [206, 208]. Table 3.6 shows the corrections for some N–H bonds which form hydrogen bonds and C–H bonds which do not. The values are compared with the values "at rest" calculated by ab-initio quantum mechanics for the isolated molecules. At 15 K, the riding and anharmonicity corrections almost cancel [186, 187]. Similar corrections could be applied to the hydrogen-bond lengths, H···A, but to the best of our knowledge this has not been done. A detailed analysis of the thermal parameters of the water molecule in a number of crystalline organic hydrates studied by neutron diffraction showed a positive correlation between H···O hydrogen-bond distances and the thermal motion parameters; the shorter the distance, the smaller the difference in the hydrogen and oxygen librational motion [209].

More complex mathematical treatment is necessary when the thermal motion is very large, as for hydrogen atoms in a room-temperature neutron structure analysis, or when it is curvilinear as in a hindered-rotor. The U_{ij} second-rank tensor does not adequately describe the nuclear or electron-scattering density when the motion is far from harmonic or when it deviates from the familiar ellipsoidal probability form. To deal with such examples, more complex mathematic expressions using Gram Charlier or Edgeworth expansions are available [210].

However, these refinements are made at a heavy cost in the observation-to-parameter ratio. The third-order terms, c_{ijk}, of the Gram Charlier expansion add ten more parameters per atom to the nine U_{ij} terms. These expressions are therefore only used when the experimental data are of exceptionally high quality, as in the neutron diffraction analysis of ice, Ih, discussed in Part IV, Chapter 21. They may also be necessary in experimental deformation density analysis, where a very precise description of the atomic thermal motion is required.

3.4 Charge Density and Hydrogen-Bond Energies

Deformation density and electrostatic potential measurements are used to study the electronic structure associated with hydrogen bonding. These methods seek to

> **Box 3.3.** Electrostatic Properties of Molecules Derived from Fourier Summations of Crystal Structure Factors, F_H [211]
>
> Electron density distribution
>
> $$\varrho(r) = (V)^{-1} \sum_H F_H \exp(-2\pi i H \cdot r)$$
>
> Electrostatic potential
>
> $$\phi(r) = -(4\pi V)^{-1} \sum_H F_H \exp(-2\pi i H \cdot r)(\sin\theta/\lambda)^{-2}$$
>
> Electric field. The gradient of the electrostatic potential. A vector with three components, e.g.,
>
> $$E_z = -(i/V) \sum_H F_H(H_z/H) \exp(-2\pi i H \cdot r)(\sin\theta/\lambda)^{-1}$$
>
> Electric field gradient. A second-rank tensor, a component of which is
>
> $$q_{zz} = -(4\pi/3V) \sum_H F_H(3H_z/H)^2 \exp(-2\pi i H \cdot r)$$
>
> For full definition of the Fourier series, see [212].

separate the electron density observed by X-ray diffraction in a crystal from that associated with an assemblage of free atoms in the same configuration (see Box 3.3). There are basically two methods, X–N and X–X.

In the X–N method, the experimental electron density is determined from the X-ray diffraction data. The free atom electron density is determined by placing theoretical free atom electron densities at the atomic nuclear positions determined by a neutron diffraction experiment on the same crystal structure at the same temperature, albeit with a crystal of larger size. In the X–N method, it is frequently found that the atomic thermal parameters determined by neutron diffraction do not agree with those determined by X-ray diffraction, even though the experiments were carried out under identical conditions. This requires the introduction of an empirical scaling factor between the two sets of data, which is effective but disconcerting.

In the X–X method, the same result is obtained as in the X–N procedure by making use of the fact that the high 2θ X-ray intensities measured with short wavelengths, i.e., AgKα = 0.560 Å, are determined only by the inner core, 1s, electrons for first row elements. This method has the obvious advantage that all measurements are made on the same crystal. The tuneable wavelength property of synchrotron X-rays is useful for this type of experiment.

The difference between the observed electron density and that from the constructed free-atom density is the deformation density or charge density. Several procedures are used for analyzing this charge density, with their associated computer programs [213–216].

From the detailed charge density distribution, the electrostatic potential and the molecular dipole moment can be obtained. This involves partitioning the charge density into a sum of pseudo-atom densities [213, 217] which consist of a Hartree-Fock core with a series of electron density deformation terms. These terms are centered on the nucleus and have variable weights (electron populations). In one method, there is a radial Slater-type function and a series of surface harmonics which are odd or even terms of Legendre functions. These functions correspond to monopoles, dipoles, quadrupoles, and octapoles, with one, three, five, and seven deformation terms per atom. A least squares refinement of $\Delta F = F_{calc} - F_{obs}$ or some function thereof then involves, in addition to the normal crystal structure analysis, the electron population parameters for each of these 16 deformation terms [217]. The parameters are then used to map the electrostatic potential in sections through the molecule [218].

Very precise experimental data are necessary for a significant charge density analysis. This is clear from the number of variable parameters involved in the least-squares refinement and the fact that the experiment seeks to measure quantities which represent the small differences between two large electron distributions; that of bonded and nonbonded atoms in the same atomic configuration. Extinction becomes a more serious source of error, particularly as it affects the low-order reflections, i.e., those sensitive to the valence electrons. More sophisticated treatment of this effect may be necessary [219, 220]. One distinct disadvantage of this method for studying the electronic structure is that it is most effective with centrosymmetrical crystal structures, where the structure factor phase angles are integers of π, whereas most biologically interesting crystal structures are chiral. The electron density distribution which is responsible for the apparent short X–H covalent bond lengths and correspondingly long H···A hydrogen-bond lengths is clearly observed in charge density maps. As shown in Fig. 3.3 these maps also show

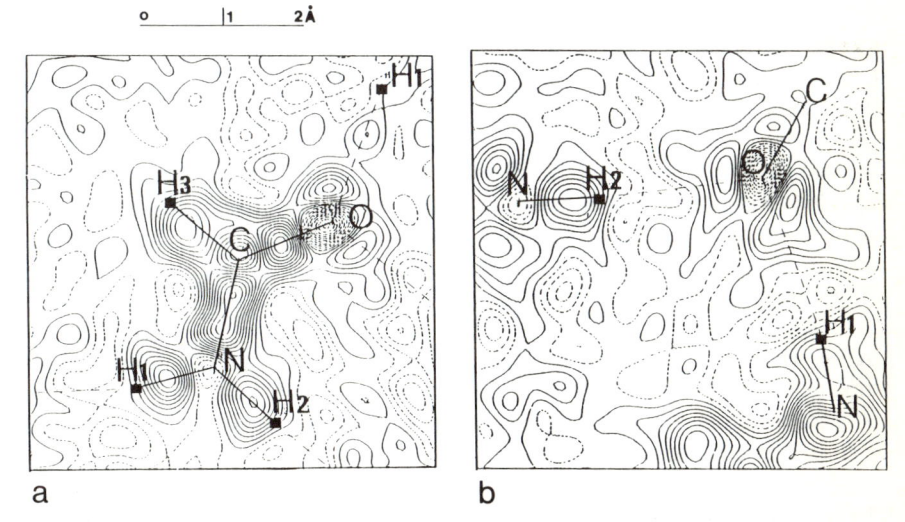

Fig. 3.3. a Experimental deformation density map for formamide in the plane of the molecule. **b** Experimental deformation density map for the hydrogen bond in the H(1), H(2) and O plane (contour at $0.01\,e\text{Å}^{-3}$, lowest contour $0.04\,e\text{Å}^{-3}$) [221]

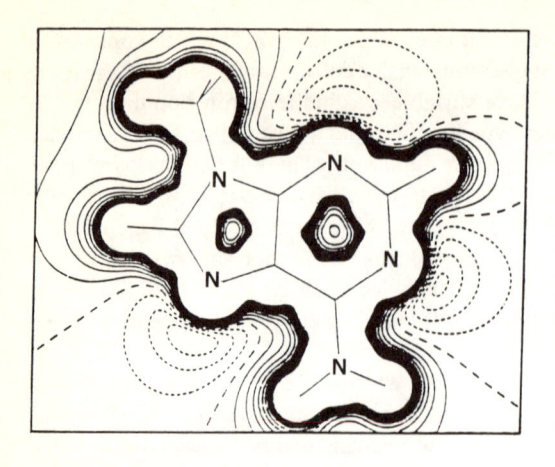

Fig. 3.4. Experimental electrostatic potential map for the molecule of 9-methyladenine removed from the lattice. Note the marked difference between the electrostatic potentials around N−H (donor), ⟩N (acceptor), and C−H (contours ±0.05 e Å^{-1}) [218]

qualitatively other expected features. The concentration of electron density in the X−H bond is accompanied by a smaller electron deficiency at the X atom and a more diffuse electron deficiency on the donor side of the hydrogen nucleus. There is pronounced deficiency of electron density at the acceptor nucleus, which is accompanied by increases in the electron density in the direction of the hydrogen bonds. All deformation density maps of crystal structures involving hydrogen bonds show these general features, to a greater or lesser degree.

The distinction between hydrogen-bond donors, i.e., N−H, and nondonors, i.e., C−H, is often not apparent in deformation density maps. This distinction appears much more clearly on the electrostatic potential maps, such as illustrated in Fig. 3.4 [218]. Such maps may therefore provide a more effective means of quantitatively analyzing the electronic differences between the different donor-acceptor hydrogen-bond combinations which is manifested by the different mean bond lengths described in Part IB, Chapter 7 [222–226]. It has been suggested that hydrogen-bond strengths can be at least qualitatively compared from the values of the electrostatic potentials at fixed distances from the donor and acceptor atoms, i.e., ~ 2.0 Å [227].

Deformation density maps have been used to examine the effects of hydrogen bonding on the electron distribution in molecules. In this method, the deformation density (or electrostatic potential) measured experimentally for the hydrogen-bonded molecule in the crystal is compared with that calculated theoretically for the isolated molecule. Since both the experiment and theory are concerned with small differences between large quantities, very high precision is necessary in both. In the case of the experiment, this requires very accurate diffraction intensity measurements at low temperature with good thermal motion corrections. In the case of theory, it requires a high level of ab-initio molecular orbital approximation, as discussed in Chapter 4.

Hydrogen-bond energies can be calculated from the results of high precision charge-density X-ray crystal structure analyses. The electron population parameters obtained from high precision low temperature X-ray and neutron Bragg diffraction measurements have been used in a model for molecular interactions to calculate the hydrogen bond energies in the crystal structures of several biological molecules; i.e., imidazole, 9-methyladenine, cytosine monohydrate and urea [93].

The model used expresses the interaction energies between two hydrogen bonds by four terms [228], $E_{total} = E_{es} + E_{pen} + E_{rep} + E_{disp}$. The first two terms are classical electrostatic. E_{es} is the electrostatic interaction energy between the atomic multipole moments of the two deformation densities and is attractive. E_{pen} is the interaction energy of the deformation multipoles of one molecule with the spherical atomic terms of the other and is repulsive. The second two terms are the short-range atom-to-atom repulsion and attraction dispersion terms which can be calculated from a Lennard-Jones or Buckingham potential. An interesting feature is that it is necessary to omit from these terms interactions involving the hydrogen atom involved in hydrogen bonding. When applied to the water dimer using a theoretical electron distribution, an equilibrium $O_W-H\cdots O$ of 1.70 Å and an energy of 25.4 kJ mole^{-1} was obtained [229], as compared with the experimental values of 2.02(1) Å and 22.6(2.9) kJ mole^{-1}. Since this model permits the calculation of bond energies from the same experimental data as is used to determine the hydrogen-bond geometry, it offers a means of direct correlation between geometry and energy. Unfortunately, even with very precise experimental measurements, the level of accuracy in the bond energies is low, with estimated standard deviations of about 30% of the energies.

3.5 Neutron Powder Diffraction

Powder neutron diffraction is a potentially useful method for studying hydrogen bond structure involving simple molecules. Powder diffraction has three advantages over single crystal analysis; single crystals are not necessary, crystal twinning is not a problem, and extinction is less. The advantage of neutron powder diffraction is that it can be used for structure analysis of simple molecules which are liquids at room temperature, where growing single crystals of neutron size and quality can be a serious obstacle. Hitherto, little use has been made of these advantages to study hydrogen bonding in organic crystal structures.

There are two types of neutron sources available for powder diffraction. One is the nuclear reactor, which provides a monochromatic beam of wavelength 1.0 Å, selected by means of a crystal monochromator from the continuous wavelength spectrum of thermalized neutrons [48]. The diffraction experiment uses the Bragg method as in X-ray single crystal diffractometry.

The other source is the continuous wavelength spectrum of neutrons produced by stopping an accelerated beam of electrons, i.e., the "spallation source". Since the electron beam is pulsed, so is the neutron beam [230]. The diffraction experiment uses the Laue method and the wavelengths are measured by their time of flight (TOF). In place of Bragg's law, $d_{hkl} = \lambda/2 \sin\theta_{hkl}$, the TOF relationship is

$$d_{hkl} = t_{hkl} \cdot \hbar/2m_n l \cdot \sin\theta_D ,$$

where $\hbar$ is Planck's constant, t is the time of flight in $\mu s^{-1} \cdot m^{-1}$, l is the flight path from source to detector and m_n is an instrumental constant, $2\theta_D$ is the angle between the incident beam and that of the detector at the sample. In convenient units,

$$d_{hkl}(A) = \frac{1.98 t_{hkl} \times 10^{-3}}{1 \cdot \sin \theta_D}.$$

Detectors are usually placed at 2θ values of $\pm 50°$, $\pm 90°$ and $\pm 150°$. Since air absorption of neutrons is negligible, the flight path can be several meters.

The TOF method has the advantage that diffraction data can be collected to higher values of $\sin \theta/\lambda$ than with the monochromatic beam method. However, with larger unit cells there is a resolution problem at short wavelengths, where the powder lines may form a continuum [230, 231].

In both methods of neutron powder diffraction, the structure analysis is by means of profile refinement. The method is known as Rietveld refinement from the name of the originator [232]. It has also been called the "born-again" method of powder diffraction analysis because of the dramatic impetus it gave to the long-established method of powder diffractometry. Because of the nature of the diffraction intensity profiles, it is mathematically simpler to apply to neutron diffraction than to X-ray diffraction. In fact, the original publication referred specifically to neutron powder diffraction. Instead of recording only the positions and intensities of the powder diffraction lines, Rietveld suggested that the analysis be based on measurements of the total scattering profile, made at small increments of the scattering angle, 2θ. The analysis of such a profile requires an analytical expression for the shape of the Bragg diffraction peak, which in the neutron diffraction case is known to be Gaussian. In X-ray diffraction, combinations of Gaussian and Lorentzian functions may have to be used, cf. [233]. In principle, this makes the X-ray powder data analysis less effective, but in practice there have been many successful applications of the method to both X-ray and neutron powder data.

The Rietveld profile refinement has been applied mainly to inorganic compounds of importance in material sciences. These analyses generally involve the refinement of the atomic parameters in crystal structures which are known from earlier analyses, or by analogy with known structures, or from stereochemical principles. The neutron diffraction method is especially useful for locating hydrogen atoms in a structure in which the nonhydrogen atomic coordinates are well-determined from X-ray methods. Ab-initio structure determination, which involves solving the phase problem, is generally beyond the scope of present facilities, but may be possible in the future with improvement of instrumental resolution [234].

The powder specimens are more easily subjected to controlled temperatures, down to 4 K, and different pressure than single crystals. Relatively large amounts of compound are required, several grams, and sample preparation to obtain a fine powder and avoid preferred orientation is critical.

For the first-row element organic molecules of primary interest in this monograph, the method has not yet achieved, and may never achieve, the level of accuracy provided by single crystal studies. However, for the location of hydrogen atoms in cases where suitable single crystals cannot be obtained, it provides an accuracy that is better than the single crystal X-ray analysis, as, for instance, in the two concurrent studies of the high pressure D_2O ice VIII [235, 236] by both monochromatic and TOF neutron powder diffraction, which are compared in Table 3.6.

Table 3.6. Comparison of refinement of structure of D_2O Ice VIII by neutron powder diffraction using constant wavelength and time-of-flight methods

	Ice VIII – Space group $I4_1/amd$	
	Bragg method [235] Constant λ, 10 K	Laue method [236] TOF at 269 K $2\theta = 60°$
Lattice parameters	a = 4.669(3) Å c = 6.810(4) Å	a = 4.6779(5) Å c = 6.8029(10) Å
Structural data: O – D (Å) O – D···O (Å) O – D···O (°) D – O – D (°) O(D)···O (Å) O···O (Å)	at 10 K 0.993(7) 1.916(10) 178.3(7) 105(1) 2.889(1) 2.710(9)	at 269 K 0.973(1) 1.920(10) 177(1) 104(2) 2.8919(3) 2.713(11)

3.6 Solid State NMR Spectroscopy

Single crystal NMR spectroscopy: A tool for the future? Since NMR chemical shifts are extremely sensitive to the electronic environment of the nuclei, they have long been used as diagnostic probes for hydrogen bonding in liquids and for examining the changes in molecular structure brought about by hydrogen bond formation. The two most common methods for biological molecules are ^1H proton and ^{13}C carbon NMR.

The proton chemical shift due to hydrogen-bond formation is one of the largest observed. This is not surprising, in view of the large asymmetry of hydrogen electron density with respect to the nucleus on bonding. As with bond energies, the only means of relating proton chemical shifts to hydrogen-bond lengths is indirect, through models and theoretical calculations. These calculations are even more complex than those relating bond lengths to energies.

The recent extension of the widely used NMR spectroscopy to solids was made possible by the technical advances of multiple-pulse and spin-enhancement methods [237, 238]. Using single crystal methods, it is possible to measure the complete chemical shift tensor, if the molecules are not too complex. From these, the isotropic equivalent chemical shifts can be derived and compared with those for the molecules in solution removed from the effects of crystal hydrogen bonding and other crystal field effects. Although the differences between the chemical shifts in solution and in the crystal are quantitatively small, they are large in terms of the sensitivity of the measurements. In consequence, a combination with diffraction methods could be very revealing if applied systematically to a group of closely related compounds.

It is also possible to measure the isotropic equivalent chemical shifts experimentally using magic angle spinning (MAS) on powder specimens. These experiments are much simpler than the single crystal measurements and require much less sample (~ 100 mg), but they are less revealing. These results can also be compared with the solution data.

From the point of view of hydrogen bonding and biological structure, the important nuclei are ^{1}H, ^{13}C, and ^{15}N. Unfortunately, nitrogen complicates the interpretation due to quadruple splitting [237], but the difficulties are not insurmountable.

Because of their importance in organic chemistry, the most commonly applied method is ^{13}C cross-polarization magic angle spinning (^{13}C-CP-MAS). This method addresses the effect of the crystal environment, including hydrogen bonding, on the electronic structure around the carbon nucleus. The combination of solution and solid-state NMR spectroscopy provides an extremely sensitive probe for studying the effects of solvation and crystal fields, including hydrogen bonding, on the structure of molecules.

Hitherto, there have been relatively few examples where solid-state NMR and crystal structure analysis have focused on the same molecule. The method can clearly identify different phases [239] and the presence of different molecules in the same phase, as in the crystal structure of lactulose, where three different configurational isomers could be identified [240]. In another carbohydrate, the presence of two symmetry-independent molecules in the crystal structure could be clearly observed by the ^{13}C-CP-MAS spectra of a powdered sample [241].

The ^{13}C-CP-MAS spectrum has also been used to complement the diffraction data relating to the crystal structure of cellulose, but hitherto the results have been inconclusive [242]. A significant amount of ^{1}H single crystal spectroscopy has been reported and has been used as a basis for correlation with neutron diffraction data and with theoretical ab-initio molecular orbital calculations [243, 244].

A critical analysis of the neutron data [245] suggested that well-designed experiments on single crystals suited for high precision in both fields could yield a very valuable correlation between proton chemical shifts and hydrogen-bond lengths.

Some general conclusions are that:

1. The shorter the hydrogen bond, the greater the chemical shift.
2. The shielding tensor is approximately symmetrical about the hydrogen-bond direction.
3. The most shielded direction is that of the hydrogen bond.
4. In a planar configuration, the least-shielded direction is perpendicular to the plane.

Unfortunately, specialists in the field of single crystal NMR spectroscopy are not specialists in single crystal neutron diffraction, and, as with solid-state infrared spectroscopy, good experiments using both methods on the same crystalline compound are relatively rare, cf. [246, 247].

Chapter 4
Theoretical Calculations of Hydrogen-Bond Geometries

4.1 Calculating Hydrogen-Bond Geometries

The calculation of reliable hydrogen-bond geometries is more difficult than for covalent bond geometries. One obvious reason for this is that the simplest relevant systems are necessarily more complex since they involve dimers, trimers, and higher polymers, instead of monomers. A second reason is that the weaker attractive forces which attenuate more slowly result in a potential energy surface near the minimum, which is much shallower than for covalent bonds. Consequently, relatively small changes in energy correspond to relatively large changes in bond lengths and bond angles. Since all methods seek the energy minimum, a higher accuracy of calculation is necessary to obtain the same accuracy in hydrogen-bond lengths and angles than can be obtained with a covalently bonded system of the same complexity.

There are three levels of theoretical calculations which can be used to predict intra- or intermolecular hydrogen bonding and its effect on molecular geometry and association.

1. Ab-initio molecular orbital calculations which use the Hartree-Fock self-consistent field theory with one-electron molecular orbitals. This method is based on the variation theorem to seek the nuclear geometry of the molecule or hydrogen-bonded complex with lowest energy [248–253]. It uses no experimental data.
2. Semi-empirical molecular orbital calculations, which are based on the same or related quantum mechanical principles, but make approximations or assumptions to simplify the computations, or include some empirical parameters based on experimental data [254–259].
3. Molecular mechanical calculations which are based on classical Newtonian mechanics [260–262], but use quantum mechanical concepts to formulate empirical equations. The parameters used are based entirely on experimental data.

From the point of view of hydrogen bonding, there is an important distinction between the first two and the third. The first two will seek structures with maximum hydrogen-bond energies, other factors being equal. The last will only include hydrogen bonding if the appropriate empirical functions and parameters are included in the methodology.

Calculations by ab-initio or semi-empirical methods on isolated molecules containing hydrogen-bond donor and acceptor groups will always predict the molecular conformations which have the maximum intramolecular bonding. For α-glucose, for example, this is a conformation with the hydroxyl groups oriented to form five intramolecular hydrogen bonds [263]. It is a hypothetical molecule, since glucose is not stable in the gas phase, or it is glucose in a nonpolar solvent in which it will not dissolve. The results from such calculations may therefore not be relevant to the chemistry or biochemistry of the molecules in solution or the prediction of molecular conformations in the solid state. Constraints may have to be applied to avoid intramolecular hydrogen bonding, when it is in competition with more favorable intermolecular bonding from adjacent molecules in the crystalline state or solvent molecules in solution. Similarly, in an amino acid, peptide, or nucleoside or nucleotide, the minimum energy conformation with maximum intramolecular hydrogen bonding may not necessarily be that found in solution or in the crystal.

The same problem applies to the third method, unless the method can be parameterized to exclude intramolecular hydrogen bonding when it is inappropriate. This presents an interesting dilemma in an oligosaccharide, for example, where parameterization for *inter-residue* intramolecular hydrogen bonding is desirable, whereas *intra-residue* intramolecular hydrogen bonding is not.

Global searching is a major problem in all methods. When these methods are used to investigate molecular structures, the atomic Cartesian coordinates are varied and the nuclear configuration of minimum energy is sought. In the general case, this involves global searching in $3(N-1)$ dimensions, where N is the number of atoms in the molecule or assemblage of molecules. In practice, chemical knowledge concerning atom connectivity is used to greatly simplify the problem. Even so, with molecules having orientational freedom involving more than three bonds, the global searching for the lowest energy minimum in the presence of a large number of minima of slightly higher energy can present a severe computational problem. All the methods used, for example, point-by-point search, steepest descents, Newton-Raphson method [264], converge on the energy well closest to that of the starting parameters. There is no other way for global searching than to start the calculations at a series of different atom coordinates. For the simpler molecules, molecular models such as the Pauling-Corey-Koltun models, or simple hard-sphere van der Waals calculations are frequently used to identify different starting points.

Global searching is more difficult for carbohydrates than for peptides, nucleosides and nucleotides. The C—OH bond has torsional freedom and therefore every C—OH···O hydrogen bond included in the model introduces an additional parameter and increases the number of local minima in the potential energy hypersurface. This is not so for $>$NH···O=C and $>$NH···N bonds, where there is no torsional freedom of the N—H bond and the hydrogen atomic position can be derived with reasonable precision from those of the nonhydrogen positions.

This problem is well illustrated by a molecular mechanics calculation on the disaccharide 6-O-β-D-glucopyranosyl-β-D-glucopyranose (β-gentiobiose) [265]. Assuming fixed standard geometry for the pyranose rings, the three inter-residue torsion angles and one linkage valence angle were used as variables to explore the possible molecular conformations. The force-field used did not include a specific hydrogen-bond potential, other than the electrostatic interaction terms based on assigned atomic charges. Twenty-four minima were reported over a calculated energy range of 30 kJ mol^{-1}. Each of the three individually observed linkage-bond torsion angles was included in the minima, but not the particular combination of the three that occur in the crystal structure, which was determined concurrently elsewhere [266]. Intramolecular hydrogen bonding is prominent in the gas-phase calculations, whereas in the crystal, intramolecular hydrogen bonding predominantly occurs as minor components of three-center bonds (see Part IB, Chap. 13). Inadequate accounting for these interactions could be the source of the problem.

In α-cyanoacetohydrazide, where there is both strong inter- and intramolecular hydrogen bonding and large dipole interactions, the conformer observed in the crystal is the eighth highest in calculated energy, with a difference of 37 kJ mol^{-1} over the minimum calculated by ab-initio molecular orbital methods [267].

As molecules become larger and more flexible, the complexity increases. In a global minimum energy study of cyclotetraglycyl, for example, 7776 original conformations (starting points) had to be examined to obtain 100 which were stereochemically reasonable [268]. The final outcome was consistent with relevant crystal structural data, but one wonders what the conclusions would have been had the crystal structural data not been available.

It is apparent from these examples that extreme caution is necessary when assessing the predictive power of the theory for complex molecules with many and varied hydrogen-bonding functional groups. To be convincing, the results must be supported by direct experimental evidence from X-ray crystallography or NMR spectroscopy.

The three methods are increasingly less demanding in computing, by orders of magnitude. As with crystal structure analysis, the effectiveness and popularity of all three methods has increased dramatically with the advances in computer technology over the past decade. However, the power of present-day super-computers is still inadequate to make the ab-initio molecular orbital method useful for the structure determination for any but a few of the simplest biologically interesting molecules. Since studies of intermolecular hydrogen-bonding complexes involve at least two molecules, this method can be applied only to very simple systems, such as the hydrogen-bonded water dimer or the formic acid and formamide dimers. A fully optimized geometry calculation for a relatively simple biological molecule, such as glucose, at an approximation better than minimal, still requires an unacceptably large amount of computing on the present-day supercomputers. Furthermore, as pointed out above, unless arbitrary assumptions are made concerning the orientation of the hydroxyl groups, the results are not applicable to glucose in a real environment, i.e., in solution or the solid state.

The third method, molecular mechanics, on the other hand, can deal with molecules of great complexity, providing the molecular configuration is known and there is not too much conformational flexibility. However, because of the global searching problem discussed above, a generally effective approach to calculating the structure of large biological macromolecules has not yet been achieved. Special procedures have been proposed, with limited success, for cyclic oligopeptides [268] and proteins [269–271].

An extension of molecular mechanics, known as molecular dynamics, requires more computer power, but promises to provide more realistic models for molecules in solution. It will undoubtedly become the theoretical method of choice for complex biological systems as super-computing becomes cheaper and more readily available [272, 273].

4.2 Ab-Initio Molecular Orbital Methods

This is the most satisfying method for calculating molecular structures, energies, and other molecular properties such as dipole moments and force constants for small molecules [253]. It requires no a-priori assumptions about the structure, although in practice the atomic connectivity is usually defined for molecules containing more than three atoms. Calculations are performed at successive levels of approximation, with each successive level requiring more computer time. The method has been programmed by a number of investigators so that a minimal knowledge of quantum mechanics is required to use them. Some of the most commonly used programs are those of the Gaussian series (Box 4.1).

Unlike experimental methods of structure analysis, there is no basis for estimating the error associated with a particular calculation. There is therefore an art, based on experience, in deciding the best level of approximation versus computing

---

Box 4.1. Acronyms for Successive Levels of Approximations Used in Ab-Initio LCAO–MO Gaussian Methodology [253]

LCAO–MO	Molecular orbitals from linear combination of atom orbitals
HF, RHF	Spin-restricted Hartree-Fock. Wave function constructed from antisymmetrized product of doubly occupied spin orbitals (UHF, spin-unrestricted Hartree-Fock calculations are used for excited states and radicals)
Basis sets	
STO-G	Slater-type orbitals, represented by a Gaussian; also referred to as minimal basis
3–21G	Split-valence or double zeta. Orbitals represented by 3 Gaussians near the nucleus and 2 away from the nucleus. Hydrogen atoms represented by one Gaussian. This is a good basis for conformational analysis of moderate-sized molecules
6–31G*, 6–31G**	As above with six and three Gaussians, respectively. * Polarization by means of d-functions on first row atoms; ** including hydrogens
Electron correlation	(Difference between HF limit and exact energy) [274]
MP1, MP2, MP3	First, second and third order Møller-Plesset perturbation theory corrections [275]
CI	Correction by means of configuration interaction calculation [253]
BSSE	Basis-set superposition error. The lowering of the calculated energy of the polymer, due to use of monomer basis functions. A mathematical consequence with no physical counterpart

Table 4.1. The effects of various levels of approximation on calculated bond lengths and angles in some first-row molecules, and comparison with experimental values. [253]

Bond	Molecule	HF/STO–3G	HF/3–21G	HF/6–31G*	MP2/6–31G*	CID/6–31G*	Exp.
C–C	C_2H_6	1.538	1.542	1.527	1.527	–	1.531
C–N	H_3CNH_2	1.486	1.471	1.453	1.465	1.460	1.471
C–O	H_3COH	1.433	1.441	1.400	1.424	1.415	1.421
C=C	C_2H_4	1.306	1.315	1.317	1.336	1.328	1.339
C=N	H_2CNH	1.273	1.256	1.250	1.282	1.268	1.273
C=O	H_2CO	1.217	1.207	1.184	1.221	1.205	1.208
C–H$_\perp$[a]	CH_3NH_2	1.089	1.083	1.084	1.092	1.091	1.099
C–H$_=$[b]	CH_3NH_2	1.093	1.090	1.091	1.100	1.098	1.099
N–H	CH_3NH_2	1.033	1.004	1.001	1.018	1.014	1.010
O–H	CH_3OH	0.991	0.966	0.946	0.970	0.963	0.963
O–H	H_2O	0.990	0.967	0.947	0.969	0.966	0.958
Angle							
C–O–H	CH_3OH	103.8	110.3	109.4	107.4	108.1	108.0
H–O–H	H_2O	100	107.6	105.5	104.0	104.3	104.5

[a] perpendicular to the $-NH_2$ plane.
[b] in the $-NH_2$ plane.

Table 4.2. Effect of offset corrections on agreement between theoretical and experimental bond lengths. Comparison of N−N and C−N bond lengths in 1,2,4-triazole, between HF/3−21G theoretical calculations and single crystal neutron diffraction analysis at 15 K. [199]. (For atomic notation see Fig. 3.2)

	Initial discrepancy Theory-Exp[a]	Offset correction[b] to MP3/6−31G	Corrected discrepancy Theory-Exp
N(1)−N(2)	+0.030	−0.027[c]	+0.003
N(1)−C(5)	+0.010	−0.005[d]	+0.005
N(4)−C(3)	+0.010	−0.006[d]	+0.004
N(2)=C(3)	−0.024	+0.019[e]	−0.005
N(4)=C(5)	−0.030	+0.019[d]	−0.011

[a] Theoretical values calculated at HF/3−21G. Experimental 15 K values corrected for thermal motion. Estimated experimental errors σ_{bond} = 0.0007 Å.
[b] Offset corrections, MP3/6−31G − HF/3−21G from
[c] planar hydrazine, H_2NNH_2,
[d] methylamine, CH_3NH_2,
[e] methylimine, $CH_2=NH$.

time for a particular result. The best approximation for molecular geometry is not necessarily the best for energies, force constants or dipole moments for any given molecule [276]. Since the approximation errors sometimes cancel, higher levels of approximation do not necessarily give results in better agreement with the experiment, as shown in Table 4.1. To minimize computing, preliminary computations using standard geometry for bond lengths and valence angles are often used to explore conformational space with minimal, or close to minimal, level of approximation before proceeding to full geometry optimization with more sophisticated basis sets. Since full geometry optimization requires much computing time, single point calculations, with or without small changes in geometry, are generally used for the highest level calculations.

Errors due to the deficiency of the basis set and neglect of electron correlation can be corrected, in part, by the Offset Method. This procedure is especially useful for calculations on molecules which are too complex for the higher level approximations shown in Table 4.1, i.e., more than 20 electrons. For the offset corrections, it is assumed that the errors due to the level of approximation are atom-pair properties and are transferable for the same type of chemical bond between different molecules. That is, for example, the error associated with a C=O bond is different from that of a C−O bond or a C=C bond and is the same irrespective of the molecule in which it occurs. With this assumption, it is possible to correct geometry optimization calculations at the HF/3−21G level for both basis-set deficiency and electron correlations.

An example is shown in Table 4.2. In the comparison of the bond lengths from the neutron diffraction analysis of 1,2,4-triazole with calculations at the HF/3−21G level, the calculated single bonds are too long and the double bonds are too short. That is, the π-electron delocalization is underestimated. The offset correction, from the data in Table 4.1, corrects for this deficiency and reduces the disagreement between the experimental and calculated bond lengths, closer to the level of the experimental errors [199].

4.3 Application to Hydrogen-Bonded Complexes

The ab-initio calculations on hydrogen-bonded dimers are necessarily limited to very simple molecules. The simplest hydrogen-bonded system of biological interest is the water dimer. Not surprisingly, this system has received intense attention from the earliest days of quantum mechanics [88, 103, 277–291]. Since it is one of the simplest systems, with two heavy atoms and four hydrogens, the calculations were pushed to the limit that the current computer technology permitted. The results of calculations at various levels of approximation, shown in Table 4.3, indicate that the hydrogen-bond length is more sensitive to the level of approximation than are the covalent bond lengths shown in Table 4.1, by an order of magnitude. There is a similar sensitivity in the various angular components of the dimer configuration. While 4–32G* gives a bond length in best agreement with experiment, 3–21G gives the best relative orientation of the two molecules, and the inclusion of electron correlation does not improve the agreement with the experimental data.

It is interesting to compare these results with those from a simple electrostatic point-charge calculation made more than 30 years ago [277] which gave a hydrogen-bond distance of 1.79 Å and a bond energy of 24.2 kJ mol^{-1}. (In that model, the O–H donor bond is represented by an unsymmetrical quadrupole and the acceptor oxygen lone pair by a dipole.) The calculated hydrogen-bond energies from the ab-initio molecular orbital calculations have ranged between 16.7 and 28.4 kJ mol^{-1}. Corrections for basis-set superposition, when more than one molecule is involved, reduces the bond energies by up to 30% with a corresponding increase in the hydrogen-bond length of 0.02 to 0.09 Å [88, 290, 291]. The best experimental value is considered to be ~ 23(3) kJ mol^{-1} [86, 292–295].

These results can be compared with those recently obtained using a much simpler chemical electrostatic theory [228, 229, 296]. In this approach, the elec-

Table 4.3. Ab-initio molecular orbital calculations on the hydrogen-bonded water dimer. Variation of geometry with level of approximation [290]

Distances (Å)	STO-3G	3–21G	6–31G*	6–311*G*	MP2/6–311G**	Experimental [294]
$O_A \cdots O_B$	2.74	2.80	2.98	3.04	2.91	2.96
$O_A - H_A$	0.99	0.97	0.95	0.95	0.96	
Angles (°)						
$O_A - H_A \cdots O_B$	180	177	175	177	175	
θ^a	124	125	118	143	136	125

[a] θ is the angle between $O_A \cdots O_B$ and the bisector of $H - O_B - H$.

Table 4.4. Some ab-initio calculations of hydrogen-bond lengths in simple dimers, polymers and water complexes (calculations at HF/STO−3G except where stated otherwise)

X−H···A	X−H (Å)	H···A (Å)	Ref.
O−H···O Methanol dimer			
linear	0.95	1.74	[91]
tandem	0.95	1.80	
O−H···O=C Formic acid			
cyclic dimer	1.01	1.53	[297]
syn-chain polymer	1.00	1.55	
anti-chain polymer	1.02	1.43	
N−H···O=C Formamide			
closed dimer (HF/3−21G)	1.02	1.82	[298, 299]
open dimer	1.02	1.91	
>N−H···N< methylene hydrazine			
closed dimer	1.02	2.06	[208]
>N−H···OH$_2$ imidazole water	1.02	1.64	[300]
HOH···N< water imidazole	0.97	1.97	[300]
formaldehyde-hydrazone dimer N(H)H···N<	1.02	2.06	[300]
>C−H···OH$_2$ glycine			
neutral	1.00	2.05	[301]
zwitterion	1.00	2.15	[301]
(C−H···OH$_2$) histidine-H$_2$O	1.00	1.85	[301]
water-formate ion linear			
HOH···O=CH (HF/4−31G) (HF/6−31G**)	0.95	1.50 / 1.75 / 1.80	[68]
water-formate ion bifurcated (chelated)	0.95	1.77 / 2.10	[68]
water-phosphate ion linear (HF/4−31G)	0.95	1.45[a] / 1.75	[68]
water-phosphate ion bifurcated (chelated) (HF/4−31G)	0.95	2.10[a] / 2.05	[68]
water-formamide cyclic dimer			

(continued on next page)

Table 4.4 (continued)

X–H···A	X–H (Å)	H···A (Å)	Ref.
O$_W$H···O=C		2.061	[302]
NH···O$_W$		2.161	
Methanol-formamide cyclic dimer			
OH···O=C		2.052	
NH···OH		2.168	

[a] Dimethylphosphate, $(CH_3O)_2PO_2^-$.

trostatic component to the energy is calculated by the interaction between the charge distribution of one water molecule with the electrostatic potential of another. The exchange and dispersion energies are represented by an atom-to-atom pair potential of r^{-6}, excluding those of the donor hydrogen to acceptor oxygen in the hydrogen bonds. This is consistent with the view that the moderate or weak biological hydrogen bonds are primarily electrostatic interactions.

The results of some theoretical calculations on other hydrogen-bonded dimers with functional groups of biological relevance are given in Table 4.4. For these dimers, comparison with experiment is not generally possible, since they are insufficiently stable in the gas phase.

Early theoretical calculations on hydrogen-bonded water polymers predicted cooperativity. Even when it was possible to carry out calculations only at relatively low approximation levels, comparison of the relative values for the water dimers, trimers, tetramers and higher polymers were very informative. They predicted the nonadditivity of hydrogen-bond energies and the phenomenon of cooperativity, discussed in Chapter 1 and elsewhere throughout this text [103–105]. One such calculation involved successive polymers, the results of which are shown in Table 4.5.

The energetic advantage of homodromic cyclic bonding was predicted. The total hydrogen-bonding energy of a pentamer forming a homodromic cycle of hydrogen bonds, for example, was shown to be greater than the binding energy of five dimers [103]. Calculations at the HF/4–31G level for six waters in a chair configuration with sequential hydrogen bonding predicted a shortening of the O···O separation of 0.14 Å relative to the hydrogen-bonded dimer [308]. In a more recent ab-initio calculation at the MP3/6–31G** level, it was found that cooperativity contributes 30% additional energy in a homodromic cyclic water tetramer relative to that of four isolated water dimers [304]. On the other hand, for the water pentamer, where a central water molecule donates two and accepts two bonds, no change in O···O separation was predicted. The other calculation compared the energies of three different trimer configurations and how they varied with hydrogen-bond length as shown in Fig. 4.1. Calculations with the minimal basis set, HF/STO–4G, gave the energy of the sequential trimer to be lower than the other two by ~8 kJ mol^{-1} at H···O = 1.8 Å, due to the cooperative effect. At H···O distances greater than 2.5 Å, the distinction disappears. This is consistent with the

Table 4.5. The non-additivity of hydrogen-bond energies in sequentially bonded H_2O polymers [103]. Fixed geometry $O-H = 0.9915$ Å; $H-O-H = 100.053°$. Calculations at HF/STO−3G level

	H···O (Å)	α (°)	β (°)	γ (°)	Binding energy per H-bond (kJ mol^{-1})	
					Cyclic	Chain
$(H_2O)_n$						
$(H_2O)_3$	1.57	62	2	19	23.5	20.4
$(H_2O)_4$	1.48	61	2	8	39.7	25.7
$(H_2O)_5$	1.45	58	2	3	44.5	34.6
$(H_2O)_6$	1.45	57	2	0	45.0	

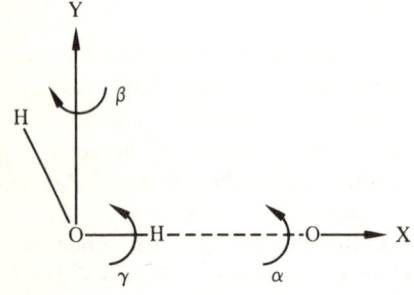

α, β and γ are successive rotations about the X, Y, Z axes performed on a water molecule with its oxygen atom located at the origin. The orientation of the other molecules in the polymer is then determined by a C_n symmetry operation.

interpretation of cooperativity as a polarization effect which attenuates as r^{-4}, i.e., faster than the electrostatic r^{-1} interactions. Theoretical calculations of open chain dimers and trimers containing water and methanol predicted that the methanol to water hydrogen bond is stronger than the water to methanol [90, 91]. This agrees with the differences in bond lengths observed in crystal structures of hydrates when the water is a donor over when it is an acceptor, as reported in Part IB, Chapter 7.

Because of the crystal field distortion, the hydrogen-bond lengths tend to be shorter in crystals than observed experimentally or predicted theoretically in hydrogen-bonded dimers. The information from gas-phase diffraction and microwave spectroscopy on hydrogen-bonded dimers is necessarily limited to the relatively few dimers which are stable in the gas phase. X-ray and neutron diffraction crystal structure analysis, in contrast, provides a plethora of structural data relating to hydrogen-bond formation, as is demonstrated in Part IB, Chapter 7 and subsequent chapters.

These data refer to molecular interactions in the crystalline state, where the many atom interactions of the crystal field forces are perturbing factors *which are unique to each crystal structure, or, at best, to each type of crystal structure* with similar molecular dimensions and the same hydrogen-bonding functional groups, cf. [305]. It might be anticipated that in statistical surveys over a large number of crystal structures for a particular type of hydrogen bond, these crystal field factors average out.

Application to Hydrogen-Bonded Complexes

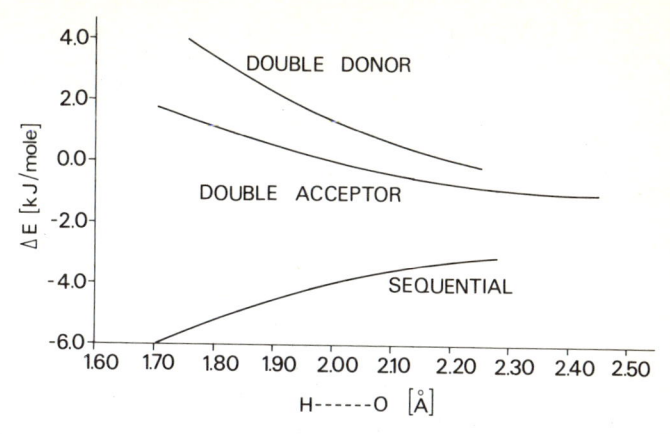

Fig. 4.1. Variation of the cooperative (nonadditivity) energy, ΔE, versus hydrogen-bond distances for the three trimer configurations [105]:

$$\text{Sequential} \quad \begin{array}{c} \text{H} \quad \text{H} \\ \text{OH} \rightarrow \text{OH} \rightarrow \text{OH}_2 \end{array}$$

$$\text{Double donor} \quad \text{H}_2\text{O} \leftarrow \text{HOH} \rightarrow \text{OH}_2$$

$$\text{Double acceptor} \quad \begin{array}{c} \text{H} \\ \text{HOH} \rightarrow \text{O} \leftarrow \text{HOH} \\ \text{H} \end{array}$$

(Fixed monomer geometry $\text{O}-\text{H} = 0.945$ Å, $\text{H}-\text{O}-\text{H} = 106°$).

$\theta = 125.3°$

This appears not to be the case. On average, the crystal field forces have a distortion effect. This was realized from some of the earliest analyses of hydrogen-bond lengths [52]. It arises from two factors; one is the influence of many-atom effects, such as cooperativity. The second is the fact that all other atom-pair interactions are striving toward the equilibrium minimum. Since hydrogen-bonded functional groups tend to protrude from molecules, this results in an overall distortion. The most obvious example of this difference is that between the values for the H···O distances of 1.7 to 1.8 Å observed in the ices and 2.0 Å for the water dimer (see Table 4.3). Similarly, the distribution for two-center $\text{O}_\text{W}\text{H}\cdots\text{O}$ bonds in the hydrates of small molecules, discussed in Part IV, has a mean value of 1.80 Å, in agreement with the ices.

To explore this difference more quantitatively, an attempt [306] was made to derive a potential energy function for the hydrogen bond that would reproduce the two-center NH···O=C and O-H···O=C bond-length distribution observed in the crystal structures of the purines and pyrimidines, described in Part IB, Chapter 7, i.e., relating the curves (a) and (b) in Fig. 2.4. Fitting the bond-length distribution curve to a Lennard-Jones type of potential energy function via a Helmholtz

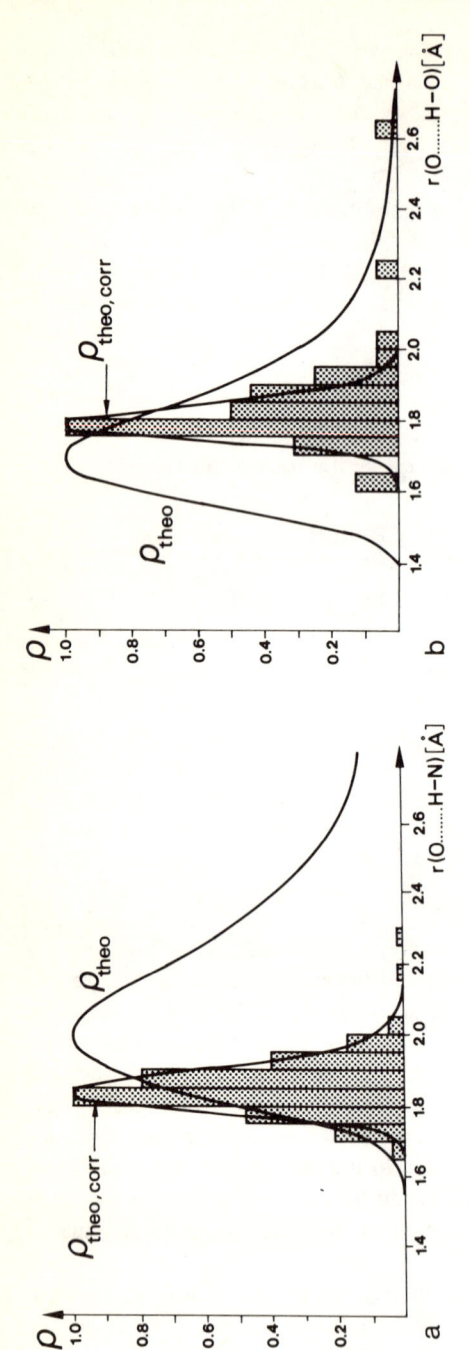

Fig. 4.2a–c. Experimental, ϱ_{exp} (dotted histograms), and theoretical, ϱ_{theo}, hydrogen-bond length distributions. **a** For $C=O\cdots H-N$, **b** for $C=O\cdots H-O$. Number of experimental data are 151 and 47 for **a** and **b**, respectively. $\varrho_{theo,corr}$ is obtained with correction of the free energy parameters, assuming a one-dimensional model of the hydrogen bond in the solid state, as indicated in **c**. A and D denote the acceptor and donor atoms, respectively. The temperature T, the Debye temperature θ and the Gruneisen constant, γ, are the macroscopic thermodynamical parameters of the ensemble. In this study, C–A is C=O = 1.22 Å, D–H is O–H = 0.97 Å and N–H = 1.03 Å [306]

Application to Hydrogen-Bonded Complexes

free energy function gave very poor agreement, as shown in Fig. 4.2, unless values were assigned to the Lennard-Jones parameters, which gave unrealistically large hydrogen-bond energies. To obtain the fit with reasonable potential energy parameters shown in Fig. 4.2, three additional terms were necessary, one of which was a simple compression term to simulate the average crystal field effect. The other terms accounted for the thermal lattice energy and the entropy and involve the Debye temperature and the Gruneisen constant [307].

Ab-initio calculations indicate that three-center hydrogen bonds are comparable in energy to two-center bonds. Ab-initio molecular orbital calculations at the HF/4−31G level on a water trimer in the three-center bond configuration with O−H···O angles of 135° and H···O = 2.15 Å and a water dimer with the linear configuration O−H···O = 180° and H···O = 1.95 Å gave comparable energies [55]. This is consistent with the widespread observation of these bonds in the crystal structures of the carbohydrates, amino acids, purines and pyrimidines, and nucleosides and nucleotides described in Part II. There is an interesting suggestion that three-center bonding plays a role in the migration of Bjerrum L-defects in ice with a calculated activation energy of about 3.5 kcal mol^{-1} [308]. It has also been proposed that the temperature dependence of the Raman spectra for water can be interpreted in terms of an increase in the proportion of three-center bonds with increase of temperature [309].

Other hydrogen-bonded dimer configurations have been studied theoretically, although they are rarely observed experimentally. These are the cyclic dimer, 1, and the bifurcated dimer, 2. The latter can be referred to as the bifurcated "donor" dimer, retaining the description bifurcated "acceptor" for the three-center bonded trimer [303].

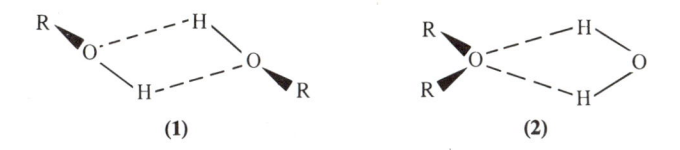

Calculations at the HF/STO−3G level indicated that the cyclic dimer of methanol was energetically unfavorable compared with the linear dimer [90]. The configuration (1) has been reported in several X-ray crystal structures, but not unequivocally confirmed. It is suspect because it involves a very short nonbonding H···H interaction. In most cases, a disordered configuration with half-occupancy of the hydrogen atoms is a preferable interpretation of the X-ray analysis. No crystal structures with this feature have been studied by neutron diffraction.

For the bifurcated water dimer, 2, the calculated energy was about half that of the linear dimer. No examples have been observed in hydrate crystal structures, except as components of three-center bonds, see Part IV, Chapter 22.

A more conceptual view of hydrogen bonding is obtained by partitioning the calculated bond energy. The ab-initio method gives the total hydrogen-bond energy. An interesting conceptual view of this interaction is obtained if this energy can be partitioned into various components, such as electrostatic (Coulombic[1]), polarization, charge-transfer, repulsion and dispersion [42]. This conceptual view is particularly useful as a guide to designing functions to include hydrogen bonding in the empirical methods described in Chapter 4.5.

Various methods have been suggested for this partitioning [89, 310–313]. When applied to $O-H\cdots O$ bonds in the water dimer [50], and the complexes of water-formamide and water-cyclopropenone [312], the electrostatic component provides 80% of the total attractive energy for $O_W-H\cdots O_W$, see Fig. 2.1, and about 75% for $O-H\cdots O=C$ bonds. This is balanced by the repulsive energy, while the polarization and charge-transfer components provide less than 5% for moderate strength bonds.

4.4 Semi-Empirical Molecular Orbital Methods

Semi-empirical methods are the only alternative to molecular mechanics for most molecules of biological interest, especially for the study of hydrogen bonding, since this necessarily involves assemblages of molecules, i.e., hydrogen-bond dimers, trimers, higher polymers and complexes between different molecules.

In semi-empirical methods, the computational complexity of the ab-initio methods is reduced by making approximations in the computational procedures or by introducing constants derived from experimental data (such as ionization potentials). These methods are more generally applied to the calculation of relative conformational energies rather than to the calculation of quantitative molecular geometries.

Two general methods which differ in their approach to approximation have been used extensively for conformational calculations on medium-sized biological molecules. These are the CNDO, INDO, MNDO series and PCILO [254–259]. CNDO, INDO, MNDO etc. simplify the computations by neglecting or partially neglecting the differentials in the energy calculations for the valence electrons, and treat the inner shell electrons as a core which screens the nuclear charge. This method has been used in fixed geometry calculations to obtain atomic charges which are subsequently used in molecular mechanics calculations.

In PCILO, the molecular orbitals are not expanded in the usual LCAO form, but in pair-wise hybrids. Perturbation theory, rather than variation theory, is then used to obtain the wave functions and calculate the energy. The method has been used for conformational analysis of small biological molecules [258], but is too computationally intensive for macromolecules. Even for small biological molecules with a significant number of variable torsion angles, the global searching problem can be severe. This is well illustrated by the applications of PCILO, MNDO and a molecular mechanics program to the conformation of N-acetyl-β-D-glucosamine (NAG) using the six variable torsion angles [258]. Nineteen minima were obtained within 7.4 kcal mol^{-1}. The lowest PCILO conformer

---

[1] This term applies to the classical interaction between charges. It is fundamentally misleading since all interactions involving the electrons and the nuclei are intrinsically electrostatic.

was the highest energy molecular mechanics conformer, and the twelfth highest MNDO conformer. Since the authors associate these differences with different energy contributions from the *intramolecular* hydrogen bonding, these results have little relevance to the conformation of the molecule in solution or in the crystal, where the hydrogen bonding will be primarily intermolecular. Had the calculations been carried out on the α-epimer, the results could have been compared directly with a crystal structure determination, but no structure analysis has been reported for the β-compound.

Another interesting example of the use of PCILO to study hydrogen bonding was a comparison of the energies of five- and six-membered (H_2O and $C-OH$) hydrogen-bonded cycles observed in the crystal structure of α-cyclodextrin hexahydrate. These results predicted that the homodromic cycles were significantly more stable than the antidromic arrangements [110], in agreement with the ab-initio calculations on the water polymers discussed earlier.

A method which would seem to have particular relevance to hydrogen-bonded systems in view of the Coulombic nature of the longer-range hydrogen-bond forces is one which evaluates long-range Coulombic interactions within the framework of the LCAO-MO method [314]. Hitherto this method has been applied to infinite polymers, where comparison with experimental structural data is not possible.

4.5 Empirical Force Field or Molecular Mechanics Methods

Molecular mechanics or empirical force field methods are the least computationally intense means of calculating molecular structures [261, 262]. These methods have their origin in the early concepts of stereochemistry [315] but their potential for application to complex biological molecules was first realized in the hardsphere calculations which formed the basis of the very influential Ramachandran plots for the conformational analysis of polypeptides [316]. In this method, the permitted regions of conformational space are defined using van der Waals radii as the criterion. This is still a useful simple procedure for the preliminary stages of conformational analysis of complex molecules, especially when combined with experimental data, such as from NMR proton coupling constants [317]. It is also used in exploring the permitted helical structures in the fibrous polysaccharides by the $n-h$ method [318], where n is the number of monomers per helix pitch and h is the rise per residue.

The more analytical methods have developed both in effectiveness and complexity as the increase in computer technology made it possible to handle multiparameter calculations. A commonly used molecular mechanics program, designed for nucleic acids and proteins [82], distinguishes between 18 different carbon atoms, 8 different nitrogens, 4 different oxygens, 6 different hydrogens, 2 different sulfurs, phosphorus, and lone pairs, and 5 different hydrogen-bond donor/acceptor combinations.

Some of the more commonly used programs have the advantages for the nonspecialist in the field that they are generally well documented and tested, and the basic principles involve classical rather than quantum mechanics. For molecules where there is no conformational ambiguity, such as the polycyclic hydrocarbons, a well-parameterized molecular mechanics program is superior to crystal structure analysis for determining molecular structure [319], since the compound need be neither synthesized or crystallized. For these reasons, molecular mechanics is the theoretical method most widely applied for predicting the structures of biological molecules ranging in size from the disaccharides, dipeptides, and nucleosides to polysaccharides, nucleic acids and proteins. The method can be applied to isolated molecules or to assemblages of molecules and is used extensively in drug design. It forms the starting point for dynamical calculations such as the Monte Carlo and molecular dynamics. These methods are used both for predicting water structure in solvated molecules, and for predicting the conformation of molecules in solution.

In the molecular mechanics methods, the energy of the molecule or assemblage of molecules is broken down into a sum of energy terms which are minimized separately, i.e.,

Intramolecular $\begin{cases} \text{bond stretch} \\ \text{bond bend} \\ \text{bond stretch/bend} \\ \text{torsion angles} \end{cases}$

Intermolecular $\begin{cases} \text{nonbonding; repulsion and attraction} \\ \text{charge or dipole interactions} \\ \text{hydrogen bonding} \end{cases}$

The calculation of the energies in molecular mechanics is based on the Newtonian laws of classical mechanics. Molecules are treated as if they consisted of an assemblage of soft rubber balls and springs. Each ball is the junction point between springs which have an equilibrium length, r_0 and interspring angle, θ_0, and torsion angle ω_0, with appropriate values of stretching, k_r, bending, k_θ, and twisting, k_ω, force constants from the equilibrium values; V_n is the torsional energy barrier with the periodicity $360°/n$. For small deviations from the equilibrium values, Hooke's law is applied.

$$E_r = \tfrac{1}{2} k_r (r - r_0)^2$$

$$E_\theta = \tfrac{1}{2} k_\theta (\theta - \theta_0)^2$$

$$E_\omega = V_n/2 k_\omega (1 + \cos n\omega) \ .$$

For large distortions from the equilibrium values, an additional anharmonicity term can be included, e.g., $[1 + k'_r (r - r_0)^2]$ or $k'_r (r - r_0)^3$, but this is rarely necessary.

The concepts of quantum mechanics are used in formulating the appropriate expressions for the nonbonding atom-to-atom interactions which are treated as sums of pairwise interactions. Box 4.2 shows the components of these atom-pair interactions which are applied, with different emphasis, depending upon the type of molecules involved.

> **Box 4.2.** Atom-Atom Interactions. Functions Used in Molecular Mechanics
>
> | | |
> |---|---|
> | Exchange repulsion | Ar_{ij}^{-12} or $A \exp{-\alpha r_{ij}}$ |
> | Dispersion | $-Br_{ij}^{-6}$ |
> | Charge transfer | $C \exp{Dr_{ij}}$ |
> | Charge-charge | $q_i q_j \cdot (\varepsilon \cdot r_{ij})^{-1}$ |
> | Dipole-dipole | $\mu_i \mu_j \cdot r_{ij}^{-3} \cdot f(\theta_i, \theta_j)$ |
> | Polarization of i by j | $-\frac{1}{2} \mu_i \alpha_j \cdot r_{ij}^{-4}$ |
>
> A, B, C, D, arbitrary constants
> r_{ij}, interatomic distances between atoms i and j
> q_i, q_j, point charges
> ε, dielectric constant
> μ_i, μ_j, dipole moments
> f, a function involving the angles θ_i, θ_j
> α_j, polarizability of atom j

All programs use the Lennard-Jones type of potential for nonbonded interactions,

$$E_{nb} = \sum_{ij} -Ar_{ij}^{-n} + Br_{ij}^{-6} , \quad \text{where} \quad n = 9 \text{ or } 12 ,$$

with the repulsion term Ar_{ij}^{-n} sometimes replaced by the two-parameters Buckingham exponential term, $C \exp(-Dr_{ij})$, which increases the flexibility at the cost of extra arbitrary parameters.

If the nonbonded potential does not include the 1,3 vicinal intramolecular atom pairs, a stretch/bend cross-term is added of the form $E_{r,\theta} = \frac{1}{2} k_{r,\theta} (r - r_0)^2 \cdot (\theta - \theta_0)^2$

The Coulombic interaction between the electrostatic potential within a molecule or between adjacent molecules is calculated by means of an atomic point-charge model. For most biological molecules, with their many hydrogen-bonding functional groups, the major component of this interaction is hydrogen bonding. These Coulombic interactions are calculated atom pairwise in the form $\sum q_i q_j (\varepsilon \cdot r_{ij})^{-1}$, where q_i and q_j are the atomic charges, r_{ij} is the atom-atom distance and ε is a dielectric constant, which is generally given a value of 1, 2 or 3. In calculations involving water solvent molecules, a distance-dependent value may be used to represent the fact that the dielectric constant of water is more than 20 times greater than those of most organic molecules.

The hydrogen-bonding energy terms from point charge interactions are not very sensitive to interatomic distances, but they are long-range interactions dependent on r^{-1}. They correspond to attractive forces which still persist when the non-bonding attractive forces have become negligible. The combination of nonbonding and Coulombic terms is referred to as a 9−6−1 or 12−6−1 potential. The results from gas-phase calculations, such as that shown in Fig. 2.1 for the water dimer, can be misleading however, if applied to molecules in the solid state or solution due to the dielectric attenuation through adjacent molecules.

Table 4.6. Atomic charges calculated by least squares fit to electric potentials calculated by ab-initio Hartree-Fock methods [329]

Molecule	Atom	Basis-set		
		STO–3G	6–31G	6–31G*
H$_2$O	O	−0.61	−0.94	−0.79
	H	+0.31	+0.47	+0.39
CH$_3$OH	O	−0.50	−0.83	−0.68
	C	+0.19	+0.42	+0.24
(Staggered)	H–(C)	−0.01	−0.01	−0.02
	H–(O)	+0.32	+0.49	+0.42
H·CONH$_2$	O	−0.44	−0.67	−0.56
	N	−0.84	−1.11	−0.88
	C	+0.68	+0.89	+0.64
	H–(C)	−0.06	−0.02	+0.03
trans	H–(N)	+0.32	+0.44	+0.37
cis	H–(N)	+0.34	+0.47	+0.40
H·COOH	O(H)	−0.55	−0.82	−0.63
	O(=C)	−0.44	−0.68	−0.57
	C	+0.70	+0.95	+0.67
	H–(C)	−0.06	0.0	+0.06
	H–(O)	+0.36	+0.54	+0.46

Unfortunately, net atomic charges are not exactly defined physical properties and they are adjustable parameters in the molecular mechanics programs. Since the interactions between biological molecules involve many of these long-range Coulombic interactions, the magnitude and distribution of these charges is an important aspect of parameterization when seeking an accurate evaluation of the conformational effects of intra- and intermolecular hydrogen bonding. Considerable attention has been directed to the problem of selecting appropriate values for the atomic point charges [320–327]. The Mulliken populations from semi-empirical calculations have been used [328], but these are no longer considered appropriate. Generally, negative charges of −0.5±0.3 e are placed on the donor and acceptor nitrogen and oxygen atoms, and positive charges of +0.5±0.3 e are placed on the hydrogen atoms. Values have been obtained from *ab-initio* calculations of small molecules which represent the fragments of larger molecules [329]. Table 4.6 shows some atomic point-charges obtained for simple molecules by least squares fitting to the molecular electrostatic potential which is calculated at various levels of approximation. Since hydrogen-bond formation will change the electronic distribution, these isolated molecule values are still only an approximation. The electrostatic induced dipole attraction can be included with a $(\alpha_j q_L^2 + \alpha_L q_j^2)r^{-4}$ term where α_L and α_j are atomic polarisabilities. The total potential is then referred to as a 12−6−4−1 potential.

In some hydration studies a polarizable dipole is added to each atom, such as that used for including the effects of cooperativity in simulating water structures [330, 331].

The electrostatic potentials can also be represented by bond dipoles, μ_{ij}, which are used in the form of the dipole interaction term

$$E = \mu_{ij}\mu_{ij} \cdot \cos\alpha - 3\cos\theta_i \cos\theta_j \cdot \varepsilon r_{ij}^{-3}$$

where α is the angle between the dipoles,
θ_i and θ_j are the angles between the dipoles and r_{ij}, where r_{ij} is the distance between the mid-points of the bonds,
and ε is a dielectric constant factor.

Lone-pairs on oxygen atoms are often included either as point charges or dipoles. This is as much an approximation as is the atomic-charge model and less effort has been directed to parameterizing this approach.

A topological approach to partitioning the electron density distribution into atomic components may ultimately provide a more realistic model [332].

Opinions differ concerning the necessity for explicit functions to account for hydrogen bonding. When the emphasis was on interpreting the spectroscopic properties of hydrogen bonds, it was generally considered necessary to have complex formulas to represent their potential energy curves [333, 334]. Now that the emphasis is more on predicting the conformation of complex organic molecules, the trend is to simplify the hydrogen-bond parameterization as much as possible. In fact, many recently developed molecular mechanics programs do not include functions specifically for hydrogen-bond interactions. It is considered that hydrogen bonding is adequately accounted for by the electrostatic and nonbonding Lennard-Jones terms, i.e., 9–6–1 or 12–6–1 potentials [335–343].

However, some investigators do supplement these interactions with specific functions relating either to the H···A distances or the X···A distances, or both [343–346]. These functions take the form of either a Morse function,

$$E_{HB} = \sum_{H\cdots A} D[\exp-2\alpha(r_{H\cdots A} - r_{H\cdots A}^0) - 2\exp\alpha(r_{H\cdots A} - r_{H\cdots A}^0)]$$

or a similar function with $r_{H\cdots A}$ replaced by $r_{X\cdots A}$, or a Lennard-Jones potential such as

$$\sum_{\text{H-Bond}} -B_{HB}r^{-6} + A_{HB}r^{-12}$$

or

$$\sum_{\text{H-Bond}} -B_{HB}r^{-10} + A_{HB}r^{-12}.$$

The problem of selecting the appropriate values, r^0, for the equilibrium hydrogen-bond lengths is avoided by using only the Lennard-Jones function.

For the stronger hydrogen bonds, angular dependence parameters may be considered necessary. These can take a very simple form such as

> **Box 4.3.** An Example of a Hydrogen-Bonding Potential Function Involving Lennard-Jones (A, B), Electrostatic (q) Morse Potentials and Directional Parameters (θ). [344]
>
> For NH$\cdots$O=C bonds
>
> $E(r, \theta) = (-A_{NO} r_{N\cdots O}^{-6} + B_{NO} r_{N\cdots O}^{-m} - A_{HO} r_{H\cdots O}^{-6} + B_{H\cdots O} r_{H\cdots O}^{-m})[1 - f(\theta)]$
> $\quad + q_N q_O r_{N\cdots O}^{-1} + q_N q_O r_{H\cdots O}^{-1} + D\{\exp[-2\alpha(r_{O\cdots H} - r^0)] - 2\exp - \alpha(r_{OH} - r^0)]\}f(\theta),$
>
> where
>
> $\quad f(\theta) = \exp[-(\theta/\theta_0)^2]$
>
> or
>
> $\quad f(\theta) = \exp[-(\theta_1/\theta_0)^2] + \exp[-(\theta_2/\theta_0)^2]$

$$f(\theta) = \cos^2\theta, \quad \text{when} \quad \theta \geq 90°$$
$$= 0, \quad \text{when} \quad \theta < 90°,$$

where θ is the X–H$\cdots$A angle, or more complex expressions such as those in Box 4.3.

The literature contains a wide choice of functions and parameters for including hydrogen bonding in molecular mechanics. A selection is given in Box 4.4. All distinguish between different types of hydrogen bonds, and some are developed for particular classes of molecules.

In a program designed for nucleic acids and proteins, the Lennard-Jones 12–10 potential is used with parameters for five different hydrogen-bond types [82]. These are included principally to prevent the hydrogen-bonded X$\cdots$A separations from becoming too short, since no H$\cdots$A interactions are included in the Lennard-Jones nonbonding component of the method. A general program for molecular mechanics and molecular dynamics calculations on macromolecules includes both a 12–10 potential and angle factors [272].

The empirical parameters used in a molecular mechanics program are derived from a mix of experimental data. These are thermochemistry, spectroscopy, crystal and gas diffraction. The results apply therefore to an *average* molecule in all three states of matter. The method is generally used for conformational analysis and exploring major differences in structure. The relatively minor changes in molecular geometry that generally occur when a molecule goes from the solution or liquid phase to the crystal or gas phase are smaller than the predictive accuracy of the method. The exception to this is where inter- or intramolecular hydrogen bonding is a significant factor in determining molecular conformation.

The key to a successful molecular mechanics program lies in the judicious choice of a consistent set of empirical parameters. Since these parameters are not necessarily transferable from one type of molecule to another, the best results are obtained with programs which are parameterized for particular classes of molecules.

A hydrogen-bond potential which is parameterized for use with a particular molecular mechanics program will not be transferable to another molecular mech-

Box 4.4. Some Hydrogen-Bond Potentials

Type of bond	Comments	Reference
General	Basic theory based on dipolar interactions	[315]
OH···O	Morse function based on $r_{X\cdots A}$ and $r_{H\cdots A}$ angle parameters. Used for interpreting spectroscopic data	[333, 334]
NH···O		
NH···N		
NH–Cl$^-$		
OH···Cl$^-$		
All types	CNDO/2 for charges with $Ar^{-10} + Br^{-12}$ term	[335–337]
C–OH···O=C	CNDO/2 for charges. Used Lennard-Jones 9–6–1 and 12–6–1. No Morse function. Tested on crystal structures of carboxylic acids, amides	[338, 339]
NH···O=C		
C–OH···O=C	CNDO/2 for charges. Morse function based on $r_{H\cdots O}$ and $\exp A(r_{O\cdots O}) - Br_{O\cdots O}^{-m}$ developed for carboxylic acids	[340]
C–OH···O=C	12–6–4–1 Potential	[341, 342]
OH···OH	Developed for use with MM1 and MM2	[345, 346, 346a]
OH···O	Morse function based on $r_{H\cdots O}$ and angle factor (Box 4.3). Combined with refinement of diffraction data for cellulose (see Part II, Chap. 13)	[344]
NH···O=C	Charges from HF/STO–3G on monomers. Lennard-Jones type 12–10 on $r_{X\cdots A}$, mainly for repulsive term. Developed for nucleic acids and proteins	[82]
NH···N		
OH···O		

anics program. It has to be reparameterized for consistency with the other components. For the relatively rigid fused-ring cyclic hydrocarbons containing only C–C, C=C, and C–H bonds, and no hydrogen-bonding functional groups, a program parameterized for alkanes only gave structural data on bond lengths, valence angles and torsion angles that agree with the crystal structural data within 0.01 Å in bond lengths and 2° in angles [319]. However, for biological molecules containing many hydrogen-bonding functional groups, the problem of obtaining the best formulas and parameters appears to be significantly more difficult.

A version of an early molecular mechanics program developed for alkanes and alcohols predicted very closely the molecular structures of the pyranose sugars in the crystalline state. A modification of MM1, referred to as MM1–CARB, was used to calculate the structures of a number of pyranoses and pyranosides for which neutron diffraction crystal structure data were available [347]. For use in conformational analysis, the agreement was very satisfactory, i.e., 0.025 Å for the C–C and C–O bond lengths, 1° in valence angles and 3° in torsion angles. These differences are the same order of magnitude as might result from the crystal field forces, and therefore give a very acceptable approximation for the molecules in solution, with no anisotropic solvent effects. No special hydrogen-bond parameterization was included and intramolecular hydrogen bonding was avoided by fixing the C–C–O–H torsion angles at the values determined by the intermolecular hydrogen bonding in the crystal. This program was not applied to disaccharides, where inter-residue intramolecular hydrogen bonding will influence conformation, but *intra-residue intramolecular bonding has to be avoided*. With such structures there is a multiplicity of energy minima in the absence of constraints on the torsion angles of the linkage bonds between the monosaccharide residues, as illustrated by the β-gentiobiose example at the beginning of this chapter.

In the determination of the crystal structures of polymers, molecular mechanics is combined with interpretation of the X-ray fiber diffraction data. Since direct methods for solving the phase problem cannot be applied to X-ray fiber diffraction data, the crystal structure is "solved" by finding the best model which gives the lowest "disagreement" R-factor. Molecular mechanics methods can be used to select the best models from the point of view of lowest conformational energy. Even combining the two approaches by minimizing the experimental disagreement factor and the theoretical energy, it is difficult to arrive at unique or even convincing solutions. This is illustrated by the example of cellulose II, which is described in Part II, Chapter 13. In that work, the combination of a Lennard-Jones and a Morse potential was used, with directional parameters, as shown in Box 4.3. Using interchain packing considerations, the number of variable parameters was reduced to nine. Thirteen solutions were reported, 11 of which had energies which differed by less than $20\,\text{kJ}\,\text{mol}^{-1}$ and all of which had "acceptable" R-factors of less than 0.21. Interestingly, the solutions with lowest experimental R-factors had the highest calculated energies.

Molecular modeling using either Monte-Carlo simulations or molecular dynamics is used to apply molecular mechanics energy minimizations to very complex systems [348]. In complex flexible molecules such as proteins or nucleic acids, the number of variable parameters, i.e., bond torsion angles, is such that the global search for energy minima becomes impossible. The same problem occurs with theoretical calculations of water structure in aqueous solutions or in heavily hydrated crystals.

In the Monte-Carlo method, which was originally proposed for a "super-computer" of 1953 (MANIAC) [348a], a walk in conformational or configurational space is generated by changing each variable parameter by $n \cdot \alpha$, where α is the maximum permitted displacement and n is a random number. The energy is calculated for each conformer or configuration generated and a structural thermodynamic average is obtained according to the Boltzmann distribution.

In molecular dynamics simulation a statistical average with time is obtained by applying initial velocities, i.e., kinetic energy, and solving the Newton equations of motion [272, 273, 349]. The kinetic energy in the simulation allows the system to pass across small energy barriers and sample all the local minima. The permitted conformations can then be derived based on a Boltzmann distribution. This method provides a simulation of the development of structure with time that may prove to have particular significance in molecular biology. Since, in both methods, an energy calculation has to be carried out for each of a very large number of configurations, only molecular mechanics can provide the necessary computational speed.

The calculations therefore necessarily contain the approximations involved in the functional form of the energy calculations, which is generally based only on Lennard-Jones type pairwise interactions [350].

In complex molecules, the bond lengths and angles are fixed and only the torsion angles are varied. In molecular assemblies, the molecules are assumed to be rigid and the variables are the coupled translational and rotational motions. Since the method is computer-intensive, its power will increase with the development of computer technology and will be increasingly capable of bringing extremely complex problems within the grasp of the theoretical methods.

In a program designed specifically to use pair potentials to model hydrogen bond interactions between donor (probe) and acceptor (target) molecules, the electrostatic $q_i q_j . r^{-1}$ modified by a geometrical term and different dielectric constants are used for the solvent and target phases. To obtain hydrogen bond geometries consistent with data from small molecule crystal structures, an 8−6 dependence term was necessary. Angular dependence terms depending upon the nature of the target atom are also included in the hydrogen bonding function.

Molecular dynamics has been applied to small molecules such as monosaccharides, mainly to test the methodology against well established experimental data [351−353] before proceeding to problems where comparison with experimental data is more difficult or even impossible [354].

The structure of liquid water has been studied both by Monte Carlo [355] and molecular dynamics [356]. In these calculations, hydrogen-bond potentials based on the water dimer model are unsatisfactory, and the simulations have to include the increased binding energy due to cooperativity to give acceptable thermodynamical results [357].

Monte-Carlo simulations have been made of the known water structure in crystalline serine monohydrate, arginine dihydrate, homoproline tetrahydrate [331], deoxycytidylyl-3',5'-guanosine-proflavine hydrate [358] and the "spine of hydration" in the minor groove of a B-type oligodeoxynucleotide duplex [359]. They resulted in "partial" agreement and demonstrated the sensitivity of the method to the pair potentials used.

A molecular dynamics simulation of the crystal structures of α-cyclodextrin hexahydrate and β-cyclodextrin dodecahydrate was more successful. In both hydrates, the carbohydrate atomic positions and two third of the water positions were reproduced within the experimental accuracy, and most of the $O-H \cdots O$ hydrogen bonds that had been determined by neutron diffraction studies (Part III, Chap. 18) also showed up in the simulation [360, 361].

Since molecular dynamics requires fast computing, progress using this method has been greatly accelerated by the development of the super-computers. It is now the best theoretical approach to understanding the hydrogen-bond structure of water and other polar solvent molecules near the surface of biological macromolecules. It also provides a method predicting the effect of solvent on the conformation of solute molecules. However, it is recognized that the proper treatment of the electrostatic interactions is one of the more difficult aspects of parameterization on the methods for energy minimization of macromolecules, either by molecular mechanics or molecular dynamics [362].

Chapter 5
Effect of Hydrogen Bonding on Molecular Structure

Hydrogen-bond formation modifies the electronic structure of the molecules involved. In very strong hydrogen bonds, the covalent X−H bond may be so much lengthened that it is barely distinguished from the H···A hydrogen bond, as shown in Part IB, Chap. 7, Table 7.2. This perturbation of the electronic structure of the donor bond and the acceptor atom will extend to the adjacent covalent bonds, depending upon the polarizability of the covalent bonding in the molecule. For this reason it is greatest in conjugated systems, where it is a structural manifestation of the *π-bond cooperative effect*, discussed in Chap. 1.

These changes in the covalent bond structure are most readily detected by the more sensitive spectroscopic methods which measure differences in the bending and stretching frequencies of the covalent bonds involved. They provided the earliest tools for recognizing hydrogen-bond formation [363, 364]. The more recent NMR methods are also very sensitive to differences in the electronic environment of the atoms in the proximity of a hydrogen bond through the measurement of chemical shifts. These observations form the basis of the spectroscopic methods for diagnosing hydrogen-bond formation in the liquid state, but they provide no direct evidence of changes in molecular geometry.

The proton chemical shift is a particularly sensitive measure of hydrogen bonding. Since it is normally observed for molecules in solution, it is not generally possible to correlate chemical shifts with changes in molecular structure. Single crystal solid-state NMR spectroscopy can provide the basis for an experimental correlation between chemical shift and bond length or valence angle in a few special cases, as discussed in Chap. 3. However, the crystal size requirements for single crystal NMR spectroscopy are even more demanding than for single crystal neutron diffraction. There are a very limited number of examples where both techniques have been applied to the same crystal structure. The examples that do exist point to a very direct linear relationship between proton chemical shift and hydrogen-bond lengths as discussed in Chapter 3, but this has not been extended to study the effect on the electronic environment of adjacent nonhydrogen atoms.

There is good evidence of a relationship between the X−H and H···A bond lengths, as shown in Fig. 5.1. However, the spread of the points in this plot is significant. It arises from a combination of the errors discussed in Chap. 3, including thermal motion effects and the fact that hydrogen-bond lengths are sensitive to the many-atom crystal field interactions. Throughout the full range of the hydrogen-

Effect of Hydrogen Bonding on Molecular Structure 95

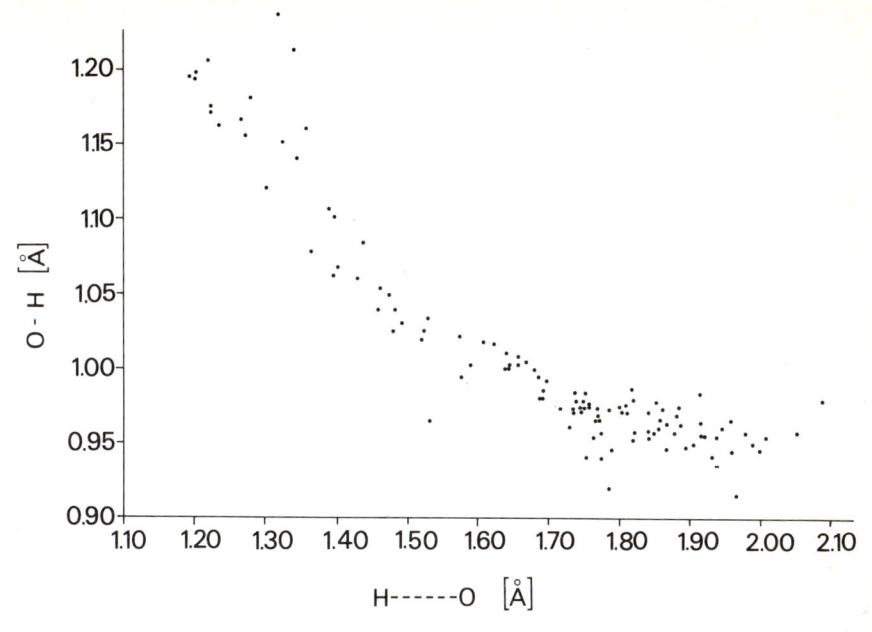

Fig. 5.1. Correlation between O–H and H···O bond lengths. These data are taken from the neutron diffraction bond lengths reported in Tables 7.2, 7.3, 7.5, 7.6 and 7.10

Table 5.1. Bond lengths in monomers and hydrogen-bonded dimers of some simple carboxylic acids from gas diffraction methods [366, 367]

	C=O	C–OH	O–H	H···O
Monomers				
Formic acid	1.217(3)	1.361(3)	0.972(5)[a]	
Acetic acid	1.214(3)	1.364(3)	1.04[b]	
Propionic acid	1.212(3)	1.396(4)	1.04	
Dimers				
Formic acid	1.220(3)	1.323(3)	1.04(2)	1.66(2)[c]
Deuterated formic acid	1.221(3)	1.333(3)	1.06(2)	1.64(1)[c]
Acetic acid	1.231(3)	1.334(4)	1.04[b]	1.64[c]
Propionic acid	1.233(6)	1.331(8)	1.04[b]	1.65[c]

[a] [368]; [b] assumed value; [c] assuming O–Ĥ···O = 180°.

bond lengths, the correlation with the X–H covalent-bond lengths is better described as a trend, rather than a relationship [365].

Changes in covalent nonhydrogen bond length as a consequence of hydrogen-bond formation is predicted theoretically and observed experimentally in particular examples. Apart from the hydrogen-bonded water polymers discussed in the preceding chapter, the simple carboxylic acids provide one of the few examples where both monomers and hydrogen-bonded dimers can be studied by the same

Table 5.2. Hydrogen-bond geometry of carboxylic acids in the crystalline state by neutron diffraction

	C=O	C–OH	O–H	H···O	O–Ĥ···O	Reference
Trichloroacetic acid[a]	1.216(4)	1.311(4)	1.013(11)	1.658(9)	177(1)	[369]
Succinic acid (77 K)[a]	1.233(2)	1.325(1)	1.005(3)	1.670(3)	174(1)	[370]
Acetic acid[a]	1.215(8)	1.330(7)	1.014(5)	1.642(13)	165(1)	[371]
Glycollic acid[a]	1.225(1)	1.323(1)	1.011(2)	1.640(2)	175.2(2)	[372]
	1.227(1)	1.329(1)	1.007(2)	1.646(2)	175.5(3)	
α-Oxalic acid 2D$_2$O	1.208(2)	1.291(2)	1.031(2)	1.493(2)	177.4(2)	[83]
α-Oxalic acid 2H$_2$O	1.212(4)	1.291(5)	1.026(7)	1.480(7)	179.3(4)	[373]
β-Oxalic acid 2D$_2$O	1.203(2)	1.291(2)	1.020(2)	1.520(2)	174.2(2)	[83]

[a] = corrected for thermal motion; standard deviations for uncorrected values.

Table 5.3. Structural data on formamide monomers, dimers, and crystal structure

	H···O	C–N	C=O	N–H(a)	N–H(b)	N–H···O	Reference
Monomer theory							
HF/3–21G		1.353	1.212	0.998	0.995		[299]
Monomer experiment							
Electron diffraction		1.367(4)	1.211(4)	1.021(9)			[374]
Dimer theory							
Closed dimer HF/3–21G	1.822	1.330	1.230	1.016	0.996	171.1	[299]
Open dimer HF/3–21G	1.914	1.343 1.345	1.219			180 (assumed)	[299]
X-ray crystal structure[a]	2.20(1)	1.319(2) 1.326(4)	1.243(4) 1.239(4)	0.89(1) 1.01(5)	0.87(1) 1.01(5)	167(1)	[376] [221]

[a] The lower values, refined with high 2θ data ($\sin\theta/\lambda > 0.85$ Å^{-1}), are more comparable to internuclear distances.

method, electron diffraction in the gas phase [175], without the added complication of crystal field effects.

The results, shown in Table 5.1, clearly demonstrate the lengthening of the acceptor C=O bond and the shortening of the donor C–OH bond on hydrogen-bond formation. This polarization of the carboxylic acid group can be represented in valence-bond resonance terms as a significant contribution from

Some high precision neutron diffraction studies of other carboxylic acids provide evidence that the shortening in the C–OH bond length is proportional to the length of the hydrogen bond, as shown in Table 5.2. The evidence relating to a cor-

Effect of Hydrogen Bonding on Molecular Structure

responding lengthening of the C=O bond, which is observed in the gas diffraction data in Table 5.1, is not apparent in the crystal structural data in Table 5.2. This is because the C=O bond length, like most multiple bonds, is much less sensitive to the changes in bond character than the single C—O bond length.

For NH···O=C hydrogen bonds, changes in the C—N and C=O bond lengths due to hydrogen-bond formation are observed experimentally and predicted theoretically. The simplest molecule containing the peptide C—N bond is formamide. The structure has been studied both experimentally as the monomer by gas-phase diffraction [374], and theoretically as the monomer and as both the cyclic and open hydrogen-bonded dimers [298, 299] **1** and **2**.

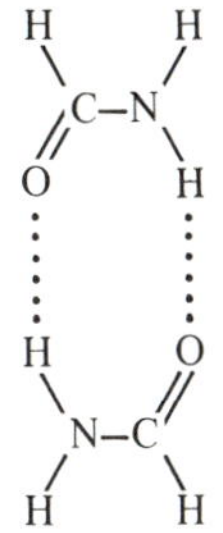

1
Closed dimer

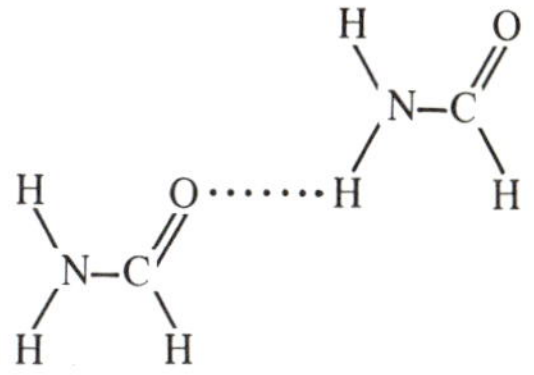

2
Open dimer

In the crystal structure [221, 375, 376], the molecules are hydrogen-bonded in a layer pattern which includes both the closed and open dimer configuration. The results of these structural studies, shown in Table 5.3, indicate that hydrogen bonding causes a shortening of the C—N bond and a lengthening of the C=O bond by an amount depending upon the number of hydrogen bonds, i.e., ~ -0.01 and $+0.01$ Å for one bond as in the open dimer, ~ -0.025 and $+0.02$ Å for two as in the closed dimer, and -0.05 and $+0.03$ Å from the experimental electron diffraction and X-ray analyses of the monomer and hydrogen-bonded polymer.

The shortening of the C—N bond implies that hydrogen bonding confers more rigidity into polypeptide chains, not only by reason of the hydrogen bonds, per se, but also because of the additional π-bond character in the peptide bond. In terms of valence-bond resonance, it increases the importance of the representation shown below:

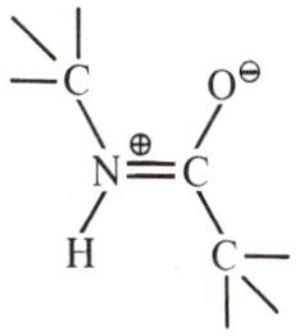

Similar shortening of the C—N bond and lengthening of the C=O bond is found when the structures of acetamide [376, 377], fluoroacetamide [299], and thioacetamide [204] in the hydrogen-bonded crystals are compared with ab-initio molecular orbital calculations for the isolated molecule, as shown in Table 5.4.

Table 5.4. Hydrogen-bonding effects on the covalent-bond lengths in formamide, acetamide, fluoroacetamide and thioacetamide. Δ (Å) refers to differences between values from the neutron diffraction analysis (corrected to *rest* structure) and the theoretical value for the isolated molecule (at rest) at the HF/3–21G level

	Formamide[a]	Acetamide	Fluoroacetamide	Thioacetamide[b]
Reference	[376]	[377]	[299]	[204]
ΔC–N	−0.027	−0.021	−0.014	0.000
				0.000
ΔC=O	+0.027	+0.034	+0.030	(−0.027)
(C=S)				(−0.022)
ΔC–C		−0.003	−0.005	0.000
				−0.003

[a] Low temperature X-ray analysis. [b] Two symmetry-independent molecules.

For hydrogen bonds weaker than OH···O=C or NH···O=C, theory predicts smaller differences in the X–H bond lengths and negligible changes in the non-hydrogen bonds. Two models have been studied theoretically with the ab-initio molecular orbital method at the HF/3–21G level of approximation. These are formaldehyde oxime cyclic dimer versus the monomer [378] and formaldehyde hydrazone cyclic dimer versus the monomer [379]. Except for the O–H and N–H bonds, these differences are too small to be detected by single crystal X-ray or neutron diffraction methods, as shown in Fig. 5.2.

Hydrogen bonding must have an effect on the electron density distribution of a molecule. In principle, this should be observed in the deformation density distributions discussed in Chapter 3. There are, in fact, two methods available. One is purely theoretical, in which the calculated deformation density for a hydrogen-bond dimer or trimer is compared with that of the isolated molecule. The other method compares the experimental deformation density of a hydrogen-bonded molecule in a crystal structure with the theoretical deformation density of the isolated molecule. Formamide has been studied by both methods [298, 380], and there appear to be significant differences in the results which are not well accounted for. Theoretical difference (dimer vs. monomer) deformation density maps have been calculated for the water dimer and the formaldehyde-water complex [312]. When those for the water dimer are decomposed into the components described in Chapter 4, a small increase in the charges on the atoms in the O–H···O bond due to the charge-transfer component is predicted [312].

Both the experimental and theoretical studies suggest that the changes in electron density distributions resulting from hydrogen-bond formation are small and subtle. On the experimental side, extremely precise X-ray analyses are necessary at low temperatures (i.e., ~10 K) combined with neutron diffraction analyses. On the theoretical side, electron correlation and basis-set superposition may have effects comparable in magnitude to the differences being sought, so that it is necessary to go beyond the Hartree-Fock limit. As remarked at a recent Sagamore meeting, it is like determining the weight of the ship's captain by weighing the ship with the captain aboard and ashore.

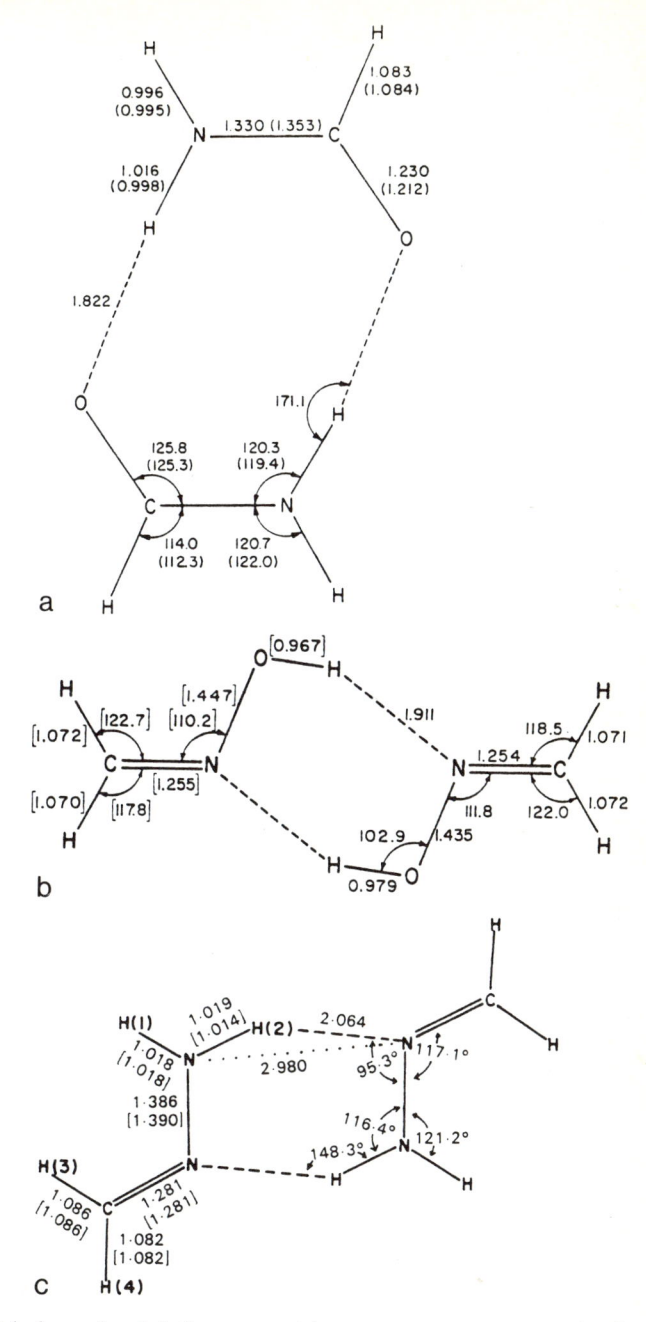

Fig. 5.2a–c. Theoretical hydrogen-bonded dimer geometries versus monomer geometries (in parentheses) in: **a** Formamide dimer calculated at HF/3–21G using Gaussian-80 [299]. **b** Formaldehyde oxime dimer calculated at HF/STO–3G using Gaussian-80 [377]. **c** Formaldehyde hydrazone calculated at HF/STO–3G using Gaussian-83 [378]

Part IB
Hydrogen-Bond Geometry

Chapter 6
The Importance of Small Molecule Structural Studies

As shown in Part IA, the term "hydrogen bond" covers bonding interactions with a wide range of geometries and energies in which the hydrogen atom is exerting a "valence" which is generally two, but can be three, and rarely, four. There is great structural versatility, with H···A distances ranging from 1.2 to beyond 3 Å and X−H···A angles from 80° to 180°.

6.1 Problems Associated with the Hydrogen-Bond Geometry

It is virtually impossible to comprehend the nature of the hydrogen bonding in a complex biological structure without some knowledge of the location of the hydrogen atoms with respect to their nearest neighbors. This structural information requires "resolution" of the atomic structure of better than 1.5 Å, which is difficult to obtain with macromolecules which have dimensions larger than 30 Å.

For the important $>$N−H and $>\overset{+}{\text{N}}$−H donor groups, it is possible to assign hydrogen atom positions from the location of the nonhydrogen atoms with a reasonable degree of certainty by assuming standard N−H covalent-bond lengths and bond angles. The same is true for −NH$_2$ if it is coplanar (conjugated) with other groups such as nucleic acid bases or amide groups. But for groups where there is rotational freedom, such as −$\overset{+}{\text{N}}$H$_3$, −NH$_2$ and −OH, the hydrogen positions can be determined only by accurate X-ray crystal structure analyses or, better, by neutron diffraction. The location of the hydrogen atoms of water molecules is often even more difficult, since these small molecules frequently exhibit both orientational and translational disorder, especially when they occur in the crystal structures of macromolecules such as the proteins and nucleic acids.

The number of crystal structure data correlates with the number of statistical analyses of hydrogen-bond geometries. As the amount of crystal structure data increased, so did the efforts to use it as a basis for statistical analyses of the geometrical properties of hydrogen bonds. Because there was only a limited amount of neutron data available and the X-ray analyses did not have the accuracy to provide reliable hydrogen atom positions, most of the earlier analyses focused on the nonhydrogen donor-acceptor distances, i.e., O···O for O−H···O bonds and N···O or N···N for N−H···O=C or N−H···N bonds [33, 53, 381−384].

The large variation in the X–H···A angles and the inability to distinguish between two- and three-center hydrogen bonds tended to obscure many of the systematic features[1] that became apparent when more accurate analyses were made accessible.

Another consequence of the limited data available was that the surveys were carried out on sets of crystal structures containing the same hydrogen bonds occurring between different donor and acceptor groups and for very different types of molecules. However, the effect of the other intra- and intermolecular forces on a particular hydrogen-bond geometry will differ from crystal structure to crystal structure. These forces tend to have an overall compression effect so that the mean hydrogen-bond length observed in crystals is generally shorter than that found in the gas phase or calculated for the equilibrium distance in isolated hydrogen-bonded distances (see Part IA, Chap. 4.3).

In order to make any statistical analyses of hydrogen-bond lengths, it is therefore necessary to assume the same *average* "crystal field" effect for all the structures included in the survey. This assumption will be more realistic if a survey extends only over molecules of the same general class, with similar shapes, sizes, and functional groups, i.e., monosaccharides or amino acids, purines and pyrimidines, nucleosides and nucleotides. If there are sufficient data available for a particular type of hydrogen bond, e.g., N–H···O=C in a particular class of compounds, like the nucleosides and nucleotides, a bond length-distribution curve can be plotted, the maximum of which is *the most probable value for that bond in that class of compounds*. We refer to this as *effective equilibrium-bond length*. Where there are not sufficient data available to make such distribution plots, as is frequently the case, we report instead a *mean value* for a particular hydrogen-bond length. Where there are sufficient data available, these distributions are approximately Gaussian, as shown in Fig. 2.5. The correspondence between the most probable and mean values is generally within the limits of precision of the experimental data.

Many of the data that are summarized in this Part IB are taken from surveys published since 1980 which have been restricted to particular classes of compounds and have used only data where reasonably reliable hydrogen atom positions are available.

6.2 The Hydrogen Bond Can Be Described Statistically

When describing the hydrogen bonding in an assemblage of molecules such as in a crystal, two aspects of the interatomic geometry have to be considered. First is the geometry of the individual hydrogen bond, and second is the structural patterns formed by these hydrogen bonds. What are the recurrent bonding geometries and what are the recurrent structural motifs, if any, and how are they related to the functional groups present and to the types of molecules involved?

---

[1] This problem is illustrated by the statement in a 1973 textbook: "The one definite fact about hydrogen bonds is that there does not appear to be any definite rules which govern their geometry" [385].

Hydrogen-bonding geometry is a statistical problem. In the liquid state, the descriptor "structure" is associated with a statistical one-dimensional probability. For intermolecular hydrogen bonding in a liquid or in solution, therefore, a dynamic rather than a static description must be sought, since the hydrogen-bonding geometries are rapidly changing.

In the crystal, the descriptor "structure" has a static conotation, and crystal structure analyses provide precise information on individual three-dimensional architectures. As described in Part IA, Chap. 2.3, hydrogen-bond lengths and angles are "soft" parameters and in any one crystal structure, a particular bond may be compressed or expanded by up to 20% of its equilibrium bond length (Part IA, Chap. 4.3).

Therefore, just as the characteristics of one human being are rarely representative of the whole of mankind, the hydrogen-bond geometry observed in one particular crystal structure is rarely characteristic of that bond. For this reason, it is necessary to examine a large distribution of bond lengths and angles for each particular variety of hydrogen bond in each class of molecules in order to determine the most probable values. This epidemiological approach to structure makes the assumption that the crystal field distortion will average out in a large sample of values. Those geometries and patterns which occur frequently in the solid state will be assumed to represent the *effective* equilibrium geometrical properties of the particular hydrogen bond. The maximum in such a bond-length distribution corresponds only qualitatively to the value at the minimum of the potential energy curve.[2] These crystal field distortions are, of course, not present in the gas-phase measurements of hydrogen-bond lengths. However, the number of hydrogen-bonded complexes that can be studied in the gas phase is very limited as there are only a few gas-phase hydrogen-bonded dimers involving those bonds which are of biological interest.

Despite this difficulty, we believe that statistical structural characteristics that are observed in the crystalline state for small biological molecules can be validly transferred to the crystalline and poorly crystalline biological macromolecular structures. At least they serve as the most useful models available for testing, using the diagnostic tools of physical chemistry. Moreover, biologically active globular proteins retain the α-helices and β-pleated sheet structures observed in their crystals. Analogous repetitive motifs are also found in the helix steps of the double helical nucleic acids. These characteristic structures are stabilized by hydrogen bonds that are intrinsically intramolecular.

To obtain significant bond length distributions, two important distinctions must be made:

---

[2] The same is true, of course, for covalent bonds, but on a different scale. In a polycyclic hydrocarbon, the C–C bond lengths may vary by up to 20% of their length, i.e., from 1.490 to 1.560 Å. This is not because of any difference in electronic character, since they are all C–C single bonds, but because of the relief of an internal strain which is distributed between the bond lengths, valence angles, and torsional angles. Within the constraint of ring connectivity, some bonds are therefore compressed, while others are expanded from their equilibrium values [386]. Covalent bonds are also expanded or contracted by crystal field effects, but these distortions are so small that they rarely exceed the precision of the experimental measurements.

Box 6.1. Hydrogen-Bond Donors and Acceptors Occurring in Biological Molecules

Donors

Hydroxyl groups

H–O$_W$–H	Water
C–O–H	Sugars, amino acid side chains, ribonucleic acid
P–O–H	Nucleic acids, prosthetic groups
N–O–H	Rarely observed

Amino and imino groups, neutral and charged

>N–H	Peptide groups, amides, amino acid side chains, nucleic acids
–N(H)–H	Amides, amino acid side chains, nucleic acids
>N$^+$–H	Amino acid side chains, nucleic acids (low pH)
>N(H)$^+$–H	Biological polyamines
–N(H$_2$)$^+$–H	Amino acids, biological polyamines
N(H$_3$)–H	Ammonium ion

Acceptors

Oxygen atoms, neutral and charged

O=C<	Peptide groups, amides, nucleic acids, carboxylic acids (free acid form)
O$_W$<^H_H	Water
O<^C_H	Sugars, amino acid side chains, ribonucleic acid
O<^C_C	Esters, ethers (most frequently pyranose and furanose ring oxygen atoms)
O<^P_H	Nucleic acids, prosthetic groups
O=C̄	Carboxylates
O=P̄	Phosphates
O=S̄	Sulfates
O=N̄	Nitrates

Nitrogen atoms

N⫽	Amino acid side chains, nucleic acids
N(H)$_2$–	Good acceptor if bound to sp^3 atom, e.g., amino sugars, prosthetic groups, *not* or very rarely in amides, nucleic acids

Salt bridges,
a term applicable if both donor and acceptor groups are charged.

---

1. Two-, three- and four-center bonds must be considered separately.
2. The effect of neighboring covalently bonded atoms on the electronic structure of the donor or acceptor groups given in Box 6.1 must be recognized. Clearly, P–OH, N–OH, C–OH and O$_W$(H)H are quite different donor groups. The same applies to N(H$_3$)H, –N(H$_2$)H, >N(H)H, >NH, >N–H, and –N(H)H. For the acceptor groups, it is necessary to distinguish between O=C<, O$_W$<^H_H, O<^C_H, O<^C_C, for example. If sufficient data are available, it has been demonstrated that even finer distinctions can be made with $^R_{R_1}$>C=O groups, accord-

ing to whether they are carboxylates, carboxylic acids, amides, ketones, etc. [75].

The distribution of hydrogen-bond lengths shows two important features. If we examine the distribution curves for particular donor-acceptor combinations under the conditions described above, we find:
− The distribution of bond lengths for a particular two-center bond in the crystal structures of a particular class of biological compounds has a well-defined maximum but is more symmetrical than might be expected from the hydrogen-bond potential energy curve (cf. Figs. 2.4 and 4.2a, b).
− The maximum and the mean value in the distribution for a particular donor/acceptor pair depends on the class of crystal structure surveyed.

As will be pointed out later, OH$\cdots$OH bonds in disaccharides tend to be longer than in monosaccharides. Comparable bonds tend to be longer in nucleosides than in purines and pyrimidines. This is because the other cohesive forces in the crystal appear to compress the hydrogen bonds, particularly those which are longer than the distribution maximum. As discussed in Part IA, Chapter 2, an attempt to fit the distribution curves for $N-H\cdots O = C$ and $N-H\cdots N\langle$ bond lengths to a generally applicable Lennard-Jones type of potential energy curve required the addition of a crystal field compression force and a thermal energy term [306].

6.3 The Problems of Measuring Hydrogen-Bond Lengths and Angles in Small Molecule Crystal Structures

Ideally, any study of hydrogen-bond geometries should be based solely on single crystal neutron diffraction studies. It is only for the amino acids [60], and the monosaccharides [58, 59] that any attempt has been made to systematically study a significant number of crystal structures of closely related biological molecules by neutron diffraction. For the peptides, disaccharides, and the nucleic acid components, no systematic neutron diffraction studies have been made, and data are available only for a few isolated examples.

For hydrogen bonds in crystal structures, statistical analyses are possible only if neutron and X-ray data are combined. To obtain the metrical information described in this chapter, it is necessary to combine the limited amount of accurate neutron diffraction data with the much more abundant, but less accurate or reliable, data from the X-ray crystal structure analyses. This combination is possible if the X−H covalent-bond lengths, which appear notoriously shortened in the X-ray analyses, are normalized to standard neutron diffraction values.

X−H covalent-bond lengths determined by X-ray diffraction methods are generally too short and the H$\cdots$A hydrogen-bond lengths are correspondingly too long. The hydrogen atom positions in the most accurate X-ray analyses are at least an order of magnitude less accurate than those from neutron analyses. X-ray analyses determine the thermally averaged maxima of the atomic electron density which, in the case of hydrogen atoms, are closer to the covalently bonded nonhydrogen atom by distances up to 0.3 Å compared with the internuclear X−H

Table 6.1. Comparison of hydrogen-bond lengths from neutron (N) and X-ray (X) diffraction studies, and X-ray normalized (X_{norm}) values[a]

Compound [REFCODE]	O—H (X)	O—H (N)	H···O (X)	H···O (N)	(X_{norm})
Methyl α-D-altropyranoside	0.81	0.971	1.90	1.736	1.74
[MALTPY]:[MALTPY01]	0.88	0.961	2.00	1.922	1.91
	0.85	0.964	2.11	2.085	2.02
			2.29	2.140[b]	2.19
	0.83	0.963	2.22	2.138	2.15
			2.29	2.185[b]	2.19
Methyl α-D-glucopyranoside	0.87	0.985	1.84	1.738	1.74
[MGLUCP]:[MGLUCP11]	0.97	0.969	1.76	1.770	1.76
	1.11	0.966	1.61	1.772	1.75
	1.05	0.923	2.19	2.328	2.24
			2.50	2.633[b]	2.45
Methyl α-D-mannopyranoside	0.80	0.976	1.98	1.810	1.81
[MEMANP]:[MEMANP11]	0.67	0.957	2.22	1.998	1.96
	0.85	0.955	2.05	1.917	2.38[c]
	0.96	0.959	2.07	2.052	2.05
Methyl β-D-galactopyranoside	0.77	0.957	1.98	1.773	1.79
[MBDGAL01]:[MBDGAL02]	0.86	0.958	1.87	1.739	1.78
	0.84	0.976	1.98	1.860	1.86
	0.86	0.971	2.31	2.24	2.24
			2.80[b]	2.75[b]	2.71[b]
Methyl α-D-galactopyranoside	1.06	0.946	1.76	1.817	1.85
monohydrate	1.06	0.971	1.66	1.747	1.75
[MGALPY]:[MGALPY01]	0.95	0.976	1.75	1.706	1.73
	0.89	0.967	1.97	1.851	1.90
	0.92	0.958	2.03	1.983	2.08
	1.04	0.938	2.07	2.128	2.03
			2.12[b]	2.210	2.15
			(X)−(N)		(X_{norm}−N)
		mean	0.102		0.033

[a] All distances are in Å. The hydrogen atom positions in the X_{norm} data have been adjusted so as to normalize the covalent O−H distance to 0.97 Å. [b] Three-center hydrogen bonds. [c] The hydrogen atom was incorrectly located in the X-ray study.

distances. *Thus O−H, N−H, and C−H covalent-bond lengths which are observed to be 0.97, 1.03, and 1.10 Å by neutron diffraction can be determined as short as 0.6, 0.7, and 0.8 Å by X-ray diffraction.*

Table 6.1 shows some comparisons of values determined by X-ray and neutron diffraction studies for O−H and H···O bond lengths for some carbohydrate crystal structures that have been analyzed by both methods. Application of the normalization reduces the discrepancy between the H···O distances obtained by X-ray and neutron diffraction experiments from 0.102 Å to 0.033 Å. Similar results have been reported for NH···O=C bonds [387], where the random errors in the X-ray N−H and H···O distances varied between 0.02 and 0.17 Å with a mean of 0.065 Å when compared with the same neutron diffraction values. Cor-

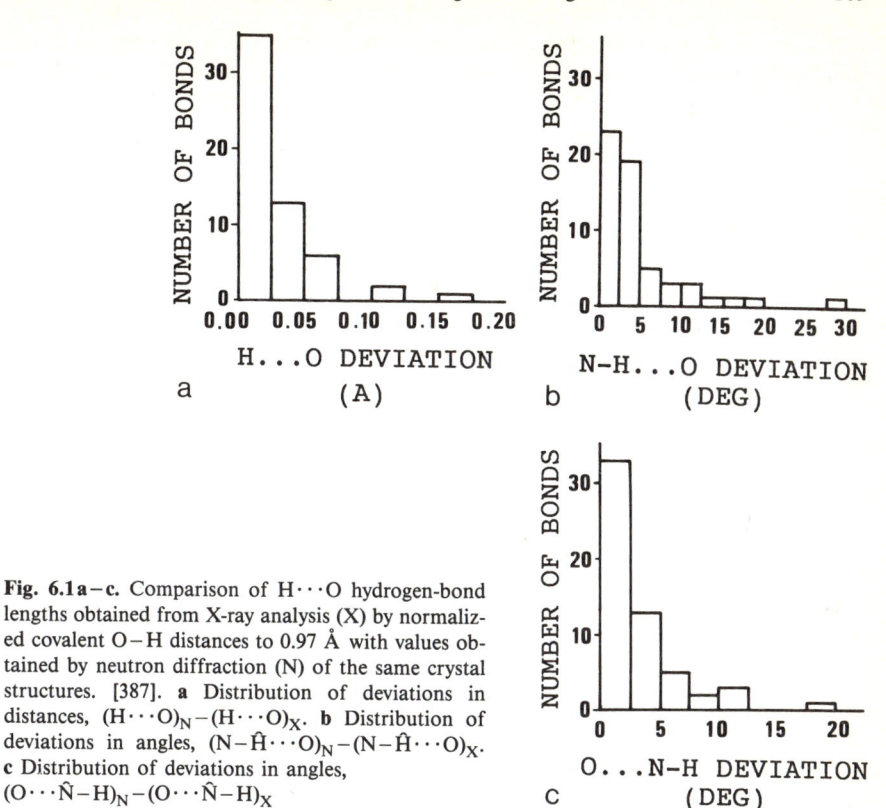

The Problems of Measuring Hydrogen-Bond Lengths and Angles

Fig. 6.1a–c. Comparison of H···O hydrogen-bond lengths obtained from X-ray analysis (X) by normalized covalent O–H distances to 0.97 Å with values obtained by neutron diffraction (N) of the same crystal structures. [387]. **a** Distribution of deviations in distances, $(H···O)_N - (H···O)_X$. **b** Distribution of deviations in angles, $(N-\hat{H}···O)_N - (N-\hat{H}···O)_X$. **c** Distribution of deviations in angles, $(O···\hat{N}-H)_N - (O···\hat{N}-H)_X$.

rections by applying the normalization procedure described below reduces these errors to an estimated standard deviation of 0.065 Å.

These analyses also compared X-ray and neutron values of the hydrogen-bond angles and conclude that the agreement $(N-\hat{H}···O)_X - (N-\hat{H}···O)_N$ is generally within 10°. The plots of these "error" distributions are shown in Fig. 6.1.

No analyses have been made which normalize the covalent C–O–H, H–O_W–H, or P–O–H angles to tetrahedral values, since such corrections would have only a second-order effect on the bond lengths.

The shortening effect arises primarily from the distribution of valence electrons in the X–H bond, leading to a significant difference in the position of the electron peak and the nucleus for the hydrogen atom. However, it is also influenced by the experimental conditions of the X-ray diffraction experiment, such as the range of scattering angle to which intensity data are collected ("resolution"), the temperature at which the crystal structure analysis is carried out, and the degree of atomic thermal motion. It will therefore vary between and within structure analyses for the same type of bond as shown in Table 6.1.

Normalization correction of X–H covalent-bond lengths is essential for deriving hydrogen-bond geometry from X-ray analyses. These "errors" in the "X-ray" hydrogen-bond distances can completely obscure any systematic trends in the bond length distributions due to the differences in the donor or acceptor groups or the patterns of bonding. They can be significantly reduced by normalizing the X–H covalent-bond lengths to "standard" neutron diffraction values, i.e., 0.97 Å for O–H[3], 1.03 Å for N–H, and 1.10 Å for C–H. These corrections are less serious for C–H bonds, where the X-ray distances are commonly within 0.1 Å of the neutron value, due to the smaller electronegativity difference between the carbon and hydrogen atoms.

The normalization is made by moving the hydrogen atom along the direction of the X–H bond to the position corresponding to the neutron standard bond length [107]. As shown in Table 6.1, this significantly reduces the systematic error in the hydrogen-bond lengths obtained by X-ray analysis. It is used on all the data quoted in this monograph.

As already pointed out, it is very important in describing metrical properties to distinguish between two-center and three-center hydrogen bonds. The distinction is made by exploring the environment of the hydrogen atom in the forward direction relative to the covalent X–H bond for electronegative acceptor atoms. In the crystal structures which we will be discussing, these first neighbor hydrogen-bond interactions always occur within 3.0 Å of the hydrogen atom. Beyond that distance, the first neighbor interactions are invariably between hydrogen atoms and with very few exceptions are considered to be of the H···H van der Waals type.

---

[3] For the stronger P–OH···O=C bonds, there is evidence that the best mean value for the O–H bond length is 1.05 Å.

Chapter 7
Metrical Aspects of Two-Center Hydrogen Bonds[1]

7.1 The Metrical Properties of O–H···O Hydrogen Bonds

The donor and acceptor groups found in biological structures which form O–H···O bonds are given in Box 6.1. The C–OH···O hydrogen bonds are the primary intermolecular cohesive force between the carbohydrate molecules, water, and carboxylic acids. They occur extensively in the structures of cyclodextrins, polysaccharides, glycolipids, and glycoproteins. The $O_W H$···O bonds are important in the ices, in hydrated crystal structures, and in the hydration shell of all biological molecules.

7.1.1 Very Strong and Strong OH···O Hydrogen Bonds Occur with Oxyanions, Acid Salts, Acid Hydrates, and Carboxylic Acids

In the wide range in strength and length, OH···O hydrogen bonds are unique amongst those found in biological structures. They range from the very short, strong bonds with hydrogen-bond lengths only about 25% longer than the covalent O–H bonds, to very weak bonds with lengths almost three times the covalent bond lengths.

A characteristic of very strong O–H···O bonds is that they are accompanied by a lengthening of the covalent O–H bond (Part I, Chap. 5). In the extreme cases, the covalent- and hydrogen-bond lengths are equal within the limits of the experimental observations. Such a symmetrical O···H···O configuration corresponds to a single potential energy well between the two oxygen atoms.

The circumstances under which a single-well potential can occur, or in which the barrier is low enough to permit proton tunneling has been a subject of interest and research since the early work of Huggins. It has been reviewed extensively elsewhere [28, 48, 49, 388, 389]. In this monograph, it will be discussed only briefly with respect to the geometry associated with very strong hydrogen bonds because

---

[1] Henceforth, crystal structure analyses of carbohydrates (class 45), amino acids (class 48), purines and pyrimidines (class 44) and nucleosides and nucleotides (class 47) are referenced by means of their Cambridge Crystallographic Data Base REFCODES. All other crystal structure analyses are referenced in the General Index.

from that point of view there is no marked discontinuity between "strong" and "normal" hydrogen bonds, which are predominant in the biological structures.

Very short O–H···O hydrogen bonds are observed in two different types of crystal structures

1. They occur as intramolecular hydrogen bonds, where a short O···O separation of between 2.4 and 2.6 Å is the result of steric constraints within the molecule. Examples of such bonds are shown in Table 7.1.
2. They are found when the hydrogen-bond functional groups are present in ions (Table 7.2). This is because the positive charge on a cation containing the donor X–H group will decrease the shielding of the proton and render it more acidic. Conversely, a negative charge in an anion containing an acceptor atom will increase its electronegativity.

These bonds are analogous to the classical example bifluoride ion [F–H–F]⁻ [48, 388]. They occur in the crystal structures with hydrated hydroxyl ions and with hydrated protons [112, 113]. These latter crystals are referred to as pseudo-

Table 7.1. Short O–H···O intramolecular hydrogen bonds observed in some organic molecules by neutron diffraction

Crystal structures		H···O (Å)	O–H (Å)	O–H···O (°)	O···O (Å)	Reference
Lithium hydrogen phthalate-methanol		1.205(5)	1.195(5)	171.3(5)	2.393(4)	[390]
		1.226(5)	1.172(5)	169.3(6)	2.388(4)	
Pyridine 2,3-dicarb-oxylic acid	(293 K)	1.238(5)	1.163(5)	174.4(4)	2.398(3)	[391]
	(100 K)	1.227(3)	1.176(3)	174.5(2)	2.400(2)	
1,3-Diphenyl-1,3-propanedione		1.360(9)	1.161(9)	154.7(7)	2.463(4)	[392]
1-Phenyl-1,3-butanedione		1.322(12)	1.238(11)	153.2(8)	2.489(5)	[393]
6-Hydroxy-fulvene-carbaldehyde		1.343(11)	1.214(12)	171.8(8)	2.557(6)	[394]
Naphthazarin C	(60 K)	1.669(4)	0.995(5)	148.0(3)	2.582(5)	[395]

Table 7.2. Some short asymmetrical O–H···O bonds observed in crystal structures of organic hydrogen anions by neutron diffraction

Crystal structures	O···O (Å)	H···O (Å)	O–H (Å)	O–H···O (°)	Reference
Imidazole H maleate	2.393(3)	1.197(5)	1.196(5)	176.8(4)	[396]
Ca H Maleate	2.424(2)	1.305(3)	1.121(3)	175.9(3)	[396]
K H Chloromaleate	2.403(3)	1.206(5)	1.199(5)	175.4(4)	[397]
K H Diformate	2.437(1)	1.270(1)	1.167(1)	179.3(1)	[398]
Na H Maleate 3H$_2$O	2.445(2)	1.367(2)	1.079(2)	176.1(2)	[399]
K H Oxydiacetate	2.476(2)	1.328(3)	1.152(3)	174.2(27)	[400]
K H Crotonate	2.488(2)	1.348(2)	1.141(2)	178.0(3)	[401]
K H Di(dichloroacetate)	2.498(3)	1.392(4)	1.107(4)	177.8(3)	[402]
K H Oxalate	2.518(3)	1.463(5)	1.054(5)	174.6(4)	[403]
Na H Oxalate H$_2$O	2.566(2)	1.531(2)	0.965(2)	176.8(2)	[404]

The Metrical Properties of O–H···O Hydrogen Bonds

Table 7.3. Some short OH···O hydrogen bonds observed in a number of hydrogen phosphates and sulfates by neutron diffraction

Compound	O···O (Å)	H···O (Å)	O–H (Å)	O–H···O (°)	Reference
$H_2PO_4^- \cdot COH \cdot (NH_2)_2^+$	2.421(3)	1.223(6)	1.207(6)	169.9(4)	[405][a]
$Na_3H(SO_4)_2$	2.432(4)	1.276(3)	1.156(3)	179.1(3)	[365]
KD_2PO_4	2.455(6)	1.397(5)	1.063(5)	172.6(5)	[406]
$CaHPO_4$	2.461(2)	1.283(3)	1.182(3)	173.5(3)	[407]
NaH_2PO_4	2.485(5)	1.459(8)	1.040(9)	167.7(7)	[408]
	2.550(5)	1.524(8)	1.026(8)	177.6(7)	
	2.591(5)	1.590(7)	1.003(7)	175.6(6)	
	2.644(5)	1.644(8)	1.001(9)	177.0(7)	
RbH_2PO_4	2.489(11)	1.430(5)	1.061(16)	175.4(25)	[409]

[a] see Fig. 7.1.

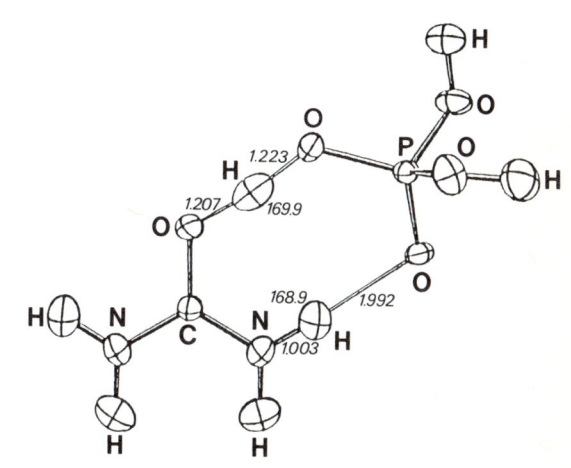

Fig. 7.1. Hydrogen bonding in urea-phosphoric acid, $[H_2PO_4]^- \cdot [(NH_2)_2COH]^+$. The standard deviations for bond distances and angles are 0.005 Å and 0.5° [405]

hydrates. They contain the hydrogen-bonded complexes $(H_2O)_nH^+$, as in the hydrates of strong acids such as $HClO_4 \cdot nH_2O$ or $H_2SO_4 \cdot nH_2O$ [113]. The O···O separations range typically from 2.4 to 2.6 Å and provide examples of both unsymmetrical and crystallographic symmetric configurations.

From the biological point of view, the important short, strong hydrogen bonds are those formed with the protonated oxyanions such as the hydrogen carboxylates and the hydrogen phosphates and sulfates, as shown in Table 7.3.

The urea trihydrogen phosphate crystal structure [405] shown in Fig. 7.1, for example, possesses one very short, almost symmetrical C–OH···Ō=P bond and one normal N(H)H···Ō=P bond. The latter hydrogen-bonding functional group is likely to be present in protein-nucleic acid interactions.

Strong hydrogen bonds can be symmetrical or unsymmetrical. In some crystal structures, the midpoint of the O···O hydrogen-bonded separation is a crystal-

Table 7.4. Some short symmetrical O···H···O hydrogen bonds observed in the crystal structures of the hydrogen carboxylate ions

Crystal structures[a]	H···O = $\frac{1}{2}$O···O	Symmetry	Reference
NH$_4$ H dichloroacetate (X)	1.216	$\bar{1}$	[410]
K H di-trifluoroacetate (N)	1.218	$\bar{1}$	[411]
K H dibenzoate (X)	1.220	$\bar{1}$	[412]
K H biphenylacetate (X)	1.221	$\bar{1}$	[413]
K H succinate (N)	1.222	2	[414]
K H di-aspirinate (N)	1.224	$\bar{1}$	[415]
Rb H oxydiacetate (N)	1.224	2	[400]
hydrazinium hydrogen oxalate (N)	1.224	$\bar{1}$	[416]
K H di-p-hydroxybenzoate hydrate (X)	1.229	$\bar{1}$	[417]
K H malonate (N)	1.229	$\bar{1}$	[418]

[a] (X) = X-ray study, (N) = neutron study.

lographic center of symmetry or a twofold axis symmetry. Some examples are given in Table 7.4. Where there is clear evidence of twofold disorder, these bonds are referred to as symmetrical double-site bonds, i.e., O−($\frac{1}{2}$)H···($\frac{1}{2}$)H−O. When the separation of the two sites is very short, the interpretation of the distribution of the proton (Fermi) nuclear density obtained from neutron diffraction experiments is ambiguous.

Since no symmetrical single-site hydrogen bonds have been found in the absence of crystallographic symmetry, the current view is that symmetrical ordered O···H···O configurations do not occur. Thus in the neutron diffraction study of rubidium hydrogen oxydiacetate a crystallographical symmetrical O···H···O bond occurs, whereas in the potassium salt, which is not isomorphous, this hydrogen bond is clearly asymmetrical [400] (see Table 7.2).

These very strong, short hydrogen bonds have energies up to ten times greater than those generally occurring in the biological systems. They are therefore too rigid and would inhibit most biological processes or at least slow them down considerably. However, they may be important in biochemical reactions incurred either by enzymatic processes or by radiation damage.

There is no evidence of a quantitative relationship between the O−H and H···O bond lengths and O···O distances. As shown for the intra- and intermolecular short hydrogen bonds in Tables 7.1 and 7.2, there is a trend for the shorter O···O distances to have correspondingly shorter hydrogen-bond lengths and longer covalent-bond lengths. It was thought at one time that there should be a smooth correlation between the O−H, H···O, and O···O lengths [419], but more recent neutron diffraction data do not support this view [365]. The intramolecular O−H···O=C bonds are shorter than their intermolecular counterparts which occur in the carboxylic acids. This suggests that these hydrogen bonds are compressed with respect to their equilibrium values and the degree of compression depends upon the steric constraints imposed by the particular molecule. The geometries of the weaker hydrogen bonds are even more affected, and it is only in the data from gas or liquid phase studies that a quantitative correlation between O−H and H···O separations might be expected.

As will be shown later, for H···O hydrogen bonds longer than 1.6 Å, any changes in the covalent-bond lengths are generally too small to be observable even by the more precise neutron diffraction analysis[2]. This is consistent with the view that the majority of the hydrogen bonds which occur in biological structures are primarily electrostatic, with relatively little covalent character.

Proton transfer occurs across short intramolecular hydrogen bonds in the gas or liquid phase but is rarely observed in crystalline state. Intra- and intermolecular keto-enol proton transfer C−OH···O=C ⇌ C=O···H−O−C occurs in the vapor and liquid states in β-diketones and in β-ketoesters where the two states are exactly symmetrically equivalent. Any configuration change which destroys this symmetry inhibits the proton transfer.

A well-explored example of this phenomenon in the crystalline state is naphthazarin C, where single crystal neutron diffraction studies indicate an order-disorder transition at 100 K. At 60 K, the hydroxyl groups are ordered, whereas at 300 K, the analysis shows a disordered O−($\frac{1}{2}$)H···($\frac{1}{2}$)H−O hydrogen-bond structure [395].

Disordered O−H···O intramolecular hydrogen bonds are not uncommon in crystal structures of molecules having cis-enol configurations, but without evidence for an order-disorder transition, they do not necessarily imply that proton transfer takes place in the crystalline state.

The OH···O hydrogen-bond lengths in the carboxylic acids and their hydrates exhibit a wide range, from very short and comparable to those described in the previous section, to relatively weak bonds with lengths of about 2.0 Å. As shown in Table 7.5, these hydrogen bonds can be divided into four distinct categories:

1. The shortest O=Ċ−OH···O=C̄ and O=Ċ−OH···O$_W$ bonds, 1.399 to 1.520 Å, which display a significant lengthening of the covalent O−H bonds.
2. The O=Ċ−OH···O=Ċ−OH bonds in the range from 1.640 to 1.699 Å. The observed lengthening of the covalent O−H bonds is barely significant[3].
3. The ≻C−OH···O=Ċ−OH and O$_W$H···O=C̄ bond lengths of about 1.75 to 1.79 Å.
4. The O$_W$H···O=Ċ−OH hydrogen bonds which are the longest, ranging from 1.895 to 2.008 Å.

The observation that *the water molecule is a stronger than normal hydrogen-bond acceptor but weaker than normal donor* is clearly demonstrated in these data and is made consistently throughout the study of small molecule hydrates reported in later chapters.

The OH···O hydrogen bonds in the amino acid crystal structures show the same characteristics as in the carboxylic acids. There are more extensive neutron data available in this category, as shown in Table 7.6. The H···O distances range

---

[2] Care has to be exercised in ascribing significance to small differences in X−H bond lengths because of the effects of thermal motion and anharmonicity discussed in Part I, Chapter 3.
[3] see Footnote[2].

Table 7.5. Hydrogen-bond lengths in the carboxylic acids, determined by neutron diffraction

Donor	Acceptor	O–H (Å)	H⋯O (Å)	O–H⋯O (°)	Structure	References
–C(=O)(OH---O)	O=C⊖(O–)	1.102(7)	1.399(7)	177.1(7)	Ammonium tetroxalate 2H$_2$O	[420]
–C(=O)(OH---O$_w$)		1.069(7)	1.403(7)	178.2(6)	Ammonium tetroxalate 2H$_2$O	[420]
		1.026(4)	1.480(7)	179.3(6)	α-Oxalic acid 2H$_2$O	[373]
		1.040(7)	1.483(7)	173.3(7)	Ammonium tetroxalate 2H$_2$O	[420]
		1.031(2)	1.493(2)	177.4(1)	Deuterated α-oxalic acid 2D$_2$O	[83]
		1.020(2)	1.520(2)	174.2(1)	Deuterated β-oxalic acid 2D$_2$O	[83]
–C(=O)(OH---O=C(OH)(C))		1.011(15)	1.642(13)	164.8(1.0)	Acetic acid	[371]
		1.001(2)	1.640(2)	175.5(3)	Glycollic acid	[372]
		1.003(2)	1.646(2)	175.2(4)		
		1.005(2)	1.670(2)	174.3(1)	Succinic acid (77 K)	[370]
		0.995(2)	1.687(3)	173.7(1)	Succinic acid (300 K)	[370]
		1.009(11)	1.658(9)	176.5(5)	Trichloro acetic acid	[369]
		0.992(7)	1.699(6)	175.2(5)	Acetic acid/phosphoric acid complex	[421]
		0.986(7)	1.694(6)	175.4(5)		
C–OH---O=C(OH)(C)		0.970(3)	1.774(3)	157.6(2)	Glycollic acid	[372]
		0.971(3)	1.753(3)	169.7(2)		
O$_w$H---O(=C⊖)(O–)		0.966(7)	1.766(7)	169.5(6)	Ammonium tetroxalate 2H$_2$O	[420]
		0.973(6)	1.787(6)	167.4(6)		
O$_w$H---O=C(OH)(C)		0.964(4)	1.917(9)	166.9(6)	α-oxalic acid 2H$_2$O	[373]
		0.956(9)	1.979(9)	156.6(4)		
		0.954(2)	1.939(2)	167.7(2)	Deuterated α-oxalic acid D$_2$O	[83]
		0.954(2)	2.008(2)	156.0(2)		
		0.944	1.960	157.2	Deuterated β-oxalic acid D$_2$O	[83]
		0.947	1.895	170.3		
		0.949(8)	1.906(7)	164.8(7)	Ammonium tetroxalate 2H$_2$O	[420]
		0.945(8)	2.000(8)	158.2(7)		

from 1.439 Å to 2.059 Å, some of the shortest being associated with the strong acceptors formed by the charged carboxylate groups which occur due to the zwitterionic character of the amino acids. The strongest bonds have O=C–OH as donors. The water molecules tend to act as donors in the weaker hydrogen bonds.

7.1.2 OH⋯O Hydrogen Bonds in the Ices and High Hydrates

The many polymorphs of ice [422, 423] and the wide variety of ice-like hydrate inclusion compounds provide a source of information about OH⋯O hydrogen bonds where the water molecules are the structure-determining molecular species

The Metrical Properties of O–H···O Hydrogen Bonds 117

Table 7.6. O–H···O Hydrogen-bond geometries in the crystal structures of the amino acids from neutron diffraction data

Type of bond	O–H (Å)	H···O (Å)	O–H···O (°)	Crystal structure	REFCODE
O=C–OH···O_W	1.085(16)	1.439(16)	179.1(8)	L-Cysteic acid·H₂O	CYSTAC01
	1.003(7)	1.658(6)	166.8(5)	Glycyl-glycine HCl·H₂O	GLCICH01
O=C–OH···O=C=O	1.050(4)	1.475(5)	172.3(3)	L-Glutamic acid (β-form)	LGLUAC11
	1.034(8)	1.530(8)	176.2(6)	N-Acetyl glycine	ACYGLY11
	1.000(15)	1.682(14)	169.7(12)	Hippuric acid	HIPPAC02
O=C–OH···O=C–OH	1.017(13)	1.624(13)	173.6(11)	L-Glutamic acid HCl	LGLUTA
O=C–OH···OH	1.018(6)	1.609(5)	170.7(4)	L-Tyrosine HCl	LTYRHC10
O=C–OH···O–S=O (O,O)	0.996(5)	1.520(5)	179.0(6)	Triglycine sulfate	TGLYSU11
	1.022(5)	1.576(5)	178.7(5)	Ammonium glycinium sulfate	AGLYSL01
C–OH···O=C=O	0.981(4)	1.689(4)	173.8(3)	L-Tyrosine	LTYROS11
	0.981(1)	1.692(2)	175.1(1)	D,L-Serine	DLSERN11
	0.958(7)	1.843(6)	166.8(4)	L-Serine·H₂O	LSERMH11
	0.971(4)	1.843(3)	164.9(3)	4 Hydroxy proline	HOPROL12
O_WH···O_W	0.957(11)	1.775(9)	167.9(6)	L-Lysine HCl·2H₂O	LYSCLH11
	0.946(7)	1.789(7)	164.9(7)	L-Arginine·2H₂O	ARGIND11
	0.909(11)	2.058(9)	117.8(7)	L-Arginine·2H₂O	ARGIND11
O_WH···O=C=O	0.954(8)	1.764(6)	170.7(6)	Glycyl-glycine HCl·H₂O	GLCICH01
	0.974(6)	1.770(6)	172.1(5)	L-Arginine·H₂O	ARGIND11
	0.950(4)	1.817(5)	174.6(4)	L-Histidine·HCl·H₂O	HISTCM12
	0.960(4)	1.857(7)	169.2(7)	L-Lysine·HCl·H₂O	LYSCLH11
	0.946(7)	1.868(6)	173.2(5)	L-Serine·H₂O	LSERMH10
	0.957(2)	1.879(4)	164.3(3)	L-Asparagine·H₂O	ASPARM02
	0.962(3)	1.888(3)	171.5(3)	L-Asparagine·H₂O	ASPARM02
	0.915(11)	1.965(9)	169.1(7)	L-Arginine·H₂O	ARGIND11
O_WH···OH	0.957(8)	1.822(7)	177.4(5)	L-Serine·H₂O	LSERMH10
	0.955(8)	1.922(7)	177.4(5)	L-Serine·H₂O	LSERMH10

Table 7.7. Hydrogen-bond geometry in the ices[a]

		O–H (Å)	H···O (Å)	O–Ĥ···O (°)	O···O (Å)	Reference
Ice Ih Neutron data, disordered[b] O–½(H)···½(H)–O	60 K 123 K 223 K	1.004(1)[a] 1.005(1) 1.008(2) 1.004(1) 1.008(4) 1.004(2)	1.746(3) 1.747(1) 1.743(4) 1.749(2) 1.751(4) 1.756(3)	180 180 180	2.750(1) 2.752(1) 2.751(1) 2.753(1) 2.759(2) 2.760(1)	[424] [425]
Ice II Neutron data, deuterated ordered, O–D···O	210 K	0.94(2) 0.96(2) 0.98(2) 1.01(2)	1.81(2) 1.84(2) 1.85(2) 1.86(2)	166(2) 166(2) 168(2) 178(2)	2.77(2) 2.80(2) 2.80(2) 2.84(2) ⟨2.80⟩	[426]
Ice IV X-ray data	110 K				2.785(2) 2.805(8) 2.880(7) 2.918(10) ⟨2.823⟩	[427]
Ice V X-ray data probably disordered	100 K				2.766(6), 2.766(6) 2.781(6), 2.819(6) 2.782(6), 2.820(6) 2.798(6), 2.867(6) ⟨2.813⟩	[428]
Ice VI Neutron powder data, deuterated, probably disordered	225 K	0.94–0.98(5) ⟨0.967⟩	1.80	157–175	2.73–2.78(3) 2.774 3.43–3.46(4) ⟨3.411⟩	[235]

Ice VII Neutron powder data, deuterated, protons and oxygens disordered		0.943(2) (best model)	1.958(2)	180	2.901(1)	[429]
Ice VIII Neutron powder data, deuterated, protons ordered	10 K	0.968(7)	1.910	178.2(7)	2.879(1) 2.743(9)[d]	[236]
Ice IX Neutron data deuterated almost completely ordered[c]	110 K	0.979(4) 0.984(3) 0.986(8)	1.789(4) 1.813(3) 1.821(7)	167.2(3) 165.3(3) 174.6(2)	2.750(3) 2.763(3) 2.793(3)	[430]
Ethylene oxide deuterohydrate Neutron data D atoms completely disordered	80 K	0.998(2) 0.993(4) 0.990(9) 0.972(5) 0.971(4) 0.973(5)	1.726(2) 1.757(9) 1.761(4) 1.797(5) 1.802(5) 1.818(5)	180 178.4(3) 177.3(5) 177.5(4) 173.6(4) 171.6(4)	2.724(2) 2.750(9) 2.750(9) 2.768(3) 2.768(3) 2.784(4)	[431]

[a] The abnormally long O–H distance has been shown to be due to additional complexities in the ice structures described in Part IV, Chapter 21.
[b] Values are corrected for thermal motion. Top line is bond along hexagonal axis. Ratio of crystallographic axes c/a = 1.6284, ideal tetrahedral = 1.633.
[c] Values corrected for thermal motion.
[d] Nonbonded.

present. These crystal structures can be considered as macromolecules in which the hydrogen bonds are the principal cohesive forces. Other interactions, such as the van der Waals attraction (which are of importance in the clathrate hydrates), play a relatively minor role in determining the hydrogen-bond geometry. This situation is distinctly different from that in the mono-, di- and trihydrates most commonly found with small biological molecules. In those crystal structures the water molecules accommodate to packing schemes and frequently serve as "space fillers" in geometrical situations dominated by the size, shape, and functional groups of the solute species present in the crystal.

It is noticeable that the monosaccharides, which are oblate molecules able to pack efficiently, form significantly less hydrates than do the disaccharides or nucleosides, which are more awkward-shaped molecules from the packing point of view.

The hydrogen-bond characteristics in the ice polymorphs [422] which have been determined by neutron diffraction data are shown in Table 7.7. The hydrogen-bond lengths extend over a range of 1.726 to 1.958 Å, which is narrower than observed in the carboxylic acid hydrates or in the amino acid hydrates, and corresponds more closely to the $H \cdots O$ distances found in the sugar hydrates (see below). Since the ices are macromolecules with three-dimensional nets of hydrogen bonds, external pressure applied to the crystals causes compression or lengthening of the hydrogen bonds from their effective equilibrium value observed under ambient conditions. The bond distances and angles distort so as to increase the macroscopic density of the ice while retaining the four-connected hydrogen-bonded network structure which is characteristic of all the crystalline ices. The atmospheric pressure ices, I_h and I_c, are very open structures, and higher densities can be obtained if the requirement of exact or almost exact tetrahedral coordination is relaxed. The $O-\hat{H} \cdots O$ bond angles are distorted from linearity by as much as 15°, while the $H \cdots O$ distances expand by up to 0.1 Å. At very high pressures the ice structures form interpenetrating lattices, in which the van der Waals $O \cdots O$ separations are smaller than the hydrogen-bond $O-H \cdots O$ separations.

These hydrogen-bond distances in ice provide a useful standard with which to compare the range of $O-H \cdots O$ distances observed in other types of crystal structures. For example, when water molecules accept hydrogen bonds from the carboxylic acid $-OH$ donors, i.e., $O=\overset{.}{C}OH \cdots O_WH_2$, the bonds are stronger than in ice. In the reverse situation, when the OH donors of the water molecules bond to carboxylic acid $C=O$ acceptors, the bonds are significantly weaker (see Table 7.5).

The hydrate inclusion compounds are a source of $OH \cdots O$ bond data but the only neutron diffraction crystal structure analysis, which has been fully reported, is that of ethylene oxide deuterate [431], $6(C_2H_4O) \cdot 46D_2O$. In that structure, the hydrogen-bond lengths range from 1.726 to 1.818 Å. The mean value is longer than that of ice Ih and the coordination at the oxygen atoms departs significantly from ideal tetrahedral. For the recent structure determination of $3.5 Xe \cdot 8 CCl_4 \cdot 136 D_2O$ at 13 K and 100 K, see [431a].

Unfortunately, none of the hydrate inclusion compounds which have less regular structures, such as the tetrabutyl and isoamylammonium salt hydrates [432] and the amine semi-clathrates [433, 434], has been studied by neutron diffraction. Furthermore, much of this early work was carried out using film methods prior to the development of single crystal diffractometers. In consequence, there is virtually no direct experimental evidence from these elaborate ice-like structures that can be related directly to the H···O bond distances or angles. Neutron single crystal analyses would provide useful data, since each structure contains many hydrogen bonds for which a wide range of hydrogen-bond distances and angles might be expected. The high hydrates of the simpler alkylammonium salts have also been studied by X-ray crystal structure analysis [125, 435–440], but none by neutron diffraction.

Water molecules are four-coordinated in the ices but may be three-coordinated in small molecule hydrates. In the ices and in the ice-like clathrate hydrate structures, the water molecules are always four-coordinated. In small molecule hydrates, they display four- and threefold coordination. They always donate two hydrogen bonds, but may accept one or two bonds, or be connected to one or two cations.

When water molecules are three-coordinated, the stereochemistry may be planar or pyramidal. A survey of the 97 hydrates formed by inorganic compounds which had been studied by neutron diffraction [441, 442] shows that in all but 15 crystal structures, the water molecules are coordinated to metal cations. The three- and four-coordinated water molecules are approximately equal in number, as is the distribution between planar and pyramidal configuration around the three-coordinated species. The 296 $O_WH \cdots O$ hydrogen bonds found in these 97 crystal structures range from 1.520 to 2.258 Å, with a weighted mean of 1.876 Å. As shown in Table 7.8, there is a definite trend for the hydrogen bonds to be shorter when the water molecule is coordinated with the higher charged cations.

In the hydrates of the small organic molecules discussed in Part IV, Chapter 22, the number of three-coordinated water molecules exceeds that of four-coordinated by a factor of 1.8. The distribution of the three-coordinated waters between planar and pyramidal is approximately equal with no marked distinction between the two configurations.

As in inorganic hydrates, the $O_WH \cdots O$ bond lengths in these organic hydrates range from 1.60 to 2.25 Å. The bonds observed in the purine and pyrimidine hydrates tend to be shorter than those in the nucleoside and nucleotide hydrates, but extend over a wide range of bond lengths, as shown in Table 7.9.

7.1.3 Carbohydrates Provide the Best Data for OH···O Hydrogen Bonds: Evidence for the Cooperative Effect

Since a relatively large body of neutron diffraction analyses is available for the carbohydrates [58], they are a good source to study OH···O hydrogen bonding. As the majority of these studies is from pyranoses and methyl pyranosides, there is little variation in the types of hydrogen bonds encountered, the donors being C–OH and the acceptors C–OH or the ring and glycosidic oxygen atoms, Oζ. This limitation to only a few well-characterized types of bonds considerably improves the significance of the statistical findings, and gives direct information on the effects of cooperativity on the OH···O bonding geometry.

Table 7.8. Effect of the cationic environment of water molecules on the O–H···O hydrogen-bond lengths [441, 442]

Class I: R–O(–H----O)(–H----O)–R'

R, R'	H, H	H, M$^+$	H, M^{2+}	H, M^{n+}	M$^+$, M$^+$	M$^+$, M^{2+}	M^{2+}, M^{2+}
No. of structures	13	14	13	2	37	18	5
No. of OH···O in survey	24	26	21	4	46	32	8
H···O min.	1.747	1.742	1.650	1.668	1.699	1.679	1.520
max.	2.099	2.258	2.034	1.742	2.242	1.952	1.827
mean	1.877	1.924	1.835	1.694	1.919	1.782	1.748

Class IIA (planar): R–O(–H----O)(–H----O)

R	H	M$^+$	M^{2+}	M^{n+}
No. of structures	8	0	26	7
No. of OH···O in survey	13	0	43	14
H···O min.	1.817	–	1.656	1.636
max.	2.059	–	2.176	2.132
mean	1.897	–	1.897	1.811

Class IIB (pyramidal): R=O(–H----O)(–H----O)

	H	M$^+$	M^{2+}	M^{n+}
	10	0	27	2
	17	0	44	4
	1.765	–	1.680	1.790
	2.137	–	2.069	1.982
	1.894	–	1.777	1.898

The Metrical Properties of O−H···O Hydrogen Bonds

Table 7.9. $O_W H \cdots O_W$ Two-center hydrogen-bond lengths in the hydrates of small molecular nucleic acid components [61, 62, 84]

	Purine and pyrimidine hydrates	Nucleoside and nucleotide hydrates
Number	26	20
Min.	1.60 Å	1.85 Å
Max.	2.25 Å	2.18 Å
Mean	1.88 Å	1.93 Å

The cooperative effect shortens and strengthens the OH···O hydrogen bonds. In Table 7.10, the hydrogen-bond geometries observed by neutron diffraction methods in the crystal structures of the pyranoses and pyranosides are arranged according to the hydrogen-bonding patterns. These are distinguished as:

1. Infinite chains: ···OH···OH···OH···OH···
2. Finite chains: OH···OH···OH···O⟨
3. Isolated bonds: OH···O⟨

Although all the H···O bond lengths extend over the same range of 1.7 to 2.0 Å, there is a significant distinction between hydrogen bonds involved in chains and those which are isolated. Based on the concept of cooperativity introduced in Part I, Chap. 1, we anticipate the hydrogen bonds to be shorter in the infinite chains than in the isolated bonds. This is consistent with an increase in hydrogen-bond strength, as calculated by ab initio methods to be in the range of 25% relative to an isolated bond [108, 110, 111]. This shortening is in fact observed with a difference of about 0.07 Å compared with the isolated bonds [106, 107]. A small, and barely significant difference of about 0.01 Å is observed in the mean covalent-bond lengths, corresponding to the shorter hydrogen bonds.

The analysis was put on a broader basis by combining 50 neutron and X-ray crystal structure analyses, with the X-ray O−H bond lengths normalized to 0.97 Å [59]. The results are illustrated in Table 7.11, where chains of hydrogen bonds ···OH···OH···OH··· are referred to as cooperative, and isolated bonds as noncooperative. From both the analyses, the H···O bond-length difference of ~0.07 Å between cooperative and noncooperative bonds is significant at the 99% probability level. There is a slight discrepancy between neutron data alone and combined X-ray and neutron data, the H···O distances in the former being 0.01 to 0.02 Å shorter than the latter. This could arise from larger errors in the C−Ô−H angles, causing a small overestimation of the H···O distances in the normalized X-ray data.

The anomeric sugar hydroxyl group has unique hydrogen-bonding properties which are apparent in the neutron diffraction data. When a mono- or oligosaccharide is a reducing sugar, it contains the so-called anomeric hydroxyl group, which is part of the hemiacetal system C−O−CH−OH. The anomeric carbon atom

Table 7.10. Hydrogen-bond geometries in the pyranoses and methyl pyranosides from neutron diffraction data [58, 59]

Compound	O–H (Å)	H···O (Å)	O–Ĥ···O (°)	REFCODE
O–H···O–H bonds in infinite chains				
Methyl α-altroside	0.971	1.736	173.0	MALTPY01
β-L-Arabinose	0.974	1.736	161.2	ABINOS01
Methyl α-glucoside	0.985	1.738	162.2	MGLUCP11
Methyl β-galactoside	0.984	1.739	159.4	MBDGAL02
α-L-Rhamnose H$_2$O	0.979	1.740	174.0	RHAMAH12
β-Fructose	0.979	1.750	154.6	FRUCTO02
β-D,L-Arabinose	0.985	1.753	171.6	ABINOR01
α-D-Glucose	0.977	1.758	169.7	GLUCSA01
α-D-Glucose	0.975	1.758	164.9	GLUCSA01
Methyl α-glucoside	0.969	1.770	162.0	MGLUCP11
Methyl α-glucoside	0.966	1.772	165.1	MGLUCP11
β-L-Arabinose	0.975	1.801	162.6	ABINOS01
α-D-Glucose	0.987	1.819	167.7	GLUCSA01
α-D-Glucose	0.980	1.821	170.1	GLUCSA01
β-D,L-Arabinose	0.978	1.858	162.4	ABINOR01
β-D,L-Arabinose	0.973	1.863	170.2	ABINOR01
Methyl β-galactoside	0.966	1.860	168.4	MBDGAL02
β-Fructose	0.963	1.869	163.2	FRUCTO02
α-L-Sorbose	0.969	1.882	157.0	SORBOL01
Methyl β-riboside	0.960	1.947	158.7	MDRIBP02
Methyl β-riboside	0.965	1.959[a]	139.0	MDRIBP02
Methyl β-riboside	0.949	1.989	147.1	MDRIBP02
Mean	0.972	1.814	169.2	
O–H···O–H bonds in finite chains				
α-L-xylose	0.961	1.731	168.8	XYLOSE01
α-L-sorbose	0.974	1.746	176.0	SORBOL01
α-L-xylose	0.975	1.751	171.5	XYLOSE01
Methyl β-xyloside	(0.920)	1.785	163.4	XYLOBM01
α-L-xylose	0.971	1.805	158.3	XYLOSE01
Methyl α-mannoside	0.976	1.810	179.8	MEMANP11
Methyl α-galactoside H$_2$O	0.957	1.851	173.9	MGALPY01
Methyl β-xyloside	0.975	1.885	160.1	XYLOBM01
Methyl α-mannoside	0.955	1.917	162.0	MEMANP11
Mean	0.968	1.809	168.2	
O–H···O< isolated bonds				
β-L-lyxose	0.974	1.718	177.0	LYXOSE01
Methyl α-Galactoside H$_2$O	0.971	1.747	169.4	MGALPY01
β-D,L-Arabinose	0.971	1.811	168.8	ABINOR01
β-L-Arabinose	0.952	1.820	170.0	ABINOS01
α-L-Xylose	0.954	1.843	165.2	XYLOSE01
α-D-Glucose	0.984	1.915	160.9	GLUCSA01
α-L-Sorbose	0.941	1.932	168.0	SORBOL01
Methyl α-mannoside	0.957	2.052	150.6	MEMANP11
Methyl β-xyloside	0.979	2.088	170.0	XYLOBM01
Mean	0.961	1.881	166.7	

[a] Intramolecular.

Table 7.11. O−H···O Hydrogen-bond lengths from neutron diffraction of carbohydrates, and from combined (normalized) X-ray and neutron diffraction data from 43 monosaccharide crystal structures

Description	Number of observations	H···O Mean value (Å)	Standard deviation σ (Å)
a) Data from neutron diffraction studies only			
H···O all cooperative bonds	59	1.805	0.069
All cooperative bonds excluding those to water molecules, anomeric oxygens, or ether oxygens	46	1.808	0.074
Noncooperative bonds	15	1.869	0.091
Noncooperative bonds excluding those to water molecules, anomeric oxygens, or ether oxygens	4	1.886	0.068
All H···O bonds in monosaccharides	44	1.812	0.077
All H···O bonds in disaccharides	13	1.845	0.052
H···O for anomeric hydroxyls	7	1.808	0.057
H···O for nonanomeric hydroxyls	67	1.819	0.080

Description	Number of observations	H···O Mean value (Å)	Range of values (Å)
b) Data from combined X-ray and neutron diffraction data			
Cooperative bonds in infinite chains	100	1.821	1.66 – 2.02
Cooperative bonds in finite chains	40	1.838	1.73 – 1.99
Termination bonds in finite chains ···O−H···O<	14	1.912	1.78 – 2.09
Noncooperative bonds O−H···O<	9	1.897	1.81 – 2.00
H···O from anomeric hydroxyls	21	1.847	1.72 – 2.02
All H···O bonds in monosaccharides	203	1.838	1.60 – 2.09
All H···O bonds in disaccharides	63	1.869	1.71 – 2.08

is unique in the pyranose or furanose ring in that it is simultaneously bonded to the ring and to the anomeric oxygen atoms. The anomeric hydroxyl can be distinguished chemically from the other hydroxyl groups by the "anomeric effect", which is also responsible for the different hydrogen-bonding characteristics.

These are illustrated by the observation that in seven out of nine examples of noncooperative (isolated) OH···O bonds in Table 7.11, the anomeric hydroxyl group is involved as donor. Therefore, as in the case of glycosyl or ring oxygen atoms, it inhibits the formation of infinite chains by acting as a *chain-stopper*.

These results are supported by a theoretical study using ab-initio molecular orbital calculations on model systems consisting of hydrogen-bonded dimers and trimers of methanediol, HO−CH$_2$−OH, versus those of methanol, CH$_3$−OH. They indicated that the anomeric hydroxyl is a stronger than normal hydrogen-bond donor, and a weaker than normal acceptor [58, 443]. In agreement with these theoretical findings is the combined experimental data set from neutron and (nor-

Table 7.12. Relative proportion in hydrogen bonds in a sample of 120 nucleoside and nucleotide crystal structure [62]

Hydrogen bonds (428 bonds)		Functional groups			
		Donors		Acceptors	
OH···OH	15%	C−OH	48	O$<^C_H$	32
OH···O=C$<$	13	−N(H)H	20	O=C$<$	21
OH···N$\ll$	9	O$_w$H	18	N$\ll$	17
O$_w$H···OH	8	$>$N$^+$ ⎫	12	O$_w$	12
N(H)H···OH	7	$>$NH ⎭		O=P̄	9
OH···O$_w$	6	P−OH	2	Cl$^-$	5
O$_w$H···N$\ll$	5			O$<$	2
$>$NH···OH	4			S=C$<$	2
$>$NH···O=C$<$	4				
OH···Cl$^-$	4				
O$_w$H···O=P̄	2				
N(H)H···O=P̄	2				
O$_w$H···O$_w$	2				
$>$NH···N$\ll$	2				
N(H)H···O=C$<$	2				
N(H)···O$_w$	2				
All others	13%				

malized) X-ray analyses [59], where of the 35 anomeric hydroxyl groups only 10 accept a hydrogen bond, and the mean value of the H···O hydrogen-bond length where anomeric hydroxyls are involved, 1.808 Å, is significantly shorter than the mean value for all other (isolated) noncooperative bonds, 1.869 Å.

In the nucleoside and nucleotide crystal structures, the OH···O hydrogen bonds are the most common. In general, these molecules do not crystallize as readily as do the carbohydrates, and in consequence, all the data, with only one exception, are from X-ray analyses. In addition, only about a quarter of the crystal structures reported are of sufficient quality to permit either direct or indirect location of all the functional hydrogen atoms. As is often the case, the larger the molecule, the less accurate are the crystal structure analyses, hence most of the data refer to compounds with less than 12 carbon atoms.

As shown in Table 7.12, the various types of OH···O hydrogen bonds constitute nearly half of all the hydrogen bonds in the nucleoside and nucleotide crystal structures[4]. The remaining half consists of a wide variety of hydrogen

---

[4] This proportion of OH···O bonds in the nucleic acids will be reduced by reason of less or no free C−OH groups (as in DNA), but this may be compensated by the high hydration leading to more O$_w$H···O$<$ and O$_w$H···O=P̄ hydrogen bonds.

Table 7.13. Metrical data on two-center O–H···O bonds in the nucleosides and nucleotides. Mean effective equilibrium bond lengths given in Å [84]

Donors		Acceptors				
		$O=\bar{P}$	O_WH_2	$O=C<$	$O<^C_H$	$O<^C_C$
$\bar{P}-OH$	No. of data	5	5		2	
	min.	1.55	1.59		1.65	
	max.	1.69	1.68		1.89	
	mean	1.58	1.64		1.77	
C–OH	No. of data	6	29	63	75	65
	min.	1.70	1.57	1.65	1.70	1.95
	max.	1.95	2.04	2.65	2.57	2.97
	mean	1.78	1.79	1.85	1.86	2.19
HO_WH	No. of data	11	12	7	39	3
	min.	1.66	1.78	1.81	1.76	1.92
	max.	2.11	2.18	2.09	2.18	2.10
	mean	1.86	1.94	1.91	1.90	2.03

bonds [with >NH and –N(H)H as donors]. For this reason, the mean effective equilibrium bond-length values are based on less reliable statistics and the systematic trends in distances and patterns are more difficult to observe than in the carbohydrate crystal structures.

The results of a survey of 120 crystal structures which provided reasonably reliable hydrogen atom positions [84] are shown in Table 7.13. Only hydrogen bonds of type C–OH···O_W, C–OH···O=C, C–OH···$O<^C_H$, and O_WH···$O<^C_H$ provide samples of sufficient number to be statistically meaningful. Nevertheless, there is a definite trend in H···O bond lengths consistent with the relative electronegativities of the donor and acceptor groups.

The sequence in H···O bond lengths is

for the donor group: $\bar{P}-OH < C-OH < O_WH$

for the acceptor group: $O=\bar{P} < O_W < O=C < O<^C_H < O<^C_C$.

As noted previously, the water molecule is a relatively strong acceptor and relatively weak donor for hydrogen-bonds.

The mean C–OH···$O<^C_H$ hydrogen bond length, 1.86 Å, is longer than that given in Table 7.11b for the H···O distances in the monosaccharides, 1.838 Å, but agrees well with the value for the disaccharides, 1.869 Å, which are molecules of size comparable to the nucleosides.

7.2 N−H···O Hydrogen Bonds

This type of hydrogen bonds includes the N−H···O=C interactions which are the most predominant hydrogen bonds in fibrous and globular proteins. Because they are responsible for the formation of the commonly occurring secondary structure elements α-helix, β-pleated sheet and β-turn, a large body of much less accurate data is available from protein crystal structures which will be analyzed in Part III, Chap. 19. The N−H···O=C type hydrogen bond is also the most common in the purine and pyrimidine crystal structures (Table 7.14), and is one of the two important bonds in the base pairing of the nucleic acids.

Donor and acceptor groups can be charged. In structures of biological molecules, the N−H···O=C bond occurs both in salt bridges with charged donor and acceptor groups and with uncharged groups (see Box 6.1). Of these bonds, those most extensively studied are the $\overset{+}{N}(H_2)H\cdots\bar{O}=C$ of the amino acid zwitterion crystal structures, and the $>$N−H···O=C$<$ in the small molecule nucleic acid components. These bonds also play an important role in the crystal structures of the oligopeptides. There are no neutron diffraction studies of the oligopeptides, but the $>$N−H hydrogen positions can be reliably deduced from those of the adjacent non-hydrogen atoms. Hitherto, there have been no systematic attempts to analyse the hydrogen bonding in other than a few dipeptides. Where hydrates are formed, a reliable assignment of hydrogen bonds becomes more difficult.

Unlike the O−H···O bonds, there are no examples of strong N−H···O hydrogen bonds. The strong and almost symmetrical N···H···O configuration which might be expected to favor the amino ⇌ imino tautomerism

$$>\!N\!-\!H\!-\!-\!-\!O\!=\!C\!< \quad \rightleftharpoons \quad >\!N\!-\!-\!-\!H\!-\!O\!-\!C\!<$$

is not observed in crystal structures. The symmetry requirement for proton tunneling referred to in the case of O−H···O bonds is lacking, which implies that proton transfer cannot occur in the ground state of these systems [389].

There have been four statistical analyses of N−H···O hydrogen-bond geometries, summarized in Tables 7.15 to 7.18:

1. The data in Table 7.15 are based on the neutron diffraction analyses of the amino acids [60]. They are mainly concerned with $-\overset{+}{N}(H_2)H\cdots\bar{O}_2C-$ bonds, since these molecules occur as zwitterions in the crystals.
2. The data in Table 7.16 reflect a broader-based analysis using mainly normalized values from the more reliable X-ray crystal structures of the purines and pyrimidines [61].
3. The data in Table 7.17 are the same type of analysis as in Table 7.16, using only the nucleoside and nucleotide crystal structures [62].
4. The data in Table 7.18 give a broad-based analysis of $\overset{+}{N}H\cdots O=\bar{C}$ and NH···O=C bonds, made irrespective of type of molecule involved [75].

Table 7.14. Distribution of hydrogen bonds in 185 crystal structures of purines and pyrimidines [61]

Hydrogen bonds total of 463		Number of different functional groups			
		Donors		Acceptors	
$>$NH$\cdots$O=C$<$	21%	$>\overset{+}{\text{N}}$H } or $>$NH	40	$>$C=O	38
$>$NH$\cdots$N$<$	12				
−N(H)H$\cdots$N$<$	12			$\searrow$N	25
−N(H)H$\cdots$O=C$<$	8	−N(H)H	22	O$_W$	16
−N(H)H$\cdots$O$_W$	6	O$_W$H	17	Cl$^-$	11
O$_W$H$\cdots$Cl$^-$	5	C−OH	8	$\bar{\text{S}}$=O	5
O$_W$H$\cdots$O$_W$	5	−$\overset{+}{\text{N}}$(H$_2$)H } or $\overset{+}{\text{N}}$(H$_3$)H	3	OH	2
O$_W$H$\cdots$O=C	5				
$>\overset{+}{\text{N}}$H$\cdots$Cl$^-$	4			C=S	2
OH$\cdots$O=C	3			$>$O	
−N(H)H$\cdots$Cl$^-$	3			−NH$_2$	
$>$NH$\cdots$O$_W$	3			Br$^-$	
−N(H)H$\cdots$O=$\bar{\text{S}}$	3			I$^-$	1
OH$\cdots$N$<$	3			$\bar{\text{N}}$=O	
Others	7			$\bar{\text{P}}$=O	

Table 7.15. $-\overset{+}{\text{N}}$(H$_2$)H$\cdots$O=$\bar{\text{C}}$ Hydrogen-bond geometries in the crystal structures of the amino acids from neutron diffraction analyses [60]

Crystal structure	$-\overset{+}{\text{N}}$(H$_2$)H$\cdots$O=$\bar{\text{C}}$		REFCODE
	(Å)	(°)	
4-Hydroxy-L-proline	1.695	167	HOPROL12
α-Glycylglycine (82 K)	1.723	162	GLYGLY04
L-Asparagine·H$_2$O	1.782	168	ASPARM02
L-Serine·H$_2$O	1.804	153	LSERMH10
D,L-Serine	1.817	161	DLSERN11
L-Alanine	1.826	164	LALNIN12
Triglycine sulfate	1.840	169	TGLYSU11
L-Lysine·HCl·2H$_2$O	1.864	148	LYSCLH11
L-Glutamic acid (β-form)	1.895	174	LGLUAC11
L-Glutamine	1.942	167	GLUTAM01
Mean	1.819	163	
	$>\overset{+}{\text{N}}$H$\cdots$O=$\bar{\text{C}}$		
	(Å)	(°)	
L-Histidine·HCl·H$_2$O	1.580	171	HISTCM12
L-Histidine	1.720	172	LHISTD13
L-Arginine·2H$_2$O [a]	1.832	177	ARGIND11
L-Arginine·2H$_2$O [a]	1.847	176	ARGIND11
L-Arginine·2H$_2$O [a]	1.943	161	ARGIND11
Hippuric acid	2.053	160	HIPPAC02
Mean	1.829	169	

[a] $>$C=$\overset{+}{\text{N}}<^{\text{H}}_{\text{H}}$

Table 7.16. NH···O Hydrogen-bond lengths observed in 224 crystal structures of purines, pyrimidines [61]

Donors		Acceptors					
		O=P̄	O$_W$H$_2$	O=C̄	O<C,H	O<C,C	O=S̄
−N̈H$_3$ or NH$_4^+$	No.		2	10			
	Min.		1.89	1.69			
	Max.		1.90	2.00			
	Mean		1.90	1.82			
>N̈H	No.	1	11	1			5
	Min.		1.61				1.59
	Max.		1.82				1.69
	Mean	1.72	1.71	1.73			1.63
				O=C<			
>NH	No.	2	19	151	1	1	2
	Min.	1.78	1.61	1.69			1.96
	Max.	1.82	1.99	2.32			2.09
	Mean	1.80	1.77	1.85	2.12	1.77	2.02
−N(H)H	No.	1	21	54	4	3	13
	Min.		1.74	1.68	1.86	2.05	1.78
	Max.		2.13	2.70	2.44	2.60	2.59
	Mean	1.82	1.94	1.99	2.11	2.32	2.01

Table 7.17. NH···O Hydrogen-bond lengths observed in the 119 crystal structures of nucleosides and nucleotides. After [62], where data for 87 crystal structures are reported

Donors		Acceptors				
		O=P̄	O$_W$H$_2$	O=C̄	O<C,H	O<C,C
>N̈H	No.	7	1		2	
	Min.	1.58			1.71	
	Max.	1.89			1.81	
	Mean	1.76	1.66		1.76	
>NH	No.	7	3	18	23	4
	Min.	1.58	1.59	1.72	1.72	1.99
	Max.	1.89	2.06	2.05	2.16	2.25
	Mean	1.76	1.84	1.90	1.88	2.10
−N(H)H	No.	12	9	13	36	6
	Min.	1.67	1.71	1.74	1.89	1.94
	Max.	2.07	2.12	2.73	2.37	2.54
	Mean	1.86	1.96	2.06	2.07	2.25

N–H···O Hydrogen Bonds

Table 7.18. Intermolecular N–H···O=C hydrogen-bond lengths (Å) observed in 889 crystal structures irrespective of molecular type [75]

Donor	Acceptor				
	O=C=O (⊖)	O=C–NH$_2$	O=C(C)(C)	O=C–OH	Weighted row mean
R$_3$NH$^+$	11 1.72	2 1.84	1 1.94	0	1.755
R$_2$NH$_2^+$	47 1.796	3 1.79	3 1.97	6 1.89	1.805
RNH$_3^+$	226 1.841	15 1.89	8 1.87	68 1.936	1.865
\>N$^+$–H	36 1.869	12 1.86	2 1.84	11 1.98	1.887
NH$_4^+$	56 1.886	4 1.99	2 1.99	13 1.92	1.900
\>N–H	74 1.928	597 1.934	38 1.970	117 2.002	1.945
Weighted column mean	1.855	1.931	2.024	1.972	

The amino acid neutron diffraction data given in Table 7.15 is very limited. In part, this is due to the high proportion of three-center bonds in these crystal structures. The data based on the nucleoside and nucleotide X-ray crystal structures shown in Tables 7.16 and 7.17 provide better statistics over a wider variety of donor-acceptor combinations. The data in Table 7.15 indicate that the $-\overset{+}{N}(H_2)H$ group is a slightly stronger donor than the $\overset{+}{\gt}NH$. This is not shown in Table 7.16, where there are only one and ten structures for comparison in two of the categories. Comparing the data from all three tables,
– there is excellent agreement for the overlap of the $\overset{+}{N}(H_2)H \cdots \bar{O}=C$ bonds in Tables 7.15 and 7.16,
– there is good qualitative agreement in the donor and acceptor sequences. For the donor groups:

$$-\overset{+}{N}(H_2)H < \overset{+}{\gt}NH < \gt NH < -N(H)H$$

and for the acceptor group:

$$O=\bar{P} < O_W < O=C < O(C)(H) < O(C)(C).$$

For the two corresponding bonds in Tables 7.15 and 7.18, those in Table 7.18 are 0.02 Å and 0.04 Å longer. This can be because the latter are primarily normalized X-ray data, or because of the heterogeneity of the types of molecules in the survey.

The analysis given in Table 7.18 is over 889 crystal structures [75]. It provides data on 1352 hydrogen bonds of type N–H···O=C and permits finer group distinctions. The statistics are better than in the other three analyses, but this study has the disadvantage that the crystal field effects are averaged over a large number of molecules with different shapes, sizes, and functional groups. There is a trend for corresponding bonds to be longer in the nucleosides and nucleotides than in the purine and pyrimidine crystal structures. This, we believe, is associated with the more efficient packing of the mainly planar molecules, versus the more awkwardly shaped nucleosides.

7.3 N–H···N Hydrogen Bonds

The N–H···N bonds constitute about a quarter of the hydrogen bonds in the purine and pyrimidine crystal structures (see Table 7.14). The proportion is much smaller in the nucleosides and nucleotides Table 7.12, where they compete with the stronger O–H···O and N–H···O interactions. In combination with N–H···O=C, the N–H···N⟨ bonds form the Watson-Crick and related base-pair configurations in purine and pyrimidine crystal structures, and in the oligonucleotides and nucleic acids.

The donor groups are ⟩N–H and –N(H)H and the acceptor group is N⟨. The aliphatic H$_2$N– very rarely acts as acceptor and the conjugated amino group found in amide groups (asparagine and glutamine) and in nucleic acid bases (adenine, guanine, cytosine) is never an acceptor.

Two surveys on bond lengths are available, based on crystal structure data from purines and pyrimidines, and from nucleosides and nucleotides [61, 62]. The results of these studies are summarized in Table 7.19.

The sequence of donor groups is

$$\text{N–H} < \text{–N(H)H} .$$

Table 7.19. Two-center ⟩NH···N⟨ bond lengths (Å) in purines, pyrimidines (column A) and nucleosides and nucleotides (column B) (Mean effective equilibrium bond length in Å [61])

Donor		A	B
⟩N–H	No.	48	9
	Min.	1.73	1.78
	Max.	2.23	1.99
	Mean	1.882	1.88
–N(H)H	No.	57	12
	Min.	1.85	1.89
	Max.	2.58	2.76
	Mean	2.050	2.13

Table 7.20. Two-center $O-H\cdots N\langle$ bond lengths (Å) in purines, pyrimidines (column A) and nucleosides and nucleotides (column B) (Mean effective equilibrium bond length in Å [61])

Donor		A	B
C–OH	No.	11	43
	Min.	1.71	1.77
	Max.	2.02	2.62
	Mean	1.82	1.890
O_WH	No.	10	13
	Min.	1.78	1.85
	Max.	2.01	2.03
	Mean	1.92	1.92

7.4 O–H···N Hydrogen Bonds

This type of hydrogen bond has the same acceptor groups as the $N-H\cdots N\langle$ bond. It is found more commonly in the nucleoside and nucleotide crystal structures, where it accounts for nearly one tenth of all the hydrogen bonds. There are no $P-OH\cdots N\langle$ hydrogen bonds, since the acceptors are invariably protonated, $\bar{P}=O\cdots H\overset{+}{N}\langle$. This is in contrast to the $\bar{P}-OH\cdots O=C$ bond, where the $O=C$ group is rarely protonated in crystal structures.

The data from the two analyses given in Table 7.20 comprise only two donors, C–OH and O_WH, and one acceptor, $N\langle$. There is qualitative agreement in the relative lengths of the hydrogen bonds, with sequence

$C-OH < O_WH$.

7.5 Sequences in Lengths of Two-Center Hydrogen Bonds

In a final summary of the hydrogen-bond lengths of the different two-center bonds discussed in this chapter, the following sequences can be given:

For the O–H···O bonds, the sequences are:

Donors $\bar{P}-OH < C-OH < O_WH$

Acceptors $O=\bar{P} < O_W < O=C \approx O\langle^C_H < O\langle^C_C$.

In the carboxylic and amino acid data, a finer differentiation is possible:

Donors $O=\overset{|}{C}-OH < C-OH < O_WH$

Acceptors $O_W < O=\overset{|}{C}-OH \approx O=\underset{\ominus}{\overset{|}{C}}=O < O\langle^C_H < O\langle^C_C$

For the N–H···O bonds, the sequence is:

Donors $>\overset{+}{N}(H)H < \hspace{2pt}\overset{+}{{\Large{>}\hspace{-8pt}}}\hspace{-4pt}\overset{+}{NH} < {>}NH < -N(H)H$

Acceptors $O=\bar{P} < O_W < O=C < O{<}^C_H < O{<}^C_C\ .$

For the N–H···O=C bonds only, the finer distinction is:

Donors ${\Large{>}\hspace{-8pt}}\hspace{-4pt}\overset{+}{NH} < {>}\overset{+}{N}(H)H < -\overset{+}{N}(H_2)H < {\Large{>}\hspace{-8pt}}\hspace{-4pt}\overset{+}{NH} < {>}\overset{+}{N}H_4 < {>}NH$

Acceptors $O=\overset{\ominus}{\underset{|}{C}}=O < O=\underset{|}{C}-NH_2 < O=C{<} < O=\underset{|}{C}-OH$

For the N–H···N{ bonds, the sequence is:

Donors ${>}NH < -N(H)H$

For the O–H···N{ bonds, the sequence is:

Donors $C-OH < O_WH$

Between different classes of molecules, the mean bond lengths for various donor-acceptor combinations do not always agree even when there are reasonably good statistics. This is because of differences in the average crystal field effect for molecules of different shape and functional groups, as discussed in Part IB, Chapter 6.

The agreement for the NH···N{ and OH···N{ hydrogen bonds in Tables 7.19 and 7.20 is better than for some other comparisons. Where there are differences, the bonds in the nucleosides and nucleotides are the longer.

7.6 H/D Isotope Effect

The substitution of hydrogen for deuterium has a very small influence on the hydrogen-bond lengths known as the isotope effect. Early work, which necessarily depended on changes in unit cell dimensions and thermal expansion coefficients, suggested that there could be small expansions or contractions of the hydrogen-bond lengths on deuterium substitution [444]. Later, crystal structure analyses reported the differences in O–H···O versus O–D···O, which ranged from zero to +0.0022 Å [54]. In none of this early X-ray crystal diffraction work were corrections made for the difference in the thermal motion of the hydrogen and the deuterium atoms. In consequence, the isotope effect could not be analyzed in terms of the actual H···O and D···O hydrogen bonds.

Theoretical work suggested that the effect should be zero or a small contraction for strong hydrogen bonds with a symmetrical (or almost symmetrical) potential well and an expansion for long bonds where there is little observable lengthening of the covalent O−H or O−D bond on hydrogen-bond formation [445].

Because of the differences in the O−H and O−D force constants, the differences in hydrogen-bond lengths are expected to vary with temperature. There is some experimental X-ray evidence to support this [446], but no neutron diffraction analyses that measure directly the changes of hydrogen-bond lengths on deuteration at different temperatures have been carried out. This appears to be a subject requiring re-investigation now that neutron diffraction structure analysis can be performed relatively easily at selected temperatures down to 10 K.

A much more promising and potentially more informative method is the NMR isotope effect. The analysis of the multiplet structure of ^{13}C NMR resonance from partially deuterated molecules is a relatively new approach to the study of hydrogen bonding. There appears to be a direct logarithmic relationship between a deuterium isotope shift effect and hydrogen-bond energy. If this can be extended to ^{13}C solid-state NMR, it would provide not only a means of comparing hydrogen bonding in solution and in the crystals, but also a direct experimental correlation between hydrogen-bond geometries and hydrogen-bond energies [447−449].

Chapter 8
Metrical Aspects of Three- and Four-Center Hydrogen Bonds

The two-center hydrogen bonds X–H···A are by far the most common in all classes of biological molecules, but there is strong evidence, based on neutron diffraction crystal structures, that the three-center bonds occur much more frequently than previously thought. Analyses of small molecule crystal structures indicate that an average of about one quarter of all the NH···O=C and OH···O hydrogen bonds are of the three-center type (Table 2.3). Because in the (zwitterionic) amino acids there is a numerical imbalance between the (three) donor properties of the $-\overset{+}{N}H_3$ and the (four) acceptor properties of the $-CO_2^-$, the proportion of three-center bonds is untypically high, amounting to almost three quarters of the total number of hydrogen bonds.

The situation is very different for the four-center bonds. They are much less common, with an average of, in general, below 1%. The amino acids are again an exceptional class of compounds, with about 15% of four-center bonds.

8.1 Three-Center Hydrogen Bonds

In the spirit of the Pimentel-McClelland definition of a hydrogen bond [28], a three-center (bifurcated) bond is defined as a configuration showing evidence of an attractive force between the covalently bonded hydrogen atom of a donor X–H group and two acceptor atoms A, A'. Since this attractive force is primarily electrostatic, attenuating as r^{-1}, the hydrogen atom environment should be explored to the first neighbor, irrespective of the distance. If there are two first neighboring electronegative atoms as shown below, a criterion for attractive forces is that the hydrogen atom lies in or close to (< 0.2 Å) the plane defined by X, A, A', or that the sum of angles $\theta + \theta' + \alpha \approx 360°$.

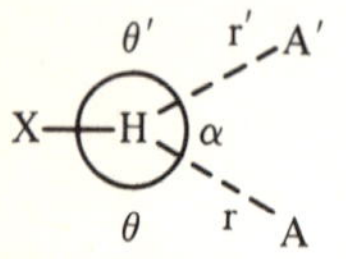

Three-Center Hydrogen Bonds

Table 8.1. Some early examples of three-center hydrogen-bond geometries by neutron diffraction

Donor X–H	Acceptors A_1, A_2	H···A_1, A_2 (Å)	X–Ĥ···A_1, A_2 (°)	A_1···Ĥ···A_2 (°)	Σ (°)
α-Glycine (1972) [65]					
N–H ⋮⋮⋮	O(2)	2.121(2)	154.0(2)	89.9(2)	359.4
	O(1)	2.365(2)	115.5(2)		
Sulfamic acid (1960) $\overset{+}{N}H_3SO_3^-$ [450]					
H(1):	·O(2)–S̄	1.95	169	89	359
	·O(3)–S̄	2.56	101		
–N̄—H(2):	·O(1)–S̄	2.45	108	88	356
	·O(3)–S̄	2.00	160		
H(3):	·O(1)–S̄	1.99	159	91	360
	·O(3)–S̄	2.45	110		

There is also a fourth oxygen atom at 2.75 Å from H(1), 2.60 Å from H(2) and 2.55 Å from H(3), making four-center configurations.

Violuric acid deuterate [67], see also Fig. 15.9h

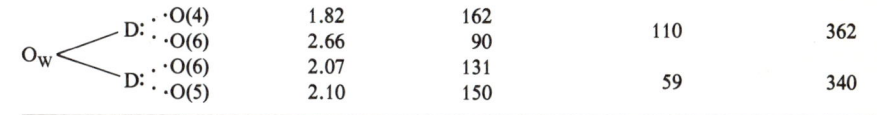

O_W ⟨ D:	·O(4)	1.82	162	110	362
	·O(6)	2.66	90		
D:	·O(6)	2.07	131	59	340
	·O(5)	2.10	150		

Table 8.2. Three- and four-center hydrogen-bond geometries in the paraelectric (room-temperature) form of ammonium sulfate. The transition to the ferroelectric (low temperature) form involves a change to two-centered bonds (nomenclature as in original publication [451]

Donor	Acceptor	Paraelectric H···O (Å)	Paraelectric N–Ĥ···O (°)	Ferroelectric H···O (Å)	Ferroelectric N–Ĥ···O (°)
N(1)–H(3) ⋮⋮	·O(2)	2.27	136.0		
	·O(4)	2.14	155.0	1.82	174.3
N(1)–H(4) ⋮⋮	·O(2)	2.27	136.0		
	·O(3)	2.14	155.0	1.92	172.7
N(2)–H(7) ⋮⋮	·O(1)	2.39	135.4		
	·O(4)	2.05	159.7	1.87	168.2
N(2)–H(8) ⋮⋮	·O(1)	2.39	135.4	1.96	164.3
	·O(3)	2.05	159.7	2.46	130.4
N(1)–H(4) ⋮⋮	·O(2)	2.27	136.0		
	·O(3)	2.14	155.0	1.92	172.7
N(1)–H(2) ⋮⋮⋮	·O(1)	2.48	139.3		
	·O(3)	2.48	124.9	1.91	160.2
	·O(3')	2.48	124.9		
N(2)–H(6) ⋮⋮⋮	·O(2)	2.36	122.2	2.28	129.0
	·O(3)	2.43	115.7	2.39	108.9
	·O(4)	2.43	115.7	2.59	111.9

The first report of a three-center (bifurcated) hydrogen bond was in 1939 in the X-ray crystal structure analysis of the amino acid glycine [63], and later confirmed by a neutron diffraction study [65]. The hydrogen-bond lengths and angles are given in Table 8.1, together with some other early studies based on neutron diffraction data.

A very interesting example of three-center hydrogen bonding is provided by the neutron diffraction analysis in 1966 of the para- and ferroelectric crystals of ammonium sulfate [451]. The transition from one to the other state involves a change in hydrogen bonding in which three of the five three-center hydrogen bonds become two-centered (see Table 8.2).

Hydrogen-bond configurations which combine the three-center bond with the bifurcated bond are observed in the crystal structures of violuric acid deuterate (Table 8.1) and the paraelectric ammonium sulfate as shown below

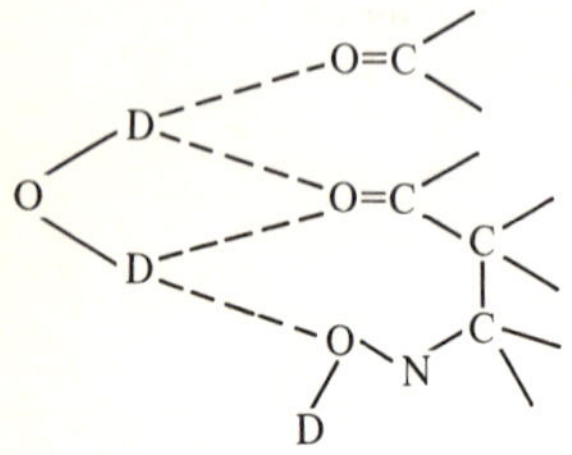

Violuric acid 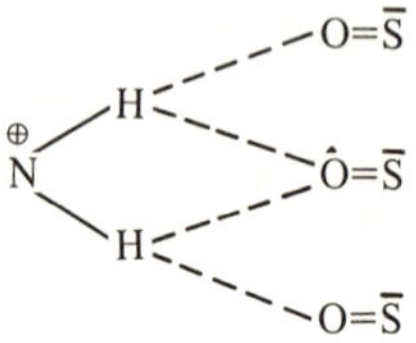   Ammonium sulfate

Three-center hydrogen bonds in the carbohydrate crystal structures. The first systematic study which suggested that three-center hydrogen bonds are more widespread than originally thought came from the examination of the results of the neutron diffraction crystal structures of the pyranose and pyranoside sugars referred to in Chapter 7 [58].

The geometrical data on these bonds are given in Table 8.3. They constitute about one quarter of the total number of $O-H\cdots O$ bonds in the sample of crystal structures surveyed. They range from almost symmetrical bonds[1] with $r \approx r' \approx 2.1$ Å and $\theta \approx \theta' \approx 135°$ to very unsymmetrical configurations where $r-r' \approx 0.6$ Å and $\theta - \theta' \approx 70°$.

As the $O-\hat{H}\cdots O$ angle in a three-center hydrogen bond approaches 90°, the $O\cdots O$ van der Waals repulsive force comes into play which gives rise to the excluded region described in Part I. The bond length of the minor component tends to increase. For an $H\cdots O$ hydrogen bond of 2.5 Å, a covalent $O-H$ bond of

---

[1] An example of a crystallographically symmetrical three-center hydrogen bond

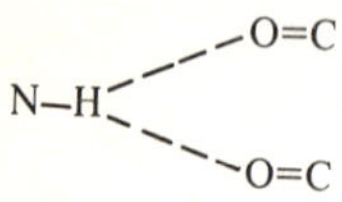

is found in the crystal structure of trans-trans diacetamide [452]

Three-Center Hydrogen Bonds

Table 8.3. Geometries of three-center hydrogen bonds from the neutron diffraction analyses of carbohydrates[a]

r (Å)	r' (Å)	θ (°)	θ' (°)	Σ (°)	REFCODE
2.085	2.140	130	140	358	MALTPY01
2.201	2.839	150	147	360	ABINOR01
1.977	2.593	169	114	358	FRUCTO02
1.983	2.636	160	124	359	MGALPY01
2.127	2.210	137	146	360	MGALPY01
1.957	2.300	147	107	360	LYXOSE01
2.114	2.634	146	123	359	LYXOSE01
1.998	2.391	153	106	360	MGLUCP11
1.981	2.608	160	107	359	RHAMAH12
1.958	2.568	139	96	360	MDRIBP02
1.989	2.495	148	135	360	MDRIBP02
1.953	2.584	163	120	356	SORBOL01
2.088	2.579	170	129	356	ABINOS01
1.927	2.613	150	97	360	MALTOS11
2.031	2.713	139	90	356	AHXGLP
2.219	2.328	165	107	358	GLUCIT01
1.908	2.506	168	94	360	SUCROS04

[a] Data recalculated from atomic coordinates.

Table 8.4. Summary of three-center hydrogen-bond geometries in crystal structures of some small biological molecules

Type of molecule	Type of bond	Symmetrical r−r' < 0.5 Å $\theta \approx \theta'$		Unsymmetrical r−r' > 0.5 Å $\theta > \theta'$	
		r (Å)	θ (°)	r (Å)	θ (°)
Carbohydrates (neutron data)	O−H···O	2.0−2.6	150−130	1.9−2.8	170−90
Purines and pyrimidines	O−H···O	2.0−2.8	150−130	1.6−3.0	170−90
(X-ray data)	N−H···O	2.0−2.5	160−100	1.7−3.0	175−90
Nucleosides and nucleotides	O−H···O	2.0−2.5	150−110	1.8−2.9	170−90
(X-ray data)	N−H···O	2.0−2.5	160−100	1.7−3.0	170−90

0.97 Å and an O−H···O angle of 90°, the O···O distance is 2.68 Å. At these distances the electrostatic attractive and the repulsive components will approximately balance, as indicated by the results of the ab initio calculation of the water dimer shown in Fig. 2.1.

Three-center hydrogen bonds in the nucleic acid constituents occur as frequently as in the carbohydrates (see Table 2.3). Because there are virtually no neutron diffraction studies available for this class of compounds, the analysis of the three-center bonds has to rely on data where the X−H bond lengths are normalized and are therefore to some extent less reliable than those of the carbohydrates (Table 8.3) which are based on neutron diffraction data.

Table 8.5. Distribution of two- (first column) and three-center (second column) hydrogen bonds[a] according to donor/acceptor type in purines, pyrimidines (top line), nucleosides, nucleotides (bottom line)

Donors	Acceptors												Totals
	O=P̄	O=C<	O_W	OH	O<^C_C	O=Ñ	O=S̄	≥N	NH_2	Cl[−]	Br[−]	S=C<	
P–OH	– 4	– 0	– 4	– 0	– –	– –	– –	– –	– –	– –	– –	– –	11 0
C–OH	1 7	16 59	11 20	3 25	1 5	4 66	3 32	– –	– –	1 –	12 36	1 18	– –	5 17	– 20	– –	– 2	257 115
O_WH	– 9	32 4	17 4	23 9	3 2	2 32	3 5	– 3	2 –	6 –	5 –	11 11	5 1	– –	24 –	1 –	4 –	3 1	3 3	172 52
>NH	2 4	151 20	27 3	18 2	8 0	2 18	– 6	– –	5 –	2 –	2 –	48 8	13 3	– –	12 –	2 –	2 –	5 2	– –	296 77
–N(H)H	1 12	54 12	14 4	21 8	11 1	9 34	3 8	– 6	1 –	6 –	8 –	50 12	33 2	– 2	5 5	1 –	2 –	6 3	1 –	254 100
≥NH[+]	– 3	1 2	5 –	11 1	1 –	– –	– –	1 1	4 –	3 –	– –	2 –	– –	14 4	1 –	– –	– –	40 18
–NH_3[+] –NH_4[+]	1 –	10 –	7 –	2 –	1 –	– –	– –	– –	– –	– –	– –	– –	– –	2 –	3 –	– –	– –	15 11
Totals	44 10	359 –	112	124 33	33	170 60	15	4	5	21	17	19	188	78	2	88	28	8	3	24	2	

[a] A three-center bond is counted once and attributed to the major acceptor.

Three-Center Hydrogen Bonds

Table 8.6. Geometry of three-center bonds in the amino acid crystal structures from neutron diffraction data

Compound	r	r'	θ	θ'	α	Σ (°)	Acceptors	REFCODE
α-Glycine	2.119	2.364	154	115	90	359	O=C̄, O=C̄	GLYCIN05
Glycylglycine·HCl·H₂O	2.070	2.361[a]	155	97	108	360	O_W, O=C̄	GLCICH01
Triglycine·SO₄	2.044	2.537	146	111	100	357	O=S̄, O=C−OH	TGLYSU11
Triglycine·SO₄	2.044	2.345	140	111	105	356	O=S̄, O=C̄	TGLYSU11
L-Glutamic acid·HCl	1.977	2.504	148	117	93	358	O=C−OH, O=C−OH	LGLUTA
NH₄, NH₃CH₂COOHSO₄	1.910	2.630	158	101	99	358	O=S̄, O=C−OH	AGLYSL01
NH₄, NH₃CH₂COOHSO₄	1.898	2.553	156	114	89	359	O=S̄, OH−C=O	AGLYSL01
L-Glutamine	1.853	2.366[a]	163	97	100	360	O=C̄, O=C	GLUTAM01
L-Glutamic acid (β)	1.844	2.578	168	96	95	359	O=C̄, O=C̄	LGLUAC11
L-Histidine	1.840	2.327[a]	160	98	100	358	O=C̄, O=C̄	LHISTD13
L-Asparagine·H₂O	1.833	2.291	157	99	100	356	O=C̄, O=C̄	ASPARM01
L-Histidine	1.786	2.520	158	107	95	360	O=C̄, O=C̄	LHISTD13
L-Glutamine	1.752	2.836	164	109	86	359	O=C̄, O=C̄	GLUTAM01
L-Histidine·HCl·H₂O	1.741	2.656	168	94	97	359	O_W, O=C̄	HISTCM12

[a] Intramolecular.

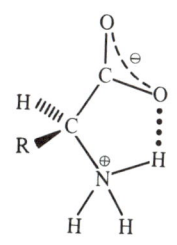

The survey on nucleic acid constituents nevertheless suggests that
- the three-center hydrogen bonds in nucleic acid constituents and in carbohydrates display comparable geometrical features and range from symmetrical to unsymmetrical, with associated changes in angles θ, θ' and in distances r, r' (Table 8.4).
- Since nucleic acid constituents exhibit a wider range of donor and acceptor types, the three-center hydrogen bonds have more variation than in the carbohydrates. It is clear from the data summarized in Table 8.5 that −N̊H₃ (and N̊H₄), O−H and −NH₂ groups tend to form more three-center bonds than O_WH and >NH groups. Similarly, N⟨, O⟨^C_H and O=C groups are more frequently involved as three-center acceptors than O_W, and the strong donor P−OH does not engage in three-center bonds.

Due to proton deficiency, crystal structures of amino acids display a much higher proportion of three-center hydrogen bonds. Their geometries, given in Tables 8.6 and 8.7, are based on neutron diffraction data, of which a relatively large number is available for this class of biological molecules.

Table 8.7. Geometry of chelated three-center bonds in the amino acid crystal structures from neutron diffraction data

Compound	r_1	r_2	θ_1	θ_2	α	Σ	Acceptors	REFCODE
Triglycine SO$_4$	2.001	2.225	161	126	69	356	O=S̄=O	TGLYSU11
NH$_4$NH$_3$CH$_2$COOHSO$_4$	1.936	2.713	166	129	60	355	O=S̄=O	AGLYSL01
NH$_4$NH$_3$CH$_2$COOHSO$_4$	1.817	2.713	177	120	60	357	O=S̄=O	AGLYSL01
L-Tyrosine	1.789	2.625	170	133	57	360	O=C̄=O	LTYROS11
L-Alanine	1.779	2.521	168	133	59	360	O=C̄=O	LALNIN12
γ-Glycine	1.763	2.641	171	132	57	360	O=C̄=O	GLYCIN15
	1.740	2.710	174	128	55	357	O=C̄=O	LYSCLH11
α-Glycine	1.728	2.648	170	131	57	358	O=C̄=O	GLYCIN05
Triglycine	1.723	2.510	177	117	66	360	O=S̄=O	TGLYSU11
L-Cysteine	1.710	2.711	174	130	55	359	O=C̄=O	LCYSTN12

The three-center bonds represent ~70% of the total number of hydrogen bonds in the crystal structures surveyed (Table 2.3). This is a significantly higher proportion than in the other biological molecules, and was attributed to proton deficiency, which occurs because the amino acids form zwitterionic crystal structures where the predominant hydrogen bonding is between the $-\overset{+}{\text{N}}\text{H}_3$ and the $\overset{|}{\overset{\text{O}}{=}}\text{C}=\text{O}$ groups. The latter have a definite preference to accept two bonds, and therefore the zwitterions pack in the crystal lattice so that each carboxylate group can, as far as possible, accept four bonds. Since the $-\overset{+}{\text{N}}\text{H}_3$ group has only three protons which can act as donors, there is a sharing of hydrogen-bond functionality, and hence a larger proportion of three-center bonds.

The three-center hydrogen bonds observed in the crystal structures of the amino acids are described in Table 8.6. They range from symmetrical to unsymmetrical, as found for the carbohydrates and nucleic acid constituents, but there is a significantly greater tendency to form more unsymmetrical bonds. A particular feature are the chelated three-center hydrogen bonds discussed below.

Chelated three-center hydrogen bonds and chelated bifurcated hydrogen bonds involve two acceptors or two donors belonging to the same molecule. They are observed in the crystal structures of the amino acids and their salts (Table 8.7), and in the crystal structures of the nucleic acid constituents (Table 8.8).

In the *amino acids*, the chelated three-center bonds have the $-\overset{+}{\text{N}}\text{H}_3$ group as donor, and they are almost invariably unsymmetrical. The primary and strong interaction is to one of the carboxylate (or sulfate) oxygen atoms, the secondary (and weaker) interaction to the other oxygen atom resulting from stereochemical constraints imposed by packing of the molecules in the crystal lattice.

In the *nucleic acid constituents*, the donors in the chelated three-center hydrogen bonds show a much greater variety with several types of charged and un-

Three-Center Hydrogen Bonds

Table 8.8. Three-center chelated hydrogen-bond configurations observed in the crystal structure of the purines, pyrimidines, nucleosides and nucleotides (normalized from X-ray data)

Donor	A_1-Y-A_2	r_1	r_2	θ_1	θ_2
$>\!\!\overset{+}{N}H$	$O=\bar{C}=O$	1.60, 1.86	2.06, 2.84	177, 114	131, 122
	$O=C-OH$	1.82	2.28	166	95
	$N-C=O$	1.81	2.73, 2.85	177, 175	135, 125
	$O=\bar{N}=O$	1.68, 1.77	2.46, 2.82	177, 174	133, 121
	$O=\bar{S}=O$	1.62, 1.70	2.73, 2.79	171, 163	123, 115
$>\!\!NH$	$O=\bar{C}=O$	1.79, 1.82	2.44, 2.78	176, 167	133, 126
	$O=C-N$	1.77	2.69	175	125
	$N-C=O$	1.82	2.92	168	141
	$O=\bar{S}=O$	1.88	2.71	170	128
OH	$O=\bar{S}=O$	1.60	2.53	163	129
O_WH	$O=\bar{C}=O$	1.84, 1.94	2.79, 2.88	170, 163	136, 114
	$O=\bar{N}=O$	1.91, 2.10	2.55, 2.88	166, 156	140, 196
	$O=\bar{S}=O$	1.84, 1.95	2.63, 2.99	179, 163	139, 116
$-N(H)H$	$O=\bar{C}=O$	1.79, 1.84	2.47, 2.99	176, 170	133, 120
	$O=\bar{C}=O$	2.67	2.70	145	155
	$O=C-N$	1.71	2.98	178	129
	$N-C=O$	2.03, 2.04	2.86, 2.92	176, 167	131, 131
	$N-C=N$	1.97	2.18	163	133
	$O=\bar{S}=O$	1.70, 1.80	2.64, 3.00	167, 161	139, 120
	$O=\bar{S}=O$	2.14, 2.33	2.08, 2.50	164, 150	140, 137
	$O=\bar{P}=O$	1.81	2.77	158	136

Table 8.9. Three-centered/bifurcated hydrogen-bond configurations in purine and pyrimidine (P & P) and nucleoside and nucleotide crystal structures (N & N)

Donor	Acceptor			Distances (°)				Angles (°)				REFCODE
	A_1	A_2	A_3	r_1	r_2	r_3	r_4	θ_1	θ_2	θ_3	θ_4	
P & P												
$-NH_2$	$O\!\!\diagup\!\!\overset{\bar{N}}{}\!\!\diagdown\!\!O$		$O=\bar{N}$	1.89	2.61	2.72	1.82	157	105	98	170	BIDRUB10
O_WH_2	$O=C$	$O\!\!\diagup\!\!\overset{\bar{N}}{}\!\!\diagdown\!\!O$		1.79	2.86	2.50	2.07	171	95	119	172	GUNPIC10
O_WH_2	$O=C$	$O=\bar{N}$	$O=\bar{N}$	1.87	2.31	2.51	2.07	148	125	109	143	THPROL
O_WH_2	Br^-	Br^-	–	2.52	3.00	2.36	–	133	95	147	–	ADENIC
N & N												
$-NH_2$	OH	$O=C$	$O\!\!<$	2.00	2.41	2.75	2.16	149	115	94	141	RBFROX
	$O=\bar{N}$	OH	$O=\bar{N}$	2.01	2.74	2.58	1.94	171	94	103	169	CYTIDN
	Cl^-	OH	Cl^-	2.35	2.79	2.49	2.68	164	88	105	145	ARFCYT10
	Cl^-	OH	OH	2.54	2.38	2.78	1.86	140	114	89	160	XFURCC10
O_WH_2	$N\!\!\diagup\!\!$	O_W	$O=\bar{P}$	1.93	2.75	2.53	1.85	152	104	120	151	AMAFAP

charged N—H groups and with hydroxyls from C—OH and O_WH (Table 8.8). The two components of the three-center bonds are again predominantly unsymmetrical, and as described above for the amino acids, they arise from the stereochemical constraints of the molecular packing.

Configurations involving simultaneously three-center and bifurcated hydrogen bonds are observed in the crystal structures of nucleic acid constituents: purines, pyrimidines, nucleosides, nucleotides.

Bifurcated (acceptor) hydrogen bonds as defined by Pimentel and McClelland [28] can be formed by $-NH_2$ and H_2O, but were never observed in the crystal structures of small biological molecules except if in combination with three-center bonds, where they adopt configurations of the type:

$$O_W \diagdown^H_H \quad \text{or} \quad -N \diagdown^{\theta_1\; H \overset{r_1}{-\!-}A_1}_{\theta_2 \overset{r_2}{\diagdown}A_2}_{\theta_3\; H \overset{r_3}{-\!-}}_{\theta_4 \;\;r_4\; \diagdown A_3}$$

The geometrical data for nine examples displaying this configuration are given in Table 8.9. In four of them, oxygen atoms of nitrate anions are engaged as acceptors. In all cases, the primary, strong bonds are with the exterior acceptor groups A_1 and A_3, and the two planes described by H, A_1, A_2 and by H, A_2, A_3 are inclined at angles of about 50°.

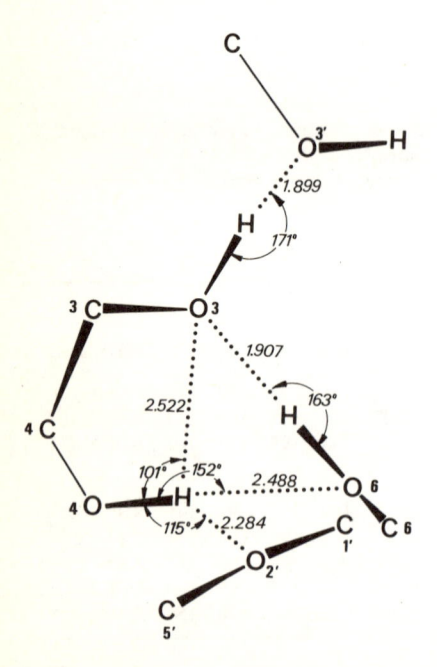

Fig. 8.1. The four-center hydrogen bond in the crystal structure of sucrose by neutron diffraction (SUCROS04)

8.2 Four-Center Hydrogen Bonds

A four-center hydrogen bond is defined as one in which the proton makes three first neighbor contacts to potential hydrogen-bond acceptor atoms in the forward direction with respect to the donor X–H bond.

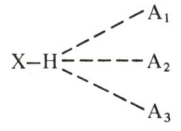

$$X-H\begin{matrix}-A_1\\---A_2\\-A_3\end{matrix}$$

The geometrical definition is less rigorous than that for three-center bonds, since it requires only that the X–Ĥ···A$_1$, A$_2$, A$_3$ angles are greater than 90°.

In the general survey of 1352 NH···O=C bonds [56], the percentage of four-center hydrogen bonds is below 1% and they occur only with N–H as donors. This may be an underestimate caused by the use of arbitrary van der Waals cut-off.

One of the well-established four-center bonds was observed by neutron diffraction in the crystal structure of sucrose (SUCROS04) shown in Fig. 8.1, in which the O(4)H of the glucopyranose residue forms two intermolecular bonds, to the fructofuranose ring oxygen O(2'), and the glucopyranose O(6), and an intramolecular hydrogen bond to O(3). All the OH···O angles are greater than 90°. Two of these bonds form an unusual triangle with the two-center bond from O(6)H to O(3).

The geometry of the four-center hydrogen bonds observed in neutron diffraction analyses of the amino acids and of the purines, pyrimidines, nucleosides,

Table 8.10. Geometry of the four-center bonds in the crystal structures of the amino acids (neutron diffraction)

Acceptors			Distances (Å)			
A$_1$	A$_2$	A$_3$	r$_1$	r$_2$	r$_3$	REFCODE
O=C̄	O=C̄^{a}	O=C̄	1.85	2.66	2.68	DLSERN11
O=C̄	O=C̄^{a}	O=C̄	1.87	2.60	2.71	LGLUAC11
O=C̄	O=C̄^{a}	O=C̄	1.91	2.52	2.71	ASPARM02
O=S̄	O=C	OH	1.97	2.74	2.79	AGLYSL01
O=S̄	O=C	O=S̄	1.99	2.46	2.64	CYSTAC01
O=C̄	Cl$^-$	O=C̄^{a}	2.22	2.59	2.26	GLYHCL
Cl$^-$	O=C̄	O=C^a	2.26	2.56	2.71	VALEHC11
Cl$^-$	O=C̄^{a}	O=C̄	2.27	2.61	2.62	HISTCM12
Cl$^-$	O=C̄	O=C̄^{a}	2.36	2.48	2.69	PHALNC01

^{a} Intramolecular.

$$\begin{matrix}&&&r_1&-A_1\\&\oplus&&&\\&N-H&-\frac{r_2}{-}&-A_2\\&&&r_3&\\&&&&-A_3\end{matrix}$$

Table 8.11. Geometry of four-center bonds in the crystal structures of the purines, pyrimidines (P & P), nucleosides and nucleotides (N & N)

P & P							
N(H₃)H⁺	O=C̃	O_W	O=C̃	2.17	2.34	2.44	AMOROT
−N(H₂)H⁺	O⟍C̃⟋O		O=C̃	1.87	2.92	2.74	TPATAA
>NH	O=Cᵃ	N⟨	N⟨	2.04	2.94	2.90	BABXAD
>NH	O=C	OH	OH	1.79	2.86	2.97	MURCAC
>NH	O⟍Ñ⟋O		O=Ñ	1.89	2.86	2.86	TCYPIC
O_WH	O_W	Br⁻	NH₂	2.00	2.33	2.65	ADENBH
N & N							
−NH₂	NH₂	N⟨ ᵃ	OH	2.53	2.79	2.88	ARBCYT10
−NH₂	Cl⁻	O=C	OH	2.54	2.57	2.74	DOCYTC
−NH₂	OH	N⟨	OH	2.50	2.51	2.56	ARBIMC10
−OH	O=C	O⟨ ᵃ	O	1.87	2.78	2.76	AFUTHU

ᵃ Intramolecular.

nucleotides are given in Tables 8.10 and 8.11. Some of these hydrogen bonds involve one intramolecular interaction. This occurs more frequently in the amino acids where the donor $-\overset{+}{N}H_3$ and acceptor $-CO_2^-$ groups are attached to the same carbon atom (C_α) and therefore in close spatial vicinity.

Chapter 9
Intramolecular Hydrogen Bonds

Intramolecular hydrogen bonds between different components of molecules stabilize conformation. They are among the most important interactions in small and in large biological molecules because they require particular molecular conformations to be formed, and when formed, they confer additional rotational stability to these conformations. They ultimately help to determine and to define the three-dimensional structures of the molecules, and are therefore involved in their functional aspects. These bonds are of major importance in the globular proteins, where all the secondary and tertiary structure hydrogen bonds are of this type. In polypeptides, the $NH \cdots O=C$ bonds between peptide bonds (n) and (n+3) give rise to β-turns and to 3_{10} helices, and between bonds (n) and (n+4) to α-helices (see Part III, Chapt. 19).

In the small biological molecules considered here, intramolecular hydrogen bonding is observed in situations where the configuration of the molecule is such that donor and acceptor groups are in close proximity[1]. As will be shown below, this interaction occurs frequently as the minor component of three-center hydrogen bonds. There are also circumstances where the molecules adopt a particular conformation to permit intramolecular bonding. Such interactions are necessarily strong and two-centered.

Intramolecular hydrogen bonds in the *amino acid crystal structures* occur as the minor components of three-center bonds in the configuration

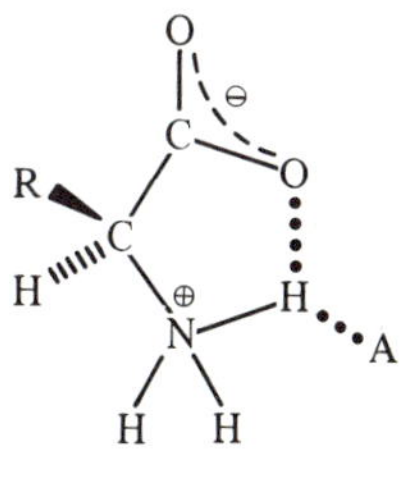

---

[1] Some examples in nonbiological molecules are given in Table 7.1. For some very short intramolecular $N-H \cdots N$ hydrogen bonds, see [453].

Some examples for this type of hydrogen bonding are given in Table 8.6. The H···O distances are 2.3 to 2.4 Å and the N–Ĥ···O angle is about 97°. The geometry in these intramolecular bonds is well defined because donor and acceptor groups are attached to the same C_α atom so that the stereochemistry is limited, i.e., it is again an intrinsic part of the molecular structure.

In the *nucleoside and nucleotide crystal structures*, intramolecular hydrogen bonding is not uncommon between ribose O(2')H···O(3') and O(3')H···O(2') hydroxyls and O(5')H···O(4' furanose). With very few exceptions, these are the minor components of three-center hydrogen bonds with H···O distances between 2.1 and 2.9 Å and O–H···O angles in the range 90° to 116°. They occur because other inter- or intramolecular forces in the crystal confer a favorable conformation to which these weak interactions add a small energetic advantage.

The interactions between sugar and base components occur in less than 10% of the crystal structures surveyed [62]. These are formed between sugar O(5')H and purine N(3) or pyrimidine O(2) if the nucleoside is in the *syn* conformation (see Part II, Chapter 17). Of particular interest are the three-center bonds, in which both major and minor components are intramolecular (shown in Table 9.1).

Since the $-NH_2$ group in adenine is conjugated and therefore coplanar with the purine heterocycle, intramolecular hydrogen bonding with the proximal N(7) atom is invariably observed as the minor component of a three-center bond. These H···N interactions are in the distance range from 2.5 to 2.9 Å, with

Table 9.1. Some important intramolecular hydrogen bonds between bases and riboside observed in nucleoside and nucleotide crystal structures

Compound	Donor	Acceptor	H···A' (Å)	X–H···A' (°)	REFCODE
6-Methyl 2'-deoxyuridine	O(5')H	O=C(4)	1.79	158	MEDOUR
		O(4')	2.38	110	
9-β-Arabinofuranosyl-8-morpholine adenine dihydrate	O(5')H	N(3)	1.78	168	ARFMAD
		O(4')	2.49	105	
8(α-Hydroxyisopropyladenosine) hydrate	O(5')	N(3)	1.82	169	HIPADS
		O(4')	2.55	104	
Inosine (orthorhombic)	O(5')···N(3)		1.95	165	INOSIN11
	O(5")···N(7)		2.61[a]	90	
5-Hydroxyuridine	O(5)H···O=C(4)		2.37[a]	107	HXURID
Xanthosine dihydrate	N(3)H	O(5')	1.86	167	XANTOS
		O(4')	2.56	113	
Oxoformycin B	N(3)H	O(5')	1.85	157	OXOFMB
		O(4')	2.37	117	

[a] Minor component of three-center bond where major component is intermolecular.

Intramolecular Hydrogen Bonds

N(6)–Ĥ···N(7) angles between 90° and 125°, and should be regarded as an intrinsic component of the molecular structure:

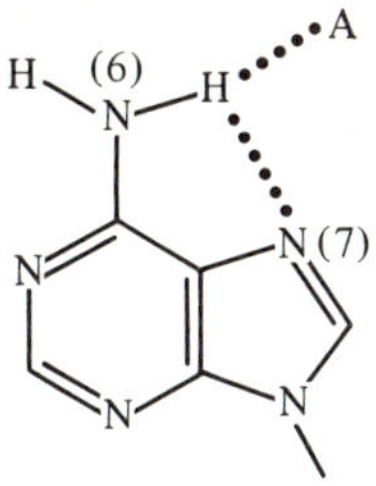

In the *pyranose sugar crystal structures*, intramolecular hydrogen bonds are formed between C(n)OH and C(n+2)OH groups when the molecular conformation is such that the C-OH bonds are in the syndiaxial orientation, i.e.:

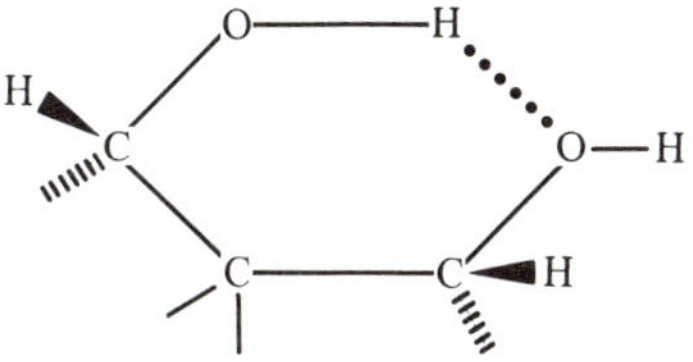

This type of bonding is not energetically favorable in most cases relative to intermolecular O–H···O bonds to adjacent molecules or solvent molecules. It is rarely observed in the crystal structures of the sugar alcohols (see Part IB, Chapt. 13). Some examples of this type of intramolecular hydrogen bonds which have been observed are described in Table 9.2.

In *di- and trisaccharide crystal structures and in cyclodextrins*, intramolecular hydrogen bonds have been observed between adjacent monosaccharides, as shown by the representative examples given in Table 9.3 and shown in Fig. 9.1. If these

Table 9.2. Intramolecular hydrogen bonds between syndiaxially oriented O–H groups in carbohydrate crystal structures

Crystal structure	Bond	H···O (Å)	O–Ĥ···O (°)	REFCODE
Methyl-β-ribopyranoside (neutron data)	O(2)H···O(4)	1.959	139	MDRIBP02
Potassium D-gluconate·H$_2$O (Form A; neutron data)	O(4)H···O(2) O(2)···O=C̄	1.737 2.121[a]	147 107	KDGLUM01
Methyl 1-thio-α-D-ribopyranoside (X-ray data)	O(2)H···O(4)	2.09	142	MTRIBP10
Methyl-1,5-dithio-α-D-ribopyranoside (X-ray data)	O(4)H···O(2) O(2)H···O(4)	2.20 2.32	129 135	MDTRPY20

[a] Component of a three-center bond.

Table 9.3. Intramolecular inter-residue OH···O hydrogen bonds in disaccharides. These hydrogen bonds are frequently three-centered. When one of the components is intermolecular, the distance is indicated by [a]. With the exception of the neutron analysis, (N), the O–H covalent bond lengths have been normalized to 0.97 Å

Compound	Bond	H···O (Å)	O–H···O (°)	REFCODE
Cellobiose	O(3′)H⟨ O(5)	1.76	151	CELLOB02
	O(6)	2.39[a]	132	
Methyl cellobioside methanolate	O(3′)H⟨ O(5)	2.02	132	MCELOB
	O(6)	2.16[a]	134	
Galabiose	O(3)H⟨ O(5′)	1.85	161	CITSIH10
	O(1′)	2.53	109	
β-Lactose	O(3′)H⟨ O(5)	1.74	175	BLACTO
	O(1)	2.39	116	
α-Lactose monohydrate	O(3′)H⟨ O(5)	1.90	156	LACTOS10
	O(4)	2.66	100	
Laminarabiose 0.20 hydrate	O(4)H···O(5′)	2.07	148	LAMBIO
Mannobiose	O(3′)H⟨ O(5)	2.02	168	DIHTUJ
	O(1)	2.60	112	
α-Maltose	O(3′)H⟨ O(2)	1.81	168	MALTOT
	O(4′)	2.47	109	
β-Maltose monohydrate (N)	O(2)H⟨ O(3′)	1.840	165.5	MALTOS11
	O(4′)	2.415	105.7	
Methyl β-maltoside monohydrate	O(2)H···O(3′)	2.00	141	MMALTS

[a] three-center bond.

molecules are hydrated, the intramolecular hydrogen bonding can also be indirect through one or more water molecules, as illustrated by the examples in Table 9.4 and discussed in more detail in Part II, Chapter 13. In the cyclodextrins, ring closure constraints orient the O(2)H and O(3)H groups of adjacent glucose units such that interglucose, intramolecular hydrogen bonds can easily form [454]. In β-cyclodextrin at room temperature they are all of the flip-flop type [O–($\frac{1}{2}$)H···H($\frac{1}{2}$)–O (Table 9.5a)] and below a phase transition temperature, all but one change to normal hydrogen bonds, which are three-centered [455] (see Table 9.5b).

There is evidence for indirect, water mediated intraresidue hydrogen bonding with water in the double helical structure of the p-nitrophenyl maltohexaose

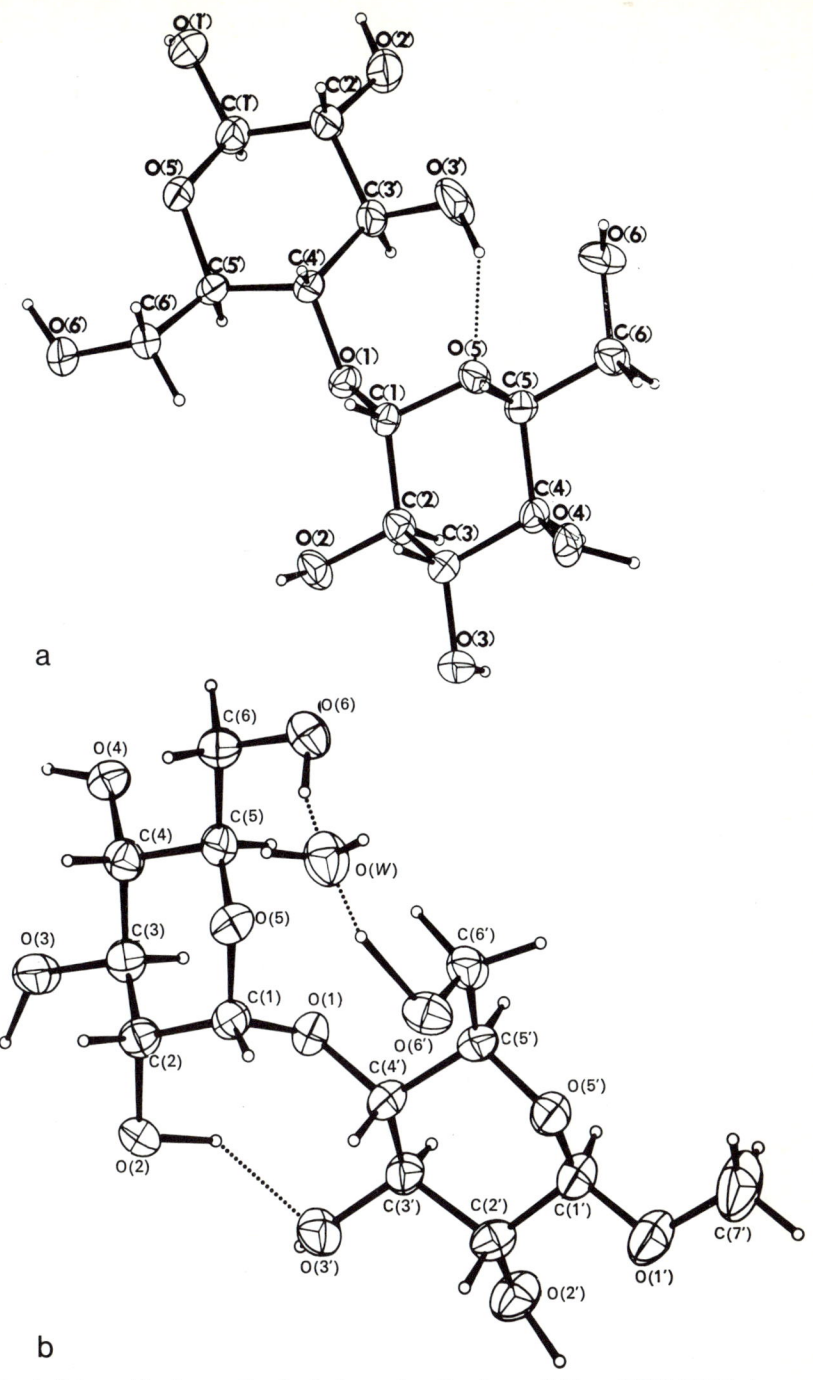

Fig. 9.1a, b. Inter-residue intramolecular hydrogen bonding in **a** cellobiose (CELLOB02), **b** methyl-β-maltopyranoside monohydrate (MALTOS11) (one is direct, one is through a water molecule)

Table 9.4. Indirect inter-residue hydrogen bonding in disaccharides[a]

Molecule	Bonds	H···O (Å)	O–H···O (°)	H'···O (Å)	O–H'···O (°)	REFCODE
Methyl β-maltoside	O(6')H···O_WH'···O(6)	1.99[b]	148	1.90	147	MMALTS
α,β-Maltose	O(6)H···O(4)H'···O(6')	1.81	176	1.79	157	MALTOT
α,α-Trehalose dihydrate	O(2)H···O_WH'···O(6')	1.77	170	1.81	167	TREHAL01
α,β-Melibiose monohydrate	O(6')H···O_WH'···O(4)	1.88[d]	172	1.85	170	MELIBM10
Sophorose monohydrate	O(2)H···O_WH···O(3)	1.92[b]	155	2.19	114[c]	SOPROS
1-Kestose	O(2)H···O(4″)H'···O(3″)	2.13[b,d]	148	1.88[b]	166	KESTOS

[a] X-ray values normalized to 0.97 Å.
[b] Major component of three-center bond.
[c] H position probably incorrect.
[d] Part of homodromic cycle.

Table 9.5a. Intramolecular interglucose hydrogen bonds of the flip-flop type $O-(\frac{1}{2})D\cdots D(\frac{1}{2})-O$ observed in β-cyclodextrin at room temperature (neutron diffraction). The occupational parameters of the oxygen atoms are 1.0, those of the deuterium atoms vary between 0.344 and 0.654. (For atom numbering scheme see Chapter 18 [454].)

Hydrogen bond	O–D (Å)	D⋯O (Å)	O–D⋯O (°)	O⋯O (Å)	D⋯D (Å)	Occupation of D-atoms (standard dev. is 0.02)	Sum of occupation
O(2)1–D⋯O(3)2 O(2)1⋯D–O(3)2	0.87 (2) 0.92 (2)	2.12 (3) 2.07 (2)	166 (2) 169 (2)	2.97 (1)	1.21 (3)	0.34 0.65	0.99
O(2)2–D⋯O(3)3 O(2)2⋯D–O(3)3	0.89 (4) 0.94 (2)	2.04 (4) 1.97 (2)	161 (4) 168 (1)	2.89 (1)	1.11 (4)	0.37 0.61	0.98
O(2)3–D⋯O(3)4 O(2)3⋯D–O(3)4	0.89 (4) 0.90 (2)	1.92 (4) 1.99 (2)	156 (2) 172 (2)	2.87 (1)	1.18 (4)	0.54 0.46	1.00
O(2)4–D⋯O(3)5 O(2)4⋯D–O(3)5	0.94 (2) 0.98 (2)	1.92 (2) 1.83 (2)	158 (2) 167 (2)	2.80 (1)	0.93 (2)	0.43 0.57	1.00
O(2)5–D⋯O(3)6 O(2)5⋯D–O(3)6	0.94 (2) 0.94 (2)	1.91 (2) 1.89 (2)	157 (2) 163 (3)	2.80 (1)	0.98 (2)	0.58 0.42	1.00
O(2)6–D⋯O(3)7 O(2)6⋯D–O(3)7	0.88 (2) 0.94 (2)	2.08 (2) 1.99 (2)	153 (2) 161 (2)	2.89 (1)	1.16 (3)	0.48 0.47	0.95
O(2)7–D⋯O(3)1 O(2)7⋯D–O(3)1	0.93 (4) 0.96 (2)	2.03 (4) 1.99 (2)	164 (2) 171 (1)	2.94 (1)	1.07 (4)	0.34 0.58	0.92
Mean values	0.92 (3)	1.98 (8)	163 (6)	2.88 (7)	1.09 (10)		0.98 (3)

Table 9.5.b. The intramolecular hydrogen bonds in β-cyclodextrin undecahydrate at 120 K are all of the three-center type, one is a flip-flop. Data taken from a neutron diffraction study [455]. (For atom numbering see Part IIB, Chap. 18)

X–D⟨A₁/A₂⟩	X···⟨A₁/A₂⟩ (Å)	D⟨A₁/A₂⟩ (Å)	X–D (Å)	X–⟨D⟩⟨A₁/A₂⟩ (°)	D⟨A₁/A₂⟩ (°)	Sum of angles around D (°)
O(2)1–D···O(3)2	2.959 (7)	2.027 (7)	0.976 (7)	158.8 (7)	79.8 (2)	347.0
···O(4)2	2.820 (7)	2.355 (7)		108.4 (6)		
O(3)1–D···O(2)7	2.991 (8)	2.023 (8)	0.975 (8)	171.6 (6)	75.0 (3)	350.7
···O(4)1	2.844 (7)	2.445 (7)		104.1 (4)		
O(3)3–D···O(2)2	2.913 (7)	1.954 (8)	0.981 (8)	165.2 (6)	77.3 (3)	352.6
···O(4)3	2.794 (7)	2.301 (8)		110.1 (5)		
···O(3)5[a]	2.754 (7)	1.84 (1)		160.0 (10)		
O(2)4–D···O(4)5	2.792 (7)	2.37 (1)	0.954 (9)	106.4 (7)	82.7 (4)	349.1
···O(2)3	2.967 (8)	2.004 (8)		176.4 (6)		
O(3)4–D···O(4)4	2.887 (7)	2.509 (7)	0.964 (8)	103.3 (5)	73.8 (2)	353.5
···O(2)4[a]	2.754 (7)	1.81 (1)		159.0 (10)		
O(3)5–D···O(4)5	2.806 (7)	2.34 (1)	0.98 (1)	107.7 (8)	83.4 (4)	350.1
···O(3)7	2.935 (7)	2.011 (7)		158.3 (4)		
O(2)6–D···O(4)7	2.746 (6)	2.228 (6)	0.971 (8)	112.2 (5)	83.0 (3)	353.5
···O(2)5	2.770 (6)	1.840 (6)		155.4 (6)		
O(3)6–D···O(4)6	2.793 (6)	2.352 (6)	0.989 (6)	106.1 (4)	80.8 (2)	342.5

[a] This is a flip-flop hydrogen bond. Occupation of hydrogen bond to O(2)4 is 0.49 (2) and to O(3)5, 0.48 (2).

Intramolecular Hydrogen Bonds

triiodide complex (see Part III, Chap. 18.9). This structure is heavily hydrated with 22 water molecules per asymmetric unit. The resolution in the X-ray analysis was such that hydrogen atoms could not be observed. The hydrogen bonding is inferred from the $O \cdots O_W$ separations. The water molecules are systematically chelated to adjacent oxygen atoms around the glucopyranose rings, forming five-membered rings as shown below.

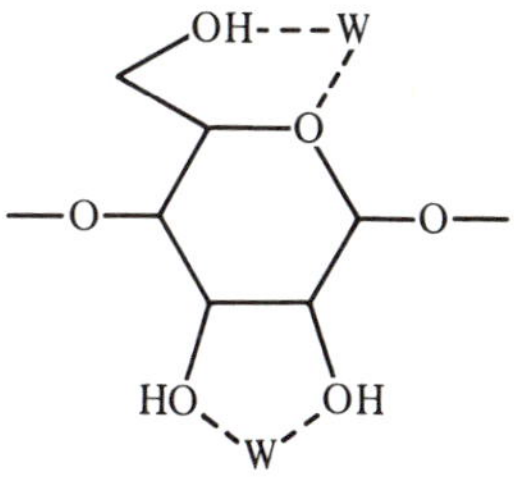

'Solvation' of this type is not observed in the hydrates of the mono-, di- and trisaccharides. It is however invoked when discussing solvation effects in solution [494].

Chapter 10
Weak Hydrogen-Bonding Interactions Formed by C–H Groups as Donors and Aromatic Rings as Acceptors

As pointed out in Part I, hydrogen bonds involving C–H groups differ from those with O–H or N–H groups in that the hydrogen atom may or may not have electronegative atoms as first neighbors in crystal structures. Unlike the stronger O–H···O or N–H···O bonds, the hydrogen bonds with C–H as donor will not have a major influence on molecular conformation. They occur when other packing forces and stereochemical constraints bring C–H bonds and electronegative acceptor atoms within the range of electrostatic interactions, about 3.0 Å.

There are, however, many C–H groups present in a biological molecule. So the circumstances when the molecular conformation or molecular packing is such that the C–H hydrogen is brought in close proximity to an electronegative hydrogen-bond acceptor, should be quite common. Also, as with N–H and O–H donors, the hydrogen bonds with C–H donors should be stronger and have a more important effect on conformational stability or molecular packing in crystals when they are formed between charged species.

The issue was first discussed in 1937 as "chemical bonds" in the light of deviations from ideal behavior of liquid mixtures of chloroform and acetone or diethyl ether [25]. The formation of C–H···O hydrogen bonds was clearly recognized by the spectroscopists [456–459] and, later, by the crystallographers [460–463], who considered inter- and intramolecular C···O distances of less than 3.3 Å as potential candidates for C–H···O bonds. In part, the reluctance to recognize C–H as hydrogen-bond donor arose from the concept that hydrogen bonds should be *co-linear* with the covalent bond. The use of sums of van der Waals radii as criteria with an arbitrary cut-off for hydrogen-bond lengths at about 3 Å also limited the consideration of these interactions. Consequently, only the shorter C–H···O distances were noted. As a result, there may be many more examples in the crystallographic literature than those that have been so identified.

Multiple bonding of C–H to electron-withdrawing groups favors C–H···A hydrogen bonds. In a 1963 review on infrared spectroscopic evidence for C–H···O bonding [458], it was concluded that C–H bonds are definitely hydrogen-bond donors to oxygens and that their donor capability decreases in the series

$$C(sp)-H > C(sp^2)-H > C(sp^3)-H .$$

In R_3CH, at least one of the R groups must be strongly electron-withdrawing for the molecule to form an interaction which can be detected by infrared spectroscopy. The effectiveness of the electron-withdrawing groups was estimated to be

$CN \gg CCl_2CCl_3 > CBr_3 > CCl_3 > Br > I > CHClCHCl_2$

$\sim Cl > F > COOCH_3 > CHBr_2 > CHCl_2$.

Geometrical data for C–H···O=C hydrogen bonding is provided by accurate X-ray and neutron analyses. A comprehensive survey of the C–H···O=C, C–H···N and C–H···Cl interactions in 113 crystal structures, irrespective of type of compound, considered intermolecular contacts with C–H···A angles greater than 90° [462]. The conclusions, although biased by being based on comparisons with the sum of the van der Waals radii ($W_H + W_A$), clearly indicated that for C–H···O=C, "the interactions are more likely to be attractive than repulsive, and can reasonably be described as hydrogen bonds". The H···O bond lengths in that analysis range from 2.04 Å to the arbitrary cut-off value of $W_H + W_A = 2.40$ Å, which is still shorter than many of the minor components of three-center N–H···O bonds discussed in the earlier chapters.

A survey based solely on the neutron diffraction analyses of 32 zwitterion amino acids [60] also suggested that there was good evidence for the existence of C–H···O=C̄ hydrogen bonds. The H···O distances range from 2.16 to 2.65 Å with a mean value of 2.45 Å, and C–Ĥ···O angles greater than 150°. It was noted that the majority of examples involves the "most acidic" C_α–H groups.

C–H···O=C hydrogen bonds may inhibit zwitterion formation in the crystalline state. This suggestion has been made on the basis of isonicotinic acid [464], and of nicotinyl glycine [465]. In these crystal structures, the observed C–H···Ō=C interactions with H···O distances from 2.35 to 2.66 Å and C–Ĥ···O angles in the range 123° to 170° appear to be preferred and suppress the formation of the more common N̄H···O=C̄ bonds. Since, however, there is no evidence of this in the amino acid crystal structures, this effect, if real, must be associated with the properties of the pyridine ring.

A few examples may help to illustrate C–H···O hydrogen-bonding interactions. In the crystal structure of acetic acid [371], there is a dimer arrangement formed by a combination of O–H···O=C and C–H···O=C hydrogen bonds:

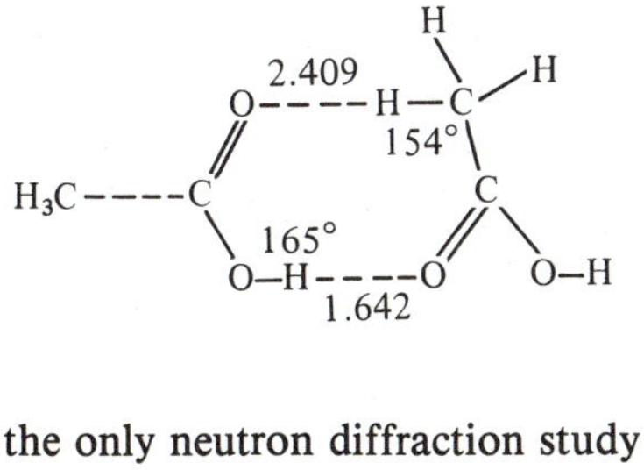

In the only neutron diffraction study of a (heavily modified) nucleoside, 5-nitro-1-(β-D-ribosyluronic acid)-uracil monohydrate [466], there is a complicated scheme of hydrogen bonds which involves two-center and three-center C–H···O

Table 10.1. CH···O hydrogen bonds observed in the crystal structure of 5-nitro-1-(β-D-ribosyl-uronic acid)-uracil monohydrate (NRURAM 11) by neutron diffraction at 80 K [466]

Bonds	H···O (Å)	C−H···O (°)
C(6)H···O⟨C(5′), H⟩	2.080 (5)[a]	155.7 (5)
C(2′)H···O=C(4)	2.310 (5)	167.7 (5)
C(5′)H⟨O−C(2′)	2.356 (5)	133.8 (5)
C(5′)H⟨O=N	2.334 (5)	132.4 (5)
C(4′)H⟨O=C(2)	2.247 (5)	145.6 (5)
C(4′)H⟨O=N	2.564 (5)	117.3 (5)

[a] Intramolecular

interactions with the geometry shown in Table 10.1. Of greater interest is the intramolecular C(6)−H···O(5′) hydrogen bond, which was also observed in other nucleosides and nucleotides, and appears to be a more general attractive force between base and sugar. This is indicated by the data in Table 10.2, obtained from accurate X-ray analyses which permitted calculation of the position of the C(6)−H hydrogen atom on the basis of the covalently bonded atoms of the pyrimidine ring system.

Another interesting example for a C−H···OH$_2$ hydrogen bond is provided by the X-ray crystal structure reported for the trihydrate of the tricyclic orthoamide shown below [467]:

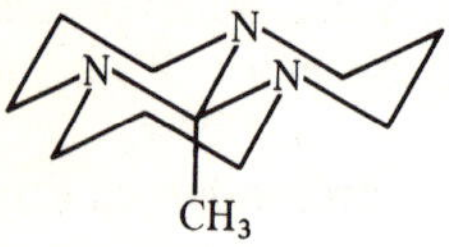

The molecule is located on a crystallographic threefold rotation axis, with the methyl group involved in C−H···OH$_2$ bonding, with an H···O distance of 2.67 Å and C−Ĥ···O angle of 170°. This interaction is strong[1] enough to force the methyl group into an eclipsed conformation with an N−C−C−H angle of only 8° which, according to quantum chemical calculations, is less favorable than the staggered conformation. This study shows that under certain circumstances the weak C−H···O interactions can be sufficient to stabilize methyl groups in other than preferred orientations.

---

[1] The energy for a CH$_4$···OH$_2$ bond is calculated at the MP4/6−311G** level to be 7.1 kJ mol^{-1} [467].

Table 10.2. Two- and three-center C–H···O hydrogen bonds in crystal structures of nucleosides and nucleotides of the pyrimidine type[a]

Compound	Bond	H···A (Å)	C–H···A (°)	REFCODE
Uridine	C(5')H···O=C(4)	2.27	148	BEURID10
3-Deaza-4-deoxyuridine	C(6)H···O(5')H	2.22[b]	166	RFURPD
Uridine-3'-monophosphate H$_2$O	C(6)H···O(5')H	2.29[b]	171	URIDMP10
	C(5')H···O=C(2)	2.41	157	
	C(4')H···O(P)(H)	2.23	149	
	C(5')H···O(P)(H)	2.54	120	
5-Hydroxyuridine	C(6)H···O=C(2)	2.36	148	HXURID
5-Methoxyuridine	C(6)H⟨O(5')H / O=C(4)⟩	2.17 / 2.37	136 / 98	MXURID01
Inosine (monoclinic)	C(8)H···O=C(10)	2.55	103	
	C(2)H⟨O(5')H / O(2')H⟩	2.25 / 2.28	131 / 139	INOSIN10
2'-O-methyl cytidine (molecule A)	C(6)H⟨O(5')H / O(4')⟩	2.25 / 2.40	166 / 101	OMCYTD20
(molecule B)	C(6)H⟨O(5')H / O(4')⟩	2.34 / 2.53	154 / 91	

[a] All O(5') hydroxyl groups are also engaged as hydrogen-bond donors.
[b] Intramolecular.

Aromatic rings may function as weak hydrogen-bond acceptors. It has been suggested on the basis of a molecular mechanics calculation that the ⟩NH donor can form a hydrogen bond directed to the center of a benzene ring [468]. In this calculation, the assumption of dipoles on the ⟩NH and C–H bonds with charges located on the nuclei gives rise to an energy minimum of 3.3 kcal mol^{-1} when the hydrogen atom is 2.4 Å from the center of the aromatic ring. A search for similar interactions in 33 highly refined protein structures leads to the conclusion that this interaction occurs more frequently than expected from a random distribution [469, 470]. A similar interaction has been noted in a hemoglobin-drug complex [471] and in pancreatic trypsin inhibitor, where an asparagine amide NH$_2$ group is in contact with a phenylalanine [472].

As with the weak C–H···O=C interactions, it is possible that this stereochemistry is energetically favorable, but is not structure-determining. It occurs only when the stronger interactions result in a favorable donor-acceptor juxtaposition, resulting in a small addition to the total hydrogen-bond energy. If this is so, some examples should occur in small molecule crystal structures where the atomic positions are known more precisely and where hydrogen atoms can be observed. A search for such structures using the Cambridge Crystallographic Data Base was unsuccessful but in a recent X-ray study on a calixarene hydrate, an $O_W H \cdots$ benzene hydrogen bonding interaction was reported [472a].

Chapter 11
Halides and Halogen Atoms as Hydrogen-Bond Acceptors

The halide ions are strong hydrogen-bond acceptors. The infrared spectroscopy data for halides show clearly the characteristic shifts associated with hydrogen-bond acceptors, in the sequence

$Cl^- > F^- > Br^- > I^-$.

There appears to be no obvious reason why the order of the spectral shifts for Cl^- and F^- are reversed relative to the hydrogen-bond strengths, which are $F^- > Cl^-$, as expected from their electronegativities. The $X-H\cdots A$ hydrogen-bond lengths are consistently shorter than for any other atom in the same row in the Periodic Table, i.e.:

$X-H\cdots F^- < X-H\cdots O < X-H\cdots N$

$X-H\cdots Cl^- < X-H\cdots S < X-H\cdots P$, etc.

In fact, $X-H\cdots Br^-$ is the only well-established hydrogen bond known to form with third-row elements [473].

Hydrogen bonds involving Cl^- are better documented than those with F^-, Br^-, I^-. There are a large number of crystal structure analyses of hydrochlorides of the small biological molecules containing $-\overset{+}{N}H_3$ or $\overset{+}{\diagdown}NH$ groups. This is because these salts crystallize more readily than do the corresponding neutral molecules to provide crystals suitable for X-ray and neutron diffraction studies. There are no crystal structures of the corresponding hydrofluorides. Hydrobromides and hydroiodides do not provide good data on hydrogen bonding because of a deterioration in the precision of the analyses due to the presence of the heavy atom.

Chloride ions are three- and four-coordinated. The neutron diffraction analyses of the amino acid hydrochlorides provided the data in Table 11.1, and the more extensive data obtained from the X-ray analyses of hydrochlorides of nucleic acid constituents are given in Table 11.2. The data indicate clearly that the chloride ion may be three- or four-coordinated. When three-coordinated, the bonds may be in planar or in pyramidal configuration, with no bimodal distribution. The ligands of four-coordinated chloride ions are only very approximately in tetrahedral con-

Table 11.1. $-\overset{+}{N}(H)_2H\cdots Cl^-$ hydrogen-bond lengths to chloride ions in the amino acid hydrochlorides from neutron diffraction data

Compound	$H\cdots Cl^-$	$X-\hat{H}\cdots Cl^-$	REFCODE
Two-center hydrogen bonds			
L-Cystine·2HCl	2.101	168	CYSTCL02
L-Glutamic acid·HCl	2.108	174	LGLUTA
Glycine·HCl	2.123	166	GLYHCL
L-Glutamic acid·HCl	2.137	172	LGLUTA
L-Phenylalanine·HCl	2.142	170	PHALNC01
L-Lysine·HCl	2.149	175	LYSCLH11
L-Valine·HCl	2.161	168	VALEHC11

Compound	$r_1{}^a$	r_2	θ_1	$+\theta_2$	$+\alpha$	$=\Sigma$	Acceptor of minor component	REFCODE
Three-center hydrogen bonds ($A=Cl^-$, $A'=Cl^-$)								
Glycine·HCl	2.160	2.686	171	79	109	359	O=COH	GLYHCL
L-Histidine·HCl·H$_2$O	2.165	2.617	169	81	109	359	O=C̄	HISTCM12
L-Lysine·HCl·2H$_2$O	2.196	2.635	172	80	107	359	O=C̄	LYSCLH11
Glycylglycine HCl·H$_2$O	2.256	2.952	151	106	101	358	Cl$^-$	GLGICH01
L-Cystine·2HCl	2.262	2.625	157	82	107	346	O=COH	CYSTCL02
Glycylglycine·HCl·H$_2$O	2.268	2.560	147	101	100	348	O=COH	GLGICH01
L-Cystine·2HCl	2.270	2.647	151	80	126	357	O=COH	CYSTCL02
L-Phenylalanine·HCl	2.302	2.496	164	87	101	352	O=COH	PHALNC01
L-Valine·HCl	2.356	2.455	169	89	100	358	O=COH	VALEHC11
L-Tyrosine·HCl	2.376	2.431	145	97	111	353	O=COH	LTYRHC10
L-Tyrosine·HCl	2.471	2.420	162	108	89	359	O=COH	LTYRHC10
L-Tyrosine·HCl	2.505	2.491	138	103	113	354	O=COH	LTYRHC10

[a] For definition, see p. 20, configuration 1.

stellation, and there is no apparent distinction in bond lengths with type of coordination.

Chloride ions and water molecules are comparable. In the three- and four-coordination, the chloride ion stereochemistry is very similar to that of the water molecules in hydrated crystals. Also, both display no strong directional properties as hydrogen-bond acceptors.

In the ice-like structures such as the quaternary ammonium clathrate hydrates, chloride or fluoride ions readily substitute for water molecules with only relatively minor changes in the ice-like lattices[1].

The $X-H\cdots Cl^-$ bond length distribution is very narrow. An interesting characteristic of the data given in Tables 11.1 and 11.2 is the relatively narrow spread of the $H\cdots Cl^-$ bond lengths compared with other donor-acceptor combi-

---

[1] Chloride ions easily replace water molecules in the butane clathrate hydrate, and counter-cations can occupy the voids in the lattice. This is one of the reasons why the hydrate crystallization method proved unsuccessful for desalination of sea water.

Table 11.2. Summary of the two-center and three-center hydrogen-bond lengths to Cl⁻ ions in the purine, pyrimidine, nucleoside and nucleotide crystal structures, normalized from X-ray data

Donors	$>\overset{+}{N}H$	$-OH^b$	$>NH^a$	O_WH^a	$-N(H)H$
Two-center					
No.	18	12	12	24	17
Min.	1.98	2.02	2.04	2.13	2.14
Max.	2.22	2.24	2.39	2.40	2.57
Mean	2.07	2.13	2.13	2.24	2.27

	Minor component	Major component
Three-center[a]		
No.	12	12
Min.	2.59	2.08
Max.	2.99	2.65
Mean	2.79	2.32

There are insufficient data to distinguish between the various donors.
[a] Purines and pyrimidines only.
[b] Nucleosides and nucleotides only.

nations. The reason for this is that the small chloride ions are relatively free from steric constraints and can more readily occupy positions of minimum energy.

In Table 11.2, the distribution of X–H···Cl⁻ bond lengths follows the same sequence for donor X–H groups as reported in Chapter 8:

$$>\overset{+}{N}-H < O-H \approx >N-H < O_W-H < N(H)H .$$

There are no reliable data on O–H···F⁻ or N–H···F⁻ bonds, and only few data on O–H···Br⁻ and N–H···Br⁻ bonds. They tend to follow the same relative sequence as the X–H···Cl⁻ bonds described above, at ~0.10 Å longer.

The fluorine atom as hydrogen-bond acceptor. From the point of view of electronegativity differences, C–F bonds might be expected to be strong hydrogen-bond acceptors. There are, however, very few references to X–H···F–C bonds to be found in the crystallographic literature. A survey of the intermolecular environment of the C–F bond in fluorinated carboxylic acids and carboxylates was inconclusive [474]. It is certain that when such bonds exist they will be weak interactions which only occur because other steric factors place the C–F bond in an appropriate position close to a strong hydrogen bond donor. In this regard C–F acceptors are perhaps similar to the C–H donors: permissive but not obligatory hydrogen-bonding functional groups.

Chapter 12
Hydrogen-Bond Acceptor Geometries

Do the functional groups which are hydrogen-bond acceptors play an important role in determining the geometry of bonding patterns in biological molecules? Since the attractive force of the hydrogen bond is primarily electrostatic, the strong directionality associated with covalent bonding is not expected. If it is to be observed with respect to the acceptor atom, it will be only for those atoms where the electron distribution in the direction of the hydrogen bond is likely to be anisotropic. For this reason, the attention directed to this question has been focused on the oxygen atom [58, 60, 475, 476]. This is understandable, since it is the hydroxyl, ether, carboxylate, carbonyl, and water oxygen which are the predominant hydrogen-bond acceptors in biological systems [477].

The lone-pair electrons of oxygen are conventionally assigned to sp^3 orbitals, which suggests some tetrahedral directionality. However, these electron distributions are diffuse and easily polarized. So this directionality, if it exists, might be dependent on the configuration of the group to which the oxygen is bonded, as between O$\langle$ and O=C, for example.

The directional properties of the oxygen atom as hydrogen-bond acceptor are weak. Two surveys [475, 476] aimed at distinguishing between the hydrogen-bond acceptor directional properties of sp^2 versus sp^3 oxygen atoms. They suggested that in the sample of crystal structures of ketones, ethers, and esters there was little difference between the various types of oxygen atoms, as the acceptor directions ranged over broad bands of ±50° from the axis of the acceptor group, O=C$\langle$ or O$\langle$.

This is consistent with the neutron diffraction studies of the amino acids, which disclosed that the O=C̄ and O=C groups most commonly accept two hydrogen bonds, sometimes only one, and sometimes three. In the more general survey of 1352 N−H···O=C hydrogen bonds from neutron and normalized X-ray data, about 75% of the neutral C=O groups accept only one hydrogen bond, the other 25% two or more, and for the charged C̄=O groups, these numbers change to 30% for one acceptor and 70% for two or more. That the number of acceptor functions is, in general, lower in this study than in the narrower survey of the amino acids might be due to the application of van der Waals cut-off criteria, which omitted some of the longer hydrogen-bonding interactions.

sp^3 Oxygen atoms are single hydrogen-bond acceptors. This follows from the neutron diffraction studies of the carbohydrates, where over 200 hydrogen bonds

Hydrogen-Bond Acceptor Geometries 165

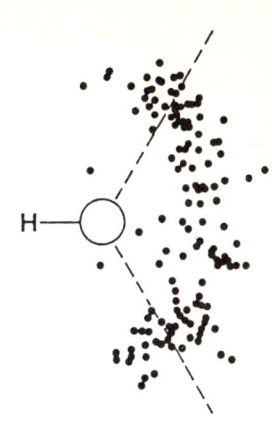

Fig. 12.1. Distributions of donor H atoms around acceptor hydroxyl O atoms in a Newman projection down the C=O bond (H−O bond for H$_2$O). *Dashed lines* lone-pair directions [476], see also Fig. 23.4a,b

of type O−H···O with acceptors O$\langle{}^C_H$ and O$\langle{}^C_C$ were analyzed. The majority of these oxygen atoms, 90%, accepted only one hydrogen bond, the remaining 10% accepted two.

Water molecules have the same characteristics as sp^3 oxygen atoms, both in the organic and inorganic small molecule hydrates.

The water oxygen atom can coordinate to one or two cations, or accept one or two hydrogen bonds, Table 7.8. When it accepts one hydrogen bond, the configuration can be pyramidal or trigonal planar, with no marked distinction in the distribution between the two. When two ligands (hydrogen bonds or metal ions) are coordinated, the configuration is sometimes tetrahedral, but can show large deviations from this ideal geometry.

Directional properties of acceptor groups are very soft or nonexistent. Just as the H···A lengths and X−Ĥ···A angles for individual hydrogen bonds are perturbed by the other intermolecular forces in the crystal, so are the directional acceptor properties of the functional groups. These properties appear to be even "softer" than the hydrogen-bond geometries, and in consequence it is not possible to extract any characteristic trends from the surveys of the type described in this chapter.

The scatterplots obtained from studies of the directionality of oxygen acceptor atoms for different donor groups suggest that the electron density distribution associated with oxygen lone pairs has only a secondary influence on the acceptor directions of the hydrogen bonds in crystal structures [475, 477] (Fig. 12.1, see also Fig. 23.4a, b). In gas-phase studies of hydrogen-bonded dimers, however, there is evidence of acceptor directionality along the axis of the lone-pair of π-bonding pair orbitals [478]. Due to this flat potential energy surface this directionality is obscured by crystal field forces in the condensed phase and possibly by solvation effects in solution. In the small molecule hydrate crystal structures discussed in Part IV, Chapter 22, three coordinated water molecules exceed four-coordinated by a factor of 1.8 and have hydrogen-bond geometries ranging from planar to pyramidal, with no suggestion of a bimodal distribution.

Part II
**Hydrogen Bonding
in Small Biological Molecules**

In the following, the hydrogen-bonding patterns in crystal structures of small biological molecules will be described in schematic form. Atom designations are given in parentheses, hydrogen-bond distances H···A are given in Å units, angles X–Ĥ···A in degrees. Infinite chains are marked by arrows, cycles by circular arrows with one head for homodromic and two heads for antidromic orientation of hydrogen bonds. Intramolecular hydrogen bonds are indicated by an asterisk given after the bond distance.

Chapter 13
Hydrogen Bonding in Carbohydrates

Carbohydrates are among the most abundant molecules in living nature, with the exception of the water molecule. They are found in wide varieties as monomers, oligomers, cyclic and linear and branched polymers. They occur alone or chemically combined with purines and pyrimidines in the nucleic acids, with polypeptides in the glycoproteins and with lipids in the glycolipids. The simpler carbohydrates resemble water molecules in one important respect. Their primary hydrogen-bonding function comes from the hydroxyl group which can have both donor and acceptor properties (Box 13.1). For this reason, their hydrogen-bonding patterns have features in common with those of the ices and the high hydrates. They differ fundamentally from those of the polypeptides or proteins, in which the majority of hydrogen-bond groups are −NH groups, which are only hydrogen-bond donors, and the C=O groups, which can only be hydrogen-bond acceptors. The hydrogen bonding in the nucleosides and nucleotides, which contain functional groups which are both donor-acceptors and donors or acceptors only, has features which resemble both the carbohydrates and the peptides.

For the carbohydrates especially, the amount of available crystal structural data decreases sharply with molecular complexity [479]. With the exception of the cyclodextrins, discussed in Part III, Chapter 18, there are less than 40 crystal structure analyses of oligosaccharides, of which less than 10 are trisaccharides, one is a tetrasaccharide, and one a hexasaccharide (Part III, Chap. 18). The majority of the basic monosaccharides that are the subunits of the polysaccharides that occur naturally have been studied; for example, the pyranose forms of β-arabinose, α-xylose, α- and β-glucose, β-fructose, α-sorbose, α-mannose, α- and β-galactose, α-fucose, α-rhamnose, N-acetyl glucosamine, and mannosamine (Box 13.2). How-

---

Box 13.1. Hydrogen-bonding functional groups in carbohydrates

Donor groups: C−O−H, H−O$_W$−H, −C(=O)O−H

Acceptor groups: O(C,H), O(C,C), $_W$O(C,H), O=C(O−H), O=C(O−C)

---

crystal structures of the furanose forms of the two important nucleic acid carbohydrates, D-ribofuranose and 2-deoxy-D-ribofuranose, have never been studied because these compounds have never been crystallized with the molecules in the furanose configuration. This is because in equilibrium in solution, the pyranose configurations predominate. In fact, relatively few crystal structures of furanoses and furanosides have been analyzed, except for those in which the ribose and deoxyribose are components of the nucleosides and nucleotides.

Of the three major components of biological structure, the proteins, nucleic acids, and polysaccharides, least is known about the polysaccharides at the secondary and tertiary level of molecular structure. This is because the polysaccharides cannot be obtained in crystals which are large enough for single crystal X-ray or neutron structure analysis. What structural information there is comes from fiber X-ray diffraction patterns. However good these diffraction patterns are, the structures derived from them will always be *model-dependent*. This is because the number of variable atomic parameters which determine the diffraction intensities exceeds the number of observed intensities.

It is likely that the most definitive structural studies of biological polysaccharides will eventually come from those glycoproteins which can be obtained as single crystals suitable for X-ray structure analysis or from enzyme crystal structures where polysaccharides are the substrates. Even there, the results have hitherto been barely at the limits of atomic resolution, because of the overall size of the protein molecules.

The subunits of biologically important carbohydrates are the pentoses and hexoses. More than 1900 crystal structure determinations of carbohydrates, nucleosides and nucleotides are listed in the 1988 Cambridge Crystallographic Data File. Those determined prior to 1980 were reviewed in critical bibliographies in Advances in Carbohydrate Chemistry and Biochemistry [479]. From the point of view of insight into hydrogen-bonding patterns in biological systems, the important structures are the cyclic aldopentoses and aldohexoses which constitute the subunits of most of the polysaccharides. These compounds are configurationally related into the four homomorphous groups shown in Box 13.2. All these molecules have been studied by crystal structure analysis, either as the α- or β-epimer, and in some cases both. Many of these compounds commonly occur naturally, or in chemical combination, as the enantiomorph (D or L). Both D and L forms have the same crystal structure except for the handedness of the molecules, and therefore have the same hydrogen-bonding patterns and structural characteristics. If D- and L-enantiomers are mixed in solution, they may crystallize separately; all D molecules in one crystal, all L molecules in another, or they may crystallize as the D,L-racemate, which will have a different crystal structure and hydrogen-bonding pattern from the D (or L) form. Since the primary functional groups are hydroxyls, a large number of hydrates might be expected. However, this is not so; relatively few monosaccharides form hydrates; less than 10% of the structures reported. More are formed by the di- and trisaccharides, and many more by the nucleosides and nucleotides, as discussed in Part IV, Chapter 22.

If relationships are expected between the molecular configurations and the hydrogen-bonding patterns observed, we should find them by comparing the

Hydrogen Bonding in Carbohydrates

Box 13.2. Configurational Relationships Between the Biologically Important Pyranoses

$R_1 = R_2 = R_3 = H$
D-xylopyranose

$R_1 = CH_2OH, R_2 = R_3 = H$
D-glucopyranose

$R_1 = R_2 = H, R_3 = CH_2OH$
D-sorbopyranose [a]

$R_1 = R_2 = R_3 = H$
D-lyxopyranose

$R_1 = CH_3, R_2 = R_3 = H$
D-rhamnopyranose [a]

$R_1 = CH_2OH, R_2 = R_3 = H$
D-mannopyranose

$R_1 = R_2 = H, R_3 = CH_2OH$
D-tagatopyranose

$R_1 = R_2 = R_3 = H$
L-arabinopyranose

$R_1 = CH_2OH, R_2 = R_3 = H$
D-galactopyranose

$R_1 = R_2 = H, R_3 = CH_2OH$
L-fructopyranose

$R_1 = CH_3, R_2 = R_3 = H$
L-fucopyranose

$R_1 = R_2 = R_3 = H$
D-ribopyranose

[a] Sorbose and rhamnose occur naturally as the L-enantiomorphs.

crystal structures within the homomorphous groups shown in Box 13.2. Hitherto, no such relationships have been discovered.

In addition to the many X-ray structure determinations of the monosaccharides and their derivatives, there are also an unusually large number, 24, neutron diffraction studies of monosaccharides [57]. These are principally pyranoses and methylpyranosides, since many of them crystallize easily and form crystals of any desired size. There are a more limited number of disaccharide X-ray crystal structure analyses, only two of which are neutron diffraction studies, those of sucrose [SUCROS 04] and α-maltose monohydrate [MALTOS 11]. Nine trisaccharide crystal structures have been determined by X-ray analysis, but only one tetrasaccharide, that of stachyose pentahydrate [STACHY 10] and one hexasaccharide, p-nitrophenyl-α-maltohexaoside (Part III, Chap. 18). Oligosaccharides, where the molecules are long enough to emulate the amyloses and celluloses, have resisted crystal structural investigation due to an inability to grow adequate single crystals. None of the interesting branch-chain oligosaccharides important in the blood sugars or the components of bacterial polysaccharides has yet been crystallized. For these more complex molecules, the combination of proton NMR spectroscopy and molecular mechanics calculations has provided a general description of the molecular conformations in solution, but no direct information concerning their hydrogen-bonding properties [480, 481].

The cyclic oligosaccharides, α-, β- and γ-cyclodextrins crystallize well. A large number of X-ray structure analyses and three neutron analyses of the hydrates and inclusion complexes of these molecules have been reported. These interesting structures, which are midway between biological small molecules and macromolecules, are discussed separately in Part III, Chapter 18.

There are no crystalline carbohydrate macromolecules analogous to the crystalline globular proteins and the transfer ribonucleic acids. The polysaccharides are either completely amorphous or semi-amorphous, as in the starches, or form fibrous structures, as in the celluloses and many of the bacterial polysaccharides. At best, the X-ray fiber diffraction patterns provide a basis for distinguishing the "most likely" conformational and packing model for molecules of known configuration, from "less likely" models [482, 483]. These models are judged on the basis of having "reasonable" hydrogen-bonding O···O distances relative to the known hydrogen-bonding behavior of their monosaccharide components. Since the hydrogen bonding involves C–OH groups, the hydrogen positions of which cannot be deduced from those of the nonhydrogen atoms, the hydrogen-bonding schemes included in the models for the crystal structure of the polysaccharides must be regarded as highly speculative. They are generally based on the simplest possible assumptions; that the hydrogen bonds are two-centered, and that the hydrogen atoms of the –OH groups lie on the line between the closest O···O separations. Some theoretical calculations on the structure of cellulose suggest that a number of energetically similar hydrogen-bonding schemes are possible which involve both two- and three-center bonds [344]. In view of the hydrogen-bonding patterns discussed later in this chapter, this is a very reasonable hypothesis.

13.1 Sugar Alcohols (Alditols) as Model Cooperative Hydrogen-Bonded Structures

The simplest carbohydrate crystal structures in which to study hydrogen bonding are the acyclic alcohols, alditols, of the general formula $CH_2OH(CHOH)_n \cdot CH_2OH$, of which the common naturally occurring compounds are the pentitols and hexitols (which all are in anhydrous form). The only hydrogen-bonding groups in these molecules are the hydroxyls, which can function both as donor and as acceptor groups. The crystal structures of three of the pentitols and six of the hexitols (which all are in the anhydrous form) have been determined [484]. Although these analyses were, with one exception, room-temperature X-ray studies, the diffraction data were excellent and most of the hydroxyl hydrogen atoms were directly located from the experimental data.

Alditols pack as rods connected by chains of hydrogen-bonds. When the alditols crystallize, the long-chain molecules pack as straight or bent rods in hex-

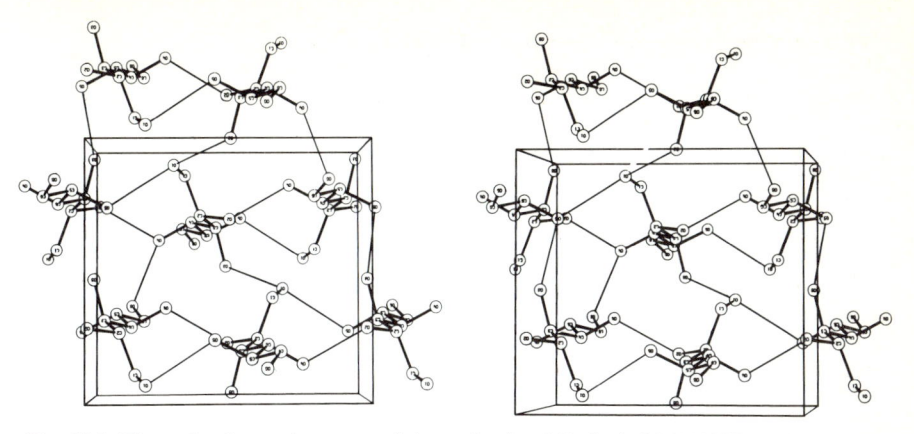

Fig. 13.1. The molecular environment of the molecule of D-glucitol [GLUCIT01]. Stereo view down the carbon chain axis in the crystal structure showing molecular packing and hydrogen bonding

agonal motifs. Irrespective of whether the carbon chains are linear, as in the crystal structures of arabinitol, D-mannitol, galactitol, or bent as in ribitol, xylitol, D-glucitol, D-iditol, allitol, D-altritol, the molecules are connected laterally by hydrogen bonds which form quite uniform patterns throughout this series of compounds. A typical example of the molecular packing and lateral hydrogen bonding is shown in Fig. 13.1.

The hydrogen bonds form infinite chains involving every hydroxyl group in the molecules [484], thereby providing the maximum hydrogen bond energy through the *cooperative effect* (Figs. 13.2a, b, c). As discussed in Part IB, Chapter 7.1.3, the mean hydrogen-bond length in the alditol crystal structures is shorter than in the cyclic monosaccharides, reflecting the influence of the cooperativity in the infinite chains. In two of these structures, allitol [ALITOL01] and D-mannitol [DMANTL01; Fig. 13.2b], there are three-center bonds, the minor components of which form weak cross-links between the infinite chains. In D-iditol [IDITOL; Fig. 13.2c], an oxygen atom is a double acceptor forming a branch to an infinite chain. In one structure, however, that of D-glucitol [GLUCIT01], the hydrogen-bond lengths are anomalous, as shown in Fig. 13.2d. In one infinite chain they are all compressed relative to a mean value of 1.8 Å, while in the other they are all expanded. This interesting feature, which was confirmed by a neutron diffraction analysis, is an extreme example of the crystal field effect, which may compress or expand the hydrogen-bond lengths from their effective equilibrium values in a particular structure [306].

The alditol crystal structures are conformationally homogeneous. The co-crystallization of different conformers of the same molecule is not common, but sufficient examples are known, especially in hydrogen-bonded crystal structures, for it to be a phenomenon with which crystallographers are familiar. It occurs when there is a population of conformers in solution, all of which have similar energies. A classical example is that of 2,3-dimethyl-2,3-butanediol (pinacol)

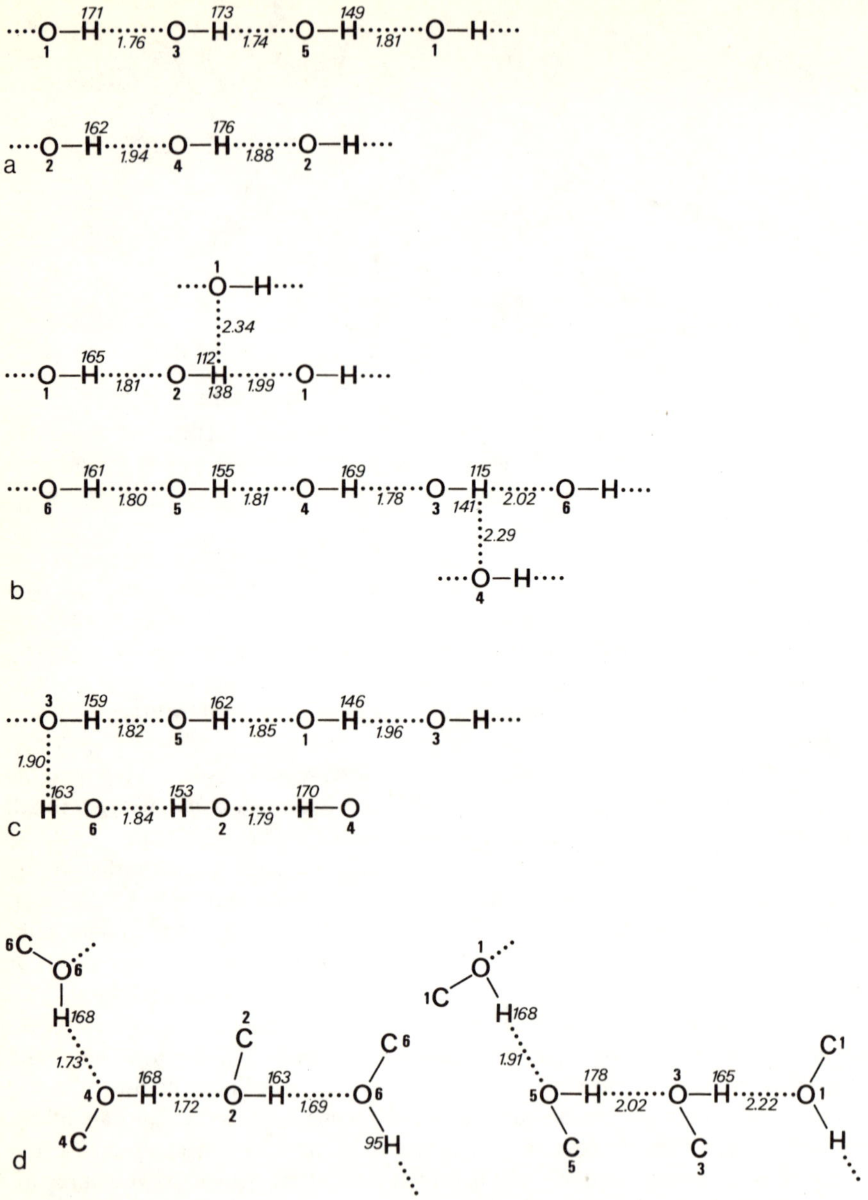

Fig. 13.2a–d. Cooperative, strong hydrogen bonds in **a** xylitol, **b** D-mannitol, **c** D-iditol, **d** the two infinite hydrogen-bonded chains in the crystal structure of D-glucitol have different H···O bond lengths, one (*left*) compressed, the other (*right*) expanded [XYLTOL, DMANTL01, IDITOL, GLUCIT01]

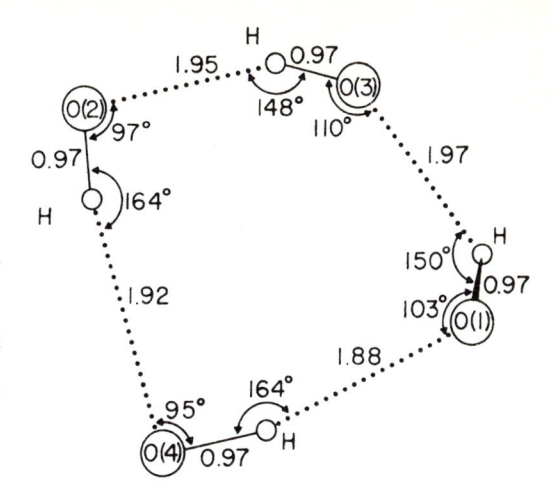

Sugar Alcohols (Alditols) as Model Cooperative Hydrogen-Bonded Structures 175

Fig. 13.3. The homodromic tetramer of hydrogen bonds in the crystal structure of pinacol (the covalent O–H bond lengths were normalized to 0.97 Å). O(1) is on molecule with symmetry $\bar{1}$. O(2)H and O(3)H are on molecules with symmetry 1. O(4)H is on molecule with symmetry 2 [485]

$(H_3C)_2C \cdot OH \cdot C \cdot OH \cdot (CH_3)_2$, in which three different conformers occur in the same crystal [485]. One is centrosymmetric, one has twofold axial symmetry, and one is asymmetric. The hydroxyl groups of all three different conformers are hydrogen-bonded to form the four-member homodromic cycle shown in Fig. 13.3.

Conformational polymorphism or co-crystallization is expected in the sugar alcohols, due to the flexibility of the carbon chains and the low optical rotations of those with enantiomorphic configurations, which are indicative of a mixed population of the conformers in solution [486]. In fact, conformational polymorphism is not observed. The reason could be that the strongly cooperative infinite chains of hydrogen bonds are conformationally very specific, and favor the crystallization of one particular conformer from the aqueous solution containing the mixture. This is equivalent to saying that the different conformers of D-glucitol or D-mannitol, which are present in aqueous solutions and account for the low optical activity, have different solubilities. If they could be separated by crystallization, they would have different crystal structures with different hydrogen-bonding schemes. When polymorphism is observed, as in D-mannitol [DMANTL; DMANTL01] the molecular conformation is the same in all three known crystal structures.

Meso configurations can have enantiomeric conformers in crystals. Ribitol and xylitol have meso configurations and are optically inactive in solution. In the crystalline state, they have the bent-chain *enantiomorphic* conformations. Xylitol crystallizes in space group $P2_12_12_1$, so each crystal contains either only the optically active left-handed or the optically active right-handed conformers. There are, of course, an equal number of left- and right-handed crystals in any one batch, which on dissolution give no optical activity.[1]

---

[1] If the left- and right-handed crystals could be separated, by the method of Pasteur, the phenomenon of "conformational mutarotation" might be observed.

Ribitol, on the other hand, crystallizes in space group $P2_1/c$, with left- and right-handed conformers in the same crystal, related by a crystallographic center of symmetry. The choice of these two alternatives depends upon which allows the greatest hydrogen-bonding energy. As yet, theory cannot make this prediction.

Intramolecular hydrogen bonding is not observed in the sugar alcohols. Straight-chain conformers are observed in the crystal structures of D,L-arabinitol, D-mannitol and galactitol, whereas in ribitol, xylitol, D-glucitol, D-iditol, allitol and D-altritol, the observed conformers have bent chains. The distinction is because in the straight chain conformers, every pair of COH bonds on alternate carbon atoms is gauche-related as in **1**. The peri, syndiaxial, or parallel orientation of alternate COH bonds, as in **2**, is in most cases avoided by adopting the bent-chain conformers such as **3**, despite the fact that conformer **2** could be stabilized by intramolecular hydrogen-bond formation **4**, as observed in potassium gluconate monohydrate, A form (Fig. 13.4).

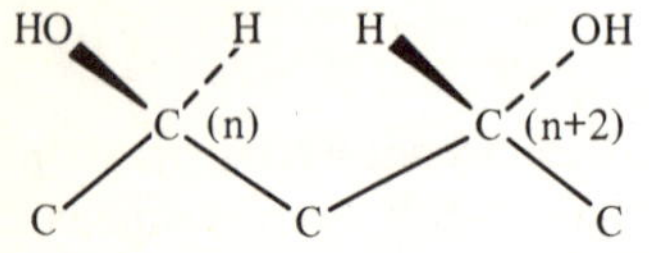

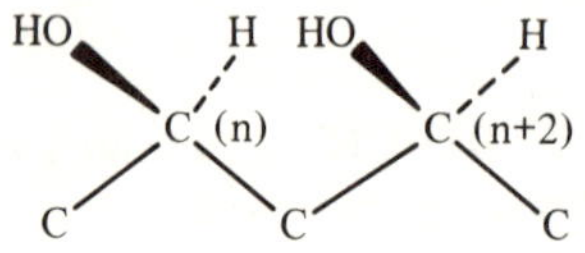

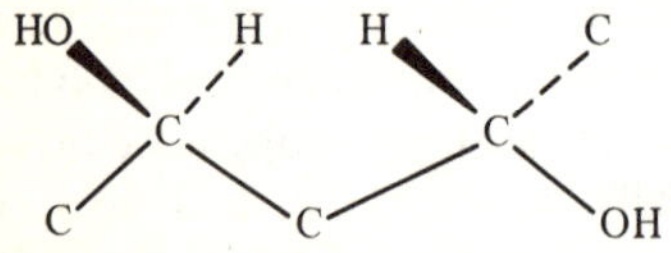

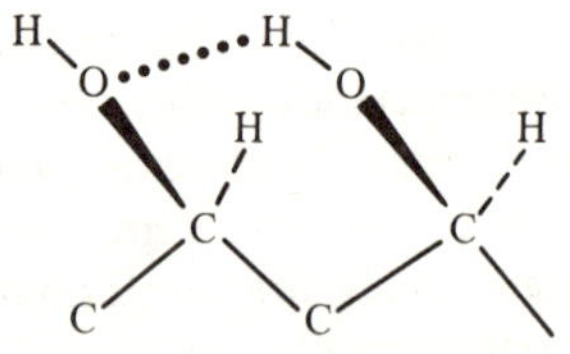

In the hydroxyl-rich environment of the molecular packing in the crystal, intramolecular bonding is sterically less favorable than the intermolecular hydrogen-bond formation that occurs in all the alditol crystal structures.

In aqueous solution, however, a sterically more favorable indirect intramolecular hydrogen bonding, as in **5**, is possible. This has not been observed crystallographically with the sugar alcohols because they do not form any hydrates.

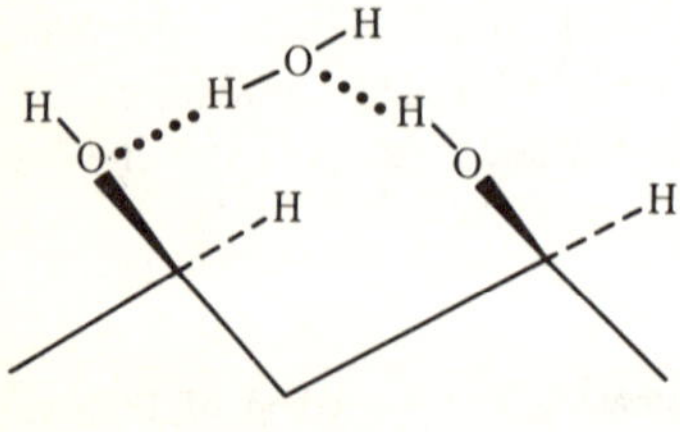

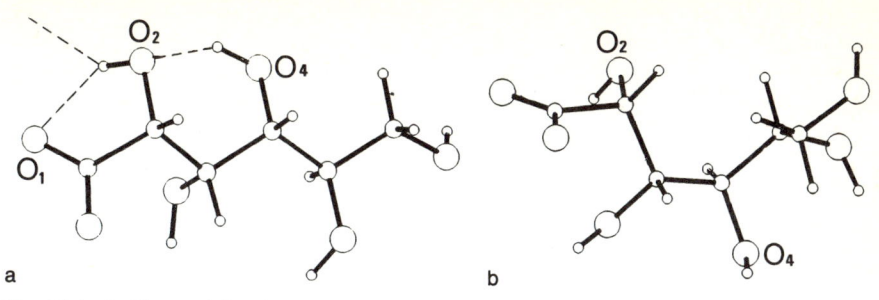

Fig. 13.4a, b. The straight chain and bent chain conformations of the gluconate ions as they appear in the A (a) and B (b) forms of potassium D-gluconate monohydrate [KDGLUM01: KGDLUM02]

Conformational polymorphism with intra- and intermolecular hydrogen bonding is observed in potassium gluconate monohydrate. Potassium gluconate monohydrate crystallizes in two forms, both of which have been studied by neutron diffraction. In form A [KDGLUM01], the molecule has the straight chain conformation shown in Fig. 13.4a, which is stabilized by an intramolecular hydrogen bond $O(4)-H \cdots O(2)$. There is also an intramolecular component of a three-center bond from $O(2)H \cdots O(1) = \bar{C}$. In the form B [KDGLUM02], shown in Fig. 13.4b, the molecule has the bent-chain conformation with no syndiaxially related C−OH bonds, and consequently no intramolecular hydrogen bonding.

The hydrogen bonding in these crystal structures shows both σ and π cooperativity. In form A, the carboxylate oxygens are strong hydrogen-bond acceptors and the cooperativity involves the bonding of the carboxylate group, as shown in Fig. 13.5. The K^+ ions are in voids between puckered sheets of strongly hydrogen-bonded anions and water molecules, and appear to have a secondary role in determining the structural pattern. The cations are as far removed as possible from the carboxylate oxygens. The eight nearest neighbor $K \cdots O$ distances include O(3), O(4) twice, O(5), O(6) and the two water molecules at distances ranging from 2.6 to 3.2 Å. In form B, the cooperativity arises from the infinite chain of

$$\overset{\overset{C}{\|}}{\to O} \leftarrow H\underset{\uparrow}{O}H \to \overset{\overset{C}{\|}}{O} \leftarrow$$

hydrogen bonds which are attached to branched finite chains, as shown in Fig. 13.5b. One of the carboxylate oxygens, O(1), is the closest neighbor to the K^+ ion with two contacts at 2.69 and 2.75 Å and O(1) is not a hydrogen-bond acceptor. The other carboxylate oxygen O(2) makes one contact to the cation at 2.75 Å, and accepts three strong hydrogen bonds.

These two neutron diffraction analyses illustrate the difficulty of identifying hydrogen-bonding patterns from $O \cdots O$ separations without information about the hydrogen atom positions. In form A, the two shortest $O \cdots O$ distances are $O(4) \cdots O(2)$ and $O(2) \cdots O(1)$, 2.610 and 2.579 Å, of which the longer corresponds to a short two-center bond, while the shorter corresponds to the longer component of a three-center bond (Fig. 13.4).

An example where hydrogen bonding causes a molecule to have a different conformation in two different solutions is trans-O-β-D-glucopyranosyl methylaceto-

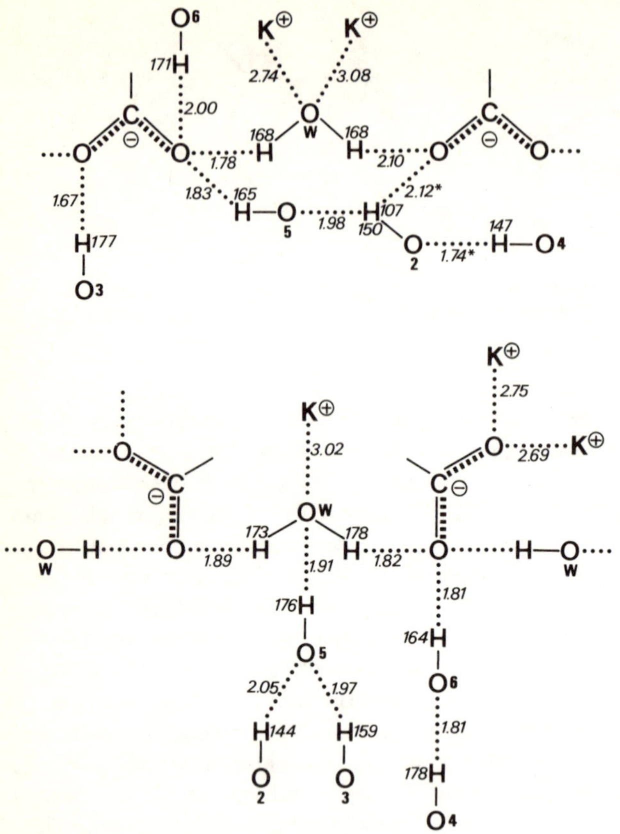

178   Hydrogen Bonding in Carbohydrates

Fig. 13.5. The hydrogen-bonding patterns in the A (top) and B (bottom) crystal structures of potassium D-gluconate monohydrate (*, intranmolecular interaction)

acetate. In 1,4-dioxane solution, the optical rotation is strongly negative ($[M]_D = -18900$ for 0.3% solution), while in aqueous solution it is strongly positive ($[M]_D = +12100$ for a 2% solution). The crystal structure analysis [GLPMAC 10] provided a basis for interpreting the necessary conformational difference in terms of intramolecular hydrogen bonding $C(6)OH \cdots O = C$ in the dioxane versus intermolecular solvent hydrogen bonding in aqueous solution.

13.2 Influence of Hydrogen Bonding on Configuration and Conformation in Cyclic Monosaccharides

Hydrogen bonding does not prevent configurational heterogeneity in crystals of reducing sugars. The cyclic monosaccharides with a free hydroxyl group at the anomeric carbon atoms, C(1), the so-called reducing sugars, have the additional complexity that different *configurations* can co-exist in equilibrium in aqueous solution [487]. For example, D-glucose can have five different configurations: open-chain, α- or β-furanose, α- or β-pyranose. This leads to the familiar phenomenon of configurational mutarotation, when one configuration in a crystal equilibrates to a mixture of configurations in solution slowly enough to be followed by changes in optical rotation of the solution. It is characteristic of the variability of the intermolecular hydrogen bonding of these molecules that the α- and β-configurations frequently co-exist in the same crystal structure. In that case, mutarotation occurs because the epimer mixture in the crystal is different from that of the equilibrium in solution. The fact that crystallization of a reducing sugar does not result in a configurationally homogeneous product has been known for a long time. To quote C. S. Hudson [488] in 1931, "... extreme caution is necessary in characterizing a homogeneous crystalline form of a reducing sugar as a pure chemical individual". This phenomenon is difficult to detect by X-ray diffraction if the proportion of the minor component is less than 10%. However, there are a number of examples where co-crystallization of the two epimers of the same reducing sugar have been detected. These are given in Table 13.1. The proportions of

Table 13.1. Examples of co-crystallization of anomeric mixtures

	Percentage proportions α	β	REFCODE
6-Deoxy-α-L-sorbofuranose	95	5	DXSORF10
Sodium α-L-guluronate dihydrate (87 K)	90	10	DUDGUE
α-Lactose monohydrate	93	7	LACTOS10
	100	0	LACTOS03[a]
Laminarabiose hemihydrate (20% H$_2$O) (3-O-α-glucosyl β-glucose; disorder between α anomer $+\frac{1}{2}$H$_2$O β anomer)	40	60	LAMBIO
α,β-Maltose	82	18	MALTOT
α,β-Melibiose monohydrate (6-O-α-D-galactopyranosyl- α,β-D-glucopyranose monohydrate)	72	28	MELIBM10
	85	15	MELIBM02
	80	20	MELIBM01
Mannobiose (O-β-D-mannopyranosyl-(1→4)-α,β-D-mannopyranose)	68	32	DIHTUJ
β-Galabiose (O-α-D-Galactopyranosyl-(1→4)-β-D-galactopyranose)	56	44	CITSIH10
Lactulose (4-O-β-D-galactopyranosyl-β-D-fructofuranose)	75	10	BOBKUY10

[a] Crystals grown from commercial samples of α-lactose monohydrate generally contain up to 10% of the β-epimer. Very slow growth from ion-free solution gives crystals with no β-epimer. The pure α crystals have a 0.9% lower unit cell volume than the α,β-mixture.

α- and β-epimers in a particular batch of crystals will depend not only upon the hydrogen-bonding of the two epimers, but also upon conditions of temperature, rate of crystallization, and probably pH of the solution. No two crystallizations necessarily have the same ratio. The differences between the independent investigations of α,β-maltose and α,β-melibiose are examples of these variations.

In every case, reasonable alternate hydrogen-bonding patterns could be postulated for the minor epimer, which could co-exist with the hydrogen-bonding pattern of the major epimer.

The structure of the laminarabiose hemihydrate [LAMBIO] is particularly interesting in this respect because it involves the co-crystallization in equal proportions of an anhydrous and monohydrate disaccharide. This structure is really a 1:1 complex between an anhydrous and monohydrate structure with different, but separate, hydrogen-bonding schemes for the anhydrous and hydrated molecules (see Fig. 13.51).

Co-crystallization of isomers with different ring configurations is also possible. The crystal structure of lactulose [BOBKUY 10], the galactose analog of sucrose, is unusual in that it involves co-crystallization of the two furanose configurations α and β, **1** and **2** with a pyranose β-configuration **3**, in the ratio 75:10:15. This novel circumstance was first detected by solid state ^{13}C CP-MAS NMR spectroscopy and confirmed by crystal structure analysis (see Fig. 13.6). The hydrogen bonding involving the galactose functional groups is the same for all three configurations and the hydrogen bonding of the fructofuranose or fructopyranose residues is flexible enough to permit this type of disorder in the crystal. It would

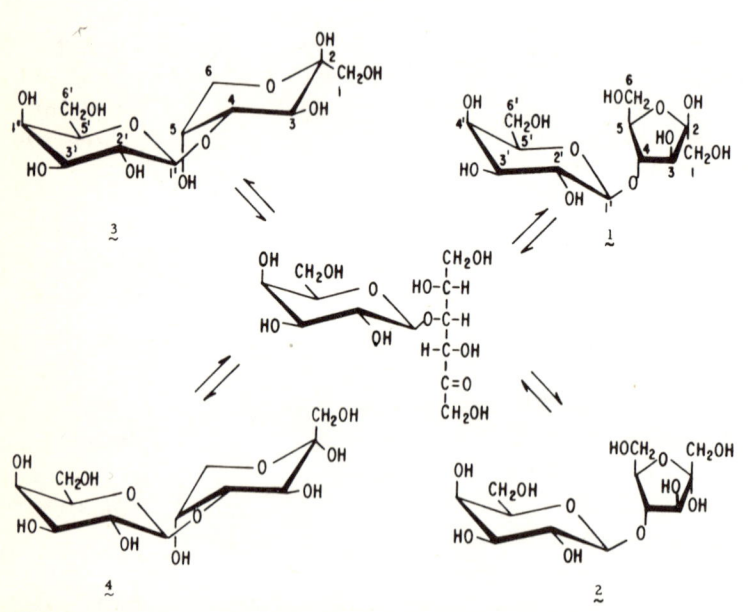

Fig. 13.6. The five configurational isomers of lactulose, 4-O-β-D-galactopyranosyl-D-fructose. The hydrogen bonding permits the co-crystallization of isomers **1**, **2**, and **3** [BOBKUY 10]

Influence of Hydrogen Bonding on Configuration and Conformation 181

Fig. 13.7. Intramolecular (∗) hydrogen-bonding pattern in methyl-β-ribopyranoside, from neutron diffraction [MDRIBP 10]

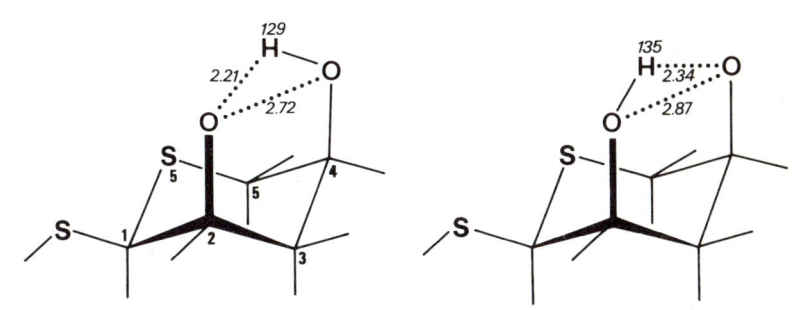

Fig. 13.8. Details of the intramolecular hydrogen bonds in the two symmetry-independent molecules in the crystal structure of methyl 1,5-dithio-α-ribopyranoside quarterhydrate [MDTRPY 20]

have been impossible to resolve the overlapping electron densities from the X-ray diffraction data without the information provided by the solid-state NMR measurements.

In some cyclic sugars, ring conformations may be stabilized by intramolecular hydrogen bonding. The configuration of the α-ribopyranose molecule is such that there is syndiaxial alignment of C–OH bonds in both the $^1C_4$ and $^4C_1$ ring conformations. In consequence, there is believed to be an equilibrium in solution as shown below.

In the isolated molecule or in nonpolar solvents, the intramolecular hydrogen bonding would be between O(2)H and O(4)H in the $^1C_4$ conformation and between O(1)H and O(3)H in the $^4C_1$ form. The α-ribopyranose molecule therefore provides an interesting example of the influence of intramolecular hydrogen bonding on ring conformation. Unfortunately, neither α- nor β-ribopyranose has been

$^4C_1-D$ $^1C_4-D$

Fig. 13.9. An example of intramolecular hydrogen bonding in the crystal structure of methyl 1-thio-α-D-ribopyranoside. The infinite O(2)H···O(4)H···O(3)H···O(2)H chains form a double ribbon linking the molecules. The —S—CH$_3$ group is omitted for clarity [MTRIBP 10]

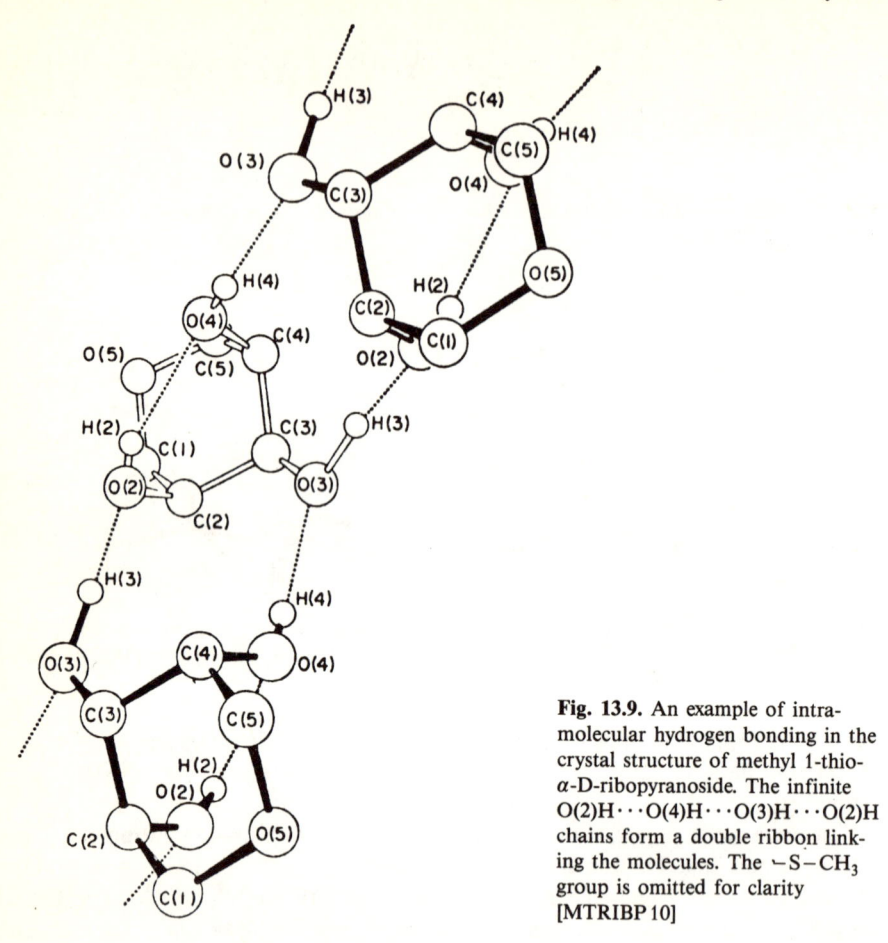

crystallized as yet. In the crystal structure of methyl-β-ribopyranoside [MDRIBP 10], the conformation is $^1C_4$-D, stabilized by a O(2)H···O(4) intramolecular hydrogen bond. The bond is the major component of a three-center bond, as shown in Fig. 13.7 from the neutron diffraction analysis.

The $^1C_4$-D chair conformation is also stabilized in the crystalline state by the formation of an intramolecular hydrogen bond in both methyl-1,5-dithio-α-ribopyranoside quarterhydrate [MDTRPY 20] and methyl-1-thio-α-ribopyranoside [MTRIBP 10]. In the former structure there are two symmetry-independent molecules with different directions for the O(2)H to O(4)H bond, as shown in Fig. 13.8. In the methyl-1-thio-α derivative, the direction is O(2)H···O(4) and the bond forms part of an infinite chain, as shown in Fig. 13.9. In the methyl-1,5-dithio-α-ribopyranoside quarterhydrate, the hydrogen-bond direction is O(4)H···O(2) in one molecule and O(2)H···O(4) in the second crystallographically independent molecule in the same crystal structure. As shown in Fig. 13.10, the intramolecular hydrogen bond is part of a cyclic arrangement of eight bonds,

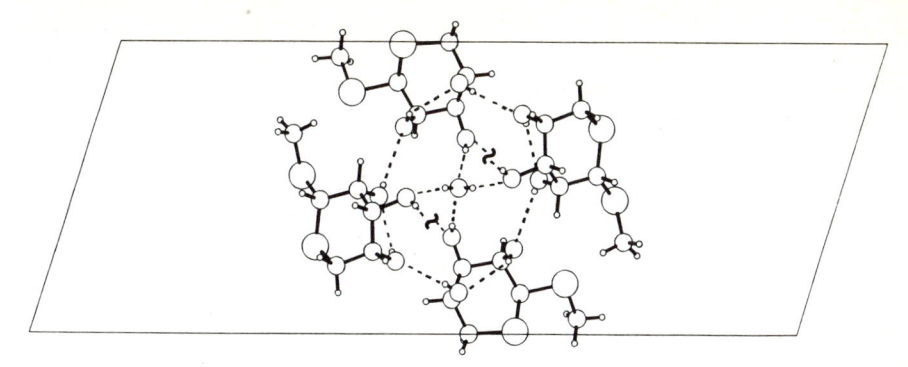

Fig. 13.10. Hydrogen-bonding pattern in the crystal structure of methyl 1,5-dithio-α-D-ribopyranoside quarterhydrate. Four molecules form a hydrogen-bonded cluster around the water molecule. The perimeter of the cluster is hydrophobic and the packing of the $4(C_6H_{12}O_3S_2) \cdot H_2O$ clusters is van der Waals. Note the eight-membered homodromic cycle and the homodromic spiral involving the water molecule. The *swivel* indicates hydrogen bonding to the next asymmetric unit [MDTRPY 20]

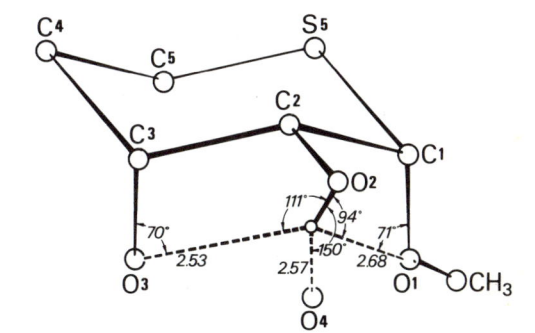

Fig. 13.11. The hydrogen bond between syndiaxial hydroxyl groups in methyl 5-thio-ribopyranoside is four-centered [METRBP 10]

in the center of which is the hydrogen-bonded water molecule. This forms a complex of four sugar molecules and one water which are internally hydrogen-bonded, and have a hydrophobic exterior. The packing of the $4(C_6H_{12}O_3S_2) \cdot H_2O$ hydrogen-bonded complex is by van der Waals interactions. This is an interesting example of hydrophilic and hydrophobic bonding in a simple crystal structure.

In contrast to these structures, the methyl-5-thio-α-ribopyranoside [METRBP 10] has the $^4C_1$ chair conformation, with no intramolecular hydrogen bonding between the syndiaxial O(3)H to O(1)CH$_3$ groups, which are separated by 2.99 Å. Instead, there is the four-centered hydrogen bond in which two of the components are intramolecular, as illustrated in Fig. 13.11. The configuration requires a relatively small distortion of the $^4C_1$ chair conformation.

Weak intramolecular hydrogen bonding can also occur between vicinal hydroxyl groups in carbohydrates. In both the pyranose and furanose sugars, weak intramolecular hydrogen bonding can occur between hydroxyl groups on adjacent carbon atoms in the rings. Irrespective of whether the hydroxyls are both equatorial, or axial and equatorial, this involves long H···O distances and donor O–Ĥ···O and acceptor H···Ô–C angles of close to or less than 90°. These bonds are weak, but can occur in crystal structures as the minor component of three-center bonding, as in α-rhamnose monohydrate [RHAMAH 12] (see Fig. 13.30).

The crystal structures of the important ribofuranoses or 2-deoxy-ribofuranoses have not been determined, since these forms are in the minority with respect to the pyranoses in solution. There are numerous examples, however, in the crystal structures of the nucleosides and nucleotides. Those that contain the ribofuranose moieties quite commonly display vicinal O(2')H···O(3') or O(3')H···O(2') intramolecular hydrogen bonding as the minor component of a three-center bond. This type of hydrogen bonding is described in Part I B, Chapter 9.

Except in these special cases, intramolecular hydrogen bonding is not commonly observed with pyranose or furanose monosaccharide molecules in the crystalline state, nor would it be expected in polar solvents when the competition is with intermolecular hydrogen bonding. Therefore, modeling carbohydrate molecules by means of theoretical calculations for comparison with observations in the crystalline state or in polar solutions, requires that all hydrogen-bond interactions are suppressed by fixing the C–OH bonds in directions appropriate for intermolecular hydrogen bonding [347, 489–493], as discussed in Part I A, Chapters 4.4 and 4.5.

In partially methylated or acetylated sugars, the opportunities for intermolecular hydrogen bonding are inhibited sterically and intramolecular bonding is to be expected when these molecules are dissolved in nonhydrolytic solvents [449, 494–497].

Hydrogen bonding determines the orientation of the primary alcohol group in hexopyranose sugars. In the pyranose hexoses and the furanose pentoses, two of the three staggered orientations of the primary alcoholic groups are energetically preferred. In the pyranose sugars, these are described by the orientation of C(6)–O(6) with reference to the C(5)–O(5) bond in the ring, as + or − *synclinal* (+or −*sc*) and *anti-periplanar (ap)* or as + or − *gauche* and *trans* [498, 499] (see Box 13.3).

In a D-hexose with the energetically preferred $^4C_1$ pyranose ring conformation, the configuration of the primary alcohol group C(6)H$_2$OH is always equatorial with respect to the ring. If the configuration of the hydroxyl on C(4) is also equatorial, as in the "gluco" configuration, the preferred orientations of the primary alcohol group are −*synclinal* (−*sc*) with O(5)–C(5)–C(6)–O(6) = −60°, or +*synclinal* (+*sc*) with O(5)–C(5)–C(6)–O(6) = +60°, since the *anti-periplanar (ap)* orientation involves a "peri" interaction between C(6)–O(6)H and C(4)–O(4)H. This could be stabilized by an intramolecular hydrogen bond, as discussed in the previous section, but this is only rarely observed. If the hydroxyl on C(4) is axial, i.e., the "galacto" configuration, the preferred orientations are +*sc* and *ap*. In the $^4C_1$-D conformers of the altro-, gluco- and mannopyranose sugars, the preferred orientations are +or −*sc*. In the $^4C_1$-D conformers of the galacto-, gulo-, ido- and talo-pyranose sugars, they are + *sc* and *ap*. This conformational factor was recognized in 1947. It is referred to as the Hassel-Ottar [501]

> **Box 13.3.** Definition of torsion and dihedral angles
>
> The conformation of a molecule is determined by angular rotations about bonds which are given by torsion angles. A torsion angle in the sequence of atoms A–B–C–D is measured by the angle which the bond A–B makes with the bond C–D when projected down B–C, see Fig. 13.12a. The angle is zero when the bonds A–B and C–D are eclipsed, and it is counted positive when C–D is rotated clockwise with respect to A–B. The terminologies used to describe angular regions are shown in Fig. 13.12b. The choice of the atoms A–B–C–D follows the recommendations given by IUPAC [500], A and D usually are substituents at atoms B and C having the highest weight.
>
> A dihedral angle is subtended by the intersecting normals to two planes, i.e. the planes defined by atoms A, B, C and B, C, D. It is the 180° complement of the torsion angle, has no sign and is now not used in describing molecular conformations. However, in many publications, dihedral angles are still incorrectly confused with torsion angles.
>
> Frequently, it suffices to indicate a torsion angle range rather than the exact torsion angle value. These ranges are indicated in Fig. 13.12. The more common nomenclature is + or − gauche and trans. In the pyranose and furanose sugar conformations, there is an exocyclic C–OH group which can almost freely rotate about the adjacent C–C bond. In hexapyranose sugar conformations, the torsion angle is defined as O(6)–C(6)–C(5)–O(5), sometimes O(6)–C(6)–C(5)–C(4) or O(6)–C(6)–C(5)–H(5). Frequently, the torsion angle ranges to both atoms O(5) and to C(4) are given so that the conformation +gauche is referred to as gauche/gauche or g/g, −gauche is referred to as gauche/trans or g/t, and trans to trans/gauche, or t/g.
>
> [Caution: this nomenclature is not standardized and sometimes the order of reference to C(5)–O(5) and C(5)–C(4) is reversed; sometimes the C(5)–H bond is used as one of the reference bonds.]

or the 1,3-diaxial effect, and is not well understood theoretically [502]. An analysis of 125 primary hydroxyl and esterified primary hydroxyl groups in 71 crystal structures gave a ratio of 3:2 for −*sc* to +*sc* with no +*sc* for the gluco configuration, and 5.8:4.2 for +*sc* to *ap* with no −*sc* for the galacto configuration [499]. The few exceptions to this rule occur only where there is an intramolecular hydrogen bond from O(6)H to O(4), as in the crystal structures of n-acetylgalactosamine [AGALAM10], 1-kestose [KESTOS] and planteose [PLANTE10].

Orientational disorder is observed in the primary alcohol group of the glucopyranose residue in the crystal structures of 1-kestose [KESTOS] and galabiose [CITSIH] and several other sugars. This disorder is not necessarily symmetrical. In galabiose and in 1-kestose, it is 0.7 −*sc*, 0.3 *ap*, the latter being the "forbidden" orientation stabilized by an intramolecular hydrogen bond. The same orientational disorder can occur in the primary alcohol group C(1)H$_2$OH of a ketopyranose sugar and is observed in the crystal structure of α-L-sorbopyranose [SORBOL01].

Both preferred orientations are believed to be in equilibrium in solution, unless there are some special solvent effects [503]. They are also observed in the crystal structures of the cyclodextrins (see Part III, Chap. 18).

A controversial aspect of one of the proposed crystal structures for the celluloses and amyloses is that it places the primary alcohol orientation in the "forbidden" *ap* position [504].

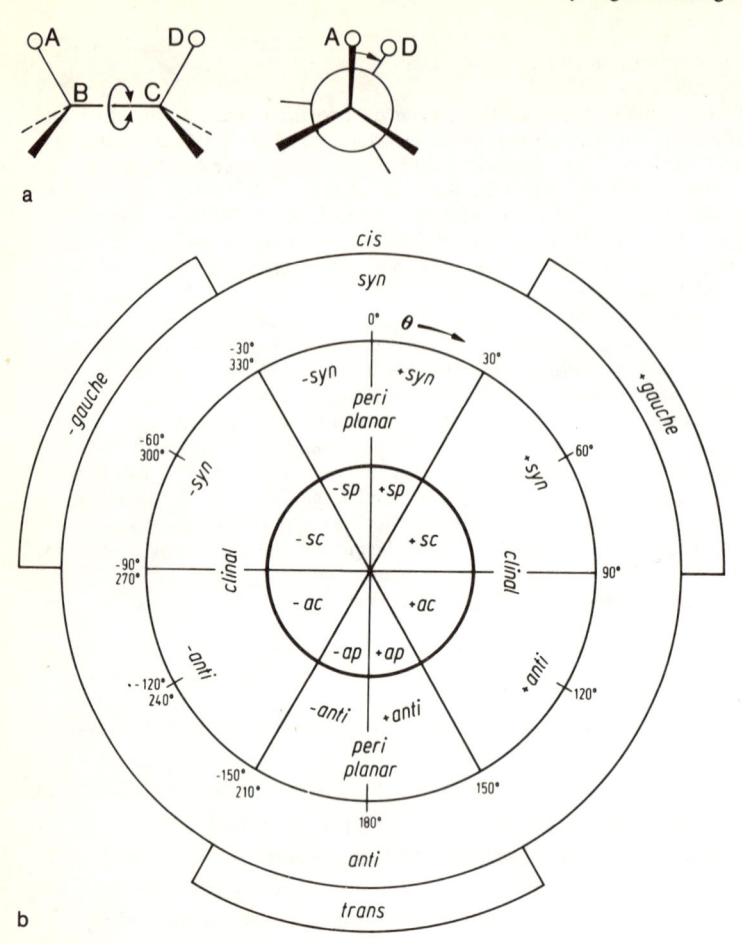

Fig. 13.12. a Definition of torsion angle $A-B-C-D$. **b** Definition of torsion angle ranges, see Box 13.3

Hydrogen bonding is more complex in cyclic monosaccharides than in the alditols. The cyclic sugars differ from the acyclic sugars by having hydrogen-bonding functional atoms which can only be acceptors. These are the ring oxygen atoms in the pyranoses and furanoses, both the ring and glycosidic oxygen atoms in the pyranosides and furanosides, and the ring and linkage oxygens in the di-, tri-, oligo- and polysaccharides.

The presence of these "acceptor only" atoms necessarily interferes with the "cooperativity", which favors the formation of the infinite $\cdots$O-H $\cdots$O-H$\cdots$O-H$\cdots$ chains which are so prominent in the crystal structures of the acyclic sugar alcohols. Either the functional character of the acetal or hemiacetal oxygen atoms is ignored, or the cooperative effect is reduced. As we will see, several different compromises are reached in which three-center hydrogen bonding plays an important part.

For the pyranoses and pyranosides, a relatively large number of high precision X-ray and neutron diffraction results are available. Both a metrical analysis [58] and a pattern recognition analysis [59] were carried out based on a combination of neutron and normalized X-ray structure data. The results of the metrical analysis of the OH···O hydrogen-bond geometry are described in Part IB, Chapters 6 and 7.

13.3 Rules to Describe Hydrogen-Bonding Patterns in Monosaccharides

The role of the three-center bond is to provide an important energetic advantage by enhancing the cooperative effect. A significant portion of hydrogen bonds in the pyranoses and pyranosides are three-centered, namely 25 out of 100 bonds, as found in 24 neutron diffraction structure analyses [58]. It was noted that this corresponds approximately to the ratio of acetal or hemi-acetal ether oxygens to the total number of ether and hydroxyl oxygens. Three-center hydrogen bonds permit the formation of extended hydrogen-bonded chains or networks between the functional groups, even when the number of hydrogen-bond acceptors exceeds the number of hydrogen-bond donors. With this hypothesis in mind, three rules were proposed for describing the hydrogen-bonding patterns in the pyranose and pyranoside crystal structures [59]. These are:

1. Make use of the cooperative effect, by forming as many infinite chains as possible, as in the alditol crystal structures.
2. Maximize the total number of hydrogen bonds per molecule by utilizing and involving as many donor hydroxyls and acceptor oxygen atoms as possible. The acceptor oxygens include not only hydroxyls but also ring and glycosidic oxygens to which there are no hydrogens attached.

   Clearly, there is an inconsistency between these two rules, because hydrogen-bond chains must necessarily terminate when they include ring or glycosidic oxygens.
3. When dealing with pyranoses, or any reducing sugars, the particular hydrogen-bonding characteristics of the anomeric hydroxyl[2] have to be considered, because as shown in Part IB, Chapter 7, they are strong donors but tend to be weak acceptors. These hydroxyls should therefore be expected to function as chain initiators, while their weak acceptor properties will, like the ether oxygens, also make them act as chain terminators.

These rules permitted the known monosaccharide crystal structures to be classified into four different hydrogen-bonding schemes. On the basis of these three rules, it was found that the 44 crystal structures for which neutron or good X-ray data were available at the time of the analysis, could be divided into four classes, which are shown in Table 13.2.

Class I, in which the cooperative effect is maximized by the formation of infinite chains or spirals of hydrogen-bonded hydroxyls. To achieve this, the acetal

---

[2] O(1)H for aldopyranoses in the standard carbohydrate nomenclature, O(2)H for ketopyranoses.

Table 13.2. Distribution of hydrogen-bonding patterns in the crystal structures of pyranoses and pyranosides. (N) indicates neutron diffraction studies; the configuration of molecules is D, unless otherwise indicated; some crystal structures appear in two groups

I Omit acetal oxygens, O(r), O(g)[a]	II Finite chains terminating at O(r), O(g)[a]	III Infinite chains with separate O—H···O(r), O(g)[a]	IV Infinite chains with 3-center bonds to acetal oxygens
2-Deoxy-2-fluoro-β-mannose [XFMANP]	Methyl 4-deoxy-4-fluoro-α-glucoside [MXFGPY]	β-Arabinose (N) [ABINOS01]	1,6-Anhydro-β-galactose [AHGALP]
α-Galactose [ADGALA01]	Methyl β-glucoside·$\frac{1}{2}$H$_2$O [MBDGPH10]	β-D,L-Arabinose (N) [ABINOR01]	1,6-Anhydro-β-mannose [AMANOF]
β-Galactose [BDGLOS01]	Methyl α-mannoside (N) [MEMANP11]	α-D,L-Fucose [ADLFUC]	2-Deoxy-β-galactose [DACHIY]
Methyl β-arabinoside [MBLARA10]	Methyl β-xyloside (N) [XYLOBM]	α-Glucose (N) [GLUCSA]	β-Fructose (N) [FRUCTO02]
Methyl β-riboside (N) [MBRIBP02]	α-Xylose (N) [XYLOSE01]	α-D,L-Mannose [ADLMAN]	β-Lyxose (N) [LYXOSE10]
Methyl α-xyloside [MXLPYR]	α-Fucose [ALFUCO]	Methyl α-arabinoside [MALARA10]	Methyl α-altroside (N) [MALTPY01]
β-Glucoheptose [BDGHEP]	β-Glucose [GLUCSE]	Methyl β-galactoside (N) [MBDGAL02]	Methyl α-galactoside H$_2$O (N) [MGALPY01]
Methyl 3,6-anhydro-α-galactoside [MANGAL]	α-Sorbose (N) [SORBOL01]	β-Digitoxose [BDDIGX]	Methyl α-glucoside (N) [MGLUCP11]
α-Sorbose (N) [SORBOL01]	α-Tagatose [TAGTOS]	α-Tagatose [TAGTOS]	α-Rhamnose·H$_2$O (N) [RHAMAH12]
α-Rhamnose H$_2$O [RHAMAH12]	α-Glucose·H$_2$O [GLUCMH]	α-Talose [ADTALO01]	1,6-Anhydro-β-allose [ANALPR]
	1,6-Anhydro-β-glucose [AHGLPY01]	β-Allose [COKBIN]	2-Deoxy-β-arabinohexose [BECGUL]

[a] O(r), ring oxygen; O(g), glycosidic oxygen.

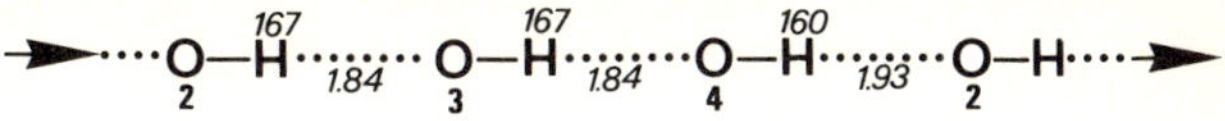

Rules to Describe Hydrogen-Bonding Patterns in Monosaccharides 189

$$\longrightarrow \cdots \underset{2}{O}-\overset{167}{H}\underset{1.84}{\cdots\cdots}\underset{3}{O}-\overset{167}{H}\underset{1.84}{\cdots\cdots}\underset{4}{O}-\overset{160}{H}\underset{1.93}{\cdots\cdots}\underset{2}{O}-H\cdots\longrightarrow$$

Fig. 13.13. Methyl-β-arabinopyranoside [MBLARA10]

$$\longrightarrow \cdots \underset{2A}{O}-\overset{175}{H}\underset{1.95}{\cdots}\underset{4A}{O}-\overset{176}{H}\underset{1.80}{\cdots}\underset{3A}{O}-\overset{173}{H}\underset{1.80}{\cdots}\underset{4B}{O}-\overset{179}{H}\underset{1.79}{\cdots}\underset{3B}{O}-\overset{180}{H}\underset{1.77}{\cdots}\underset{2B}{O}-\overset{177}{H}\underset{2.07}{\cdots}\underset{2A}{O}-H\cdots\longrightarrow$$

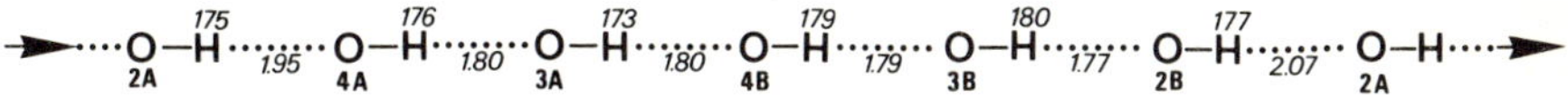

Fig. 13.14. Methyl-α-xylopyranoside [MXLPYR]

$$\underset{1}{O}-\overset{158}{H}\underset{1.81}{\cdots}\underset{2}{O}-\overset{172}{H}\underset{1.75}{\cdots}\underset{4}{O}-\overset{169}{H}\underset{1.73}{\cdots}\underset{3}{O}-\overset{165}{H}\underset{1.84}{\cdots}\underset{5}{O}$$

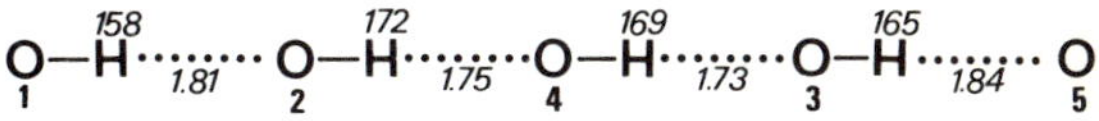

Fig. 13.15. α-Xylopyranoside [XYLOSE02]

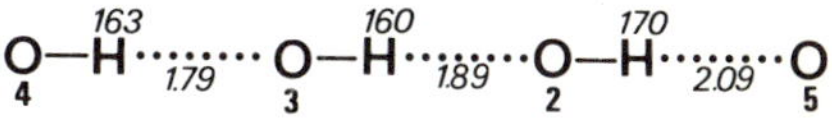

$$\underset{4}{O}-\overset{163}{H}\underset{1.79}{\cdots}\underset{3}{O}-\overset{160}{H}\underset{1.89}{\cdots}\underset{2}{O}-\overset{170}{H}\underset{2.09}{\cdots}\underset{5}{O}$$

Fig. 13.16. Methyl-β-xylopyranoside [XYLOBM01]

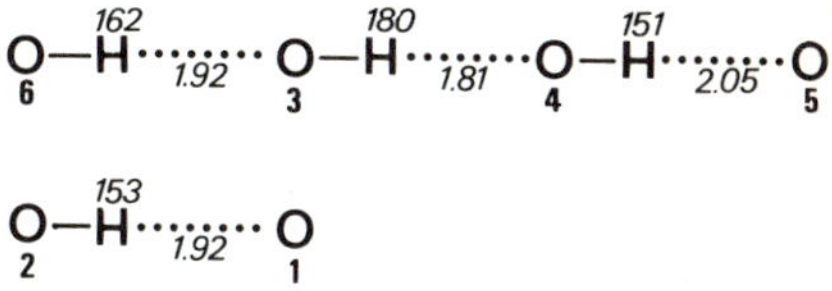

$$\underset{6}{O}-\overset{162}{H}\underset{1.92}{\cdots}\underset{3}{O}-\overset{180}{H}\underset{1.81}{\cdots}\underset{4}{O}-\overset{151}{H}\underset{2.05}{\cdots}\underset{5}{O}$$

$$\underset{2}{O}-\overset{153}{H}\underset{1.92}{\cdots}\underset{1}{O}$$

Fig. 13.17. Methyl-α-mannopyranoside [MEMANP11]

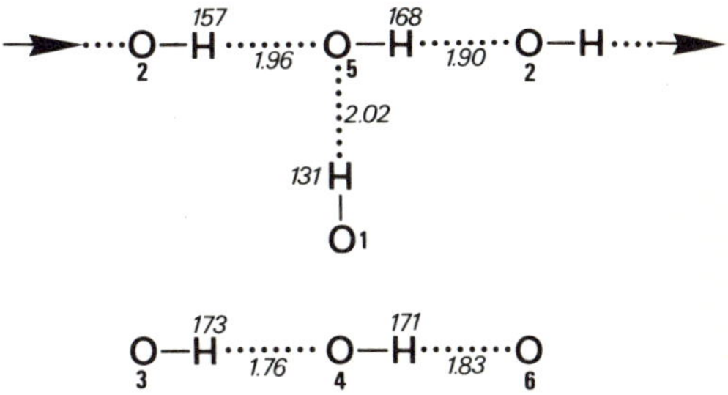

$$\longrightarrow \cdots \underset{2}{O}-\overset{157}{H}\underset{1.96}{\cdots}\underset{5}{O}-\overset{168}{H}\underset{1.90}{\cdots}\underset{2}{O}-H\cdots\longrightarrow$$

$$\vdots 2.02$$

$$\underset{O_1}{\overset{131}{H}}$$

Fig. 13.18. α-D-Tagatose [TAGTOS] $\underset{3}{O}-\overset{173}{H}\underset{1.76}{\cdots}\underset{4}{O}-\overset{171}{H}\underset{1.83}{\cdots}\underset{6}{O}$

ring or glycosidic oxygens are omitted from the hydrogen bonding. Examples are methyl-β-arabinopyranoside [MBLARA10] (Fig. 13.13) and methyl-α-xylopyranoside [MXLPYR] (Fig. 13.14), which has two symmetry-independent molecules **A** and **B** in the crystal structure.

Class II, in which a more limited cooperative effect is obtained by means of finite chains with the acetal oxygens terminating the chains. In three of the seven reducing sugars in this class, the finite chain originates with the anomeric hydroxyl

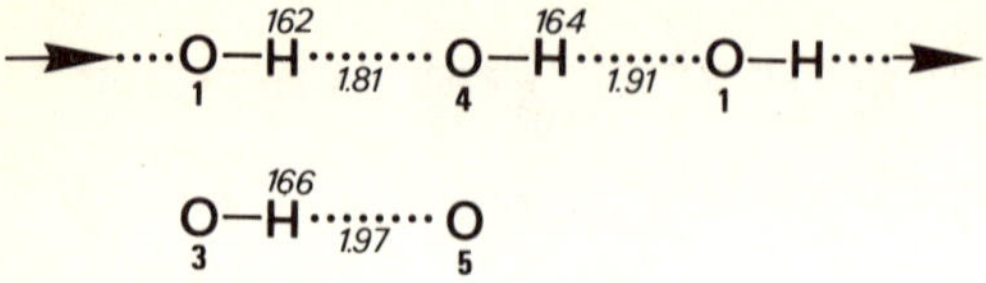

Fig. 13.19. β-D-Digitoxose [BDDIGX]

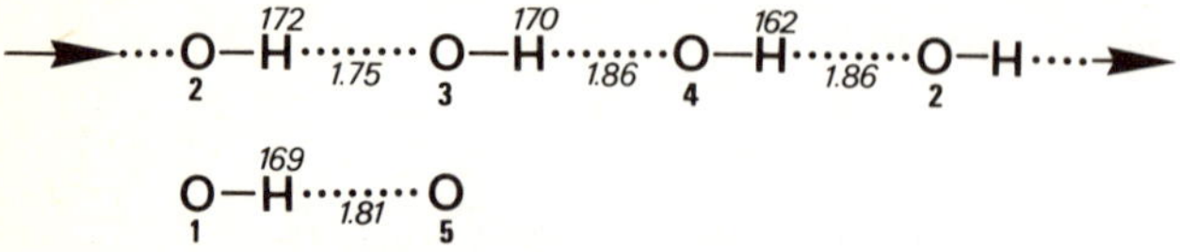

Fig. 13.20. β-D,L-Arabinose [ABINOR 01]

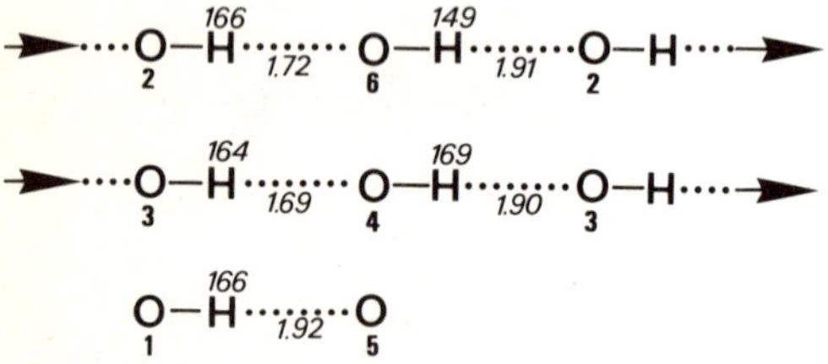

Fig. 13.21. α-D,L-Mannopyranose [ADLMAN]

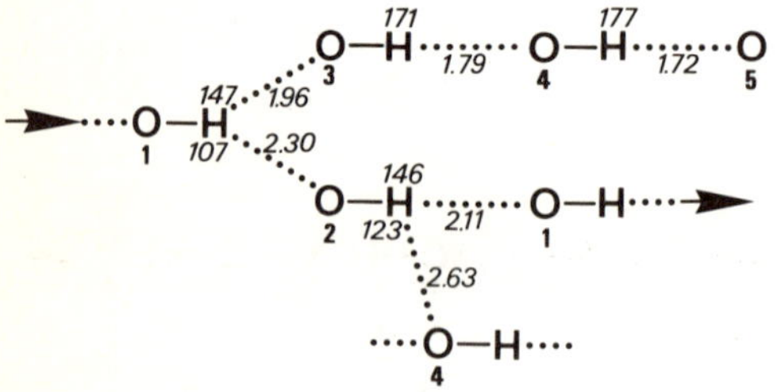

Fig. 13.22. β-Lyxose [LYXOSE 01]

group, thereby fulfilling the strong-donor/poor-acceptor properties of this hydroxyl group. Examples are α-xylopyranoside [XYLOSE 02] (Fig. 13.15) methyl-β-xylopyranoside [XYLOBM 01] (Fig. 13.16) where the glycosidic oxygen is excluded, and methyl-α-mannopyranoside [MEMANP 11] (Fig. 13.17) where both glycosidic and ring oxygen are included.

Class III, in which the maximum cooperative effect is achieved with all the hydroxyl groups except one, which forms a separate two- or three-centered hydrogen bond to a ring or glycosidic oxygen. In seven of the ten reducing sugars, the separate bond is from the anomeric hydroxyl to a ring oxygen. In α-D-tagatose [TAGTOS] (Fig. 13.18), there is an infinite chain and a separate short finite chain of two hydrogen bonds which terminate at the ring oxygen atom. Other examples

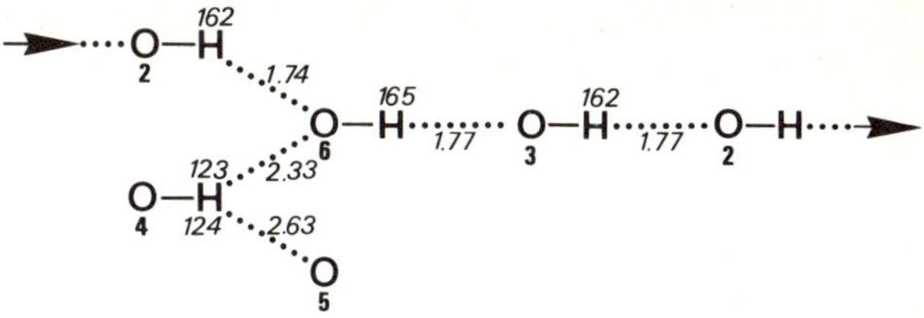

Rules to Describe Hydrogen-Bonding Patterns in Monosaccharides

Fig. 13.23. Methyl-α-glucopyranoside [MGLUCP 11]

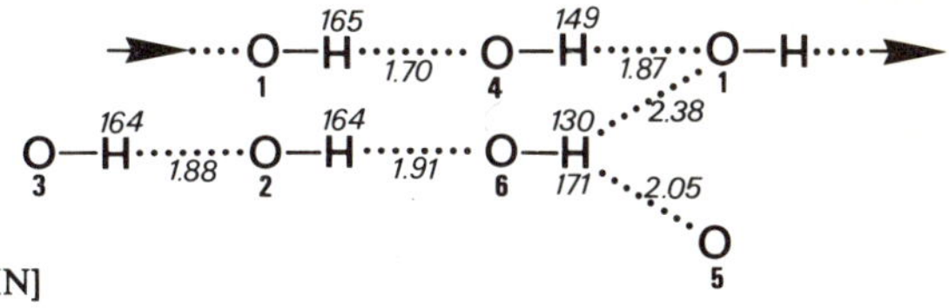

Fig. 13.24. β-Allopyranose [COKBIN]

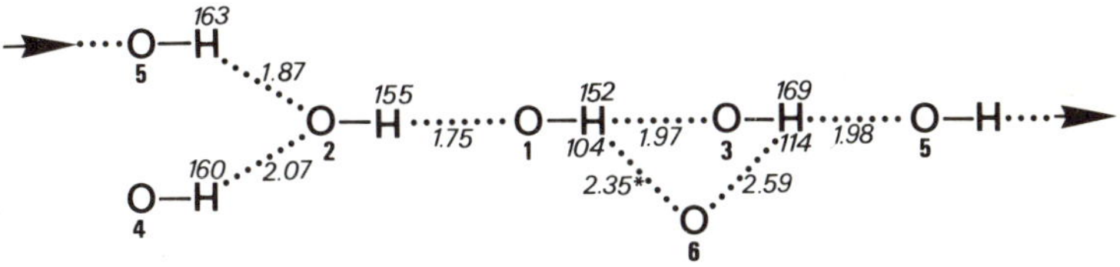

Fig. 13.25. β-Fructopyranose [FRUCTO 02]. *intramolecular

are β-D-digitoxose (2,6-dideoxy-D-ribopyranose) [BDDIGX] (Fig. 13.19), β-D,L-arabinose [ABINOR 01] (Fig. 13.20), and α-D,L-mannopyranose [ADLMAN] (Fig. 13.21).

Class IV. In this type of hydrogen-bonding pattern, the infinite chains are retained and the ring and glycosidic oxygens are included in the hydrogen-bonding scheme by means of three-center bonds. Examples are β-lyxose [LYXOSE 01] (Fig. 13.22); methyl-α-glucopyranoside [MGLUCP 11] (Fig. 13.23); β-allopyranose [COKBIN] (Fig. 13.24); and β-fructopyranose [FRUCTO 02] (Fig. 13.25).

Of the 43 crystal structures for which adequate data were available, the distribution was 10, 11, 11, 11 in the four classes, as shown in Table 13.2. Clearly, they must all be of comparable energy. The determining factor for any particular structure is from the intermolecular forces other than hydrogen bonding, which in turn depend on the shapes of the molecules.

As discussed in Part I B, Chapter 7, the stronger binding energy due to the cooperative effect is evident in the hydrogen-bond lengths, which are summarized by the mean values given below:

The variation in H···O hydrogen-bond length with pattern type is:

100	infinite chain bonds	⟨1.822 Å⟩
40	finite chain bonds	⟨1.839 Å⟩
54	finite chain bonds including terminal O–H···O	⟨1.860 Å⟩
9	single links	⟨1.897 Å⟩.

13.4 The Water Molecules Link Hydrogen-Bond Chains into Nets in the Hydrated Monosaccharide Crystal Structures

In the carbohydrates, the water molecules most commonly have fourfold, nearly tetrahedral coordinations, which function as crossing-points for infinite or finite chains, thereby forming network structures. In this sense, the inclusion of water molecules in carbohydrate structures makes the hydrogen bonding more *ice-like*.

Fig. 13.26. α-Glucose [GLUCSA01]

Fig. 13.27. β-Glucose [GLUCSE01]

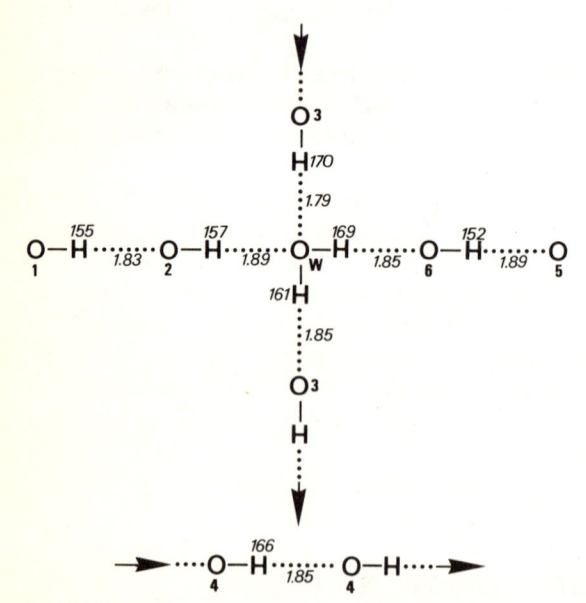

Fig. 13.28. α-Glucose monohydrate [GLUCMH11]

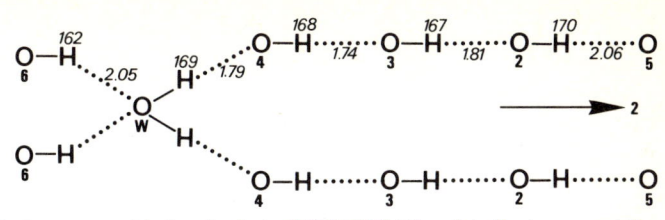

Fig. 13.29. Methyl-β-D-glucopyranoside hemihydrate [MBDGPH 10]. →2 indicates a crystallographic twofold rotation axis

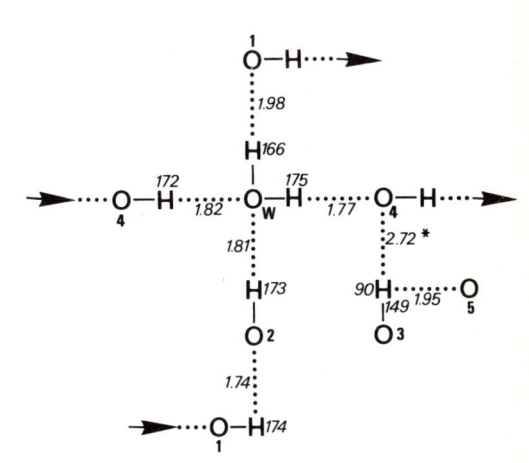

Fig. 13.30. α-Rhamnose monohydrate [RHAMAH 12]; *, intramolecular

Fig. 13.31. Methyl-α-galactoside monohydrate [MGALPY 01]

Fig. 13.32. 2-Keto-L-gulonic acid monohydrate [KGULAM]

Fig. 13.33. N-Acetylglucosamine [ACGLUA 11]. *intramolecular

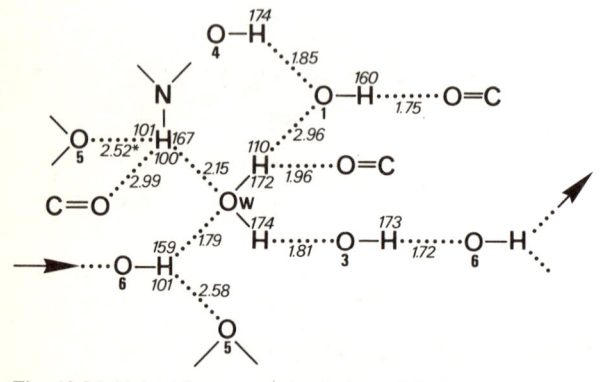

Fig. 13.34. N-Acetylmannosamine hydrate [NACMAN 10]. *intramolecular

Fig. 13.35. N-Acetylgalactosamine [AGALAM 01]. *intramolecular

This is illustrated by comparing the hydrogen-bonding schemes in the crystal structures of α- and β-glucose [GLUCSA 01, GLUCSE 01] (Figs. 13.26, 13.27), with that in α-glucose monohydrate [GLUCMH 11] (Fig. 13.28). In the anhydrous crystals, the hydrogen bonds form infinite or finite chains. In the α-epimer, the anomeric hydroxyl O(1)H forms an isolated bond to the ring oxygen O(5). In the β-epimer, a finite chain is initiated at the anomeric hydroxyl and terminates at the ring oxygen. One hydroxyl, O(4)H, is linked to the chain by a weak two-center bond. In the hydrate, finite chains similar to those in the anhydrous β-form intersect with an infinite chain at the water molecules to form a three-dimensional network of hydrogen bonds. As in the anhydrous α-form, there is an infinite chain of bonded O(4)H groups.

In methyl-β-D-glucopyranoside hemihydrate [MBDGPH 10] (Fig. 13.29), the water molecule is on a twofold axis of symmetry so that the two finite chains which intersect at the water molecule are symmetry-related and are identical. A similar network structure is observed in α-rhamnose monohydrate [RHAMAH 12] (Fig. 13.30), where two infinite chains intersect at the water molecule. In that structure, a OH···O< bond is linked to one of the infinite chains by the minor component of a three-center bond. In methyl-α-galactoside monohydrate [MGALPY 01] (Fig. 13.31), however, the water molecule is a single acceptor. In consequence, a finite chain and an infinite chain join at the water molecule. In 2-keto-L-gulonic acid monohydrate [KGULAM] (Fig. 13.32), the water molecule is a double acceptor, at the intersection of two finite chains of strong hydrogen bonds. The π-bond connectivity of the carboxylate group links two of the finite chains.

NH···O=C hydrogen bonding is not very strong in the crystal structures of amino sugars. In N-acetylglucosamine [ACGLUA 01] and mannosamine hydrate [NACMAN 10] (Figs. 13.33, 13.34), the N—H groups form weak three- and four-center bonds, one component of which is intramolecular. In N-acetylgalactosamine [AGALAM 01], an NH···OH bond initiates a finite chain of four bonds, which terminates in an O—H···O=C bond (Fig. 13.35).

13.5 The Disaccharide Crystal Structures Provide an Important Source of Data About Hydrogen-Bonding Patterns in Polysaccharides

In general, disaccharides crystallize well and a significant number have been studied by crystal structure analysis, of which two are by neutron diffraction. These molecules contain between 6 and 8 donor/acceptor hydroxyl groups and between 3 and 4 ether oxygens per molecule which are only acceptors. Like the monosaccharides, the crystal structures are extensively hydrogen-bonded, forming hard crystals with relatively high melting points.

The fact that hydrates are more common in the disaccharides provides an opportunity to study the way in which the inclusion of water molecules may influence the hydrogen-bonding patterns of oligo- and polysaccharides. The crystal structures of several compounds have been studied in both the anhydrous and hydrated forms.

The hydrogen-bonding patterns in disaccharides are different in some respects from those in the monosaccharides. Only two of the crystal structure analyses were by neutron diffraction, i.e., sucrose [SUCROS 04], and β-maltose monohydrate [MALTOS 11], the crystals generally give good X-ray diffraction data and all the hydroxyl hydrogen atoms were located in the X-ray analyses. The hydrogen bonding in these structures differs from that of the monosaccharides in two respects. One is the higher proportion of intramolecular hydrogen bonds which are

Table 13.3. Summary of Hydrogen Bonding in Disaccharide Crystal Structures

Compound	Intramolecular inter-residue bonds	Finite or infinite chains	Number of bonds 2-center : 3-center	Comments
Cellobiose [CELLOB02]	O(3')H···O(5)	2-infinite 1-finite	5:3	Infinite chains intersect at O(6)
Methyl cellobioside MeOH [MCELOB]	O(3')H⟨O(6) / O(5)	1-infinite 1-finite	7:1	Chains linked by 3-center bond
β-Galabiose [CITSIH]	O(3')H···O(5)	1-finite O(1')H···O(5) single link	5:3	Infinite chain contains 4-membered cycle
α-Lactose H₂O [LACTOS10]	O(3')H⟨O(5) / O(1)	2-infinite 2-finite	8:2	Finite chain intersects infinite chain at O(W) and finite chain at O(2')
β-Lactose [BLACTO]	O(3')H···O(5) / O(4')	2-finite	8:0	O(4)H forms weak 2-center bond
α-Maltose [MALTOT]	O(3')H⟨O(2) / O(4') O(6)H···O(4')H···O(6')	1-finite 1 4-bond cycle	6:2	Antidromic 4-bond cycle with separate O(1')H···O(5) bond
β-Maltose H₂O [MALTOS11]	O(2)H⟨O(1) / O(3')	2-infinite 1-finite	6:3	Infinite chains intersect with finite chain at O(6) and O(W). Ring and glycosidic oxygen included by 3-center bonds
Methyl β-maltoside H₂O [MMALTS]	O(2)H⟨O(1) / O(3') O(6)H···O(W)···HO(6')	4-infinite	9:2	Two of infinite chains intersect at O(W)
Laminarabiose 0.25 H₂O [LAMBIO]	O(4')H···O(5)	2-infinite 2-finite	16:0	In hemihydrate structure, 2 infinite chains intersect at O(W)

Hydrogen Bonding in the Disaccharide Crystal Structures

Compound	Bonding	Chain type	Ratio	Notes
4-O-β-Galactosyl rhamnose [GAPRHM10]	none	1-infinite 1-finite	7:1	Chains do not join
4-O-β-Glucosyl glucitol [BDGPGL]	none	2-infinite	7:3	Infinite chains linked by minor component of 3-center bond
4-O-α-Glucosyl glucitol [BIZHIB]	none	2-finite		Two finite chains join at O(1')
Sucrose [SUCROS04]	O(1')H⋯O(1) ⋱O(2)	1-infinite (8-membered cycle)	6:2	Contains a 4-center bond from O(4)H
	O(6)H⋯O(5) ⋱O(3)			
Isomaltulose H$_2$O [IMATUL]	O(2)H⋯O(2') ⋱O(6')	2-infinite	9:1	Chains do not intersect at O(W)
Turanose [TURANS01]	Weak minor component O(4')H⋯O(5) ⋱O(2) O(1)	1-infinite 2-finite	3:4	O(2)H is non-bonding. O(4)H forms a weak 2-center bond. O(4')H forms a 4-center bond
α,α'-Trehalose [DEKYEX]	None	1-finite 1-infinite	4:4	Infinite and finite chains are linked by minor components of 3-center bonds
α,α'-Trehalose 2H$_2$O [TREHAL01]	O(2)H⋯O(W)H⋯O(6')	1-infinite 2-finite	11:1	Finite and infinite chains intersect at 3- and 4-coordinated waters
Gentiobiose [GENTBS]	None	1-infinite	7:2	Infinite chain contains short O(3)H⋯O(6') bond, 1.65 Å
Melibiose H$_2$O [MELIBM10]	O(6')H⋯O(W)H⋯O(4)	5-member homodromic cycle 2-finite	9:1	Finite chains join cycle at O(W)
Sophorose H$_2$O [SOPROS]	O(2)H⋯O(W)H⋯O(3')	3-infinite 2-finite	8:2	2 infinite chains intersect at O(W)

between the monosaccharide residues as shown in Table 13.3. The second is the higher proportion of hydrates, which occur in about one-third of the crystal structures.

The hydrogen-bonding schemes fit less easily into the patterns discussed in the previous section for maximizing the hydrogen-bond energy. In consequence, these crystal structures provide some features which are not observed or are rare in the monosaccharides. For example, a non-hydrogen-bonding hydroxyl, in turanose [TURANS01], and several examples of very weakly bonding hydroxyls, in sucrose [SUCROS04], β-lactose [BLACTO], and gentiobiose [GENTBS01]. It is a reasonable presumption that the greater the stereochemical complexity of the molecules involved, the more likely is the occurrence of nonbonding or weakly hydrogen-bonding functional groups.

Intramolecular hydrogen bonding is more common in disaccharides than in the monosaccharides, because there are less stereochemical restrictions to forming nearly linear bonds between donor and acceptor groups on the two different monosaccharide residues. Inter-residue intramolecular hydrogen bonding does not have the stereochemical constraints of the intramolecular hydrogen bonding in the monosaccharides. Examples of this direct intramolecular hydrogen bonding between residues are shown in Table 9.3. This type of hydrogen bond confers additional rigidity to the disaccharide molecule, converting it from a two-ring to a three-ring system, as in β-cellobiose [CELLOB02], or a four-ring system as in sucrose [SUCROS04]. In a number of structures, the inter-residue bond is the major component of a three-center bond.

In 1→4-linked pyranosyl-pyranose disaccharides, there are two common forms of inter-residue hydrogen bonds. The $O(3')H \cdots O(5)$ bond occurs in the β-linked disaccharides, cellobiose, galabiose, methyl-β-cellobioside methanolate, β-lactose, α-lactose monohydrate, mannobiose, and mannotriose trihydrate. The $O(4')H \cdots O(5)$ bond occurs only in laminarabiose quarter-hydrate. Both these bonds orient the pyranose ring oxygens to opposite sides of the long molecular axis. The $O(3')H \cdots O(2)$ or $O(2')H \cdots O(3)$ bond in the α-linked disaccharides α-maltose, α-maltose monohydrate, methyl-β-maltoside and phenyl-α-maltoside orients the pyranose rings so that the ring oxygens tend to be on the same side of the long molecular axis (see, for example, Fig. 9.1).

The hydrogen bonding in cellobiose and methyl cellobioside as models for that in the celluloses. In the absence of crystal structure analyses of higher oligomers of 1→4-linked β-glucopyranose, cellobiose is frequently used as a model for interpreting the X-ray fiber diffraction patterns of the celluloses, especially since the cellobiose unit is considered to be the repeating unit of the polysaccharide chain.

The hydrogen bonding in the crystal structure of cellobiose [CELLOB02] (Fig. 13.36) consists of two infinite chains which are cross-linked at $O(6)-H$ through the minor component of a three-center bond; the major component is the inter-residue $O(3')H \cdots O(5)$ bond. There is also a finite three-link chain originating in the anomeric hydroxyl $O(1)-H$ and terminating in the ring oxygen $O(5')$. All hydroxyl and both ring oxygens $O(5)$, $O(5')$, and the glycoside link $O(4)$, are included in the hydrogen bonding through the use of three-center bonds.

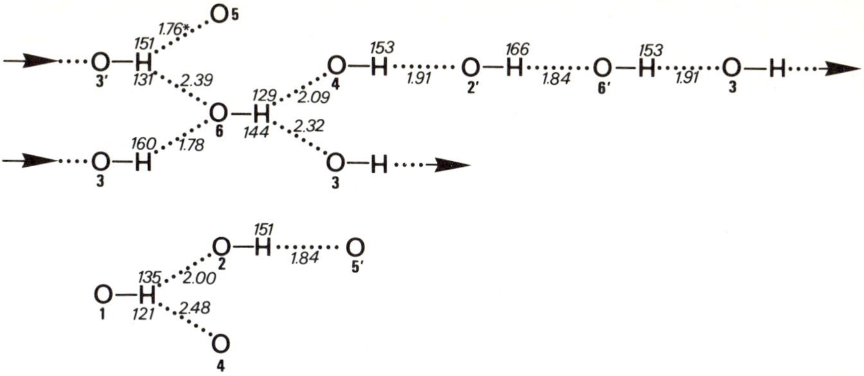

Hydrogen Bonding in the Disaccharide Crystal Structures 199

Fig. 13.36. Cellobiose [CELLOB02]. *intramolecular

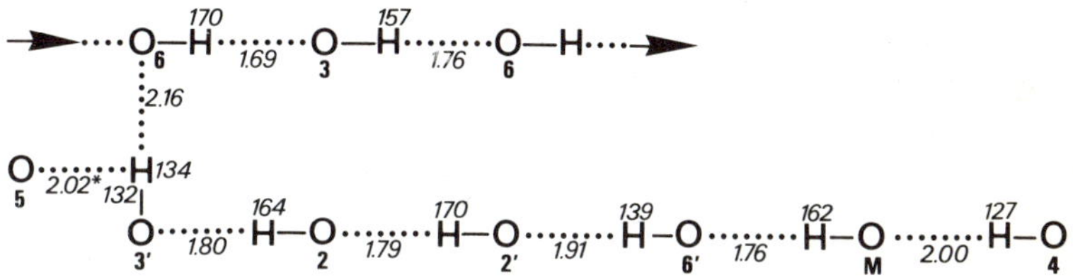

Fig. 13.37. Methyl-β-cellobioside methanolate [MCELOB]. *intramolecular

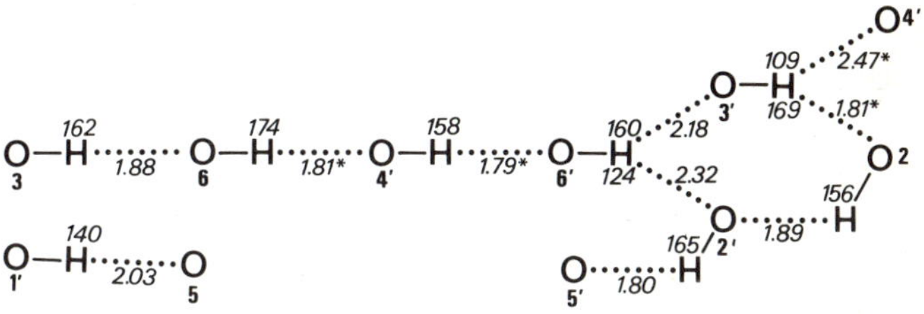

Fig. 13.38. α-Maltose [MALTOT]. *intramolecular

Methyl-β-cellobioside methanolate [MCELOB] (Fig. 13.37), has a simpler hydrogen-bond structure. This disaccharide is perhaps a better model for the hydrogen bonding in the polymers, since it does not include the anomeric hydroxyl O(1′)H. The hydrogen bonding is simpler, consisting of an infinite chain involving ···O(3)H···O(6)H··· to which is attached at O(6) a six-link finite chain originating at O(4)H, by means of a three-center bond. The major component of this bond is intra-residue to the ring oxygen O(5). The ring oxygen O(5′) and the linkage oxygen O(1) are excluded from the hydrogen-bonding scheme.

Hydrogen bonding in maltose, maltose monohydrate, and methyl maltoside provide models for that in the amyloses. Again the disaccharides are used as

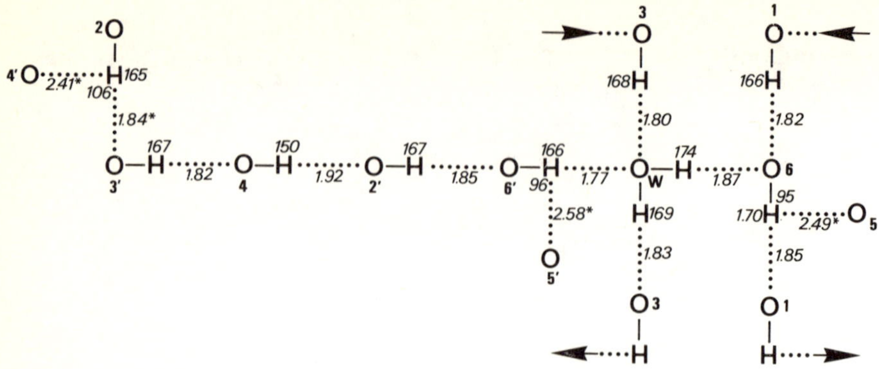

Fig. 13.39. β-Maltose monohydrate [MALTOS11]. *intramolecular

models for the α-linked 1,4 polymers in the absence of crystal structure analyses of higher oligomers. α-Maltose, 4-O-(α-D-glucopyranosyl)-α (β)-D-glucopyranose [MALTOT] (Fig. 13.38), forms anhydrous crystals and a monohydrate, of which the latter is the more common form. Crystalline anhydrous maltose is a mixture of α- and β-epimers in the proportion 0.82 α, 0.18 β. There was no evidence from the neutron diffraction analysis of an epimeric mixture in the monohydrate.

In the anhydrous crystal structure, there is an interesting intramolecular three-center hydrogen bond, the major component of which is the inter-residue O(3′)H···O(2) between the glucose residues, while the minor component is to the linkage oxygen O(4′) between the two residues. This places both ring oxygens on the same side of the molecular axis. Both ring oxygens are also involved in the hydrogen bonding, which consists of a finite chain and a separate O(1′)H···O(5) link. The finite chain originates in O(3)H and branches at a three-center bond at the primary alcohol group O(6′)H to form an antidromic four-membered ring.

In the β-maltose monohydrate [MALTOS11] (Fig. 13.39) the same inter-residue hydrogen bond occurs, but the direction is reversed, i.e., O(2)H···O(3′). In this structure, the water molecule links infinite chains with finite chains to form an irregular network structure. The two ring oxygens O(5) and O(5′) and the linkage oxygen O(4′) are incorporated in the hydrogen bonding through the minor intramolecular components of three-center bonds.

Direct and indirect inter-residue hydrogen bonds occur in the crystal structure of methyl β-maltoside monohydrate [MMALTS] (Fig. 13.40). This is a better model for the 1,4-linked α-glucose polysaccharides, since the anomeric hydroxyl is substituted. As in maltose and maltose monohydrate, there is an inter-residue intramolecular hydrogen bond, O(2)H···O(3′), but in the maltoside, there is also an indirect inter-residue bond through a water molecule O(6)H···O(W)···HO(6′) linking the two primary alcohol groups. The separation of the primary alcohol groups of 1→4-linked hexoaldose sugars is too great to be linked by direct bonding, but their separation is well suited to the insertion of a water molecule. This indirect hydrogen bonding does not occur in the maltose monohydrate, where the

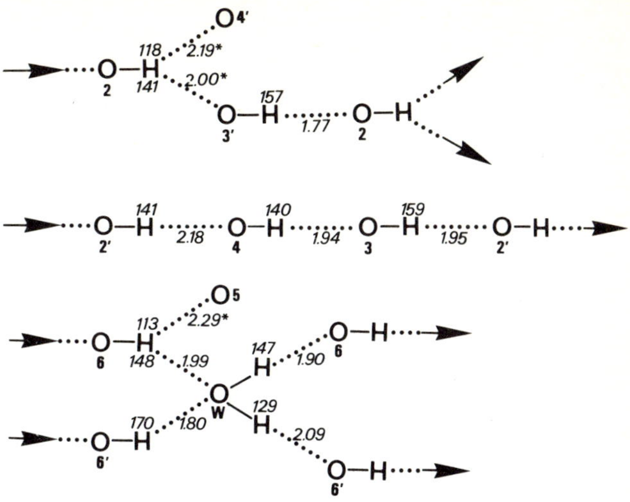

Hydrogen Bonding in the Disaccharide Crystal Structures

Fig. 13.40. Methyl-β-maltoside monohydrate [MMALTS]. *intramolecular

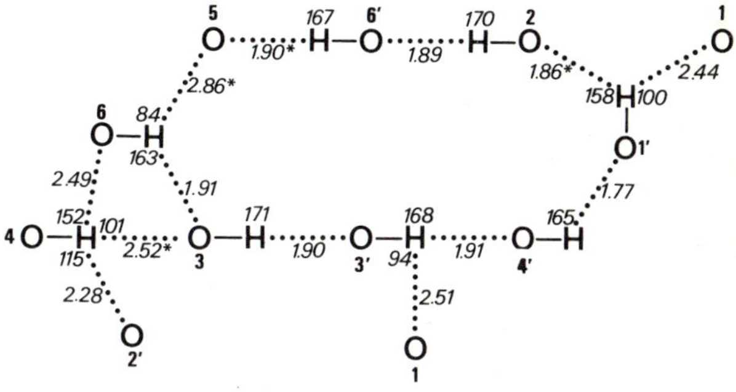

Fig. 13.41. Sucrose [SUCROS04]. *intramolecular

water molecules are hydrogen-bonded together in pairs. As in the maltose structures, the O(2)H···O(3') bond places both ring oxygens on the same side of the molecular axis, although the direction of the hydrogen bond is reversed between the two structures.

In this crystal structure, the hydrogen bonding consists of four infinite chains. Two of them are nonintersecting and the other two, which are formed by the primary alcohol hydroxyls, intersect at the water molecule. The linkage oxygen O(4') is included through the minor intramolecular component of a three-center bond from O(2)H. The ring oxygen O(5) is included through the minor intramolecular component of a three-center bond from O(6)H. The ring oxygen, O(5'), is excluded from the hydrogen bonding (Fig. 13.40).

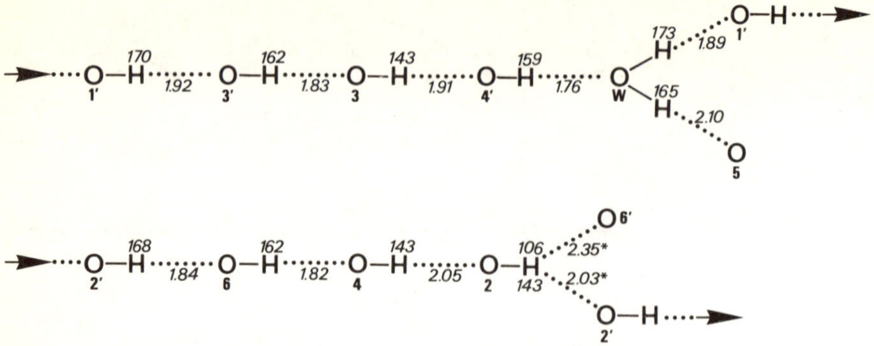

Fig. 13.42 Isomaltulose [IMATUL]. *intramolecular

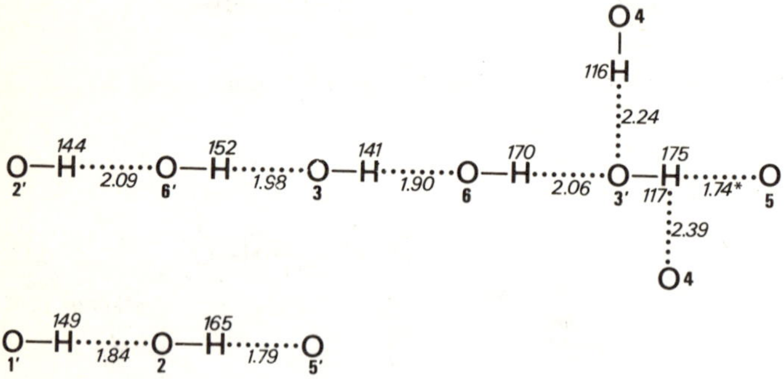

Fig. 13.43. β-Lactose [BLACTO]. *intramolecular

The hydrogen bonding in sucrose includes two inter-residue bonds, and an unusual four-center bond. The hydrogen bonding in the crystal structure of sucrose [SUCROS04], α-D-glucopyranosyl-β-D-fructofuranoside, has been determined by neutron diffraction (Fig. 13.41). It includes two inter-residue intramolecular hydrogen bonds, O(1')H···O(2) and O(6')H···O(5), both from the fructofuranose to the glucopyranose residue. This places the ring oxygens on the same side of the mid-line of the molecule. These bonds form part of an eight-membered cycle which includes the minor intramolecular component of a three-center bond from the glucose primary alcohol group O(6)H to the ring oxygen O(5), which accepts two intramolecular bonds, one intra- the other inter-residue. The glucose O(4)H forms a four-center bond as shown in Fig. 8.1. While all the oxygen atoms in the molecule are involved in hydrogen bonding, this crystal structure contains an unusually large number of weak hydrogen bonds, there being six with H···O distances greater than 2.2 Å, and only one less than 1.8 Å.

Intramolecular hydrogen bonding in sucrose may be retained in aqueous solution. Empirical force field calculations suggest that the molecule of sucrose retains its crystal structure conformation in aqueous solution with the loss of one of the

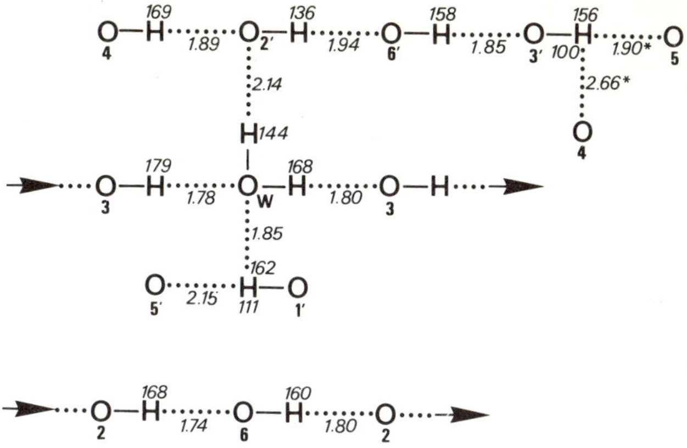

$$\underset{4}{O}-H\underset{1.89}{\overset{169}{\cdots\cdots}}\underset{2'}{O}-H\underset{1.94}{\overset{136}{\cdots\cdots}}\underset{6'}{O}-H\underset{1.85}{\overset{158}{\cdots\cdots}}\underset{3'}{\overset{156}{O-H}}\underset{1.90*}{\cdots\cdots}\underset{5}{O}$$

(vertical: 2.14 down to H 144; 2.66* down to O 4)

$$\longrightarrow\cdots\underset{3}{O}-H\underset{1.78}{\overset{179}{\cdots\cdots}}\underset{W}{O}-H\underset{1.80}{\overset{168}{\cdots\cdots}}\underset{3}{O}-H\cdots\longrightarrow$$

(vertical from W: 1.85 down to 162 then to O 5' ··2.15·· H 111 — O 1')

$$\longrightarrow\cdots\underset{2}{O}-H\underset{1.74}{\overset{168}{\cdots\cdots}}\underset{6}{O}-H\underset{1.80}{\overset{160}{\cdots\cdots}}\underset{2}{O}\cdots\longrightarrow$$

Fig. 13.44. α-Lactose monohydrate [LACTOS 10]. *intramolecular

two intramolecular hydrogen bonds [493]. ^{13}C-NMR spin lattice relaxation measurements support this hypothesis [449].

Isomaltulose monohydrate is an analog of sucrose. Isomaltulose, 6-O-α-D-glucopyranosyl fructofuranose [IMATUL] (Fig. 13.42), has the same residues as sucrose, with a 1→6 rather than 1→4 inter-residue linkage and crystallizes as the monohydrate. Unlike sucrose, there is only one inter-residue intramolecular hydrogen bond, O(2)H···O(2′), which is relatively weak (2.03 Å). The hydrogen bonding consists of two infinite chains which do not intersect at the water molecule, as is more usual with carbohydrates. The water molecule in this crystal structure accepts only one bond. It donates one strong bond to O(1′)H, which is part of the chain, and a second weaker bond to the glucopyranose ring oxygen atom O(5). The linkage oxygen, O(6′), is included as the minor component of a three-center bond, but the fructose ring oxygen is excluded.

Both epimers of lactose have been studied. In the crystal structures of β-lactose, 4-O-β-D-galactopyranosyl-β-D-glucopyranose [BLACTO], and α-lactose monohydrate [LACTOS 10] there is a strong inter-residue intramolecular hydrogen bond from O(3′)H to the ring oxygen O(5). This places the ring oxygens on opposite sides of the mid-lines through the molecules (Figs. 13.43, 13.44).

The hydrogen bonding in the anhydrous β-lactose structure (Fig. 13.43), consists of five-link and two-link finite chains, terminating in the ring oxygens. The linkage oxygen O(4) is included through the minor component of a three-center bond. One hydroxyl, O(4)H, forms only a very weak bond, H···O = 2.24 Å, with an O−H···O angle of 116°. The position of this hydrogen atom is questionable and requires confirmation by neutron diffraction.

In the α-lactose monohydrate (Fig. 13.44), the hydrogen bonding consists of two infinite and two finite chains. One of the infinite chains and one of the finite chains intersect at the water molecule. Two of the finite chains intersect at O(2′), which is a double acceptor. There is an intramolecular, inter-residue O(3′)H to O(5) bond, which is the major component of the three-center bond. The minor com-

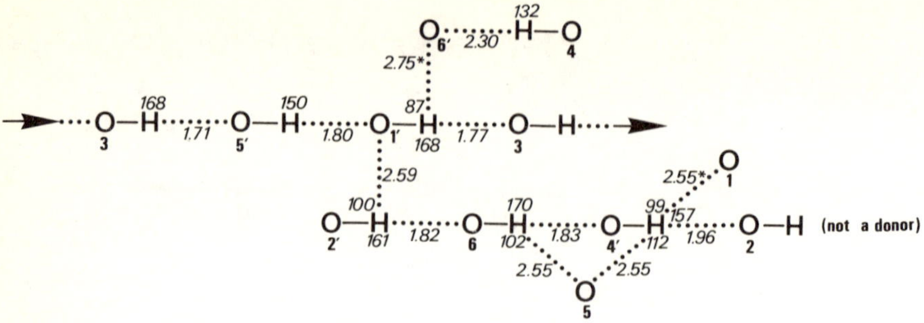

Fig. 13.45. Turanose [TURANS01]. *intramolecular

ponent is also intermolecular to a linkage oxygen O(4). The other ring oxygen O(5′) is also bonded by the minor component of a three-center bond.

A nonbonding hydroxyl group occurs in turanose. This crystal structure of 3-O-α-D-glucopyranosyl-β-D-fructopyranose [TURANS01] has an interesting hydrogen-bond structure (Fig. 13.45) because it provides a rare example of a hydroxyl which is almost nonbonding. The O(2)H hydrogen atom of the fructopyranose residue is 2.64 Å from the O(3) of the same molecule, making an O−H···O angle of 97°. All other H···O distances involving this hydrogen are greater than 2.5 Å. The nonhydrogen-bonding hydroxyl is supported by the appearance of a sharp absorption at 3000 cm^{-1} in the solid-state infrared spectrum. The second sharp peak at 3550 cm^{-1} is associated with the O(4′)H hydroxyl, which forms a four-center bond with H···O distances of 1.96, 2.55 and 2.55 Å.

The similar distinct absorption peak shown for β-D-fructopyranose is associated with a weakly bonding O(4)H group forming a two-center H···O bond of 2.07 Å (Fig. 13.25). A similar separate absorption peak should also be observed for the very weakly bonded O(4)H group in the β-D-glucopyranose (Fig. 13.27) and β-lactose (Fig. 13.43) crystal structures, but hitherto these have not been reported.

In turanose, the hydrogen bonding consists of an infinite chain and two finite chains linked by the weak components of three-center bonds. Both ring oxygens O(5), O(6′) and the linkage oxygen O(1) are included through the weak components of three-center bonds.

The two forms of trehalose have been studied both by X-ray diffraction and solid-state NMR. α,α′-Trehalose is an interesting disaccharide, since it occurs in bacteria, fungi, algae, and insects. It is the sugar component of the mycollic esters, known as *cord factors* [505, 506], which are said to have immuno-stimulant and anti-tumor properties [507]. The common crystalline form of α-D-glucopyranosyl-α-D-glucopyranoside is a *dihydrate*, but crystals of the anhydrous form can be obtained, with difficulty, and have been studied. Both investigations were by X-ray diffraction. That of the dihydrate was published almost simultaneously by four investigators, and provides a good check of accuracy versus precision in the X-ray analysis of a disaccharide. The anhydrous study was a low-temperature (−160°C), X-ray crystal structure analysis.

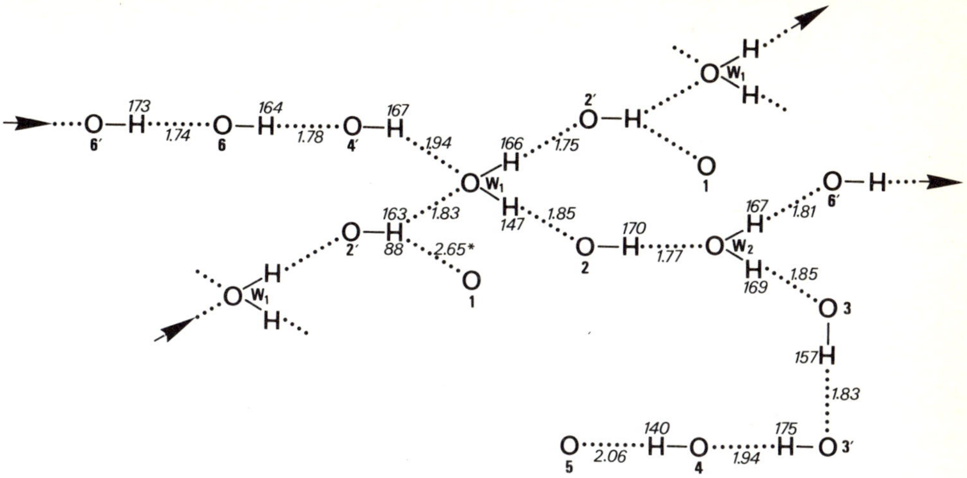

Fig. 13.46. Trehalose dihydrate [TREHAL 01]. *intramolecular

In the dihydrate [TREHAL 10] (Fig. 13.46), there is an indirect inter-residue hydrogen bond through a water molecule linking O(2)H···O(W2)H···O(6'), which forms part of an infinite chain of strong bonds. The hydrogen bonding involving O(W1)H and O(2')H forms a helical arrangement about a twofold axis. The water molecules form crossing and branch points for the two infinite chains and one finite chain, which terminates at a ring oxygen O(5). The structure is unusual in containing only one three-center bond, the minor component of which is a weak intramolecular interaction to the glycosidic linkage oxygen O(1). Excluding the bond to the ring oxygen O(5), the hydrogen bonds are unusually strong, ranging from 1.75 to 1.94 Å, with a mean value of 1.825 Å. This explains why it is difficult to form the anhydrous crystals from solution.

In anhydrous α,α'-trehalose [DEKYEX] (Fig. 13.47), the structure consists of a long eight-link finite chain weakly bound to form a network structure by the minor components of three-center bonds. Both ring oxygens are included, but the linkage oxygen is not. These crystals have been studied by solid-state C^{13} NMR spectroscopy [293].

Gentiobiose has a very weakly bonded hydroxyl. In the crystal structure of 6-O-β-D-glucopyranosyl-β-D-glucopyranose [GENTOS01] (Fig. 13.48), all but one of the hydroxyls, O(2')H, form an infinite chain of relatively strong bonds which include an unusually short O(3)H···O(6') bond of 1.65 Å. The hydroxyl O(2')H does not accept a hydrogen bond and forms a weak three-center bond, one member of which is intramolecular inter-residue. The ring oxygen O(5') is included through a three-center bond, but the other ring oxygen and the linkage oxygen are not.

The hydrogen bonding in melibiose monohydrate includes a homodromic cycle. Cyclic hydrogen-bonding schemes are not common in mono- and disaccharides, but in the crystal structure of the epimeric α,β-melibiose monohydrate, 6-O-α-D-galactopyranosyl-α,β-D-glucopyranose-H$_2$O [MELIBM01] (Fig. 13.49),

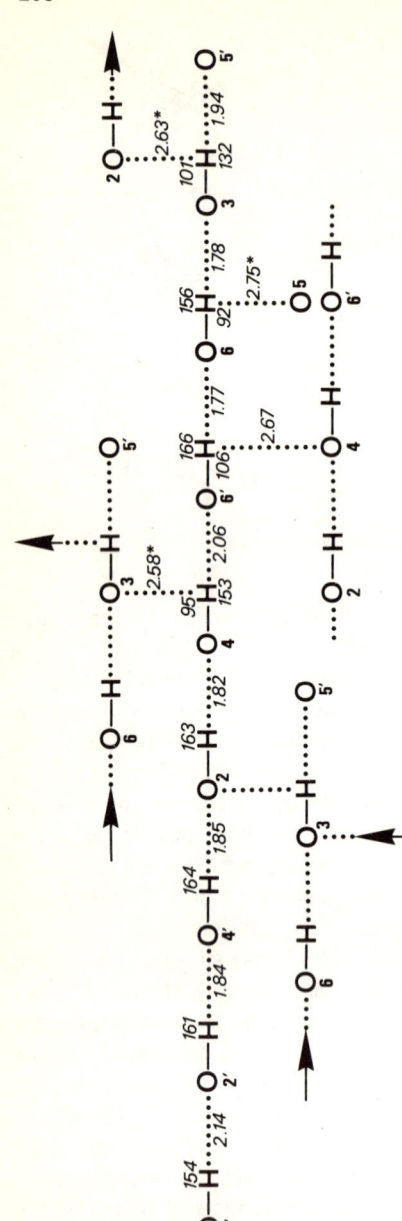

Fig. 13.47. α,α-Trehalose [DEKYEX]. *intramolecular

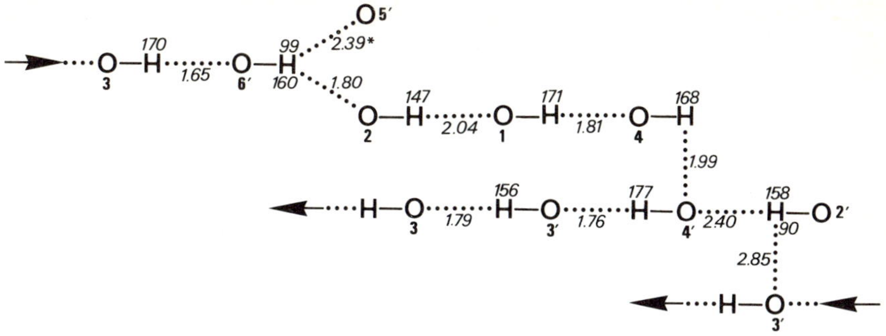

Fig. 13.48. Gentiobiose [GENTOS01]. *intramolecular

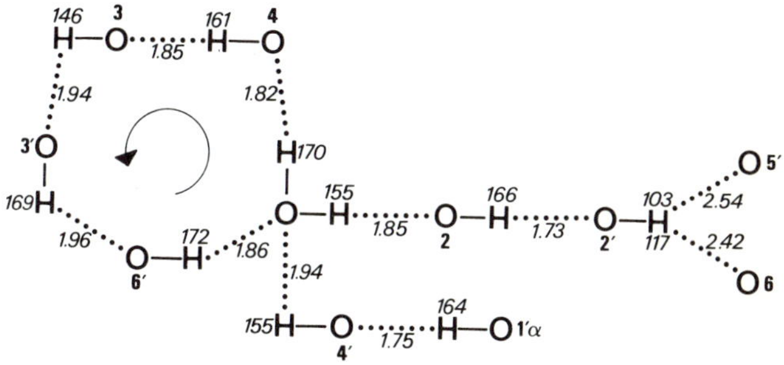

Fig. 13.49. Melibiose [MELIBM01]

there is a homodromic cycle of five hydrogen bonds. It includes the water molecule which accepts two bonds and donates two, thereby forming an intersection of the cycle with a finite chain. The minor β-epimer forms a separate hydrogen bond from the anomeric hydroxyl O(1') to a ring oxygen.

The hydrogen-bonding pattern in sophorose monohydrate exhibits three types of hydrogen-bonding chains. In the crystal structure of sophorose, 2-O-β-D-glucopyranosyl-α-D-glucopyranose monohydrate [SOPROS] (Fig. 13.50), all three types of hydrogen-bond chains are represented. There is a finite three-link chain, originating at O(4')H and terminating at a ring oxygen O(5'). There is an infinite chain involving O(3)H and O(6)H, to which there is attached a three-center bond from the anomeric hydroxyl O(1')H to O(6) in the chain and the ring oxygen O(5). There are two infinite chains which intersect at the water molecule. The linkage oxygen O(1) is incorporated through the minor intramolecular components of two three-center bonds. There is an indirect inter-residue bond from O(2)H ···O(W)H···O(3'). All the oxygen atoms are included in the hydrogen-bonding scheme. However, except for the infinite chain, all the hydrogen-bond lengths are longer than average, exceeding 1.9 Å.

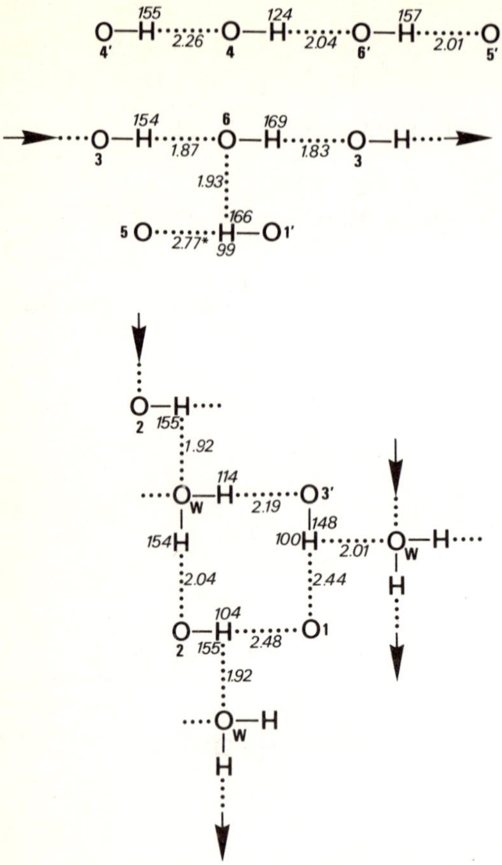

Fig. 13.50. Sophorose monohydrate [SOPROS]. *intramolecular

Laminarabiose hemihydrate is a mixture of a β-anomer and a hydrated α-anomer. The crystal composition of laminarabiose is 3-O-α-D-glucopyranosyl-α/β-D-glucopyranose hemihydrate [LAMBIO] (Fig. 13.51). The crystals studied had the ratio 2α to 3β, with 0.19 H$_2$O molecules located on a twofold axis of symmetry, as in methyl-β-D-glucopyranoside hemihydrate. The water molecules are only associated with the α-anomer. The structure is therefore a mixture of the β-anomer with the α-anomer hemihydrate. Both anomers have an inter-residue intramolecular hydrogen bond from O(4)H to the ring oxygen O(5), thereby orientating the pyranose rings so that the ring oxygens are on opposite sides of the long molecular axis. The hydrogen-bonding pattern consists of two infinite chains involving only pairs of hydroxyls, and finite chains linked at the water molecule. The contrast between the strongly bonded infinite chains and the weak hydrogen bonding donated by the water molecule is striking.

The crystal structures of five disaccharides with an acyclic residue have been analyzed. The hydrogen-bonding patterns show the same tendency to form infinite chains as was observed in the simple sugar alcohols.

Hydrogen Bonding in the Disaccharide Crystal Structures

Fig. 13.51. Laminarabiose hemihydrate [LAMBIO]. *intramolecular, →2 is a crystallographic twofold rotation axis

Fig. 13.52. 4-O-β-Galactopyranosyl-rhamnitol [GAPRHM 10]. *intramolecular

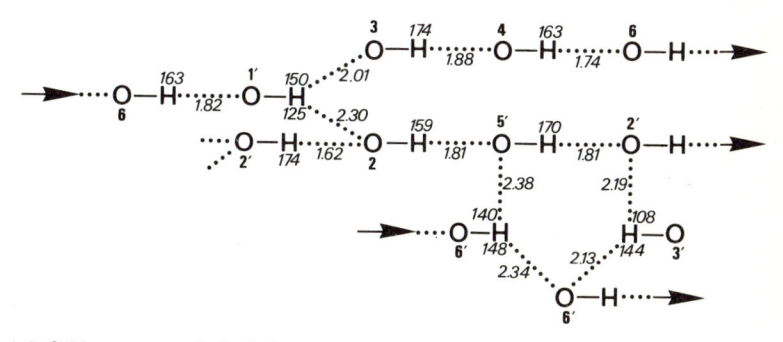

Fig. 13.53. 4-O-β-Glucopyranosyl-glucitol [BDGPGL]

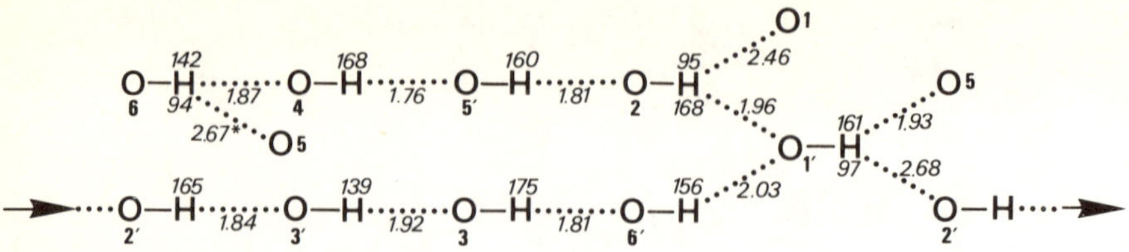

Fig. 13.54. 4-O-α-Glucopyranosyl-glucitol [BIZHIB]. *intramolecular

In the 4-O-β-galactopyranosyl-L-rhamnitol [GAPRHM 10] (Fig. 13.52), the hydrogen bonding is an infinite chain, with a weak intramolecular three-center component to the linkage oxygen O(1), and a finite chain originating in the anomeric hydroxyl O(1)H and terminating in the ring oxygen O(5).

In the 4-O-β-glucopyranosyl-D-glucitol [BDGPGL] (Fig. 13.53), there are three infinite chains, linked by the minor components of three-center bonds.

In 4-O-α-glucopyranosyl-D-glucitol (maltitol) [BIZHIB] (Fig. 13.54), there are two chains which join at O(1')H, one finite chain terminating at O(5) and an infinite chain through the minor component of a three-center bond, O(1')−H···O(2').

In 6-O-α-D-glucopyranosyl-D-glucitol [BAVCAC], there are three infinite chains linked by the minor components of three-center bonds. In 1-O-α-D-glucopyranosyl-D-mannositol dihydrate [BAGZEO], a hydrogen-bonded water dimer links the hydroxyls on O(2)H and O(5)H of the same mannositol residue.

13.6 Hydrogen Bonding in the Tri- and Tetrasaccharides Is More Complex and Less Well Defined

With the exception of the cyclodextrins, which are discussed in Part III, Chapter 18, there is much less information about hydrogen bonding in the higher members of the oligosaccharide series. The crystal structures of one anhydrous trisaccharide, 1-kestose, and four hydrates, mannotriose trihydrate, melizitose monohydrate, planteose dihydrate, and raffinose pentahydrate, have been determined. One tetrasaccharide, stachyose pentahydrate, has been studied, and the Ba(I$_3$)$_2$ complex of a derivative of maltohexaose (see Part III, Chap. 18.9). All were room-temperature X-ray studies. Although the overall conformations of the sugar molecules were well determined, the positions of the hydroxyl hydrogen atoms were not, and in some cases the hydrogen atoms were not observed.

1-Kestose contains a cyclic hydrogen-bonding system. In the crystal structure, the molecule of 1-kestose, O-α-D-glucopyranosyl-(1→2)-O-β-D-fructofuranosyl-(1→2)-β-fructofuranoside [KESTOS], has the curled conformation which may be stabilized by a homodromic cycle of hydrogen bonds, shown in Fig. 13.55. The hydrogen atoms attached to one of the primary alcohol groups, O(6), and a secondary group, O(3'), were not observed and their location is unknown. This may be due to orientational disorder of the primary alcohol hydroxyls. The hydrogen-

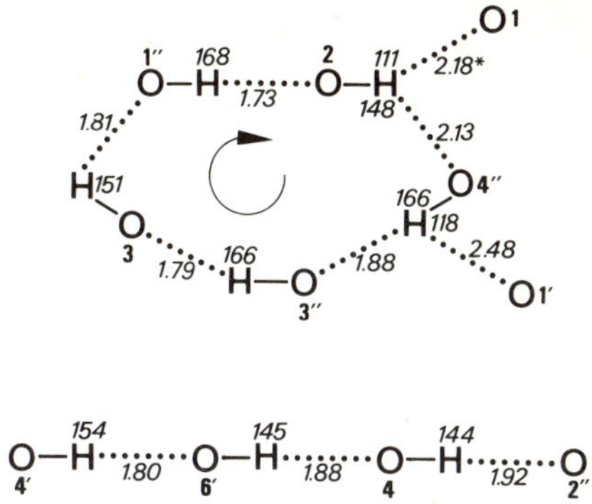

Hydrogen Bonding in the Tri- and Tetrasaccharides

$$\underset{1''}{O}\underset{1.73}{-H}\cdots\underset{168}{\underset{}{}}\underset{2}{O}\underset{148}{-H}\cdots\underset{111}{\underset{2.18*}{}}O1$$

Fig. 13.55. 1-Kestose [KESTOS]. *intramolecular

bond structure as far as it is known consists of a finite three-link chain terminating in the ring oxygen O(2″) of the fructofuranose residue, together with a pentagonal homodromic system of two three-center and three two-center bonds.

In mannotriose trihydrate [COFMEP 10] the hydrogen bonding consists of infinite chains cross-linked to themselves and finite chains by three-centered hydrogen bonds shown in Fig. 13.56. The water molecule hydrogen bonding is unusual in that all three are single acceptors. One of the water molecules, W(2), donates very weak bonds, while another, W(3), appears to donate only one bond. The two finite chains terminate in intramolecular inter-residue bonds to the ring oxygens. All oxygens, except the two linkage oxygens O4(1), O4(2), are involved in the hydrogen bonding. The glycosidic group O(1′)H forms a three-center bond, the shorter component of which is to the ring oxygen on the same mannose residue.

Infinite chains of hydrogen bonds occur in the crystal structure of melizitose monohydrate II. The molecule of melizitose, O-α-D-glucopyranosyl-(1→3)-β-D-fructofuranosyl-α-D-glucopyranoside, [MELEZT] [MELEZT 02], like 1-kestose, has a central furanoside residue and its conformation is curled. The monohydrate has two crystalline forms; only form II has been analyzed. All the hydrogen atoms were located and there was no disorder. Unlike 1-kestose, the hydrogen-bonding pattern shown in Fig. 13.57 has no cyclic component. It consists of a complex system of infinite chains cross-linked through the water molecule, and weak three-center bonds. The water molecule accepts one bond and donates two bonds, resulting in chain branching. There is one four-center bond from a primary alcohol hydroxyl, O(6) – H.

The hydrogen bonding in planteose dihydrate is poorly defined. In the crystal structure of planteose dihydrate [PLANTE 10], not all the hydroxyl hydrogen atoms are located and some appear in unlikely positions. The pattern is similar to that of melizitose hydrate, with infinite and finite chains cross-linked by the water molecules and three-center bonds. One of the water molecules, O(W 1), accepts one strong and one weak bond and donates a strong bond and a weak three-center bond. This includes the interesting chelated configuration shown in Fig. 13.58.

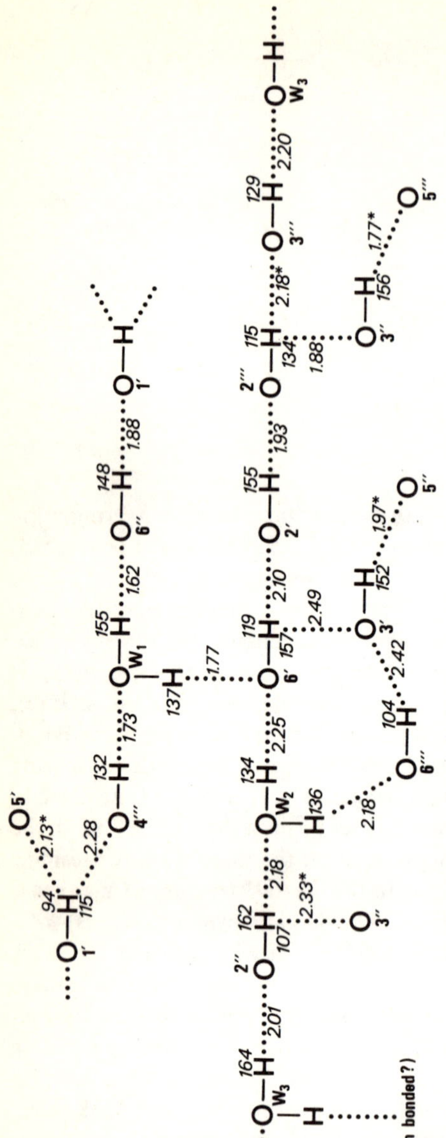

Fig. 13.56. Mannotriose trihydrate [COFMEP10]. *intramolecular

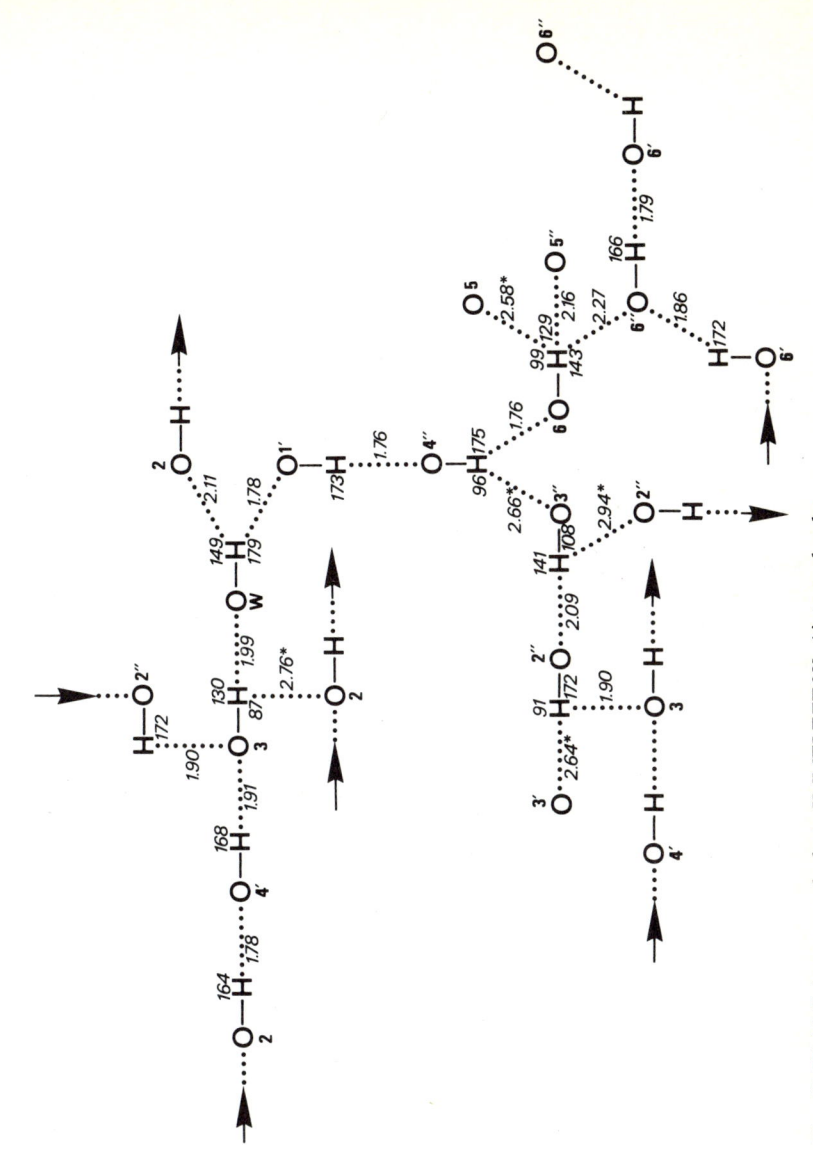

Fig. 13.57. Melizitose monohydrate II [MELEZT02]. *intramolecular

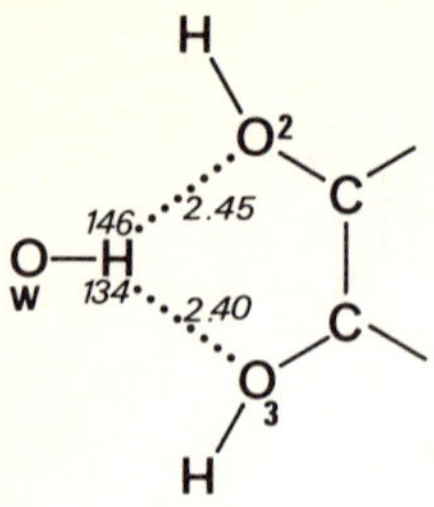

Fig. 13.58. Planteose dihydrate [PLANTE 10]

The other water molecule, W(2), receives and donates well-defined bonds, but the position allocated to one of the water hydrogens seems unlikely.

Raffinose pentahydrate has interesting infinite chains of hydrogen bonds. The hydrogen-bonding pattern is composed of four infinite chains, three of which extend in the direction of the crystallographic axes as shown in Fig. 13.59. These chains intersect at the water molecules to form an ordered homodromic network. The five water molecules are linked together. Four are double acceptors and one (W 1) is a single acceptor [507a].

The only tetrasaccharide for which there is a crystal structure analysis is stachyose pentahydrate, O-α-D-galactopyranosyl-(1→6)-O-α-D-galactopyranosyl-(1→6)-O-α-D-glucopyranosyl-(1→2)-β-D-fructofuranoside [STACHY 10]. It is an interesting molecule, because it incorporates the configurations of the trisaccharides, mannotriose and raffinose, and the disaccharide, sucrose. The hydration is uncertain, since seven water oxygen sites were located, all of which refined with partial occupancies and none could be omitted without a deterioration in the agreement factor. The occupancy factors ranged from 0.34 to 0.70.

13.7 The Hydrogen Bonding in Polysaccharide Fiber Structures Is Poorly Defined

The experimental data for the crystal structure analyses of polysaccharides is from X-ray fiber diffraction patterns. No neutron diffraction fiber patterns have been reported and it is doubtful whether these would help to resolve some of the ambiguities in interpretation. The "fiber diagrams" are the equivalent of single crystal rotation photographs with the problems of overlapping reflections made worse by the diffuseness of the diffraction spectra. In some cases, additional information can be obtained using precession camera methods and by inducing higher order in stretched films. The diffraction spectra are not only very diffuse relative to single crystal spectra, but it is rare that more than 50 independent structure amplitude measurements can be made [483, 484]. The number of parameters exceeds the number of observations unless many a-priori assumptions are made concerning the structure of the polymer. Even then, the data are insufficient to establish unequivocal crystal structures. Model structures are proposed and tested by their consistency with known stereochemical principles and how well they fit the obser-

The Hydrogen Bonding in Polysaccharide Fiber Structures Is Poorly Defined

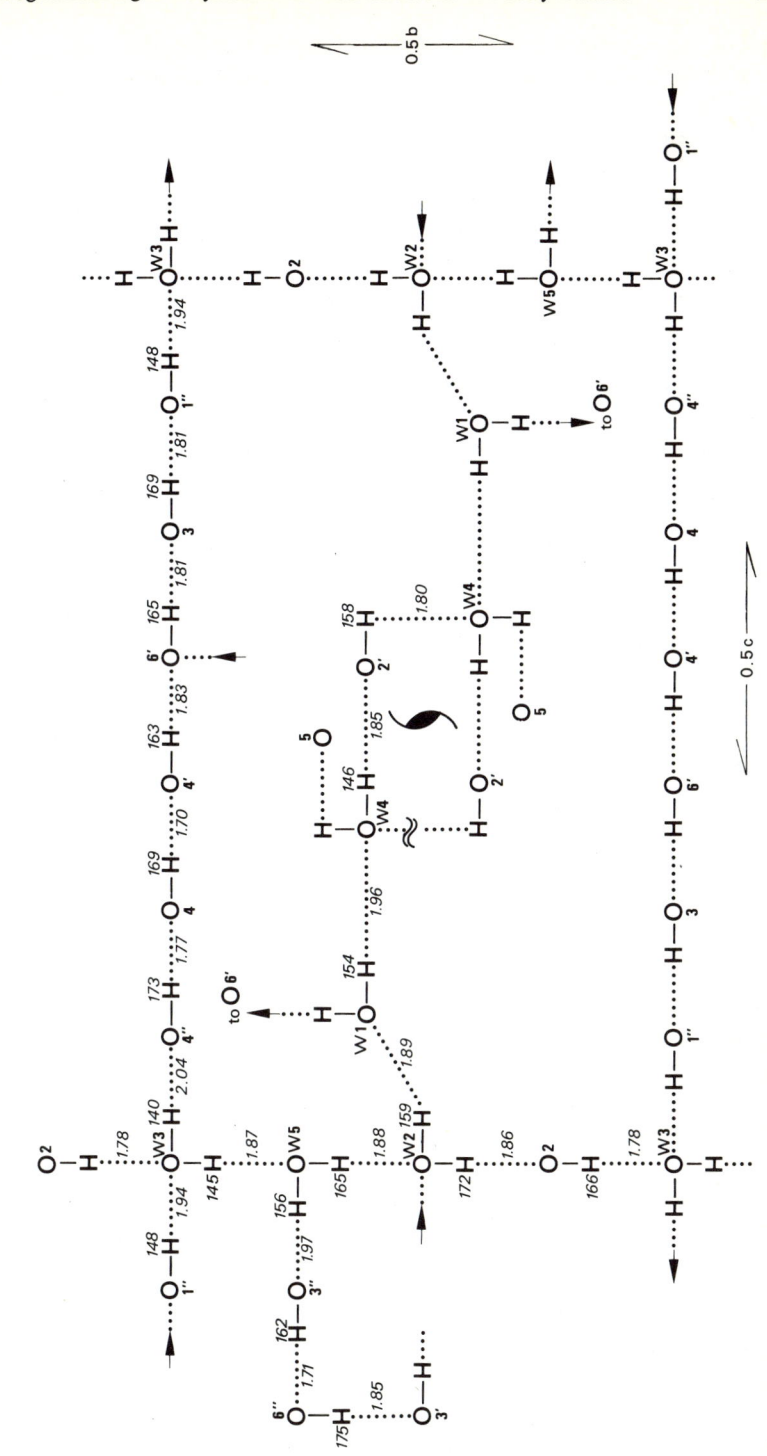

Fig. 13.59. Raffinose pentahydrate [507a]

vations, i.e., reduce the value of the disagreement factors. These model structures have to be consistent with the unit cell dimensions, space group symmetry or choice of space groups if this is ambiguous, and acceptable molecular geometries and intermolecular separations. It is considered axiomatic that the values for the bond lengths, valence angles, torsion angles, hydrogen-bonded and nonbonded interatomic distances lie within the range of values observed in the crystal structures of relevant mono- or disaccharide molecular structures.

The most advanced interpretative methods combine a least-squares fitting to the experimental structure amplitudes with energy minimization calculations by means of molecular mechanics. Since the limited amount of observations places severe constraints on the number of parameters that can be varied, it is common practice to use standard geometries for the monomer units [508], at least until the final stages of the refinement calculations.

In order to discuss the hydrogen bonding in polysaccharide crystal structures, two interrelated problems have to be resolved. One is to determine the intrachain hydrogen bonding, which may be inter- or intramonomer residue. This bonding must be consistent with fitting the polymer repeat unit to the length of the fiber axis, which is usually a well-defined quantity. When the fiber axis coincides with the axis of the polymer, the conformation of the polymer repeat unit must fit the symmetry of the fiber axis.

The second problem is to determine the interchain hydrogen bonding. This must be consistent with acceptable interchain atom-pair nonbonding separations, the lateral unit cell dimensions and the space group symmetry. It is a reasonable assumption from small molecule crystal structures that all hydrogen-bonding donor functional groups will form hydrogen bonds. Models that do not meet this criterion are viewed with disfavor. It is desirable, but not necessary, that all potential hydrogen-bond acceptor atoms are acceptors. Small molecule crystal structural studies indicate that linkage oxygens are not commonly acceptors, except for the minor components of three-center bonds. Ring oxygens are acceptors in about half of the small molecule carbohydrate crystal structures. Hydrogen bonds are identified by the O···O separations, since there is no information from the diffraction patterns concerning the orientation of the hydroxyl groups. It is customary to assume that all the hydrogen bonds are two-centered and to place the hydrogen atoms on the line of centers of the shortest O···O separations. Evidence from small molecule studies suggested that this will be incorrect for about one third of the bonds.

The much-studied example of cellulose illustrates these problems. Cellulose I and II may or may not have the same intrachain hydrogen bonding, although the fiber axis lengths of 10.36 Å correspond to a cellobiose repeat structure in both forms. All proposed models for cellulose I and II do include the inter-residue intrachain O(3')H···O(5) hydrogen bond, which is a feature of the crystal structures of cellobiose and methyl cellobioside methanolate. This places the ring oxygens alternately on opposite sides along the chain axis. It also requires that the mean planes of the pyranose rings are approximately coplanar and form a relatively rigid fused-ring conformation.

The favored polymer chain model for cellulose I places the crystallographic 2_1 axis through the polymer chain (Fig. 13.60). This permits the formation of two in-

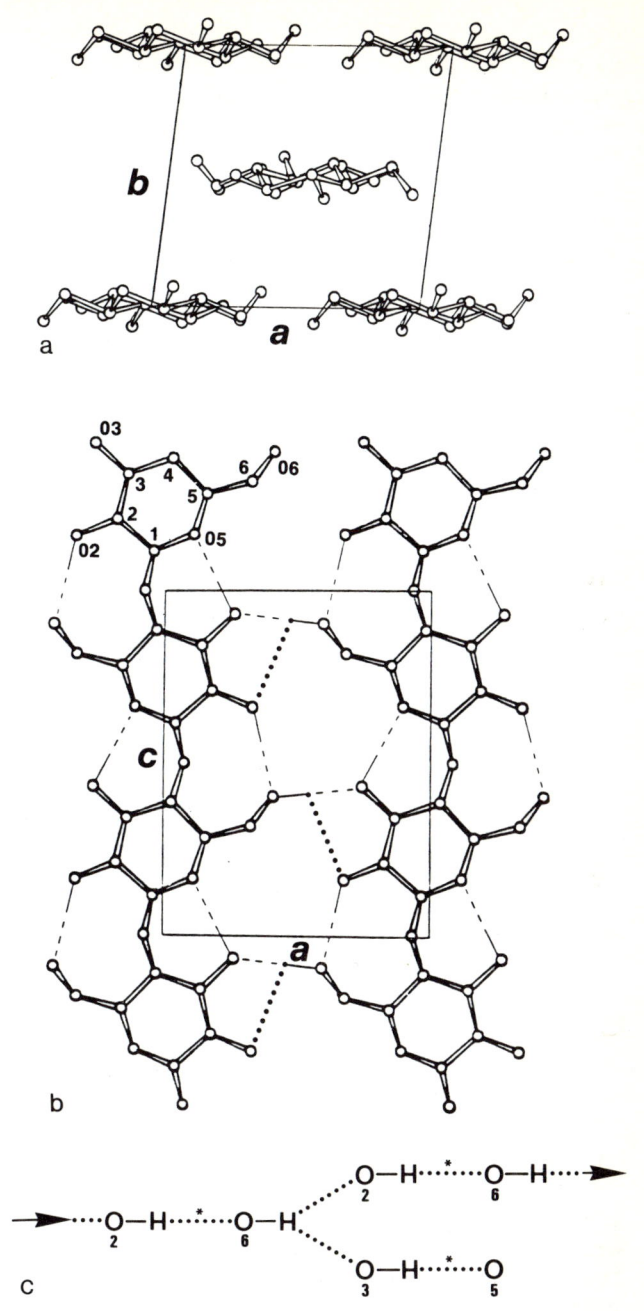

Fig. 13.60a–c. Crystal structure of cellulose I. **a** View along the crystallographic c-axis. **b** View on the a,c-plane. The 2_1 screw axis is along c. O–H groups indicated by O– [242]. *Dashed lines*, hydrogen bonds as proposed on the basis of X-ray fiber diffraction and molecular packing arguments; *dotted lines*, three-center hydrogen bonds as proposed in **c**. In **c**, *indicate intramolecular hydrogen bonds

tra-chain inter-residue hydrogen bonds O(3)H···O(5′) and O(2)H···O(6′), to give an even more rigid four-ring layer configuration. There are three ways of arranging these chains in the unit cell, which are almost indistinguishable when viewed in projection down the fiber axis. These are parallel or antiparallel up or down, with displacements of c/4 for the center chain [242, 509–514]. The distinction rests in a difference in disagreement factors (R-values) that could easily be reversed by new structure amplitude measurements [482]. In these structures there is an interchain hydrogen bond O(6)H···O(3) linking the layers laterally, but no bonding between layers. All three donor hydroxyls form hydrogen bonds, and by assuming a three-center bond from a primary alcohol hydroxyl, a very reasonable hydrogen-bond scheme can be postulated, as shown in Fig. 13.60c.

The cellulose II structure is more complex to unravel, since it involves two symmetry-independent chains. Polymer chain structures have been proposed that involve six different intrachain hydrogen-bond arrangements [344]:

1. One inter-residue bond, O(3′)H···O(5);
2. Two inter-residue-bonds, O(3′)H···O(5) and O(2′)H···O(6), as in cellulose I;
3. Two inter-residue bonds, O(3′)H···O(5) and O(6)H···O(2′);
4. Two inter-residue bonds, O(3′)H···O(5), O(2)H···O(6′) and an intra-residue bond, O(6′)H···O(4′);
5. One two-center inter-residue bond O(3′)H···O(5) and one three-center inter-residue bond O(6)H⟨O(3′) / O(5);
6. One three-center inter-residue bond O(3′)H⟨O(5) / O(6).

Models (1), (5) and (6) leave O(2)H and O(6)H available for interchain hydrogen bonding; (2) and (3) leave O(2)H or O(6)H available and (4) has no interchain hydrogen bonding, unless it is through the minor component of a three-center bond.

The favored models are anti-parallel or parallel-up. In both models, one chain has two intramolecular inter-residue hydrogen bonds as in cellulose I, while the other chain has only the O(3′)H···O(5′) observed in cellobiose. All hydroxyls are utilized. The parallel-down models appear to be inconsistent with this criterion; however, this does not necessarily exclude them from consideration [515].

A molecular mechanics energy optimization, which included a Morse potential for the hydrogen bond, combined with a disagreement R-factor minimization, has been reported [344]. This computation involved 15 geometrical factors, a weighting factor and an average isotropic temperature factor. It resulted in a large number of local minima, of which 12 were documented within a range of about 40 kJ mol^{-1}. Of these, all but one involved antiparallel chains. Two of the antiparallel chain models were unusual in not having the cellobiose O(3′)H···O(5) inter-residue bond. This was apparently compensated by the reorientation of the O(3)H group to form interchain hydrogen bonds to O(6) in one case and a three-center bond to two O(6)s on adjacent residues in the other. Similar ambiguities are in the analysis of cellulose III. It seems unlikely that an unambiguous solution to the structures of cellulose II and III will come from further calculations. It clearly requires either more precise and more extensive X-ray data or evidence from complementary methods.

Models for the crystal structures of the many crystalline forms of amylose [516], mannan [517], chitin [515] and many of the bacterial polysaccharides [518] have been proposed. All included the inter-residue $O(3')H \cdots O(5)$ intrachain hydrogen bond when stereochemically feasible. For mannan, a three-center bond has been proposed on the basis of infrared measurements [519]. Intrachain interresidue hydrogen bonds also feature in the structures proposed for the bacterial polysaccharides whenever stereochemically possible.

Even if an unequivocal crystal structure is established for a polysaccharide structure, the hydrogen bonding is likely to remain ambiguous in the absence of direct evidence relating to the orientation of the hydroxyl groups. As in the mono- and disaccharide crystal structures, the possibility of forming two-center and three-center bonds of comparable energy with similar $O \cdots O$ separations increases the number of choices. It is conceivable that in structures such as the celluloses, there is orientational disorder of the hydroxyls and the primary alcohol C–OH bonds, leading to a multiplicity of hydrogen-bond structures. Precedents for these disordered structures are found in the carbohydrate crystal structures described earlier in this chapter, and there is no evidence from the fiber diffraction patterns which excludes this complexity.

Chapter 14
Hydrogen Bonding in Amino Acids and Peptides: Predominance of Zwitterions

Amino acids of the general form, **1**, are the monomeric molecules which are condensed to form the polypeptide chains of the fibrous and globular proteins. The naturally occurring molecules are the L-enantiomers, shown in **1**; for chemical formulae see Fig. 19.1. D-amino acids can be synthesized and the individual L- or D-amino acids or the D,L-racemates can be crystallized. All the common amino acids have been studied by neutron or X-ray crystal structure analysis (see Table 14.1), in the anhydrous or hydrate forms, as hydrochlorides or hydrochloride hydrates.

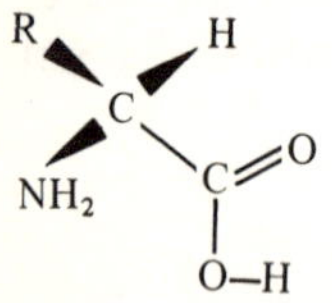

1

As with the carbohydrates, the crystal structures of the D,L-mixtures are different from those of the enantiomorphs, and their crystal structures may or may not be centrosymmetrical. As shown in Table 14.1, more than 50 amino acids and their salts or hydrates have been analyzed as the enantiomorph or the racemate, or both.

In solution, the several charged species, **2, 3, 4**, can occur in equilibrium, depending on the pH of the solution. In the crystalline anhydrous and hydrated amino acids, however, only the zwitterion configuration, **3**, is observed. The acidic configuration, **2**, occurs in the hydrochlorides and hydrated hydrochlorides, but there are no crystal structures containing the anionic form **4**. Polymorphism occurs in the amino acids, but it is relatively rare and certainly no more common than

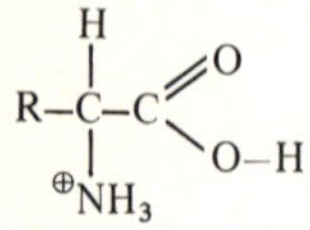

2 Acidic

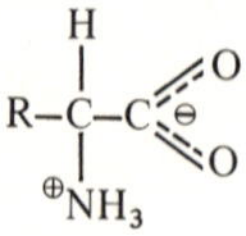

3 Neutral zwitterion

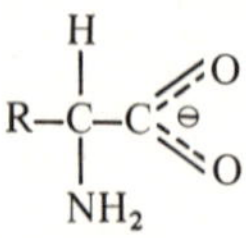

4 Basic

Table 14.1. REFCODES for crystal structure analyses of commonly occurring amino acids and their hydrochlorides and hydrates (n = neutron analysis)

Amino acid	Anhydrous	Hydrate	Hydrochloride	Hydrochloride hydrate
Alanine	L (N) [LALNIN12] D,L [DLALNI]	—	L–HCl [ALAHCL]	—
Arginine	—	L–2H$_2$O (N) [ARGIND11] L–H$_2$O (N) [ASPARM03,05]	L–HCl [LARGIN]	L–HCl H$_2$O [ARGHCL10]
Asparagine	—		—	—
Aspartic acid	L [LASPRT] D,L [DLASPA10]	—	D,L–HCl [ASPART10]	—
Cysteic acid	L [CYSTEA]	L–H$_2$O (N) [CYSTAC01]	—	—
Cysteine	L (mono) (N) [LCYSTN] L (ortho) [LCYSTN21, 12]	—	—	L–HCl H$_2$O [CYSCLM10]
Cystine	L (tetra) LCYSTI11 L (hex) LCYSTI10	—	L–2HCl (N) [CYSTLC01, 02]	—
Glutamic acid A form	L (N) [LGLUAC03]	D,L–H$_2$O [CADVUY]	L–HCl (N) [LGLUTA]	—
B form	L (N) [LGLUAC11]	—	D,L–HCl [DLGLAC]	—
Glutamine	L (N) [GLUTAM01]	—	L–HCl [GLUTAN]	—
α-Glycine	(N) [GLYCIN03,05] [GLYCIN]	—	HCl (N) [GLYHCL]	—
β-Glycine	(N) [GLYCIN 15,16,17,18]	—	½HCl [DGLYHC]	—
γ-Glycine	L (ortho) (N) [LHISTD13]	—	—	—
Histidine	L (mono) [LHISTD02] D,L [DLHIST]	—	L–2HCl [HISTDC10]	L–HCl H$_2$O (N) [HISTCM12] D,L–HCl 2H$_2$O [DCHIST]
Hydroxyproline	L (N) [HOPROL12]	—	—	—
Isoleucine	L [LISLEU] D,L [DLILEU]	—	—	L–HCl H$_2$O [LILEUC10]
Leucine	L [LEUCIN01] D,L [DLLEUC]	—	L–½HCl [DLEUHC]	—
Lysine	—	—	D,L–HCl [DLLYSC10]	L–HCl 2H$_2$O (N) [LYSCLH02,11]

(continued on next page)

Table 14.1 (continued)

Amino acid	Anhydrous	Hydrate	Hydrochloride	Hydrochloride hydrate
Methionine	L [LMETON10] D,L (α-form) [DLMETA02] D,L (β-form) [DLMETA03]	—	L – HCl [METHCL]	—
Phenylalanine		—	L – HCl (N) [PHALNC01]	—
Proline	L [PROLIN]	D,L H$_2$O [DLPROM]	D,L – HCl [DLPROL] D,L – $\frac{1}{2}$HCl [PROLNH]	—
Serine	L [LSERIN01,10] D,L (N) [DLSERN11]	L – H$_2$O (N) [LSERMH10]	—	—
Threonine	L$_s$ (N) [LTHREO02]	—	—	—
Tryptophane	D,L [QQQBTP01]	—	L – HCl (N) [TRYPTC01]	—
Tyrosine	L (N) [LTYROS11] D,L [DLTYRS]	—	L – HCl (N) [LTYRHC10]	—
Valine	L [LVALIN] D,L [VALIDL]	—	L – HCl (N) [VALEHC11] D,L – HCl (N) [DLVALC]	L – HCl H$_2$O [VALHCL10]

in other classes of biological monomers such as the carbohydrates. There are, however, a greater number of hydrated crystals, 30% out of a sample of 53 structures, compared with less than 10% for a comparable sample of monosaccharides. Some crystals of amino acids have been studied only as their hydrates or hydrochlorides, i.e., arginine, lysine, and phenylalanine. In addition to the zwitterion groups $\overset{+}{N}H_3$ and $-CO_2^-$, many other hydrogen-bonding functional groups are provided by the amino acid side chains (see Box 14.1 and Fig. 19.1).

Box 14.1. Hydrogen-bonding functional groups in amino acids and their salts

Donor groups: $-\overset{\oplus}{N}(H_2)-H$, $>\overset{\oplus}{N}(H)-H$, $>\overset{\oplus}{N}-H$, $-N(H)H$, $>N-H$,

$H-O_W-H$, $-C\overset{\diagup O}{\diagdown O-H}$, $C-O-H$, $-S-H$

Acceptor groups: $O=C\overset{\diagup}{\diagdown O^\ominus}$, $O=C\overset{\diagup}{\diagdown O-H}$, $O\overset{\diagup H}{\diagdown C}$, $N\overset{\diagup}{\diagdown}$, $N(H_2)$, $Cl^\ominus$, $_WO\overset{\diagup H}{\diagdown H}$, $O=\tilde{S}$

The unusually large number of neutron diffraction analyses, 25, and of high precision X-ray analyses for this class of compounds has provided a source of data for an in-depth survey of the hydrogen-bonding geometry associated with this type of molecule [60, 74, 75]. The metrical aspects of this geometry are described in Part IB. In this chapter, we will be emphasizing the pattern recognition aspects of the hydrogen bonding, which, however, have not received the detailed attention that has been directed toward the mono- and disaccharides. This is possibly because the zwitterion ion $\overset{+}{N}H\cdots O=\bar{C}$ hydrogen bonding which dominates the amino acid crystal structures is not the most important bonding in the biologically important polypeptides.

The zwitterion bridges $\overset{+}{N}H\cdots O=\bar{C}$ determine the hydrogen-bond patterns in the amino acid crystal structures. The amino acids are zwitterions in the crystalline state, or in the case of the hydrochlorides or hydrobromides, they are acid cations. In consequence, the predominant hydrogen-bond donor is the $\overset{+}{N}H_3$ and the predominant acceptors $O=\bar{C}$ or Cl^-. There is a numerical imbalance of hydrogen-bond donor and acceptor capability; the $\overset{+}{N}H_3$ group can provide only three hydrogen bonds, while both oxygens of the carboxylate group prefer to accept two bonds each, and the chloride anion prefers a tetrahedral environment of four bonds. This imbalance is partially alleviated by hydration, since water molecules have two donor protons and frequently accept only one bond. This provides a rationalization of why the amino acids form an unusually large number of hydrates, compared with the carbohydrates, for example.

As pointed out in Part IA, the imbalance between donor and acceptor properties of the hydrogen-bonding functional groups leads to proton deficiency for two-center hydrogen bonding and a larger than usual proportion of three-center bonds. In addition, the configuration of the amino acids is particularly favorable for intramolecular hydrogen bonding of the type shown below. When the orientation of the $\overset{+}{N}H_3$ group is such that a hydrogen lies in or close to the plane of the

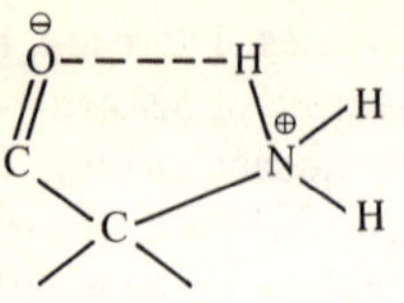

N–C–C=O group, the H···O distance is about 2.5 Å. This intramolecular interaction is frequently the minor component of a three-center bond with H···O distances ranging from 2.45 to 2.75 Å.

Even two-center $\overset{+}{N}H\cdots O=\overset{-}{C}$ bonds are very bent. Unlike the C–OH group, the position of a proton in a $-\overset{+}{N}H_3$ group is constrained by that of the other two protons. In some circumstances, not all three protons can lie in positions of optimum hydrogen-bond energy. For this reason, the amino acid crystal structures provide a number of examples of unusually *bent* hydrogen bonds. These are given in Table 14.2. In every case where the $\overset{+}{N}-H\cdots O$ angle is less than 140°, the H···O distance is greater than 1.9 Å.

These examples are distinguished from being described as a component of a three-center bond, in that there are no electronegative atoms in the direction of the extension of the bisector of the N–H···O angles within 3.0 Å.

Sometimes these bent bonds are associated with a particular configuration, as in D,L-tyrosine, where the bent two-center bonds form part of the cyclic trimer geometry shown below.

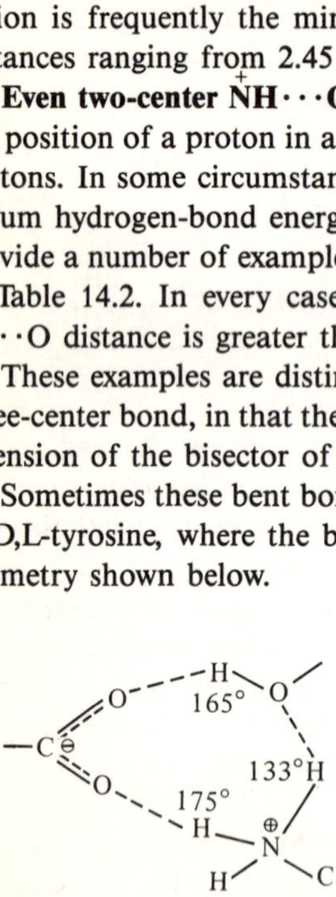

The zwitterion ammonium group forms three classes of hydrogen-bonding patterns. Inspection of the hydrogen-bonding patterns in the crystal structures of the amino acids revealed three distinct categories [74]. These classes I, II, and III,

Table 14.2. Bent two-center bonds in the amino acid crystal structures

Amino acid	Bond	H···A	N–Ĥ···A	REFCODE
L-Glutamine	$\overset{+}{N}-H\cdots O=\overset{-}{C}$	1.92 Å	129°	GLUTAM01
L-Serine H$_2$O	N–H···O$_W$	2.13	132	LSERMH10
Glycylglycine phosphate H$_2$O	O$_W$–H···O=$\overset{-}{C}$	2.16	121	GLYGLP
Glycyl-L-tyrosine 2H$_2$O	O$_W$–H···O=$\overset{-}{C}$	2.29	110	GLTLYR10
L-Arginine phosphate H$_2$O	$\overset{+}{N}-H\cdots O=\overset{-}{C}$	2.53	139	LARGPH01
D,L-Tyrosine	$\overset{+}{N}-H\cdots OH$	2.09	133	DLTYRS
L-Lysine HCl 2H$_2$O	$\overset{+}{N}-H\cdots O=\overset{-}{C}$	2.08	134	LYSCLH11
D,L-Aspartic acid HCl	$\overset{+}{N}-H\cdots O=\overset{-}{C}$	2.04	137	ASPART10

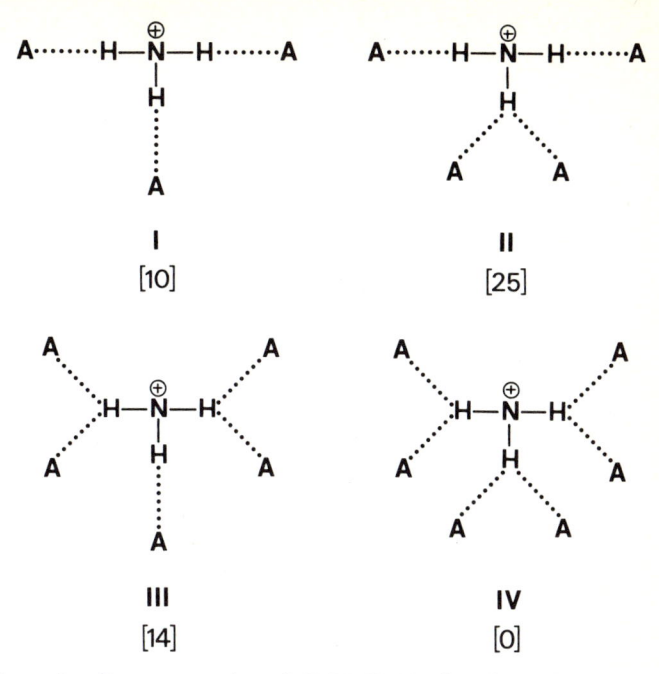

Hydrogen Bonding in Amino Acids and Peptides: Predominance of Zwitterions 225

Fig. 14.1. Scheme for hydrogen-bonding structure classes I, II, III, IV. Number of crystal structure analyses in each class is given in parentheses

Fig. 14.2. L-Cysteine [LCYSTN12, neutron]

which arise from the four possible first neighbor hydrogen-bond patterns that can be assigned to a $-\overset{+}{N}H_3$ group, are shown in Fig. 14.1.

In a survey of a representative group of 49 amino acid crystal structures which were determined by neutron diffraction or good precision X-ray diffraction analyses, there were 10 in class I, 25 in class II, 14 in class III and none in class IV. The absence of a representative of configuration IV is likely to be due to steric repulsions arising from molecular overcrowding. Examples of the three classes are illustrated in the following [74].

Class I. Of the ten amino acid crystal structures belonging to this class, five have carboxylate or hydroxyl oxygens as acceptor groups A, as shown in Fig. 14.2 for L-cysteine [LCYSTN12] and in Fig. 14.3 for L-lysine-L-aspartate [LYSASP]. In two of the cases, A is a carboxylate or water oxygen or chloride ion, as shown in

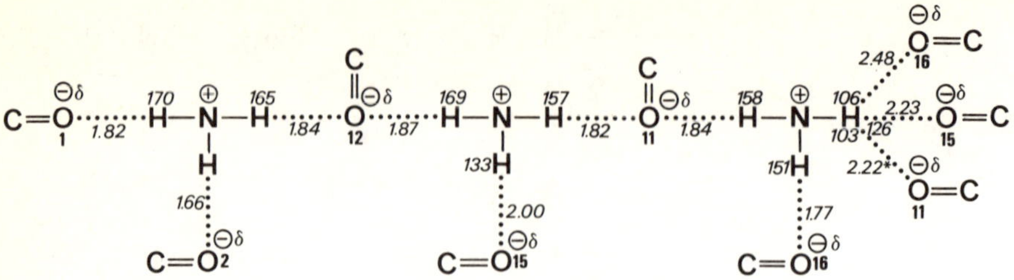

Fig. 14.3. L-Lysine-L-aspartate [LYSASP]; *, intramolecular

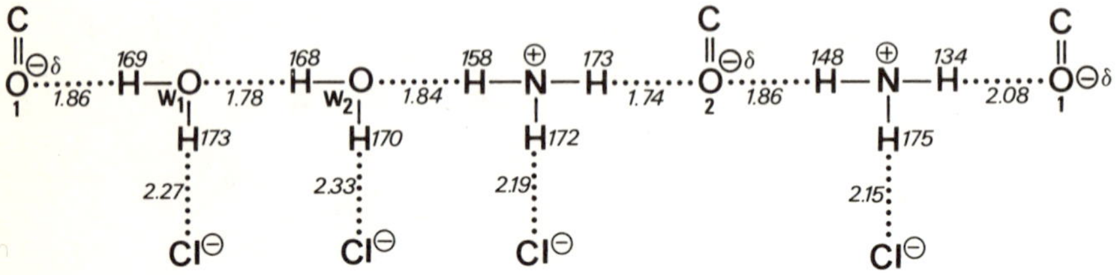

Fig. 14.4. L-Lysine·HCl·2H$_2$O [LYSCLH11, neutron]

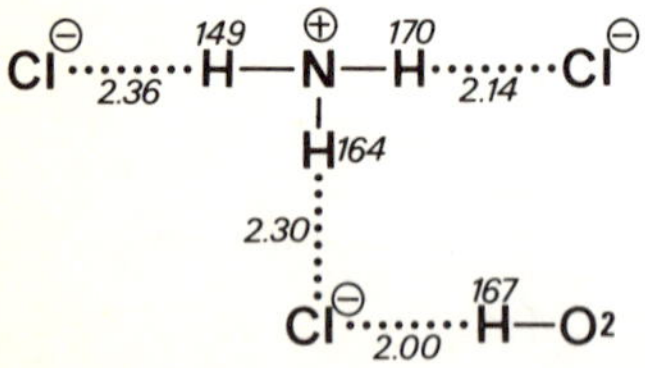

Fig. 14.5. L-Phenylalanine·HCl [PHALNC01, neutron]

Fig. 14.4 for L-lysine·HCl·2H$_2$O [LYSCLH11]. A is chloride in only three amino acid hydrochloride salts, see L-phenylalanine·HCl [PHALNC01] in Fig. 14.5.

Class II. Most of the amino acid crystal structures belong to this class. The hydrogen-bond acceptor groups are oxygen atoms from carboxylate, hydroxyl or water, or nitrogen atoms, or chloride ions. In ammonium glycinium sulfate [AGLYSL01] the ammonium group forms Class II hydrogen bonding and the primary ammonium group belongs to Class III (Fig. 14.6). The phosphate group and the arginine guanidinium group give rise to a complicated hydrogen-bonding pattern in L-arginine phosphate·H$_2$O [LARGPH01] (Fig. 14.7). Two of the acceptor groups are chloride ions in glycine·HCl [GLYHCL] (Fig. 14.8), and three are chloride ions in L-valine·HCl [VALEHC11] (Fig. 14.9).

Class III. The majority of crystal structures in this class has hydrogen bonds $\overset{+}{N}-H\cdots O$ with carboxylate or hydroxyl oxygen atoms as acceptors, and in only four are the acceptors mixed, chloride ions or oxygen from water and carboxylate

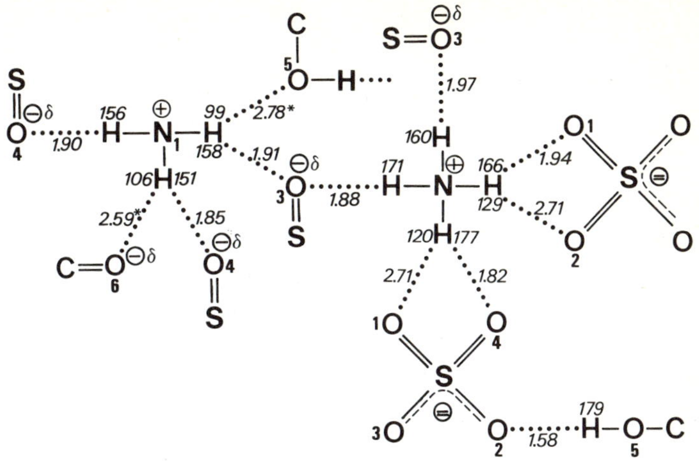

Hydrogen Bonding in Amino Acids and Peptides: Predominance of Zwitterions 227

Fig. 14.6. Ammonium glycinium sulfate [AGLYSL01]; *, intramolecular

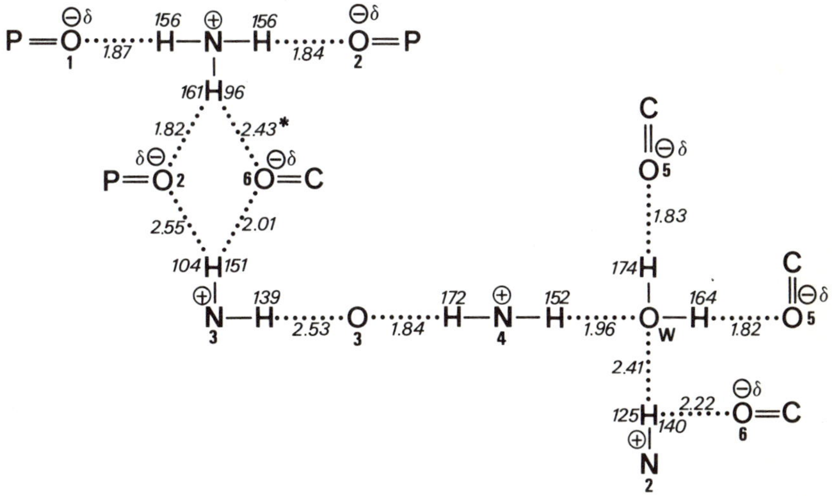

Fig. 14.7. L-Arginine phosphate·H$_2$O [LARGPH01]; N(2), N(3), N(4) are in the guanidinium group; *, intramolecular

groups. Frequently one of the hydrogen bonds is of the intramolecular type (indicated by an asterisk).

D,L-serine [DLSERN11] (Fig. 14.10) and L-glutamic acid [LGLUAC11] (Fig. 14.11), are good examples to illustrate how proton deficiency promotes the formation of three-center hydrogen bonds. In the more complex crystal structure of glycylglycine·H$_2$O·HCl [GLCICH01] (Fig. 14.12), oxygen atoms and chloride ions are involved as hydrogen-bond acceptors.

There are three related pattern II′, III′ and III″ involving other three- and four-center hydrogen bonds (Fig. 14.13). II′ occurs in L-glutamic acid·HCl [LGLUTA]

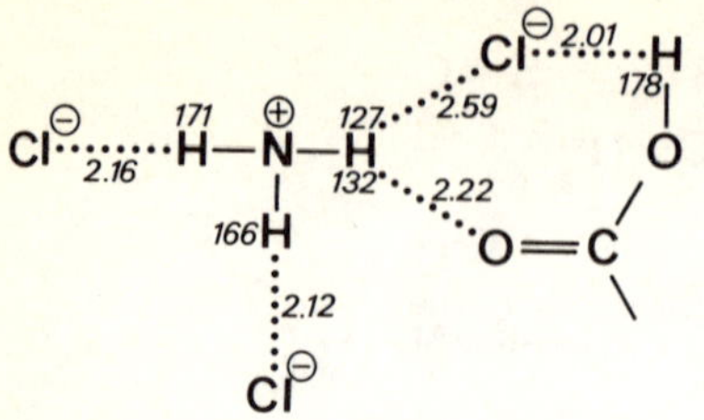

Fig. 14.8. Glycine·HCl [GLYHCL, neutron]

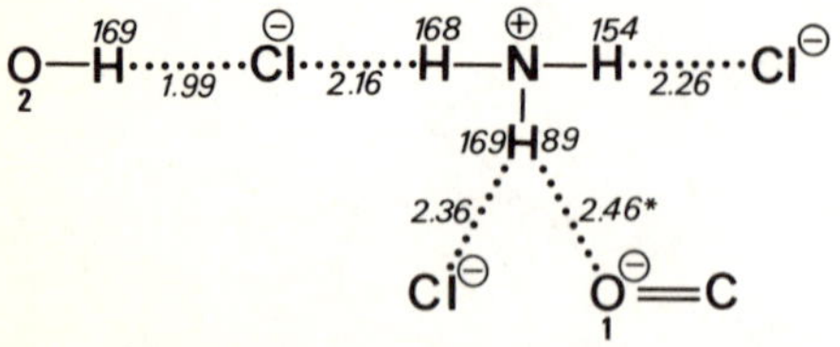

Fig. 14.9. L-Valine·HCl [VALEHC11, neutron]; *, intramolecular

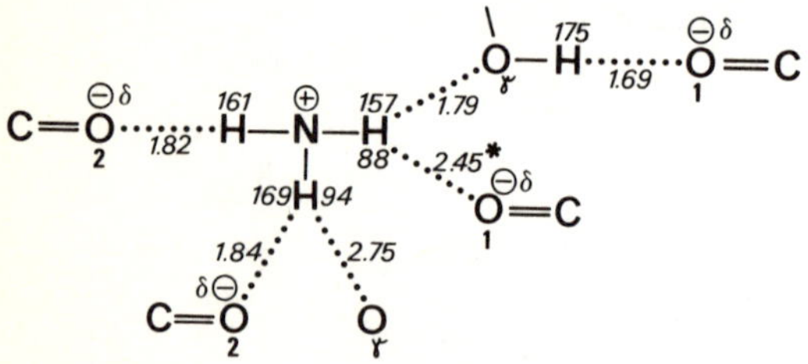

Fig. 14.10. D,L-Serine [DLSERN11, neutron]; *, intramolecular

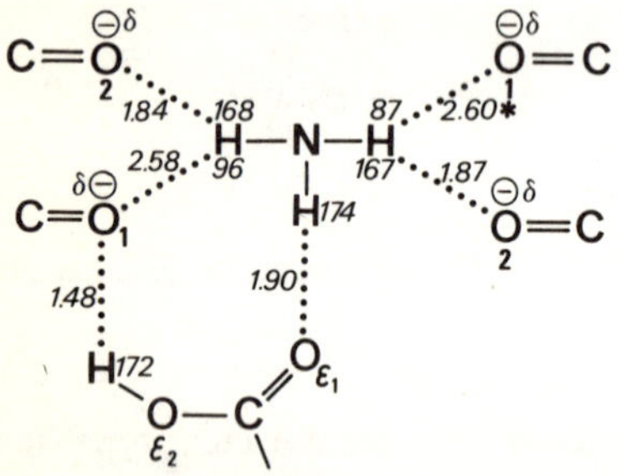

Fig. 14.11. L-Glutamic acid β-form [LGLUAC11, neutron]; *, intramolecular

and in L-lysine-L-aspartate [LYSASP] (Fig. 14.3). In the former, the N—H···O distances in the four-center bond are 1.98, 2.50, and 2.61 Å; in the latter they are 2.22, 2.23, and 2.48 Å. The configuration III' is observed in L-histidine·HCl ·H$_2$O [HISTCM12] (Fig. 14.14), and in L-cysteine dimethyl ester·2HCl·H$_2$O [CYSMEC], and III'' is found in L-cysteic acid·H$_2$O [CYSTAC01] (Fig. 14.15).

Hydrogen Bonding in Amino Acids and Peptides: Predominance of Zwitterions 229

Fig. 14.12. Glycylglycine·H₂O·HCl [GLCICH01, neutron]; *, intramolecular

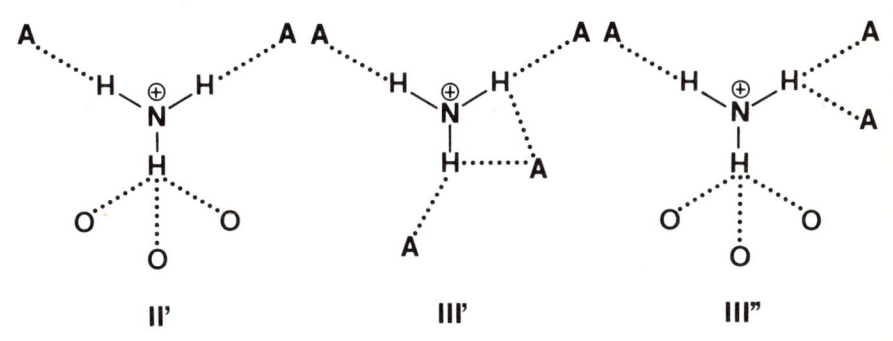

Fig. 14.13. Scheme for hydrogen-bonding structure classes II′, III′, III″

Fig. 14.14. L-Histidine·HCl·H₂O [HISTCM12, neutron]; *, intramolecular

Data on NH$_2^+$ hydrogen bonding are more limited. There are only three amino acid crystal structure analyses containing NH$_2^+$ groups, one of which is a neutron diffraction study. They are all based on the amino acid proline:

Prolylsarcosine hydrate [PRSARH], in which there is one two-center and one three-center NH$^+\cdots$O bond, the minor component of which is intramolecular (Fig. 14.16).

Allo-4-hydroxyl-L-proline dihydrate [AHLPRO] contains two infinite chains formed by two hydrogen-bonded water molecules (Fig. 14.17), and in 4-hydroxyl

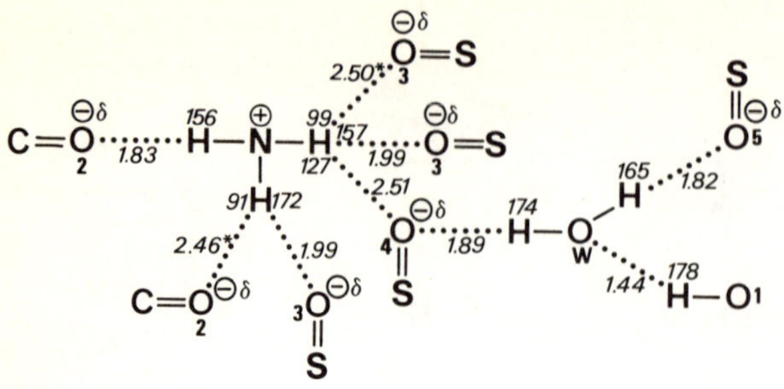

230 Hydrogen Bonding in Amino Acids and Peptides: Predominance of Zwitterions

Fig. 14.15. L-Cysteic acid·H$_2$O [CYSTAC01, neutron]; *, intramolecular

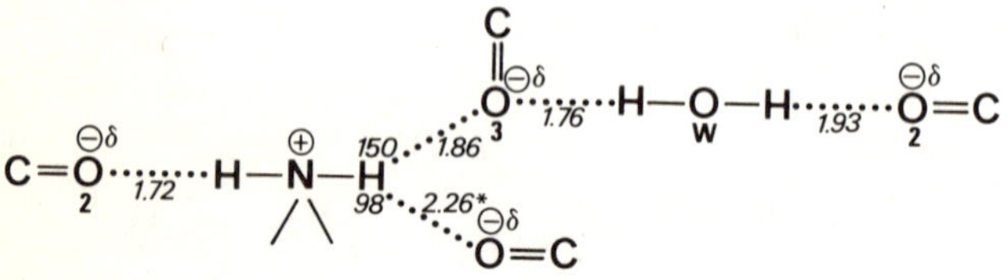

Fig. 14.16. Prolylsarcosine·H$_2$O [PRSARH]; *, intramolecular

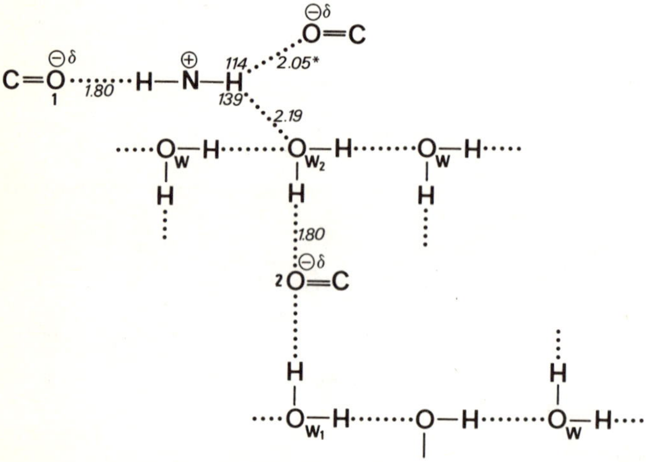

Fig. 14.17. Allo-4-hydroxyl-L-proline dihydrate [AHLPRO]; *, intramolecular

proline [HOPROL12], there is one two-center and one three-center hydrogen bond (Fig. 14.18).

Some general observations can be made from these hydrogen-bond geometries and patterns.

Three-center bonds tend to shorten two-center bonds. There is a systematic trend for the two-center bonds to become shorter with an increase in the number of three-center bonds. The mean two-center $\overset{+}{N}-H\cdots O=\bar{C}$ bond lengths in

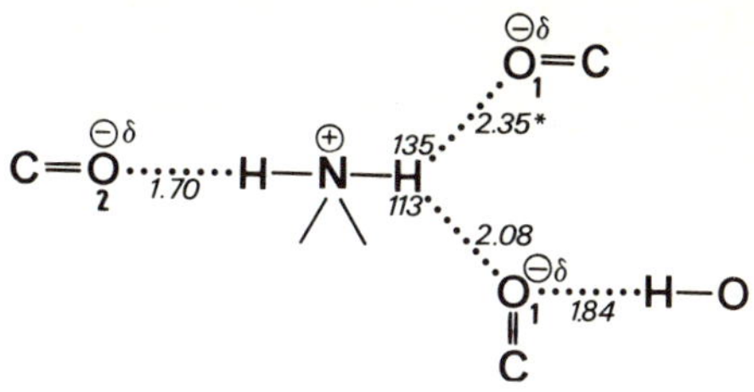

Fig. 14.18. 4-Hydroxyl proline [HOPROL12]; *, intramolecular

classes I, II, and III are, respectively, 1.899, 1.880, and 1.789 Å. This is consistent with the mean value of 1.841 Å for the 226 N–H···O=C two-center bonds, irrespective of class of bonding.

Several interesting cyclic configurations are observed. The very short carboxylic acid hydrogen bond C–OH...O=C̄ of 1.475 Å occurs in L-glutamic acid B-form, as part of a trimer system. Cyclic trimers involving the components of three-center bonds are also observed in the A-form of glutamic acid. In this pattern, there is also a chelated three-center bond of the type $-OH\cdots\overset{O}{\underset{O}{\gtrless}}\bar{C}-$. Other cyclic trimers occur in L-tyrosine [LTYROS11], Ca-di-glutamine 4H$_2$O [LGLUCA], and glycine HCl [GLYHCL]. Cyclic tetramers are observed in L-lysine-L-asparate [LYSASP], L-asparagine H$_2$O [ASPARM03], and L-lysine HCl 2H$_2$O [LYSCLH02], the latter involving a water dimer.

Hydrogen-bonding patterns in the linear and cyclic peptides have not been studied systematically in the same way as in the di- and trisaccharides and in the nucleosides and nucleotides. Although there many crystal structure analyses reported for the linear and cyclic oligopeptides, they are not generally as precise as for the di- and trisaccharides and nucleosides. Few include hydrogen atom coordinates, particularly those of the hydrogen-bonding groups. There are no neutron diffraction analyses and few low-temperature X-ray crystal studies.

On the other hand, the hydrogen atom positions of the principle donor ⟩N–H groups can be calculated with reasonable reliability. The cyclic peptides, in particular, have conformational constraints that can make hydrogen bonding sterically difficult or impossible. A wider range of ⟩N–H...O=C bond lengths is anticipated, with a significant number of nonbounded ⟩N–H or O=C groups. A systematic analysis of the hydrogen bonding in the oligopeptides would provide an interesting comparison with that in the protein structures discussed in Part III, Chapter 19.

Chapter 15
Purines and Pyrimidines

As a class, the compounds derived from purine and pyrimidine bases are of considerable biochemical interest, particularly if they are the constituents of the nucleic acids. With their many hydrogen-bonding donor and acceptor functions, these molecules can interact specifically with the enzymes involved in the metabolism of the nucleic acids and with regulatory proteins. Purines, pyrimidines, and their derivatives have been used most successfully in the investigations of biological processes and have found application in pharmacology [520, 521]. The same is true for the barbiturates, where the main emphasis is in pharmaceutical industry.

For these reasons, a wide variety of derivatives of the purine and pyrimidine bases and of the barbiturates have been synthesized. Many of them were crystallized and investigated by X-ray diffraction techniques [522].

Hydrogen atom positions were not always determined. Many of the X-ray analyses of purine and pyrimidine derivatives were carried out prior to 1970, and the crystals were generally poor. In consequence, hydrogen atomic positions were not reported for the majority of these crystal structures, and there are very few neutron diffraction studies. This is not a serious problem for $>$N−H and exocyclic, π-conjugated NH$_2$ groups because hydrogen atom positions can be calculated on the basis of the covalently bonded atoms. However, it can give rise to uncertainty concerning keto/enol and amino/imino tautomerism of the bases. It also makes ambiguous the assignments of bonds originating at donor groups with rotational freedom, such as −O−H, nonconjugated −NH$_2$, and water molecules.

15.1 Bases Are Planar and Each Contains Several Different Hydrogen-Bonding Donor and Acceptor Groups

The more important molecules in the pyrimidine and purine class of heterocyclic compounds are the naturally occurring and chemically modified constituents of the nucleic acids, and the barbiturates.

1. *The four principal bases of the nucleic acids* are uracil and cytosine, which are derivatives of pyrimidine, and adenine and guanine, which are derived from the purine heterocycle (Fig. 15.1). In the nucleic acids, ribose (in ribonucleic acid,

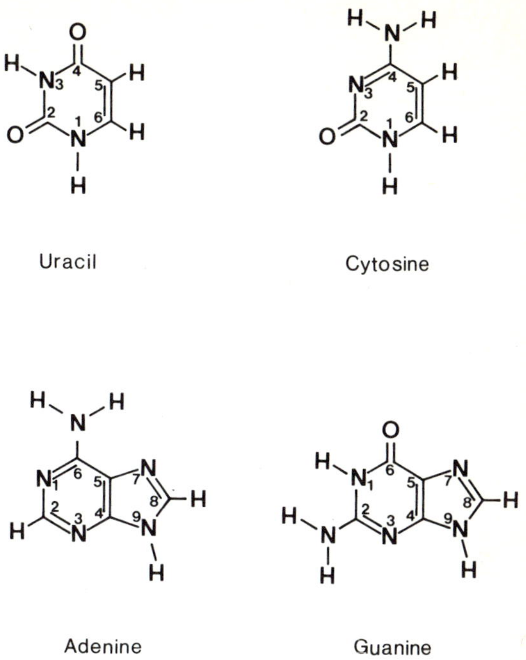

Uracil

Cytosine

Adenine

Guanine

Fig. 15.1. The four bases found in the nucleic acids. Uracil occurs in RNA and is substituted by the analogous thymine (5-methyluracil) in DNA. Uracil is the keto tautomer of 2,4-dihydroxy pyrimidine with two donors N(1)H, N(3)H and two acceptors O(2), O(4). Cytosine is the keto tautomer of 4-amino, 2-hydroxy pyrimidine with three donors N(4)H$_2$, N(1)H and two acceptors O(2), N(3). Adenine is 6-aminopurine with three donors N(6)H$_2$, N(9)H and three acceptors N(1), N(3), N(7). Guanine is the keto tautomer of 2-amino, 6-hydroxy purine with three donors N(2)H$_2$, N(9)H and three acceptors, N(3), N(7), O(6). In the nucleosides, pyrimidine N(1) and purine N(9) are substituted by ribose or deoxyribose (see Fig. 17.1)

RNA) and 2'-deoxyribose (in deoxyribonucleic acid, DNA) are attached at pyrimidine N(1) and at purine N(9) positions.

2. *Modified nucleic acid bases.* These are derived from the purines and pyrimidines given in Fig. 15.1. They either occur naturally, mostly as rare (minor) components of the transfer RNA's [523], or they are synthesized as mimics of or substitutes for the nucleic acid bases. They are of considerable biochemical and pharmaceutical interest, but we can consider only some of the more important compounds (Fig. 15.2).

3. *The barbiturates* differ from the pyrimidines in having the conjugated ring system with sp^2 carbon and nitrogen atoms interrupted by a sp^3 carbon atom at position 5 (see Fig. 15.3). These compounds have drug activity when the total number of carbon atoms in R and R' is four or more. The simplest barbiturate is barbital (veronal) with R=R'=C$_2$H$_5$.

All these bases derived from pyrimidine and purine heterocycles are planar because the ring systems are conjugated. The exocyclic carbonyl and amino groups

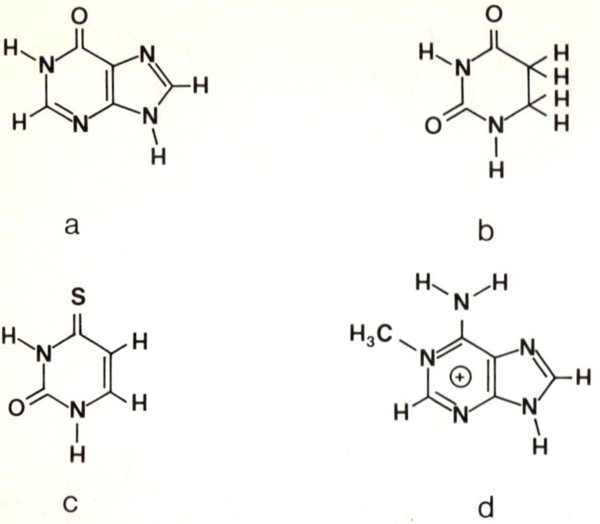

Fig. 15.2a–d. Formulas of some modified bases occurring naturally in transfer RNA's. **a** Hypoxanthine, a guanine derivative lacking the 2-amino group; **b** dihydrouracil and **c** 4-thiouracil, two derivatives of uracil; **d** 1-methyladenine, a positively charged derivative of adenine

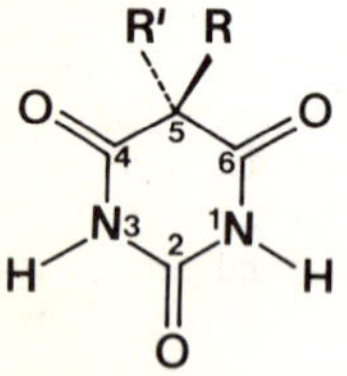

Fig. 15.3. Structural formula of barbiturates

are also involved in this conjugation. In consequence, the C=O and C–NH$_2$ bond lengths have the characteristics of partial double bond order. This is consistent with the resonance forms according to the principles of organic chemistry [522].

The conjugated amino group is *not* a hydrogen bond or proton acceptor. The double bond character of the C–NH$_2$ groups is important, because it restricts rotation of the NH$_2$ group. The NH$_2$ group is a good hydrogen-bond donor, with negligible acceptor properties. This poor basic character of the base amino groups is also reflected in the protonation behavior. The amino groups are *never* protonated (except for *very* extreme acidic conditions), but the free ring nitrogen atoms are: N(3) in cytosine, N(1) in adenine and N(7) in guanine. The N(3) of adenine and of guanine are also (weak) hydrogen-bond acceptors and, in general, they are not protonated.

15.2 Many Tautomeric Forms Are Feasible But Not Observed

Theoretically, the purine- and pyrimidine-based nucleic acid constituents and the barbiturates have the potential to occur in several tautomeric forms of the keto/enol and amino/imino type where the aromatic character of the six-membered pyrimidine ring is fully or, as in the barbiturates, partially retained, as illustrated in Fig. 15.4. In these molecular species, which are all feasible on the basis of organic chemical considerations, the hydrogen-bonding donor/acceptor properties of the functional amino, imino, enol and keto groups vary considerably, being donor in one form and acceptor in the other.

There are some additional possibilities for proton tautomerism with adenine, guanine, and cytosine (Fig. 15.5), where one of the protons can move from one ring nitrogen atom to another one. This type of tautomerism is restricted to the free bases and cannot occur in the nucleosides where pyrimidine N(1) and purine N(9) are substituted by the furanose moiety.

The bases occur only in the canonical tautomeric forms [524, 525]. In both polar and nonpolar environments in the solid and solution states, the canonical forms of the bases illustrated in Fig. 15.1 are the *only* ones that have been observed. There is some spectroscopic NMR evidence for the occurrence of the nucleic acid bases in other tautomeric forms in solution, but only at or below the 0.01% level. This concentration is very difficult to measure, if at all possible [123, 526].

The keto/enol or amino/imino equilibrium can be shifted, however, if the bases are chemically modified as in isoguanine (2-hydroxy,4-amino-purine). In this molecule, the tautomeric equilibrium depends largely on the dielectric constant of the solvent [525].

In consequence, the donor/acceptor sites of nucleic acid bases are clearly defined and not influenced by tautomeric changes. It is one of the fundamental principles in biology that molecules which are involved in recognition processes have clearly defined hydrogen-bonding donor/acceptor properties. It is with this understanding that nature has selected the purine and pyrimidine derivatives adenine, guanine, cytosine, uracil, and the analogous thymine as suitable molecules to carry the genetic information inscribed in the nucleotide sequence in DNA, as described in Part III, Chapter 20. *Hydrogen bonding is the key feature in the biological information transfer mechanisms by the nucleic acids.*

15.3 π-Bond Cooperativity Enhances Hydrogen-Bonding Forces

In the pyrimidines and purines, the major hydrogen-bonding donor groups are $-NH_2$ and $>NH$, and the major acceptor groups are $C=O$ and $>N$. The $-NH_2$ and $>NH$ groups invariably function as donors in crystal structures. There are a few examples where $>N$ atoms are not acceptors, and even less when $>C=O$ groups are not engaged in hydrogen bonding. Because all of these groups are part of π-conjugated systems, π-bond polarization occurs upon hydrogen-bond formation. As discussed in Part I A, Chapter 2.6, this is a characteristic feature of this

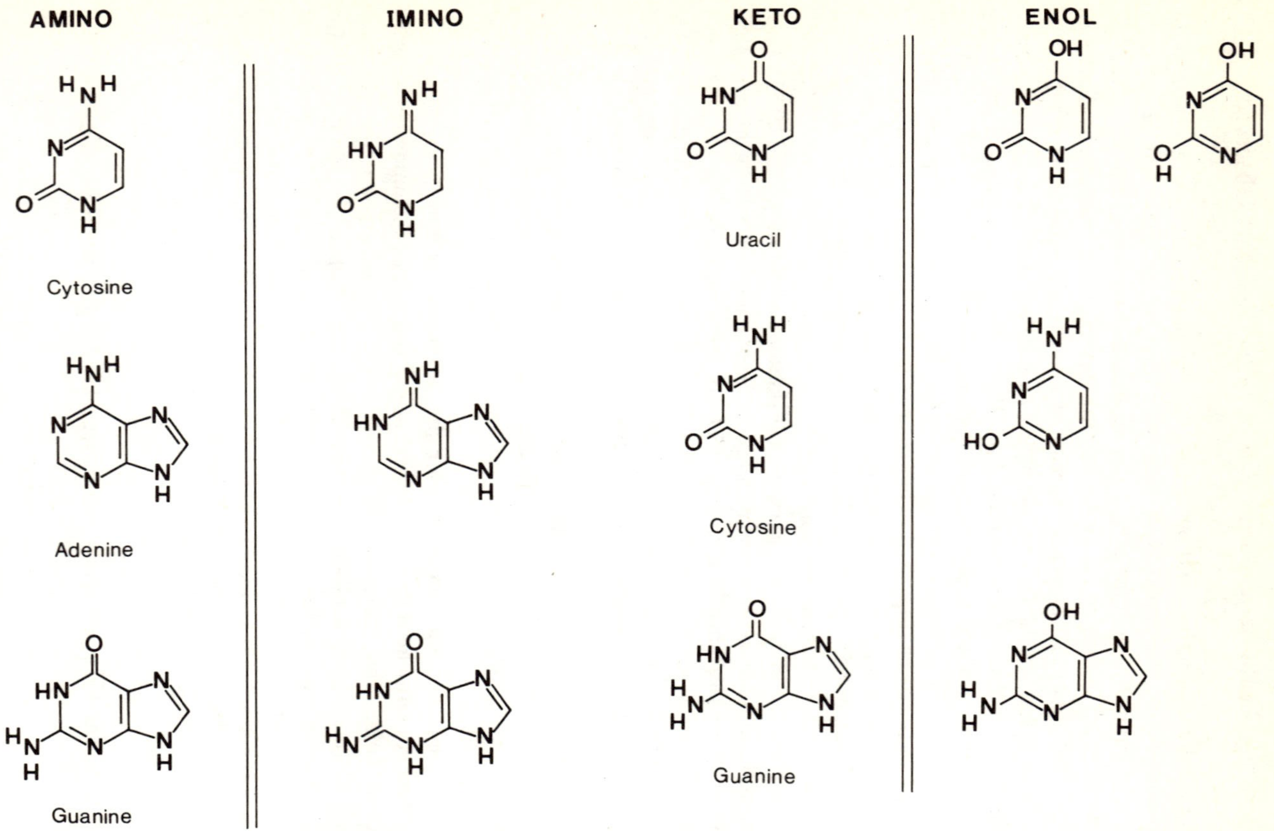

Fig. 15.4. Keto/enol and amino/imino tautomeric forms of uracil, cytosine, adenine, guanine. The keto and amino forms predominate, the enol and imino forms are only present in equilibrium at 0.01% or less

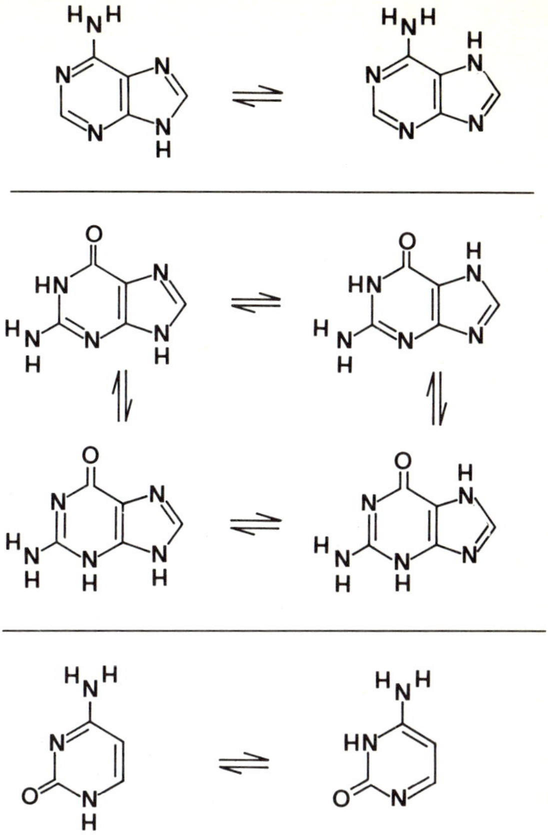

Fig. 15.5. Forms of prototropic tautomerism in (*top*) adenine, (*center*) guanine, (*bottom*) cytosine

class of compounds and gives rise to π-cooperativity, which enhances the stability of hydrogen bonds. It leads, if no other molecules than the bases are involved, to cyclic or catenary configurations, which are similar to those of the carboxylic acids and the peptides, as shown in Figs. 15.6 and 2.7.

Where cyclic hydrogen-bond configurations occur, they are referred to as *base pairs*. They can be of the *homo* type, if the bases self-associate, or of the *hetero* type, if different bases interact with each other. The hydrogen-bonded base pairs can form isolated dimers and in many cases this dimerization is extended to two-dimensional ribbon patterns or to three-dimensional networks of hydrogen bonds (see Chap. 16).

15.4 General, Non-Base-Pairing Hydrogen Bonds

There is a considerable variety of hydrogen bonds in the crystal structures of purine and pyrimidine bases, far more than in the carbohydrates and amino acids.

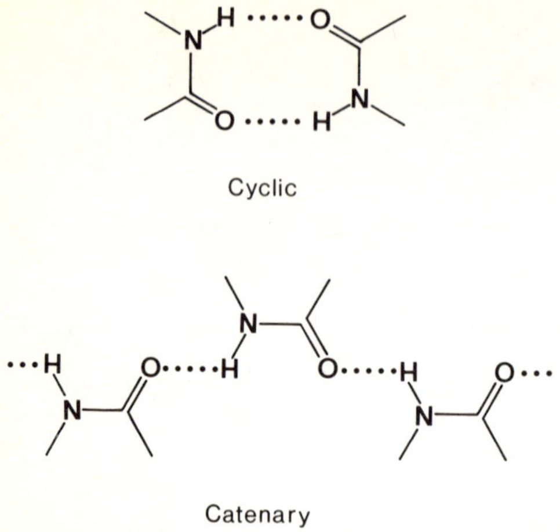

Cyclic

Catenary

Fig. 15.6

The percentage occurrence of different hydrogen bonds in a sample of 463 hydrogen bonds from 185 crystal structures is given in Table 7.14; their metrical characteristics are described in Part IB, Chapter 7.

The proportion of three-center bonds in the pyrimidine, purine, and barbiturate crystal structures of about 30% is comparable to the number observed in the carbohydrate crystal structures (Table 2.3). There is also a small proportion, about 1%, of four-center bonds. A particular feature of the three-center bonds in these crystal structures is the occurrence of *chelation*, where both acceptor atoms are covalently bonded to the same atom(s), as in Fig. 15.7.

These three-center bonds are invariably unsymmetrical with one strong bond (H···O, 1.9 to 2.4 Å; N–Ĥ···A, 145–170°) and one weak bond (H···O, 2.4 to 2.8 Å; N–Ĥ···A, 105–135°). These unsymmetrical chelated structures are more a property of the molecular configuration and crystal packing than of the hydrogen bonds. In the absence of crystal packing forces, i.e., in solution, we doubt whether such chelation would be energetically favorable.

The $-NH_2$ group, like the water molecule, has the potential to form bifurcated hydrogen bonds (i.e., Fig. 15.8). This configuration is never observed except in combination with three-center bonds. In these, the two three-center components are always unsymmetrical, with the bifurcated pair being the longer and sometimes close to or even beyond van der Waals contact distances. The two three-center configurations are not coplanar, and make angles of ~50°. Some examples are shown in Fig. 15.9 with long, yet stereochemically acceptable distances.

Intramolecular three-center hydrogen bonding is a characteristic feature in the crystal structures of the majority of the molecules containing the adenine moiety. Because the amino group is π-conjugated with the purine ring system, all atoms are coplanar. This requires that the N(6)H···N(7) distance in adenine is in the range 2.5 to 2.9 Å with an angle of 90° to 125° (Fig. 15.10 and Part IB, Chapter 9). This interaction, although forced by the conjugation, can be considered as the invariable long (minor) component of an intramolecular three-center bond.

General, Non-Base-Pairing Hydrogen Bonds

Fig. 15.7

Fig. 15.8

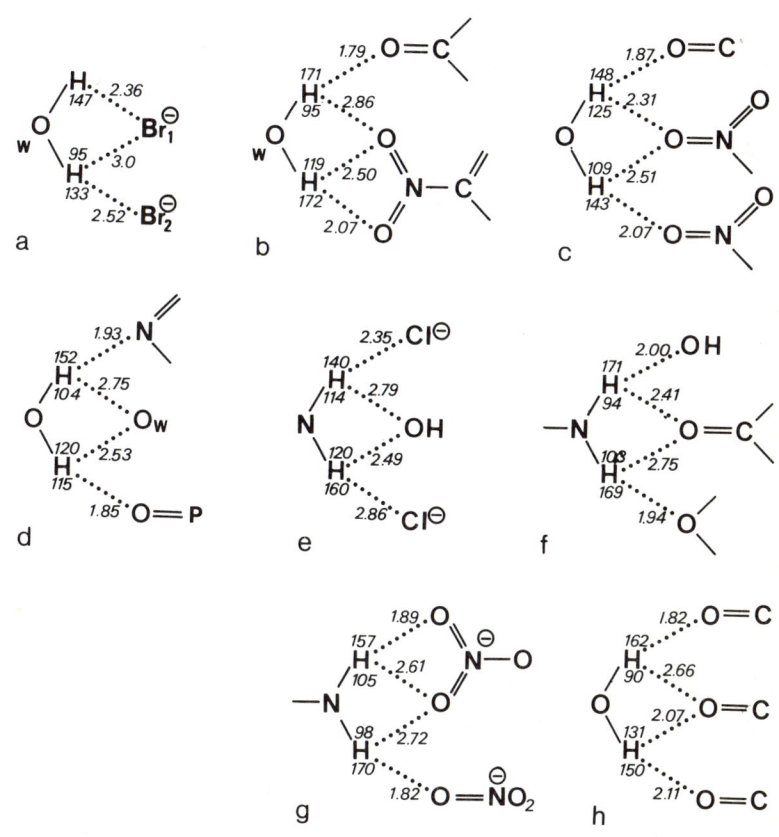

Fig. 15.9a–h. Some hydrogen-bonding configurations involving two three-center bonds. **a** N-[3-(aden-9-yl)propyl]-3-carbamoyl-pyridinium bromide hydrobromide dihydrate [ADENIC]; **b** guanine picrate monohydrate [GUNPIC10]; **c** thiamine picronololate dihydrate [THPROL]; **d** 8,2′-anhydro-8-mercapto-β-D-arabinofuranosyladenine-5′-monophosphate trihydrate [AMAFAP]; **e** α-D-xylofuranosyl cytosine·HCl [XFURCC10]; **f** 3-amino-6-(β-D-ribofuranosyl)-6H-1,2,6-thiadiazine-1,1-dioxide [RBFROX]; **g** adeninium dinitrate [BIDRUB10]; **h** perdeuterated violuric acid [67]; ^{2}H are actually D, see also Table 8.1

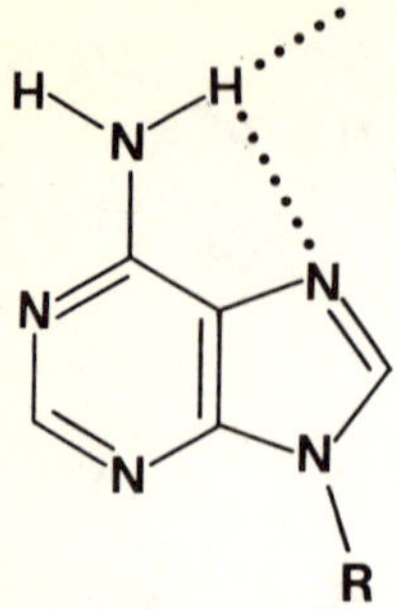

Fig. 15.10. The intramolecular hydrogen bond in adenine is due to the coplanarity of amino group and purine heterocycle

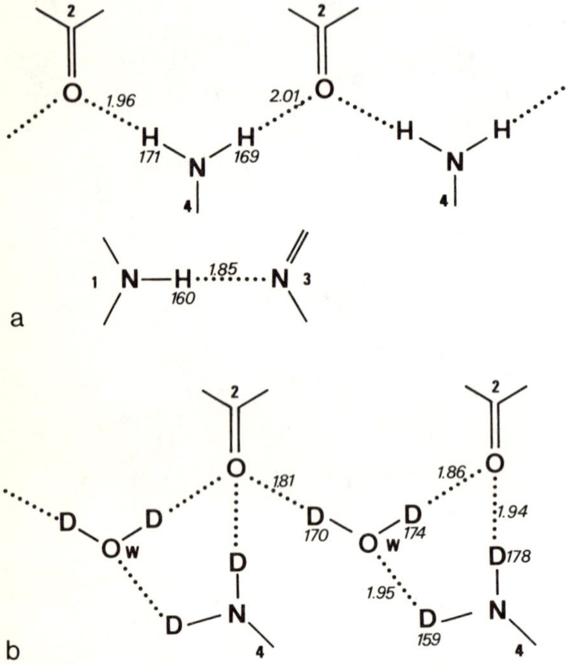

Fig. 15.11. a Cytosine [CYTSIN01]; **b** deuterated cytosine·D$_2$O [CYTOSH]

Hydrogen-bonding patterns occurring in crystal structures of bases are described below, excluding base-pairing hydrogen bonds, which will be discussed in Chapter 16. From a large sample of structural data, we have selected those which display more characteristic motifs. They include localized, finite chains of hydrogen bonds, cyclic arrangements, and infinite chains. The latter appear only when water molecules or halide ions are present, and they involve both π- and σ-bond cooperativity.

General, Non-Base-Pairing Hydrogen Bonds

Fig. 15.12. N-4-amino cytosine [AMCYTS]

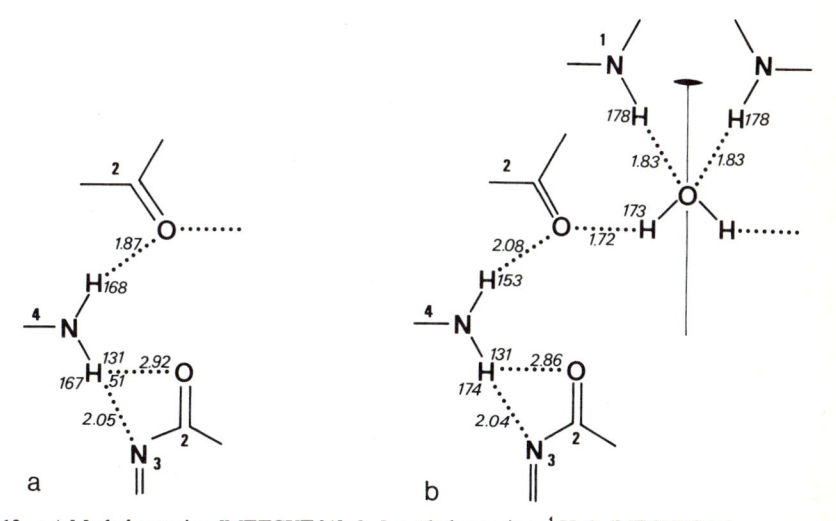

Fig. 15.13. a 1-Methyl cytosine [METCYT01]; **b** 5-methyl cytosine·$\frac{1}{2}$H$_2$O [MECYTS01]

The crystal structure of *cytosine* [CYTSIN01] exhibits a typical catenated hydrogen-bond structure linking the NH$_2$ and C=O groups. The NH···N$\big\langle$ bond forms a separate link, the cytosine molecules are thereby self-associated to form a ribbon pattern (Fig. 15.11a).

A more complex variation of the structure is observed in *cytosine monohydrate* [CYTOSH], which was studied as the deuterated compound by neutron diffraction. It contains cyclic trimers linking −ND$_2$, D$_2$O and C=O groups (Fig. 15.11b).

In *N-4-amino cytosine* [AMCYTS], the hydrogen bonding is quite different. It is centered around the triple acceptor property of the C=O group with no base-

Fig. 15.14. 3-Methyl cytosine·½H$_2$O [MCYTSH]

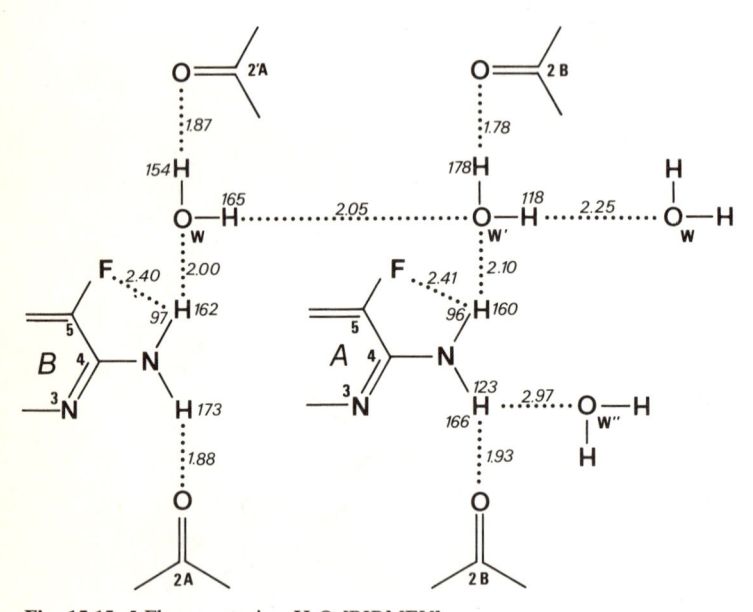

Fig. 15.15. 5-Fluoro cytosine·H$_2$O [BIRMEU]

base self-association. Both hydrogen bonds from the NH$_2$ group are very weak (Fig. 15.12).

Both *1-methyl cytosine* [METCYT01] (Fig. 15.13a) and *5-methyl cytosine hemihydrate* [MECYTS01] (Fig. 15.13b) have a chelated, unsymmetrical three-center bond and very similar crystal structures. The molecules are self-associated to form ribbon structures. The water molecule of the hemihydrate lies on a crystallographic twofold axis.

A related pattern as in 5-methyl cytosine hemihydrate is observed in *3-methyl cytosine hemihydrate* [MCYTSH] (Fig. 15.14). There is no three-center chelated bond because N(3) is methylated.

In *5-fluoro cytosine monohydrate* [BIRMEU] (Fig. 15.15), there are two symmetry-independent molecules which have almost identical hydrogen bonding. The

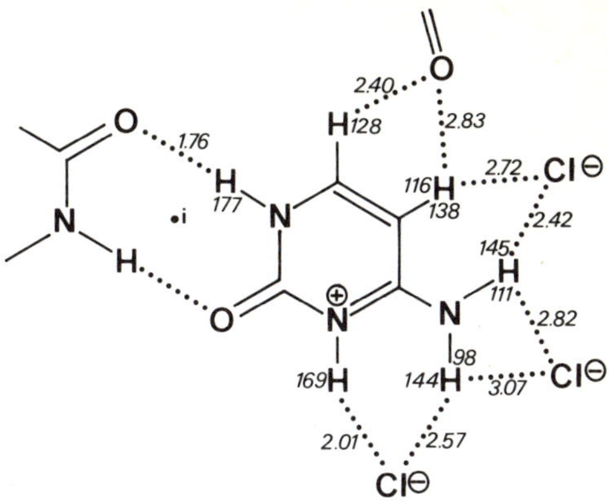

General, Non-Base-Pairing Hydrogen Bonds   243

Fig. 15.16. Cytosine·HCl [CYTOSC]; i, inversion center

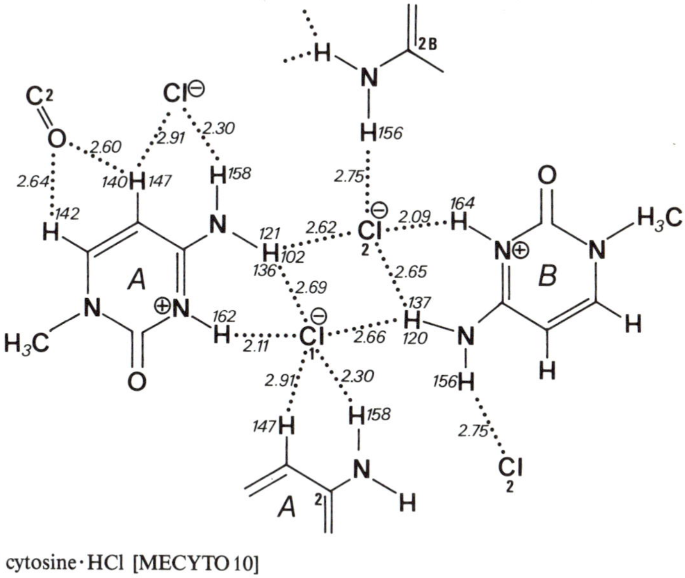

Fig. 15.17. 1-Methyl cytosine·HCl [MECYTO10]

water molecules are arranged in hydrogen-bonded infinite homodromic chains which intersect with short finite chains formed by the functional groups of the organic molecules. The intramolecular H···F distance of 2.41 Å might be regarded as a weak hydrogen-bond interaction to a covalently bonded fluorine atom.

In *cytosine hydrochloride* [CYTOSC] (Fig. 15.16), the four-coordinated chloride ions dominate the structure. The NH$_2$ group forms a bifurcated three-center hydrogen bond to three chloride ions, similar to those in Fig. 15.9. There is evidence for chelating hydrogen bonding from the C(5)–H and C(6)–H groups to the carbonyl C(2)=O oxygen, and C(5)–H forms additionally a three-center bond to a chloride ion. There are separate N–H···O=C cyclic hydrogen bonds.

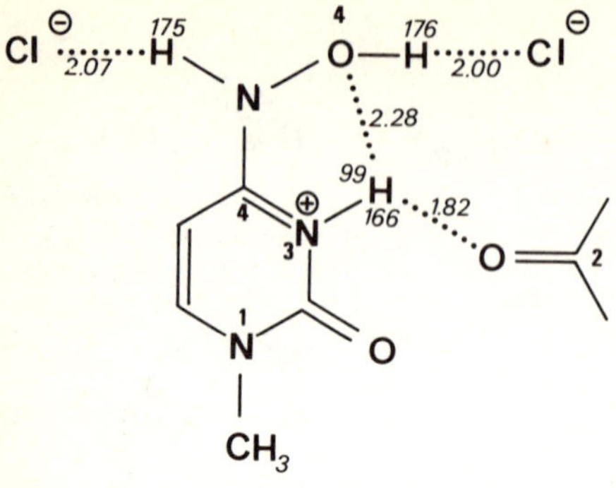

Fig. 15.18. 1-Methyl N-4-hydroxy cytosine ·HCl [MHCYTC]

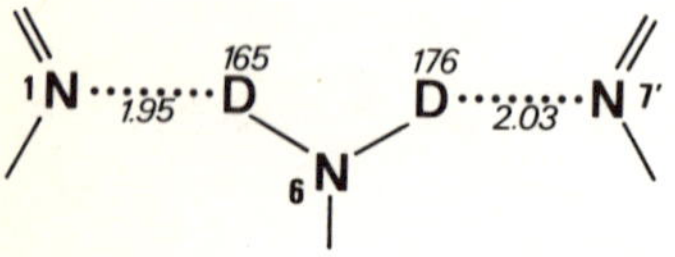

Fig. 15.19. 9-Methyladenine [MEADEN 02]

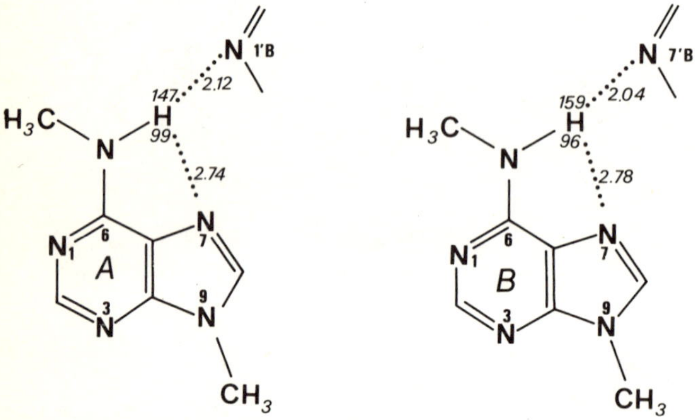

Fig. 15.20. Dimethyladenine [DMADEN 10]

In the *1-methyl cytosine hydrochloride* [MECYTO 10] (Fig. 15.17) crystal structure, there are two symmetry-independent molecules in the unit cell with almost identical hydrogen-bond environments. The NH$_2$ groups form a two-center and a three-center bond to the chloride ions. There is a chelating hydrogen bonding from the adjacent C(5)–H and C(6)–H groups to a carbonyl oxygen C(2)=O which is very similar to that in CYTOSC (Fig. 15.16).

The chloride ion is two-coordinated only in *1-methyl N-4-hydroxy cytosine hydrochloride* [MHCYTC] (Fig. 15.18), and there is an intramolecular

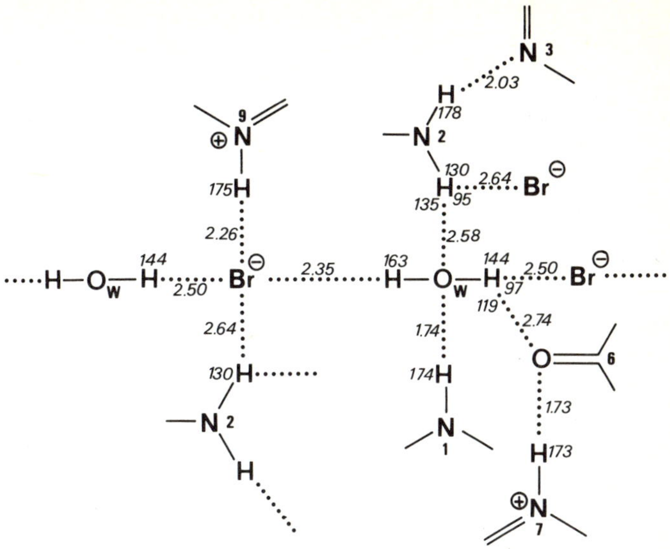

General, Non-Base-Pairing Hydrogen Bonds 245

Fig. 15.21. Guanine·HBr·H₂O [GUANBM]

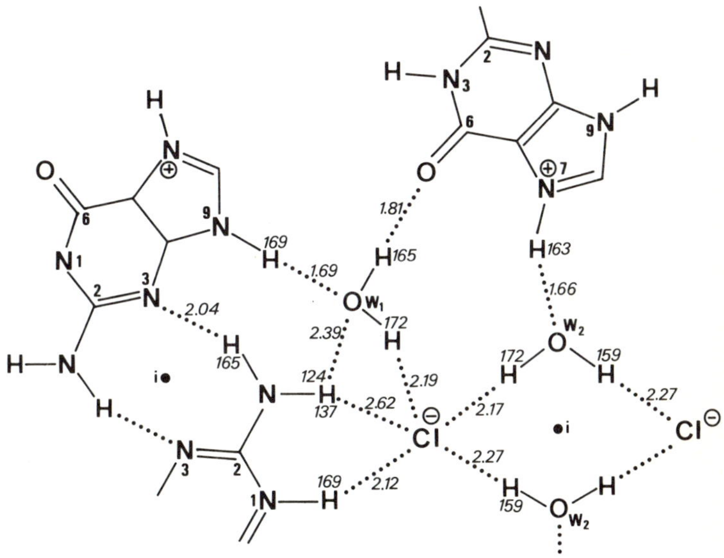

Fig. 15.22. Guanine·HCl·2H₂O [GUANCD]; i, inversion center

$\overset{+}{\underset{\diagup}{N}}$(3)H···O(4) hydrogen bond which stabilizes the hydroxylamino group *cis* to the $\overset{+}{N}$(3)H group.

9-Methyl adenine [MEADEN02] (Fig. 15.19) was studied by neutron diffraction. In this unusually simple hydrogen-bond structure, there is a possible

C(9)H···(N)3 bond with a long H···N distance of 2.52 Å, otherwise N(3) is not an acceptor.

There are two symmetry-independent molecules in *N(6), N(9) dimethyl adenine* [DMADEN10] (Fig. 15.20) with very similar, simple hydrogen bonding. Both have the N(6) methyl group *cis* to N(1), and this conformation gives rise to an intramolecular NH···N(7)⟨ bond as minor componnt of a three-center bond. As with the preceding structure, ⟩N(3) is not an acceptor.

The hydrogen-bonding structure in *guanine hydrobromide monohydrate* [GUANBM] (Fig. 15.21) is a three-dimensional net with both the water molecules and bromide ions four-coordinated. There is a fifth interaction to the bromide ion from C(8)−H, which corresponds to weak hydrogen bonding. It is questionable whether the positive charge is located on N(9)H (as assigned by the authors), or distributed between N(9)H and N(7)H.

In *guanine hydrochloride dihydrate* [GUANCD] (Fig. 15.22), the hydrogen-bond structure is a three-dimensional net. The positive charge appears to be located on N(7). The Cl$^-$ is four-coordinated, and five-coordinated when the minor component from the N(2) amino group is included. One water molecule is three-coordinated, the other four-coordinated.

Chapter 16
Base Pairing in the Purine and Pyrimidine Crystal Structures

As discussed in Chapter 15, the arrangement of the functional groups on adjacent ring positions in the purine and pyrimidine bases is such as to increase the strength of hydrogen bonding due to π-cooperativity. This favors the formation of base-paired dimers and in some cases of trimers and tetramers. These can occur between the same bases (homo- or self-association) and between different bases (hetero-association).

The hetero-base pairs are of the greatest biological importance because they can be of the complementary Watson-Crick type [27, 527]. They have a key role in the information transfer processes discussed in Part III, Chapter 20. To describe the base-pairing geometries more systematically, we consider first the possible combinations between like (homo) bases. They are of some significance in the tertiary interactions of transfer RNA and are observed in the self-aggregation of homopolynucleotides like polyuridylic acid [poly(U)], polycytidylic acid [poly(C)], polyadenylic acid [poly(A)] and polyguanylic acid [poly(G)]. The resulting complexes form double helices as in poly(U)·poly(U) [528, 529], in poly(C)·poly(CH$^+$) [530, 531], where cytosine in the homo base pairs is statistically half protonated at N(3), and in poly(AH$^+$)·poly(AH$^+$) [532], where both strands are fully protonated at N(1). They form triple helices as in poly(A)·2poly(U) or in poly(dA)·2poly(dT) [533], and even quadruple helices as in [poly(G)]$_4$ [534]. The helical geometries of these aggregates will be considered in more detail in Part III, Chapter 20.

There are many possible homo- and hetero-base-pair combinations. They depend on the sequence of the functional groups with donor and acceptor properties at adjacent positions around the heterocyclic ring systems. The possible configurations can be illustrated in matrix form where the donor/acceptor groups of one of the interacting molecules are given in the rows, and of the other in the columns. Base pairs with only one hydrogen bond, although feasible and probably occurring transiently, are not considered in the following. We have also omitted base pairs with sterically improbable hydrogen-bond geometry. Moreover, we have restricted the discussion to the four naturally occurring bases, uracil (thymine), cytosine, adenine, guanine.

Both donor functions of amino groups are never involved in base-pair dimers. In no case has a base pair been observed, either in crystal structures of monomers or in helical complexes of the polynucleotides, where the amino group utilizes *both* its hydrogen donor functionalities to form two hydrogen bonds with *one* associat-

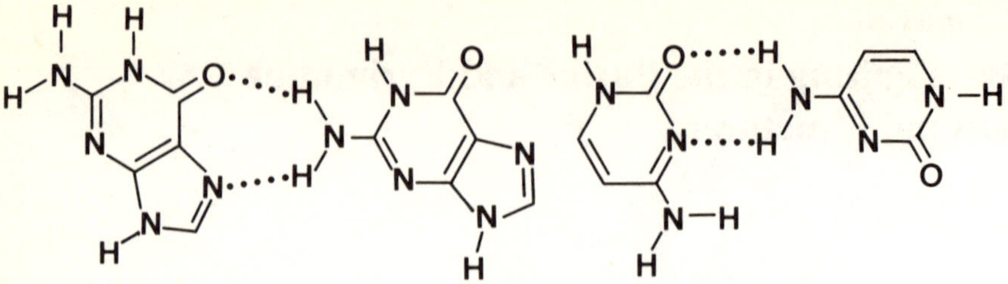

Fig. 16.1. Homo- and hetero-base pairs where the amino group of one base donates both hydrogen bonds to the partner base have not been observed and are unlikely to occur

ed base, as shown in Fig. 16.1. This is probably because in these configurations, the π-cooperativity does not contribute to the hydrogen-bonding energy, and the other configurations where π-cooperativity is involved are favored and formed preferentially. Consequently, base pairs of this type are very unlikely and not considered here.

16.1 Base-Pair Configurations with Purine and Pyrimidine Homo-Association

A large number of base-pair combinations are possible between the four naturally occurring nucleic acid bases in their free, unsubstituted form (see Fig. 16.2). In the following matrix notation, the homo base pairs are described, and for a more complete survey we have also included barbiturate self-association. The base pairs are denoted in a four-character code where the first two letters give the bases involved, U for uracil, C for cytosine, A for adenine, G for guanine, followed by a sequential number M and by an index 2 or 3, describing whether two or three hydrogen bonds are formed in the base pair. The functional groups of base 1 are given in column 1, and of base 2 in columns 3 and higher. In nucleosides, nucleotides, and the nucleic acids, the bases are substituted at N(1) of the pyrimidines and N(9) of the purines; the permitted base-pair combinations are indicated in italics.

Uracil (or *thymine*) has hydrogen-bonding functions on four adjacent positions with two donors (D) N(1)H, N(3)H and two acceptors (A) O(2), O(4). Hydrogen-bonded dimers can be formed in six possible configurations involving two bonds each denoted **UUM²**.

Base-pair configuration[a]			$UU1^2$	$UU2^2$	$UU3^2$	$UU4^2$	$UU5^2$	$UU6^2$
Base 1	Hydrogen bond	Base 2						
N(1)–H	D–A		O=C(2)	O=C(2)	O=C(4)			
C(2)=O	A–D		H–N(1)	H–N(3)	H–N(3)	H–N(3)	H–N(3)	
N(3)–H	D–A					O=C(2)	O=C(4)	O=C(4)
C(4)=O	A–D							H–N(3)

[a] Configurations in italics can occur in the nucleic acids.

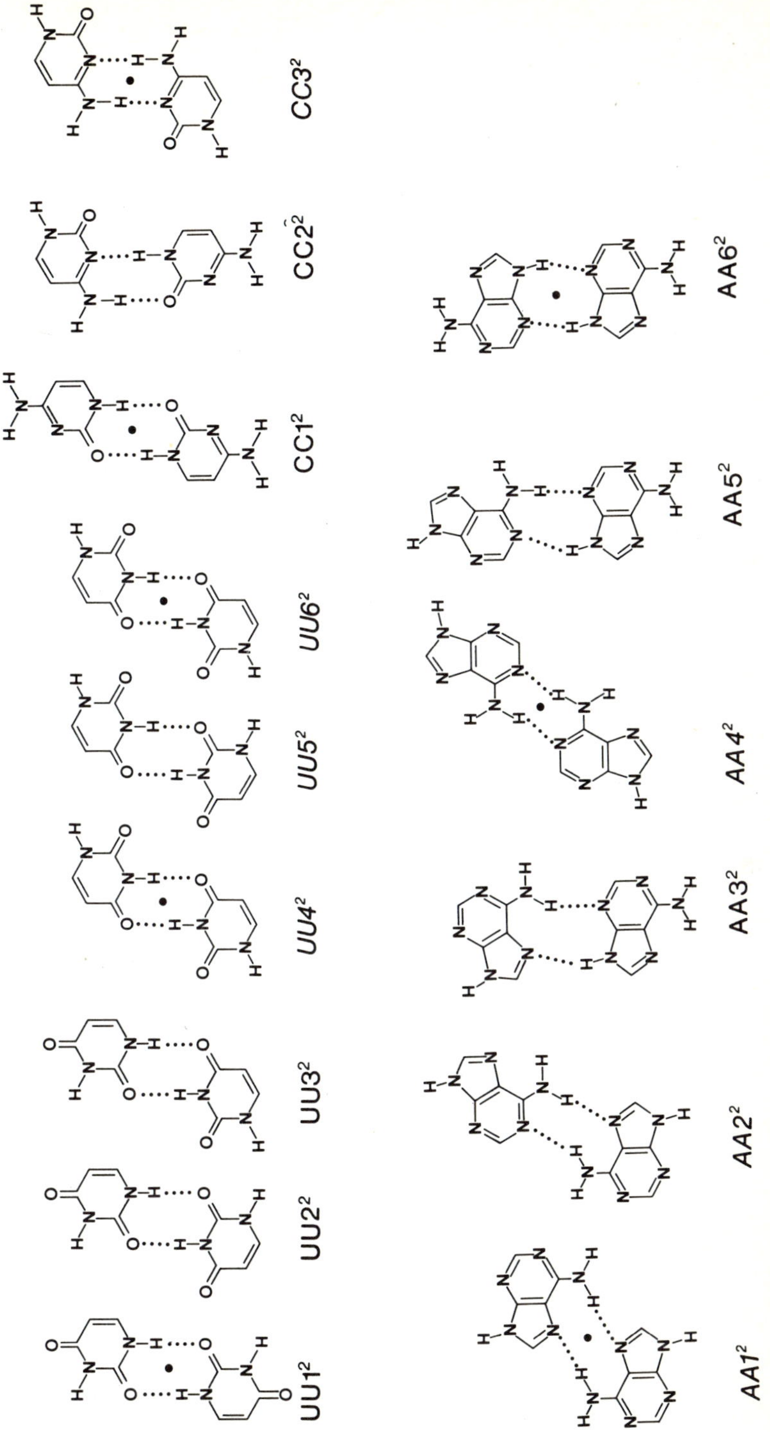

Base-Pair Configurations with Purine and Pyrimidine Homo-Association 249

Fig. 16.2. (continued on next page)

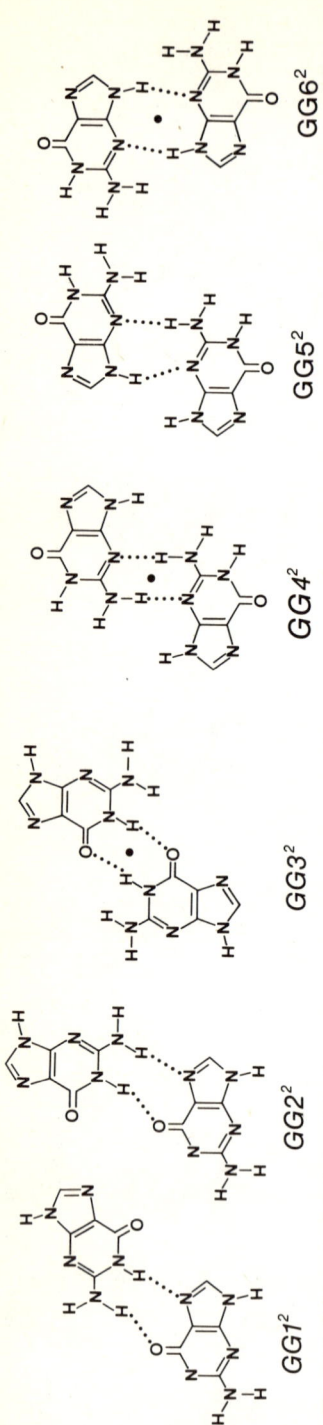

Fig. 16.2. Schematic drawings of the base pairs formed between like (homo) bases, numbered according to the matrix notation. Base pairs in *italics* can occur with nucleosides, nucleotides, and nucleic acids. Centrosymmetric base pairs are indicated by a *dot* at the inversion center

Of these, **UU1^2**, **UU4^2**, and **UU6^2** are centrosymmetrical configurations (see Fig. 16.2).

Cytosine has hydrogen-bonding groups on four adjacent ring positions with three donor and two acceptor functionalities, N(1)H, N(4)H, N(4′)H and O(2), N(3). Hydrogen-bonded dimers can be formed with three different configurations involving two bonds.

Base-pair configuration[a]			CC1^2	CC2^2	CC3^2
Base 1	Hydrogen bond	Base 2			
N(1)–H	D–A	O=C(2)	N(3)		
C(2)=O	A–D	H–N(1)	H–N(4)		
N(3)	A–D			H–N(4)	
N(4)⟨H, H′	D–A, D–A				N(3)

Of these, **CC1^2** and **CC3^2** are centrosymmetrical (Fig. 16.2). There is no three-bond configuration for the cytosine dimer having good stereochemistry except for a centrosymmetrical, hemiprotonated base pair where the proton is shared between the N(3) groups, and hydrogen bonds are: N(4)–H···O(2), N(3)···H$^+$···N(3), O(2)···H–(4)N (Fig. 16.3).

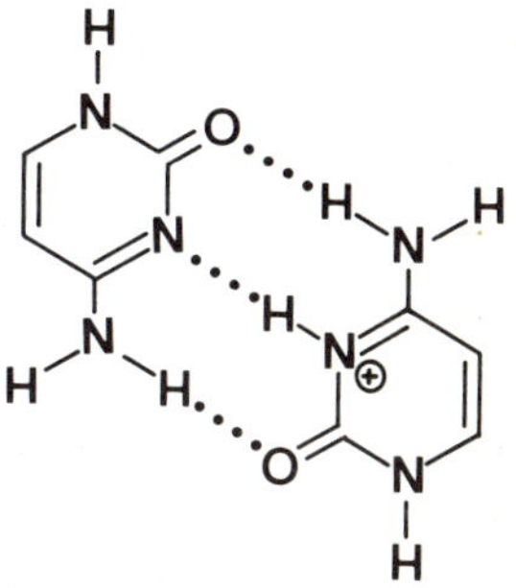

Fig. 16.3. The hemi-protonated base pair formed by cytosine has three hydrogen bonds

Adenine has three donors N(6)H, N(6)H′, N(9)H and three acceptors N(1), N(3), N(7). It cannot form any three-bonded dimers with acceptable geometry. There are six possible two-bond dimer configurations.

Base-pair configuration[a]			$AA1^2$	$AA2^2$	$AA3^2$	$AA4^2$	$AA5^2$	$AA6^2$
Base 1		Hydrogen bond	Base 2					
N(7)		A–D	H–N(6)	H'–N(6)	H–N(9)			
H		D–A	N(7)	N(1)	N(3)			
N(6)〈								
H'		D–A				N(1)	N(3)	
N(1)		A–D				H–N(6)	H–N(9)	
N(3)		A–D						H–N(9)
N(9)–H		D–A						N(3)

Of these, $AA1^2$, $AA4^4$ and $AA6^2$ are centrosymmetrical configurations (see Fig. 16.2).

  Guanine is potentially the most versatile of the bases in its hydrogen-bond configurations. It has four donor and three acceptor functions, with donors N(1)H, N(2)H, N(2)H', N(9)H, and acceptors N(3), O(6), N(7). There is no base-pair configuration with three hydrogen bonds, but six dimer configurations with two-bond interactions.

Base-pair configuration[a]			$GG1^2$	$GG2^2$	$GG3^2$	$GG4^2$	$GG5^2$	$GG6^2$
Base 1		Hydrogen bond	Base 2					
N(7)		A–D	H–N(1)	H–N(2)				
C(6)=O		A–D	H–N(2)	H–N(1)	H–N(1)			
N(1)–H		D–A			O=C(6)			
H		D–A						
N(2)〈								
H'		D–A				N(3)	N(3)	
N(3)		A–D				H'–N(2)	H–N(9)	H–N(9)
N(9)–H		D–A						N(3)

Of these $GG3^2$, $GG4^2$, $GG6^2$ are centrosymmetrical configurations (see Fig. 16.2). The configuration N(1)H···N(3), N(2)H···O(6) is not possible because of interference between N(2)H$_2$ and N(9)H.

  The barbiturates can form many self-associated structures (see p. 253). Because the donor/acceptor capabilities alternate, acceptor O(6), donor N(1)H, acceptor O(2), donor N(3)H, acceptor O(4), there are ten different configurations, of which **BB1^2, BB5^2, BB8^2, BB10^2** are centrosymmetrical.

  Some of the hydrogen-bonded dimer configurations described in the matrices are observed in the crystal structures of the purines and pyrimidines, but this is by no means the general rule. This may be because the number of crystal structures in which self-base pairing occurs is relatively small compared with the number of possibilities. Certain arrangements appear to be particularly favored for reasons

Base-pair configuration	BB1²	BB2²	BB3²	BB4²	BB5²	BB6²	BB7²	BB8²	BB9²	BB10²
Base 1	Hydrogen bond									
C(6)=O	A−D	H−N(1)	H−N(1)	H−N(3)	H−N(3)					
N(1)−H	D−A	O=C(6)	O=C(2)	O=C(2)	O=C(4)					
C(2)=O	A−D					O=C(2)	O=C(2)	O=C(4)		
N(3)−H	D−A					H−N(1)	H−N(3)	H−N(3)	H−N(3)	H−N(3)
C(4)=O	A−D								O=C(2)	O=C(4)
									O=C(4)	H−N(3)

Table 16.1. Experimentally determined and calculated dipole moments of bases

Compound	Dipole moment		
	Calculated		Measured dipole (D)[b]
	Magnitude (D)	Direction (°)[a]	
Adenine	2.9	64°	3.0
Uracil	4.6	36°	3.9
Guanine	7.5	−31°	
Cytosine	7.6	102°	

[a] Directions of calculated dipole moments are from + to −, with the angle to the axis N(1)−C(4) in pyrimidines and C(4)−C(5) in purines counted counterclockwise if structural formulas are drawn such that atom C(2) is left with regard to N(1) in pyrimidines and N(9) in purines. After [535].
[b] For 9-methyladenine and 1,3-dimethyluracil.

of crystal packing that are not obvious merely by observing the hydrogen-bond geometry of the dimerization. When dimerization occurs, it is frequently integrated in a ribbon or layer arrangement of hydrogen bonds which is part of a more extended two- or three-dimensional framework. In the hydrates or hydrochlorides, which contain three- or four-coordinated water molecules or chloride ions, the dimers are generally part of a three-dimensional network structure.

In the crystal structures, homo base pairs with centrosymmetrical configuration are favored. This general statement is true for all the four different bases of pyrimidine and purine type. The individual bases exhibit considerable dipole moments, as shown in Table 16.1, and we assume that the preference for base pairs with a center of symmetry is due to the antiparallel orientation of the dipole moments in this particular arrangement. This results in a favorable cancellation of the total electric field over the crystal volume.

Of the six possible *uracil-uracil (thymine-thymine)* configurations, five have been observed experimentally.

The centrosymmetrical $UU1^2$ configuration (Fig. 16.2) occurs frequently as in thymine [THYMIN]; thymine monohydrate [THYMMH]; Dihydrouracil [DHURAC10]; dihydrothymidine [DHTHYD10]; 6-methyluracil-5-acetic acid [MURCAC]; 5,6-dihydro-2-thiouracil [DHTURC]; 6-methyl-5,6-dihydrouracil [MDHURC10]; 2,4-dithiouracil [DTURAC].

This is equivalent to the corresponding $CC1^2$ configuration in cytosine (Fig. 16.2), which is least commonly observed with this base. It is also interesting to note that the same type of molecular association $UU1^2$ occurs in 1,3-dimethyl uracil [DMURAC] (Fig. 16.4), where the NH $\cdots$ O=C bonds are replaced by an unusual

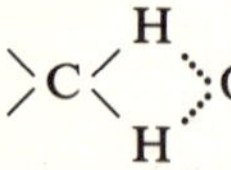

$$\begin{matrix} H \\ >C< \\ H \end{matrix} \cdots O(2)$$

bifurcated hydrogen-bond configuration.

The $UU2^2$ configuration is found in the 1:1 complex of thymine and p-benzoquinone [THYBNZ10], and $UU3^2$ occurs in 5-fluorouracil [FURACL].

The centrosymmetrical $UU4^2$ configuration is frequently observed as in thymine [THYMIN]; thymine monohydrate [THYMMH]; 1-methylthymine [METHYM01] (Fig. 16.5); dihydrouracil [DHURAC10] (together with the $UU1^2$ base pair); 1-methyl-4-thio-uracil [MTURAC]; 1-methylthymine trans-anti dimer [MTHYMD]; 5-nitro-6-methyl uracil [NIMURC10]; 1-methyl-5-fluoro-5-methoxy-5,6-dihydrouracil [MFXHUR]; 1-ethyl-5-bromouracil (form II) [EBURCL11]; 1-methyl-5,6-dihydro-uracil [MHURAC].

The uracil-uracil base-pair configuration $UU5^2$ (Fig. 16.2) is not observed with the monomer bases, but in the modified nucleoside 5-dimethylaminouridine [DMURID], and it has been proposed to occur in the double helix formed by the

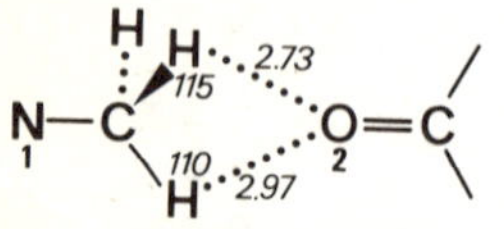

Fig. 16.4. The crystal structure of 1,3-dimethyluracil (DMURAC) is not stabilized by N–H $\cdots$ O=C hydrogen bonds but by C–H $\cdots$ O=C interactions

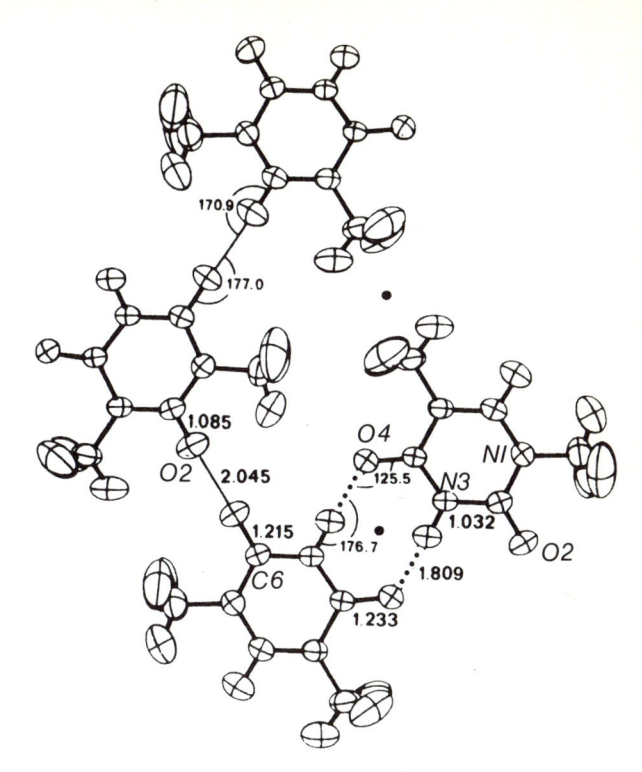

Fig. 16.5. The geometry of the centrosymmetrical base pair of type $UU4^2$ in the crystal structure of 1-methylthymine is based on neutron diffraction data [METHYM01]. Besides $N(3)-H\cdots O(4)$ hydrogen bonds (*dotted*), there is a $C(6)-H\cdots O(2)$ hydrogen bond (*thin line*). Relevant distances given in (Å), angles in (°), standard deviations are 0.003 Å for distances, 0.3° for angles, $X\cdots A$ distances are $N(3)\cdots O(4)$, 2.841(2) Å; $C(6)\cdots O(2)$, 3.121(2) Å. The *dot* indicates the center of symmetry

polymers polyuridylic acid [poly(U)·poly(U)] and poly-2-thiouridylic acid [poly(s²U)·poly(s²U)] [528, 529]. In contrast, the centrosymmetrical $UU6^2$ base pair is frequently found as in uracil [URACIL]; 1-methylthymine and its Ag complex [AGMTHY]; 5-ethyl-6-methyl-uracil [EMURAC]; 1-methyl-4-thiouracil [MTURAC]; 2,4-dithiouracil [DTURAC]; 1-ethyl-5-bromouracil (form I) [EBURCL10]; 1-(2-carboxyethyl)-uracil [BIYRIK]; potassium thyminate trihydrate [KTHYMT].

The hydrogen-bonded *cytosine-cytosine* dimers illustrated in Fig. 16.2 are observed in the following compounds:

The centrosymmetrical $CC1^2$ base pair occurs in cytosine hydrochloride ([CYTOSC]; N(3) is protonated and bound to the chloride ion), and in 5-bromocytosine-N-tosyl-L-glutamic acid [BCYTGA].

The configuration $CC2^2$ is observed in cytosine [CYTSIN01]; cytosine monohydrate [CYTOSM11]; thiocytosine [THCYTO10]; 5-bromocytosine-hemi-dioxane solvate [BRCYTS].

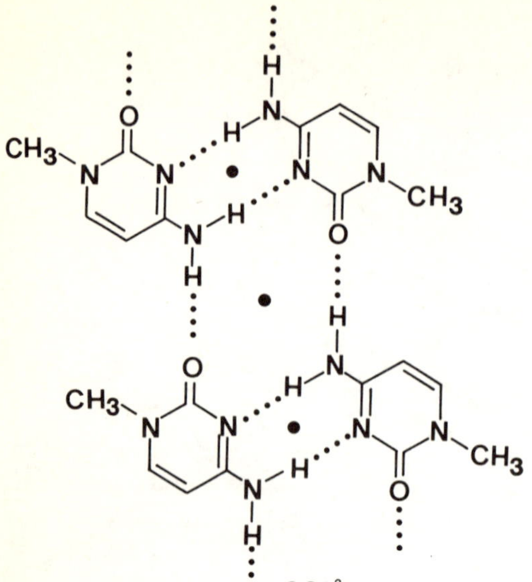

Fig. 16.6. The base pairs of type $CC3^2$ in the crystal structure of 1-methylcytosine [METCYT01] have a center of symmetry (see *dots*), and are part of an extended two-dimensional ribbon

The centrosymmetrical $CC3^2$ base pair is found as in 1-methylcytosine [METCYT01]; 5-methyl-cytosine hemihydrate [MECYTS01]; 2-aminopyrimidine [AMPYRM01]; 5-bromocytosine-phthaloyl-D,L-glutamic acid [BRCPDG] and in 5-bromocytosine-N-tosyl-L-glutamic acid [BCYTGA].

In 1-methylcytosine [METCYT01], the centrosymmetrical $CC3^2$ dimers are further associated to form the ribbon structure, shown in Fig. 16.6.

The crystal structure of isocytosine [ICYTIN] is interesting since it is a 1:1 complex of the 1,4- and 3,4-dihydro tautomers and has both triple and double hydrogen-bond self-association (Fig. 16.7). As outlined in Chapter 15, this tautomerism is *not* observed in the naturally occurring nucleic acid bases.

If cytosine is hemiprotonated, it can form a centrosymmetrical base pair with three hydrogen bonds (Fig. 16.3), as observed in the crystal structures of cytosine-5-acetic acid [CYACET], of bis 1-methylcytosine hydroiodide ([MCYTIM10] and [530]), and in the polynucleotide duplex poly(C) poly(CH$^+$) [531].

In *adenine-adenine* base pairs, four of the six possible configurations are realized in crystal structures (Fig. 16.2):

The centrosymmetrical $AA1^2$ base pair is frequently found as in adenine hydrochloride [ADENCH02]; 1,9-dimethyl-adeninium chloride [MADEND]; the 1:1 complex of 9-ethyl-8-bromoadenine with 1-methyl-5-bromouracil [BURBAD]; poly(AH$^+$)·poly(AH$^+$) and adenylyl-3′,5′-adenylyl-3′,5′-adenosine hexahydrate [APAPAD10] where adenine is protonated at N(1) [532]. The unsymmetrical $AA2^2$ base pair occurs in 3-adenin-9-yl-propionamide [ADPROP] and in 9-ethyladenine in complex with indole [EADIND].

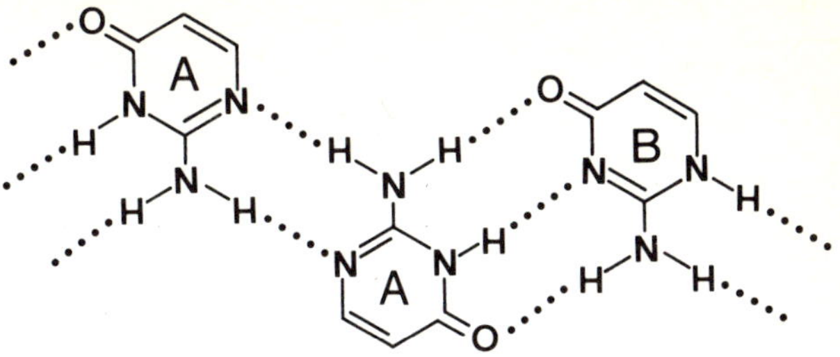

Fig. 16.7. In the crystal structure of isocytosine [ICYTIN], the base adopts two different tautomeric forms *A*, *B*. This cannot occur with the natural bases found in the nucleic acids

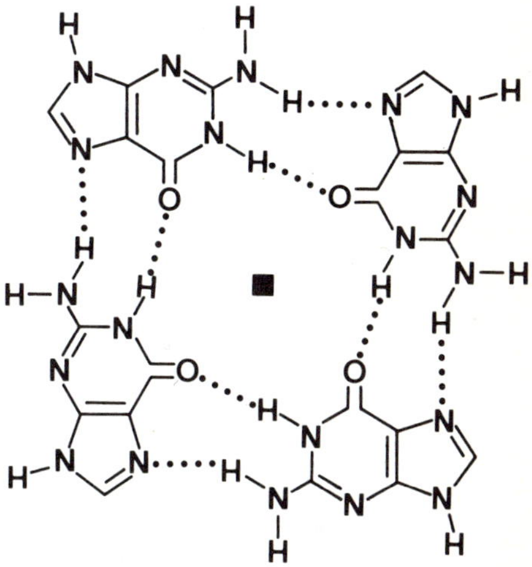

4 x GG2^2

Fig. 16.8. Schematic description of the *GG2*^{2}-type guanine base tetramer formed in the quadruple helix [poly(G)]$_4$. The bases are related by a fourfold rotation axis indicated by the black square

The centrosymmetrical *AA4*^{2} base pair is mostly observed in complexes: 9-ethyl-8-bromo-adenine with 9-ethyl-8-bromo-hypoxanthine [EBAEBH]; 9-methyladenine with 1-methyl-4-thiouracil [SURMAD10] and with 2-thiohydantoin [BIFYOE]; 9-ethyladenine with parabanic acid [EADPBA]; 3-(adenin-9-yl)propiontryptamide [ADPRTR].

The centrosymmetrical **AA6**^{2} base pair is found in crystal structures where adenine is protonated at N(1): adeninium hemisulfate monohydrate [ADESUL]; adeninium phosphate [ADENPH]; bis-adeninium dinitrate monohydrate [ADENOH10], and with N^6-(Δ^2-isopentenyl)-2-methylthioadenine [PMTADN10].

There is no obvious reason for the absence of examples of configurations **AA3**^{2} or **AA5**^{2} except, as noted above, that they are not centrosymmetrical and will have a net dipole moment.

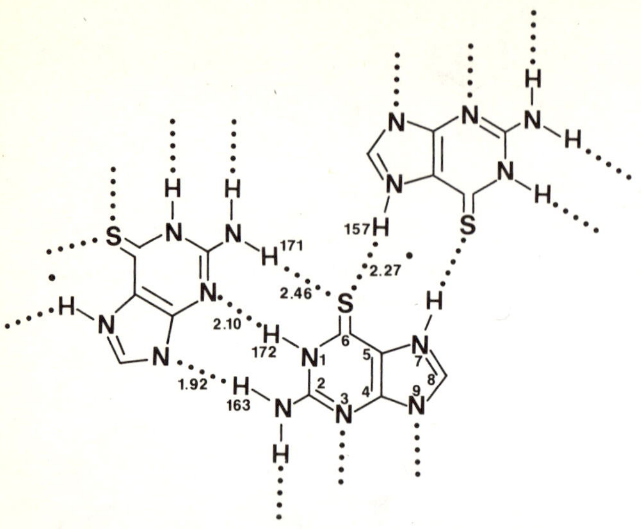

Fig. 16.9. In the crystal structure [THGUAN 10], the unusual N(7)H, N(9) tautomer of 6-thioguanine forms simultaneously two base pairs with two and three hydrogen bonds respectively. Relevant distances (Å) and angles (°) are given for the hydrogen bonds. The associated X···A distances are: N(1)···N(3), 3.053 Å; N(2)···N(9), 2.973 Å; N(2)···S(6), 3.327 Å; N(7)···S(6), 3.303 Å

Although guanine has the greatest versatility for hydrogen-bond self-association, only four of the six possible guanine-guanine base pair configurations have been observed (Fig. 16.2). This may be because relatively few crystal structures of guanine and its derivatives have been studied, i.e., less than 20, as compared with the more than 60 for adenine derivatives.

The observed guanine-guanine base-pair configurations are the unsymmetrical base pair *GG1²* which occurs in 9-ethylguanine [ETGUAN]; the two centrosymmetric pairs *GG4²* and *GG6²* in guanine monohydrate [GUANMH 10] and guanosine dihydrate [GUANSH 10]; the latter base pair is formed together with the *GG5²* configuration.

All other configurations were not hitherto observed except for *GG2²*, which was proposed to occur in the quadruple helix formed by polyguanylic acid [poly(G)]₄ [533]. The base quadruple is of the form depicted in Fig. 16.8.

In 6-thioguanine [THGUAN 10], the unusual N(7)H,N(9) tautomer occurs. It gives rise to a three-bond and a two-bond self-association (Fig. 16.9) not described in the matrix form because in the naturally occurring bases considered there, N(7) is acceptor and N(9)H donor.

The same two-bond association is found together with the *GG4²* configuration in guanine hydrochloride monohydrate and in guanine hydrobromide monohydrate [GUANBM], which are also protonated at N(7).

In guanine hydrochloride dihydrate [GUANCD], water molecules interrupt the bonding of N(7)H···O(6), and the self-association is that described for the *GG4²* configuration (Fig. 16.2).

16.2 Base-Pair Configurations with Purine-Pyrimidine Hetero-Association: the Watson-Crick Base Pairs

The most important hetero base pairs are the adenine-uracil A−U, or thymine A−T, and guanine-cytosine G−C combinations proposed by Watson and Crick in 1953 [27, 527]. Their biological implications are discussed in connection with the nucleic acids in Part III, Chapter 20.

The discovery of these base pairs, which imply the double helical structure of DNA, stimulated a series of crystal structural studies not only of complexes of purines and pyrimidines, but of other complexes involving related molecules and their derivatives. Although we can formulate a large number of possible hetero-combinations in matrix form, as shown below; these complexes are reluctant to crystallize even when there is spectroscopic evidence of hetero-complex formation in solution. This is presumably because self-(homo)-association is energetically more favorable and only in rare cases were crystals of hetero complexes actually formed. Because of their three hydrogen bonds, G−C complexes form and crystallize more readily. There have been many attempts to crystallize the Watson-Crick A−U base pair, but none was successful and it only formed when the dinucleoside phosphate adenylyl-3′,5′-uridine (ApU; [536]) or higher oligomers were crystallized (see Part III, Chapter 20).

As in the homo-base pair series, most hetero-base pair associations involve two hydrogen bonds. Only the Watson-Crick G−C base pair and two guanine-uracil pairs of no biological importance form three hydrogen bonds. In the following, we summarize the possible base-base combinations in matrix notation, with those possible with nucleic acids [where pyrimidine N(1) and purine N(9) are substituted] indicated in italic letters, as before.

Adenine-uracil. There are nine possible configurations (see scheme page 260):
Combination $AU6^2$ is the Watson-Crick base pair, $AU5^2$ is called "reversed" Watson-Crick (Fig. 16.10, left). $AU3^2$ is the base pair discovered by Hoogsteen, and $AU2^2$ is the "reversed" Hoogsteen base pair (Fig. 16.10, right).

Guanine-cytosine. Guanine has an acceptor-donor-donor sequence which is complementary to the donor-acceptor-acceptor sequence of cytosine, so that the three-bond Watson-Crick base pair $GC2^3$ can form (Fig. 16.11). In addition, there are seven more two-bond configurations (see scheme 1, page 262).

Of these 16 possible A−U and G−C combinations, only five have been observed in crystal structures of monomeric bases, mainly due to the difficulty in crystallizing the appropriate complexes referred to above.

For the A−U or A−T combination (Fig. 16.10), the reversed Hoogsteen configuration $AU2^2$ is observed in the crystal structures of 9-methyladenine·1-methyl-4-thiouracil [SURMAD10] and 9-methyladenine·1-methyl-5-bromouracil [MBUMAD10]; with 6% of $AU3^2$, in 9-ethyladenine·1-methyl-5-fluorouracil [ETABFU] and in the nucleoside pair adenosine·5-bromouridine [ADBURM]. The Hoogsteen $AU3^2$ configuration is found in 9-ethyladenine·1-methylthymine [MTHMAD11; neutron diffraction] (Fig. 16.12) with a minor component of $AU2^2$, and the reversed Watson-Crick base pair $AU5^2$ occurs in 9-ethyl-8-bromoadenine·1-methyl-5-bromouracil [BURBAD]. Thus far, the Watson-Crick, A−U,

260 Base Pairing in the Purine and Pyrimidine Crystal Structures

Base-pair configuration[a]		AU1[2]	AU2[2]	AU3[2]	AU4[2]	AU5[2]	AU6[2]	AU7[2]	AU8[2]	AU9[2]
Adenine	Hydrogen bond	Uracil								
N(7)	A–D	H–N(1)	H–N(3)	H–N(3)						
	D–A	O=C(2)	O=C(2)	O=C(4)						
N(6)\H	D–A				O=C(2)	O=C(2)	O=C(4)			
N(1)	A–D				H–N(1)	H–N(3)	H–N(3)			
N(3)	A–D							H–N(1)	H–N(3)	H–N(3)
N(9)–H	D–A							O=C(2)	O=C(2)	O=C(4)

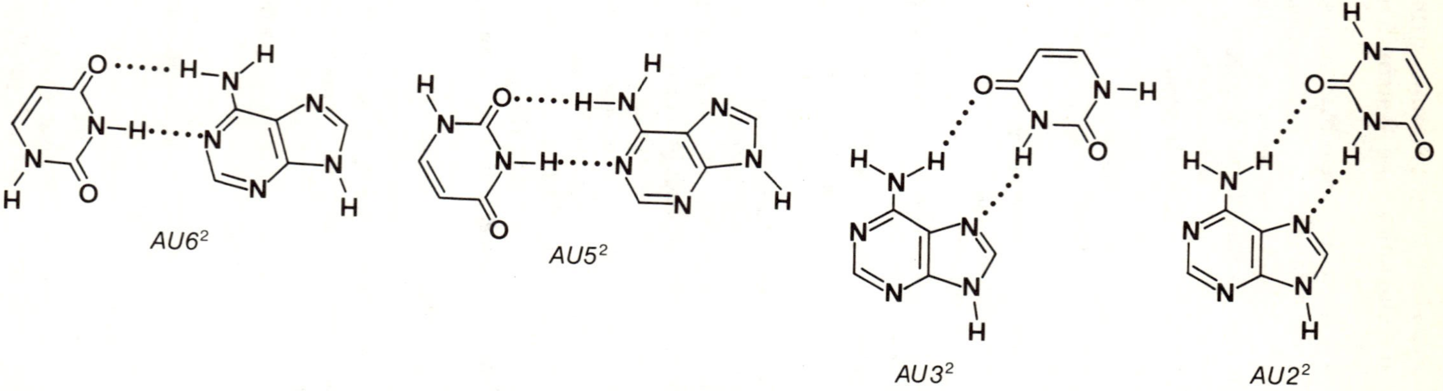

Fig. 16.10. Schematic description of A–U base pairs found in crystals and in double helical nucleic acids. *AU6[2]* Watson-Crick; *AU3[2]* Hoogsteen; *AU2[2]* reversed Hoogsteen; *AU5[2]* reversed Watson-Crick.

Base-Pair Configurations with Purine-Pyrimidine Hetero-Association 261

Fig. 16.11. Watson-Crick $GC2^3$ and reversed Watson-Crick $GC4^2$ base pairs formed by guanine and cytosine

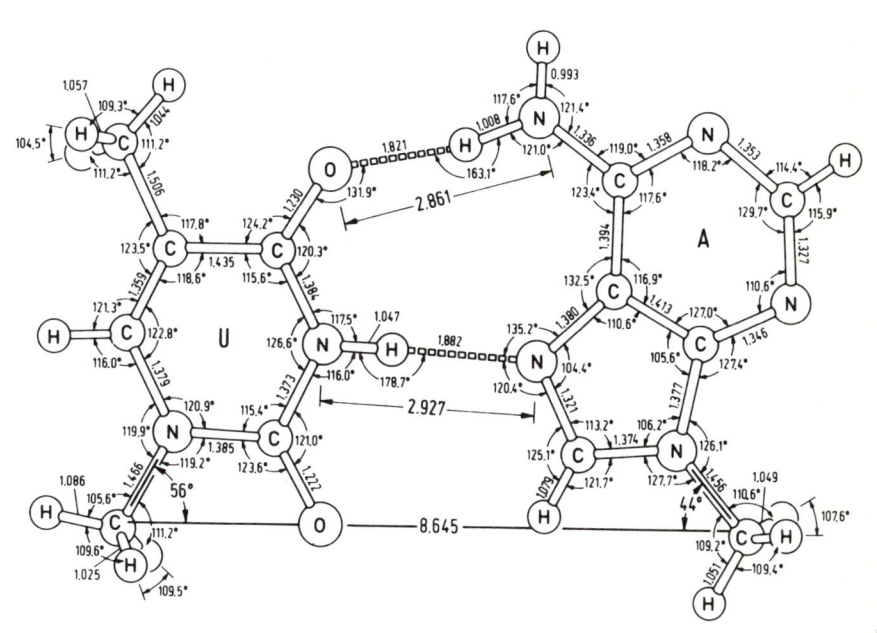

Fig. 16.12. Neutron diffraction analysis of the 9-methyl-adenine·1-methylthymine complex [MTHMAD11] forming a Hoogsteen-type base pair with configuration $AU3^2$. Distances given in (Å), angles in (°). After [537] taken from [522]

base pairing $AU6^2$ has not been observed in an adenine-uracil complex crystal structure, but has been verified in several crystalline oligonucleotide complexes with double helical structure (Fig. 16.13) ([536] and Part III, Chap. 20).

In contrast, the G–C combination occurs *only* in the Watson-Crick configuration $GC2^3$ (Fig. 16.13). This is observed in 9-ethyl-guanine·1-methylcytosine [EGMCYT10], and in 9-ethylguanine·1-methyl-5-bromocytosine [EGUMBC].

262   Base Pairing in the Purine and Pyrimidine Crystal Structures

Scheme 1

Base-pair configuration[a]		GC1²	GC2³	GC3²	GC4²	GC5²	GC6²	GC7²	GC8²
Guanine	Hydrogen bond	Cytosine							
N(7)	A–D	N(4)							
C(6)=O	A–D	H–N(1)	H–N(4)						
N(1)–H	D–A	O=C(2)	N(3)	O=C(2)					
H	D–A		O=C(2)	N(3)					
N(2)〈					N(3)	O=C(2)			H–N(4)
H	D–A				H–N(4)	H–N(1)	H–N(2)	H–N(1)	N(3)
N(3)	A–D						N(3)	O=C(2)	
N(9)–H	D–A								

Scheme 2

Base-pair configuration[a]		GU1²	GU2²	GU3²	GU4²	GU5²	GU6³	GU7³	
Guanine	Hydrogen bond	Uracil							
N(7)	A–D								
C(6)=O	A–D	H–N(1)	H–N(3)	H–N(3)			O=C(4)	O=C(2)	
N(1)–H	D–A	O=C(2)	O=C(2)	O=C(4)			H–N(3)	H–N(3)	
H	D–A						O=C(2)	O=C(4)	
N(2)〈					O=C(2)	H–N(1)			
H	D–A				H–N(1)	O=C(2)			
N(3)	A–D								
N(9)–H	D–A								

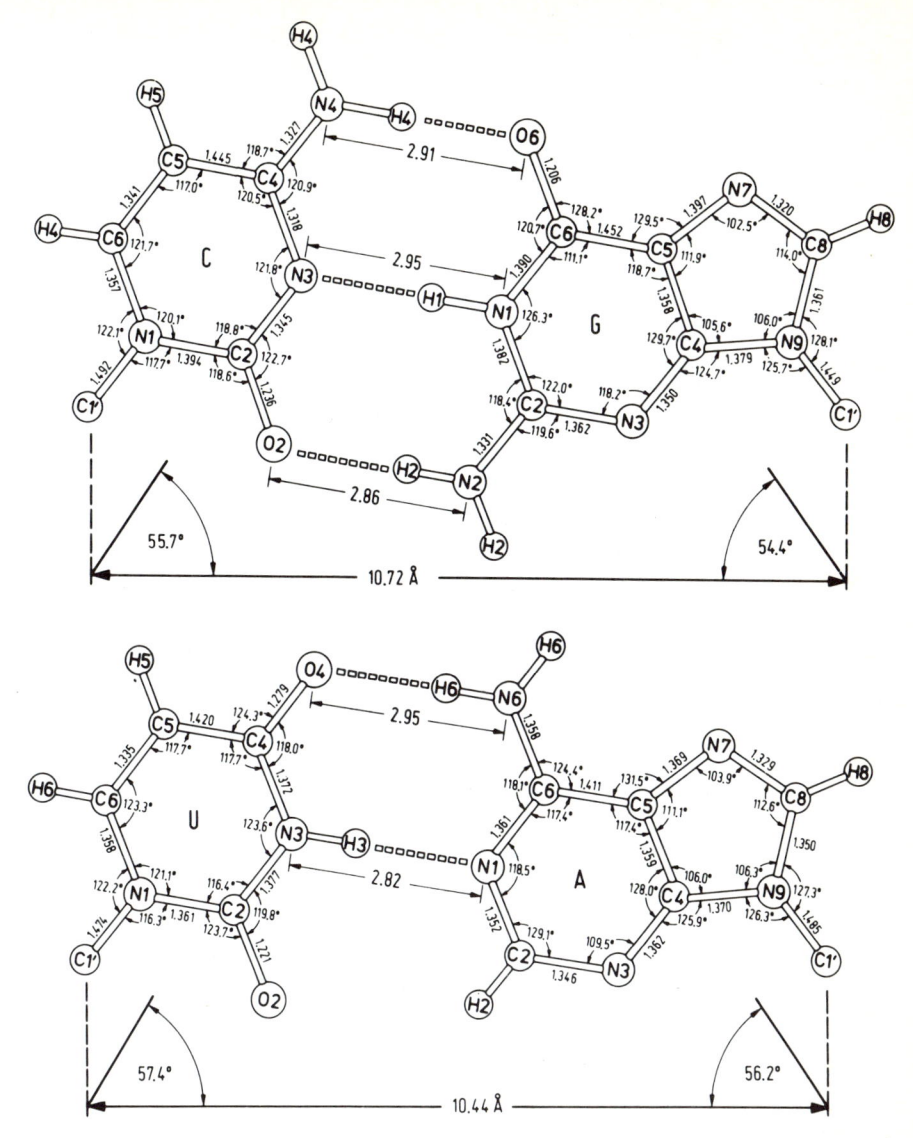

Fig. 16.13. The two Watson-Crick base pairs C–G (*top*) from guanylyl-3',5'-cytidine [538] and U–A (*bottom*) from adenylyl-3',5'-uridine [536]. Distances (Å) between and angles (°) formed by glycosyl N–C(1') links are indicated to illustrate the isomorphism of the two base pairs [522]

None of the two-bond base pairs has been found in a guanine-cytosine complex except for the configuration *GC3²* in transfer RNA (Part III, Chap. 20, Fig. 20.9). Presumably this is because configurations with only two hydrogen bonds are less stable than those with three hydrogen bonds.

The base pair combination *adenine-cytosine* has six possible configurations with two hydrogen bonds and none with three bonds.

Base-pair configuration[a]		ACl[2]	AC2[2]	AC3[3]	AC4[2]	AC5[2]	AC6[2]
Adenine	Hydrogen bond	Cytosine					
N(7)	A–D	H–N(1)	H–N(4)				
N(6)⟨H H	D–A	O=C(2)	N(3)				
	D–A			O=C(2)	N(3)		
N(1)	A–D			H–N(1)	H–N(4)		
N(3)	A–D					H–N(1)	H–N(4)
N(9)–H	D–A					O=C(2)	N(3)

The adenine-cytosine base pair was not observed experimentally except in a mismatch oligonucleotide (Part III, Chap. 20) and probably has no significance in biology except in mutation events.

The combination *guanine-uracil* has two possible configurations with three hydrogen bonds in good stereochemistry, and five configurations with two hydrogen bonds (see scheme 2, page 262).

Of these combinations, only *GU2*[2] and *GU3*[2] can occur in the nucleic acids because of substitution at N(1) of uracil and N(9) of guanine. The configuration *GU2*[2] (Fig. 16.14) was found in transfer ribonucleic acid. This base pair is important as "wobble" base pair in interactions between tRNA anticodon and messenger RNA codon (see Part III, Chap. 20). A variation of the Watson-Crick *GC2*[3] configuration (Fig. 16.11) is observed in the crystal structures of the complexes of 9-ethyl-2-aminopurine with 1-methyl-5-fluorouracil [FUREAP] and with 1-methyl-5-bromo-uracil [BUREAP], where all hydrogen bonds are formed except the cytosine N(4)–H···O(6) guanine, because these groups are missing.

The minor base hypoxanthine (X) occurs in some transfer RNA's. It is comparable to guanine but lacks the N(2) amino group. Therefore, all hydrogen bonds can be formed with cytosine and uracil where this group is not essential. For the *hypoxanthine-cytosine* interaction, we can write the matrix:

Base-pair configuration[a]		XC1[2]	XC2[2]	XC3[2]	XC4[2]
Hypoxanthine	Hydrogen bond	Cytosine			
N(7)	A–D				
C(6)=O	A–D	H–N(1)	H–N(4)		
N(1)–H	D–A	O=C(2)	N(3)		
N(3)	A–D			H–N(2)	H–N(1)
N(9)–H	D–A			N(3)	O=C(2)

In the nucleic acids, only *XC2*[2] is possible due to substitution. This base pair corresponds to the guanine-cytosine Watson-Crick combination *GC2*[3] (Fig. 16.11).

Similarly, we can write the interaction matrix for the possible *hypoxanthine-uracil* base pairs:

Base-pair configuration[a]			$XU1^2$	$XU2^2$	$XU3^2$	$XU4^2$	$XU5^2$	$XU6^2$
Hypoxanthine	Hydrogen bonds	Uracil						
N(7)	A–D							
C(6)=O	A–D		H–N(1)	H–N(3)	H–N(3)			
N(1)–H	D–A		O=C(2)	O=C(2)	O=C(4)			
N(3)	A–D					H–N(1)	H–N(3)	H–N(3)
N(9)–H	D–A					O=C(2)	O=C(2)	O=C(4)

There are no configurations possible with three hydrogen bonds. Of the six base pairs with two hydrogen bonds, only two ($XU2^2$, $XU3^2$) can occur if hypoxanthine N(9) and uracil N(1) are substituted, as in the nucleic acids.

Configuration $XU2^2$ (comparable to $GU2^2$, Fig. 16.14) is a "wobble" base pair and of importance for some transfer RNA's with hypoxanthine in the anticodon sequence (see Chap. 20).

For the purine-purine combination *hypoxanthine-adenine*, there are six possible base pairs with two hydrogen bonds. Of these, only two can occur in the nucleic acids, $XA1^2$ and $XA2^2$, with $XA1^2$ representing a "wobble" pair which is of some biological importance in transfer RNA, and comparable to the guanine-uracil "wobble" illustrated in Fig. 16.14.

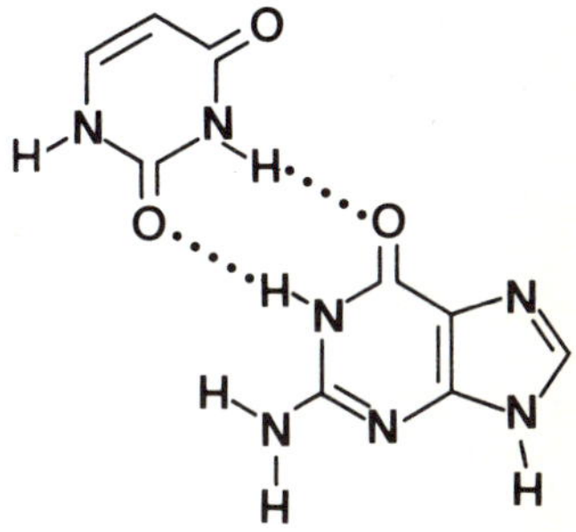

$GU2^2$

Fig. 16.14. The "wobble" G–U base pair $GU2^2$ formed by guanine and uracil is of importance in the codon-anticodon interaction between messenger RNA and transfer RNA

Base-pair configuration[a]		$XA1^2$	$XA2^2$	$XA3^2$	$XA4^2$	$XA5^2$	$XA6^2$
Hypoxanthine	Hydrogen bonds	Adenine					
N(7)	A – D						
C(6)=O	A – D	H – N(6)	H – N(6)	H – N(9)			
N(1) – H	D – A	N(1)	N(7)	N(3)			
N(3)	A – D				H – N(6)	H – N(6)	H – N(9)
N(9) – H	D – A				N(7)	N(1)	N(3)

Uracil-cytosine base pairs are mixed pyrimidine-pyrimidine complexes, which are biologically less or even not at all important. There are six two-bond configurations, of which only $CU5^2$, $CU6^2$ could occur in nucleic acids.

Base-pair configuration[a]		$CU1^2$	$CU2^2$	$CU3^2$	$CU4^2$	$CU5^2$	$CU6^2$
Cytosine	Hydrogen bond	Uracil					
N(1) – H	D – A	O=C(2)	O=C(4)	O=C(2)			
C(2)=O	A – D	H – N(1)	H – N(3)	H – N(3)			
N(3)	A – D				H – N(1)	H – N(3)	H – N(3)
N(4)⟨H / H	D – A				O=C(2)	O=C(4)	O=C(2)

The $CU4^2$ configuration is observed in the complex of 1-methylcytosine·5-fluorouracil [FURMCY], where it forms part of a tetrameric arrangement (Fig. 16.15).

16.3 Base Pairs Can Combine to Form Triplets and Quadruplets

In a base pair, only two or three hydrogen bonds are linking the two partners, and in general several hydrogen-bonding acceptor and/or donor sites are still free to interact. In the crystal lattice, these sites can give rise to the formation of more extended two- or three-dimensional sheets or networks, as illustrated for 1-methylcytosine in Fig. 16.6. In this sense, a base pair occurring in a crystal is only one (specific) hydrogen-bonding interaction among others which stabilize the crystal lattice.

In polymeric nucleic acids, the Watson-Crick base pairs also have the potential to form additional hydrogen bonds. These will generally be saturated by water of hydration molecules. They are, in fact, necessary for specific recognition of a particular nucleic acid base sequence by a particular protein such as, for example,

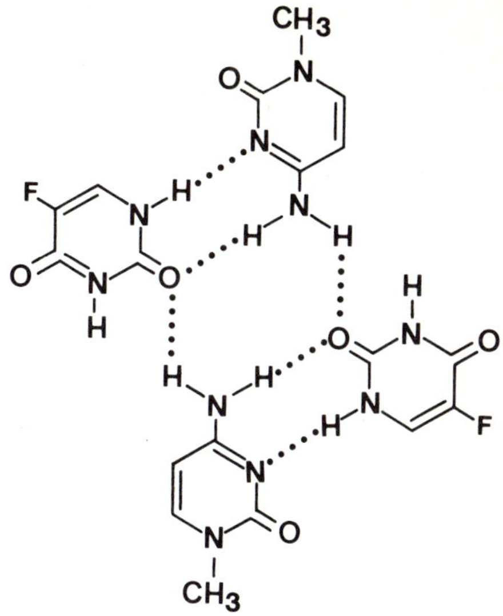

Fig. 16.15. The **CU4²**-type base quadruple in 1-methylcytosine·5-fluorouracil [FURMCY]

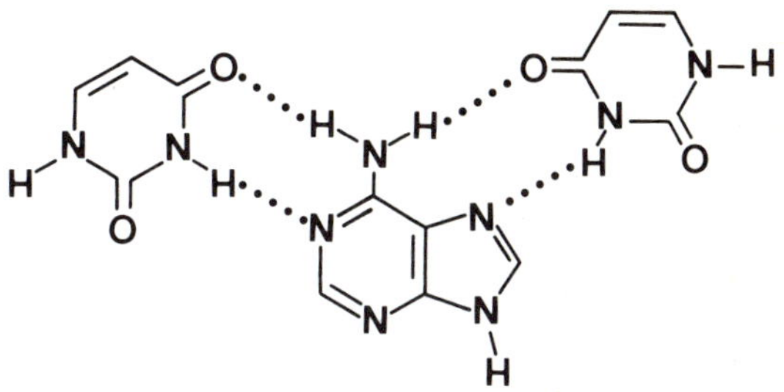

AU6² + AU3²

Fig. 16.16. The U–A–U base triple observed in poly(U)·2poly(A) is composed of a Watson-Crick *AU6²* and a Hoogsteen base pair *AU3²*. A corresponding C–G–C triple is not observed. It would only be possible if the C of the Hoogsteen base pair were protonated

repressors, restriction endonucleases, or methyl transferases (discussed in Part III, Chap. 20).

The extra hydrogen-bonding sites can, however, also form further base pairs, as observed for poly(A)·poly(U). In the Watson-Crick A–U pair, one of the adenine N(6)–H donors and the N(7) acceptor are still free and can form another A–U pair of the Hoogsteen type (Fig. 16.16). This is in fact observed; if the Watson-Crick duplex poly(A)·poly(U) is exposed to high salt conditions, it disproportionates to form a triplex poly(A)·2poly(U) and a single strand poly(A) [539]:

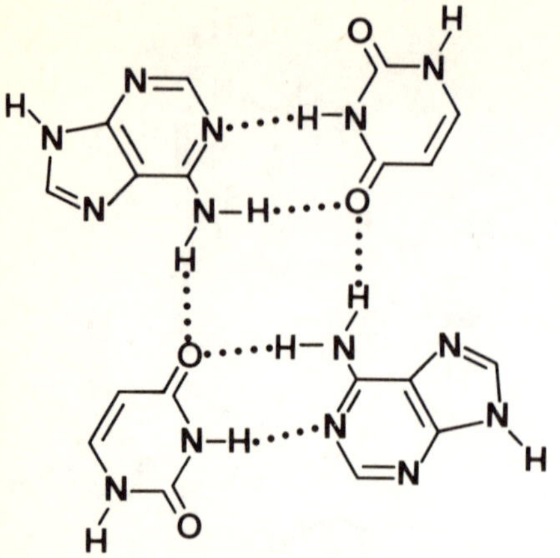

$AU6^2 + AU6^2$

Fig. 16.17. The base quadruplets formed by two Watson-Crick base pairs have not yet been directly observed experimentally

Antiparallel Watson-Crick	Dispropor- tionation	Antiparallel Watson-Crick	Parallel Hoogsteen
2 [poly(A)·poly(U)]		poly(U)·poly(A)·poly(U)	+ poly(A)
Double helix		Triple helix	

Because the glycosyl links and the associated sugar-phosphate backbone are related, in the Watson-Crick duplex, by a pseudodyad perpendicular to the helix axis, the two strands are oriented *antiparallel.* They are *parallel* in the Hoogsteen duplex because the glycosyl links are related by rotation about the helix axis.

The double helix formed by poly(G)·poly(C) does not disproportionate. The reason is that it forms an extremely stable Watson-Crick duplex and that the Hoogsteen base pair could only form if poly(C) were protonated (legend to Fig. 16.16).

Although the base triplets are of only minor importance in double-stranded nucleic acids, they have a structural role in determining and stabilizing the tertiary structure of transfer RNA, as discussed in Chapter 20. Base quadruplets where two Watson-Crick base pairs are associated as shown in Fig. 16.17 have been invoked to play a role in DNA-DNA aggregation and DNA recombination, but there is no direct evidence for their occurrence.

A homo-base quadruple has been actually observed for [poly(G)]$_4$ and for [poly(I)]$_4$ where the amino group in position 2 is lacking. It has the form depicted in Fig. 16.8. Since the glycosyl links are related by the fourfold helix axis, the polynucleotide strands are parallel to each other.

Chapter 17
Hydrogen Bonding in the Crystal Structures of the Nucleosides and Nucleotides

All the important hydrogen-bonding functional groups which occur in biological macromolecules are found in the nucleosides and nucleotides (Box 17.1). From the point of view of hydrogen bonding, they are the most complex of the small biological molecules, since they contain the greatest variety of hydrogen-bonding functional groups. To the $-NH_2$, $>NH$, $O=C$ and $>N$ groups which are predominant in the purine and pyrimidine bases are added the $-OH$ and $>O$ groups of the ribo- or deoxyribofuranosyl moieties, and for the nucleotides, the $P-OH$ and $\bar{P}=O$ groups of the phosphoric acid components (Box 17.2).

Box 17.1. Hydrogen-bonding donor and acceptor groups in the nucleosides and nucleotides

Donor groups: $C-O-H$, $\overset{\ominus}{P}-O-H$, $-N(H)-H$, $>N-H$, $\overset{\oplus}{>N}-H$, $H-O_W-H$

Acceptor groups: $O{<}^H_C$, $O{<}^C_C$, $O{<}^H_P$, $O=\bar{P}$, $O=C{<}$, $N{\ll}$, $S=C{<}$, $O{<}^H_H$

Box 17.2. Nucleoside and Nucleotide Nomenclature

There are eight basic nucleosides which are the constituents of RNA and DNA. They consist of the four bases displayed in Fig. 15.1 which are substituted, at pyrimidine N(1) and at purine N(9), by a D-ribose in RNA and by a 2′-deoxy-D-ribose in DNA where the O(2′)H of the ribonucleosides is replaced by hydrogen; both sugars are in the furanose configuration and the glycosidic link is of the β-form (Fig. 17.1); the α-anomers frequently have antibiotic properties. The *nucleosides* are called uridine, cytidine, adenosine, guanosine in the ribo series and thymidine, deoxycytidine, deoxyadenosine, deoxyguanosine in the deoxyribo series. Sugar atoms are designated with a "prime", base atoms without a "prime".

In the *nucleotides*, the O(3′) or O(5′) hydroxyls (or both) are esterified by phosphoric acid (Fig. 17.2). At and below neutral pH, the phosphate group carries one negative charge, it is protonated (neutral) at pH<2, and it is double negatively charged above neutral pH.

There are a large number of nucleoside and nucleotide derivatives with modifications at the base (see Fig. 15.2) and at the sugar, as, for instance, the naturally occurring arabinonucleosides, which display antibiotic activity (see Fig. 17.1).

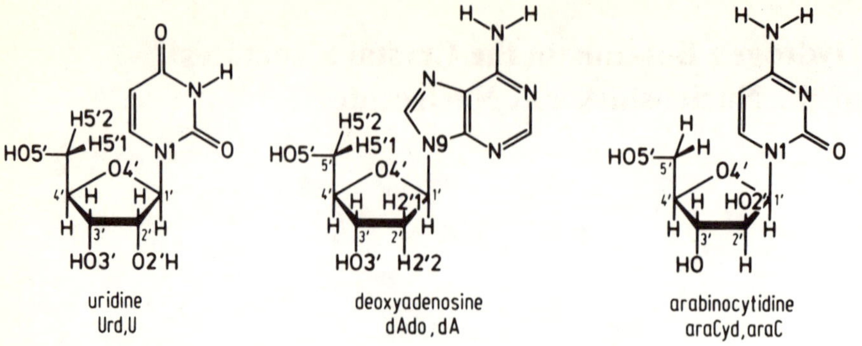

Fig. 17.1. Chemical structures of and atomic numbering scheme in the nucleosides, shown for uridine, deoxyadenosine, and the antibiotic arabinocytidine

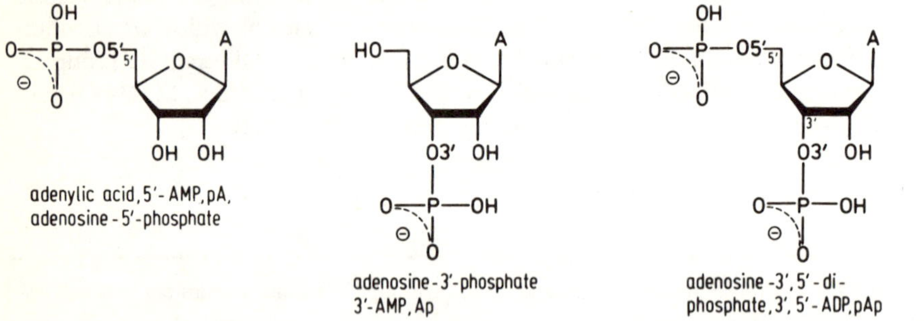

Fig. 17.2. Chemical structures of some nucleotides: adenosine-5'-phosphate or 5'-adenylic acid; adenosine-3'-phosphate or 3'-adenylic acid; adenosine-3',5'-disphosphate

The nucleosides and nucleotides do not form good diffraction quality crystals as readily as do the amino acids, carbohydrates, or the purines and pyrimidines. Therefore, the number of X-ray structure analyses providing reliable hydrogen atom positions for the groups with orientational freedom are relatively small and there is only one neutron diffraction analysis, which is of a rather complex nucleoside derivative [466]. Of the over 600 crystal structure analyses reported in the Cambridge Data Base, only about one fifth provides hydrogen atomic positions that are sufficiently reliable to be useful in a survey of the hydrogen bonding. This is particularly true for the hydrate structures where the water molecules can be disordered and the positions of the hydrogen atoms are frequently uncertain.

17.1 Conformational and Hydrogen-Bonding Characteristics of the Nucleosides and Nucleotides

As with the corresponding macromolecules, the nucleosides and nucleotides are generally hydrated in the crystalline state; about 45% of their crystal structures re-

Conformational and Hydrogen-Bonding Characteristics of the Nucleosides and Nucleotides   271

Table 17.1. Proportion of hydrogen bonds in a selection of nucleoside and nucleotide crystal structures versus the purines and pyrimidines [61, 62]; see also Tables 7.12 and 7.14

Nucleosides and nucleotides		Purines and pyrimidines	
OH···OH	15%	NH···O=C	21%
OH···O=C	13%	NH···N	12%
OH···N	9%	N(H)H···O=C	8%
O_WH···OH	8%	N(H)H···N	12%
N(H)H···OH	7%	O_WH···O=C	5%
OH···O_W	6%	N(H)H···O_W	6%
O_WH···N	5%	Others	36%
Others	37%		

ported in the Cambridge Crystallographic Data Base are hydrates. This compares with 30% for the purines and pyrimidines, 20% for the amino acids, and 20% for the carbohydrates. The nucleoside and nucleotide structures therefore provide a useful portfolio of the different stereochemical configurations that water molecules can adopt in conjunction with the functional groups of biological molecules. This is discussed in Part IV, Chapter 21.

Hydrogen bonds involving OH groups are important. The significance of the sugar, phosphate, and water OH groups in the crystal structures of the nucleosides and nucleotides is shown by the difference between the proportion of hydrogen bonds in these crystal structures and those of the purines and pyrimidines, where the weaker NH···O=C and NH···N$\langle$ hydrogen bonds dominate (see Table 17.1).

However, it is important to note that in the nucleic acids, the number of hydroxyl groups per monomer unit is reduced from three (in the ribonucleoside) to one in RNA, and to none in DNA. The proportion of O−H···O−H and O−H···O=C, O−H···N$\langle$ and O−H···O=$\bar{P}$ bonds would be greatly reduced from those shown in Table 17.1 were it not that the nucleic acids are heavily hydrated and therefore must have a high proportion of O_W−H···O=C, O_W−H···N$\langle$, and O_W−H···O=$\bar{P}$ interactions.

The metrical analysis of the hydrogen-bond geometries in this class of molecules is reported in Chapter 7. It shows similar distributions of bond lengths and angles for the same donor/acceptor combinations as observed in the carbohydrates, amino acids, purines and pyrimidines.

For the two-center bonds, the sequence of hydrogen-bond lengths for the donor groups is [61, 62]:

$$\bar{P}-OH < C-OH < NH < O_WH < N(H)H < C-H \ .$$

For the acceptor groups it is

$$O=\bar{P} < O_WH_2 < O=C < N\langle\ < Cl^\ominus < O\langle\ < S=C \ .$$

Three-center hydrogen bonds occur frequently but do not involve donor $\bar{P}$−OH. The proportion of the three-center bonds present in the crystal structures of the nucleosides and nucleotides is about 30% and comparable to that observed in the crystal structures of the other biological small molecules. This is to be expected, since the number of acceptor groups exceeds the number of donors by about the same proportion. As in the carbohydrate crystal structures, the weak components of the three-center bonds frequently link different sections of the hydrogen-bond pattern into a more extensive cooperative network. The chelated three-center bonds, which are a special feature of the purine and pyrimidine structures, are less common. There are a few examples of four-center bonds and of three-center/bifurcated bonds.

The strong $\bar{P}$–OH group is never the donor in three-center bonding, whereas the weak N(H)H donor group is most frequently involved. There are some examples of $PO_2^\ominus$ and $P(OR)O^\ominus$ groups functioning as three-center chelating acceptors, with one strong and one weak interaction $P{\displaystyle{\overset{\ominus}{\underset{O\cdots}{O\cdots}}}}\!:H-X$. There are also some examples of the chelating acceptor $-C{\displaystyle{\overset{O\cdots}{\underset{N}{}}}}\!:H-X$. Interestingly, the H$\cdots$N bond is the shorter of the components of the three-center bond [61, 62].

A significant number of intramolecular hydrogen bonds are observed. Those between the purine or pyrimidine bases and the sugar moieties have received particular attention because they can impart additional rigidity to the conformation of the nucleosides and nucleotides (Part I B, Chap. 9). Moreover, in the ribo series, there are frequently hydrogen bonds as the minor component of the three-center type between the vicinal hydroxyl groups.

The intramolecular base$\cdots$sugar hydrogen bonds require the exocyclic O(5') in its preferred orientation "over" the sugar, with torsion angle γ in the +sc range (see Box 17.3). The O(5')H group can act as a donor in nucleosides or as an acceptor in nucleosides and in nucleotides, depending on whether the orientation of the base is in the syn or anti form (for nomenclature see Box 17.3). If the base is in the less common syn orientation, O(2) of pyrimidine or N(3) of purine is located "over" the sugar and close to O(5')H so that an intramolecular hydrogen bond O(5')–H$\cdots$O(2)/N(3) can form. This interaction contributes to the stability of the syn conformation as shown in Fig. 17.5 for 8-bromoguanosine, but is not a necessary requirement, as illustrated by several crystal structures of syn-nucleosides where such hydrogen bonds are not observed. A good example is 6-methyluridine [546], where there are two independent molecules in the syn form in the crystal structure. One displays an O(5')H$\cdots$O(2) intramolecular hydrogen bond which is three-centered with the minor intramolecular component to the ribose ring oxygen O(4'). The O(5')H in the other molecule also forms a three-centered hydrogen bond, but both components are intermolecular.

If the bases are in anti orientation, the C(8)H of purine and C(6)H of pyrimidine are located "over" the sugar and close enough to O(5') that a (base) C–H$\cdots$O(5') hydrogen bond can form, Fig. 17.6. Because the C(8)H and C(6)H hydrogen atoms have slightly acidic character as shown by H/D exchange experi-

Box 17.3. Nomenclature of Nucleotide and Nucleic Acid Conformation

Nucleosides and Nucleotides occur in a limited number of preferred conformations. Except for the planar purine or pyrimidine base, the nucleotide component of the nucleic acids contains only single bonds (see Fig. 17.1). Rotation about bonds is restrained by the requirements of the furanose ring closure, and for the acyclic bonds, rotation is restricted by electronic factors and is limited by intramolecular interactions. Therefore only a few conformations have to be considered if nucleosides and nucleotides are described in their three-dimensional shapes.

In the literature, there are different and confusing nomenclatures. In this text, we follow the IUPAC rules [540].

As illustrated and defined in Fig. 17.3, torsion angles along the sugar-phosphate chain are denoted in alphabetical order α to ζ [starting with P–O(5′)], those in the furanose ring are called v_0 to v_4, and the orientation about the glycosidic link C(1′)–N is described by χ.

Furanose five-membered rings are never planar. They adopt a puckered form for which a family of pseudorotational conformers [541] are possible. The main conformations are the envelope and twist forms denoted by atom names, i.e., C(2′) and C(3′), and descriptors endo and exo. They refer to the displacement of these atoms from the mean plane through the other atoms of the five-membered ring, with reference to the exocyclic C(4′)–C(5′) bond (Fig. 17.4).

Relatively small differences in the puckering of the ribofuranose ring are amplified in the orientation of the O(3′) and O(5′) atoms to which, in the nucleic acids, the phosphate groups are bound. For this reason, extensive analyses have been made of the relationship between ring conformation and the overall stereochemistry of the nucleic acid molecules.

Since the energy barrier between the two main pucker families C(2′)-endo and C(3′)-endo is small, 4 to 6 kcal mol^{-1} [542], the transition occurs easily along a pseudo-rotational pathway in DNA [541–544] but is sterically restricted to C(3′)-endo in RNA.

The bases are in syn or anti orientation. The orientation of the base relative to the sugar moiety is defined by torsion angle χ which is constrained by steric interactions, and by the anomeric effect. The main conformations are referred to as syn and anti (see Fig. 17.5). In syn, χ is close to 0° (sp; for torsion angle definition, see Box 13.3; Fig. 13.12), and the base is oriented "above" the ribose ring causing steric interactions which in the anti conformation with χ close to 180° (ap) are avoided. The anti conformation is therefore preferred and is the only form observed in double-helical DNA and RNA. An exception is the left-handed Z-DNA with alternating purine/pyrimidine nucleotide sequence where the purines are in the syn conformation.

In helical arrangements, the torsion angles β, ε about C–O bonds are ap and α, ζ about P–O ester bond, are –sc. In nucleotides and the double helical oligo- and polynucleotides, the other acyclic torsion angles also have preferred ranges. In general, β is in the ap range, whereas ε is more shifted to –ac. For the P–O ester bond torsion angles α and ζ, only the –sc range is consistent with right-handed helical arrangement. Combinations where one or both P–O torsion angles are in the ap or +sc range give rise to the nonhelical conformations that are observed in the loop and interstem regions of transfer RNA.

The "rigid nucleotide" as building block in the nucleic acids. In this discussion, which is primarily concerned with aspects of hydrogen bonding, the simplified "rigid nucleotide" concept will be sufficient in the description of nucleotide and nucleic acid conformation. According to this concept [545], *there are two basic nucleotide units which differ mainly in sugar pucker:*

Sugar pucker	Base orientation	Torsion angle					
		α	β	γ	δ	ε	ζ
C(2′)-endo	anti	–sc	ap	+sc	ap	ap to –ac	–sc
C(3′)-endo	anti	–sc	ap	+sc	+ac	ap to –ac	–sc

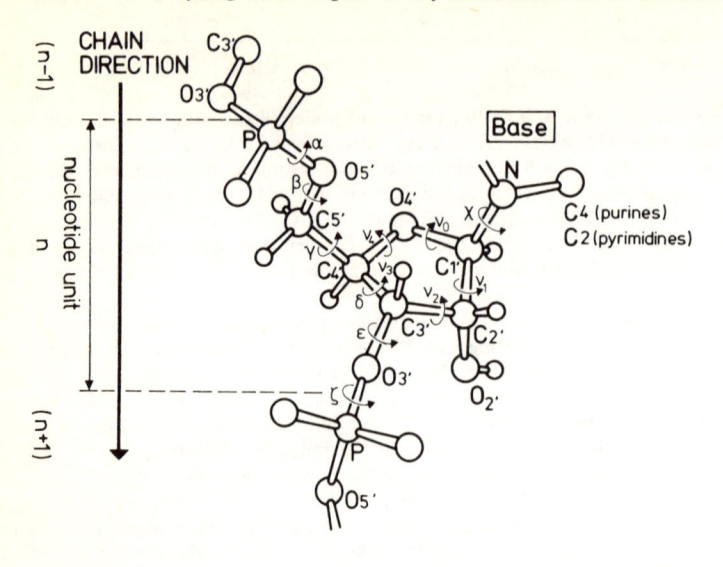

Torsion angle	Atoms involved[1]
α	$_{(n-1)}$O(3')−P−O(5')−C(5')
β	P−O(5')−C(5')−C(4')
γ	O(5')−C(5')−C(4')−C(3')
δ	C(5')−C(4')−C(3')−O(3')
ε	C(4')−C(3')−O(3')−P
ζ	C(3')−O(3')−P−O(5')$_{(n+1)}$
χ	O(4')−C(1')−N(1)−C(2) (pyrimidines)
	O(4')−C(1')−N(9)−C(4) (purines)
v_0	C(4')−O(4')−C(1')−C(2')
v_1	O(4')−C(1')−C(2')−C(3')
v_2	C(1')−C(2')−C(3')−C(4')
v_3	C(2')−C(3')−C(4')−O(4')
v_4	C(3')−C(4')−O(4')−C(1')

[1]) Atoms designated (n−1) and (n+1) belong to adjacent units.

Fig. 17.3. Definition of torsion angles in the polynucleotide chain [522]

ments [548, 549], this intramolecular interaction is further augmented. It is nevertheless weak and in the range 1.1 to 2.7 kcal mol^{-1}, as assessed by CNDO/2 calculations [463].

In the survey of nucleoside and nucleotide crystal structures [62], there are five clear examples of C(6)H···O(5') hydrogen bonds of which three are three-centered (Table 10.2). All are in pyrimidine nucleosides. For the purine nucleosides and nucleotides, there are insufficient structure determinations with hydrogen atoms located from X-ray data to draw any conclusions. However, in many nucleoside and nucleotide crystal structures, the nonhydrogen atom C···O(5') distances are

Fig. 17.4. Description of the main sugar conformations in nucleosides, nucleotides, and nucleic acids [522]

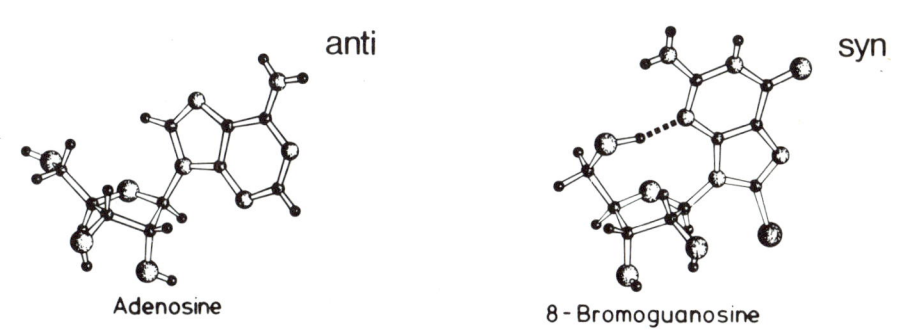

Fig. 17.5. Definition of syn and anti orientation of the base relative to the sugar in nucleosides and in nucleotides [522]

in the range 3.0 and 3.2 Å, suggesting H···O(5') hydrogen bonds between 2.0 and 2.2 Å which correspond to the shorter of those reported for C−H···O interactions.

The other type of intramolecular hydrogen bond is mostly through the minor component of three-center bonds and involves vicinal O(2')H and O(3')H groups of the ribose moiety. It occurs in about 20% of the structures examined. The O(2')H···O(3') interaction is three times more common than the O(3')H···O(2') interaction, H···O distances range from 2.29 to 2.58 Å with O−Ĥ···O angles between 90° and 112°. Unusual two-center O(2')···O(3') hydrogen bonds are observed in the crystal structures of 5,6-dimethyl-1-(α-D-ribofuranosyl)benzimidazole [MRFBZI10], O(2')H···O(3') with H···O 1.96 Å, O−Ĥ···O 127°; xanthosine dihydrate [XANTOS], O(2')H···O(3') with O···O, 2.60 Å, H···O, 2.03 Å, O−Ĥ···O, 105° (Fig. 17.7); in 6-amino-10-(β-D-ribofuranosylamino) pyrimido[5,4-D] pyrimidine [RPPYPY20] with O(3')H···O(2') O···O, 2.58 Å, H···O, 1.78 Å; O−Ĥ···O; 154°.

276 Hydrogen Bonding in the Crystal Structures of the Nucleosides and Nucleotides

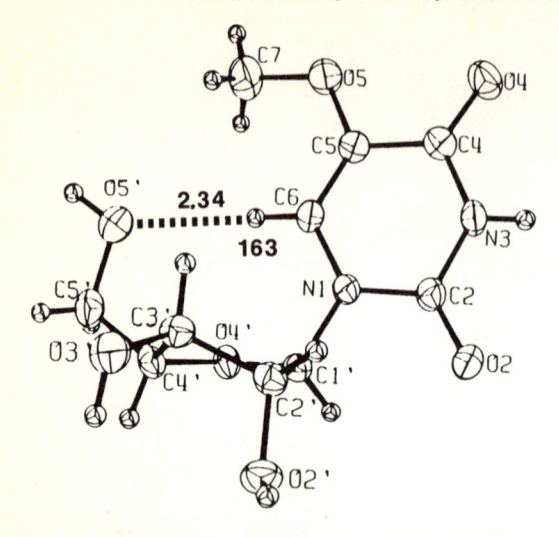

Fig. 17.6. The (base) C–H···O(5') intramolecular hydrogen bond, shown here for 5-methoxyuridine, stabilizes the anti conformation [MXURID01]; sugar pucker is C(3')-endo

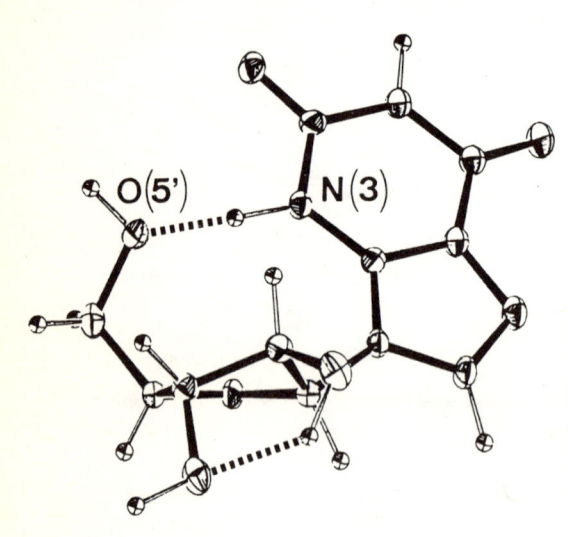

Fig. 17.7. Xanthosine, a derivative of guanosine. It displays a rare example of an intramolecular O(2')–H···O(3') hydrogen bond and its syn conformation is stabilized by an intramolecular N(3)–H···O(5') bond [XANTOS]; sugar pucker is C(2')-endo

17.2 A Selection of Cyclic Hydrogen-Bonding Patterns Formed in Nucleoside and Nucleotide Crystal Structures

Hydrogen-bonding patterns in the nucleosides and nucleotides resemble more closely those of the carbohydrates and carboxylic acids than the purines and pyrimidines because of the predominance of the OH···OH hydrogen bonding, see Table 1. As in the carbohydrate crystal structures, weak three-center components involving sugar and water hydroxyl groups frequently link different sections of the hydrogen-bond pattern into a more extensive cooperative network. Patterns based on the self-association of the purine and pyrimidine bases are less common,

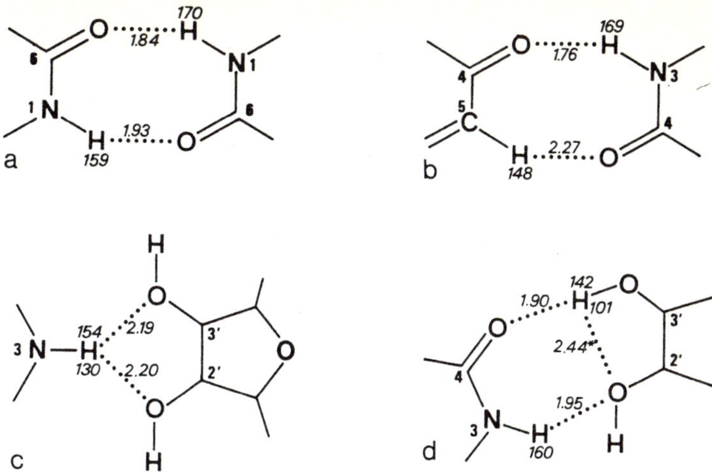

A Selection of Cyclic Hydrogen-Bonding Patterns 277

Fig. 17.8. **a** 8-Bromoinosine [BRINOS10]; **b** uridine [BEURID]; **c** 5-carboxymethyluridine [CXMURD]; **d** 5-methoxyuridine [MXURID01]

although there are a number of cyclic systems which utilize the π-cooperative aspect of bonding to adjacent groups in the purine or pyrimidine rings. Examples of these cyclic dimer, trimer, tetramer, and pentamer systems which are part of the more extended hydrogen-bonding patterns in the crystal structures are described below.

Cyclic Dimer Configurations. In 8-bromoinosine [BRINOS10] the inosine residues are self-associated (Fig. 17.8a). This is a rare example of self-association in the crystal structures of the nucleosides. Surprisingly, in uridine [BEURID10] (Fig. 17.8b) the CH$\cdots$O=C hydrogen bond plays the same role as the NH$\cdots$O=C interaction in 8-bromoinosine.

In 5-carboxylmethyluridine [CXMURD] (Fig. 17.8c) there is an interesting chelation of the ribosyl hydroxyl groups by the uracil N–H, and a comparable chelated base-sugar association is observed in 5-methoxy-uridine [MXURID01] (Fig. 17.8d).

Cyclic Trimer Configuration. The α-cytidine [ACYTID] crystal structure contains two cyclic trimer configurations (Fig. 17.32). In one, N(H)H is engaged in a three-center bond, the other is a homodromic cycle stabilized by σ- and π-cooperativity. A comparable situation is found in the crystal structure of α-pseudouridine monohydrate [APSURD] (Fig. 17.13). The glycosyl link is to uracil C(6) instead of N(1), and therefore the N(1)H group is free to donate a hydrogen bond in a homodromic cycle.

The 5-amino-1-β-D-ribofuranosylimidazole-4-carboxamide [ARBIMC10] structure contains both a cyclic trimer and a cyclic pentamer (Fig. 17.9a). The amide NH$_2$ group is engaged in a four-centered hydrogen bond with long H$\cdots$A distances between 2.50 and 2.56 Å. There are two intramolecular components indicated by (∗).

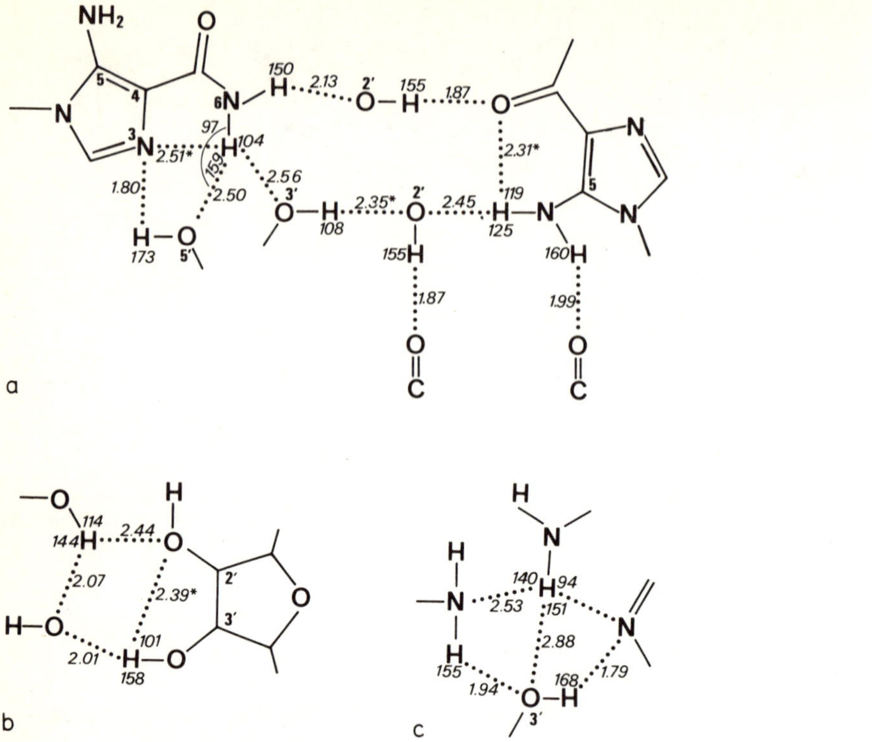

Fig. 17.9. a 5-amino-1-β-D-ribofuranosylimidazole-4-carboxamide [ARBIMC10]; ∗, intramolecular; **b** 6-chloropurine-D-riboside [CLPURB]; ∗, intramolecular; **c** 9-β-D-arabinofuranosyladenine [ARADEN10]; ∗, intramolecular

In the crystal structure of 6-chloropurine-D-riboside [CLPURB] illustrated in Fig. 17.9b, the bonding involves the vicinal ribosyl hydroxyls which are linked by an intramolecular hydrogen bond (∗), and is similar to the trimer configurations reported above. In the four-membered cycle formed by hydrogen bonds, two O−H groups in opposite corners act as double acceptors, and the other two are donors of three-center hydrogen bonds. In 9-β-D-arabinofuranosyladenine [ARADEN10], there are two trimers arranged around a four-centered N(H)H hydrogen bond (Fig. 17.9c).

In 5-aminouridine [AMURID], the weak three-center bond from O(3′)H includes an intramolecular component (∗) to O(2′) and a rare example of an NH$_2$ acceptor (Fig. 17.19). This group, however, is not conjugated as, for example, the amino group in adenine, guanine, and cytosine, but *only* in conjugation with the C(5)=C(6) double bond. We assume that it is still pyramidal with the lone electron pair located on the N atom, and therefore serves as hydrogen-bond acceptor.

In 3′-O-methyl-1-β-D-arabinofuranosyl cytosine [MARAFC] (Fig. 17.39) there is a chelated three-center hydrogen bond donated by O(5′)−H, with one very short (1.78 Å), and one very long (2.61 Å) component. A CH group donating a symmetrical three-center hydrogen bond is in inosine [INOSIN10] (Fig. 17.58).

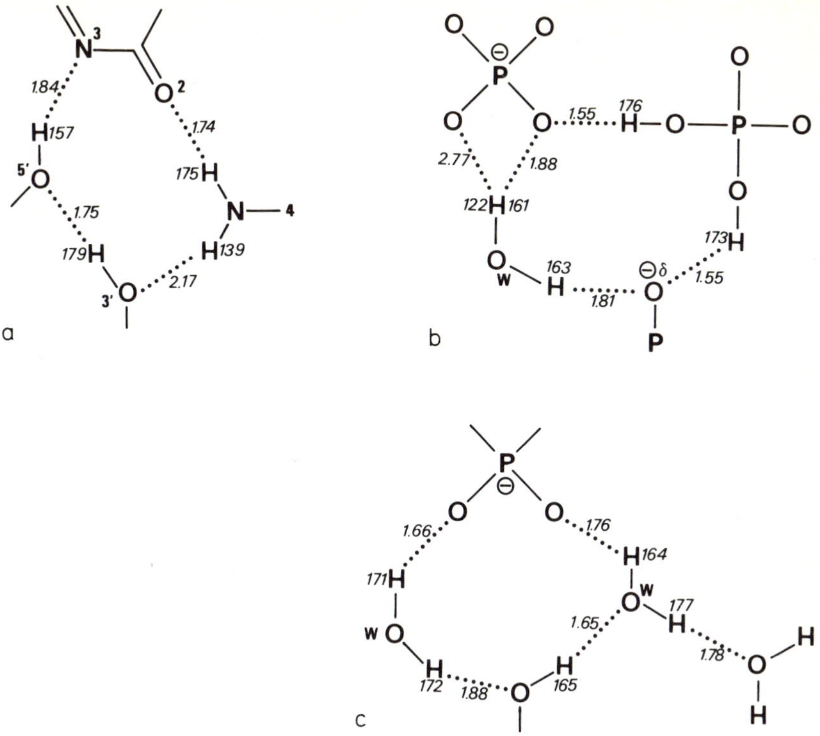

Fig. 17.10. **a** 6,2′-anhydro-1-α-D-arabinofuranosyl-6-hydroxycytosine [ARFHCY]; **b** 2,2′-anhydro-1-β-D-arabinofuranosylcytosine-3′,5′-diphosphate [AFCYDP]; **c** guanosine 5′-phosphate·3H$_2$O [GUANPH01]; see also Fig. 17.56

Cyclic Tetramers. In the cyclic tetramers given in Fig. 17.10, amino groups or water molecules act as double donors and the cycles are consequently of the antidromic type. This type of hydrogen-bonding structure is found in 2′-deoxycytidine [DXCYTD] (Fig. 17.34), where the cycle is pseudosymmetrical. It also occurs in the crystal structure of a nucleoside in β-configuration with additional covalent sugar-base link between sugar C(2′) and base C(6), 6,2′-anhydro-1-β-D-arabinofuranosyl-6-hydroxycytosine ([ARFHCY]; Fig. 17.10a). A chloride ion is part of the four-member ring in adenosine hydrochloride (ADOSHC, Fig. 17.51).

Cyclic configurations involving the phosphate or phosphoric acid $\bar{P}=O$ and $\bar{P}-OH$ groups are observed in some nucleotide crystal structures. In the nucleotide with covalent link between sugar C(2′) and cytosine O(2), 2,2′-anhydro-1-β-D-arabinofuranosylcytosine-3′,5′-diphosphate [AFCYDP] there is a strongly bonded cyclic tetramer configuration involving $\bar{P}(OH)_2$, $\bar{P}=O$, $\bar{P}O_2$ and H$_2$O (Fig. 17.10b). There is also a very unsymmetrical chelated three-center bond from the water hydrogen to the phosphate group.

In adenosine-3′-phosphate dihydrate (ADPOSD), there is a cyclic tetramer configuration involving $\bar{P}O_2$, H$_2$O, NH$_2$ and $\overset{\backslash\backslash}{>}$N groups, with an unsymmetrical chelated three-center bond from the >NH to the phosphate group (Fig. 17.54).

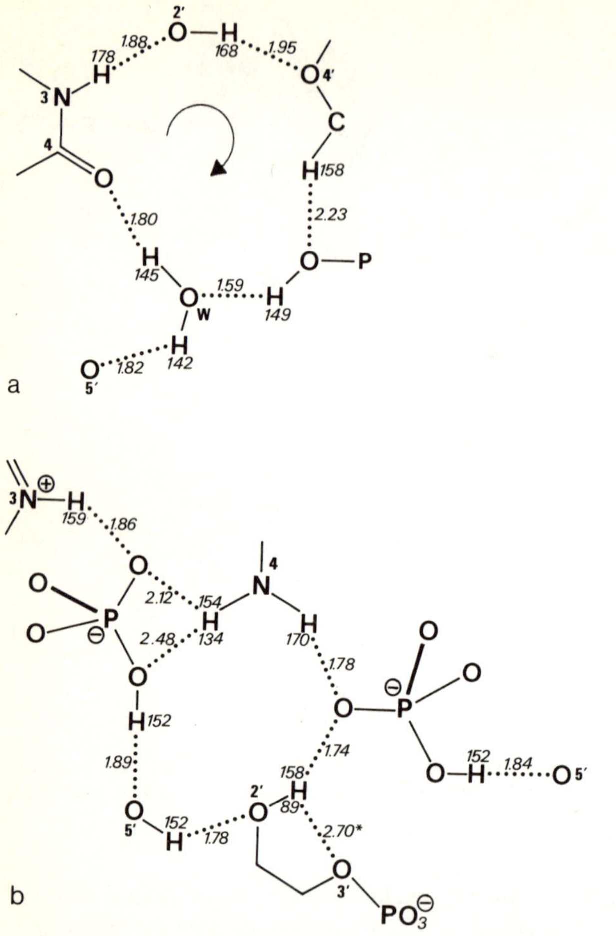

Fig. 17.11. a Uridine-3'-phosphate·H$_2$O [URIDMP10]; **b** cytidine-3'-phosphate (monoclinic) [CYTIAC01], *, intramolecular

In guanosine 5'-phosphate trihydrate [GUANPH01], a tetramer configuration is stabilized by short H···O hydrogen bonds involves $\bar{P}O_2$, OH and two H$_2$O, (Fig. 17.10c).

Cyclic Pentamers. In uridine-3'-phosphate monohydrate [URIDMP10], the cyclic pentamer configuration involves >NH···OH, OH···O<, CH···O(H)P, P–OH···O$_W$H and O$_W$H···O=C hydrogen bonds, all of which are short (Fig. 17.11a). It is a homodromic cycle if we include the polarizability of the ribose-O(4')-CH-component, and probably stabilized by the strong donor/acceptor P–OH of the neutral phosphate group.

In cytidine-3'-phosphate (monoclinic) [CYTIAC01], there is an antidromic cyclic pentamer configuration involving $\bar{P}$=O, NH$_2$, and three OH groups (Fig.

17.11 b). The structure contains a chelated three-center bond from the −N(H)H to the phosphate group.

17.3 General Hydrogen-Bonding Patterns in Nucleoside and Nucleotide Crystal Structures

If the crystal structures and hydrogen-bonding patterns of uridine and thymidine and their simpler derivatives are compared with those of cytidine, it becomes obvious that the −NH$_2$ functional group has a characteristic influence. The nucleosides without the −NH$_2$ groups have less than average hydrated crystals, although the one nucleotide represented, uridine-3′-monophosphate, crystallizes as a monohydrate.

In the uridines and thymidines, there is a tendency to form long finite sequences of hydrogen bonds, and there are a few examples of infinite chains. In the cytidines, the NH$_2$ group does not function as a hydrogen-bond acceptor, and is most frequently the starting point of two shorter finite chains.

The purine nucleoside crystal structures adenosine, inosine, and guanosine provide less reliable data from the hydrogen-bonding point of view. This is particularly true of the guanosines, where only four structures can be reported.

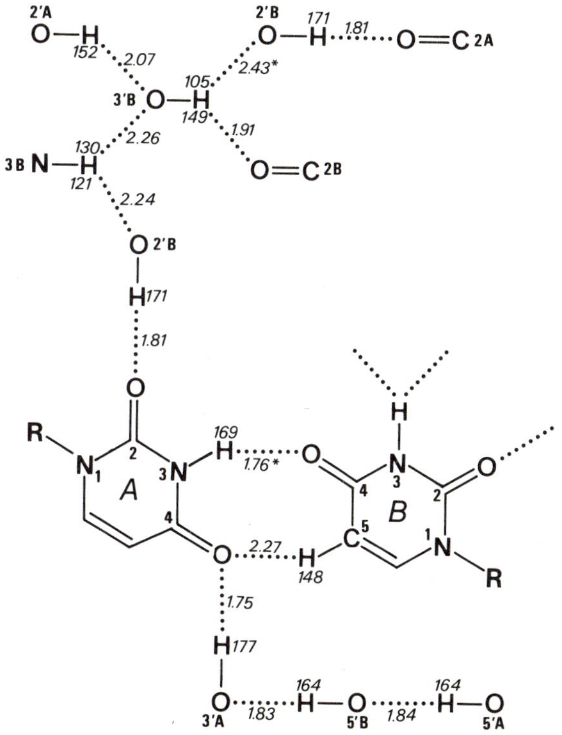

Fig. 17.12. Uridine [BEURID10]; ∗, intramolecular

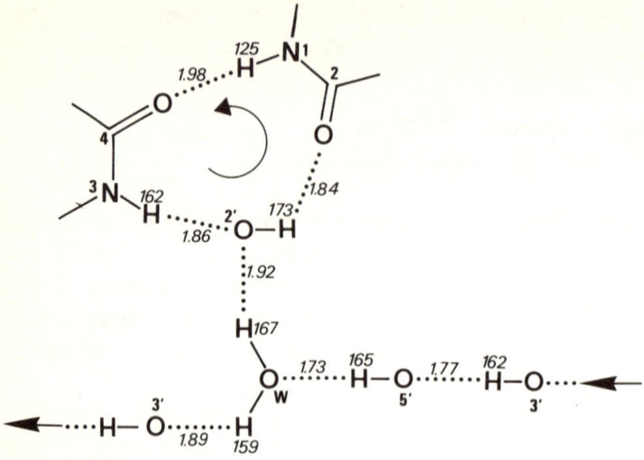

Fig. 17.13. α-Pseudouridine·H₂O [APSURD]

In the crystal structures formed by uridine and its derivatives, a recurrent motif is N(3)H···(OH···OH)···O=C, with many variations.

All the functional groups of the two molecules (**A, B**) in the asymmetric unit of the crystal structure of *uridine* [BEURID10] are linked into a hydrogen-bonding network (Fig. 17.12). As noted before in Fig. 17.8b, this is an interesting example of a structure where a C–H···O=C bond forms a cyclic uracil dimer configuration. The >N(3B)–H group forms an almost symmetrical three-center bond, and there is an intramolecular O(3')–H···O(2') hydrogen bond (∗) in molecule **B**.

The crystal structure of *α-pseudouridine monohydrate* [APSURD] (Fig. 17.13) has a particularly strong hydrogen-bond system consisting of a three-bond homodromic cycle stabilized by π- and σ-cooperativity (see Fig. 17.9b) linked to an infinite –O–H···O–H···O–H··· chain through the water molecule.

The crystal structure of *6-methyluridine* [MEYRID] contains two molecules (**A, B**) in the asymmetric unit (Fig. 17.14). The hydrogen bonding forms an infinite chain through the intramolecular O(3'B)-H···O(2'B) bond (∗) which is the minor component of a three-center bond. There are also two short finite chains linked through the O(3'A)–H···O(2'A) intramolecular bond of the other molecule. One of these chains includes three-center intramolecular bonding from O(5'A)–H to both the ring oxygen and the uracil O=C(2A), which is possible because the base is in the syn orientation.

The hydrogen-bonding structure of *5-methyluridine hemihydrate* [MEURID] (Fig. 17.15) consists, essentially, of one finite homodromic chain. It starts at O(2')–H as donor, involves O(3')–H, O(5')–H, a water molecule, and uracil O=C(4)–N(3)–H as π-cooperative group, and ends at O=C(2). Because the water molecule is located on a twofold axis, it connects two such finite chains in tetrahedral coordination.

The crystal structure of *6-methyl-2'-deoxyuridine* [MEDOUR] has a very simple finite chain (Fig. 17.16). Because the nucleoside is in the syn form as 6-methyl-

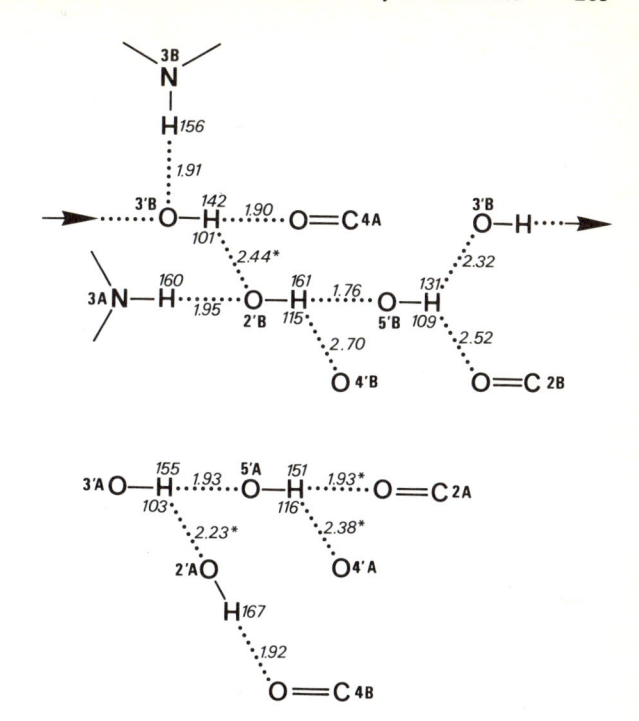

Fig. 17.14. 6-Methyluridine [MEYRID]; *, intramolecular

Fig. 17.15. 5-Methyluridine·½H₂O [MEURID]

Fig. 17.16. 6-Methyl-2'-deoxyuridine [MEDOUR]; *, intramolecular

Fig. 17.17. 5-Methoxyuridine [MXURID10]; *, intramolecular; O(2') and O(3') are in the same molecule, see Fig. 17.8

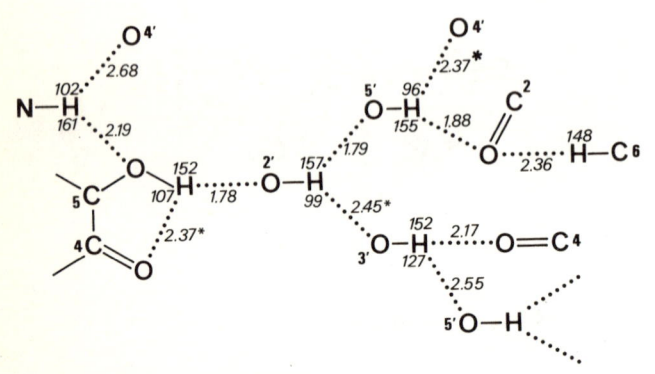

Fig. 17.18. 5-Hydroxyuridine [HXURID]; *, intramolecular

uridine [MEYRID] (Fig. 17.14), the O(5')–H makes a three-center bond, both components of which are intramolecular and indicated by *.

There is good evidence for an intramolecular C(6)–H···O(5') hydrogen bond (*) of 2.17 Å and 162° in the crystal structure of *5-methoxyuridine* [MXURID10] (Fig. 17.17). The hydrogen bonding consists of two finite chains originating at N(3)–H as a three-center double donor (see also Fig. 17.8).

The rather complex structure of *5-hydroxyuridine* [HXURID] includes three intramolecular hydrogen bonds (*) O(5)–H···O(4), O(2')–H···O(3'), O(5')–H···O(4') which are the minor components of three-center bonds (Fig. 17.18). It consists of three finite chains which are joined at the O(2')–H. There is evidence for a C(6)–H···O=C bond of 2.36 Å and 148°.

The hydrogen-bonding scheme of *5-aminouridine* [AMURID] (Fig. 17.19) contains a number of interesting features. There are three intramolecular hydrogen bonds (*). Two are components of three-center bonds, N(H)–H···O=C(4) in the uracil and O(3')–H···O(2') in the ribosyl moieties, and the third is a C(6)–H···O(5') bond of 2.54 Å and 151° between the uracil and the primary alcohol O(5'). The O(3')–H group donates a three-center bond with its weak component accepted by the –NH$_2$ group. This amino group is *not* really conjugated and is therefore assumed to be pyramidal (see remarks on p. 278).

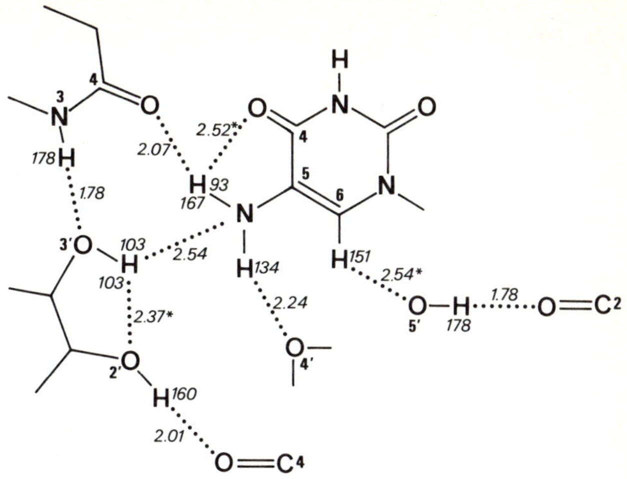

General Hydrogen-Bonding Patterns in Nucleoside and Nucleotide Crystal Structures 285

Fig. 17.19. 5-Aminouridine [AMURID]; *, intramolecular

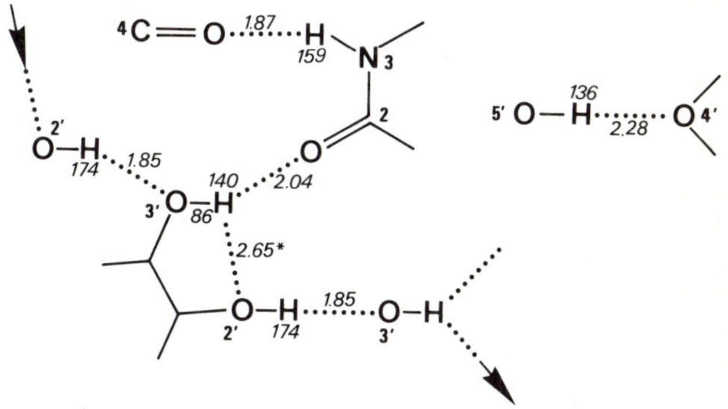

Fig. 17.20. 5-Chlorouridine [CLURID01]; *, intramolecular

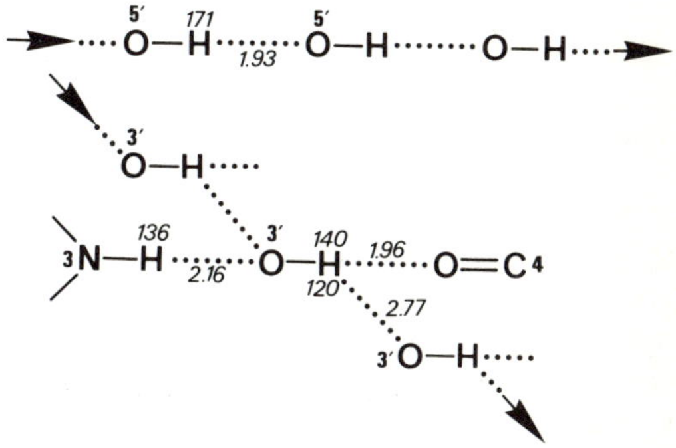

Fig. 17.21. 5-Chloro-2′-deoxyuridine [CLDOUR]

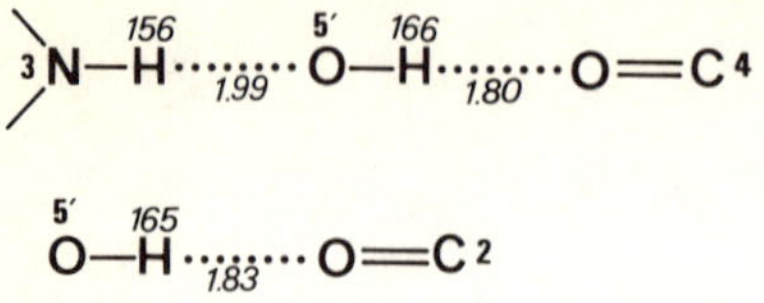

```
      156       5'  166
 \              
 3N—H······O—H······O=C 4
 /    1.99      1.80


  5'  165
  O—H······O=C 2
      1.83
```

Fig. 17.22. 2'-Chloro-2'-deoxyuridine [CDURID]

```
         169              175              168            154            169
    ···O—H······O—H······O—H······O—H······O—H······→
       3'A 1.75   5'A 1.86  5'B 1.74  3'B      3'A
                                          1.96
                                    2.35**.
                                         2'B O—H······O=C
                                               1.98      4A
                                             :2.10
      \   135                          142.:
   3AN—H······O 4'A                2'A O—H
      /  2.02                       132'.
                                         .2.71
      \   124
   3BN—H······O 4'B                  O=C 4B
      /  2.14
```

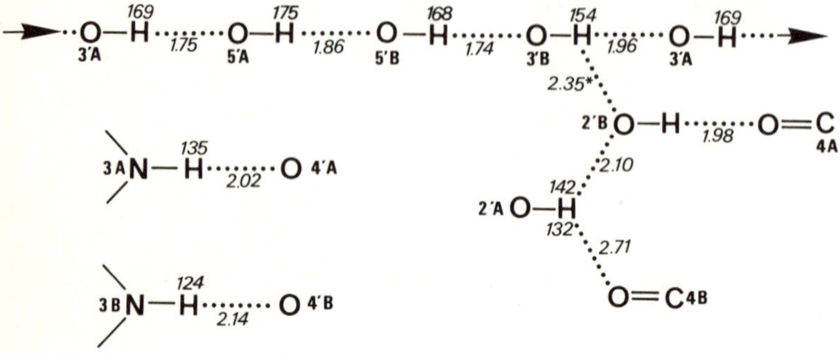

Fig. 17.23. 6-Azauridine [AZURID10]; *, intra-molecular

In the crystal structure of *5-chlorouridine* [CLURID01], the inter- and intramolecular (*) O(2')–H···O(3')–H···O(2')–H bonds form an infinite chain (Fig. 17.20). The primary alcohol O(5')–H forms only a weak two-center bond to a ribose ring oxygen which is not part of the major pattern.

The crystal structure of *5-chloro-2'-deoxyuridine* [CLDOUR] has a very simple pattern with two infinite chains (Fig. 17.21). One is through the O(5')–H groups, and the other through the minor component of a three-center bond formed by O(3')–H. It strongly resembles a carbohydrate hydrogen-bonding pattern. A very simple scheme is also formed by *2'-chloro-2'-deoxyuridine* [CDURID] with a short finite chain and a separate link from the primary alcohol (Fig. 17.22).

The crystal structure of *6-azauridine* [AZURID10] contains two symmetry-independent molecules **A** and **B** (Fig. 17.23). The O(3')–H and O(5')–H of both molecules form an infinite chain, to which is linked a finite chain through the minor O(3')–H···O(2') intramolecular component (*) of a three-center bond in molecule **B**. The aza-nitrogen N(6) atoms are *not* hydrogen-bond acceptors in either molecule.

The interesting feature of the hydrogen-bonding pattern in the crystal structure of *3-deaza-4-deoxyuridine* [RFURPD] (Fig. 17.24) is that it involves only O–H, >C=O, O< and C–H groups, because there is no N atom except for the glycosyl N(1). The O(2')–H donates a three-center chelate-type bond to both vicinal O(2'), O(3') atoms on an adjacent molecule, which form an intramolecular (*) O(3')–H···O(2') component of another three-center bond completing the triangle of hydrogen bonds. There is good evidence of a base···sugar C(6)–H···O(5') intramolecular bond (*) at 2.22 Å, 166°.

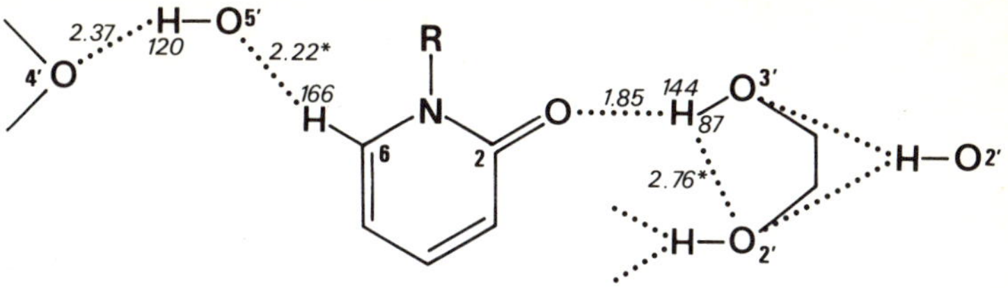

General Hydrogen-Bonding Patterns in Nucleoside and Nucleotide Crystal Structures 287

Fig. 17.24. 3-Deaza-4-deoxyuridine [RFURPD]; *, intramolecular

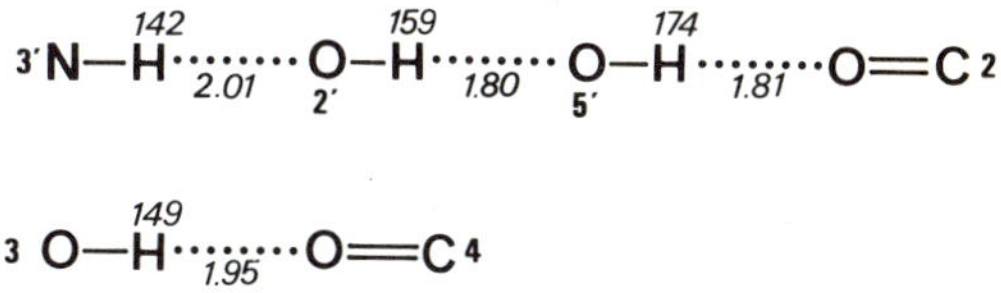

Fig. 17.25. 1-β-D-arabinofuranosyluracil [URARAF01]

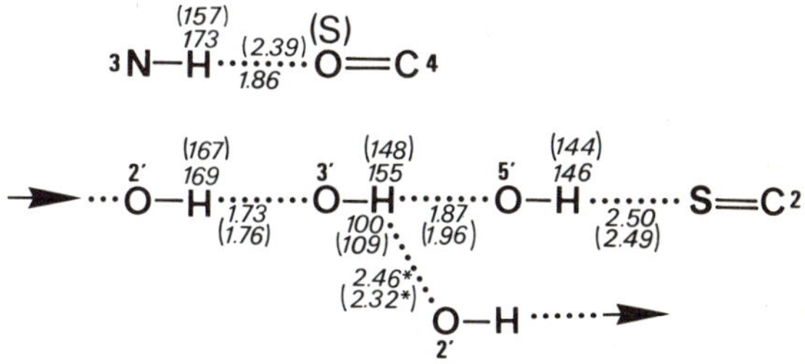

Fig. 17.26. 2-Thiouridine [TURIDN10] and 5,6-dihydro-2,4-dithiouridine [HDTURD10] (in parentheses); *, intramolecular

The crystal structure of *1-β-D-arabinofuranosyl uracil* [URARAF01] has a very simple system of two-center bonds which form one short finite chain and an isolated bond (Fig. 17.25).

In *2-thiouridine* [TURIDN10], the weak intramolecular (*) three-center bond component between O(3')−H···O(2') links the finite chains into an infinite chain (Fig. 17.26). The N−H···O=C bond forms a separate link. A very similar pattern is observed in 5,6-dihydro-2,4-dithiouridine [HDTURD10], with the substitution of C(4)=S for C(4)=O. Values for bond distances and angles of the latter crystal structure are given in parentheses in Fig. 17.26.

The hydrogen-bonding structure in *uridine-3'-phosphate·H_2O* [URIDMP10] contains a homodromic cyclic pentamer arrangement (Fig. 17.27). The two P−O−H groups form strong bonds, P−O−H···O=P and P−O−H···O_W (see also Fig. 17.11a). There is evidence of two C−H···O bonds − one intramolecular (*) C(6)−H···O(5'), at 2.29 Å and 171°, the other inter- and intramolecular from the ribosyl C(2')−H···O=C(4), at 2.56 Å and 156° and (*) to O(5') at 2.55 Å and 111°.

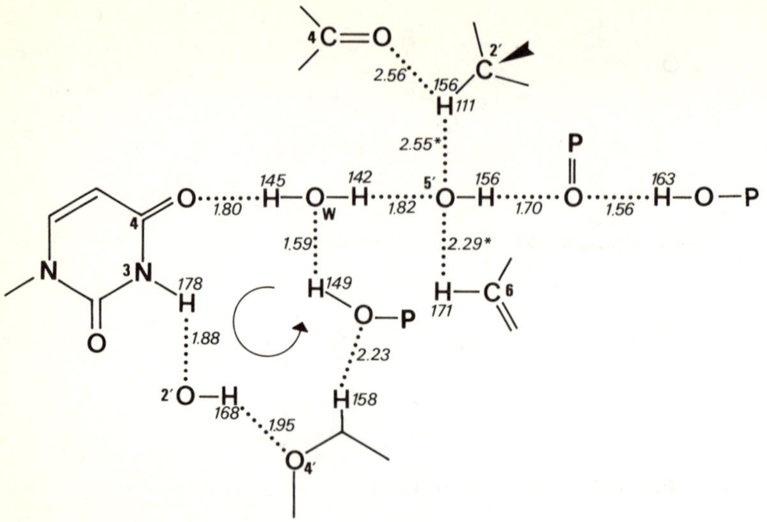

Fig. 17.27. Uridine-3'-phosphate·H₂O [URIDMP10]; ∗, intramolecular

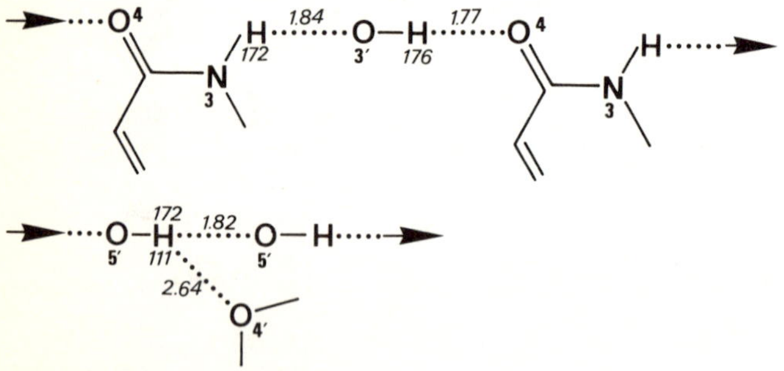

Fig. 17.28. Thymidine [THYDIN]

Selected crystal structures of thymidine and its derivatives. The hydrogen-bonding patterns of thymidine and its derivatives are comparable to those of the uridine derivatives, but relatively simple and they have a uniformity which is rare. In three of the four structures, infinite chains involving σ- and π-cooperativity occur through the adjacent donor and acceptor groups of the thymine ring.

The hydrogen bonding in the crystal structure of *thymidine* [THYDIN] consists of two infinite chains (Fig. 17.28), one with both σ- and π-cooperativity. An unusual feature is that the C(2)=O is not included in the hydrogen bonding. A similar scheme is found in *6-azathymidine* [AZTYMD] (Fig. 17.29) except that both O(3')–H and O(5')–H are included in one infinite chain.

The crystal structure of *thymidine 5'-carboxylic acid* [TYMCXA] is unusual in that it has two molecules in the asymmetric unit with almost identical hydrogen-bonding schemes related by a noncrystallographic pseudo-twofold axis of symmetry. In Fig. 17.30, data for molecule **A** given without, those for molecule **B** with

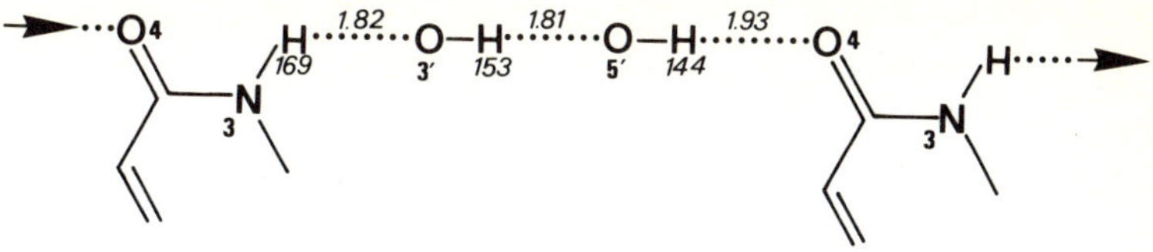

Fig. 17.29. 6-Azathymidine [AZTYMD]

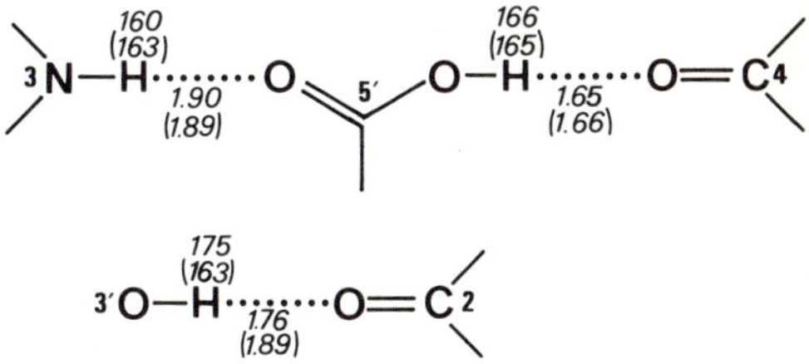

Fig. 17.30. Thymidine-5'-carboxylic acid [TYMCXA]

parentheses. It contains the same infinite chain structure through the N–H···O=C groups with the carboxylic acid group with π-cooperativity in place of the one or two hydroxyls with σ-cooperativity in the preceding structures (Figs. 17.28, 17.29). In this case the C(2)=O group is included in a separate bond from O(3')–H.

Selected crystal structures of cytidine and its derivatives. The cytidine-related nucleoside and nucleotide crystal structures for which reliable hydrogen-bond patterns can be reported include five nucleotides. Amongst the nucleotides there is more variety than with the uridine compounds. There are three hydrates, four hydrochlorides, and a nitrate. The conjugated NH_2 group is always involved as a donor, but never as an acceptor. Several different configurations are observed:

with eleven examples

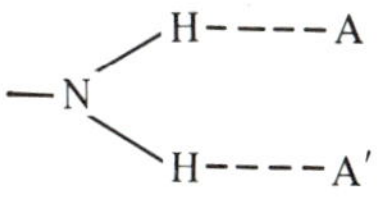

with five examples

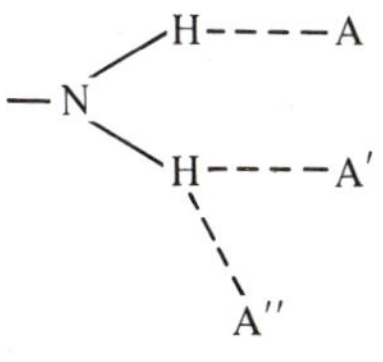

with three examples

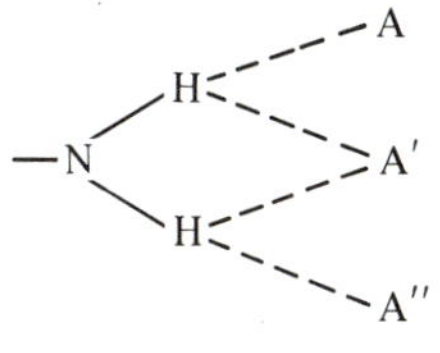

and one example of

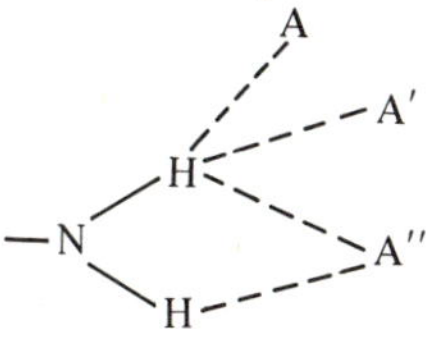

It is interesting to note that the same four configurations are observed with water molecules, as described in Part IV, Chapter 21.

There are two examples of chelated three-center bonds:

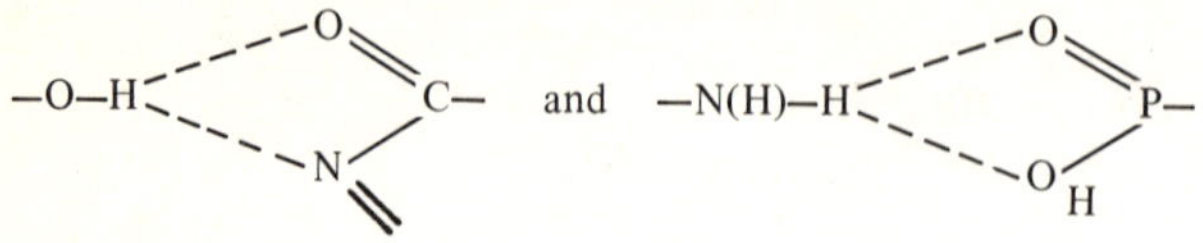

and an unusual trimer configuration:

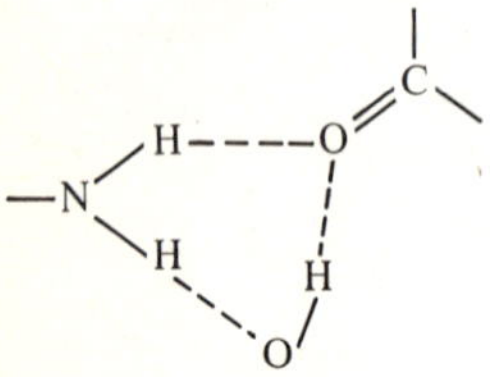

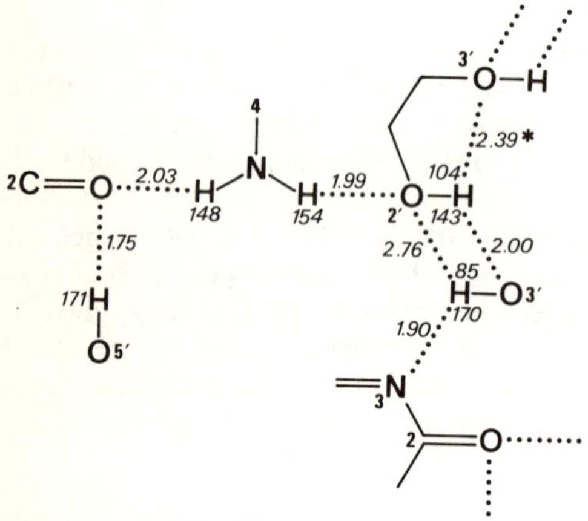

Fig. 17.31. Cytidine [CYTIDI10]; *, intramolecular

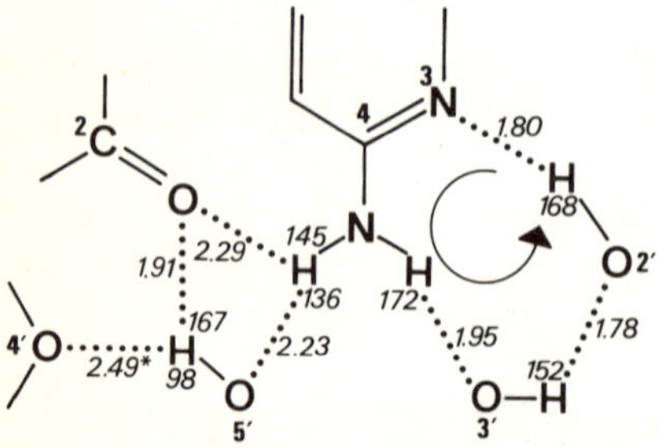

Fig. 17.32. α-Cytidine [ACYTID]; *, intramolecular

The hydrogen bonding in the crystal structure of *cytidine* [CYTIDI10] forms an infinite chain through the adjacent groups of the cytosine ring (Fig. 17.31). All functional groups are involved except the ring oxygen O(4'). This structure is reported to have a three-center hydrogen bond with O(2')–H···O(3') as minor intramolecular compound (*), and an unusual tandem arrangement of weak O–H···O–H bonds. This 1965 analysis requires confirmation by modern methods, preferably by neutron diffraction.

The hydrogen-bond pattern in *α-cytidine* [ACYTID] contains two cyclic trimer systems (Fig. 17.32). One is homodromic and involves π- and σ-cooperativity, the other has a weak intramolecular bond (*) from O(5')–H to the ring oxygen O(4').

The hydrogen-bond structure of *cytidinium nitrate* [CYTIDN] (Fig. 17.33) contains the unusual bifurcated/three-center bond configuration, with strong bonds to the nitro oxygens (2.01 Å and 1.94 Å) and weak bonds (2.74 Å and 2.58 Å) to the hydroxyl oxygen O(3'). There is evidence of a three-center intramolecular bond (*) with C(6)–H as donor and O(4'),O(5') as acceptors, similar to that observed in O(2')-methylcytidine (Fig. 17.36), and there is another three-center bond with the intramolecular O(2')–H···O(3') as minor component (*). The cytosine moiety is protonated at N(3).

With two independent molecules **A** and **B** in the asymmetric unit (Fig. 17.34), the hydrogen-bond structure of *2'-deoxycytidine* [DXCYTD] contains an interesting, almost symmetrical, cyclic tetramer, to which is attached a finite chain of three hydroxyl groups.

In *3-deazacytidine* [DAZCYT10], there is an infinite chain of O–H···O–H···O–H bonds through the intramolecular O(3')–H···O(2') bond (*) to which are linked two short finite chains (Fig. 17.35).

A complex hydrogen-bonding system is found in *O(2')-methylcytidine* [OMCYTD20] with two independent molecules **A**, **B** in the asymmetric unit (Fig. 17.36). It includes a chelated three-center bond, O–H⋯ $\begin{smallmatrix}O\\\\N\end{smallmatrix}$ C, an intramolecular O(3')–H···O(2') bond (*), and evidence for a three-center C–H···O bond in which both components are intramolecular (*) similar to that observed in cytidinium nitrate (Fig. 17.33).

The crystal structure of *1-β-D-arabinofuranosyl cytosine* [ARBCYT10] contains hydrogen-bonded NH$_2$, C=O, O–H trimers, which are linked together by the major component of the N(H)H three-center bond (Fig. 17.37).

In *1-α-D-xylofuranosyl cytosine* [XYFCYT10], there is an antidromic infinite chain structure through the minor component of the O–H···O=C three-center bond donated by O(2')–H (Fig. 17.38).

The crystal structure of *O(3')-methyl-1-β-D-arabinofuranosyl cytosine* [MARAFC] (Fig. 17.39) contains a very compact trimer involving a chelated three-center bond formed by O(5')–H as donor and cytosine N(3),O(2) as acceptors, with two external bonds, from the O(2')–H and the amino group.

The hydrogen bonding in *2'-deoxycytidine·HCl* [DOCYTC] centers around the four-coordinated chloride ion (Fig. 17.40). It contains an unusual bifurcated bond

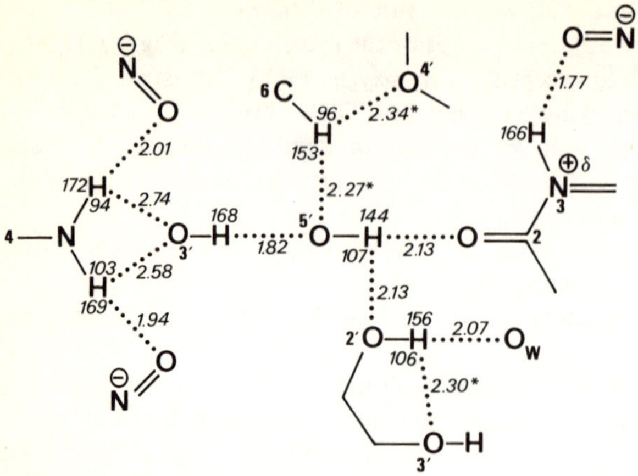

Fig. 17.33. Cytidinium nitrate [CYTIDN]; ∗, intramolecular

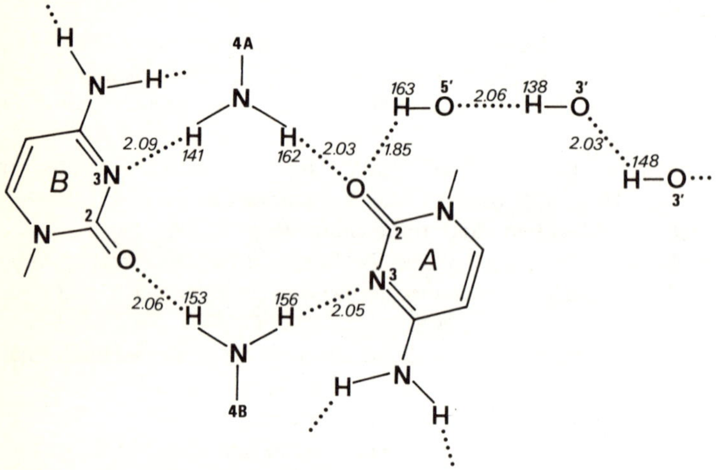

Fig. 17.34. 2′-Deoxycytidine [DXCYTD]

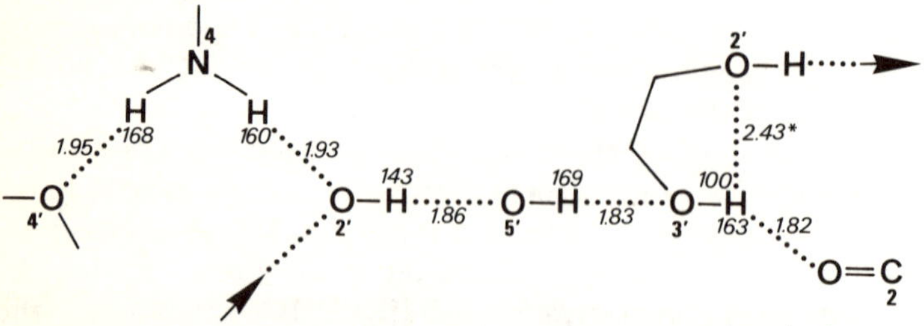

Fig. 17.35. 3-Deazacytidine [DAZCYT10]; ∗, intramolecular

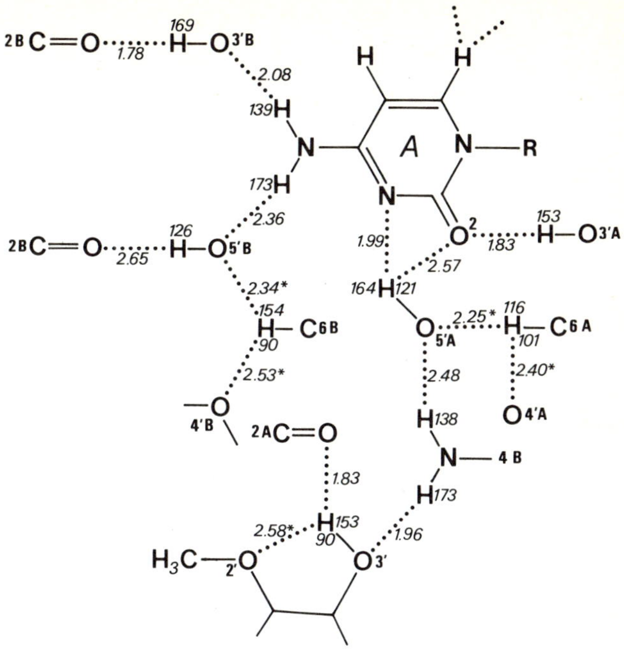

General Hydrogen-Bonding Patterns in Nucleoside and Nucleotide Crystal Structures 293

Fig. 17.36. O(2')-methylcytidine [OMCYTD20]; *, intramolecular

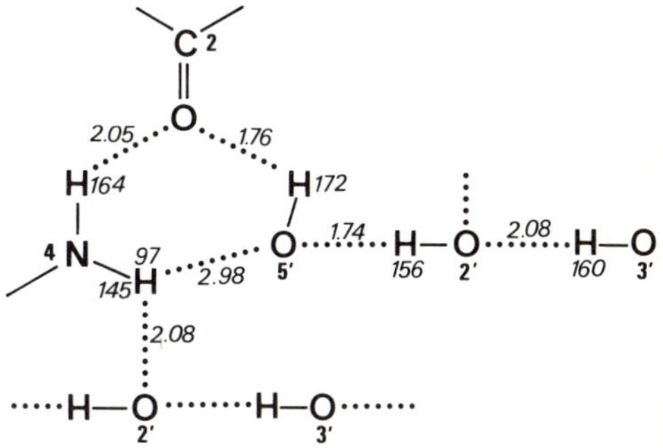

Fig. 17.37. 1-β-D-Arabinofuranosylcytosine [ARBCYT10]

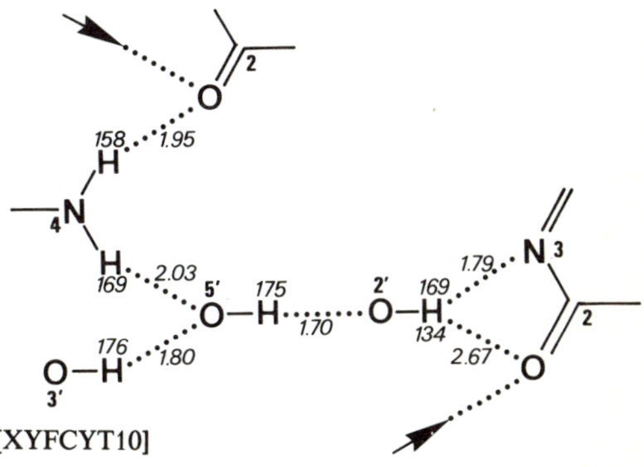

Fig. 17.38. 1-α-D-xylofuranosylcytosine [XYFCYT10]

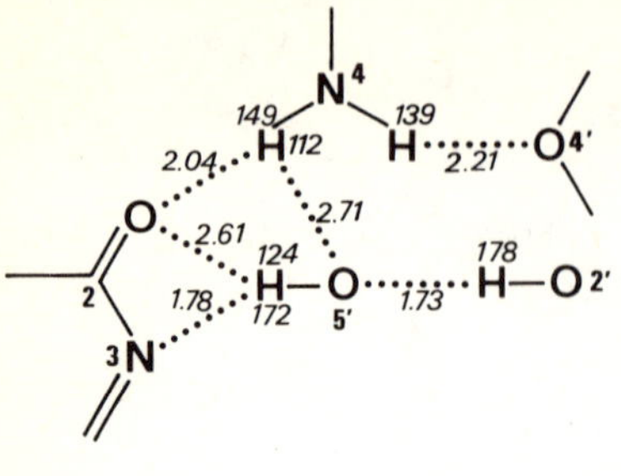

Fig. 17.39. O(3′)-methyl-1-β-D-arabinofuranosylcytosine [MARAFC]

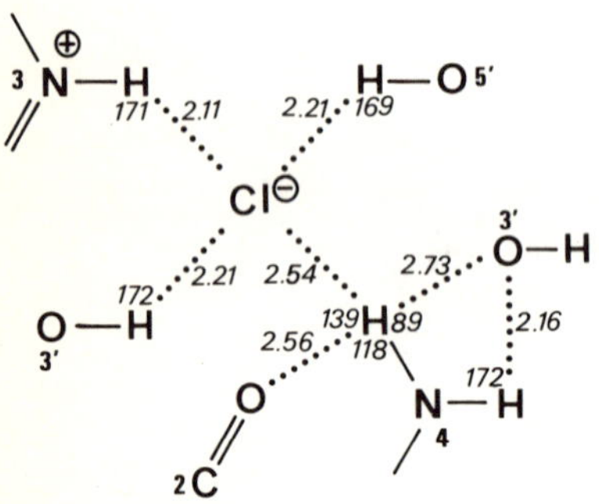

Fig. 17.40. 2′-Deoxycytidine·HCl [DOCYTC]

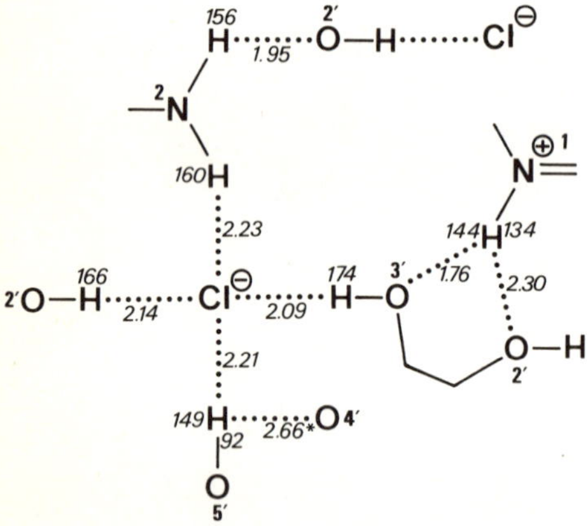

Fig. 17.41. Pseudo-isocytidine·HCl [PSCYTD]; *, intramolecular

from the amino group to the O(3′)–H oxygen, in combination with a four-center bond involving one of the two amino hydrogens.

The Cl$^\ominus$, NH$_2$ and O(2′)–H groups in *pseudo-isocytidine·HCl* [PSCYTD] form infinite chains (Fig. 17.41). The positive charge is formally on N(1)–H, but neither of the N–H groups bonds directly to the four-coordinated chloride ions.

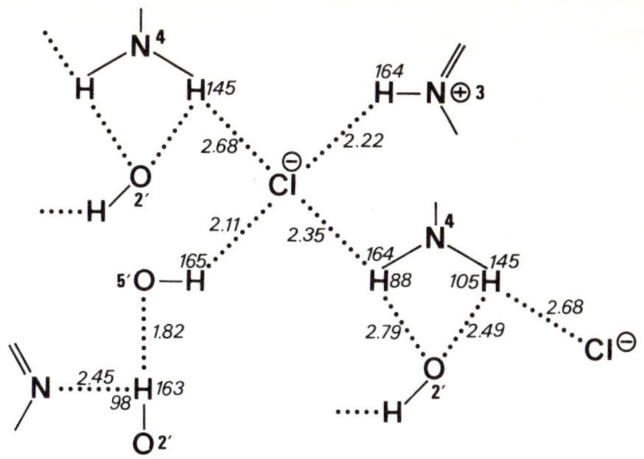

General Hydrogen-Bonding Patterns in Nucleoside and Nucleotide Crystal Structures 295

Fig. 17.42. 1-β-D-arabinofuranosylcytosine·HCl [ARFCYT10]; *, intramolecular

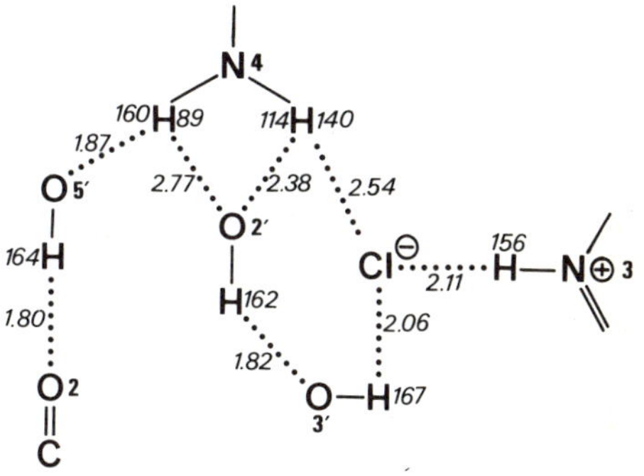

Fig. 17.43. 1-α-D-xylofuranosylcytosine ·HCl [XFURCC10]

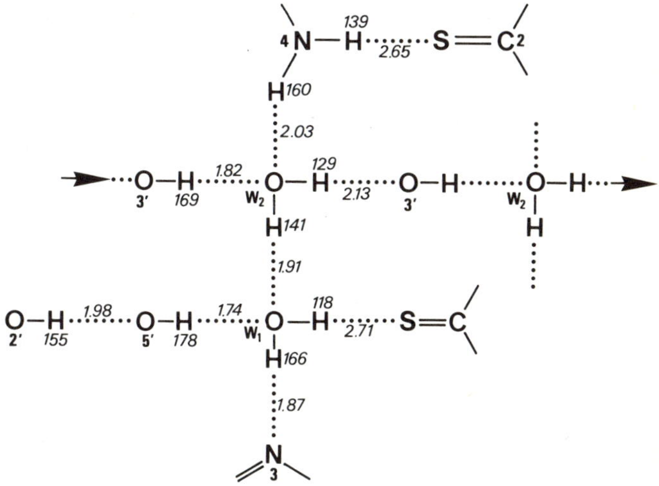

Fig. 17.44. 2-Thiocytidine·2H$_2$O [TCYTDH]

The hydrogen-bonding pattern in *1-β-D-arabinofuranosyl cytosine·HCl* [ARFCYT10] (Fig. 17.42) centers around the four-coordinated chloride ion. It contains a bifurcated/three-center bond configuration from the NH$_2$ to the oxygen atom of O(2′)–H and to two chloride ions. The $\overset{\oplus}{N}$(3)–H group binds directly to Cl$^\ominus$.

Like the arabinosyl hydrochloride, the structure of *1-α-D-xylofuranosylcytosine·HCl* [XFURCC10] (Fig. 17.43) also contains a bifurcated/three-center bond system. The chloride ion is three-coordinated and directly hydrogen bonded to $\overset{\oplus}{N}$(3)–H.

The presence of two water molecules in the structure of *2-thiocytidine dihydrate* [TCYTDH] results in a hydrogen-bonding pattern (Fig. 17.44) very similar to that observed in the carbohydrate hydrates. It consists of an infinite chain crosslinked to two finite chains through the four-coordinated water molecules.

In the zwitterionic cytidylic acids and their derivatives, the hydrogen-bonding structures are remarkably similar. They contain as recurrent motif phosphate groups linked by amino groups, with some variations due to O(2′) and O(3′) hydroxyls and water molecules.

The crystal structure of *cytidine-3′-phosphate (monoclinic)* [CYTIAC01] (Fig. 17.11 b) contains a chelated three-center N–H 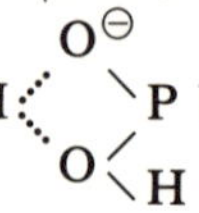 bond with O$^\ominus$ and O–H to P, and a strongly bonded cyclic pentamer involving the NH$_2$, P=O and three O–H groups. The ring oxygen O(4′) is not involved in the hydrogen bonding. The hydrogen-bonding structure of the *orthorhombic form* [CYTIAC] is simpler (Fig. 17.45) than that of the monoclinic form. In both structures there is an intramolecular O(2′)–H···O(3′) bond (∗) as the minor component of a three-center bond.

The hydrogen-bonding schemes of the three nucleotides *cytidine-2′,3′-cyclic phosphate* [CYCYPH10], *deoxycytidine-5′-phosphate·H$_2$O* [DOCYPO], *arabinosylcytosine-2′,5′-cyclic phosphate* [ARACYP] are very similar with a sequence $\overset{\oplus}{N}$(3)–H/phosphate/amino group/phosphate/water or hydroxyl O–H. The simplest structure is that of cytidine-2′,3′-cyclic phosphate, (Fig. 17.46). In deoxycytidine-5′-phosphate·H$_2$O, the water molecule is three-coordinated and links a short side chain. Because the phosphate group is protonated in this monoester, there is an infinite, homodromic hydrogen-bond chain utilizing the π-cooperativi-

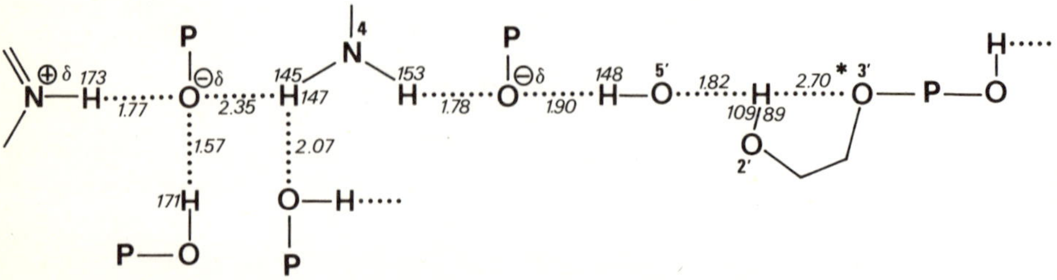

Fig. 17.45. Cytidine-3′-phosphate (orthorhombic) [CYTIAC]; ∗, intramolecular

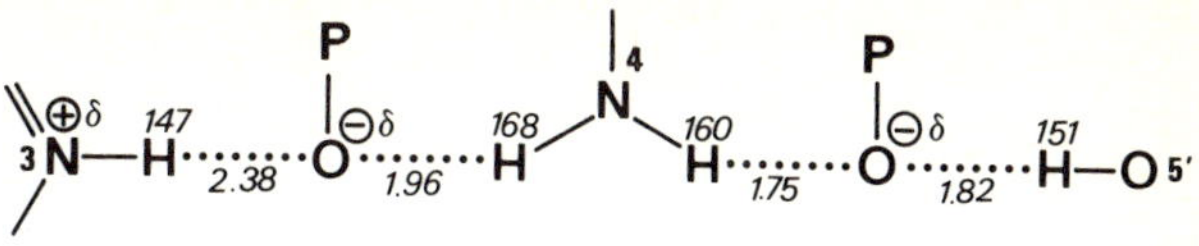

Fig. 17.46. Cytidine-2′,3′-cyclic phophate [CYCYPH10]

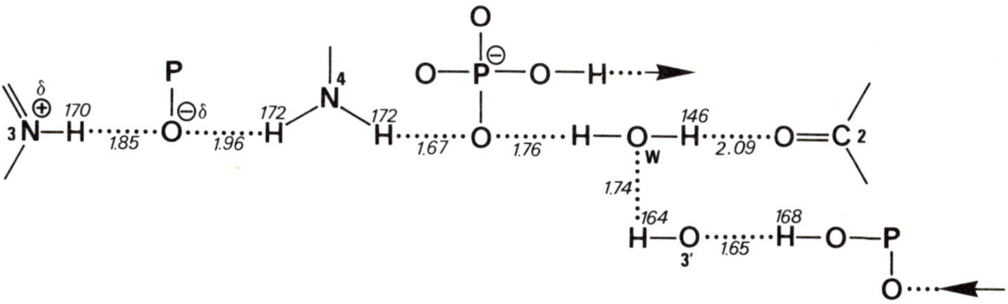

Fig. 17.47. Deoxycytidine-5′-phosphate·H₂O [DOCYPO]

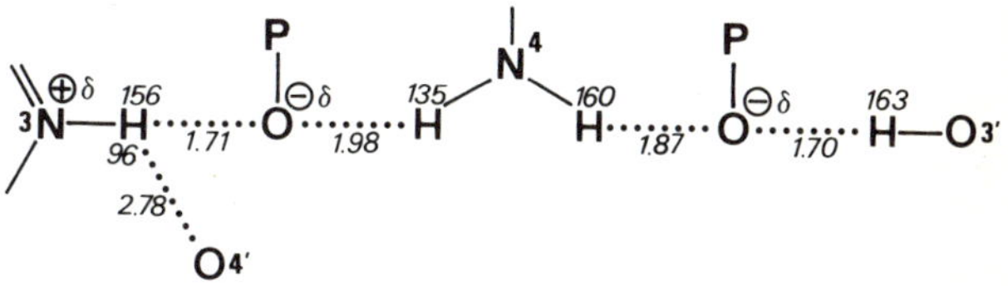

Fig. 17.48. Arabinosylcytosine-2′,5′-cyclic phosphate [ARACYP]

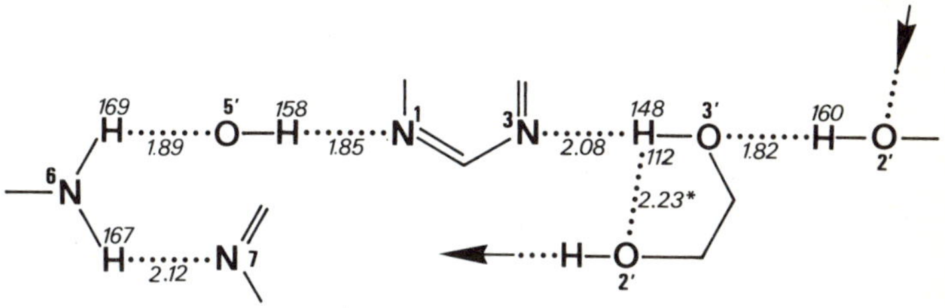

Fig. 17.49. Adenosine [ADENOS10]; *, intramolecular

ty of the phosphate group (Fig. 17.47). In arabinosylcytosine-2′,5′-cyclic phosphate, the N(3)–H group forms a three-center bond to the ribose ring oxygen (Fig. 17.48).

Compared with the pyrimidine series, there is less hydrogen bond data available for the purine-derived nucleosides and nucleotides. The structures which provide hydrogen-bond data are discussed in the following.

Selected crystal structures of adenosine, guanosine, inosine and their derivatives. In *adenosine* [ADENOS10], the hydrogen-bond structure involves all func-

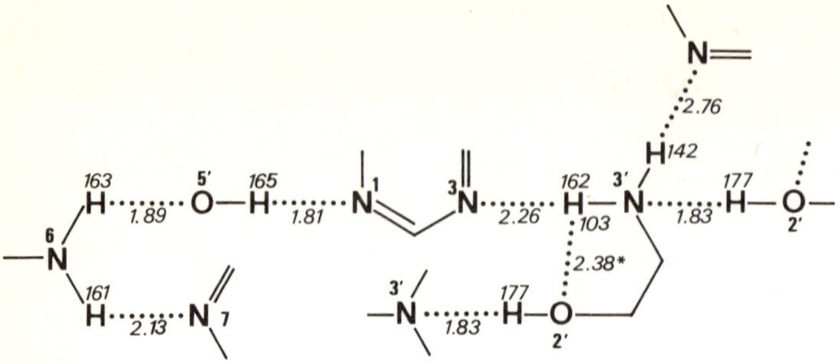

Fig. 17.50. 3'-Amino-3'-deoxyadenosine [AMOADA]; *, intramolecular

tional groups except the ring oxygen O(4') (Fig. 17.49). There are three short finite chains, donated by the N(6)H$_2$ group and the O(2')–H hydroxyl and accepted by adenine nitrogens N(1), N(3), N(7). A homodromic infinite chain is formed by O(2')–H and O(3')–H hydroxyls linked by the minor compound of an intramolecular hydrogen bond (*) between the vicinal ribose hydroxyls.

In *3'-amino-3'-deoxyadenosine* [AMOADA], a very similar hydrogen-bonding pattern is observed, with N(3')–H replacing the O(3')–H group (see Fig. 17.50). It provides an example where an amino group accepts a hydrogen bond. However, this is an aliphatic amino group and should not be confused with the π-conjugated amino groups of the cytosine, adenine, and guanine bases which do not act as hydrogen-bond or proton acceptors.

The crystal structure of *adenosine HCl* [ADOSHC] (Fig. 17.51) contains a hydrogen-bonded tetramer involving NH$_2$, O(2')–H, O(5')–H and ·Cl$^\ominus$. The charge is on N(1)–H. Neither N(3) nor N(7) are hydrogen-bond acceptors.

The formal charge in *9-β-D-arabinofuranosyl adenine HCl* [ARFUAD01] is on N(1)–H, which is not directly hydrogen-bonded to the three-coordinated chloride ion (Fig. 17.52).

The structure of *8-thiooxoadenosine hydrate* [TOADEN] (Fig. 17.53) contains an infinite chain of –O–H···O$_W$–H···O–H–. There is an interesting chelation involving the N(6) amino group as double donor and the ribose O(2') and O(3') hydroxyl groups as acceptors. In the modified adenine heterocycle, the 8-thio group is in the keto form and accepts a hydrogen bond, and the N(7) bears a hydrogen and acts as donor.

In the crystal structure of *adenosine-3'-phosphate·2H$_2$O* [ADPOSD] (Fig. 17.54), there is a complex network of hydrogen bonds which includes a cyclic tetramer formed by P̄=O, –NH$_2$, ⟩N, H$_2$O and a chelated three-center bond involving N(1)–H and P̄O$_2$. The hydrogen-bonding network around the phosphate and amino groups is reminiscent of the patterns found in the zwitterionic cytidylic acids (see Fig. 17.11 b, and 17.45 to 17.48). As in deoxycytidine-5'-phosphate·H$_2$O (Fig. 17.47), the phosphate monoester group is protonated. It forms a homodromic infinite hydrogen-bonded chain with water molecule W1 and contributes to the stability by π-cooperativity.

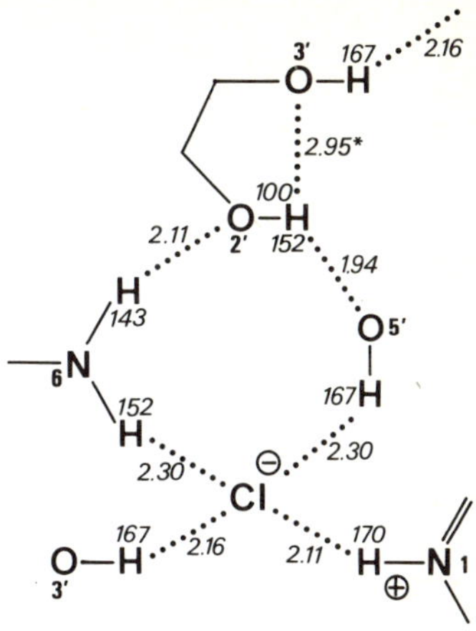

General Hydrogen-Bonding Patterns in Nucleoside and Nucleotide Crystal Structures 299

Fig. 17.51. Adenosine·HCl [ADOSHC]; *, intramolecular

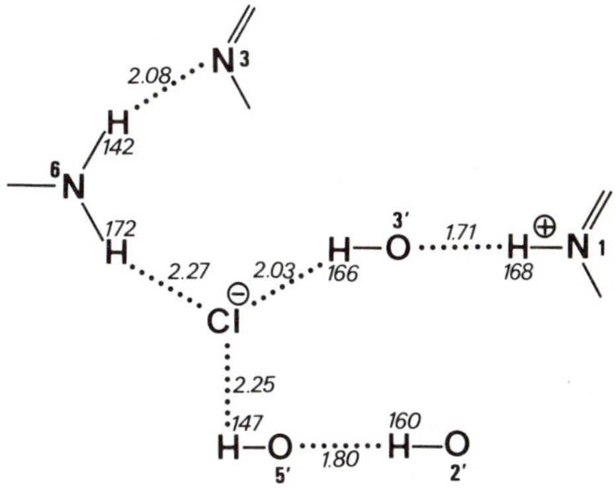

Fig. 17.52. 9-β-D-Arabinofuranosyladenine·HCl [ARFUAD01]

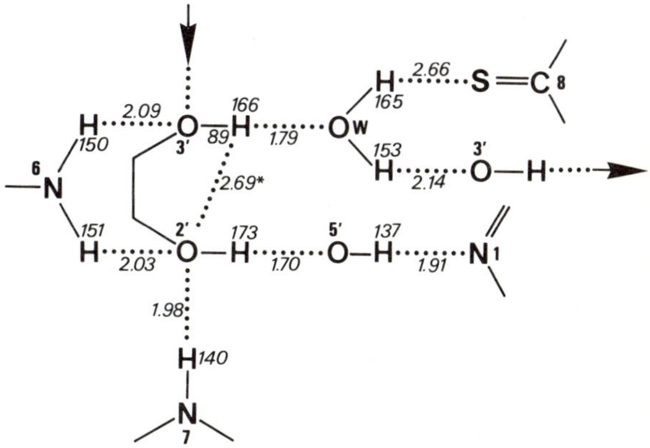

Fig. 17.53. 8-Thiooxoadenosine·H$_2$O [TOADEN]; *, intramolecular

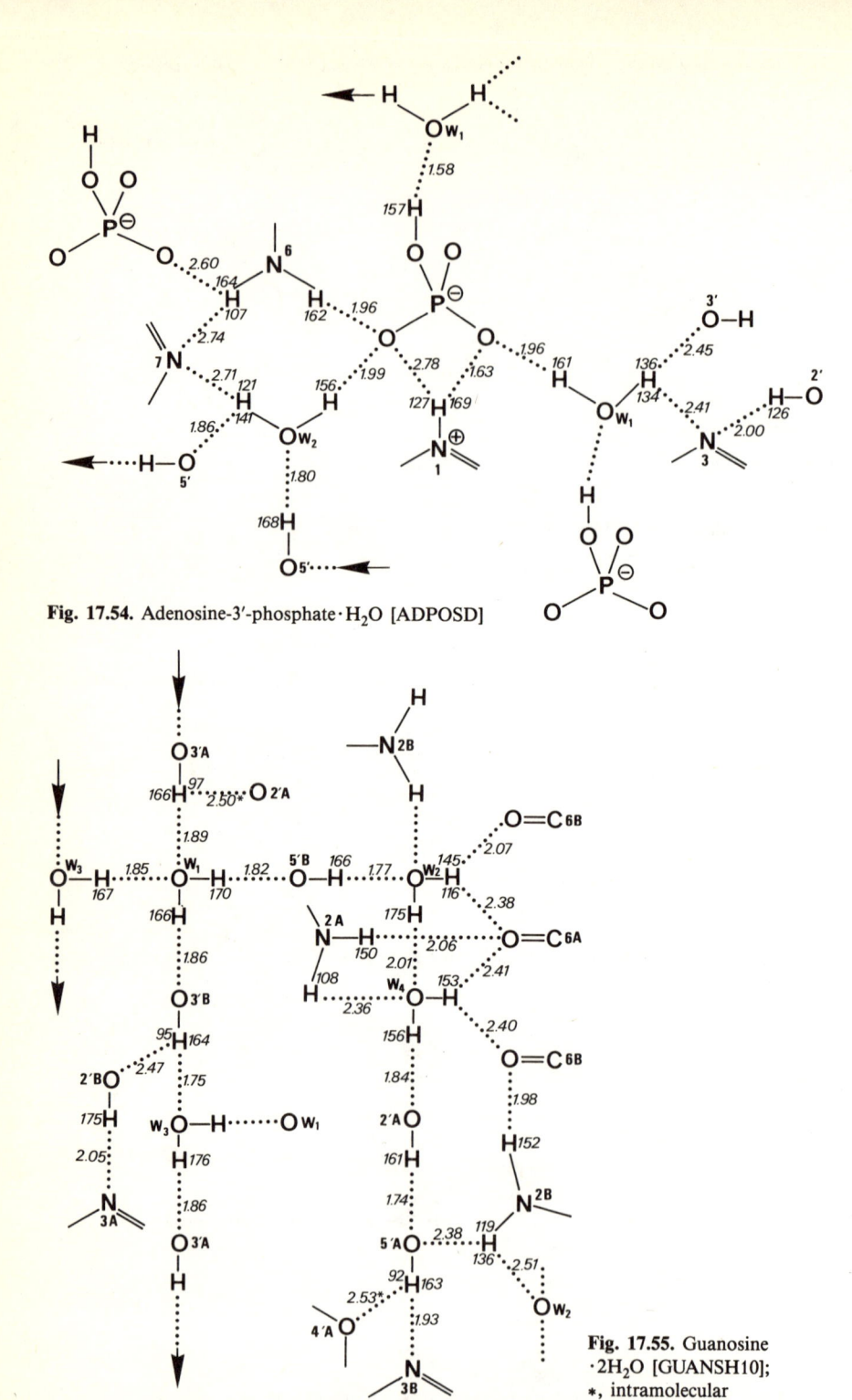

Fig. 17.54. Adenosine-3'-phosphate·H$_2$O [ADPOSD]

Fig. 17.55. Guanosine ·2H$_2$O [GUANSH10]; *, intramolecular

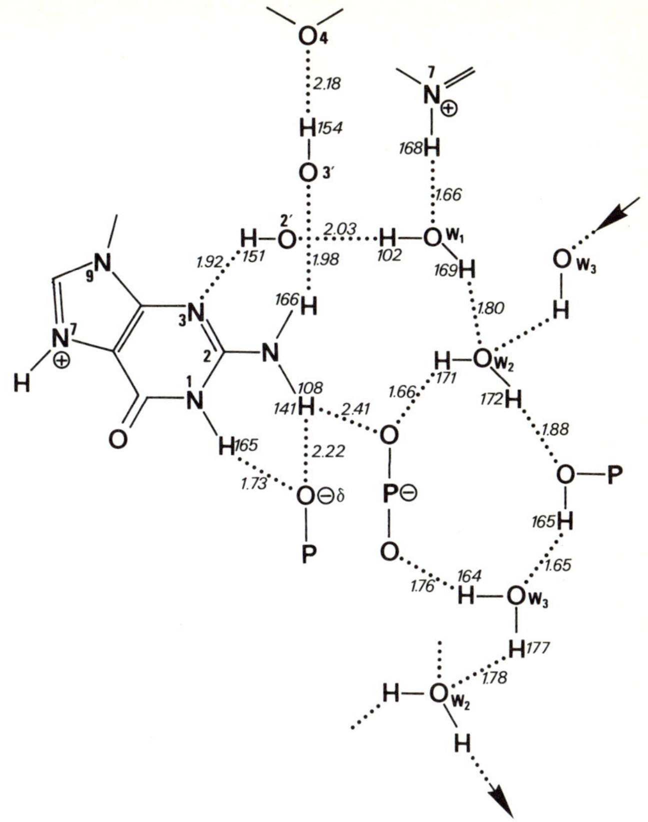

Fig. 17.56. Guanosine-5′-phosphate·3H$_2$O [GUANPH01]

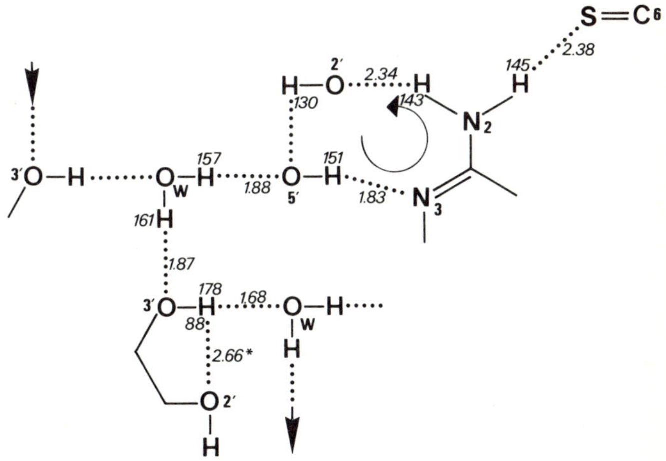

Fig. 17.57. 6-Thioguanosine·H$_2$O [TGUANS10]; ∗, intramolecular

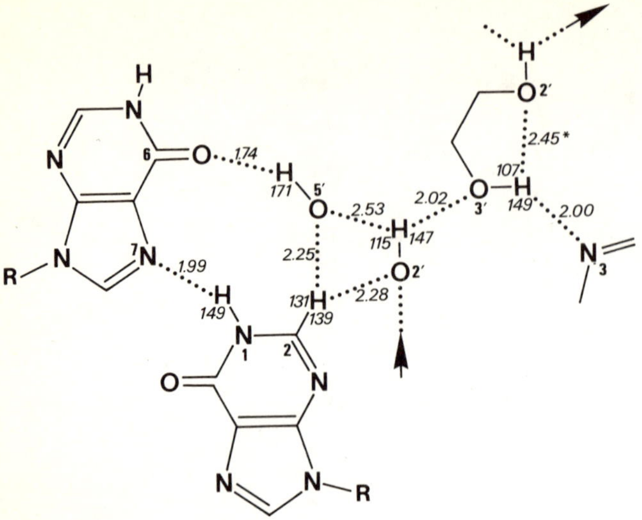

Fig. 17.58. Inosine (monoclinic) [INOSIN10]; *, intramolecular

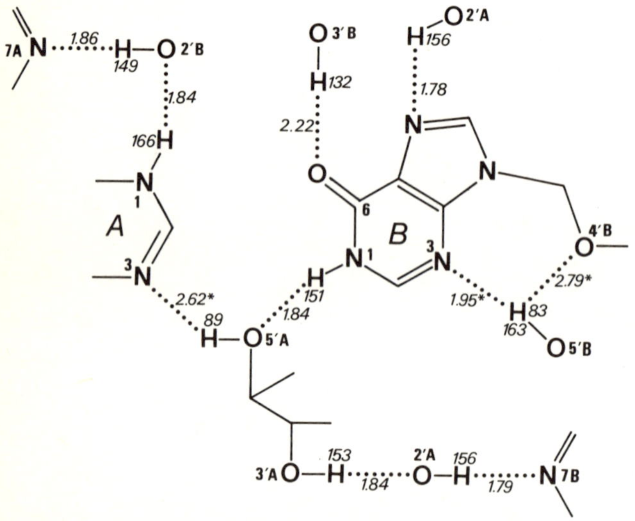

Fig. 17.59. Inosine (orthorhombic) [INOSIN11]; *, intramolecular

The three crystal structures of guanosine and its 6-mercapto derivative for which reliable data are available are all hydrated. The hydrogen-bond patterns of guanosine dihydrate and guanosine 5'-monophosphate hydrate are very complicated.

A complex hydrogen-bonding structure is observed in *guanosine dihydrate* [GUANSH10] (Fig. 17.55) with two nucleoside molecules and four waters in the

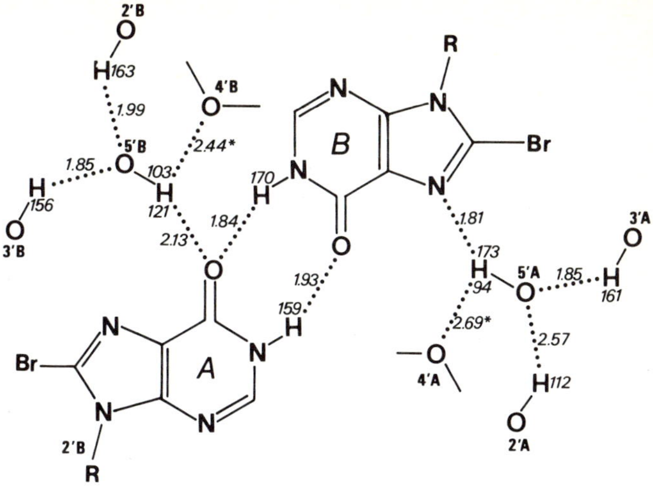

Fig. 17.60. 8-Bromoinosine [BRINOS10]; *, intramolecular

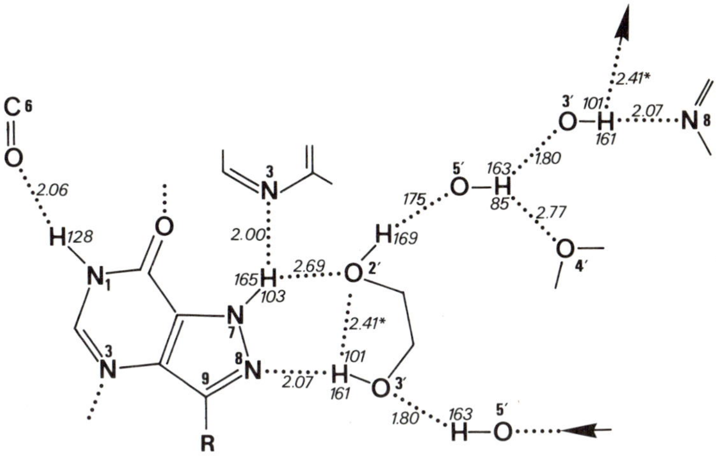

Fig. 17.61. 8-Aza-9-deazainosine [FORMYB01]; *, intramolecular

asymmetric unit. It consists of homodromic infinite and finite chains, with O(4'), O(6), and N(3), but not N(7), acting as chain terminators.

In the crystal structure of *guanosine 5'-phosphate·3H₂O* [GUANPH01] (Fig. 17.56) the free acid of the nucleotide is in the zwitterion form with N(7) protonated. There is a four-membered cycle formed by two water molecules and two phosphate groups which is part of an infinite, homodromic chain. The N(1)–H and amino N(2)–H form a chelated bifurcated hydrogen bond with a phosphate oxygen, and N(2)–H is further involved in a three-center bond with another phosphate oxygen atom.

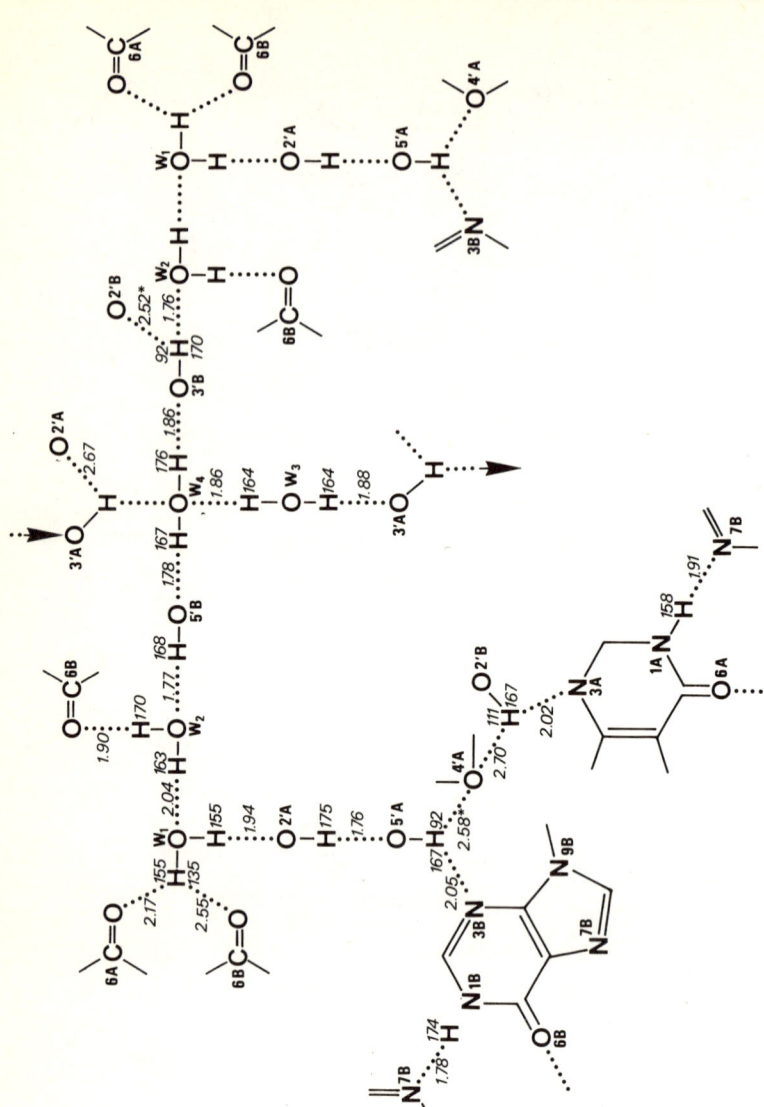

Fig. 17.62. Inosine·2H$_2$O [INOSND01]; *, intramolecular

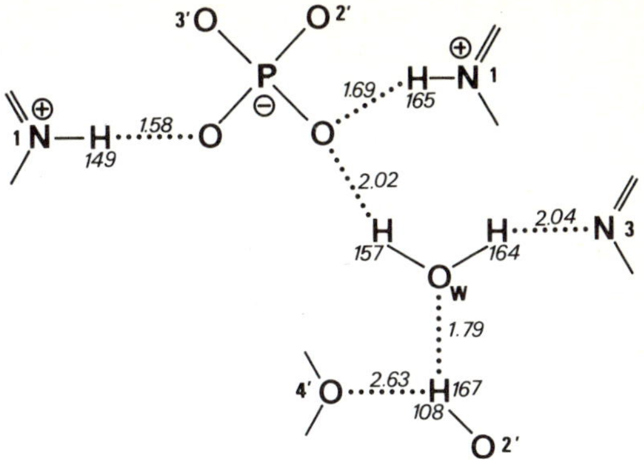

Fig. 17.63. Inosine-3′,5′-cyclic phosphate·H$_2$O [BEPRAP]

In *6-thioguanosine·H$_2$O* [TGUANS10], a homodromic infinite chain of O(3′)–H···O$_W$–H··· bonds is linked to a homodromic trimer loop with σ- and π-cooperativity (Fig. 17.57). O(3′)–H is engaged in a three-center bond with the minor component forming an intramolecular O(3′)–H···O(2′) hydrogen bond indicated by *.

In inosine, the NH$_2$ at C(2) on guanine is removed, thereby reducing the hydrogen-bonding capability by a donor group with two functional hydrogens. 8-Bromoinosine is a rare example where the self-association of the purine residues occurs in the crystal structure of a nucleoside.

All the functional groups are engaged in hydrogen bonding in *inosine (monoclinic)* [INOSIN10], with O(6), N(3), N(7) acting as chain terminators. An infinite homodromic chain formed by O(2′)–H and O(3′)–H groups involves the minor, intramolecular (*) component of a three-center bond donated by O(3′)–H. The most interesting feature of this hydrogen-bonding pattern is an almost symmetrical three-center interaction donated by the C(2)–H group (see Fig. 17.58).

There are two molecules **A**, **B** in the crystal asymmetric unit of *inosine (orthorhombic)* [INOSIN11] (Fig. 17.59). They are both in the syn conformation which is stabilized by intramolecular hydrogen bonds O(5′)–H···N(3) indicated by (*). Oxygen atoms O(4′) and O(6) of molecule A and the O(5′)–H of molecule B are not engaged as hydrogen-bond acceptors.

The crystal structure of *8-bromoinosine* [BRINOS10] with two molecules **A**, **B** in the asymmetric unit (Fig. 17.60) contains a centrosymmetric base-pair configuration. It has an interesting overall pseudosymmetry, linking the functional ribosyl groups of one molecule to the purine of the other.

8-Aza-9-deazainosine [FORMYB01] is the antibiotic formycin which contains a glycosyl C(1′)–C(9) link. The =N(8)–N(7)–H group is engaged in hydrogen bonding with ribose O(2′)–H, O(3′)–H hydroxyls involving a three-center interaction with the minor component intramolecular (*). It is part of an infinite, homodromic chain formed by all the O–H groups (Fig. 17.61).

All functional groups in two molecules **A**, **B** and four waters per crystal asymmetric unit are hydrogen bonded in *inosine dihydrate* [INOSND01] except for the ring oxygen O(4'B) (Fig. 17.62). The structure has an infinite antidromic chain (marked by arrow) intersected at a four-coordinated water molecule (W4) with a complex finite chain. This pattern is reminiscent of that of guanosine dihydrate (Fig. 17.55).

The nucleotide *inosine 3',5'-cyclic phosphate·H$_2$O* [BEPRAP] crystallizes in the zwitterionic form. The free oxygen atoms of the phosphate accept two strong, salt-type hydrogen bonds from two $\overset{\oplus}{N}$(1)–H groups. O(6) and N(7) are not involved in hydrogen bonding (Fig. 17.63).

Part III
Hydrogen Bonding in Biological Macromolecules

Chapter 18
O−H···O Hydrogen Bonding in Crystal Structures of Cyclic and Linear Oligoamyloses: Cyclodextrins, Maltotriose, and Maltohexaose

For the study of the hydrogen bonding, the cyclodextrins are unique in being molecules in the 1000- to 1300-Da range which provide excellent quality crystals for both X-ray and neutron diffraction experiments. Their hydrogen-bonding functionality is relatively simple in that they contain only hydroxyl groups and water molecules as donors and acceptors with the sugar ring and glycosidic linkage oxygen atoms as additional acceptors.[1] As will be shown, the crystal structures of cyclodextrin hydrates provide remarkable examples of the importance of σ-bond cooperativity in determining the hydrogen-bond structures. There are relatively few three-center bonds except for those intramolecularly to the glycosidic oxygen atoms. In the hydrates of the β- and γ-cyclodextrins, water molecules and hydroxyl groups display positional and/or rotational disorder and provide examples of the flip-flop type hydrogen-bond disorder.

In the crystal structures of the linear cyclodextrin analogs maltotriose and maltohexaose, the O−H hydrogen atoms could not be located. Because the hydrogen-bonding scheme remains obscure, these crystal structures will be discussed only in more general terms. They are, nevertheless, of great interest, since the two linear oligosaccharides form single- and double-stranded left-handed helical structures which relate to the structure of amylose.

18.1 The Cyclodextrins and Their Inclusion Complexes

The soluble fraction of starch consists of the linear amyloses which are composed of $\alpha(1-4)$-linked D-glucose units and adopt helical conformations. If hydrolyzing enzymes called glucanotransferases act upon the amyloses, individual turns of the helices are cleaved off and resealed to produce cyclically closed oligosaccharides. Because the glucanotransferases are not very specific with respect to their cutting sites, a family of oligosaccharides is obtained with 6, 7, and 8 glucoses per annulus

---

[1] Each glucose in a cyclodextrin has two acceptor functions: −O(4)− (which, in fact, rarely accepts a hydrogen bond), −O(5)−, and three hydroxyl groups. In addition, there are water of hydration molecules which act as double donors and as single or double acceptors (see Part IV, Chap. 21).

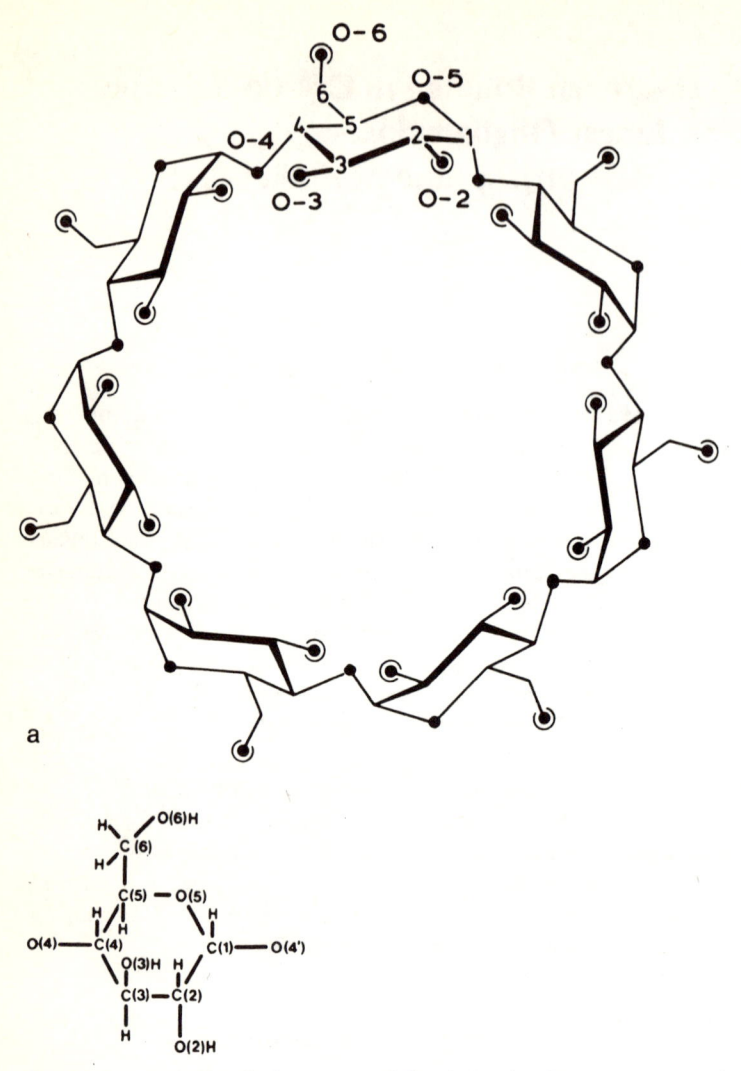

Fig. 18.1. a (top) Chemical structure of β-cyclodextrin. Oxygen atoms are indicated by *filled circles*, hydroxyl groups by ⊙, (bottom) the glucose numbering scheme. In the text, a *lower index* indicates to which glucose in the cyclodextrin ring an atom belongs; e.g., $O(2)_6$ is atom O(2) in glucose 6. **b** (see p. 311)

(Fig. 18.1). These are called the cyclohexa-, -hepta-, octa-amyloses, or α-, β-, γ-cyclodextrins [551–558].

Cyclodextrins are macrocycles with limited flexibility. In all the crystal structure analyses of cyclodextrins, the glucose units have the D-pyranose configuration with the normal $^4C_1$ chair conformation [556, 558, 559]. This chair form is nearly rigid, as illustrated by only small variations (<7°) of the intrapyranose ring tor-

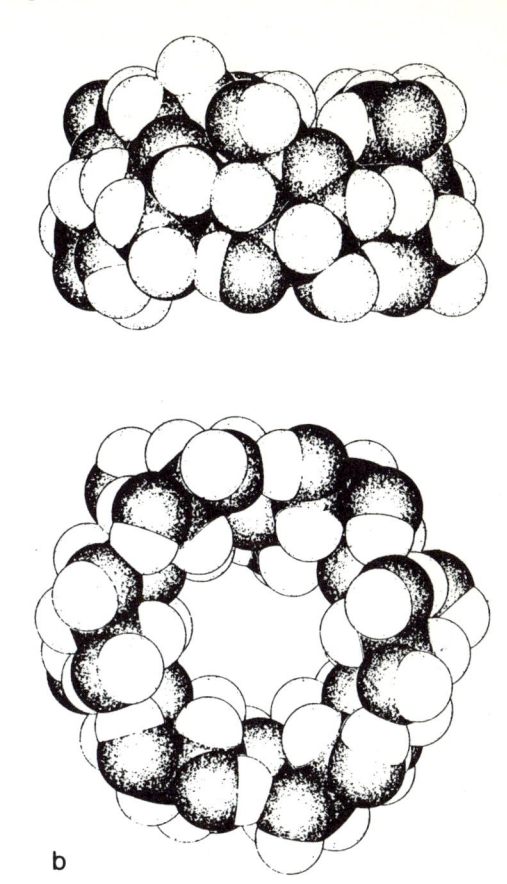

Fig. 18.1b. Space-filling plot in top and side views of α-cyclodextrin [558], drawn with atomic coordinates of the methanol inclusion complex (see also Fig. 18.8; all guest and water atoms are omitted). Note the conical form of α-cyclodextrin (top) and the annular shape (bottom) with the cavity lined by C–H hydrogen atoms. All O–H groups are on both rims of the cone

sion angles in individual glucose units. The glucoses can rotate relative to their neighbors about the α(1–4) glucosidic linkage bonds, but these rotations are also limited so that the torsion angles comprising this link, i.e., O(5)–C(1)–O(4′)–C(4′) and C(1)–O(4′)–C(4′)–C(5′), rarely vary by more than 30° (for atomic numbering see Fig. 18.1; primed and unprimed atoms belong to adjacent glucose units).

In contrast, the primary O(6)–H hydroxyl group can rotate about the exocyclic C(5)–C(6) bond. As with hexopyranoses discussed in Part II, Chapter 13, there are two preferred conformations +sc and −sc for the O(6)–C(6)–C(5)–O(5) torsion angle which are determined by the perio- or 1,3-diaxial or Hassel-Ottar effect [499, 501, 502, 560]. The ap conformation has never been found in the cyclodextrin

Table 18.1. Selected physical parameters of the cyclodextrin molecules [557]

Properties	Cyclodextrin		
	α	β	γ
Number of glucoses	6	7	8
Molecular weight	972	1135	1297
Cavity diameter (Å)	5.0 ± 0.2	6.2 ± 0.2	7.9 ± 0.4
Outer diameter (Å)	14.6 ± 0.4	15.4 ± 0.4	17.5 ± 0.4
Height of torus (Å)	8.0	8.0	8.0
Solubility in water (g/100 ml solution)	14.5	1.85	23.2
Average interglucose $O(2)\cdots O(3')$ distance (Å)	3.0	2.85	2.80

Box 18.1. Cyclodextrins as Models to Study Molecular Inclusion Phenomena and Enzymatic Reactions

The inclusion capability of the cyclodextrins is of special interest for the study of weak interactions. Because of their central cavity, the cyclodextrins act as "host", to form inclusion compounds with "guest" molecules. The only requirement for inclusion is that the guest fits into the cyclodextrin cavity. Other factors such as hydrophobic, hydrophilic, molecular, or ionic character are less important. Since the cyclodextrin cavity is a property of the covalently bonded molecule, it exists also in solution. Therefore the inclusion process can be investigated both by solid state and solution methods. This contrasts with inclusion complexes where the clathration occurs exclusively in the solid, crystalline state as a result of intermolecular hydrogen bonding or ionic forces which stabilize the host lattice. Examples of such solid state host compounds are provided by urea, clathrate hydrates (Part IV, Chap. 21), choleic acids, the nickel thiocyanates.

As in the other inclusion complexes, the cyclodextrin cavity is filled by guest species, including water. The cohesive forces between the host structure and the guest molecules can be van der Waals interactions or hydrogen bonds, and hydrophobic forces may also contribute to inclusion formation. Because these weak interactions are so important in biological systems and difficult to investigate with the large proteins or nucleic acids, the cyclodextrins provide suitable models for their study [551–554, 556, 561, 562].

Cyclodextrins are excellent enzyme models: Catalysis and induced fit. Due to their cavities, which are able to accommodate guest (substrate) molecules, and due to the many hydroxyl groups lining this cavity, cyclodextrins can act catalytically in a variety of chemical reactions and they therefore serve as good model enzymes. Thus, benzoic acid esters are hydrolyzed in aqueous solution by factors up to 100 times faster if cyclodextrins are added. The reaction involves an acylated cyclodextrin as intermediate which is hydrolyzed in a second step of the reaction, a mechanism reminiscent of the enzyme chymotrypsin. The catalytic efficiency can be further enhanced if the cyclodextrins are suitably modified chemically so that a whole range of artificial enzymes have been synthesized [551–555, 556, 563, 564].

Whereas β- and γ-cyclodextrins adopt a "round" shape in aqueous solution which does not significantly change upon complex formation, α-cyclodextrin is somewhat collapsed and opens to a "round" shape when a guest molecule enters. This is comparable to the "induced fit" proposed for enzyme-substrate complexation and is discussed in some detail in Box 18.2.

crystal structures. The −sc orientation where the O(6)−H group points away from the cyclodextrin cavity is generally favored. The +sc orientation where the primary alcohol group points toward the cavity (for definition of torsion angle ranges, see Box 13.3), occurs only if O(6)−H is engaged in hydrogen bonding to an enclosed guest molecule (vide infra).

The cyclodextrin macrocycle is best described as a short, hollow, truncated cone. Its length of 8 Å is determined by the width of the glucose unit, and the diameter of the central cavity (and of the macrocycle) depends on the number of glucose units per cyclodextrin (Table 18.1). The glucose rings are always oriented in the same direction, so that the narrow rim of the cone is formed by the primary O(6)H alcohol groups, and the secondary O(2)H, O(3)H hydroxyls occupy the wide rim. It is these peripheral hydroxyl groups that give the cyclodextrins their hydrophilic surface.

In contrast to the periphery of the cyclodextrins, the internal cavities, with diameters of 5 to 8 Å (Table 18.1), have hydrophobic character because they are lined by the methylene C−H groups and by the ether-like O(4) and O(5) oxygen atoms. The distribution of hydrophilic and hydrophobic surfaces, together with the annular shapes of the cyclodextrins, gives rise to the "microheterogeneous environment" [555] which is the reason for some of their most interesting properties (Box 18.1).

18.2 Crystal Packing Patterns of Cyclodextrins Are Determined by Hydrogen Bonding

Cyclodextrins can be crystallized as hydrates[2] *or as inclusion compounds with appropriate guest molecules.* There is a clear indication that the packing of α-, β-, and γ-cyclodextrin molecules in their various crystal structures is determined by three main factors: the type of cyclodextrin, the size and nature of the enclosed guest molecules, and the host-guest and host-host hydrogen bonding. The interstices formed between cyclodextrin molecules are occupied by water molecules which are incorporated into the overall hydrogen-bonding structure.

The cyclodextrins crystallize in patterns which form cages and channels illustrated in Fig. 18.2 [558, 565]. These patterns can be classified into the two cages of types **A** and **B**, where the cavity of each cyclodextrin is closed on both ends by adjacent molecules, and the channel type **C**, where cavities of adjacent cyclodextrins merge into infinite tube-like cavities.

In *cage-type A packing motifs*, cyclodextrin molecules are arranged cross-wise in a herringbone mode. This packing is found for α-, β-, and γ-cyclodextrins if

---

[2] Because in the cyclodextrin hydrates the cavity is occupied by water molecules, they can be considered as the inverse of the clathrate hydrates discussed in Part IV, Chapter 21. In these, the water molecules form the host structure and the organic molecule is the guest.

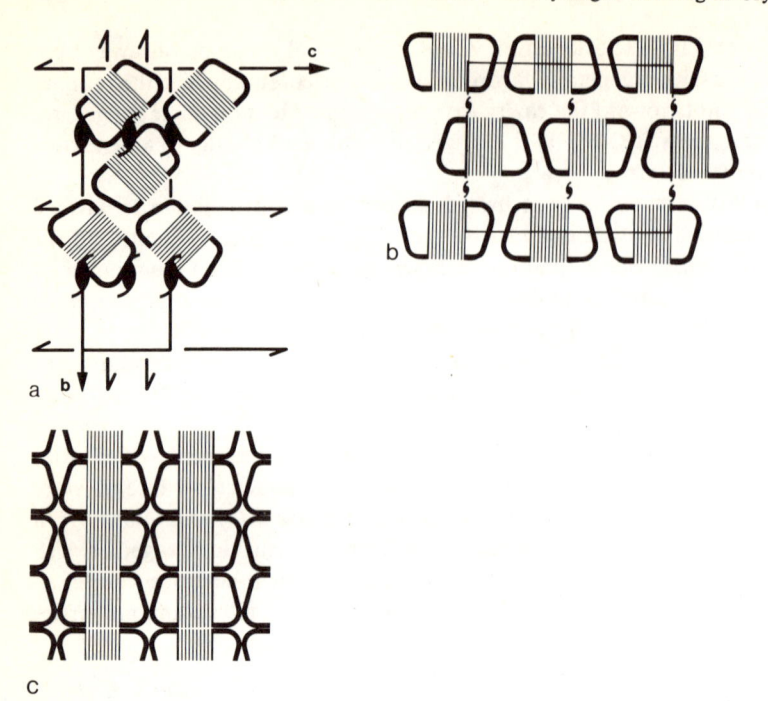

Fig. 18.2a–c. Packing of cyclodextrin molecules in crystal lattices. There are two *cage* packing patterns, in herringbone (**a**) and in brick-wall-type arrangement (**b**). In the *channel*-type packing (**c**), cyclodextrins are stacked like coins in a roll. In the illustrations, cyclodextrins are seen *from the side*, cavities are indicated by *shading* [565]

small molecular guests such as water, methanol or ethanol are included. In the *cage-type B motifs* formed by α- and β-cyclodextrins, the molecules are arranged side by side so that extended layers are formed. Adjacent layers are stacked with some lateral displacement, thereby closing the cavities of one layer by the cyclodextrin molecules in the next layer, and the overall appearance is that of a brick wall.

If the guest molecules are too long to fit the cavity of an individual α-, β-, or γ-cyclodextrin or if they are of ionic character, the cyclodextrin molecules arrange like coins in a roll and the cavities line up in *cage-type C channel motifs*. The cyclodextrins can be oriented head-to-head or head-to-tail, with extensive hydrogen bonding between the O(2), O(3) and O(6) hydroxyls.

The individual packing motifs are a molecular property and characteristic for type of cyclodextrin and type of included guest molecule.

For α-cyclodextrin, the packing motifs can be predicted because they depend largely on size and ionic or molecular character of the guest molecules [565]. With small molecular guests of dimensions comparable to the host cavity, type **A** cage structures are formed. If the guest molecules are longer or if they are of ionic character, type **C** channel structures predominate (Fig. 18.3). These are stabilized by intermolecular hydrogen bonds between O(2), O(3) and O(6) hydroxyl groups (see Fig. 18.3b) [566]. In the stacks, the α-cyclodextrin molecules can be oriented head to head (as in Fig. 18.3) or in head-to-tail mode. With small aromatic guest molecules, the preferred packing mode is the brick-wall motif type **B**.

The situation is different for β- and γ-cyclodextrins, which appear to aggregate very easily by rim-to-rim hydrogen bonding between O(2) and O(3) hydroxyl groups to form head-to-head dimers, similar to that shown in Fig. 18.3 for α-cyclodextrin. This occurs even when the guest molecules, such as n-propanol, are so small that they could be accommodated in cavities provided by the monomeric cyclodextrins in the cage-type **A** herringbone arrangement [565–570].

In β-cyclodextrin, the dimers can either stack linearly to produce type **C** channels or they are displaced laterally so that the cavity of each dimer is closed on both ends in a brick-wall type **B** motif. In the latter, direct $O(6)H \cdots O(6')$ hydrogen bonding between the dimers is difficult due to their separation, and water molecules are added at the dimer interface to form the indirect $O(6) \cdots H_2O \cdots O(6)$ interactions.

In γ-cyclodextrin with eight glucoses in the macrocycle, the preferred packing for more extended guest molecules, such as n-propanol and larger, is in a tetragonal space group where the molecular (and channel) axis coincides with the fourfold crystallographic axis (see Fg. 18.4) [568, 569]. The packing in this tetragonal space group is peculiar because head-to-head and head-to-tail arrangement alternate so that the asymmetric unit consists of a stack of three fourths of the γ-cyclodextrin molecules. Again, a large number of hydrogen bonds form at the interfaces between the γ-cyclodextrin molecules, and there are hydrogen bonds to and between the water molecules located in interstices between the stacks.

The guest molecules located in the channel with fourfold symmetry are, in general, so disordered that the atomic positions cannot be determined from crystal structure analyses, except when the guest molecule itself possesses fourfold symmetry [569].

18.3 Cyclodextrins as Model Compounds to Study Hydrogen-Bonding Networks

The property of cyclodextrins of primary interest in this text is the formation of hydrogen-bonding networks in the crystalline state. These networks are extensive because the cyclodextrin crystal structures contain not only the three donor hydroxyl groups and five potential acceptor oxygen atoms characteristic of the glucose unit, but also many waters of crystallization. The total number of donors and acceptors respectively is 18 and 30 for α-cyclodextrin, 21 and 35 for β-cyclodextrin, and 24 and 40 for γ-cyclodextrin, to which is added the double donor and double acceptor function of each of the water molecules.[3]

In only a few of the many publications on X-ray studies of the cyclodextrin inclusion compounds and of the cyclodextrin hydrates have the hydrogen atoms of

---

[3] As in the small molecule crystal structures the water molecules are three- and four-coordinated (see Part IV, Chap. 21).

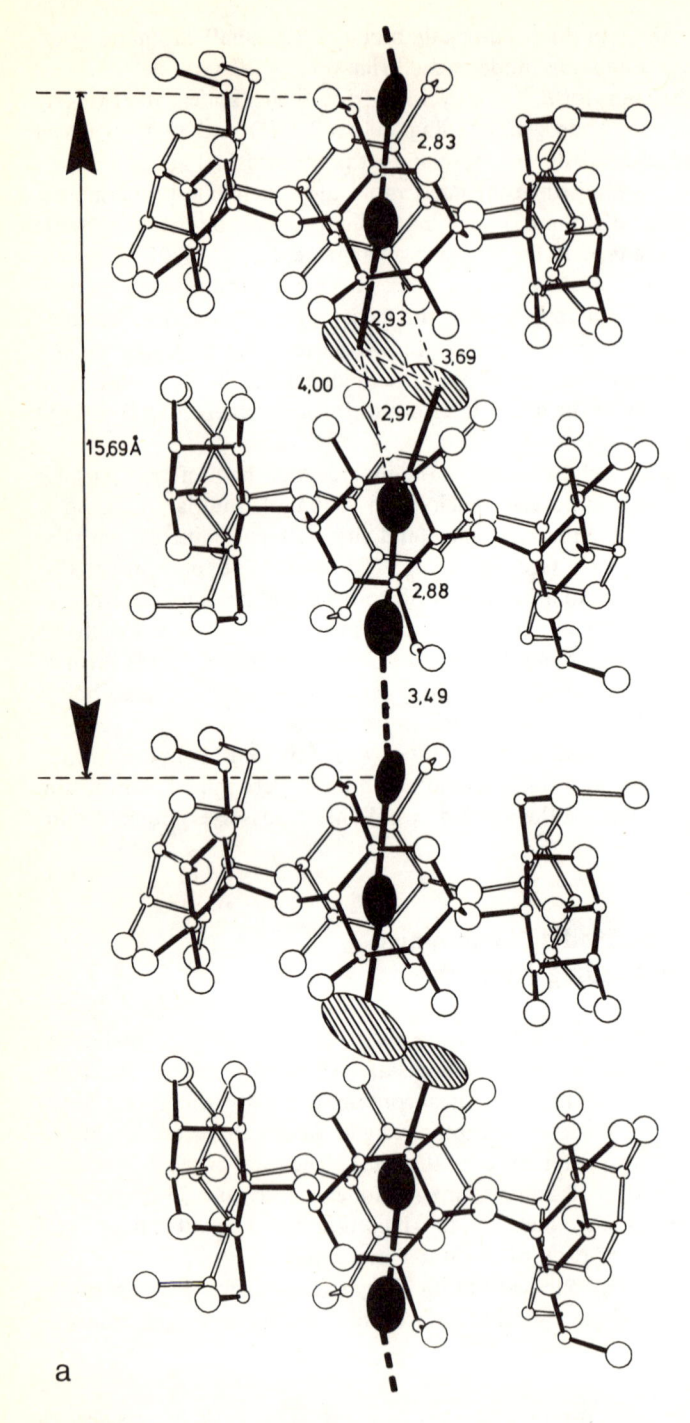

Fig. 18.3a

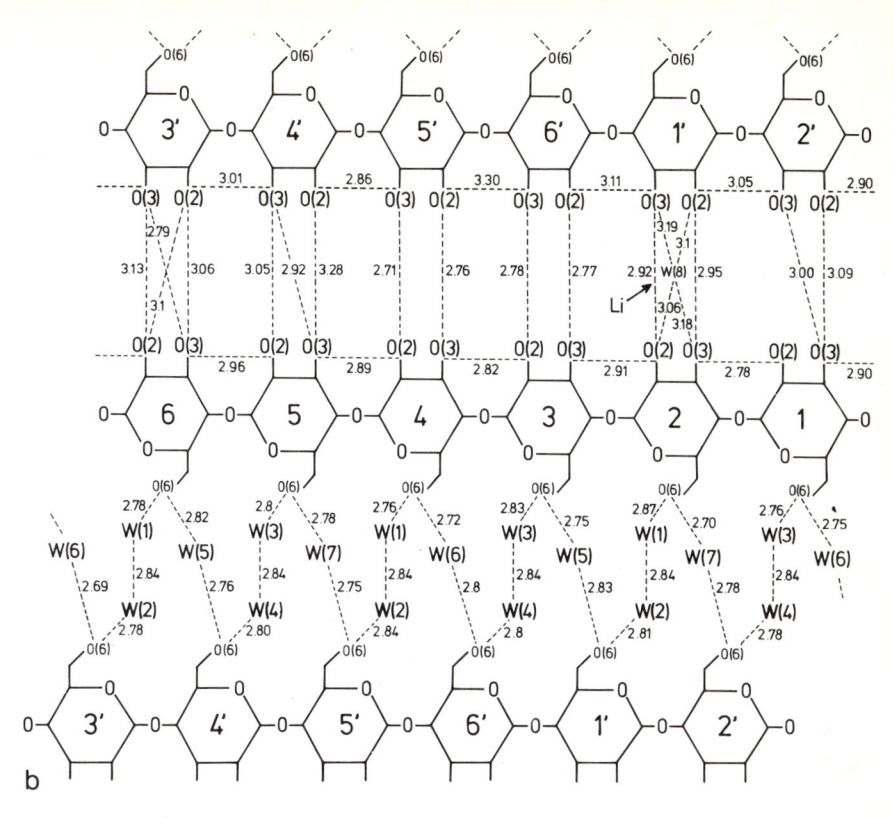

Fig. 18.3. a Schematic view of the channel-type structure formed by head-to-head stacked α-cyclodextrins in the presence of lithiumiodide and iodine, with formula (α-cyclodextrin)$_2$ · LiI$_3$ · I$_2$ · 8H$_2$O. Water and Li$^+$ are located in voids between the α-cyclodextrin stacks, the cavities are filled by I$_3^-$ and I$_2$ (*black*), with one of the iodines (*shaded*) disordered over two sites (I · · · I distances given in Å units). **b** There is extensive intra- and intermolecular hydrogen bonding between O(2), O(3) hydroxyl groups, and hydrogen bonding between O(6) groups is mediated by water molecules W. Li$^+$ is coordinated by O(2), O(3) hydroxyls from glucoses 1' and 2 and by water molecule W8 in tetragonal pyramidal form, with W8 as apex. Hydrogen-bond O · · · O contact distances given in Å units [566]

the hydroxyl groups and of the water molecules been experimentally located [556, 558, 566–570]. However, the results of three neutron diffraction analyses are presently available (see Table 18.2).

Intramolecular hydrogen bonding stabilizes the macrocycle. As in α-maltose and in β-maltose monohydrate, the separation of the O(2)H and O(3)H groups on adjacent glucose residues in the cyclodextrin molecules is such that intramolecular interglucose hydrogen bonds can be formed.

They have average O · · · O separations of about 3.0 Å for α-cyclodextrin, of 2.85 Å for β-cyclodextrin, and of 2.80 Å for γ-cyclodextrin (see Table 18.1). The intramolecular hydrogen bonds are either O(2)H · · · O(3') or O(3)H · · · O(2'), or disordered as in β-cyclodextrin · 11 H$_2$O, O(2)–($\frac{1}{2}$H) · · · ($\frac{1}{2}$H)–O(3'). There is

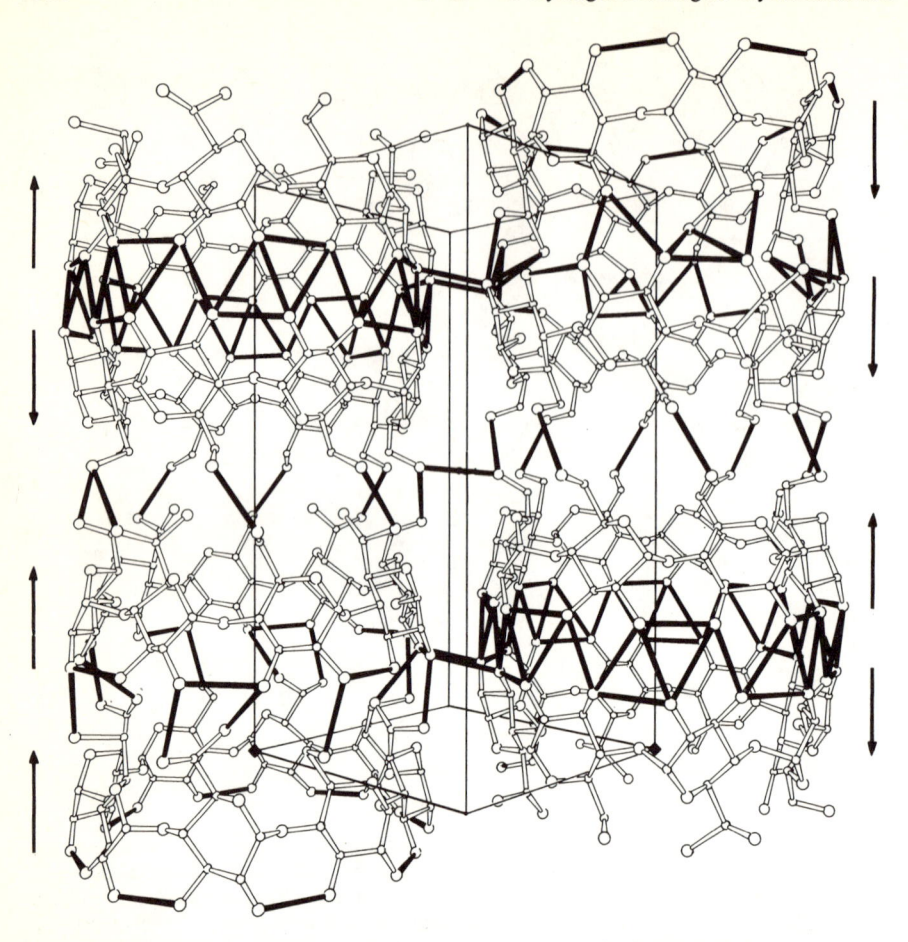

Fig. 18.4. Packing of γ-cyclodextrin molecules in the crystal structure of γ-cyclodextrin·x (n-propanol)·26H$_2$O. The γ-cyclodextrins are stacked along fourfold rotation axes in head-to-tail and head-to-head mode, indicated by the *arrows*. Some of the O(6)H hydroxyles are twofold disordered. Hydrogen-bonding contacts are drawn in *solid lines* [567]

spectroscopic evidence that these hydrogen bonds persist in solution where they stabilize the annular conformation and contribute to the rigidity of the macrocycles [571–573].

In addition to these intramolecular hydrogen bonds, a large number of intermolecular bonds O–H···O exist in the crystalline state. Since all known cyclodextrin crystal structures contain water of crystallization, extensive intermolecular hydrogen-bonding networks are formed. If the guest molecules have hydrogen-bond donor or acceptor properties, they will also be incorporated in the hydrogen-bonding scheme. As with the mono- and disaccharide crystal structures, the formation of three-center bonds and of cooperative chains of O–H···O hydrogen bonds can be expected. In highly hydrated structures such as β-cyclodex-

Table 18.2. Crystallographic data for cyclodextrin hydrates

	Cyclodextrin						
	α Form I	Form II	Form III	β Form I	Form II	γ Form I	
Composition of hydrate	6H$_2$O	6H$_2$O	7.5 7H$_2$O	11H$_2$O room temp.	12H$_2$O 120K	18H$_2$O	
Space group	P2$_1$2$_1$2$_1$	P2$_1$2$_1$2$_1$	P2$_1$2$_1$2$_1$	P2$_1$	P2$_1$	P2$_1$	
Unit cell constant a (Å)	14.858(3)	13.70(1)	14.356(5)	21.261(6)	21.617(4)	22.520(4)[c]	
b (Å)	34.038(7)	29.35(2)	37.538(12)	10.306(3)	10.026(2)	11.197(3)	
c (Å)	9.529(2)	11.92(1)	9.400(4)	15.123(4)	14.891(4)	16.810(4)	
β	—	—	—	112.3(5)	112.52(2)	105.23(2)	
O–H···O hydrogen bonds per unit cell	124	152	136	106[a]	94[b]	—	
X-ray (X) or neutron (N) data	N	X	X	N	N	X	
References	[574]	[575]	[78]	[118]	[455]	[570]	

[a] 70 O–H···O hydrogen bonds and 36 O–H···H–O flip flops.
[b] 90 O–H···O hydrogen bonds and 4 O–H···H–O flip flops.
[c] V. Zabel, unpublished data from neutron diffraction of γ-cyclodextrin deuterate.

trin·11H$_2$O and γ-cyclodextrin·18H$_2$O, most of the water molecules located in cyclodextrin cavities and in intermolecular voids are disordered. This is associated with the disordered hydrogen bonds O−($\frac{1}{2}$H)···($\frac{1}{2}$H)−O, which are a characteristic feature of the ices and of the clathrates hydrates (Part IV, Chap. 21).

Hydrogen bonding in the cyclodextrin hydrates is relevant for biological structures. The most interesting cyclodextrin crystal structures, which can serve as models for biological hydrogen-bonding systems and for hydration of biological macromolecules, are those in which the hydrophobic cavities and the hydrophilic interstices between cyclodextrins are occupied by water molecules.

Because of the complexity of these crystal structures, unambiguous assignment of hydrogen bonds is rarely possible from X-ray analyses. Much of the information described in the next paragraph is therefore from the three neutron diffraction analyses (Table 18.2). As the large amount of hydrogen atoms in these crystal structures gives rise to incoherent neutron scattering and impairs the signal-to-noise ratio, the cyclodextrins were crystallized from D$_2$O instead of H$_2$O. Full substitution of hydroxyl hydrogen for deuterium is not easy to obtain, and this adds to the complexity of the interpretation of the hydrogen bonding. In the following discussion, no distinction is made between D and H.

18.4 Cooperative, Homodromic, and Antidromic Hydrogen-Bonding Patterns in the α-Cyclodextrin Hydrates

The hydrate of α-cyclodextrin crystallizes in three forms with almost identical type **A** cage packing: α-cyclodextrin·6H$_2$O, forms I [574] and II [575], and α-cyclodextrin·7.57H$_2$O [78] (Table 18.2). These crystals were studied by X-ray diffraction and the form I hexahydrate by neutron diffraction. Because the X-ray data were of unusually good quality, the positions of the hydrogen atoms could be determined in these crystal structures and the hydrogen-bonding schemes can be discussed with confidence.

In α-cyclodextrin hydrates, the packing of the host molecules is similar but hydrogen-bonding structures are different. Although the general packing modes of α-cyclodextrin molecules in the hexahydrates form I and II and in the 7.57 hydrate are comparable because the crystal structures are almost isomorphous, the water molecules are in different positions and therefore the hydrogen-bonding patterns are different (Fig. 18.5a, b, c). In the form I hexahydrate two ordered water molecules are located in the α-cyclodextrin cavity and four are in intermolecular voids (Fig. 18.6a). In form II crystals, one of the enclosed waters is substituted by the O(6) hydroxyl of an adjacent α-cyclodextrin molecule, thus forming a self-complex, and five water molecules are in intermolecular voids. In these two hydrates the α-cyclodextrin macrocycle is to some extent "collapsed" (see Chap. 18.6). This contrasts with the 7.57 hydrate, where the α-cyclodextrin macrocycle adopts a "round" shape because its cavity accommodates 2.57 water molecules which are statistically disordered over four sites, thereby filling the "round", more open cavity (Fig. 18.6b; vide infra and Chap. 18.6). Of the remaining five water molecules located in voids between the α-cyclodextrin macrocycles, four are ordered and one is disordered over two sites.

The cooperative, infinite chains and cycles formed by O–H···O hydrogen bonds in the α-cyclodextrin hydrates are a characteristic structural motif [109]. As with the simpler carbohydrate crystal structures described in Part II, Chapter 13, the hydrogen bonds can be traced from donor to acceptor in the cyclodextrin hydrate crystal structures. Networks of O–H···O–H···O–H··· interactions are observed in which the distribution of hydrogen bonds patterns with two characteristic motifs. One are the "infinite" chains which run through the whole crystal lattice, and the others are the loops or cyclically closed patterns (a special case of the "infinite" chains). As in the small molecule hydrates, such as α-maltose monohydrate, the chains and cycles are interconnected at the water molecules to form the complex three-dimensional networks illustrated schematically in Fig. 18.5, with some sections shown in more detail in Fig. 18.7a, b, c.

α-Cyclodextrin·6H$_2$O, form I [574]. The principal motif in the hydrogen-bonding structure of this crystal form of the α-cyclodextrin hydrate are three cycles which share water molecules W(1) and W(4) (see Figs. 18.5a, 18.7a). There is one homodromic, six-membered cycle (I) linked with an antidromic, five-membered cycle (II). In the latter cycle, W(4) acts as a double hydrogen-bond donor and O(3)$_1$ as a double acceptor. The third cycle (III), of antidromic type, has W(3) as double donor and W(1) as double acceptor. There is one weak hydrogen bond, W(3)H···O(6)$_6$, which is the minor component of a three-center bond with O(6)$_{1B}$ as second acceptor group (for atom numbering, see legend to Fig. 18.1).

This cluster of cycles is part of two infinite chains. One runs in b direction,

$$\cdots[O(3)_4H\cdots O(2)_3H\cdots W(2)H\cdots O(3)_6H]_\infty\cdots O(3)_4H\cdots$$

and the other chain runs diagonally across the unit cell, with the two water molecules W(A) and W(B) located in the α-cyclodextrin cavity (Fig. 18.6a):

$$\cdots[W(A)H\cdots W(B)H\cdots O(3)_5H\cdots O(6)_4H\cdots O(2)_2H\cdots W(4)H\cdots O(6)_2H$$

$$\cdots O(3)_1H\cdots O(6)_5H]_\infty\cdots W(A)H\cdots .$$

The spatial distribution of hydroxyl or water oxygen atoms around water molecules approximately satisfies tetrahedral geometry. The most regular tetrahedra with O···W···O angles close to 109° are around W(2) and W(4), while the tetrahedron formed with W(1) as center is very distorted. The oxygen atoms around W(3), O(2)$_1$ and O(6)$_3$ are at normal hydrogen-bonding distances, whereas O(6)$_6$ and O(6)$_{1B}$ (at 8% occupancy) are "loosely bonded" and the O···W(3)···O angles vary between 60° and 171°. The waters W(A) and W(B), which are included in the α-cyclodextrin cavity, have only three hydrogen-bonded partners. For W(A), the angle O(6)$_5$···W(A)···O(6)$_{1A}$ of 77° is determined by geometrical restrictions imposed by the orientation of the O(6)H hydroxyl groups while the other two angles involving W(B), 109° and 120°, are roughly tetrahedral. For W(B), the

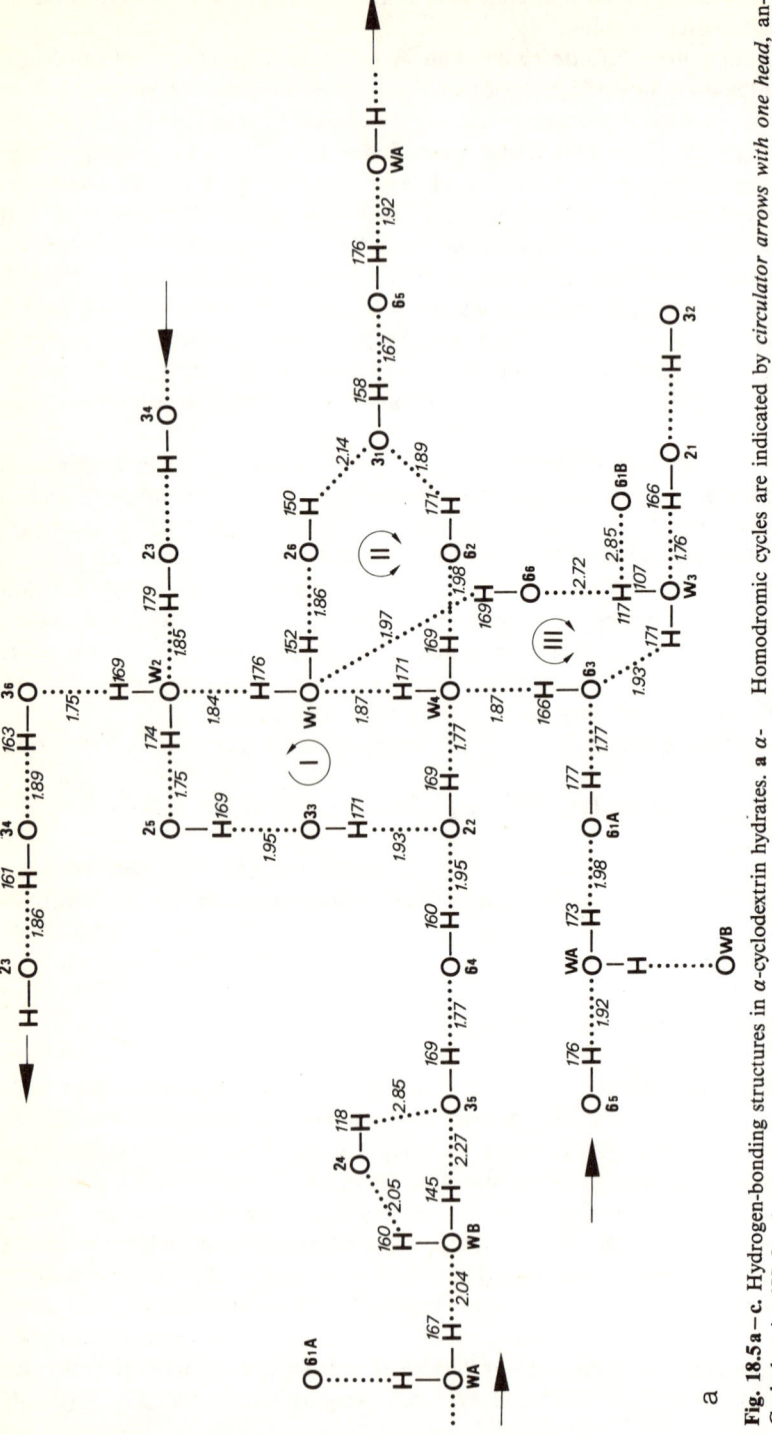

Fig. 18.5a–c. Hydrogen-bonding structures in α-cyclodextrin hydrates. **a** α-Cyclodextrin·6H₂O, form I; **b** α-cyclodextrin·6H₂O, form II; **c** α-cyclodextrin·7.57 H₂O; *, intramolecular. Data in **a** are from neutron and in **b, c** from X-ray analyses. In **c** covalent O–H bonds are normalized to 0.97 Å. Homodromic cycles are indicated by *circulator arrows with one head*, antidromic cycles by *arrows with two heads*. Atom designation, 2₆ means O(2) of glucose number 6

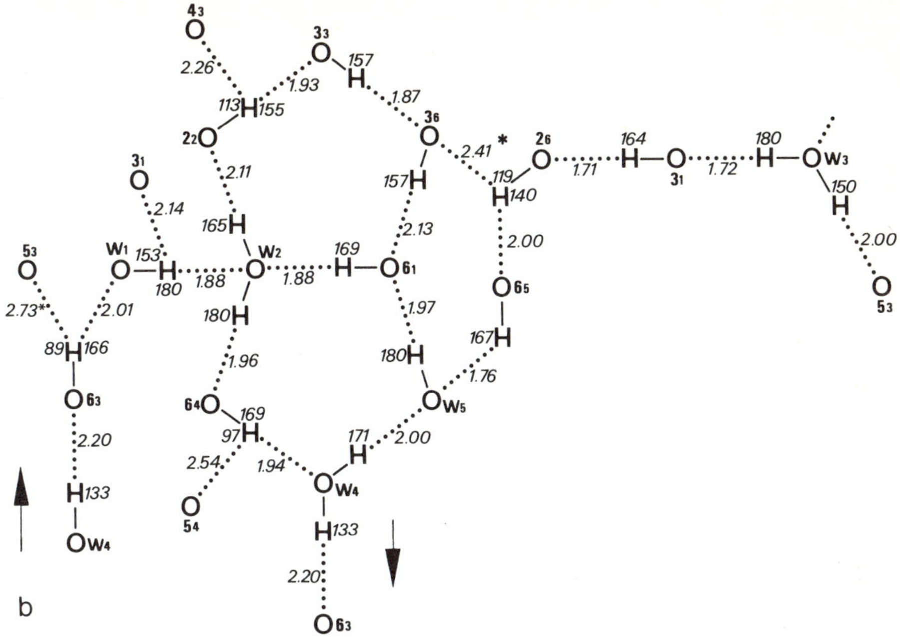

Cooperative, Homodromic, and Antidromic Hydrogen-Bonding Patterns 323

Fig. 18.5b

sum of angles between the three ligands O(2)$_4$, O(3)$_5$ and W(A) is 360°, indicating that this is a trigonal planar coordination.

The geometry of the α-cyclodextrin molecule with the two enclosed water molecules W(A), W(B) is shown in Fig. 18.6a. The interglucose, intramolecular hydrogen bonds O(2)···O(3') are all formed except to glucose number 5, which is rotated out of alignment with the other five glucoses. As a result, both hydrogen bonds O(6)$_5$H···W(A) and W(A)H···O(6)$_1$ can form, and the cavity diameter of 5.0 Å is diminished so that water molecule W(A), with van der Waals width of 3.8 Å, is tightly held. This water is not located in the center of the cavity. It is displaced toward glucoses 2, 3 and 4, making nearest neighbor contacts to the hydrogen atoms C(3)H and C(5)H which are at the interior of the cavity (see Fig. 18.6a).

α-Cyclodextrin·6H$_2$O, form II [575]. The hydrogen-bonding structure in the crystals of this α-cyclodextrin hexahydrate (Figs. 18.5b, 18.7b) reflects some features of the scheme described for crystal form I, since there are also three connected cycles. However, the arrangement of water and hydroxyl groups is different and the three cycles are all five-membered. Two of the cycles (**I, II**) are of the homodromic type, and one is antidromic (**III**), with the O(2)$_6$H group donating a three-center bond, both components of which are part of the cycle. This three-center bond is particular because the O(2)$_6$H···O(3)$_6$ interaction, marked

Fig. 18.5c

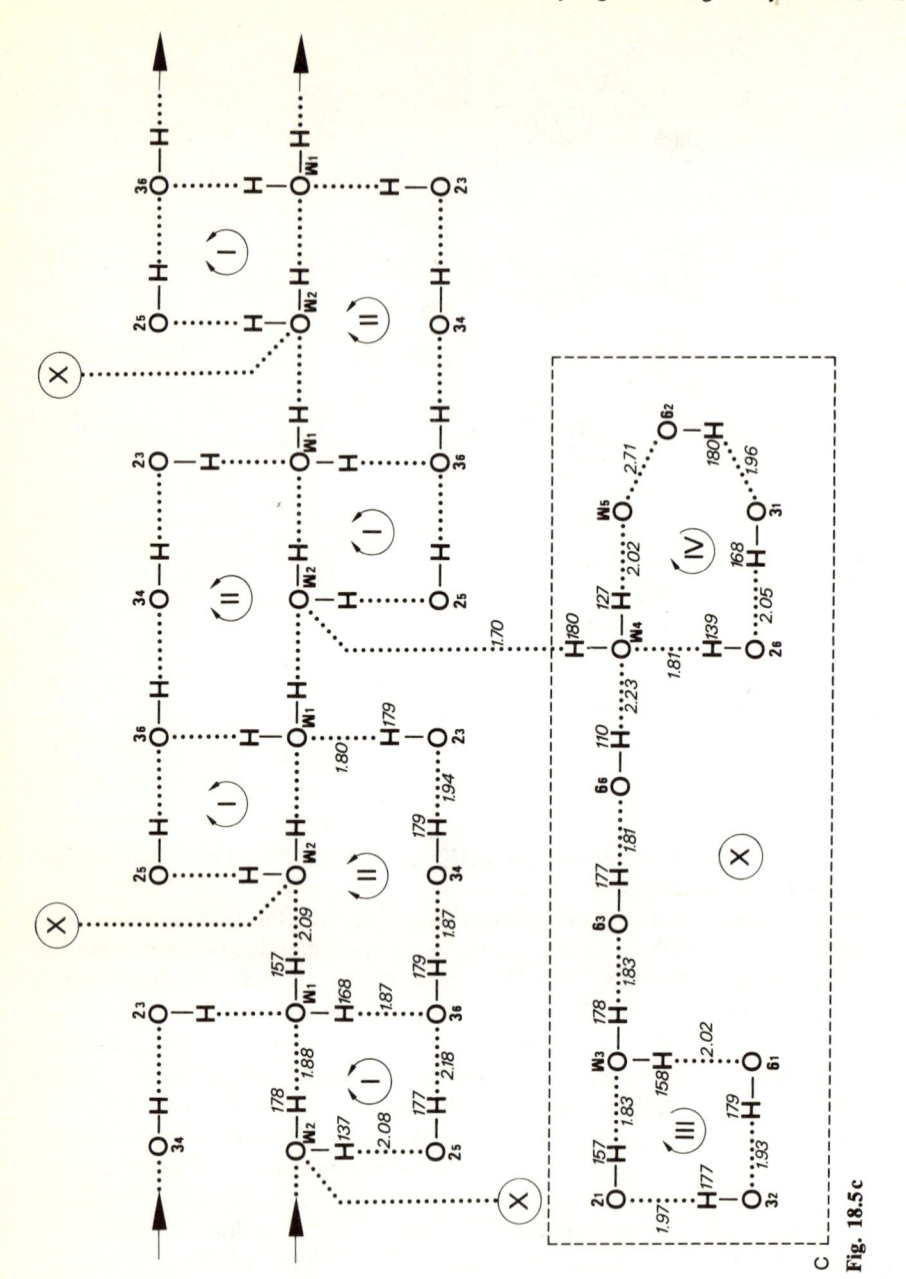

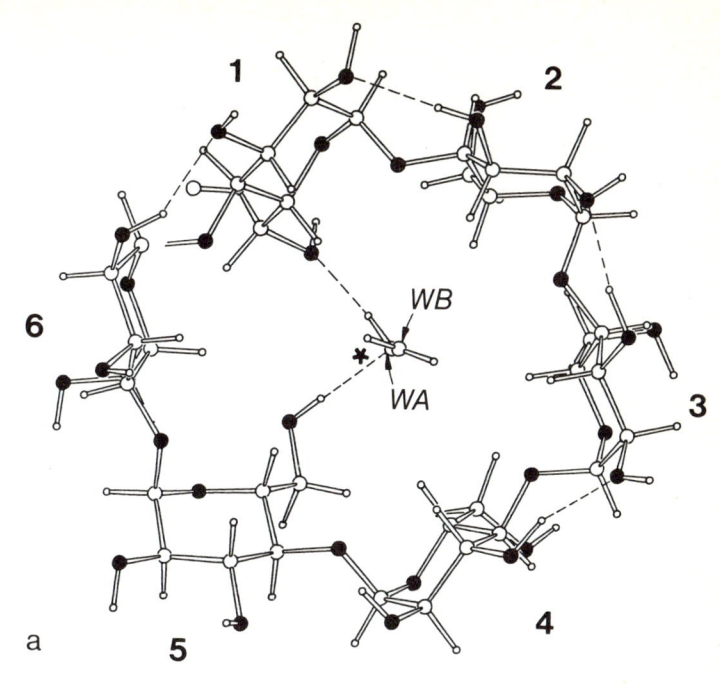

Fig. 18.6a, b. Structures of α-cyclodextrin in **a** hexahydrate form I; **b** 7.57 hydrate. H and C atoms indicated by *small* and *larger circles* (C–H hydrogen atoms are omitted in **b**), O atoms by *filled circles*, and hydrogen bonds are drawn with *dashed lines*. Corresponding glucoses in the nearly isomorphous crystal structures are numbered *1* to *6*. **a** Two water molecules are enclosed in the cavity, the "lower" one (WA) accepting a hydrogen bond from O(6)$_5$ and donating two bonds to O(6)$_1$ and to WB (not shown). Glucose 5 is rotated relative to the other glucoses so that the hydrogen bond to WA can form, but O(2)···O(3') contacts to glucose 5 are broken. The enclosed water molecules are not at the center of the cavity (marked by *) but moved to contact the wall of the cavity. **b** The enclosed 2.57 water molecules are statistically disordered over four sites and the α-cyclodextrin macrocycle adopts a "round" shape with all six O(2)···O(3') hydrogen bonds formed. All O(5)–C(5)–C(6)–O(6) torsion angles are in the −sc range whereas in **a**, those engaged in hydrogen bonding to the enclosed water are in +sc range [574, 575]

by an asterisk in Fig. 18.5b, is one of the rare intramolecular hydrogen bonds formed between vicinal glucose hydroxyls. The cluster of three cyclic systems is part of an infinite, homodromic chain:

$$\cdots[\text{W}(4)\text{H}\cdots\text{O}(6)_3\text{H}\cdots\text{W}(1)\text{H}\cdots\text{W}(2)\text{H}\cdots\text{O}(6)_4\text{H}]_\infty\cdots\text{W}(4)\text{H}\cdots$$

α-Cyclodextrin·7.57H$_2$O [78]. This hydrate was obtained when a solution of α-cyclodextrin was crystallized in the presence of 1.2 M BaCl$_2$ aimed at producing a barium complex of α-cyclodextrin. However, there is no BaCl$_2$ in the crystals. Even so, the presence of the salt must have had a strong influence on the crystallization behavior of α-cyclodextrin because the unit cell constants of this hydrate differ significantly from those of the other two hydrate forms (see Table 18.2), and the water content per formula unit is higher.

Fig. 18.6b

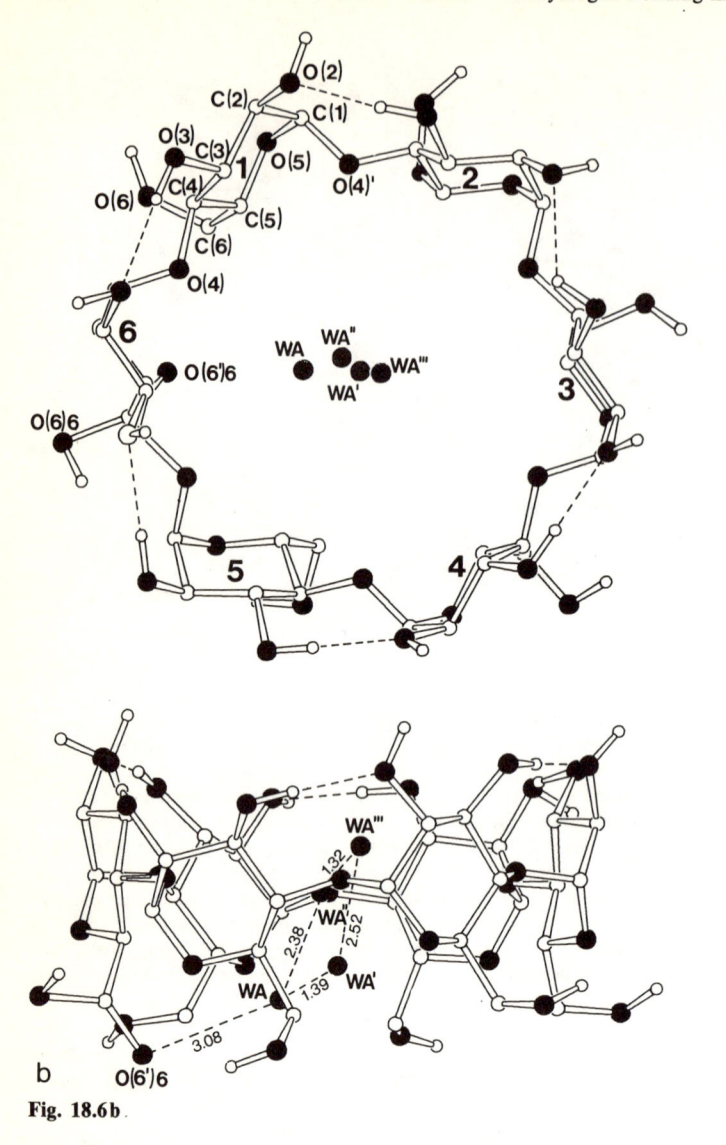

Of the 7.57 water molecules per asymmetric unit, four are fully ordered and one, W(5), shows two positions at 0.64 and 0.36 occupation. These water molecules are located in intermolecular voids, and the remaining 2.57 waters are statistically distributed over four sites in the α-cyclodextrin cavity (see Fig. 18.6b). Associated with this disorder in the α-cyclodextrin cavity is a "round" shape of the macrocycle, where in contrast to the other two α-cyclodextrin hydrate forms, all six intramolecular interglucose O(2)···O(3') hydrogen bonds are formed. Because hydrogen atoms attached to the disordered water molecules in the 7.57 hydrate could

Cooperative, Homodromic, and Antidromic Hydrogen-Bonding Patterns

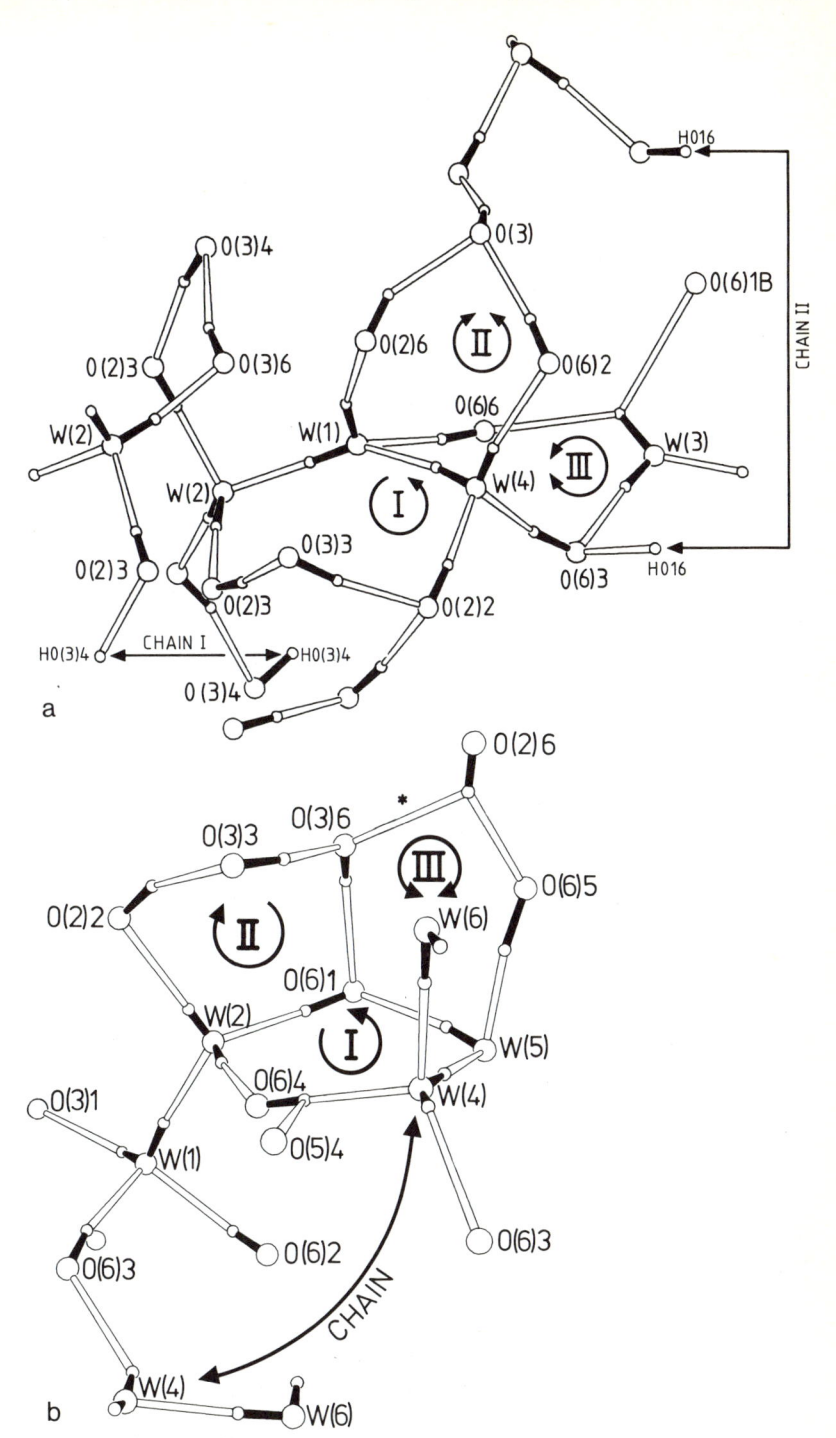

Fig. 18.7a,b (Legend see p. 329)

Fig. 18.7 c, d (Legend see p. 329)

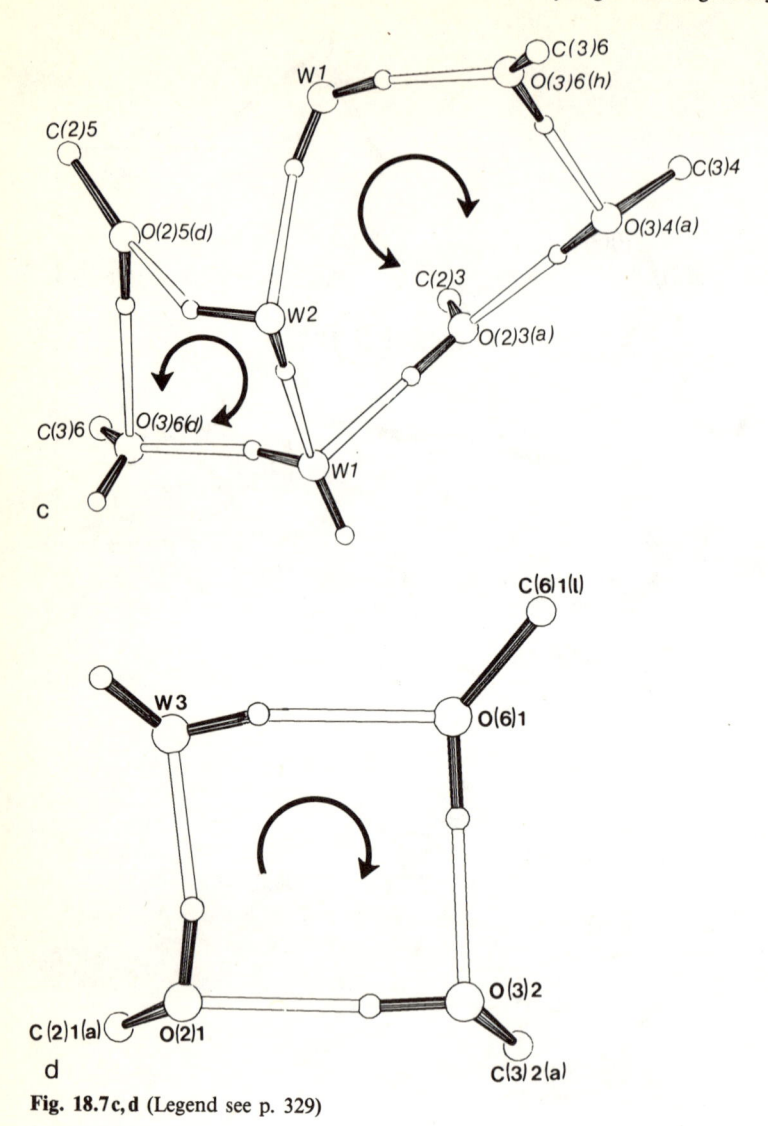

not be located, the hydrogen-bonding scheme can only be described for the five well ordered waters and the α-cyclodextrin hydroxyl groups.

A predominant motif in the hydrogen-bonding structure in α-cyclodextrin·7.57H$_2$O is a ribbon of fused four- and six-membered antidromic cycles, shown in Figs. 18.5c and 18.7c, d. This is generated by the symmetry operation of a screw axis in the b direction. It contains an infinite chain of water molecules,

$$\cdots W(1)-H\cdots W(2)-H\cdots W(1)-H\cdots W(2)-H\cdots ,$$

Cooperative, Homodromic, and Antidromic Hydrogen-Bonding Patterns 329

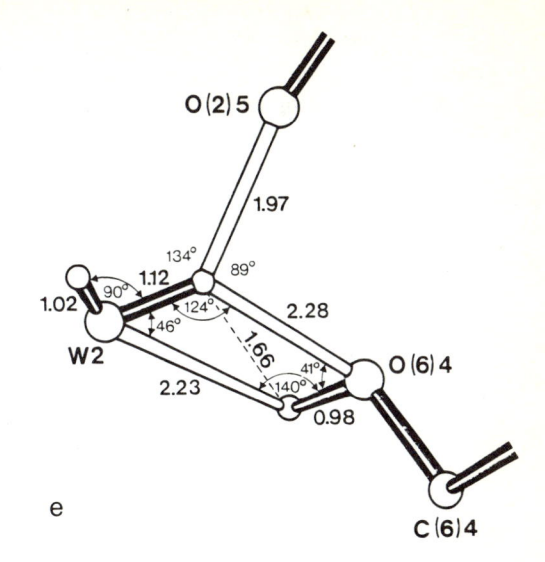

Fig. 18.7a–e. Cyclic and chain-like hydrogen-bonding patterns O–H···O observed in **a** α-cyclodextrin·6H$_2$O, form I; **b** α-cyclodextrin·6H$_2$O, form II; *, intramolecular; **c,d,e** α-cyclodextrin·7.57 H$_2$O. Infinite homodromic chains are marked *chain*, and cycles are indicated by *round arrows* with one (homodromic) and two (antidromic) heads. Only atoms of water molecules and of hydroxyl groups are drawn, the former denoted by *W* and the latter by the nomenclature described in Fig. 18.1. Each of the pictures **a** to **e** shows only a section of the respective crystal structure [78, 109, 575]

which intersects at W(1) with another infinite chain shown below which includes the hydroxyls O(3)$_6$, O(3)$_4$, O(2)$_3$.

$$\cdots [\text{W}(1)-\text{H}\cdots\text{O}(3)_6-\text{H}\cdots\text{O}(3)_4-\text{H}\cdots\text{O}(2)_3-\text{H}]_\infty \cdots \text{W}(1)-\text{H}$$

These intersections result in the formation of the four- and six-membered antidromic cycles **I** and **II**, Fig. 18.5c.

The ribbon is linked to another component of the hydrogen-bond structure (**X** in Fig. 18.5c) through a bond which connects W(4)–H···W(2). This component consists of four- and five-membered homodromic cycles **III** and **IV** (Fig. 18.7d) linked by a chain of three hydrogen bonds ···O–H···O–H···O–H···.

All the hydrogen bonds in the ribbon are of normal H···O bond length from 1.80 to 2.18 Å, and angles are in the range 157° to 179° except for W(2)H···O(2)$_5$, 137°. This water molecule forms a three-center bond to O(2)$_5$, with the minor component to O(6)$_4$, shown in Fig. 18.7e. The three-center bond includes the tandem configuration with an H···H separation of 1.66 Å, implying that there is twofold disorder involving these hydrogen atoms (see Part IA, Chap. 2.2).

The water W(5) in cycle **IV** (Fig. 18.5c) is disordered and the hydrogen atoms could not be located. Since W(5)···O(6)$_2$ is 2.7 Å, a hydrogen bond is likely to complete the homodromic cycle. The hydrogen-bond lengths and angles are normal in this section (**X**), with significant, but not unusual deviations from tetrahedral coordination about the water W(4).

Hydrogen-bonding patterns in crystal structures of the cyclodextrins and the simpler carbohydrates differ. The infinite, homodromic chains are common both in the low molecular-weight carbohydrates and in the cyclodextrins. The principal difference lies in the frequency of occurrence of the homodromic and antidromic cycles, which are common in the cyclodextrin crystal structures and rare in the mono-, di-, and trisaccharides. The cyclic patterns are the rule in the clathrate hydrates and in the ices. From this point of view, the hydrogen-bonding patterns of the hydrated cyclodextrins lie between those of the simpler hydrated carbohydrates and those of the hydrate inclusion compounds, discussed in Part IV, Chapter 21.

Cycles are four-, five-, and six-membered. The smallest cycle thus far observed contains four hydroxyl groups (cf. Figs. 1.3 and 18.7d) and the most common ring sizes are four-, five-, and six-membered. The data available thus far suggest that the five-membered cycles are formed more frequently in the α-cyclodextrin hydrates and are approximately planar. They are therefore very similar to the pentagonal faces of the clathrate hydrate dodecahedron illustrated in Fig. 21.3, in which the $O \cdots \hat{O} \cdots O$ angles, $\sim 108°$, are close to the $105°$ observed for the $H - \hat{O} - H$ angle in the water molecule. The four-membered rings described thus far are also approximately planar with $H \cdots \hat{O} - H$ angles closer to $90°$ (see Fig. 18.7d). The six- and higher-membered rings are puckered but more irregular than the chain and boat configurations which occur in the ices. This is also a feature of the less symmetrical clathrate water structure of the "semi-clathrates" formed by amine hydrates, see Part IV, Chap. 21.2.

In the ices and gas hydrates, the cyclic arrangements are constrained by space group symmetry. As shown in Part IV, Chapter 21, the hydrogen-bonding schemes in the ices and in the gas hydrates can be decomposed into patterns of three-dimensionally arranged cyclic systems. These cycles, however, are geometrically constrained by the requirements of the high-symmetry space groups with mirror and glide planes and inversion centers which provide the lowest lattice energy of these simple molecules.

Such symmetry elements are not possible in crystals of cyclodextrins, proteins, and nucleic acids, which all contain asymmetric carbon atoms, and therefore cannot be organized into highly symmetrical three-dimensional patterns. Natural product molecules of this sort crystallize in space groups in which the cyclic hydrogen-bonding motifs are formed without symmetry limitations. In this respect they resemble the lower symmetry amine hydrates described in Part IV, Chapter 21.

18.5 Homodromic and Antidromic $O - H \cdots O$ Hydrogen-Bonding Systems Analyzed Theoretically

The effect of cooperativity in stabilizing cyclic homodromic systems of hydrogen bonds has a sound theoretical basis in the ab-initio quantum mechanical calculations of water polymers discussed in Part IA, Chapter 5. The cyclic and chain geometries experimentally observed in the crystal structures of α-cyclodextrin hydrates are too complex for ab-initio methods at a useful level of approximation with presently available computers. Therefore, calculations were carried out at the semi-empirical level using PCILO [110] with coordinates of cyclic and chain-like structures as obtained from neutron diffraction analyses. In addition, ab-initio studies were performed on four-membered cyclic structures with fully optimized geometrical parameters [111].

Some quantitative numbers for the effect of cooperativity in cyclic and chain-like structures. Based on the O–H···O configurations displayed in Fig. 18.7a, b, hydrogen-bonding energies (ΔE^{tot}) were calculated for the cyclic structures **I** to **III** in Fig. 18.7a and **I** in Fig. 18.7b, and for one of the infinite chains. The oxygen atom positions were as found in the crystallographic study, O–H distances were normalized at 0.97 Å, and C–O–H groups were replaced by H–O–H. These calculations carried out with the semi-empirical PCILO method are given in Table 18.3 and suggest that homodromic cycles and the homodromic chain are stabilized by between 6.8% and 9.8% of total energy with respect to isolated O–H···O hydrogen bonds, whereas antidromic cyclic arrangements provide only marginal stabilization below 1%. These results are consistent with those obtained for the sequential, double donor and double acceptor models for the water trimer [105], described in Part IA, Chapter 5.

Homodromic cycles have smaller dipole moments compared with antidromic cycles and infinite chains. The semi-empirical PCILO calculations also indicate that the cyclic arrangements of hydrogen bonds will have significant total dipole moments, those of the homodromic cycles being less than those of the antidromic cycles or of the homodromic chain arrangements [110]. Macroscopically, the dipoles must cancel for the whole crystals, either by antiparallel arrangements in the crystal unit cells or by antiparallel alignment of domains in the crystals.

These calculations also predict that the formation of chains or cycles of O–H···O hydrogen bonds causes polarization such that the charges on oxygen become more negative while those on the hydrogen atoms become more positive (Table 18.3). The polarization is calculated to be greater for homodromic than for antidromic arrangements.

The cooperative effect in homodromic O–H···O hydrogen bonds contributes 26% more energy if the geometry is optimized. In the PCILO study, the oxygen atom positions were those derived from crystallographic data, and therefore the effects of crystal packing are included. In order to determine the contribution due to the cooperative effect in a free system, ab-initio methods were applied on the water monomer, dimer, and a cyclic tetramer with homodromic hydrogen bonds, as shown in Fig. 18.7d. Starting from a theoretical arrangement for the water dimer and a cyclic tetramer with S_4 symmetry, the geometry was optimized with respect to the total energy. As the calculations indicated, the cooperative effect strengthens the O–H···O bond by 26% relative to the isolated bond in the dimer [304], and the charges on hydrogen and oxygen atoms are changed significantly by polarization (see Table 18.3).

Table 18.3. Calculation of the contribution of the cooperative effect in the stabilisation of cyclic and chain-like $O-H\cdots O-H\cdots O-H\cdots$ hydrogen bonds in homo- and antidromic arrangement using PCILO methods. E^{tot} = total stabilisation interaction, E^{hb} = energy of all n individual hydrogen bonds. Interaction energies are always negative and given in kcal mol^{-1}. Δq_O and Δq_H give mean changes in electronic charges measured in $|e|$ of oxygen and hydrogen atoms relative to isolated molecules, $\mu(D)$ is the total dipole moment [110]

System	Cooperative effect (%)				Δq_O	Δq_H	$\mu(D)$	Anti- or homodromic
	ΔE_{tot}	$\dfrac{\Delta E^{tot}}{n}$	ΔE^{hb}	$\dfrac{\Delta E^{tot}-\Delta E^{hb}}{\Delta E^{tot}}$				
Cycle I[a]	28.42	4.74	26.43	7.3	−0.032	0.034	2.7	Homo
Cycle I[c]	20.96	4.19	19.53	6.8	−0.028	0.031	3.3	Homo
Chain[b]	14.88	4.96	13.42	9.8	−0.027	0.025	6.0	Homo
Cycle II[a]	19.56	3.91	19.44	0.6	−0.026	0.023	7.7	Anti
Cycle III[a]	17.09	3.42	16.93	0.9	−0.020	0.018	8.6	Anti

[a] See Fig. 18.7a.
[b] Chain I in Fig. 18.7a.
[c] See Fig. 18.7b.

18.6 Intramolecular Hydrogen Bonds in the α-Cyclodextrin Molecule Are Variable — the Induced-Fit Hypothesis

There is evidence that the α-cyclodextrin macrocycle can adjust its cavity diameter, within certain limits, to accommodate either two hydrate water molecules in the "collapsed" or "tense" form, or a nonaqueous guest molecule in the "round" or "relaxed" form. This is the concept of the "induced fit" [576] which has been applied to comparable conformational changes observed with several enzymes upon substrate binding.

The "tense" and "relaxed" forms of α-cyclodextrin mimic the "induced-fit" model proposed for enzymes. In the inclusion complexes of α-cyclodextrin, the macrocycle adopts a "round, relaxed" shape with all the O(2),O(3) hydroxyls engaged in intramolecular $O(2)\cdots(3')$ hydrogen bonds (Fig. 18.8) [577]. The average $O\cdots O'$ distance of around 3.0 Å is consistent with an $H\cdots O$ separation of about 2.0 Å, indicating relatively weak interactions. This form of α-cyclodextrin persists even when an included guest molecule, such as krypton (3.8 Å diameter) is too small to fully occupy the 5.0 Å cavity and is disordered over several sites [578], analogous to the gas hydrate structures discussed in Part IV, Chapter 21.

A comparable situation is found in α-cyclodextrin·7.57 H$_2$O, where 2.57 water molecules are disordered over four sites in the cavity [78] (see Fig. 18.6b). In the hexahydrates I and II, however, the two enclosed water molecules (in I), and the enclosed water molecule and hydroxyl group (in II) are well ordered. The water is hydrogen-bonded to the O(6) hydroxyl of glucose 5 (see Fig. 18.6a), which is rotated inwards in order to form the hydrogen bond and to reduce the cavity size so that the water molecule with diameter of approximately 4 Å is more tightly bound. The rotation of glucose 5 requires that the hydrogen bonds involving its

O(2) and O(3) hydroxyls are broken. This allows the remaining four intramolecular hydrogen bonds to become stronger with a mean O···O distance of 2.85 Å, or about 1.8 Å for the H···O bond length. The described changes in the host structure upon binding of a guest molecule are consistent with the induced-fit concept described in Box 18.2.

In contrast, β-cyclodextrin and γ-cyclodextrin do not display conformational changes upon complex formation. Due to the variations in ring size, the interresidue O(2)···O(3') hydrogen-bond distances in α-cyclodextrin, about 3.0 Å, are longer and therefore weaker than those observed in β- and γ-cyclodextrins, where average O(2)···O(3') distances are 2.85 Å and 2.80 Å respectively (see Table 18.1). The molecular shapes in β- and γ-cyclodextrins are therefore more stable and rigid compared to the α-analog, and no induced-fit type conformational change is observed when complex formation occurs. These results obtained from crystal structure analyses are also in agreement with circular dichroism measurements in aqueous solution, which clearly indicated conformational flexibility for α-cyclodextrin but not for β-cyclodextrin [573].

18.7 Flip-Flop Hydrogen Bonds in β-Cyclodextrin · 11 H$_2$O

In the crystal structure of β-cyclodextrin · 11 H$_2$O, the hydrogen bonding is even more complicated than in the α-cyclodextrin hydrates [454]. There are not only more hydroxyl groups and water molecules in the crystal asymmetric unit, but they also display considerable positional disorder giving rise to more complex hydrogen bonding patterns.

Disorder in the β-cyclodextrin · 11 H$_2$O crystal structure. Within the β-cyclodextrin cavity of 6.2 Å diameter (Table 18.1), there are 6.13 water molecules statistically disordered over eight sites, of which only one is fully occupied and the other seven are statistically filled to an average of 73%. The remaining 4.87 water molecules are shared by eight sites in the interstitial regions between the β-cyclodextrin molecules. Of these eight sites, two are fully occupied and the other six are filled only to an average of 48%.

This distribution suggests that positions of water molecules located in the hydrophobic cavity are more densely populated and better ordered than those in the interstices bounded by the hydrophilic surface of the cyclodextrin molecules. In the interstices, the occurrence of many potential hydrogen-bonding donor and acceptor sites appears to provide several different alternative hydrogen-bonding patterns of comparable energy and hence more opportunity for disorder of the water molecules.

In the crystal structure of β-cyclodextrin · 11 H$_2$O the hydroxyl groups and water molecules form extended arrangements of cyclic and chain-like O−H···O hydrogen-bond patterns. In the following, however, our attention is focused on the statistically distributed O−H groups and water molecules which give rise to disorder in hydrogen bonds and lead to 18 interactions of the form O−($\frac{1}{2}$H)···($\frac{1}{2}$H)−O out of a total of 53 hydrogen-bonding interactions per asymmetric unit [118, 454].

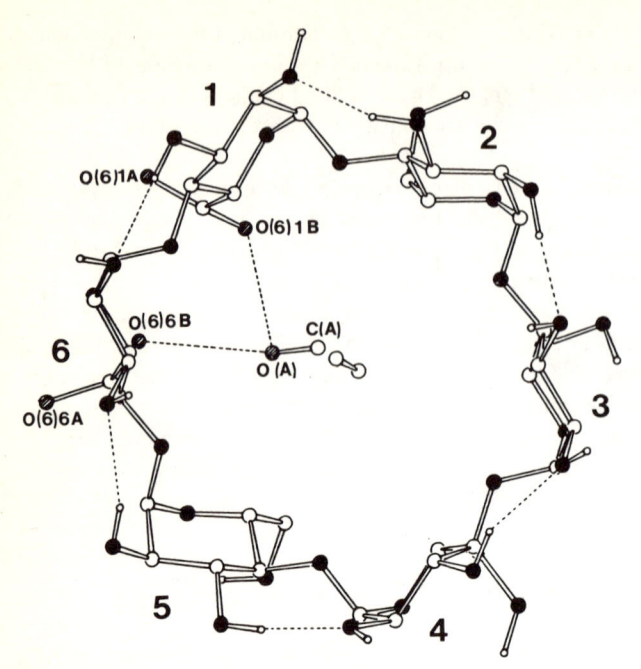

Fig. 18.8. Structure of α-cyclodextrin in the complex with methanol, α-cyclodextrin·MeOH·5H$_2$O; the water molecules are omitted for sake of clarity. The methanol molecule is twofold disordered, the α-cyclodextrin macrocycle adopts the "round" structure typical for inclusion complexes [577]

Box 18.2. Schematic Description of the Induced Fit Mechanism of α-Cyclodextrin Inclusion Formation

The "tense" and "relaxed" structures observed for α-cyclodextrin in the free and complexed state suggested the scheme described in Fig. 18.9 [579].

When α-cyclodextrin is "free" in aqueous solution, it adopts the "tense" form found in the two hexahydrate crystal structures (Fig. 18.6a): one glucose unit is rotated inwards to reduce the size of the α-cyclodextrin cavity so that the enclosed water molecules are held more tightly. The two intramolecular O(2)···O(3') hydrogen bonds to this glucose unit are broken, the other four remain intact.

Complex formation with substrate (S) can proceed directly, by route **A**, to yield a "relaxed" α-cyclodextrin with all six O(2)···O(3') hydrogen bonds engaged (as in the α-cyclodextrin methanol complex, Fig. 18.8), or the macrocycle can first open up to a "relaxed" form, route **B**, with the enclosed water molecules disordered over several sites so as to fill, statistically, the 5 Å diameter α-cyclodextrin cavity (as observed in the α-cyclodextrin·7.57H$_2$O crystal structure, Fig. 18.6b). The water is now in an "activated" form and can be replaced directly by the substrate. In a third possible mechanism, route **C**, the substrate aggregates first at the periphery of "tense" α-cyclodextrin, and in a second step replaces the two enclosed water molecules.

This scheme is adopted from enzyme-substrate complex formation where comparable processes are believed to take place [576]. We assume that routes **A** and **B** are nearly equivalent and more probable than **C**, but experimental evidence is still lacking.

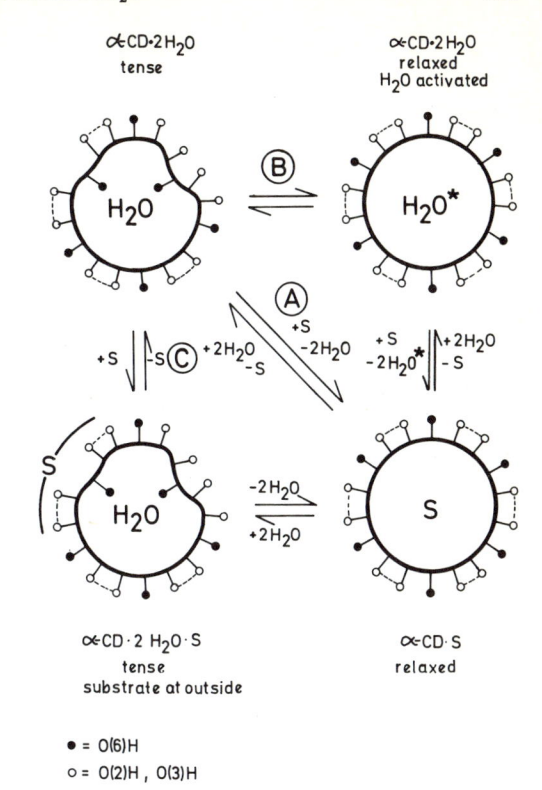

Fig. 18.9. Schematic description of the "induced-fit" type inclusion formation of α-cyclodextrin. S substrate (guest molecule); H_2O^* activated water (as observed in the α-cyclodextrin 7.57 H$_2$O crystal structure, Fig. 18.6b), ──● O(6) hydroxyl groups, ──○ O(2) and O(3) hydroxyl groups. Hydrogen bonds are indicated by *dashed lines* [579]

$O-(\frac{1}{2}H)\cdots(\frac{1}{2}H)-O$ **interactions are a statistical average over** $O-H\cdots O$ **and** $O\cdots H-O$**: the disordered flip-flop hydrogen bond.** In all the 18 hydrogen bonds of type $O-(\frac{1}{2}H)\cdots(\frac{1}{2}H)-O$ observed in the crystal structure of β-cyclodextrin·11H$_2$O, covalent O−H and hydrogen-bond H···O distances and O···O separations are as commonly observed for O−H···O hydrogen bonds (see Table 9.5a). The hydrogen atom positions are only half occupied and the H···H distance of about 1 Å is so much shorter than expected from the sum of van der Waals radii, 2.4 Å, that they are mutually exclusive (see Fig. 18.10a). This was interpreted as a dynamic disorder:

The nuclear density in the difference Fourier maps is clearly in agreement with this suggestion, because hydrogen atom positions are as expected half occupied, oxygen atom positions fully occupied, and the covalent and hydrogen-bond distances are consistent with those in commonly observed O−H···O hydrogen bonds.

The disorder in the hydrogen bonds can be described by two different mechanisms, as discussed in Part I, Chapter 2.8. In one, the configurational change, covalent O–H bonds are broken and the hydrogen atom is transferred (by tunneling) across the O–H···O bond. In the other, the conformational change, covalent bonds remain intact and hydroxyl groups are rotated from one hydrogen-bonding position into the other.

The net result is the same: two different states with nearly identical geometry and, presumably, nearly identical energy. This is favorable for entropical reasons and we expect considerable entropic stabilization of the disordered flip-flop hydrogen bonds.

In the medium strong to weak O–H···O hydrogen bonds observed in biological molecules and in water, the conformational mechanism is more likely to occur. The reason is that it requires only rotation of the O–H group after breaking the (weak) H···O hydrogen bond. In contrast, we would anticipate a large energy barrier for the configurational mechanism which requires breaking of the covalent O–H bond and transfer of the hydrogen atom across a distance of about 0.8 Å corresponding to the difference between hydrogen bonding H···O, 1.8 Å, and covalent O–H bond lengths, 0.97 Å. In the case of very strong, nearly symmetrical hydrogen bonds, where the covalent and hydrogen-bonding distances are nearly equal, O–H ≈ H···O (Part IIA, Chap. 7.1.1), the configurational mechanism will be preferred because the covalent bond is already weakened by lengthening from the standard 0.97 Å to about 1.2 Å.

An experimental distinction between the two mechanisms is lacking. We have to assume that the transition from one to the other kind of mechanism will be continuous and we expect a certain narrow range where the O–H···O hydrogen bond is so strong that both mechanisms occur simultaneously, see Chapter 2.8.

The O–($\frac{1}{2}$H)···($\frac{1}{2}$H)–O interactions are not isolated but are connected to form larger systems. As displayed in Fig. 18.10b, which gives a section of the crystal structure of β-cyclodextrin·11H$_2$O, these systems can be deconvoluted into two patterns with homodromic, ordered chains in which the hydrogen bonds run in either one or the other direction, with all H-atom positions **A** *or* **B** filled. Simultaneous occupation of **A** *and* **B** positions in any of the O–($\frac{1}{2}$H)···($\frac{1}{2}$H)–O systems is forbidden for sterical reasons.

There are two arrangements of disordered hydrogen bonds which are of particular interest. These are (a) an infinite chain, and (b) inter-residue O(2)···O(3') hydrogen bonds.

a) A flip-flop infinite chain in the crystal structure of β-cyclodextrin·11H$_2$O, is shown in Fig. 18.11. The chain is formed by the water molecule W(1) and two hydroxyls O(2), O(3) of adjacent residues 6 and 7 of the same β-cyclodextrin molecule. These atoms are close to a crystallographic 2$_1$ screw axis, the symmetry operation of which produces an infinite chain structure. In this flip-flop chain, the hydrogen-bonding potential of hydroxyls O(2)$_6$, O(3)$_7$ is fully satisfied and water W(1) is engaged in another short flip-flop chain comprising four hydroxyls and one water W(4) (not shown in Fig. 18.11).

b) Intramolecular, interglucose flip-flops stabilize the β-cyclodextrin macrocycle. As shown in Fig. 18.12, which displays a view of the β-cyclodextrin molecule occurring in the β-cyclodextrin·11H$_2$O crystal structure, all the seven glucoses are linked by flip-flop type hydrogen bonds (see also Table 9.5a). This might be a

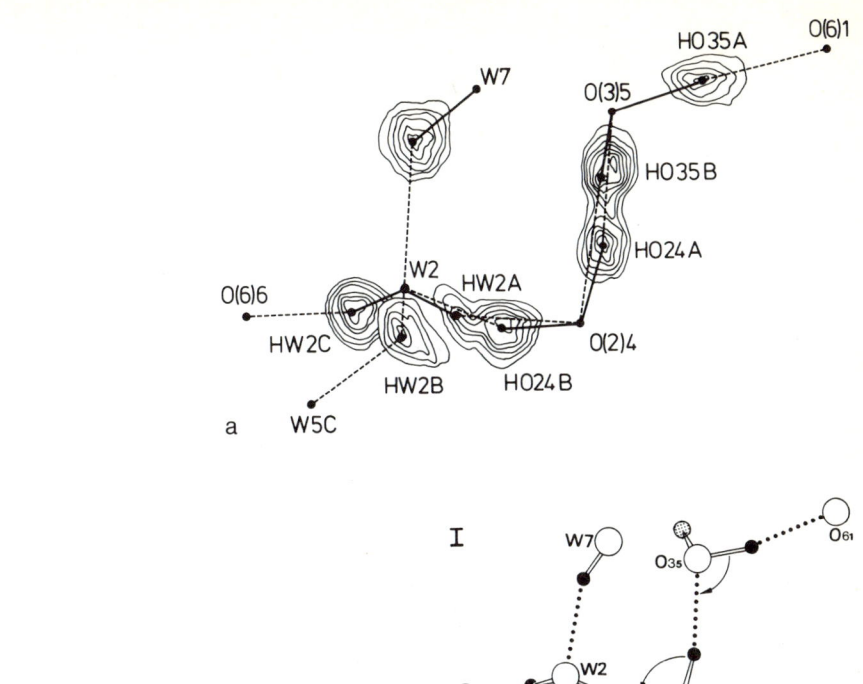

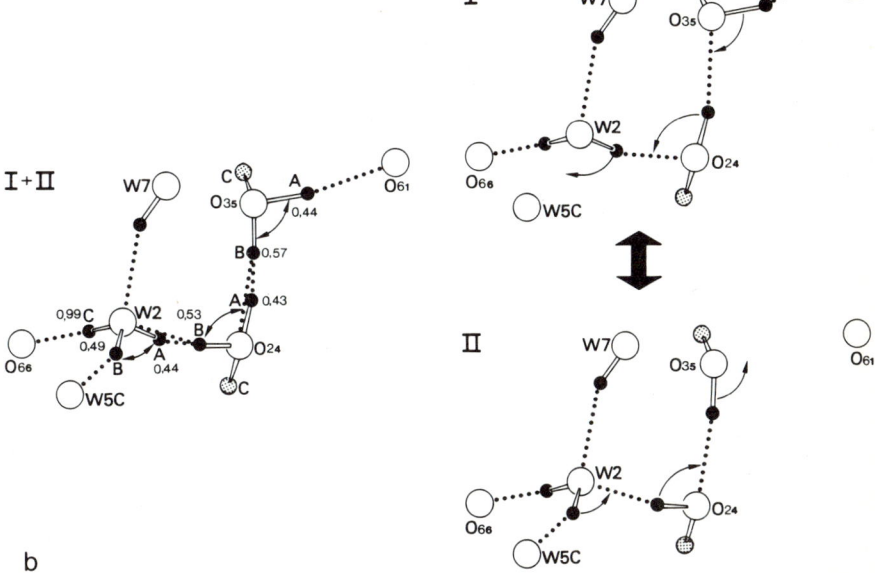

Fig. 18.10a, b. Section of a difference Fourier map (**a**), calculated from neutron diffraction data of β-cyclodextrin $\cdot$ 11 H$_2$O. Contours give H-atom positions. **b** Interpretation of **a**, with the flip flop O$-(\frac{1}{2})$H$\cdots(\frac{1}{2}$H$)-$O hydrogen bonds (*left*) deconvoluted into the two states (*I*) and (*II*) having normal O$\cdots$H$-$O and O$-$H$\cdots$O hydrogen bonds in homodromic arrangement (*right*). Small (*black*), medium (*stippled*) and large (*open*) circles represent H,C,O atoms, hydrogen bonds drawn as *dotted lines*. Occupational parameters of hydrogen atoms given in the left picture, all other atom positions are fully occupied. *Curved arrows* indicate how O$-$H groups rotate to go from one flip-flop state to the other. Only hydrogen atom positions **A** *or* **B** can be filled and a change **A** ⇌ **B** has to proceed in a concerted, cooperative rotation

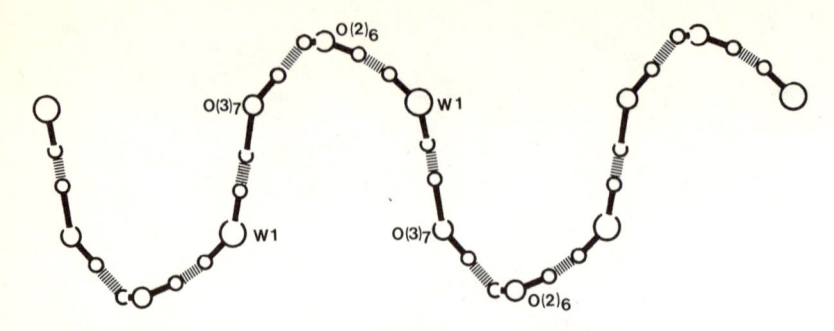

Fig. 18.11. The infinite flip-flop chain in β-cyclodextrin·11H$_2$O is composed of water molecule *W1* and hydroxyl groups *O(3)$_7$* and *O(2)$_6$* which belong to the same β-cyclodextrin molecule (see also Fig. 18.12). The operation of a 2$_1$ screw axis produces the infinite chain which is also anchored in the crystal lattice by many hydrogen bonds to and from adjacent hydroxyl groups (not shown) [454]

special feature of the β-cyclodextrin (and probably also of the γ-cyclodextrin) molecule where the intramolecular O(2)···O(3') distances are about 2.85 Å. It explains the conformational rigidity, whereby β- and γ-cyclodextrin in the "empty" hydrate and in the inclusion complexes with guest molecules have the same shape, in contrast to α-cyclodextrin, where "induced-fit" type changes are observed (Box 18.2).

The intramolecular flip-flop hydrogen bonds are of the three-center type. All the hydrogen atoms involved in intramolecular flip-flops have intramolecular contacts to adjacent glycosyl oxygens O(4) in the range 2.23 to 2.58 Å. This geometry is consistent with the formation of unsymmetric three-center bonds similar to those observed in the carbohydrate crystal structures (Part II, Chapter 13).

$$O(2)-H \cdots O(3') \quad \text{and} \quad O(3)-H \cdots O(2')$$
$$\cdot\cdot \qquad\qquad \cdot\cdot$$
$$O(4') \qquad\qquad O(4)$$

Do the flip-flop hydrogen bonds represent dynamical or statical disorder? Although we assumed dynamical disorder of the flip-flop hydrogen bonds in the previous discussion, from a crystallographer's view it could also be static. Static disorder means that the hydrogen bonds are ordered one way in one unit cell (or domain) in the crystal, and in opposite way in another. The diffraction experiment cannot distinguish between static and dynamic disorder because it observes the space and time average of a statistical distribution of hydrogen bonds. In dynamic disorder the conformational changes may require energies comparable to energies in the crystal at room temperature, and if so, they may become ordered at lower temperatures. Thus temperature-dependent experiments can distinguish between static and dynamic disorder.

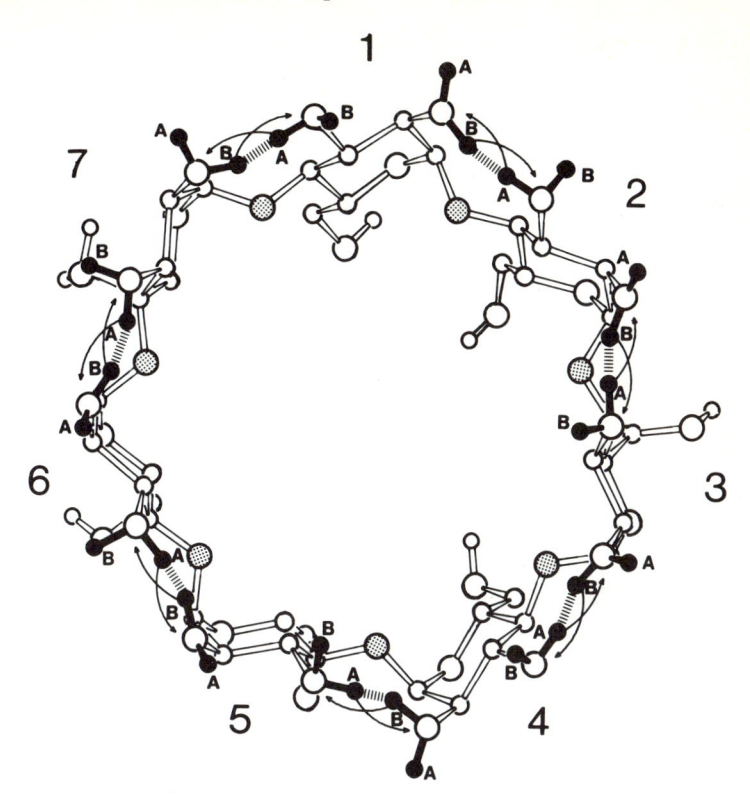

Fig. 18.12. Molecular structure of β-cyclodextrin in the crystal structure of the undecahydrate. Hydrogen atoms on O(6) drawn as *small open circles*, on O(2),O(3) as *small black circles*, all other hydrogens omitted for the sake of clarity; oxygen and carbon atoms are symbolized by *large-* and *medium-sized circles*, glycosidic oxygens by *dotted circles*. Intramolecular flip-flop hydrogen bonds involve disordered O(2), O(3) hydroxyls of adjacent glucose units. Only hydrogen atom positions **A** or **B** can be filled at one time because the H···H separation in each flip-flop, indicated by ‖‖‖‖, is too short (~1 Å) to allow simultaneous occupation. O–H···O bonds between hydrogen atoms (**A**, **B**) and oxygen atoms are indicated by *bent arrows*; hydrogen atoms **A**, **B** that are not involved in *intra*molecular flip-flop bonds are hydrogen-bonded intermolecular to hydroxyl groups or water molecules

Evidence that the dynamic disorder occurs was provided by a quasielastic incoherent neutron scattering experiment carried out with β-cyclodextrin · 11 H$_2$O. It verified that several protons jump, with mean residence time around 10^{-11} s [579a]. A differential scanning thermogram showed that an exothermic transition takes place at 227 K [580, 481], which was interpreted as that temperature where the dynamic process ceases and the hydrogen bonds become ordered. When β-cyclodextrin was crystallized from D$_2$O instead of H$_2$O, the transition temperature changed to 205 K, indicating an H isotope effect [455]. Conclusive evidence for dynamic disorder was, however, provided by a neutron diffraction study carried out below the transition temperature which is described in the next paragraph.

18.8 From Flip-Flop Disorder to Ordered Homodromic Arrangements at Low Temperature: The Importance of the Cooperative Effect

Below the transition temperature, water molecules become ordered and change the hydrogen-bonding pattern. At room temperature, only 3 of the 16 water sites observed in β-cyclodextrin·11H$_2$O are fully occupied. A neutron diffraction study carried out at 120 K, well below the transition temperature [455], showed that these sites are changed to 11 fully occupied and only one, water W8, remains partially occupied at 65%. At the reduced temperature, the water molecules have moved into positions of lower energy. Associated with these changes in water structure are rotations of water molecules and of β-cyclodextrin hydroxyl groups into ordered O–H$\cdots$O hydrogen bonds which are incorporated in homodromic systems (see Fig. 18.13). This has occurred at the expense of disordered bonds which have disappeared, except for one newly formed four-membered flip-flop cycle which is not present at room temperature (see Figs. 18.13a, b, 18.16). The observed changes are:

a) The infinite flip-flop chain existing at room temperature (Fig. 18.11) has changed into a homodromic chain (Fig. 18.14), which is anchored to β-cyclodextrin molecules through short hydrogen-bonding chains in homodromic arrangement.

b) The seven inter-residue, intramolecular flip-flops in the β-cyclodextrin macrocycle (Fig. 18.12) have disappeared and are now of the ordered O–H$\cdots$O type except for one four-membered flip-flop cycle (see Fig. 18.15). All these inter-residue hydrogen bonds are of the three-center type and involve a longer minor component to the glycosyl oxygen atoms O(4) (see Table 9.5b).

c) The quadrilateral flip-flop cycle (Figs. 18.15a and 18.16), is the most striking feature of the low-temperature β-cyclodextrin·11H$_2$O crystal structure. Although all the oxygen atoms involved in this cycle are in comparable positions at room temperature, there are only two of the four flip-flops present, W(2)–($\frac{1}{2}$H)$\cdots$($\frac{1}{2}$H)–O(2)$_4$–($\frac{1}{2}$H)$\cdots$($\frac{1}{2}$H)–O(3)$_5$. Hydrogen atom positions have changed upon cooling and give rise to the flip-flop cycle with disordered clockwise and anticlockwise orientation of the four intracyclic hydrogen bonds, as illustrated in Fig. 18.16.

The reason for the formation of the flip-flop cycle can be seen in its function as versatile junction for several homodromic chains which intersect in this place. These are three infinite chains, one in crystallographic c and two in b direction (see Figs. 18.13 and 18.17). Moreover, a short homodromic chain formed by W(13) and W(14) connects opposite corners of the flip-flop cycle and therefore gives rise to a homodromic cycle, Fig. 18.13b. If the flip-flop cycle consisted of ordered O–H$\cdots$O bonds, other homodromic chains would have to be interrupted.

d) The disorder in the β-cyclodextrin cavity observed at room temperature is considerably reduced at 120 K. Five of the six enclosed water sites are rearranged such that a pentagonal homodromic cycle is formed, which also includes the

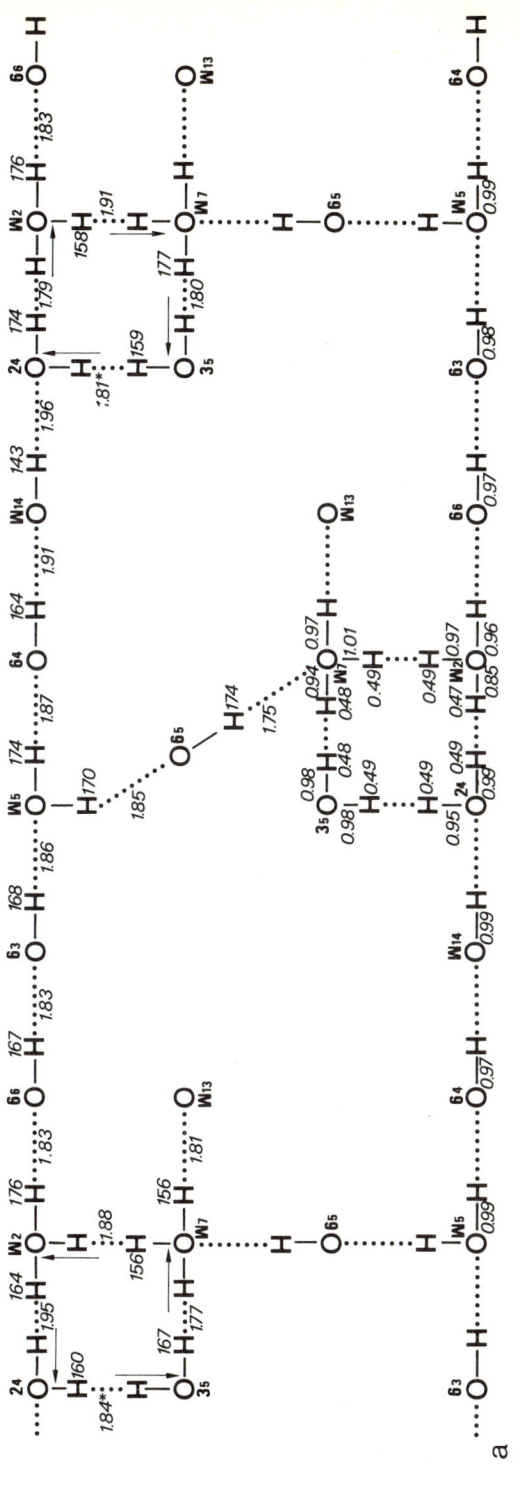

Fig. 18.13a, b. Shown are hydrogen-bond patterns occurring in the crystal structure of β-cyclodextrin·11H$_2$O at 120 K. Because the total scheme is too complicated to be shown in one picture, it is broken up into two different schemes. In these, water molecules are denoted by W, and hydroxyl groups by two numbers, e.g., O(6)$_4$, hydroxyl O(6) of glucose number 4 in the β-cyclodextrin macrocycle. **a,** Hydrogen-bonding details for flip-flop quadrilateral (see also Fig. 18.16) in anticlockwise (*top left*) and clockwise (*top right*) direction, and occupancies and O·H covalent bond lengths (*bottom center*). **b,** A is the flip-flop quadrilateral, B the homodromic pentagonal cycle in the β-cyclodextrin cavity, see also Fig. 18.17a. Water molecules W7 and W13 occur in both schemes and are the points where the hydrogen-bonding systems are jointed to form a complex three-dimensional structure

Fig. 18.13b

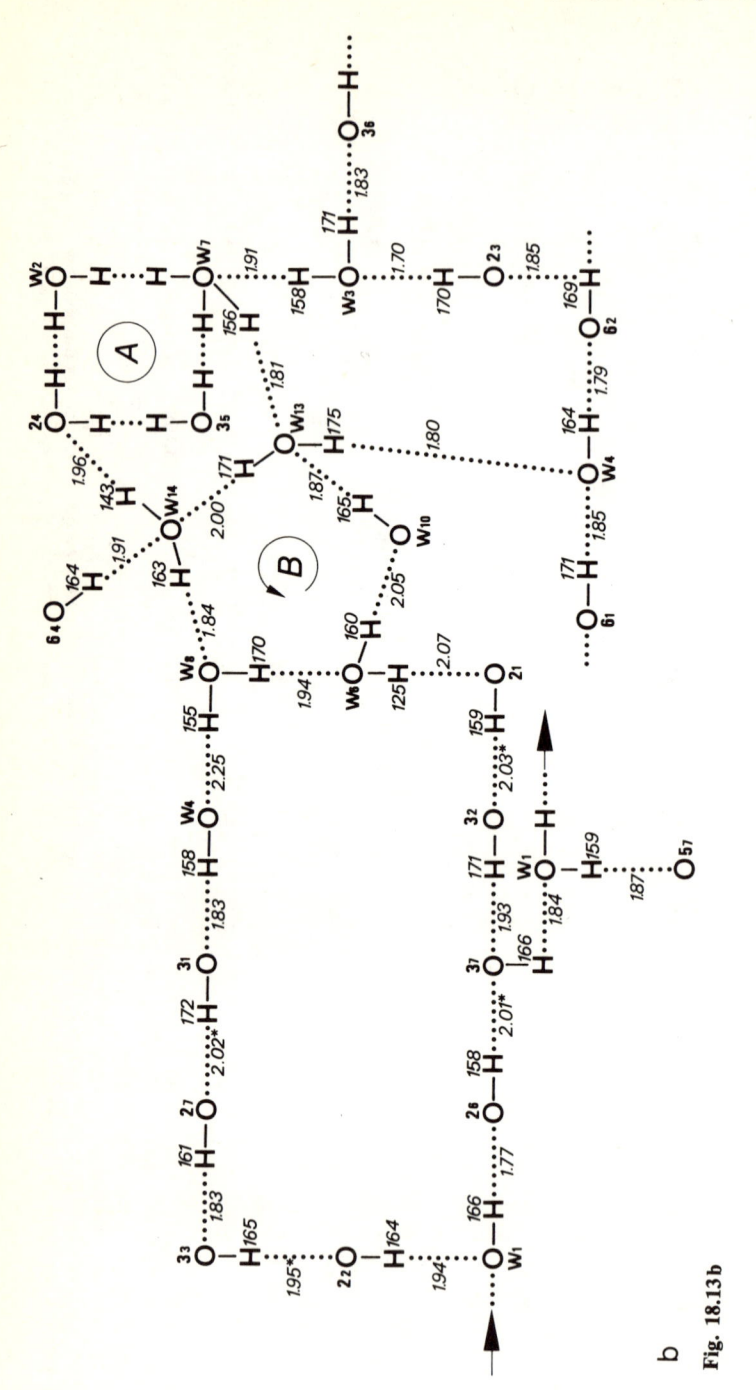

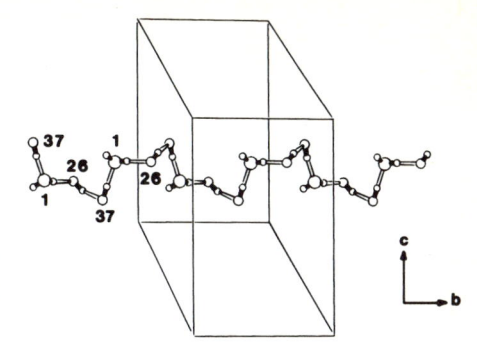

Fig. 18.14. The flip-flop chain occurring at room temperature (Fig. 18.11) changes to a homodromic chain at 120 K. 26 and 37 denote $O(2)_2$ and $O(3)_7$; 1 is water W(1)

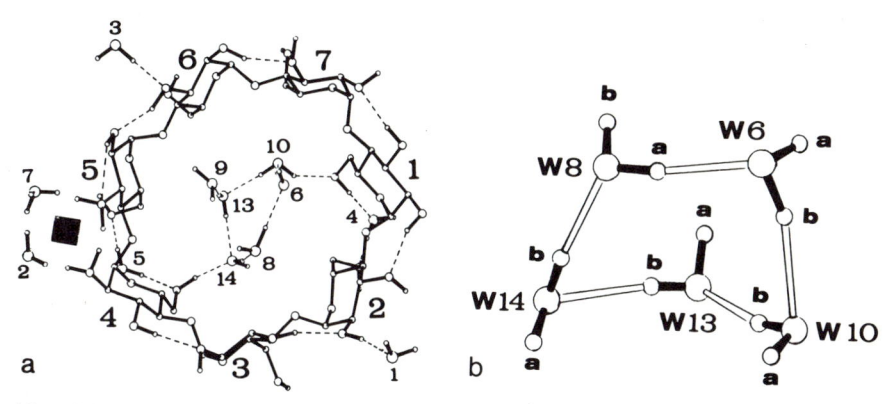

Fig. 18.15. a Stereo view of β-cyclodextrin·11 H₂O at 120 K, with water molecules W1 to W14. Only one $O(2)\cdots O(3')$ flip flop has remained which is part of the flip-flop quadrilateral marked by a *solid square*. **b** View of the homodromic cycle formed in the cavity of β-cyclodextrin at 120 K. At room temperature, these five water molecules are statistically disordered over eight sites

only partially occupied (65%) water molecule W(8) (see Fig. 18.15b). The cycle is anchored in the β-cyclodextrin cavity by virtue of hydrogen bonds to two O(6) hydroxyls (Fig. 18.15a) and between the water molecules and the ether-like O(4),O(5) oxygen atoms (not shown), which function less frequently as acceptors in oligosacchride crystal structures.

e) An infinite homodromic network occurs in the crystallographic b, c plane (Figs. 18.13, 18.17b, c). It consists of chains running in c direction, and individual chains related by translation in b are connected by short homodromic chains. Therefore, transition from one to the other chain is homodromic, and the chains and their connections form an infinite network, with individual meshes consisting of 22 O−H groups in antidromic arrangement.

The cooperative effect determines the formation of extended O−H···O patterns. The conclusions that can be drawn from a comparison of the crystal structures of β-cyclodextrin·11 H₂O above and below the transition temperature (−46 °C) are twofold. First, the flip-flop disorder is dynamic. Second, on reducing

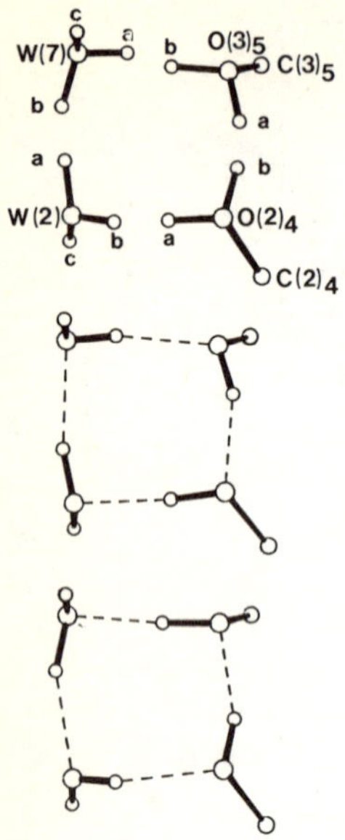

Fig. 18.16. The quadrilateral flip-flop cycle and its deconvolution into two cycles with homodromic O—H···O hydrogens bond in clockwise and anticlockwise orientation; hydrogen atom positions **a, b** are only half occupied

the temperature, the disordered water molecules move into lower energy, fully occupied positions and the hydroxyl groups rotate to form as many ordered hydrogen bonds in homodromic patterns as possible. The entropy lost from the ordering of the hydrogen-bond protons is gained by additional energy from the new cooperative arrangement of the hydrogen bonds.

18.9 Maltohexaose Polyiodide and Maltotriose — Double and Single Left-Handed Helices With and Without Intramolecular O(2)···O(3') Hydrogen Bonds

The crystallization of oligosaccharides beyond the trisaccharides is difficult. Whereas the cyclodextrins show modest to good solubility in water (Table 18.1), their linear analogs are extremely water-soluble and tend to form syrups and glasses. These different properties are probably due to the inherent flexibility of the linear oligosaccharides, contrasting the more rigidly defined structures of the cyclically closed cyclodextrins. As a consequence, the crystallization of higher oligosaccharides was unsuccessful except for the tetrasaccharide stachyose pentahydrate described in Part II, Chapter 13.

In addition to the flexibility, epimerisation at the reducing group of oligomers contributes to configurational heterogeneity in solution. To avoid this, the α-glucosides of maltooligomers were crystallized. The extreme solubility in water was reduced by employing a mixture of n-butanol, ethanol, water (4:1:1 by volume) for methyl-α-maltotrioside [582], and by complex formation with barium-polyiodide for p-nitrophenyl-α-maltohexaoside [583].

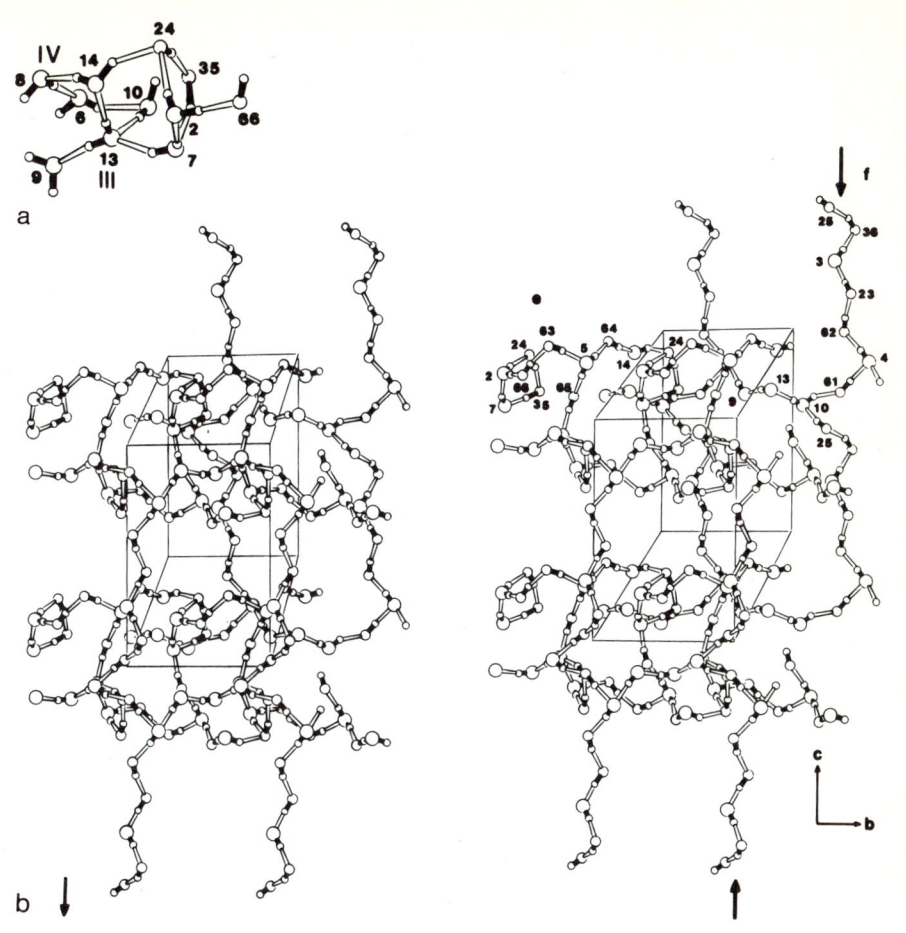

Fig. 18.17a–c. Hydrogen-bonding scheme in β-cyclodextrin·11H$_2$O at 120 K, showing only O–H groups and water molecules. **a** The cluster formed between flip-flop cycle (atoms O(2)$_4$, O(3)$_5$, W(2), W(7)) and homodromic pentagon (water molecules W(6), W(10), W(13), W(14), W(8)). **b** Stereo view of the hydrogen-bonding pattern in the b, c plane. **c** Schematic diagram of parts of the pattern given in **b** [455]

The crystals obtained are highly hydrated, with compositions methyl-α-maltotrioside·4H$_2$O and (p-nitrophenyl-α-maltohexaoside)·Ba(I$_3$)$_2$·27H$_2$O. The hydration appears to be important for the crystallization because it contributes to the intermolecular hydrogen bonding, stabilizing the crystal structures so that direct intermolecular contacts between the maltooligomers, which might influence their conformations through packing forces, are largely reduced. Unfortunately, in both crystal structures, the O–H hydrogen atoms, which are required to describe the hydrogen-bonding schemes unambiguously, could not be located from difference Fourier maps due to experimental difficulties. For this reason, the discussion on hydrogen-bonding schemes is limited and can only be based on O···O distance criteria.

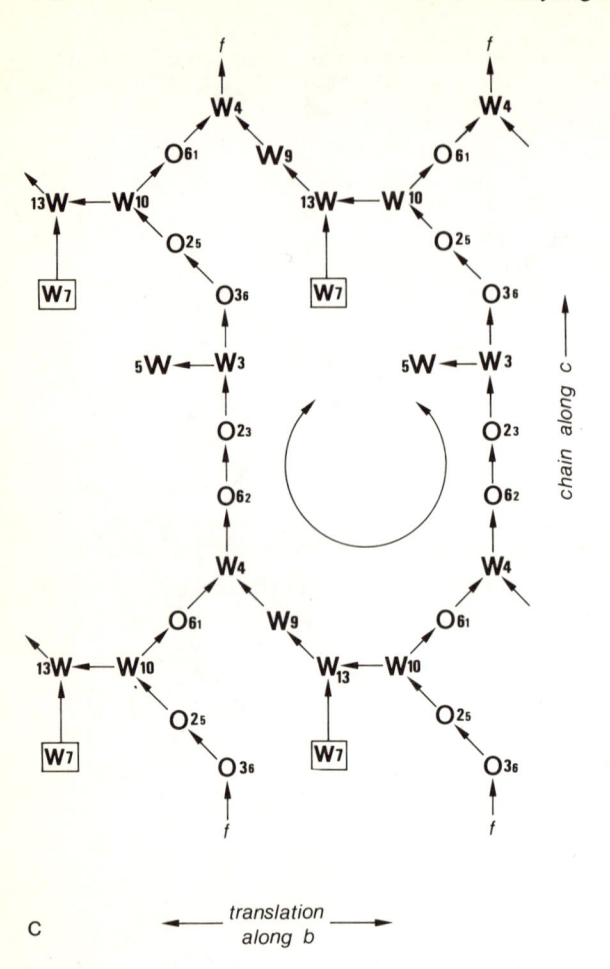

Fig. 18.17c

The two maltooligomers form left-handed helices with different molecular parameters. As shown in Table 18.4, the structural parameters in methyl-α-maltotrioside and in p-nitrophenyl-α-maltohexaoside are largely comparable. The glucoses are all in $^4C_1$ chair conformation as in the cyclodextrins. They are similar in dimensions, as reflected by the $O_4 \ldots O_{4'}$ virtual bond distances, and the average angle at the glucosidic link, $C(1)-O(4')-C(4')$, 115°, is identical. The torsion angles Φ, ψ which describe the relative rotation of adjacent glucoses about this link could all be different for different pairs of glucoses if the molecular structures were "random" coils, or they could be identical in case of more ordered, helical structures. The latter is true, with the negative Φ, ψ torsion angles characteristic for left-handed helices. There is, however, a clear distinction in the magnitude of Φ, ψ, which gives rise to different helices for the methyl-α-maltotrioside·$4H_2O$ and for the complex (p-nitrophenyl-α-maltohexaoside)$_2$·Ba(I$_3$)$_2$·27H$_2$O.

Table 18.4. Some average geometrical parameters of the left-handed single and double helices formed by methyl-α-maltotrioside·4H$_2$O and by (p-nitrophenyl-α-maltohexaoside)$_2$·Ba(I$_3$)$_2$·27H$_2$O, respectively [582, 583]

	Intramolecular angles and distances					Helical parameters	
	Φ^a	ψ^a	O$_4$···O$_{4'}$	C$_1$–O$_{4'}$–C$_{4'}$	O$_2$···O$_{3'}$	Rise per residue	Residues per turn
Methyl-α-maltotrioside	–38°	–31°	4.40 Å	115.0°	≥3.63 Å	3.6 Å	5
p-Nitrophenyl-α-maltohexaoside	–19°	–12°	4.52 Å	115.0°	2.89 Å	2.33	8

[a] Torsion angles are defined as Φ, H$_1$–C$_1$–O$_{4'}$–C$_{4'}$ and ψ, C$_1$–O$_{4'}$–C$_{4'}$–H$_{4'}$, where the positions of H atoms are calculated on the basis of the C,O skeleton.

Methyl-α-maltotrioside has no intramolecular O(2)···O(3′) hydrogen bonds and is reminiscent of A-starch. The Φ, ψ angles of methyl-α-maltotrioside are greater (–38°, –31°) than those of p-nitrophenyl-α-maltohexaoside (–19°, –12°), indicating that adjacent glucose residues are more rotated relative to each other. This gives rise to wide inter-residue O(2)···O(3′) separations of greater than 3.63 Å which is not evidence of hydrogen bonds as in the maltose disaccharides.

Moreover, the O(2) and O(3) hydroxyl groups are all engaged in *inter*molecular hydrogen-bonding contacts with O···O distances smaller than 3.05 Å to hydroxyl groups of symmetry-related molecules or to water molecules. Besides these interactions, hydrogen bonds are formed to two of the three O(5) oxygens but not to the glycosidic links O(4) (see Fig. 18.18a).

Because the Φ, ψ angles at the two glycosidic links in methyl-α-maltotrioside are so similar, the geometry of the trimer could be extended to that of a polymer. The resulting left-handed amylose single helix described in Fig. 18.18b has a rise per residue of 3.6 Å and 5 glucoses per turn resulting in a pitch height of 18.0 Å [582]. Only slight modifications in Φ, ψ torsion angles would be needed to obtain a left-handed helix with six residues in a pitch of 21.04 Å as derived from X-ray fiber diagrams on A-amylose [584]. Based on electron diffraction studies on microcrystals of amylose fractions with a degree of polymerization (DP) of 15, combined with the crystallographic information on methyl-α-maltotrioside, a model was constructed in which two left-handed amylose helices in parallel orientation intertwine to form an A-type double helix [585]. Since there are no O(2)···O(3′) intramolecular hydrogen bonds, the two strands are associated mainly by hydrophobic forces and by one intermolecular hydrogen bond, O(6)···O(2′).

This model differs from the right-handed parallel double helix proposed on the basis of combined X-ray fiber diffraction data and of energy minimisations [584]. From an energetical point of view, the two models are nearly identical [585], and a clear distinction can be made only if more (crystallographic) data become available.

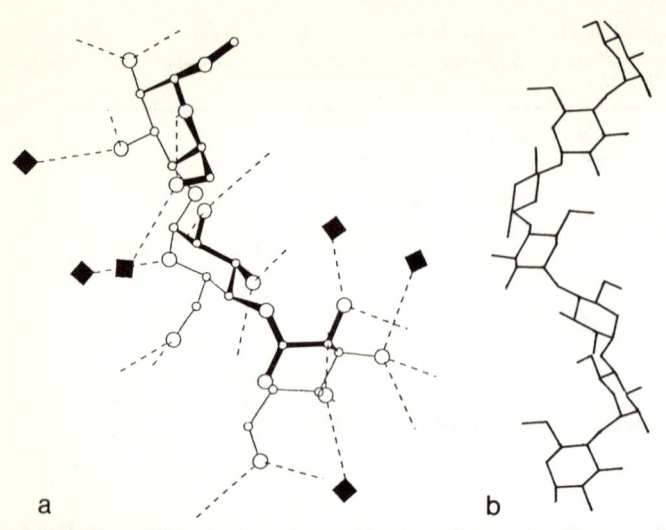

Fig. 18.18. a Molecular structure of methyl-α-maltotrioside. Hydrogen-bonding contacts are indicated by *dashed lines*, water molecules are marked by *solid squares*, C and O atoms are represented by *small* and *large spheres*. Note that there are no intramolecular O(2)···O(3') hydrogen bonds. **b** View of the left-handed single helix with five residues per turn derived from the structure of methyl-α-maltotrioside [582]

Intra- and intermolecular hydrogen bonds stabilize the left-handed, antiparallel double helix in the maltohexaoside·polyiodide complex. Since the p-nitrophenyl-α-maltohexaoside molecules enclose the polyiodide formed by slightly zigzagged $I_3^- \cdot I_3^- \cdot I_3^- \cdots$ units, their conformation differs strikingly from that of the methyl-α-maltotrioside (Table 18.4). The Φ, ψ angles are small, average $-19°$, $-12°$, so that the relative rotations between adjacent glucose units are also small and intramolecular O(2)···O(3') hydrogen bonds can form (see Fig. 18.19). They give rise to a left-handed helical configuration with 2.33 Å rise per residue, and 8 glucoses in a pitch of 18.6 Å. The O(2),O(3) hydroxyl groups are in well-defined positions due to the stereochemical constraints of the glucose chair form and of the O(2)···O(3') intramolecular hydrogen bonds, thereby facilitating the energetically favorable arrangement of two strands in antiparallel mode, stabilized by intermolecular O(2)···O(2') and O(3)···O(3') hydrogen bonds (Figs. 18.19, 18.20).

This antiparallel double helix has a central cavity which is filled by the polyiodide chain or, in other terms, the polyiodide chain serves as a matrix around which the p-nitrophenyl-α-maltohexaoside molecules are arranged. The complex is stabilized by the inter- and intramolecular hydrogen bonds, the coordination of Ba^{2+} to hydroxyl groups and O(5) atoms, and by the stacking interactions between p-nitrophenyl groups [583, 583a].

Why are the maltooligomer helices so different? The main difference is the complex formation of p-nitrophenyl-α-maltohexaoside with polyiodide, whereas the methyl-α-maltotrioside crystallized with water molecules. In the polyiodide complex, the polyiodide is enclosed by the p-nitrophenyl-α-maltohexaoside

Maltohexaose Polyiodide and Maltotriose 349

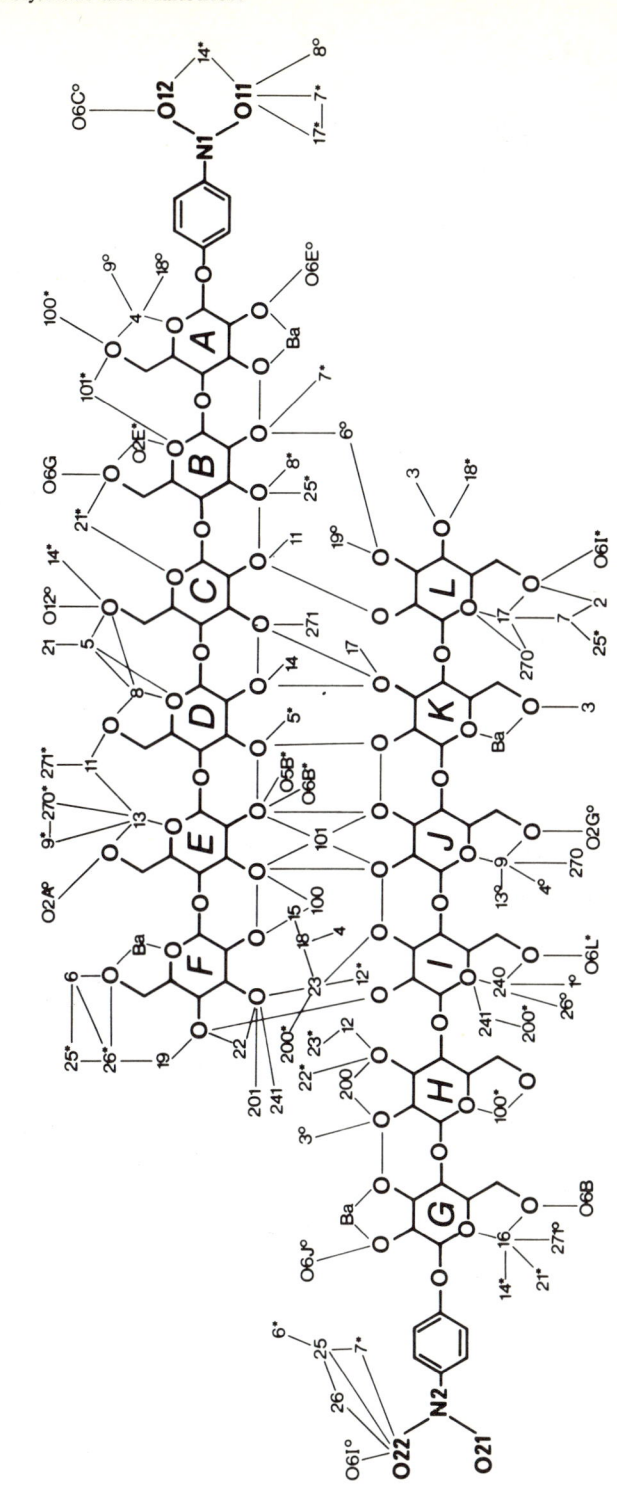

Fig. 18.19. Hydrogen-bonding scheme and Ba^{2+} coordination in the complex (p-nitrophenyl-α-maltohexaoside)$_2$·Ba(I$_3$)$_2$·27H$_2$O. The lines indicate O···O distances <3.5 Å, or Ba···O coordination. Water oxygen atoms are only given by their numbers, glucose hydroxyls are differentiated by "0", i.e.

O6G is O(6) of glucose G. Contacts marked "○" are within the "infinite" double helix in the crystal, "*" are those between symmetry related helices. Note the water (or Ba^{2+})-mediated intraglucose hydrogen bonding (coordination) involving O(5), O(6) in all glucoses except G, see Part IB, Chapter 9, p. 155

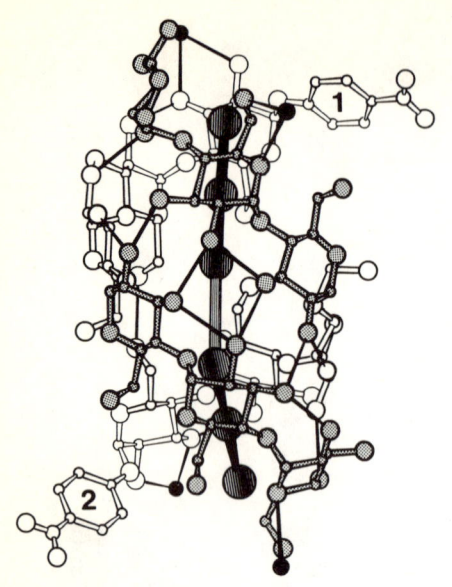

Fig. 18.20. View of the molecular complex (p-nitrophenyl-α-maltohexaoside)$_2$·Ba(I$_3$)$_2$·27H$_2$O, omitting water molecules. Hydrogen-bonding contacts, interiodine covalent bonds, and the Ba−O coordinative bonds are drawn *black*. Molecules *1* and *2* [see Fig. 18.19, glucoses A−F (*1*) and H−M (*2*)] are indicated by *numbers* in the phenyl rings. Parts of the complex closer to the viewer are *stippled*. C, Ba, N, O, I atoms drawn with *increasing radii*, Ba filled *black* [583]

molecule just as iodine or polyiodide is enclosed by α-cyclodextrin [566]. If we open the α-cyclodextrin ring and let it relax, a smooth, lockwasher-type structure will result, reminiscent of the actually observed helix. The double helical arrangement of p-nitrophenyl-α-maltohexaoside in the presence of polyiodide is directly comparable to the packing patterns of the cyclodextrins (Sect. 18.2 and Fig. 18.3). In their dimeric head-to-head arrangements, they have similar intermolecular hydrogen-bonding schemes involving O(2)···O(2'), O(3)···O(3'), O(6)···O(6').

If there is no central matrix around which the maltooligomers can arrange, they adopt a more extended conformation where intramolecular O(2)···O(3') hydrogen bonds are absent. It appears that, instead, the conformational energy around the glucosidic bond is minimized and the hydrogen-bonding potential is satisfied by intermolecular interactions to other oligosaccharide molecules and, notably, to water of crystallization as in Fig. 18.18. We must assume that these different helical forms and, probably, other forms which are possible through slight modifications of Φ, ψ angles, are in equilibrium.

Chapter 19
Hydrogen Bonding in Proteins

In proteins, the 20 different α-L-amino acids (Box 19.1) are linked in specific sequence by peptide bonds to form linear polypeptides of molecular weight in the range of a few thousand to several hundred thousand. If a protein contains cysteines, these can cross-link by oxidation to form disulfide bridges. Besides these covalent bonds, the main stabilization of the very complex three-dimensional (tertiary) structure, which is characteristic for each protein, is by hydrophobic and van der Waals forces and, even more importantly, by hydrogen bonds [133, 134, 586–589].

Four levels of structural organization are defined in globular proteins. The amino acid sequence or *primary* structure is, above all, responsible for the higher-level structure and biological function of a protein. The *secondary* structure defines the conformation of the polypeptide backbone (the "main-chain") with the typical repetitive elements, α-helix and β-pleated sheet, which were recognized 40 years ago by Astbury, Huggins, Pauling, and Corey as the essential folding motifs in wool and silk [17, 21, 26, 590]. In addition, there are bends in the polypeptide chain called *β*-turns, which give rise to sharp hairpin-like folds [591]. Because the different amino acids have tendencies to occur preferentially in certain secondary structure elements, these can often be predicted with (modest) success from the primary sequence of a protein [586, 588, 592].

The arrangement in three-dimensional space of the regions with secondary structure elements and the regions of irregular polypeptide conformation is called *tertiary* structure (Box 19.1). This gives the globular proteins their characteristic shape. It describes the folding of the polypeptide chains into functionally active sites and is, above all, essential for the specific biological properties of a protein.

In multisubunit proteins, the individual subunits are combined in certain arrangements to form *quaternary* structures. They represent the highest level of organization in protein assembly and, per definition, are not held together by covalent forces.

Secondary, tertiary, and quaternary structures of a protein are mainly stabilized by hydrophobic interactions, van der Waals forces, and by hydrogen bonds. The latter will be considered in the following. The secondary structure is stabilized only by *main-chain* to *main-chain* inter-peptide $N-H \cdots O=C$ hydrogen bonds. Tertiary and quaternary structures are held together by hydrogen bonds of the type *main-chain* to *main-chain*, *main-chain* to *side-chain*, *side-chain* to *side-chain*. Besides these, we have to consider interactions between *water* and *main-chain* or

Box 19.1. Amino Acids, Peptide Link, and Structural Organization of a Protein

The different amino acids occurring in the proteins are depicted in Fig. 19.1. Only one, glycine, has no side-chain; the side-chains of the other amino acids are of hydrophobic to hydrophilic character. Those of aspartic and glutamic acids and of lysine, arginine and (below pH ~7) histidine are charged. The side-chain of proline is cyclically closed so that the peptide nitrogen cannot donate a hydrogen bond. The amino acids are linked head-to-tail by peptide bonds to form the primary structure. Counting of the amino acids in a polypeptide chain is sequentially from the positively charged amino (N) terminus to the negatively charged carboxy (C) terminus, as illustrated for the enzyme ribonuclease T_1 in Fig. 19.2.

The C–N bond in the peptide groups has double bond character due to the π-conjugation. It adopts a *trans* conformation,[1] so that its hydrogen-bond donor $-\overset{H}{\underset{|}{N}}-$ and acceptor $-\underset{\underset{O}{\|}}{C}-$ groups are diametrically opposite, as are the C_α atoms to which the amino acid side-chains are attached:

This limits the flexibility of the polypeptide chain to rotations about the $N-C_\alpha$ and $C_\alpha-C$ bonds with torsion angles Φ (C–N–C_α–C) and Ψ (N–C_α–C–N) respectively. As recognized by Ramachandran [593], the rotations about these bonds are also sterically restricted, with energy minima in Φ, Ψ ranges typical of the *secondary structure* elements α-helix and β-pleated sheet, with a small region for left-helical arrangement (Fig. 19.3). The secondary structure is stabilized by hydrogen bonds formed exclusively between main-chain peptide groups (see Fig. 19.4). The folding of the secondary structure elements and of unstructured chain segments in three-dimensional space is called *tertiary structure*. Because the protein molecules are too complicated to be illustrated with all their atoms, they are represented as cartoons, as shown in Fig. 19.5 for the enzyme ribonuclease T_1 in complex with a specific inhibitor, guanosine-2'-phosphate [594].

[1] Some rare cases of cis conformation were found in X-ray structure analyses of proteins.

side-chain. The side-chains have a variety of functional groups which can act as hydrogen-bond donors and acceptors (see Fig. 19.1).

19.1 Geometry of the Secondary-Structure Elements: Helix, Pleated Sheet, and Turn

The secondary structure elements are stabilized by hydrogen bonds formed exclusively between peptide groups. Because the peptide groups are simultaneously donors and acceptors, cooperativity comes into play.

Hydrogen bonding of the peptide groups is cooperative. As indicated in Box 19.1, the donor and acceptor functions of the peptide group $-\overset{H}{\underset{|}{N}}-\underset{\underset{O}{\|}}{C}-$ are in

Geometry of the Secondary-Structure Elements: Helix, Pleated Sheet, and Turn 353

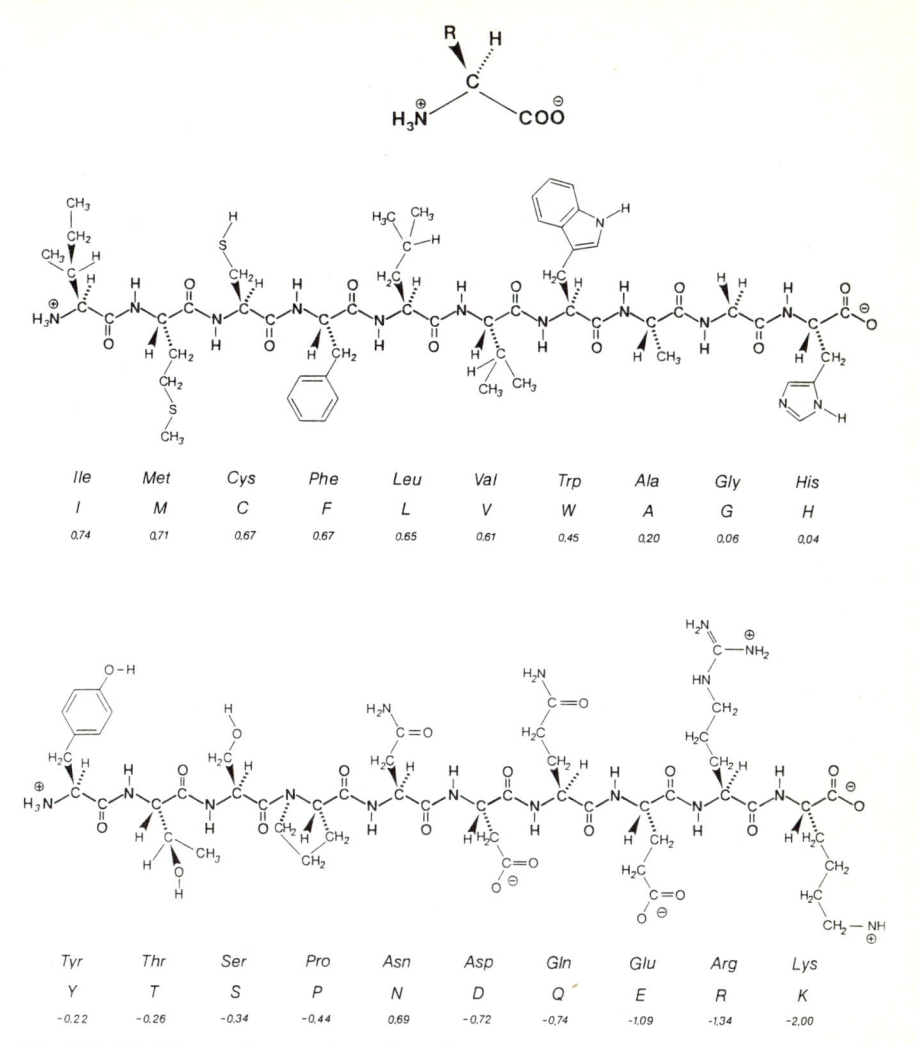

Ile	Met	Cys	Phe	Leu	Val	Trp	Ala	Gly	His
I	M	C	F	L	V	W	A	G	H
0.74	0.71	0.67	0.67	0.65	0.61	0.45	0.20	0.06	0.04

Tyr	Thr	Ser	Pro	Asn	Asp	Gln	Glu	Arg	Lys
Y	T	S	P	N	D	Q	E	R	K
-0.22	-0.26	-0.34	-0.44	0.69	-0.72	-0.74	-1.09	-1.34	-2.00

Fig. 19.1. The 20 naturally occurring amino acids drawn in the form of two polypeptide chains, one (*top*) containing the hydrophobic and the other (*bottom*) the hydrophilic amino acids. In each chain the hydrophilic character increases *from left to right*. Below each amino acid are given three-letter and one-letter codes and a hydrophobicity index (most positive = most hydrophobic; most negative = least hydrophobic) [595]. The main-chain (backbone) atoms are drawn more heavily

trans configuration and π-conjugated, i.e., the central N–C bond displays partial double-bond character. If a hydrogen bond N–H$\cdots$O=C to or from a peptide group is formed, the latter is polarized and consequently (1) its central N–C bond is shortened, implying more double-bond character and greater rigidity, and (2) its hydrogen-bonding potential is increased due to π-cooperativity.

The topologies of helices and pleated sheets involve fundamentally different intra- and interchain interactions. The α-helix is formed by a *continuous* segment

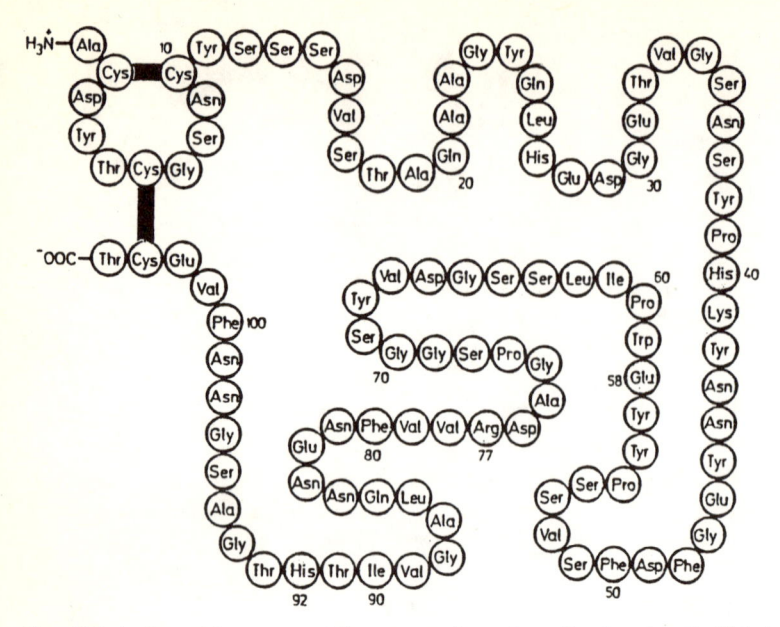

Fig. 19.2. Amino acid sequence of the enzyme ribonuclease T_1. Cysteine disulfide bridges which cross-link the linear polypeptide chain are indicated by *heavy lines* [593a]

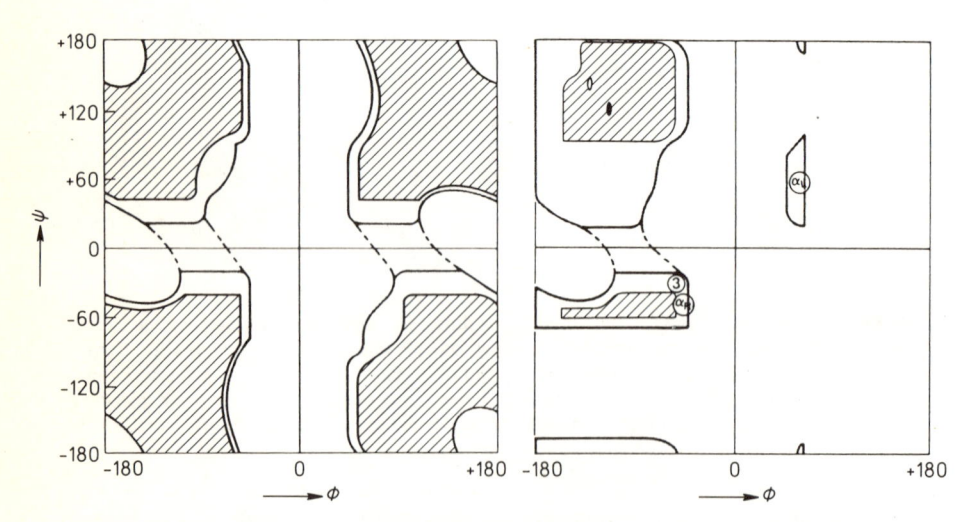

Fig. 19.3. Ramachandran plots illustrating the sterically allowed regions for Φ, ψ for glycine (*left*) and for the other 19 amino acids (*right*). The fully allowed regions are indicated by *shading*, the partially allowed regions by *thick lines*; the connecting region enclosed by the *dashed lines* is permissible with slight bond angle distortion. Secondary-structure conformations are α_R, α-helix; 3, 3_{10} helix; α_L, left-handed α-helix; () antiparallel and ● parallel β-sheet [586]

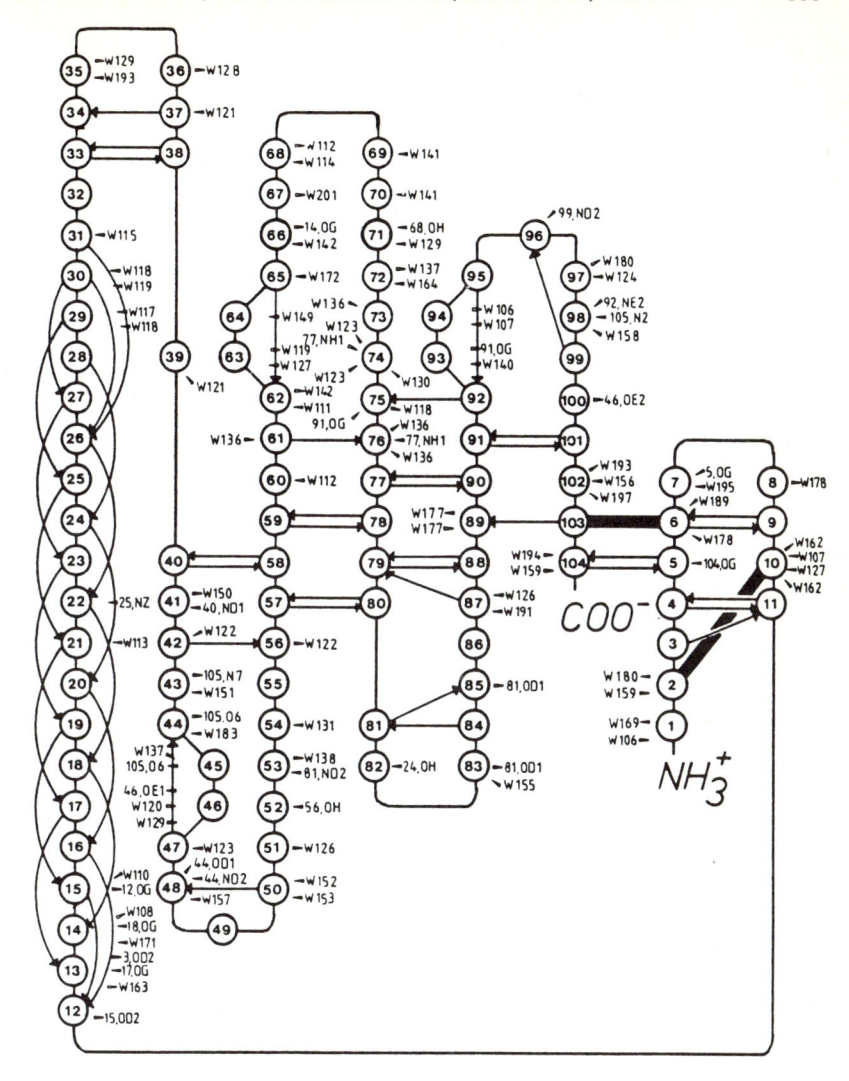

Geometry of the Secondary-Structure Elements: Helix, Pleated Sheet, and Turn 355

Fig. 19.4. Secondary-structure hydrogen bonding in ribonuclease T_1. *Arrows* ($\rightarrow$) point in the direction $N-H\cdots O=C$. Note regions of α-helix and of antiparallel β-pleated sheet. Also indicated are hydrogen-bonding contacts from main-chain atoms to side-chains and water molecules [594]

of the polypeptide chain and is stabilized by a series of hydrogen bonds $(n+4)N-H\cdots O=C(n)$. These local interactions can be described as intrachain and compared to intramolecular hydrogen bonds in small molecules.

In contrast, the pleated sheet is formed by *different segments* of the polypeptide. These can be adjacent and fold back like a hairpin, or the segments can be located anywhere along the devious polypeptide chain. The pleated sheets can consist of only two strands in parallel or antiparallel orientation. Since the sheet has

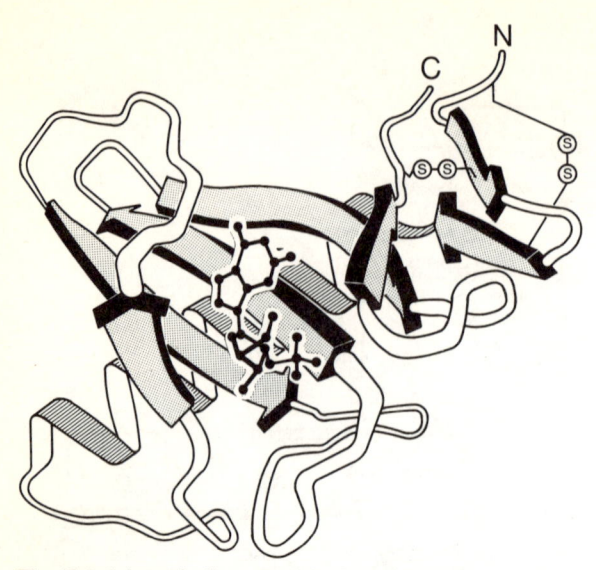

Fig. 19.5. Schematic diagram of the three-dimensional folding of the polypeptide chain (Fig. 19.2) of ribonuclease T_1 in a complex with the specific inhibitor guanosine-2'-phosphate [594]; S–S, disulfide bridges; thick arrows indicate β-strands

free hydrogen-bonding groups at both edges, more strands can be added to form a multitude of strands. The sheets can coil to form cylinder-like structures referred to as β-barrels which, as a ring, are "infinite" β-sheets. The hydrogen bonds in the β-sheets can be described as interchain and are comparable to the intermolecular hydrogen bonds in small molecule crystal structures.

Pleated sheet structures are parallel or antiparallel. In the local minimum in the Ramachandran φ/ψ plot (Fig. 19.3) of β-pleated sheet structures, two configurations are possible, with parallel and antiparallel orientation of the polypeptide strands (Fig. 19.6). The strands are linked by *inter*chain N–H···O=C hydrogen bonds, which run both ways between the strands and produce a characteristically different pattern in parallel and antiparallel sheets. It is a particular stereochemical feature of the β-pleated sheets that amino acid side-chains point alternately up and down, and adjacent side-chains interact sterically to produce a right-handed twist [597, 598] (see Fig. 19.7a). The regular pattern of a β-sheet can be interrupted locally by insertion of an extra amino acid, giving rise to a so-called β-bulge [599].

A family of polypeptide helices is stereochemically feasible. The helices are stabilized by *intra*chain N–H···O=C hydrogen bonds between different peptide units, depending on type of helix. To describe the helices, two numbers are used, X_n. The X (not necessarily integral) denotes the number of amino acids in one turn; n is the number of covalent bonds (including N–H) which are in the main-chain segment and bridged by each of the N–H···O=C hydrogen bonds in the helix.

The tightest and widest possible helices, 2_7 and 4.4_{16} (π) are rarely found in globular proteins, because the intermediate 3.6_{13} and 3_{10} helices are more stable

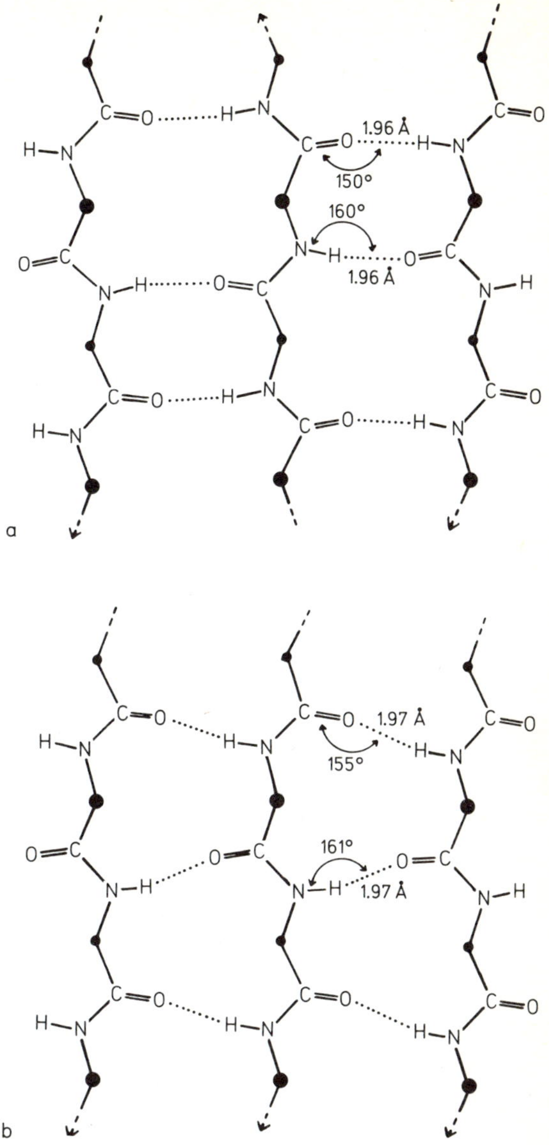

Geometry of the Secondary-Structure Elements: Helix, Pleated Sheet, and Turn   357

Fig. 19.6a, b. Geometry of **a** antiparallel and **b** parallel β-pleated sheets, with average hydrogen-bond distances and angles drawn in. "Bending" of hydrogen bonds at the oxygen atoms is different, toward C_α in **a** and away from C_α in **b**. C_α atoms are indicated by *dots*; if side-chains point up (toward the viewer), the *dots are large*, otherwise they *point down* [596]

and therefore form more readily. By far the most frequently observed is the 3.6_{13} or α-helix. The $C=O \cdots H-N$ hydrogen bonds are between the carbonyl C=O of the n-th peptide unit and the N–H of the (n+4)th peptide along the chain, often referred in shorthand notation as 1←5. The hydrogen bonds are nearly parallel to the helix axis and the C=O groups are tilted slightly and displaced away from the helix axis, as shown in Fig. 19.7b, c.

The α-helices have been observed to extend over long stretches of the polypeptide chain. In contrast, the 3_{10} helix is found only in short segments. It occurs fre-

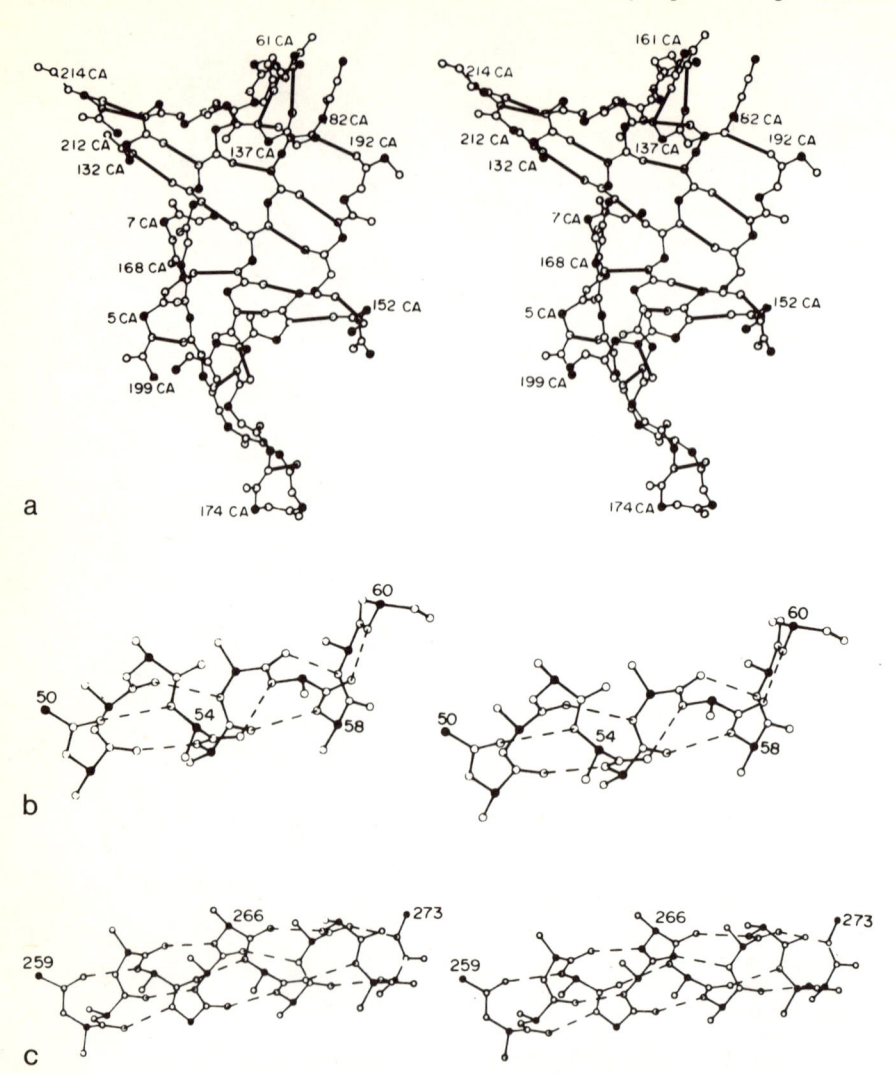

Fig. 19.7 a–c. Some examples of secondary structure in proteins. **a** The antiparallel β-sheet in actinidin showing a β-turn (residues 172–175) and a β-bulge (residues 167–168) is twisted in an overall right-handed form. **b** An α-helix in ribonuclease A is terminated by a piece of 3$_{10}$-helix with 1←4 hydrogen bonds (residues 55–60). Compared with the α-helix in **c**, the hydrogen bonds in the 3$_{10}$-helix are less linear and the C=O groups are more tilted. In **a** only C$_\alpha$ atoms are drawn (*dots*) and hydrogen bonds are indicated by solid lines, in **b** and **c** the C$_\beta$ atoms are indicated by *open circles*, and hydrogen bonds by *dashed lines* [596]

quently as one turn either at the carboxy terminus of an α-helix or as a sharp turn which reverses the direction of the polypeptide chain. The 3$_{10}$ helix is more tightly twisted than the α-helix, and therefore the C=O groups are more tilted away from the helix axis (see Fig. 19.7b). This gives rise to a less favorable geometry for the

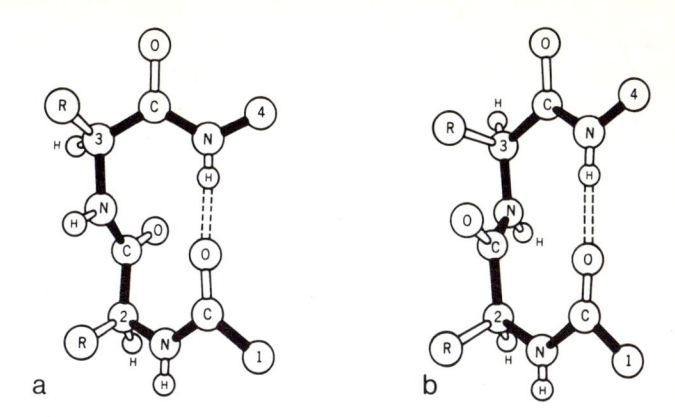

Fig. 19.8a, b. Geometry of β-turns of types I(a), II(b). The β-turn of type III corresponds to one turn of the 3_{10}-helix [602]

1←4 hydrogen bonds formed between peptide C=O of residue (n) and N−H of residue (n+3) and makes 3_{10} helices less stable relative to the α-helix.

In helices all main-chain atomic groups and hydrogen bonds are oriented in one direction along the helix axis. Therefore, a significant dipole moment is created, with positive pole in the direction of the amino terminus. The potential at about 5 Å distance from the N-terminus of an α-helix of 10 Å length (about 2 turns) was estimated to be about 0.5 eV [600, 601].

Reversal of chain direction occurs by well-defined turns of the 3_{10} or β-type. A single turn of the 3_{10} helix in globular proteins is frequently used to give rise to a chain reversal. It is stabilized by a main-chain 1←4 hydrogen bond. A similar interaction between four consecutive amino acids is also observed in a structurally related series of so-called β-turns [591]. Depending on the torsion angles at residues 2 and 3, three different types of β-turns, I, II, III, have been identified.[1] Figure 19.8 shows only types I and II, type III is very similar to one turn of a 3_{10} helix.

Other kinds of turns which are much more rarely encountered are the α-turn with 1←5 hydrogen bond which represents one single turn of an α-helix, or the three-residue γ-turn with a 1←3 hydrogen bond. This latter turn produces a kink rather than a hairpin-like reversal in a polypeptide chain [596, 603], see Table 19.5.

19.2 Hydrogen-Bond Analysis in Protein Crystal Structures

A deficiency in all the protein crystal structures investigated thus far at high resolution <1.8 Å is that the data/parameter ratio is insufficient to permit least-squares

---

[1] In fact, six different types of turns were characterized, but this is of no importance in our discussion (see [588] and Table 19.5).

refinement of the individual atomic positions and of thermal parameters for the C, N, O, and S atoms. Therefore, bond lengths and angles have to be constrained to assumed values derived from the more accurate analyses on the amino acids and oligopeptides. Similarly, hydrogen-bond and nonbonded (steric) interactions are also constrained to assumed values. These constraints are generally more relaxed than those for the covalent bond lengths and angles and for the planar peptide bonds. In no protein X-ray study hitherto has the position of a hydrogen atom been derived from the electron density map. Therefore, in the following analysis, peptide, amide, imidazole, guanidinium and indole hydrogen atom positions are calculated on the basis of the C, N, O skeleton employing an N−H distance of 1.0 Å [596].

This procedure is comparable to that used to analyze the N−H···O=C interactions in small molecule crystal structures described in Part IA, Chapter 2.3 [75]. For those groups with rotational freedom such as the −OH in serine, threonine, and tyrosine, the S−H in cysteine, and the −$^+$NH$_3$ in lysine and at the amino terminus, it is more difficult to guess the likely position of the hydrogen atoms. In these instances, the assumptions are made that the hydrogen bonds are of the two-center type when the R−$\hat{X}$···A angle is close to 110°. The bonds are assumed to be linear with X−$\hat{H}$···A angles of 180° (a better approximation would be to take the most commonly observed angle of ~160°).

The evidence for three-center bonds is large deviations from the R−$\hat{X}$···A angles of 110°. They may be as small as 80° and as large as 150°, as shown in the survey of intermolecular three-center N−H⋯(O=C, A) hydrogen bonds occurring in small molecule crystal structures [56].

Main-chain atoms are better defined than side-chain atoms. In the electron density map of a protein, main-chain atoms are in general better defined than side-chain atoms, especially if they are involved in secondary and tertiary structure hydrogen bonding. Likewise, atomic groups in the core of the protein are more tightly bound and therefore better defined than those at the periphery. The latter are more free to move, which is expressed in larger temperature factors due to higher thermal motion or disorder. Because solvent molecules (water) are frequently associated with the peripheral groups of the proteins, they are only as well or even less clearly defined and often not observed.

The identification of hydrogen bonds in proteins is, as in small molecule X-ray crystal structures, more reliable in cases where the C₂N···O angles and N···O distances are close to ideal values, and more difficult when the angles are far from 110° and the distances are large.

In the following, the geometrical data obtained from a number of protein crystal structures determined at high resolution are analyzed in terms of hydrogen-bonding patterns, and then the data are compared metrically. This procedure and the data analysis are taken mainly from a review of Baker and Hubbard [596], who considered in detail the protein structures listed in Table 19.1.

Table 19.1. Proteins used by Baker and Hubbard [596] in the analysis of hydrogen-bonding geometry[a]

Protein	Structural type	Number of residues	Number of solvent molecules	Resolution (Å)	R factor	σ(bonds) (Å)	Refinement method	Reference
Actinidin	$\alpha+\beta$	218	272	1.7	0.165	0.013	FFT	[604]
Carboxypeptidase A	α/β	307	279	1.54	0.174	0.027	RLS	[605]
Cytochrome c (Tuna, ferro)	Small[b]	104	54	1.5	0.173	0.018	EREF	[606]
Dihydrofolate reductase (*L. casei*)	α/β	162	264	1.7	0.152	0.070	RLS	[607]
Erythrocruorin	All-α	136	94	1.4	0.183	0.023	Real sp.	[608]
Insulin (2-Zn)	Small[b]	2×51	325	1.5	0.179	0.017	FFT	[609]
Lysozyme (human)	$\alpha+\beta$	130	140	1.5	0.187	0.024	RLS	[610]
Myoglobin (sperm whale, deoxy)	All-α	152	388	1.4	0.16	0.048	EREF	[611]
Penicillopepsin	All-β	323	—	1.8	0.136	0.019	RLS	[612]
Phospholipase A2 (bovine)	All-α	123	99	1.7	0.171	0.009	FFT	[613]
Plastocyanin	All-β	99	44	1.6	0.17	0.021	FFT/RLS	[614]
Ribonuclease A	$\alpha+\beta$	124	78	1.45	0.26	0.036	LS	[615]
Streptomyces griseus proteus A	All-β	181	—	1.5	0.121	0.020	RLS	[616]
Thermolysin	$\alpha+\beta$	316	173	1.6	0.213	0.026	RLS	[617]
Trypsin (bovine)	All-β	223	154	1.34	0.157	0.054	Diff.	[618]

[a] "Structural type" describes α-helix and β-sheet content: all-α or all-β, and types where α and β are clearly separated ($\alpha+\beta$), and cases where α-helices and β-strands are mixed or alternate (α/β). Refinement methods are FFT (fast Fourier transform least squares), Real sp (real space least squares); RLS (restrained least squares where bond angles and distances are restrained), and EREF (where energy minimization of the molecule is included), Diff (difference Fourier methods with constraints on molecular geometry).

[b] These proteins contain only small segments of secondary structure.

In this survey [596], hydrogen bonds were considered with X–Ĥ···A angles in the range 90° to 180° so that three-center bonds are included, with cut-off distances of 2.5 Å for H···O and 3.5 Å for O···O(N) interactions. For those hydrogen bonds where the hydrogen atom positions could not be derived from the C, N, O skeleton as in hydroxyl and amino or ammonium groups, O···O(N) distances of 3.5 Å and a minimum C–Ô···O(N) angle of 90° were used as cut-off values. The effect of these cut-off criteria, which are unavoidable in the absence of direct knowledge of the hydrogen atom positions, would be to identify some unsymmetric three-center bonds as two-center because the minor components are not recognized as such. For instance, an H···O cut-off of 2.5 Å in the survey on amino acids neutron diffraction studies would have identified more than half of the three-center bonds as two-centered. Based on the neutron diffraction data, a cut-off of H···O <2.7 Å would have been more appropriate.

19.3 Hydrogen-Bonding Patterns in the Secondary-Structure Elements

Because main-chain to main-chain hydrogen bonds in the secondary-structure elements, α-helix, β-pleated sheet and β-turn, are comparable in character and geometry, they are considered together in the following.

Most of the main-chain C=O and N–H groups are involved in secondary-structure hydrogen bonding. In the analysis of the 10 protein structures listed in Table 19.1, only 194 out of 1733 main-chain C=O groups (or 11.2%), and 209 out of 1679 N–H groups (or 12.4%) are not in any obvious hydrogen-bonding contact within the cut-off limits described in Section 19.2, as indicated in Table 19.2c. This is expected from the small molecule crystal structure survey discussed in Part IA, Chapter 2.3 because a nonhydrogen bonding 〉NH group is rarely, if ever, observed. Obviously, the long polypeptide chains place much more severe conformational constraints on the positioning of the functional groups. It is not unreasonable, therefore, that about 10% of the 〉NH groups in the protein structures were not located in a hydrogen-bonding acceptor environment.

The nonbonding of the 〉C=O groups may be overestimated, since some may be involved in long three-center bonds. The presence of unidentified disordered water molecules as donors may also reduce this statistic.

The hydrogen bonds accepted and donated by the main-chain C=O and N–H groups are further analyzed in Table 19.2a. Main-chain to main-chain interactions N–H···O=C predominate, with 1040 of the total 1532 (or 68%) donated by N–H and 1064 of the total 2292 (or 46%) accepted by C=O. These numbers will be reduced to 55% N–H and 40% C=O if the "all α-helix" proteins myoglobin and erythrocruorin having 70% of all peptide groups involved in main-chain to main-chain hydrogen bonding are excluded from the analysis. A much smaller number of main-chain to side-chain hydrogen bonds are identified. As shown in Table 19.2a, only 11% are to main-chain C=O and a comparable number to N–H groups.

Table 19.2. Hydrogen bonding by main-chain C=O and N–H groups in ten well-refined proteins[a] [596]

Total number of C=O groups 1733			Total number of N–H groups 1679		
a) C=O accept hydrogen bonds from			**N–H donate hydrogen bonds to**		
Main-chain N–H	1064	(46.4%)	Main-chain C=O	1040	(67.9%)
Ser/Thr O–H	88		Ser/Thr O–H	38	
Tyr O–H	17		Tyr O–H	4	(10.9%)
Arg $\overset{+}{\text{N}}$–H, NH$_2$	39		Side-chain amide C=O	35	
Lys $\overset{+}{\text{NH}}_3$	19	(10.7%)	Asp, Glu, Terminus COO$^-$	90	
Side-chain amide N–H	55		Water	325	(21.2%)
His N–H	18		Total	1532	(100%)
Trp N–H	8				
Water	984	(42.9%)			
Total	2292	(100%)			
b) C=O accept hydrogen bonds from			**N–H donate hydrogen bonds to**		
Only one main-chain N–H	621	(41.1%)	Only one main-chain C=O	1012	(76.1%)
Only water	406	(26.8%)	Only water	293	(22.0%)
2 main-chain N–H	80	(5.3%)	2 main-chain C=O	25	(1.9%)
Main-chain N–H + side-chain	101	(6.6%)			
Main-chain N–H + water	240	(15.8%)	Total	1330	(100%)
Side-chain + water	66	(4.4%)			
Total	1514	(100%)			
c) C=O groups with			**N–H groups with**		
No hydrogen bond	194	(11.2%)	No hydrogen bond	209	(12.4%)
1 hydrogen bond	924	(53.3%)	1 hydrogen bond	1400	(83.4%)
2 hydrogen bonds	489	(28.2%)	≥2 hydrogen bonds	70	(4.2%)
>2 hydrogen bonds	126	(7.3%)	Total	1679	(100%)
Total	1733	(100%)			
d) C=O groups hydrogen bonded in			**N–H groups hydrogen bonded in**		
Helices[b]	563	(63.4%)	Helices[b]	586	(62.8%)
Pleated sheets	204	(22.9%)	Pleated sheets	209	(22.4%)
Double turns[c]	50	(5.6%)	Double turns[c]	55	(5.9%)
Single turns[d]	72	(8.1%)	Single turns[d]	83	(8.9%)
Total	889	(100%)	Total	933	(100%)

[a] Proteins are: actinidin, carboxypeptidase A, dihydrofolate reductase, eythrocruorin, insulin, lysozyme, myoglobin, phospholipase A2, plastocyanin, thermolysin.
[b] >2 consecutive bonds.
[c] 2 consecutive bonds.
[d] Single hydrogen bond.

Main-chain to water hydrogen bonding plays a significant role in proteins, as the numbers in Table 19.2a indicate, 43% for C=O and 21% for N−H groups. Because the water molecules are predominantly at the periphery of the protein molecules with a few enclosed in the interior, this means that main-chain to main-chain hydrogen bonding (or secondary structure) predominates in the core of the proteins. Secondary structure is less well defined at the periphery where the peptide C=O and N−H groups are more free to interact with water molecules.

C=O groups are more frequently involved in multiple hydrogen bonds than N−H (Table 19.2c). When the main-chain to main-chain N−H···O=C interactions are analyzed, 53% of C=O accept one hydrogen bond, 28% two bonds and 7% are engaged in three or more interactions. For the donor N−H groups, this distribution is very different; the majority of 83% forms one hydrogen bond and only 4% are donating to more than one acceptor, and are therefore identified as three-center bonds.

This number is significantly smaller than the ~20% observed in small molecule crystal structures (Part IA, Chap. 2.3), suggesting that some of the longer hydrogen-bond interactions may not have been recognized as such. It is tempting to suggest that a significant proportion of the 53% single acceptor C=O groups are actually involved in long three-center bonds with H···O distances which exceed the cut-off value of 2.5 Å. The same argument would increase the number of N−H donors closer to the 20% observed in the N−H···O=C small molecule survey [56].

C=O often accepts hydrogen bonds simultaneously from main-chain N−H and water molecules. This happens in 240 of the 615 cases (or 39%) listed in Table 19.2b, c, where C=O accepts two or more than two hydrogen bonds. Of these, 146 are in α-helices, only 33 are at the ends of β-sheets, and 37 are in β-turns, the remaining 24 being not involved in secondary structure. Although in the proteins analyzed there are a higher proportion of α-helices than β-sheets, these numbers still reflect the greater tendency of C=O groups in α-helices to bind to external water molecules. The same holds for C=O groups in β-turns which are found predominantly at the periphery of the proteins and are as frequently associated with water molecules as the C=O groups are in α-helices. Similarly, of all the peptide C=O groups accepting hydrogen bonds simultaneously from main-chain N−H and from side-chain groups, about 60% are in α-helices, 20% in β-turns and only about 10% in β-sheets [596]. These numbers suggest that the peptide groups in the β-sheet regions are more "shielded" compared with α-helices and β-turns, in agreement also with H/D exchange experiments described in Section 19.9. This is probably associated with the tighter packing of amino acid side-chains in β-sheets which, in general, are more hydrophobic in character than those in the other secondary structure elements [596].

Looking more closely at the ability of peptide C=O groups in α-helices to accept more than one hydrogen bond, it was found that 36% are engaged in one extra interaction from external water or from side-chains 3 to 4 residues (~1 turn) apart in sequence (n+3 or n+4). The side-chains most frequently encountered are Ser and Thr, with His, Gln, Asn, Arg, Lys, Tyr contributing less often. These side-chain to main-chain contacts are discussed later.

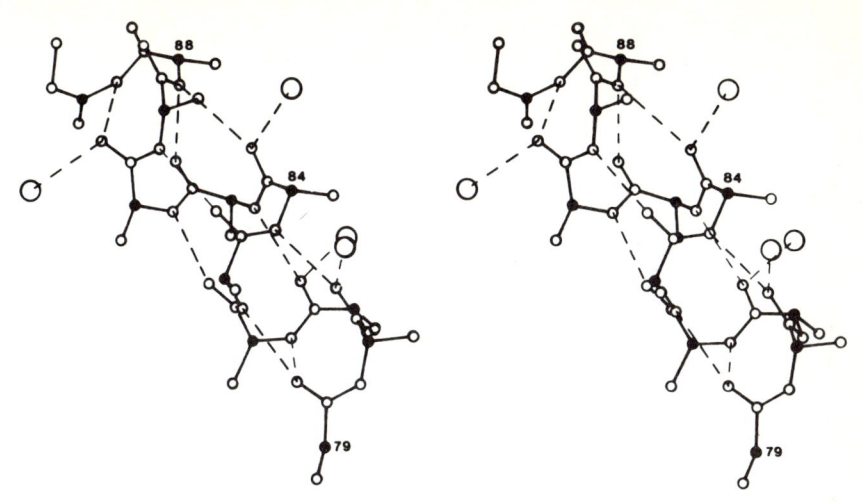

Fig. 19.9. In α-helices, the β-carbon atoms of amino acid side-chains in positions n, (n+3), (n+4) form "pockets" which can harbor a water molecule hydrogen bonded to the peptide C=O in position n. The peptide oxygen of residue 79 accepts two hydrogen bonds from the peptide N−H of residues 82, 83 in a bifurcated "α_N distortion" geometry and the peptide N−H of residue 88 donates a three-center hydrogen bond in the "α_{C_1} distortion" [596]

Hydrogen bonding of water molecules to α-helices. The helix structure has "pockets" formed by the β-carbons of residues (n), (n+3) and (n+4) (Fig. 19.9), which can accommodate water molecules. These hydrogen-bond to the C=O group of residue (n) with almost ideal C=Ô···O_W angles of ~120°. This hydration occurs preferentially at solvent exposed sides of α-helices, and distorts the C=O group away from ideal intrahelical hydrogen-bonding geometry. As a consequence, the helices bend locally. Instead of water, side-chain donor groups can also hydrogen bond to the C=O groups in the α-helices and induce a similar bending [619].

Hydrogen-bonding geometry at termini of helices is often more irregular. Three main types of helix distortions are observed [596]. At the N-terminus, α-helices may display a 3_{10} helical turn (α_N distortion), providing a conformation where the C=O oxygen of residue 1 can accept two hydrogen bonds from N−H groups of residues 3 and 4. At the C-terminus, two kinds of distortion are observed (see Fig. 19.9). In one (α_{C1}), the α-type turn with 1←5 hydrogen bond changes to the narrower 3_{10} with 1←4 bond. In the other (α_{C2}), the wider π-helix turn with 1←6 interaction is formed.

19.4 Hydrogen-Bonding Patterns Involving Side-Chains

Hydrogen bonds to amino acid side-chains are classified according to the distance of the respective partner groups along the polypeptide chain. "Local" side-chain to main-chain interactions are formed if a side-chain is folded such that it hydrogen bonds with a peptide within 4 residues along the chain, and "long-range" hydrogen bonds occur when the side-chain is bonded to a sequentially more distant

Table 19.3. Intermolecular and intramolecular (long-range, local) hydrogen bonds between main-chain and side-chain atoms in the 15 proteins listed in Table 19.1 [596]

Side-chains donating hydrogen bonds to main-chain C=O

Side-chain	Total	Long-range interactions			Local interactions								
		Intermolecular	>10 Residues	5–10 Residues	Back-folding of side-chain n to C=O of residue						
					n−4	n−3	n−2	n	n+1	n+2	n+3	n+4
Ser	75	8	23	4	14	21	1	–	1	3	–	–
Thr	71	4	14	8	32	9	–	–	2	1	–	1
Asn	52	8	29	5	3	1	1	2	1	–	1	1
Gln	53	8	21	5	4	3	3	3	4	1	1	–
Asp	15	2	7	–	–	1	1	–	–	3	–	1
Glu	12	3	3	3	–	1	–	–	1	1	–	–
Lys	32	12	16	–	1	9	1	3	–	–	1	1
Arg	60	10	27	7	–	–	2	–	2	–	1	–
His	34	5	18	2	5	–	2	–	–	–	–	–
Tyr	20	1	16	1	1	–	1	–	–	1	1	–
Trp	14	2	8	–	1	1	–	–	–	–	–	–
	63	182	35	61	46	12	8	11	10	6	4
Total	438	280			158								

Table 19.3. (continued)

Side-chains accepting hydrogen bonds from main-chain N–H

Side-chain	Total	Long-range interactions			Local interactions						
		Intermolecular	>10 Residues	5–10 Residues	Back-folding of side-chain in position (n) to N–H of residue						
					n−4	n−3	n−2	n	n+1	n+2	n+3	n+4
Ser	41	–	4	3	–	1	–	1	3	14	14	1
Thr	25	–	4	4	–	1	–	–	5	6	5	–
Asn	51	2	9	4	–	–	–	3	4	21	7	1
Gln	16	1	7	1	–	4	–	3	–	–	–	–
Asp	88	3	23	2	2	4	–	9	4	29	10	2
Glu	44	1	21	1	1	9	1	10	–	–	–	–
Lys	2	–	2	–	–	–	–	–	–	–	–	–
Arg	–	–	–	–	–	–	–	1	–	–	–	–
His	4	–	–	1	–	–	–	–	–	2	–	–
Tyr	7	–	4	1	–	–	–	–	–	–	–	2
Tryp	–	–	–	–	–	–	–	–	–	–	–	–
		7	74	17	3	19	1	27	16	72	36	6
Total	278		98					180				

main-chain peptide. Side-chain to side-chain interactions, which can also be "local" or "long range", contribute to the overall structure stabilization of a protein.

In long-range interactions, side-chains donate hydrogen bonds more frequently to peptide C=O than they accept bonds from N−H groups. As shown by the numbers provided by the sample of the 15 proteins listed in Table 19.1, 280 contacts are to main-chain C=O and 98 to N−H groups (see Table 19.3). These data illustrate again that the C=O oxygen is more accessible as an acceptor for hydrogen bonding than N−H is as a donor. This is not true for the local side-chain to main-chain interactions.

In local interactions, side-chains hydrogen bond equally well to peptide C=O and N−H groups. The entries in Table 19.3 clearly illustrate that the number of side-chains binding to main-chain C=O and N−H groups are about equal, 158 and 180, respectively. Local folding of side-chains to bind with main-chain N−H is most common for Ser, Thr, Asn, Asp in position (n) to N−H in forward positions (n+2) and (n+3), i.e., toward the carboxy terminus of the polypeptide chain. Another common form of side-chain bonding is that in which the side-chains Ser, Thr in position (n), backfold to hydrogen bond with C=O in positions (n−3) and (n−4). This configuration is a recurrent motif in α-helices. It has been proposed for membrane-bound proteins where Ser, Thr are frequently located in α-helical regions which are embedded in the hydrophobic membrane [620].

Local side-chain to main-chain hydrogen bonds are not evenly distributed. If the geometry of these interactions is analyzed in more detail (Table 19.3), it becomes clear that the contacts between the side-chain donor in position (n) are preferentially to the main-chain acceptor C=O in positions (n−3) and (n−4) toward the N-terminus whereas the main-chain donor N−H is more preferred if it is 2 to 3 residues (n+2, n+3) toward the C-terminus of the main-chain. There is no case of an interaction between side-chain (n) and main-chain residue (n−1), and folding back of side-chain (n) on its own peptide is mainly with Asx, Glx hydrogen bonding to N−H (n) and with Asn, Gln, Arg binding to C=O (n).

The side-chains of Asp and Glu are probably protonated when they are hydrogen-bonded to main-chain C=O. According to Table 19.3, there are a number of long-range and local interactions between peptide C=O oxygens and the carboxylic acid groups of Asp, Glu. The latter have to be protonated because if they were carboxylate groups, the C=O functions would repel each other on the basis of their (partial) negative charges. The reason that the Asp, Glu carboxylate groups are protonated is because their pK is raised due to favorable steric interactions, so that (Asp, Glu)COOH···O=C hydrogen bonds can form [596]. A similar suggestion was made to explain interactions between carboxylate groups in penicillopepsin [612].

Local side-chain to main-chain N−H hydrogen bonds stabilize N-termini of α-helices and of β-turns (Figs. 19.10, 19.11). Of the 71 α-helices in the sample set (Table 19.1), 46 display hydrogen bonds formed locally at the amino termini between main-chain N−H groups and side-chains. Most of these interactions involve the side-chains of Asp, Asn, Glu, Gln, Ser, Thr at position (n) which hydrogen bond with N−H of peptide (n+3), or, in less frequent arrangements, with peptide

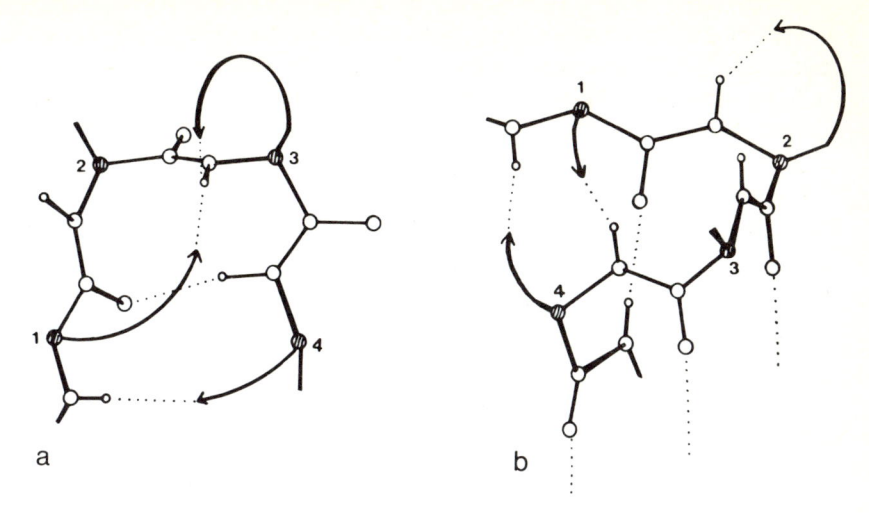

Hydrogen-Bonding Patterns Involving Side-Chains

Fig. 19.10 a, b. Schematic representation of main classes of local (back-folding) side-chain to main-chain (NH) hydrogen bonds. **a** In β-turns of type I, interactions are (n) to (n) (see side-chain at residue 3), and (n) to (n+2) (see side-chain at residue 1), and (n) to (n−3) (see side-chain at residue 4). **b** At α-helix N-termini, interactions are (n) to (n) (at any position, here shown for residue 2), (n) to (n+3) (see side-chain at residue 1), and (n) to (n−3) (see side-chain at residue 4) [596]

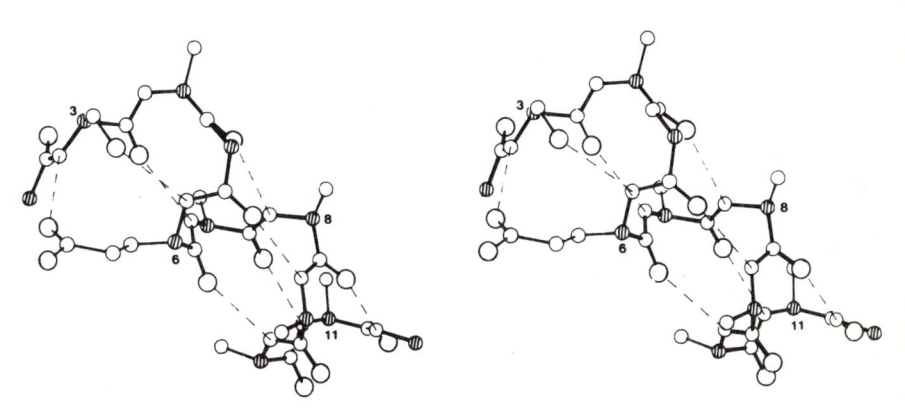

Fig. 19.11. Stereo view showing two of the situations given in Fig. 19.10b. In myoglobin, the side-chain of Ser3 at the N-terminus of helix A hydrogen bonds with the free NH of residue 6, (n) to NH(n+3), and the carboxylate of Glu6 interacts with NH of residue 3, (n) to NH(n−3). C_α atoms drawn hatched, all side-chains (except those mentioned) indicated only with their C_β atoms [596]

N−H at positions (n) and (n−3). As illustrated in Fig. 19.11, these hydrogen bonds often occur as clusters.

The turns of the 3_{10} or β-type can be considered as fragments of helices, and it is therefore not surprising to find them involved in side-chain to main-chain hydrogen bonds [589, 604]. Interactions predominate where side-chains (n) are in contact with main-chain N−H in position (n+2) (Fig. 19.10a). The amino acid residues Asp, Asn, Ser are those most frequently involved.

Most of the side-chains are hydrogen bonded. In the sample of proteins considered (Table 19.1), 85% of the side-chains are involved in hydrogen bonding [596]. In reality, this number is likely to be close to 100%. Many of the side-chains which are not in obvious hydrogen-bonding contact have high thermal motion or disorder, so that an assignment of individual atomic positions is questionable.

Because side-chains are more exposed on the protein surface than main-chain atoms, they are more frequently bonded to water (52%) than the main-chain peptide C=O (43%) and N−H (21%) groups. Again, this number is likely to be an underestimate.

Side-chains tend to fully satisfy their hydrogen-bonding acceptor and donor potentials. Crystal structures on small molecules disclose that ether and hydroxyl oxygens commonly accept one hydrogen bond, and sometimes two. Oxygens of carboxylate groups, on the other hand, are more generally involved in two hydrogen bonds. The same trend is also observed with amino acid side-chains in proteins, although the relative number of bound ligands per side-chain functional group is lower than in small molecule crystal structures because of the uncertainties in locating water molecules and side-chain atoms at the periphery of a protein [596]. For water hydrogen bonding, see Part IV, Chapter 23.

The donor functions of the His and Trp side-chains are exceptional because they are nearly 100% and 85% respectively engaged in donating hydrogen bonds. The reason is that His and Trp are mostly located in the core of proteins, and His is a good hydrogen-bond former with N−H···N interactions in D,L-histidine, 1.88 Å [DLHIST], and in L-histidine, 1.90 Å [LHISTD13], within the range of 1.73 to 2.12 Å (maximum at 1.88 Å) observed for purine-pyrimidine hydrogen bonds [84].

The potential for accepting hydrogen bonds is met to about 85% for Asp, to 77% for Gln, and 68% for Tyr. For all other side-chains, only about 50% of their hydrogen-bonding accepting/donating potential appears to be satisfied, but again this is certainly an underestimate [596].

Side-chains display a much stronger tendency to act as donors than as acceptors of hydrogen bonds. This is observed in particular for the O−H groups of Ser, Thr, and for the imidazole group of His. The residues Asn and Gln (in those cases where amide C=O and NH_2 could be distinguished), show no definite preference.

Proteins contain, on average, more acceptor than donor sites [596]. Similar proton deficiency in the amino acid zwitterion crystal structures results in the formation of three-center hydrogen bonds rather than in unsatisfied acceptor sites [74] (Part IA, Chap. 2.6). There is less flexibility in the orientation of hydrogen-bond donor and acceptor groups in proteins which would lead to a relatively larger number of unsatisfied acceptor sites. Some of the more unsymmetrical three-center bonds which might have been missed in the survey [596] because of the 3.5 Å X···A cut-off limit will also contribute to reduce the number of unsatisfied acceptors in side-chains.

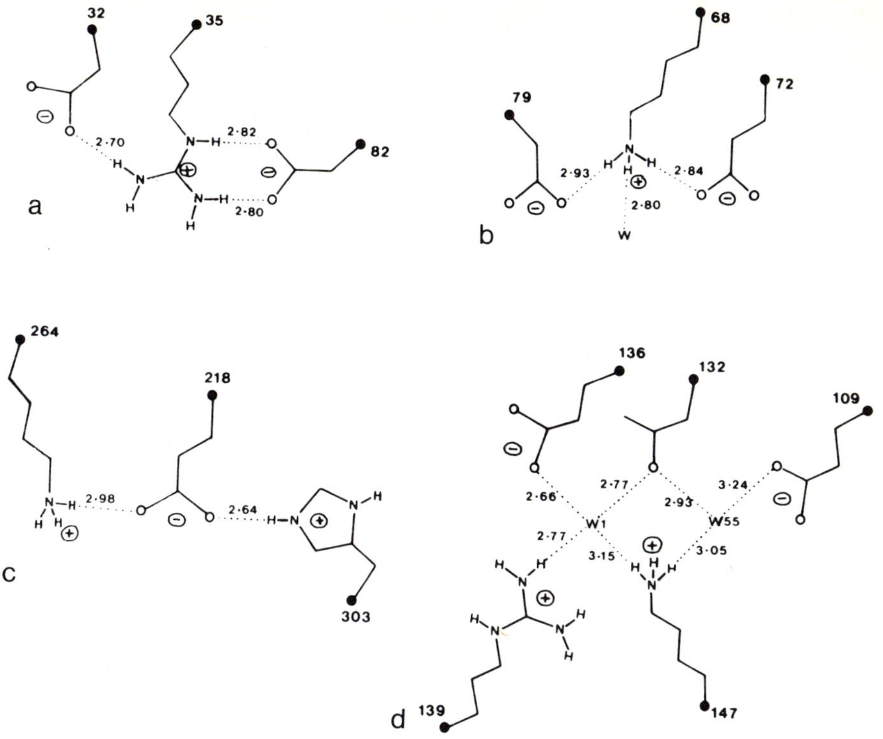

Fig. 19.12a–d. Schematic illustrations of salt bridges (ion-pair interactions) in **a** thermolysin; **b** erythrocruorin; **c** carboxypeptidase; **d** myoglobin. O···N/O distances given in Å units [596]

Strong hydrogen bonds are formed by charged side-chains and by main-chain termini. These interactions are of the general hydrogen-bonded "salt-bridge" types

$$N^+ - H \cdots O = C$$
$$N^+ - H \cdots O = \bar{C}$$
$$N - H \cdots O = \bar{C}.$$

Some examples are illustrated in Fig. 19.12. In the proteins given in Table 19.1, such hydrogen bonds account for the following proportions of the total hydrogen bonds involving side-chains respectively: Lys 19%, Arg 29%, His 25%, Asp 26%, Glu 33%. These salt bridges, with one or both partners charged, are predominantly intramolecular and therefore stabilize tertiary structure. They are often engaged in multiple contacts in which the Lys and Arg side-chains bridge pairs of Asp, Glu carboxylate groups, or a carboxylate forms cyclic hydrogen bonds with two N–H groups of Arg, or they are found to bridge two positively charged side-chains. There are also many cases where the charged side-chains are not in direct contact, but are bridged by water molecules.

19.5 Internal Water Molecules as Integral Part of Protein Structures

Without water, proteins lose their three-dimensional structure. They denature and their biological functions disappear. Water molecules not only hydrate the surface of proteins and occupy the volume between the protein molecules in crystals, as discussed in Part IV, but they are also found in specific locations within their interior. Since water molecules are so essential, their interactions with main-chain and side-chain atoms of a protein deserve special attention [621, 622].

Internal water is an integral part of a protein structure. Even globular proteins of small size are observed to contain "buried" water molecules. Examples are pancreatic trypsin inhibitor, with 58 amino acid residues and 4 interior water molecules, lysozyme with 129 residues and 4 waters, and larger proteins like actinidin with 218 residues which may contain 10 to 20 molecules of internal water (Fig. 19.13). These water molecules can be buried deep inside the globular proteins or located in cavities near the surface. In some cases, therefore, the distinction internal/external water can be ambiguous.

Internal water molecules may be located in the same sites in related proteins. In many proteins, internal waters are an essential part of the three-dimensional structure and their positions are typical for a whole family of proteins. Thus, in the three related plant cysteine proteinases, actinidin, papain, and calotropin D1, 15 of the 16 internal waters (8 of which are illustrated in Fig. 19.13) are found in

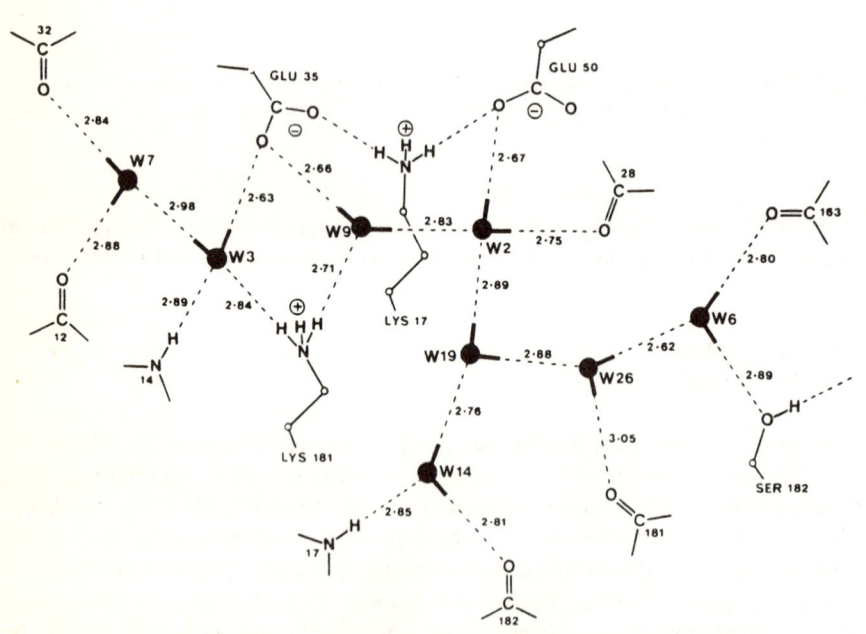

Fig. 19.13. A cluster of 8 internal water molecules in the structure of actinidin. All water hydrogen positions were derived on the basis of "logical" $N-H \cdots O$ and $O-H \cdots O$ hydrogen bonds. Note charged side-chains Lys181, Glu35 bridged by two water molecules W3, W9. $O \cdots O/N$ distances in Å units [596]

nearly the same positions. The root-mean square difference in these water oxygen atom positions in the three proteins is only 0.48 Å [596]. Likewise in the serine proteases elastase, trypsin, and chymotrypsin, 14 water molecules are in homologous positions [623, 624], and human egg white and tortoise lysozymes feature the same 4 internal water molecules [625].

The conservation of internal water structure suggests that the hydrogen-bonding networks involved in their binding are identical or nearly so within the families. This implies that any changes in amino acid sequence (mutations) in the proximity of the water molecules are such that the networks are not affected and are retained during evolution.

Internal water molecules are essential for structure stabilization. In proteins, internal water not only fills the structural cavities, but is necessary to stabilize the three-dimensional folding. This is achieved by hydrogen bonding to otherwise unsatisfied donor or acceptor sites or by mediating charges. In penicillopepsin, for instance, several internal sites (7 C=O and 9 N−H main-chain groups, and 2 polar side-chains) were without (intramolecular) hydrogen-bond partners, and in lysozyme, the same would happen with 3 internal C=O and 3 N−H main-chain groups if there were no internal water molecules. In carboxypeptidase, the internal protonated Glu 108 would be in a highly unfavorable apolar environment if a water did not share the proton, or at least associate with the protonated Glu carboxylate. In many cases, internal water molecules bridge charged side-chains and prevent the structural collapse of oppositely charged residues [621]. This is found in carboxypeptidase (Asp 104-water-Arg 59) and in actinidin (Glu 35-water-Lys 181; see Fig. 19.13).

Because internal water molecules are in mostly apolar environments, their hydrogen bonds are often strong and well defined. The average O···O distance is 2.89±0.21 Å for the internal water molecules in lysozyme, carboxypeptidase, cytochrome c, actinidin, and penicillopepsin. As with the small molecule crystal structures, the water molecules are involved in three or four hydrogen bonds, with 48% engaged in three and 37% in four interactions.

Internal water molecules tend to form clusters. In general, internal water molecules in protein structures are not found isolated but are assembled in clusters. Their hydrogen-bonding scheme could be derived in actinidin (Fig. 19.13), in lysozyme, and in penicillopepsin, based on the assumption that water molecules act as double donors and acceptors. In some of the protein structures, which have been analyzed in greater detail, an internal water is associated with three acceptor sites indicating three-center bonding as observed in the amino acid zwitterion crystal structures (see Part I B, Chap. 8).

An alternate interpretation is that hydrogen bonding is dynamic and that the water molecule is satisfying all acceptor sites by rotating into two positions [596]. A distinction between these two hypotheses is difficult for small molecule crystal structures, and impossible for proteins.

Hydrogen-bond partners for internal water molecules. In the five proteins lysozyme, carboxypeptidase, cytochrome c, actinidin and penicillopepsin (Table 19.1), the protein groups (numbers in parentheses) which are bonded to internal water molecules are the main-chain C=O (75), N−H (38), and side-chain atoms

(45), which are Asp and Glu (16), Asn and Gln (10), Ser and Thr (9), Tyr (4), Lys and Arg (6). The internal clusters also exhibit 46 water-water contacts. All the 38 main-chain N−H bind only to water, whereas 33 of the 75 C=O are simultaneously hydrogen-bonded to other protein groups, again showing that C=O can be in contact with more partners than N−H.

19.6 Metrical Analysis of Hydrogen Bonds in Proteins

In protein X-ray structure determinations hydrogen atoms cannot be located, and their positions have to be calculated from the known coordinates of the C, N, O atoms. Their positions are an order of magnitude less well defined than with small molecule crystal structures. Nevertheless, general trends can be derived and yield valuable information.

In their analysis of hydrogen bonds in proteins [596], Baker and Hubbard located the hydrogen atoms on the basis of the covalently bonded atoms with an N−H distance of 1.0 Å. For the main-chain to main-chain hydrogen bonding N−H$\cdots$O=C, the geometry is analyzed in terms of the distances H$\cdots$O and N$\cdots$O, and of angle N−Ĥ$\cdots$O. In addition, the angle at the carbonyl oxygen is given, C−Ô$\cdots$N, and the peptide plane is used to define out-of-plane (β) and in-plane (γ) angular deviations which are defined in Fig. 19.14 [610].

Main-chain to main-chain hydrogen bonds in α-helices are relatively uniform and well defined, with only small deviations from average values (Table 19.4). This regularity is also reflected in a narrow spread of torsion angles Φ, Ψ which cluster around −63°, −40°.

Compared with N−H$\cdots$O=C hydrogen bonds in small molecule crystal structures, the N−Ĥ$\cdots$O angle in α-helices, 155(11)°, is slightly less linear, and the out-of-plane component β, 27(8)°, illustrated in Fig. 19.15a, is in the range usually observed [75, 382, 475]. The in-plane component γ, −18(9)°, however, is considerably smaller than the 30° to 80°[2] found in small molecule crystal structures, and this deviation is also reflected in the C=Ô$\cdots$H angle, 147(9)°, which should be close to 120° as expected for an sp^2 geometry around carbonyl oxygen. These data indicate that the N−H$\cdots$O=C hydrogen bonds in the α-helices are distorted from their equilibrium configuration.

Another nonstandard feature in α-helix hydrogen bonding concerns the H$\cdots$O distance of 2.06(16) Å, which is longer than the average of 1.95 Å in small molecule crystal structures and in β-sheets (Table 19.4b). The distribution of H$\cdots$O distances in the range of 1.7 Å to 2.5 Å is broad and consistent with the softness of hydrogen bonds (Part IA, Chap. 2.1). It suggests that there are local irregularities in the α-helix structures, which are caused by hydrogen bonding of

---

[2] In the peptide bond, γ can have positive and negative values because the C$_\alpha$ atoms distinguish between each side of the peptide plane. This is not so in the small molecule survey where a number of different types of molecules were considered. Therefore, absolute values should be compared, 18(9)° in proteins and 30° to 80° in small molecule crystal structures.

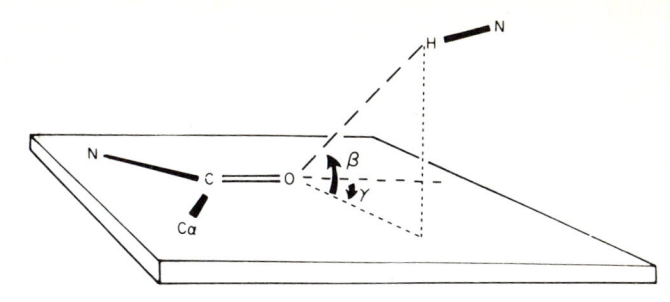

Fig. 19.14. Definition of out-of-plane (β) and in-plane (γ) hydrogen-bond angles as given in [596, 610]

Table 19.4a. Mean hydrogen-bond geometry in α-helices[a] ([596], where more details are given)

Number of hydrogen bonds	C=Ô···H (°)	N–Ĥ···O (°)	H···O (Å)	N···O Å
577	147(9)	155(11)	2.06(16)	2.99(14)

Distortions α_N, α_{C1}, α_{C2} at the ends of α-helices:

a) N-Terminus (α_N)

C=O···H–N 1 4	105(9)	126 (7)	2.28(12)
C=O···H–N 1 5	150(6)	156 (7)	2.21(15)

b) C-Terminus (α_{C1})

C=O···H–N 1 5	141(7)	139 (8)	2.37(10)
C=O···H–N 2 5	97(5)	128 (7)	2.39 (8)

c) C-Terminus (α_{C2})

C=O···H–N 1 6	153(9)	156(13)	2.11(11)
C=O···H–N 2 5	103(6)	148 (8)	2.16(13)

[a] Average data from proteins listed in Table 19.1 (except all-β proteins).

b. Mean hydrogen-bond geometry in β-sheets[a]

	Number of bonds	C=Ô···H	N–Ĥ···O	O···H	O···N
Overall	516	151(12)	160(10)	1.96(16)	2.91(14)
Parallel β-sheets	96	155(11)	161(9)	1.97(15)	2.92(14)
Antiparallel β-sheets	420	150(12)	160(10)	1.96(16)	2.91(14)

[a] See legend to **a**, data from proteins except all-α.

side-chains or water molecules to carbonyl oxygen atoms. This also gives rise to helix bending or curvature, as outlined in Section 19.3.

Main-chain to main-chain hydrogen bonds in β-pleated sheets. The most striking features of these N–H···O=C hydrogen bonds are that (1) their metrical

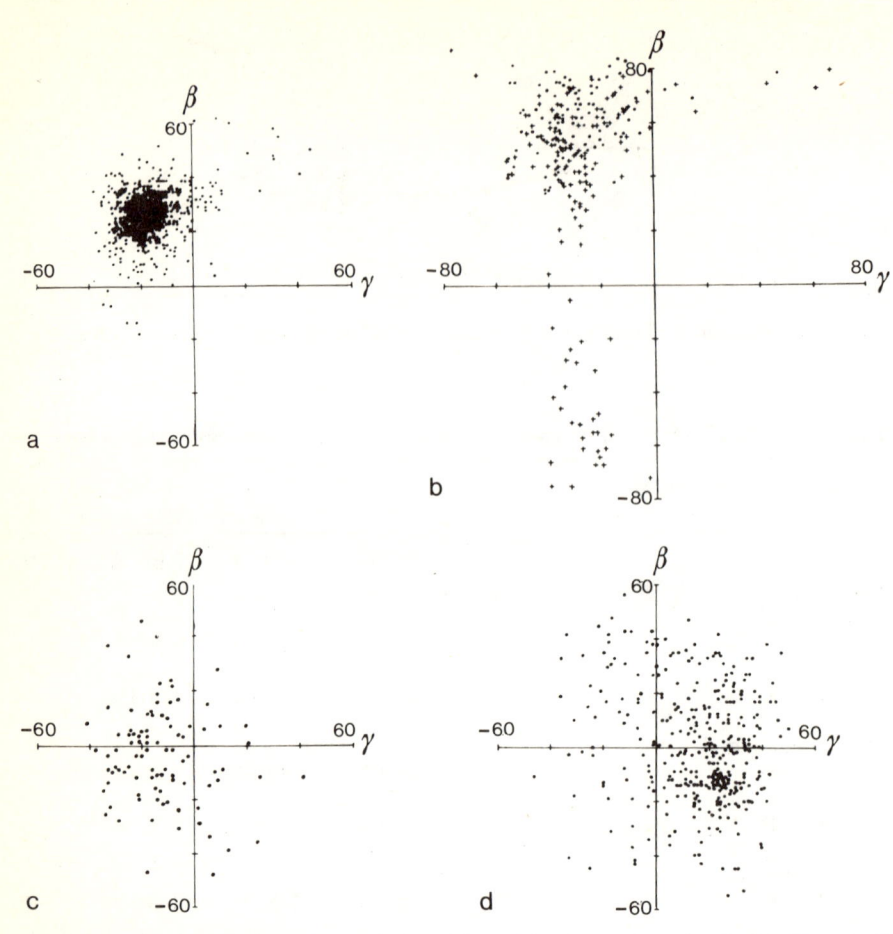

Fig. 19.15a–d. Distribution of out-of-plane (β) and in-plane (γ) angles for N–H$\cdots$O=C hydrogen bonds in **a** α-helices, **b** β-turns (*crosses*) and 3_{10}-helices (*dots*), **c** parallel and **d** antiparallel β-pleated sheets [596]

properties are similar in all proteins analyzed, and the mean values agree with those observed in small molecule crystal structures [75]; (2) their geometry is close to that calculated for a minimum energy β-sheet [598]; (3) the geometry of hydrogen bonds in parallel and antiparallel β-sheets is almost the same. In general, hydrogen bonds in β-sheets tend to be shorter and closer to being linear than those in α-helices [596].

Why do antiparallel β-sheets occur much more frequently than parallel ones [589], despite the observation that the geometry of the N–H$\cdots$O=C hydrogen bonds is virtually the same in both types? This could be due to more unfavorable van der Waals interactions in parallel sheets, which reduces their flexibility so that they are less suited to adapt to the local steric requirements in the globular proteins. This has also been suggested on the basis of a theoretical study [597].

Metrical Analysis of Hydrogen Bonds in Proteins 377

Table 19.5. Mean geometry of 3_{10}-type 1←4 hydrogen bonds in β-turns[a], and of 1←3 hydrogen bonds in γ-turns[b] [596]

Number of hydrogen bonds in 1←4 turns		Angles and distances			
		C=Ô···H (°)	N-Ĥ···O (°)	H···O (Å)	N···O (Å)
Overall:	295	116(15)	149(14)	2.17(18)	3.05(15)
In 3_{10} helices:	102	114(10)	153(10)	2.17(16)	3.09(14)
In turns of type					
I:	59	116(10)	155 (9)	2.12(16)	3.05(13)
II:	24	131(11)	157 (7)	2.12(13)	3.06(10)
III:	30	115 (9)	155 (9)	2.15(13)	3.08(11)
I':	8	116 (8)	159 (5)	2.03(12)	2.98 (9)
II':	9	139 (8)	149 (9)	2.11(18)	3.01(15)
III':	7	116 (5)	152(11)	2.17(13)	3.09(12)
In all β-turns	137	122(13)	154 (9)	2.13(15)	3.06(12)
In γ-turns:	20	103 (7)	137(10)	2.19(18)	

[a] Sample of 15 proteins given in Table 19.1.
[b] Sample of 10 proteins given in Table 19.2 footnote a.

The differences in hydrogen bonding in parallel and antiparallel β-sheets are more obvious if we compare the associated β-, γ-values (Fig. 19.15 c, d). The average γ value is −15° for parallel sheets and 20° for antiparallel sheets, i.e., the in-plane component of the N−H···O=C bond changes sign. The average out-of-plane β-angle, on the other hand, is ~0° for both types of β-sheet, with a wide spread in the −60°<β<+60° range. This suggests that, although β-sheets display an overall right-handed twist, individual residues show local right- or left-handed out-of-plane distortions [626] which cancel on average, in contrast to the average out-of-plane component β of 27(8)° derived for α-helices (Fig. 19.15a).

Hydrogen bonds are longer and less regular in β-turns and 3_{10} helices. Table 19.5 gives the geometrical data for β-turns, isolated 3_{10} turns and short pieces of 3_{10} helices obtained from the 15 proteins listed in Table 19.1. Also included are the 1←4 hydrogen bonds in which one C=O oxygen accepts *two* hydrogen bonds simultaneously from the 1←4 and 1←5 type turns in distorted α-helices.

If geometrical data of hydrogen bonds in β-turns I, II, III and their "mirror" images I', II', III' in Table 19.5 are compared with 3_{10} turns, there is no significant difference between the 3_{10}, I and III varieties. Turns II and II', however, deviate because they display a wider C=Ô···H angle of 131(11)° and 139(8)°.

As the data suggest, the hydrogen bonds in these turns are less regular than those in α-helices and in β-sheets, with longer H···O distances of 2.17(18) Å and less linear N−Ĥ···O angles of 149(14)°. The C=Ô···H angles, 116(15)°, are much more in agreement with the 120° required for hydrogen bonding to sp^2 carbonyl oxygen as acceptor than, for comparison in α-helices, 147(9)° and in β-sheets, 151(12)°.

If out-of-plane and in-plane components β and γ of N−H···O=C hydrogen bonds in β and 3_{10} turns are analyzed (Figs. 19.15b and 19.16), angles for 3_{10} and

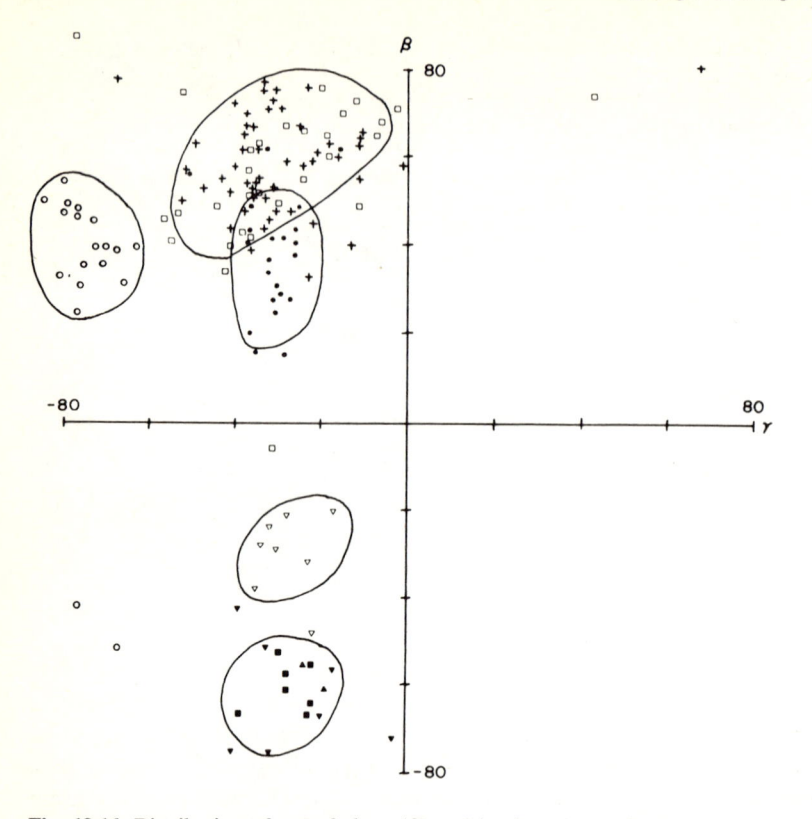

Fig. 19.16. Distribution of out-of-plane (β) and in-plane (γ) angles for 1←4 hydrogen bonds N−H···O=C in different classes of turns, and for 1←3 hydrogen bonds. Characteristic regions are *circled* for β-turns of type I (+), type II (●) and type III (□), and for the related type I' (▼), type II' (▽) and type III' (■). 1←3 hydrogen bonds in γ-turns are marked (○) [596]

type I, III turns are nearly identical and around $\beta = 60°$, $\gamma = -30°$. The value $\beta = 60°$ is larger than the average 30° to 40° found for small molecule N−H···O=C hydrogen bonds [75, 382]. It is also in conflict with the almost ideal C=Ô···H angle of 114(10)° in 3_{10} helices and 122(13)° in β-turns, and gives rise to the relatively long H···O distance of 2.17(16) Å and 2.13(15) Å. The large out-of-plane deviation suggests that N−H···O=C bonds in β-turns are not very stable, in agreement with the observation that in many cases these turns do not have any hydrogen bonds [627], and that the 3_{10} helices are very short.

The type II turns, with β-, γ-clustering at more favorable values of $\beta = 30°$, $\gamma = -30°$, and the II' turns with $\beta = -30°$, $\gamma = -25°$, appear to display a more normal hydrogen-bond geometry because the out-of-plane angles β are considerably reduced. There are, however, steric limitations requiring Gly to be in position 3 of the turn which limits their frequencies [591].

γ-Turns with 1←3 hydrogen bonds are rare. In the analysis of 15 proteins [596], only 20 γ-turns were observed with hydrogen bonds between C=O of residue (n) and N–H of residue (n+3). The average data for these interactions are presented in Table 19.5.

The H···O distances of 2.19(18) Å are greater than those of the α-helices, and the N–Ĥ···O angles, 137(10)°, are far from linear. This suggests that three-center hydrogen bonding occurs in which the γ-turn is the shorter of the two components. In terms of β- and γ-values, the γ-turns cluster at $\beta = 40°$, $\gamma = -75°$ and at the mirror-related $\beta = -40°$, $\gamma = -75°$ (Fig. 19.16), i.e., the deviation from linearity is not as much as in the type I and III β-turns. Since the in-plane angle is $\gamma = -75°$, the N–H bond points directly toward an sp^2 lone pair orbital of oxygen, while the other orbital is frequently engaged in a second hydrogen bond with water or a donor group from the protein molecule.

The γ- or 1←3 turn occurs in a variety of situations which is expressed in a wide spread of torsion angles Φ, Ψ of residues 1 and 3. In the central residue 2, the Φ-, Ψ-values cluster more around $\Phi = -90°$, $\Psi = 70°$, with one exception in which $\Phi = 75°$, $\Psi = -54°$ (a γ-turn in thermolysin). In fact, the γ-turn is not a sharp chain reversal, as is the β-turn or 3$_{10}$ helix, but is better described as a kink in a polypeptide chain [596].

19.7 Nonsecondary-Structure Hydrogen-Bond Geometry Between Main-Chains, Side-Chains and Water Molecules

As we have seen in the previous Section 19.6, secondary-structure hydrogen-bonding geometry in the α-helices, β-pleated sheets and β-turns is constrained by the requirements of polypeptide chain folding. When the hydrogen bonds which are not involved in secondary-structure interactions are examined, their geometry is in better agreement to that observed in the small molecule crystal structures discussed in Part I B, Chapter 7. In most cases, therefore, the discussion can be limited to situations where deviations occur [596].

Main-chain to side-chain hydrogen bonding displays, at the main-chain carbonyl oxygen, a C=Ô···X (X=N, O) angle which is consistently about 135(17)°, with a wide spread of 110° to 160° (Table 19.6a). This is more obtuse than the 126° for C=Ô···N and 128° for C=Ô···O in small molecule crystal structures [382, 384]. If main-chain N–H groups are involved in side-chain or water hydrogen bonding, N–Ĥ···O angles are more linear, average 151(15)°, and the distribution of 140° to 170° is much narrower than for angles C=Ô···X (Table 19.6b).

In comparison with hydrogen-bond angles, the distances display a wider spread and are more difficult to evaluate. This is because there are many weak interactions with long hydrogen bonds, a large proportion of which are probably three-centered. In most instances, there is no distinction between (main-chain) N–H···O=C (side-chain) and (side-chain) N–H···O=C (main-chain) separations, with mean O···H distances ranging from 1.88 Å to 2.16 Å, and O···N/O distances from 2.80 Å to 3.07 Å, see Table 19.6. Both ranges are similar to those obtained from small molecule crystal structure analyses. Histidine is frequently buried internally in the protein structures. In consequence, the hydrogen bonds to the imidazole ring are better defined and shorter, with H···O at 1.88(26) Å and O···N at

Table 19.6. Hydrogen bonds formed by main-chain C=O and N–H groups[a] to side-chains and to water molecules [596]

Side-chain	Number of hydrogen bonds	C=Ô···X[b]	O···X[b]	C=Ô···H	O···H	N–Ĥ···O
a) Involving main-chain C=O						
Ser, Thr	146	134(17)	2.93(28)	–	–	–
Asn, Gln	105	–	2.95(17)	132(16)	2.05(20)	152(15)
Lys	32	140(20)	2.92(22)	–	–	–
Arg	60	–	2.93(23)	135(19)	2.09(24)	144(16)
Tyr	27	132(16)	2.80(26)	–	–	–
His	29	–	2.80(21)	135(15)	1.88(26)	150(18)
Trp	14	–	2.96(15)	135(19)	2.16(20)	142(23)
b) Involving main-chain N–H						
Ser, Thr	66		3.02(17)	120(16)	2.14(17)	150(17)
Asn, Gln	68		2.94(13)	128(17)	2.03(14)	153(14)
Asp, Glu	125		2.92(17)	125(19)	2.03(19)	151(13)
Tyr	7		3.07(19)	120(11)	2.13(24)	160(15)
c) (Main-chain)C=O···water						
Overall	1139	129(16)	2.94(29)			
C=O bound only to one H$_2$O	308	133(14)	2.91(26)			
C=O bound to two H$_2$O	258	130(16)	2.94(33)			
d) (Main-chain)N–H···water						
Overall	389		2.97(21)		2.06(20)	156(15)

[a] For all proteins listed in Table 19.1.
[b] X is N or O.

2.80(21) Å (Table 19.6a). Likewise, hydrogen bonds from main-chain N–H to carboxylate oxygens of aspartic and glutamic acids and of C-termini are short, with mean H···O distances at 2.03 (19) Å and O···N at 2.92(17) Å (see Table 19.6b).

If side-chains (n) fold back on the protein main-chain to form hydrogen bonds with N–H groups in downstream (toward the carboxy terminus) positions, i.e., (n) to (n+2) or (n+3) or with upstream (toward the amino terminus) positions (n–3), or (n–4), there are obviously no steric limitations and geometries are normal, with distances involving Ser, Thr being 0.04 to 0.15 Å longer in general [596]. For hydrogen bonding between side-chain (n) and upstream C=O carbonyls in positions (n–3) and (n–4), a slight trend is observed, suggesting a small steric effect. Thus, bond angles C=Ô···X at main-chain carbonyl are 130 (19)° for (n–3) and wider 141(12)° for (n–4), whereas the reverse is observed for C=Ô···X at side-chain carbonyl, 115(12)° for (n–3) and narrower, 104(9)° for (n–4).

If a side-chain (n) folds back on its own peptide N–H (n), the H···O distance of 2.00 Å is short for Glu and Gln, and the angle N–Ĥ···O is 143°. These data are less favorable for Asp and Asn with a shorter side-chain and indicative of steric strain, with H···O widened to 2.25 Å and N–Ĥ···O at 112° [596].

Main-chain to water hydrogen bonds (see also Part IV, Chapter 23). With water molecules, there are fewer restrictions than in the main-chain to side-chain interactions just discussed because the water molecules are free to orient in the most favorable positions without any geometrical constraints. The problem with water molecules is, however, that their oxygen atoms O_W are generally located less precisely than the side-chain atoms. They frequently have larger thermal parameters which could be due to disorder, and the analysis of the hydrogen bond is more ambiguous.

The mean angle at the main-chain carbonyl oxygen, $C=\hat{O}\cdots O_W$, is 129(16)° and close to the values given for main-chain to main-chain interactions (see Table 19.6c). The distribution of angles is broad (110° to 150°), yet symmetrical, and only 4% of the total display a more linear hydrogen bond with an angle greater than 160°. This can be taken as a clear indication that the preferred angle at carbonyl oxygen is around 130°, and agrees also with data from small molecule crystal structures [384] with one salient difference: if, in small molecules, a $C=O$ group accepts only one hydrogen bond, the $C=\hat{O}\cdots H$ angle is, in general, around 180° instead of 120°. This trend is *not* followed in proteins, where all $C=O\cdots H$ angles are around 120°, no matter whether one or two groups act as hydrogen-bond donors [596].

As observed for the $N-H\cdots O$ interactions between main-chain and side-chain groups, the $N-\hat{H}\cdots O_W$ angles are almost linear, 156(15)°, if the "all-α-helix" proteins are omitted from the sampling, see Table 19.6d. The distribution of these angles is 140° to 180° for 90% of the data, and consistent also with small molecule crystal structures [75, 382, 475]. The spread of hydrogen-bond distances is broad, and wider for $C=O$ than for $N-H$, probably because the $C=O$ oxygen is more readily accessible and multiple $C=O\cdots HO_W$ interactions are frequently observed.

These findings are generally supported by out-of-plane and in-plane deviations of $C=O\cdots O_W$ interactions defined by the angles β and γ. There is a marginal clustering at $\beta = -40°$ to $+40°$ and $\gamma = 30°$ to 60°, and $\gamma = -30°$ to $-50°$, indicating a weak preference of water molecules to bind to carbonyl oxygen sp^2 lone pair orbitals. By and large, however, a wide scatter with radius 50° to 60° exists around the origin, suggesting that for water molecules the tendency to bind in the plane of the $\rangle C=O$ group is weak, see also Fig. 23.4a, Part IV.

Hydrogen bonds involving side-chains often display considerable angular spread, about the values expected from the small molecule crystal structures. The angles $C-\hat{O}\cdots X$ for Ser, Thr hydroxyls are smaller, 113(14)° and close to tetrahedral (109.5°) if they act as donors and greater, 122(15)° if they accept hydrogen bonds. A comparable yet weaker trend is observed for the hydroxyl group of Tyr. In Asp, Glu the carboxylate acts as acceptor with $C-\hat{O}\cdots X$ angles of 124(17)° as expected (120°). Side-chains with $N-H$ as donors such as Asn, Gln, Arg, His display $N-\hat{H}\cdots X$ angles near 150°, close to the value found in small molecule structures. In general, this wide distribution of angles reflects the variation in hydrogen-bond donors and acceptors comprising side-chains, main-chains, and water molecules.

In contrast to small molecule crystal structures, where the variation in most $H\cdots A$ distances falls within 0.3 Å, hydrogen-bonding distances involving side-

chain groups are more widely distributed. Distances involving His and Tyr are smaller than the average because, as already outlined, these side chains are located mostly in the protein interior and have a strong influence on the stabilization of tertiary structure. If the N−H···O hydrogen distances are compared, a trend is found:
(His)N−H···O<(Asp, Glu)COO⁻···H−N; (Arg)$\overset{+}{N}$−H···O; (Lys)$\overset{+}{N}$−H···O <(Asn, Gln)N−H···O<(Ser, Thr)O···H−N. The shorter distances are N···O = 2.76 Å, H···O = 1.83 Å, and the longer N···O = 2.95 Å, H···O = 2.00 Å [596].

Binding of water molecules to side-chains follows an even broader distribution because many side-chains are flexible and mobile and the waters are poorly ordered. Many water molecules are located outside the "inner", better-defined hydration sphere. The O⁻···O$_W$ distance distribution of water molecules around the carboxylate groups of Asp, Glu, and of C-termini is fairly clear, 2.5 to 3.1 Å with mean value near 2.77 Å, whereas Ser, Thr, Tyr, Asn, Gln display wider distributions with a mean value at 2.80 Å [596]; for more details, see Part IV, Chap. 23.

19.8 Three-Center (Bifurcated) Bonds in Proteins

The analysis of hydrogen bonds in 15 proteins recognized a number of three-center hydrogen bonds, of which 70 involve main-chain N−H, and 40 are associated with the side-chain N−H of His, Arg, Asn, Gln. Other feasible three-center bonds could be associated with hydroxyl groups of Ser, Thr, Tyr but hydrogen atom positions could not be defined in these cases.

Ninety percent of the three-center hydrogen bonds formed by main-chain N−H groups involve at least one protein acceptor atom; the other acceptor being either a water molecule (53%) or another protein atom (47%). With side-chain N−H groups, one half interact with water molecules, the other half with one or more protein acceptors.

As already described for three-centered, N−H⟨^O_O⟩ hydrogen bonds in small molecule crystal structures, interactions with equal or nearly equal H···O distances are in the minority. In proteins, they amount to 35% of the total, having H···O distances at 2.3 to 2.5 Å and N−Ĥ···O angles in the range 120° to 140°. By far the greater proportion (65%) of three-center bonds in proteins are unsymmetrical, with N−Ĥ···O angles differing by 40° and in the range 90° to 120° for one bond, and 140° to 160° for the other which is associated with the shorter H···O distance. If angles between the hydrogen atoms (as vertex) and the two acceptors are analyzed, N−H⟨^O_O⟩, they have a mean value of 82(17)°. The mean increases to 91(9)° if both acceptors are protein C=O oxygens and it reduces to 75(18)° if one partner is water. Because water is a small molecule, the steric interference is much less. In these respects, there is a good correlation between the small molecule and the protein crystal data.

More three-center hydrogen bonds are expected in protein structures. Compared to the N−H···O=C hydrogen bonds in small molecule crystal structures where 304 out of 1509 (or 20%) are three-centered and 6 are four-centered [56], the ratio of three-center bonds in proteins is very small, <10% [596]. This number appears unrealistic in view of the many potential hydrogen-bonding contacts which main-chain and side-chain acceptor and donor groups can form among themselves and with water molecules. We therefore have to assume that many of the existing three-center bonds in the protein crystal structures were not recognized as such because (1) they involved water molecules in disordered sites, (2) the X−H and A groups were not sufficiently well defined for proper assignment of the hydrogen positions, (3) the long H···A component fell beyond the cut-off limit of 2.5 Å, and (4) the analyses of the hydrogen bonding in the protein structures were carried out with the preconceived idea that "they (the three-center bonds) are not common" [596]. A new analysis with more relaxed geometrical parameters has shown that three-center bonds occur more frequently in proteins than previously assumed. They are found systematically in α-helices with major (n)N−H···O=C(N−4) and minor (n)N−H···O=C(n−3) components, and in β-sheets with major (n)N−H···O=C and minor (n)N−H···O−C(n) components [627a].

In future high-resolution crystal structure studies of proteins, hydrogen bonding has to be analyzed in terms of two-center *and* three-center interactions. It should be stressed that three-center bonds may be especially important in protein dynamics because they can be regarded as transition intermediate from one two-center bond to another (see Part IV, Chap. 25).

19.9 Neutron Diffraction Studies on Proteins Give Insight into Local Hydrogen-Bonding Flexibility

In recent years, several protein crystal structures have been investigated by neutron diffraction methods [628]: oxy-, met-, and carbonmonoxymyoglobin [629–631], ribonuclease A [632], and its complex with the transition state analog uridine-vanadate [633], trypsin [634, 635], and basic pancreatic trypsin inhibitor [636], lysozyme [637] and crambin [638]. The original objective of this work was to better define the hydrogen bonding by directly locating the hydrogen atoms which have a neutron-scattering power comparable to the nonhydrogen atoms (Part IA, Chap. 3.2).

In order to reduce the high incoherent background scattering which is due to the abundance of hydrogen atoms in protein and mother liquor, the crystals grown from H_2O were soaked for several months in D_2O. This proved fortuitous because H/D exchange of functional O−H and N−H groups within the protein molecule is observed. In fact, the research has focused on the interpretation of the finding that some hydrogen-bond functional hydrogen atoms exchange fully, while others are more protected, as this provides insight into protein flexibility.

The H/D exchange occurs when a proton involved in hydrogen bonding is accessible to solvent (D_2O), and is irreversible due to the large excess of D_2O in the sample:

$$N-H\cdots O \rightleftharpoons N-H+O \xrightarrow{D_2O} N-D+O \rightleftharpoons N-D\cdots O.$$

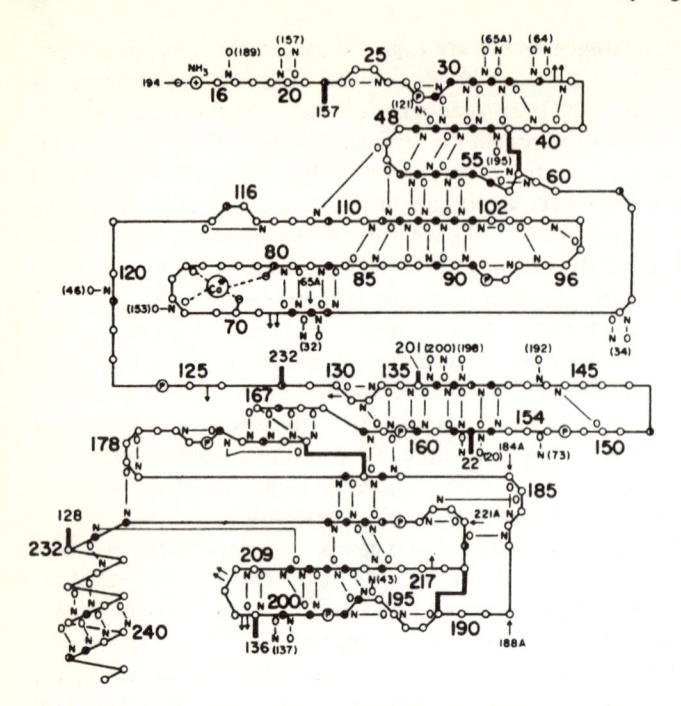

Fig. 19.17. Schematic drawing illustrating the secondary structure of trypsin and giving H/D exchange at each peptide group. ○ = full, ◐ = partial, ● = no exchange. P = proline. Peptide NH and carbonyl C=O are shown when hydrogen bonded. Disulfide bridges in *broad lines*. Note that H/D exchange in β-pleated sheet regions is largely reduced or does not take place, whereas in the α-helix (residues 233–243), nearly all protons are exchanged [634]

Because H and D are chemically equivalent and have neutron scattering lengths of opposite sign (see Part I A, Table 3.1), their exchange in the protein molecule is easily detected in Fermi density maps. Since, however, the resolution is such that individual hydrogen atoms are barely observed, functional donor groups can disappear and sometimes are not seen at all in the Fermi density (see Part I A, Table 3.4). In general, interpretation of these maps therefore requires a careful analysis of *both* the X-ray electron density and neutron Fermi maps, and the crystal structure has to be very well refined. If the analysis is successful, this provides an elegant tool for exploring the accessibility of the functional groups to bulk solvent.

H/D exchange is variable in proteins. In a protein in solution or in the crystalline state, all sterically unprotected peripheral side-chain N−H and O−H protons are exchanged. Of more interest for the flexibility of a protein are the peptide hydrogen atoms as a measure of secondary-structure hydrogen bonding because their exchange/unexchange pattern relates directly to conformational rigidity in these regions [628, 632, 634, 635, 637].

The H/D exchange pattern for peptide N−H groups of crystalline trypsin, displayed schematically in Fig. 19.17, shows clearly that the H/D exchange is not evenly distributed among secondary N−H···O=C hydrogen bonds. Some are fully

Table 19.7. Distribution of H/D exchange in five different crystalline proteins

Protein		RNase A	Trypsin	Oxymyogl.	Crambin
Total number of peptide bonds		120	215	149	46
Fully	⎫	65%	68%	71%	22%
Partially	⎬ exchanged N–H	25%	8%	21%	24%
Not	⎭	10%	24%	8%	53%

exchanged, some partly and some not at all. This is summarized in Table 19.7 for trypsin, ribonuclease A, oxymyoglobin, and for the very small protein, crambin.

Of the proteins listed in Table 19.7, crambin is unusual, having a 50% or larger ratio of unexchanged N–H protons. It appears from the crystal structure that this is due to a particular tight intramolecular packing of the polypeptide chains which prevents access for exchange of the secondary structure hydrogen bonds. For the other proteins, the comparable H/D exchange patterns reflect the relative stabilities of α-helices and β-pleated sheets.

α-Helices are more flexible than β-sheets. The H/D exchange patterns of crystalline proteins indicate that peptide N–H protons in α-helices are more commonly and more completely exchanged than in β-sheets, except in regions where they are in contact with hydrophobic side-chains or with β-sheet structure. In contrast, β-sheets are more resistant to H/D exchange, indicative of conformational stability and stiffness.

The reason for these differences in H/D exchange behavior of the two secondary-structure elements could be that β-sheets are, in general, more hydrophobic in character than α-helices because they are preferentially formed by hydrophobic amino acids [627]. On the other hand, it appears that multiple-stranded β-sheets are additionally stabilized for topological reasons because many polypeptide chains are linked together. If one N–H···O=C bond breaks, a distortion occurs, which is counteracted by the other chains in the pleated sheet. In contrast, in an α-helix, breaking of one N–H···O=C hydrogen-bond will not dramatically affect the geometry of the neighboring hydrogen bonds and therefore can take place more easily.

A case of special interest in the trypsin H/D exchange experiment is the Ser 54 hydroxyl bridged N–H···O contact between peptide groups 55 N–H and 43 C=O:

```
55N-H----O-D-----O=C43
         |
         CH₂
           \Ser54
```

It is surprising to find that 55 N–H is *not* exchanged, whereas the (Ser 54)O–H is *fully* exchanged. If we assume that the (Ser 54)O–D is incorporated into the β-sheet without major distortion, the H/D exchange suggests that, in fact, β-sheets are not as static as they appear. In addition, the N–H/N–D exchange is kinetically slow compared with O–H/O–D, a result supported by spectroscopic data [639, 640].

Table 19.8. % H/D exchange in lysozyme in the crystalline state (neutron diffraction) and in solution (NMR; measured after 6 weeks)[a] [641]

Position	Crystal	Solution
Peptide N–H of		
Thr 43	76	100
Asn 44	4	34
Thr 51	95	100
Asp 52	4	5
Tyr 53	4	12
Asn 59	100	100
Ser 60	10	1
Cys 64	11	50
Asn 65	25	100
Side-chain of Tryp 26 (indole NH)	7	1

[a] Errors in the experimental data ~20%.

Exchange of individual protein hydrogens in crystalline state and in solution is comparable. In case of hen egg white lysozyme, the H/D exchange in D_2O solution at pH 3.8 was monitored by NMR techniques, and a neutron diffraction analysis was carried out on triclinic crystals grown from 2% $NaNO_3$ and 0.025 M sodium acetate at pH 4.2. Comparison of these experiments suggests that the difference in ionic strength and pH between the solution and crystalline state has no significant influence on H/D exchange [641].

The H/D exchange pattern for the indole hydrogen of Trp 26 and for some of the clearly NMR-assignable peptide hydrogen atoms is given in Table 19.8. Although the errors in the experimental data are about 20%, it is obvious that the correspondence in the two data sets is good. The indole hydrogen of Trp 26 is not exchanged because it is buried in a hydrophobic cavity and hydrogen-bonded to peptide C=O of Tyr 23. For the H/D exchange of the peptide groups, the ratios obtained from the crystal data are in some cases lower than those from the solution data (see Asn 44, Cys 64 and Asn 65), which might reflect the expected higher mobility and flexibility of some groups when the protein molecule is free in solution. In general, however, the data suggest clearly that the structures of lysozyme in solution and in the crystalline state are essentially identical, but that the structure in the crystal is somewhat tighter, so that exchange is slower or more inhibited [641].

For H/D exchange, two mechanisms have been proposed. In one, referred to as the "cooperative or local unfolding mechanism", a segment of secondary structure opens cooperatively and extrudes transiently into the bulk solvent, where exchange takes place by normal water chemistry [640, 642, 643]. In the other, called the "penetration mechanism", reacting solvent (D_2O) molecules diffuse into the core of the protein, the path being opened by local atomic fluctuations and/or mobile defects in protein packing of the tightly folded polypeptide chain [644]. It is likely that both mechanisms occur and a clear distinction between the two cannot be made. It does appear that the latter is more probable than the former and, if in combination, the latter is predominant [632].

Whereas NMR can be used to study exchange kinetics, neutron diffraction can provide only long-term information because the soaking of crystals in D_2O and data collection takes of the order of months. In a recent diffraction study where certain neutron diffraction intensities were monitored, the D_2O/H_2O exchange with time could be followed directly [637]. While this provides insight into the *overall* exchange in a crystal, individual H/D exchange rates cannot be observed. For this, NMR methods have to be applied.

Is there any correlation between H/D exchange ratio and distance of peptide group into the molecule away from the molecular surface? To answer this question, the distance of each peptide proton in trypsin to the nearest bulk solvent interface was calculated [635]. The average distance was found at 3.8 Å. More than 90% of the unexchanged peptide hydrogens are located at 4 Å or more from the surface, i.e., they are buried deeper than average in the core of the enzyme. However, the average depth of the 11 fully exchanged peptide protons in the β-sheet structures is also about 4 Å. While this is closer to the 3.8 Å average depth of the total sheet population, it is clear that distance from the surface of a protein does not provide a complete and satisfying explanation for the observed H/D exchange pattern. The departure from equilibrium geometry of the hydrogen-bonding interactions or some special topological characteristics of the surroundings (like hydrophobic or large aromatic groups) of a peptide proton must play a major role in the exchange mechanism. This holds also for the peptide protons in non-β-sheet arrangements.

In fact, in trypsin 27% (36 sites) of the fully exchanged peptide hydrogens are buried more than 4 Å from the surface, and 15% (20 sites) are even 5 Å or deeper. H is replaced by D in these peptide groups, because there are internal, fully D_2O-exchanged water molecules in trypsin, and the average distance between them and the exchanged peptide H's is around 2.5 Å.

H/D exchange correlates strongly with density of atom packing and of hydrogen bonding [635]. If exchange characteristics do not necessarily correlate with distance from solvent boundaries because internal water molecules come into play and promote exchange, maybe the packing density, calculated as the number of atoms within a sphere of 7 Å radius drawn around each peptide bond as center, give a better measure to assess and understand H/D exchange. Also the number of hydrogen bonds surrounding a peptide group in a 7 Å sphere could be of importance, since they are indicative of conformational and structural rigidity.

The results of such calculations carried out for trypsin suggest that the sites of unexchanged peptide hydrogen atoms correspond to regions of highest atomic or hydrogen-bond density. Conversely, fully or partially exchanged hydrogen atoms correlate with low density regions in the protein structure. Expressed in numbers, 78% of the unexchanged sites and only 5% of the exchanged sites have a packing density of >160 atoms/7 Å sphere. Or, in terms of hydrogen bonds, the average exchanged site has 6.2 hydrogen bonds within a sphere of 7 Å radius, whereas the unexchanged site has 9.1 hydrogen bonds as neighbors [635].

Hydrogen-bonding geometry is virtually identical from X-ray and neutron studies except for H···O distances. In a comparison of crystallographic data obtained for trypsin by both diffraction methods, it was found [596] that the angles involving hydrogen bonds in secondary-structure elements, i.e., $N-\hat{H}\cdots O$ and $C=\hat{O}\cdots H$, are nearly identical in X-ray and neutron diffraction analyses, not differing by more than 4°. However, the hydrogen-bond distances $H\cdots O$ and $N\cdots O$ are consistently larger by about 0.1 Å in the X-ray study except for (main-chain)$N-H\cdots$water interactions, where the reverse is found. This cannot be due to the positioning of hydrogen atoms along the X–H bonds. The average N–H distance from the neutron data in deutero-trypsin is 1.013(0.02) Å, while in the X-ray hydrogen-bond analysis, the N–H distances are normalized by assuming a comparable 1.0 Å. The reason for this discrepancy, which is not observed with small molecule crystal structures [387], remains unclear [596].

19.10 Site-Directed Mutagenesis Gives New Insight into Protein Thermal Stability and Strength of Hydrogen Bonds

In a molecular structure as complicated as a protein, one could argue that one out of several hundred hydrogen bonds would have only a small effect, if any, on the overall structure and on function. Since genetic engineering has provided us with the tools to systematically and specifically mutate any amino acid of a protein into one of the other 19 naturally occurring amino acids, we have at hand an excellent possibility of probing subtle differences in hydrogen bonding. In this context, we shall describe three studies, of which two, concerned with the enzymes subtilisin and phage T4 lysozyme, were aimed at aspects of thermal stability, whereas the third investigated the involvement of hydrogen bonding in substrate and transition state binding of the enzyme tyrosyl-tRNA synthetase.

In a protein, mutation of a single amino acid involved in hydrogen bonding can influence the thermal melting: subtilisin and phage T4 lysozyme. With *subtilisin*, the proteins expressed from bisulfite-treated clones were screened for thermostable mutants. The best mutant contained a single amino acid change, Asn→Ser 218, with an increase in melting point temperature from $T_m = 78.3 \pm 0.1°$ for wild type to $80.7 \pm 0.1°$ for the mutant, measured under identical conditions [645].

The crystal structures of wild-type and mutant subtilisins are practically identical at a high resolution of 1.8 Å except in the vicinity of the mutation site, amino acid 218. The wild-type Ser 204 $N-H\cdots O\delta$ Asn 218 hydrogen bond is not optimal because the acceptor oxygen atom $O\delta$ of the Asn 218 side-chain is displaced from the plane of the donor peptide group. In contrast, the $O\gamma H$ group of the mutant Ser 218 is *in* this plane and therefore more favorably positioned for strong hydrogen bonding Ser 204 $N-H\cdots O\gamma$ Ser 218. Moreover, the shorter side-chain of Ser 218 permits closer packing and consequently all the hydrogen bonds in this area, which is part of a β-pleated sheet, are shortened by ~0.1 Å and strengthened (see Fig. 19.18).

In order to study the influence of a mutation in position 218 further, Asn 218 was changed genetically to Asp, Cys, Ala. The relative thermal stabilities of these variants are Ser > Asp > Asn > Cys > Ala, as expected: Ser forms the optimum hydrogen bonds; Asp is better than Asn because of the negative charge, but not

Site-Directed Mutagenesis Gives New Insight 389

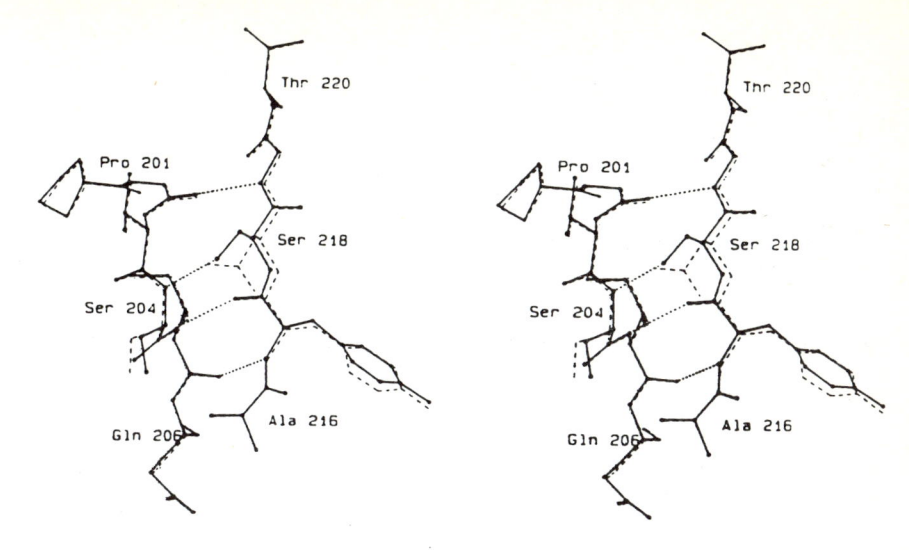

Fig. 19.18. Stereo view of a section of the tertiary structure of subtilisin, showing the environment of the mutation site (wild type, *dashed lines*) Asn 218→(mutant, *solid lines*) Ser218. As a result of the mutation, the β-pleated sheet hydrogen bonds 219N–H···O=C202 and 205N–H···O=C217 are shortened from 2.86 Å to 2.76 Å and from 2.93 Å to 2.77 Å, respectively, for N···O distances. The hydrogen bond between Ser204N–H and the side-chain oxygen is longer (2.87 Å) for Ser218 than for Asn218 (2.77 Å), but exhibits better overall geometry [645]

Table 19.9. Thermodynamic stability[a] of phage T4 lysozymes with different amino acids at position 157 [646]

Amino acid at position 157	ΔT_m (°C)	$\Delta\Delta G$ (kcal mol^{-1})
Thr (wild type)	–	–
Asn	–1.7	0.45
Ser	–2.5	0.66
Asp	–4.2	1.1
Gly	–4.2	1.1
Cys	–4.9	1.3
Leu, Arg, Ala, Glu, Val	–5.0 to –6.0	1.3 to 1.6
His	–7.9	2.1
Phe	–9.2	2.4
Ile	–11.0	2.9

[a] ΔT_m is the change in melting temperature relative to the wild-type enzyme [T_m = 42.0(5) °C]. $\Delta\Delta G$ is the corresponding change in free energy at 42 °C.

as good as Ser; Cys is poor because the –SH group is not a good hydrogen-bond acceptor, and Ala with no acceptor at all is even worse [645].

In contrast, in *phage T4 lysozyme* a thermolabile mutant was found in which one single amino acid was exchanged Thr→Ile 157, and the thermal melting point dropped from 42 °C for the wild-type enzyme to 31 °C for the mutant. Systematic exchange of the amino acid in position 157 by site-directed mutagenesis led to the

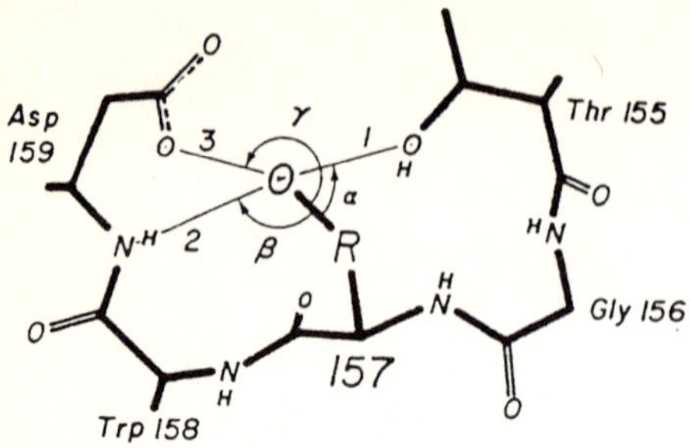

Fig. 19.19. Schematic description of the hydrogen-bonding geometry formed with different amino acids at position 157 in phage T4 lysozyme. The potential hydrogen-bond lengths 1, 2, 3 and associated angles α, β, γ are:

Amino acid 157	Atom	Hydrogen bonding distance X···A(Å)			Angle (°)			Temperature factor B (Å^2)[a]
		1	2	3	α	β	γ	
Thr	Oγ	2.8	3.2	3.2	119	103	106	20
Ser	Oγ	2.8	3.4	3.2	119	107	114	30
Asn	Oδ	3.1	3.1	3.2	97	130	101	30
Asp	Oδ	2.9	3.1	3.1	100	125	94	–
Gly	H$_2$O	2.9	3.1	2.8	–	–	–	29

[a] Of the central hydrogen-bonded oxygen atom [646].

lysozymes given in Table 19.9. X-ray structure analyses were carried out at 1.7 Å resolution for all the mutants [646]. They show that for the most thermostable mutants and the wild type, a characteristic hydrogen-bonding pattern is satisfied around the hydrogen-bonding functionality of side-chain 157. This pattern involves the donor functions of the peptide N–H of Asp 159 and of OγH of Thr 155, and the strong acceptor properties of the side-chain Oδ of Asp 159 (see Fig. 19.19). The hydrogen-bonding scheme is essentially retained even with the mutant Gly 157 where a water molecule occupies the position otherwise filled by the side-chain functional group of amino acid 157. The water molecule is especially tightly bound with a temperature factor of 29 Å^2, well below the average of about 48 Å^2 for other ordered water molecules in the crystal structure.

It becomes clear from this study that hydrogen bonding is of prime importance, although other effects due to van der Waals interactions and steric repulsion contribute when amino acid 157 in phage T4 lysozyme is mutated. The deleterious influence of Ile 157 in the thermosensitive mutant can be explained on this basis, as well as the effects of mutations where Thr 157 is substituted by Phe, His, Val (see Table 19.9). In these cases, all hydrogen bonds in the pocket around the side-chain of amino acid 157 are broken except for one to a water molecule, Thr 155 Oγ···O$_W$.

Energy of individual hydrogen bonds in a protein; tyrosyl-tRNA synthetase as a test case. When an enzyme interacts with its substrate, the bound water molecules

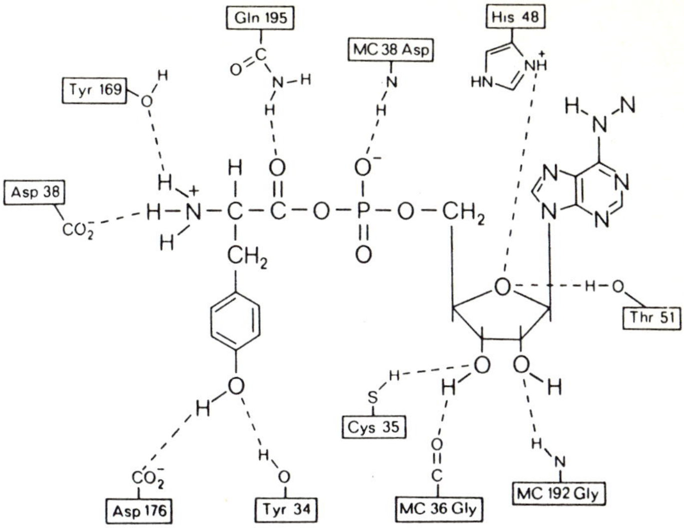

Fig. 19.20. Schematic drawing of hydrogen-bonding interactions between the enzyme tyrosyl-tRNA synthetase and the substrate analog tyrosyl adenylate. The enzyme groups interacting with tyrosyl adenylate are boxed, MC indicates main-chain C=O or N–H groups [647]

covering the interface between enzyme and substrate must be stripped off before direct enzyme-substrate contacts can form which are stabilized by hydrophobic, van der Waals and hydrogen-bonding forces. Consequently, hydrogen bonding is the *difference* in energy between molecules in solution interacting with water and molecules in a complex interacting with each other [647, 648]. If we consider a substrate with an acceptor group S–A, and an enzyme with a donor group E–X–H, we can formulate:

$$S-A\cdots HOH + H_2O\cdots H-X-E \rightleftharpoons [S-A\cdots H-X-E] + H_2O\cdots HOH \ .$$

Since the number of hydrogen bonds on both sides of the equation is the same, the reaction is essentially isoenthalpic. It is driven entropically because two bound water molecules are released into the bulk solvent.

The tRNA synthetases are enzymes which are responsible for the "charging" of tRNA with their cognate amino acids. They recognize specifically the tRNA and the amino acid and link them by an ester bond between the 3'-OH of the terminal adenosine in tRNA (see Chap. 20.4) and the carboxylate of the amino acid. The enzyme of *Bacillus stearothermophilus* has been co-crystallized with the substrate-analog tyrosyl-adenylate, and the structure of the complex was determined and refined using 2.0 Å resolution X-ray data [649]. A schematic diagram of the individual hydrogen-bonding interactions which occur between the substrate analog and the active site of tyrosyl-tRNA synthetase is given in Fig. 19.20. These can be classified according to (1) high specificity of the enzyme for tyrosine rather than phenylalanine (side-chain of Tyr 34), (2) binding of the terminal adenosine of

Table 19.10. Comparison of energies of individual hydrogen bonds in vacuo and in enzyme-substrate complexes [647, 648]

Type of bond	In vacuo: energy of formation	In enzyme-substrate: change in energy upon deletion
	Both partners uncharged	
HOH···OH$_2$	-6.4 kcal mol^{-1}	
CH$_3$SH···OH$_2$	-3.2 kcal mol^{-1}	0.5 – 1.8 kcal mol^{-1}
CH$_3$(H)S···HOH	-3.1 kcal mol^{-1}	(factor ~10 loss in specificity)
	One partner charged	
Imidazolium/H$_2$O	-14.0 kcal mol^{-1}	3 – 6 kcal mol^{-1}
CH$_3$COO$^-$···HOH	-19.0 kcal mol^{-1}	(factor ~1000 loss in specificity)

tRNA (side-chains of Cys 35, His 48, Thr 51), and of tyrosine (side-chain of Tyr 169), and (3) a bond closer to the seat of reaction (Gln 195).

These amino acids were point-mutated by genetic engineering into such amino acids that altered the hydrogen-bonding scheme one by one. The following observations were made [647, 648]:

1. The deletion of a side-chain which interacts with an uncharged group on the substrate, e.g., the mutation Tyr→Phe 34 doubles K_m, results in a low apparent bond energy change of 0.5 kcal mol^{-1}, and drops the specificity for tyrosine by a factor of 15. For the larger Cys→Gly 35 exchange, the change in bond energy is 1.1 kcal mol^{-1}, again a relatively small value. In general, deletion of an uncharged or charged (His 48) side-chain which interacts with an uncharged group on the substrate, reduces the binding energy by about 0.5 to 1.8 kcal mol^{-1}. This value is ~25% of the binding energy measured for a comparable hydrogen bond in vacuo (see Table 19.10).

2. Deletion of a side-chain that interacts with a charged group on the substrate (as in Tyr→Phe 169) results in a larger binding energy loss of about 3.7 kcal mol^{-1}. It is 4.5 kcal mol^{-1} in the case of Gln→Gly 195, where the substrate group is not C=O but C–O$^-$ in the transition state. Energies of about 4 kcal mol^{-1} correspond to a factor of 1000 in specificity.

A comparable effect is found if a group on the substrate is replaced that interacts with a charged side-chain of the enzyme. In general, we have to expect changes in hydrogen-bonding energy in the range 3 to 6 kcal mol^{-1} if one of the partners is charged, again much less compared with the in vacuo results (Table 19.10).

3. If the hydrogen bond formed between enzyme and substrate is expanded beyond its normal equilibrium value, i.e., too long H···A distance, total deletion of this interaction may lead to an increase in binding energy. An example is given by the Thr→Ala 51 exchange, which increases the binding energy by 0.44 kcal mol^{-1}.

In general, the data obtained from point mutations of tyrosyl-tRNA synthetase demonstrate that hydrogen bonding contributes significantly to the specificity of enzyme-substrate interactions, especially if charged groups are involved. As expected, steric factors are even more important because of the strong dependence of van der Waals energies (which change as r^{-6}) at shorter interatomic distances.

Genetic engineering will provide new insight in protein hydrogen bonding. The examples described in this section have shown that genetic engineering is a very subtle and powerful tool which enables us to study certain individual hydrogen bonds in a large protein molecule. This opens new perspectives, and it is hoped that our understanding of hydrogen-bonding interactions in the functional mechanism and in the stabilization of the three-dimensional structure of proteins will increase rapidly as a consequence of the new methodology.

Chapter 20
The Role of Hydrogen Bonding in the Structure and Function of the Nucleic Acids

In the form of deoxyribonucleic acid (DNA), the macromolecular nucleic acids are of prime importance in biology because they carry the building plan for each living individual. They are identically reduplicated and inherited from one generation to the next, be it bacterium, plant, animal, or man. The information about every feature of and about every molecule contained in a living being is encoded in the nucleotide sequence of its DNA, which is read out and translated into the amino acid sequences of its proteins. In the many different steps involved in this protein biosynthesis, information transfer takes place which would be impossible without the weak hydrogen bonds. Because they can easily and rapidly be formed and broken, they are ideally suited for these dynamic processes which are so important for life.

20.1 Hydrogen Bonding in the Nucleic Acids is Essential for Life

There are two types of nucleic acids – the genetic material deoxyribonucleic acid (DNA), and the ribonucleic acids (RNA) –, which are involved in protein biosynthesis. DNA and RNA have very different functions in biology, but are similar in chemical terms. As illustrated in Fig. 20.1, both nucleic acids are linear polymers, DNA being formed by the four deoxyribonucleosides (see Fig. 17.1) and RNA by the four ribonucleosides. The nucleosides are connected at their 3'- and 5'-hydroxyl groups by phosphodiester linkages. DNA, which is 10^7 to several 10^9 nucleotides long, can be compared with computer magnetic tapes: the tape material, a plastic ribbon, is the "inert" sugar-phosphate backbone, and the information contained on the tapes in the form of bits (letters) and bytes (words) is encoded in the nucleic acids in the sequence of the four bases (bits), with three of them (a triplet) being the equivalent of the byte.

Base-pair hydrogen bonding of the Watson-Crick type is fundamental in all biological processes where nucleic acids are involved. These processes, which are chiefly DNA replication and protein biosynthesis [650, 651], were understood only at the molecular level when Watson and Crick discovered the three-dimensional structure of DNA [27, 527]. This structure consists of two polynucleotide chains running in opposite directions (antiparallel), and twisted into a right-handed double helix. The hydrophobic purine and pyrimidine bases are stacked in the center

Hydrogen Bonding in the Nucleic Acids is Essential for Life

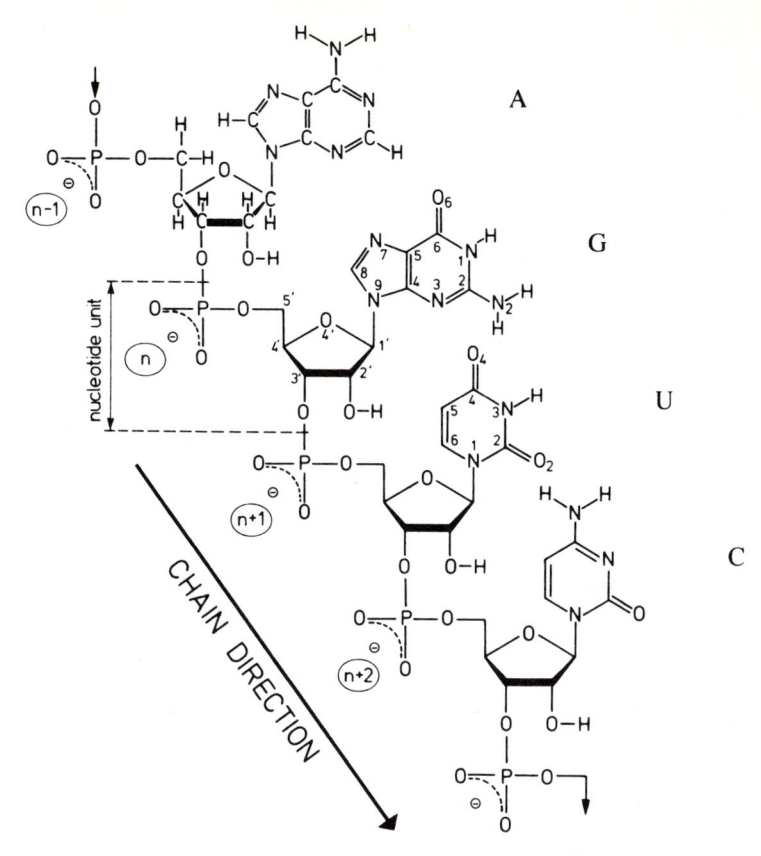

Fig. 20.1. Chemical structure of RNA with sequence ...pApGpUpCp... or, in short, AGUC. All hydrogen atoms are drawn in adenosine, and only functional hydrogen atoms are given in the other nucleotide units. In DNA, hydroxyl groups in sugar 2'-position are replaced by hydrogen atoms, and uridine is methylated in 5-position and called thymidine [522]

of the helix and the hydrophilic sugar-phosphate backbones are at the periphery. The bases are linked by hydrogen bonds in the base pairs such that adenine (A) in one strand opposes thymine (T) in the other strand, and guanine (G) opposes cytosine (C), so that one strand of DNA is said to be *complementary* to the other (Part II, Chap. 16, Figs. 16.10, 16.11, 16.13).

For *reduplication*, the chains are separated and on each a new, complementary strand is synthesized by enzymes called DNA polymerases [652]. For *protein biosynthesis*, the DNA is copied (transcribed) into the "messenger" ribonucleic acid (mRNA) by the enzyme RNA polymerase (Fig. 20.2) where, in contrast to DNA, the deoxyribose is replaced by ribose and thymine by the equivalent uracil. Here again, the Watson-Crick base pair plays the crucial role so that the mRNA sequence is complementary to the DNA sequence.

In order to read the sequence of nucleotides in the mRNA and to *translate* it into the protein sequence, the complicated multi-protein-subunit machinery called

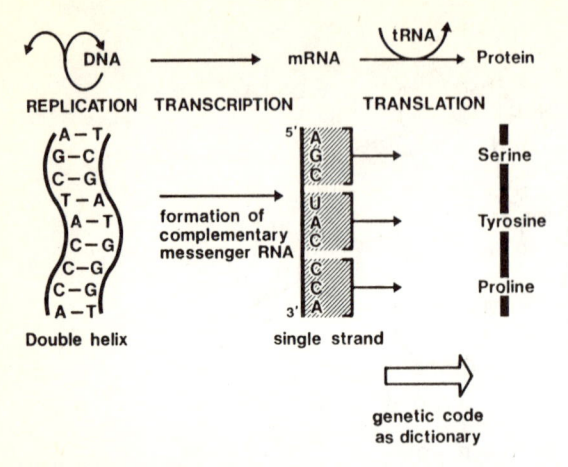

Fig. 20.2. Simplified scheme describing the "central dogma" in molecular biology. DNA is *replicated* and passed from one generation to the next. For protein biosynthesis, DNA sequence is first *transcribed* into complementary messenger RNA (mRNA) sequence which, by means of the "adapter molecule" transfer RNA (tRNA), is *translated* into protein sequence. The translation follows the genetic code where a nucleotide triplet (e.g., AGC) codes for an amino acid (e.g., serine) [522]

ribosome is needed. The necessary link between nucleic acid and protein is provided by adapter molecules. These are small RNA's with about 70 nucleotides called *transfer RNA* (tRNA). For each of the different 20 naturally occurring amino acids, there is at least one specific tRNA. Specific means that each tRNA contains an "anticodon" region consisting of three consecutive nucleotides, and the sequence of this triplet determines which amino acid is covalently linked by an ester bond at the 3'-OH terminus of that tRNA (*genetic code*).

At the ribosome, which travels along the mRNA, the tRNA molecule is bound such that its anticodon can interact with a nucleotide triplet on mRNA (the codon). If the anticodon is complementary to a codon triplet on the mRNA, the amino acid attached at the 3'-terminus of the tRNA is transferred to the amino terminus of the growing polypeptide chain; if it is not complementanty, the tRNA is rejected and another one is checked for complementary. The whole process is repeated until the synthesis of the protein is completed. It is initiated, as well as terminated, by specific codons regulating this translation.

In all these events, the hydrogen-bonded Watson-Crick base pair is operative and is responsible for DNA reduplication, transcription, and translation. Since mispairing can occur, all these processes are checked for fidelity by several enzymes which can correct for errors [652]. At this and all other levels of DNA reduplication and protein biosynthesis, intermolecular hydrogen bonds between nucleic acids and between nucleic acids and proteins are responsible for recognition, interaction, and, finally, for information transfer.

20.2 The Structure of DNA and RNA Double Helices is Determined by Watson-Crick Base-Pair Geometry

Much of the structural information for the nucleic acids was originally derived from X-ray fiber diffraction patterns [522, 653, 654]. While they provide sufficient data to establish the conformation of an average nucleotide in a double helical arrangement, they cannot distinguish between the A:T, T:A, G:C and C:G base pairs. To determine the fine structure of each individual nucleotide in an oligo- or polymeric assembly, single crystal diffraction analyses are necessary [654–658]. In recent years, single crystal studies of transfer RNA and of several oligonucleotides have provided such detailed information on nucleotide structures [659, 660] and culminated in the discovery of left-handed Z-DNA [661]. Since, however, the crystal quality is often inadequate to carry an analysis to a resolution beyond 2 Å (except for Z-DNA oligonucleotides), direct location of hydrogen atoms is not possible and the details of hydrogen bonding may still be obscure. As in the proteins, the positions of the hydrogen atoms on $>$N−H groups and on the conjugated, coplanar −NH$_2$ groups of the bases can be inferred from a knowledge of the nonhydrogen geometry. Those on the C−OH groups of the ribose residues in the RNA's can be obtained only by assuming colinear two-center X−H$\cdots$A bonds.

Of all the hydrogen-bonded base pairs displayed in Part II, Chapter 16, the Watson-Crick A:U (or T) and G:C pairs are best suited for the formation of smooth double helices with both chains in antiparallel orientation. The reasons are purely geometrical.

A:U (or T) and G:C base pairs are approximately isostructural. Although the A:U (or T) and G:C base pairs differ in having a single and double dimer hydrogen-bonding system respectively, they have comparable geometries (see Fig. 16.13). This is because the distances between the glycosyl links in the two base pairs are nearly identical, 10.72 Å versus 10.44 Å, as are the angles formed by these bonds and the line connecting the C(1') atoms. In addition, the two glycosyl links in a base-pair are related by a twofold axis lying midway between the glycosyl bonds and in the base-pair plane. These structural features of the Watson-Crick base pairs are maintained in the double helices and have several structural consequences:

− A:U(T) and G:C base pairs can substitute for each other without major distortions of the DNA or RNA double helix, because they are approximately isostructural;

− A:U(T) corresponds structurally to U(T):A, and G:C to C:G due to the pseudo-twofold axis. This means that the double helix can have a regular structure irrespective of nucleotide sequence;

− The sugar-phosphate chains attached to atoms C(1') are in opposite directions because of the pseudo-twofold axis of the base pairs. The two strands in DNA or RNA double helices with Watson-Crick-type base pairing are therefore always antiparallel.

Both right- and left-handed double helices occur in the nucleic acids [522, 653, 662]. In the double helical DNA structure proposed by Watson and Crick in 1953,

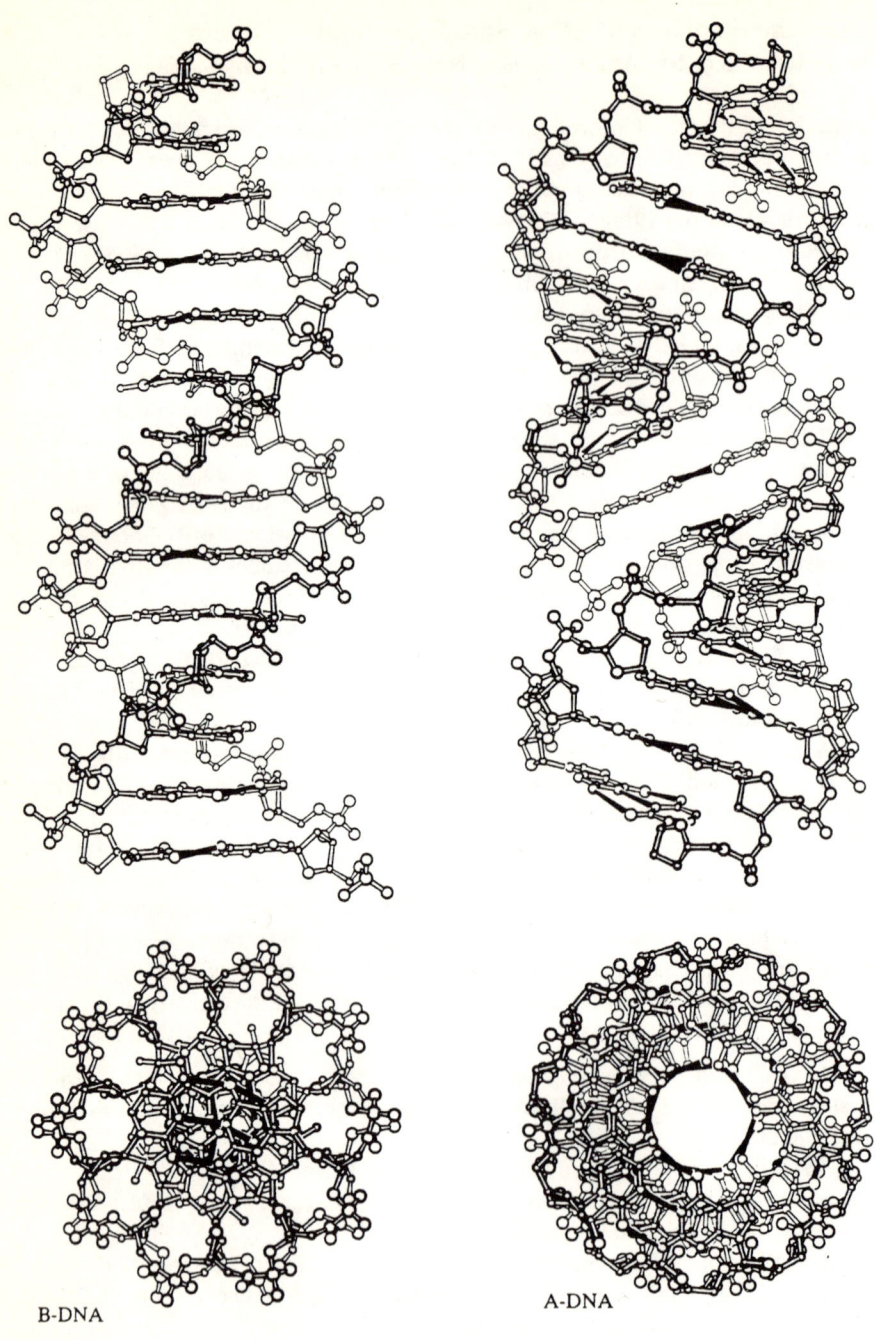

398   The Role of Hydrogen Bonding in the Structure and Function of the Nucleic Acids

B-DNA

A-DNA

Fig. 20.3 (Legend see next page)

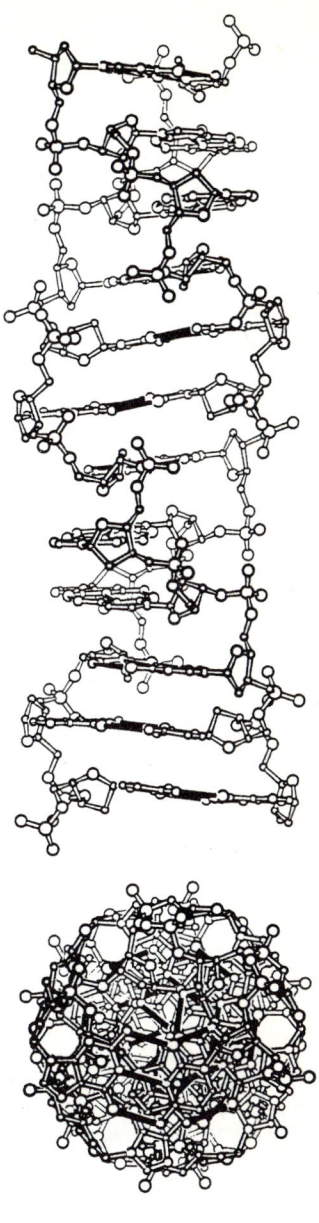

Z-DNA

Fig. 20.3. Structures of A-, B-, and Z-DNA in side and top views. The helix axis is located in the major groove side (Fig. 20.15) of the base pairs in A-DNA, it pierces the base pairs in B-DNA and it is located in the minor groove side (Fig. 20.15) in Z-DNA. Base pairs are tilted toward helix axis in A-DNA but nearly perpendicular to it in B- and Z-DNA. In A- and B-DNA, the sugar-phosphate backbone describes a smooth helix and the repeat unit is one nucleotide. In the left-handed Z-DNA, the sugar-phosphate backbone is zigzagged and the repeat unit is a dinucleotide. A-RNA has a structure similar to A-DNA [522]

and now verified by a number of different methods including, notably, crystallography, the base pairs are stacked along the helix axis. The sugar-phosphate backbones of the two strands are antiparallel and wind around the base pair stack in a right-handed twist (Fig. 20.3). Depending on salt concentration (and humidity), which control the hydration of the phosphate groups (see Part IV, Chap. 24), DNA can adopt several double helical forms with the main representatives, A and B, having respectively 11 and 10 base pairs per helical turn of 28 Å and 34 Å pitch height (Table 20.1 and Box 20.1). These conformational changes between the different forms of DNA take place without any evidence of major distortions in the hydrogen bonding of the base pairs. This again points to the characteristic soft stretching and bending properties of hydrogen bonds discussed in Part I.

Box 20.1. Polymorphism of DNA (A-, B-, Z-) and Structural Rigidity of RNA

In all the naturally occurring DNA and RNA double helices, the base pairs are of the Watson-Crick type A:U(T) and G:C. If DNA with the four nucleotides in "random" sequence [and not in alternating purine-pyrimidine sequence as in poly(dG-dC)] is exposed to different salt concentrations (and metal ions), it adopts two principal secondary structures A and B.

Differences in A- and B-DNA double helices are caused by differences in the sugar puckering modes. The B-DNA form occurring at low salt conditions is believed to be the "native", biologically active DNA found in chromatin. Base pairs are located on the helix axis and nearly perpendicular to it with an average tilt angle of only −5.9°. There are 10 base pairs per right-handed turn of 34 Å pitch, the sugar pucker is C(2')-endo (for definition see Box 17.3). The major and minor grooves differ in width but are comparable in depth (Fig. 20.3 and Table 20.1) [522, 653, 654, 661 – 663].

If the salt concentration is raised or alcohol is added to the solution, the phosphate groups become less hydrated (see Part IV, Chap. 24), and transformation to the A-DNA form takes place. The helix axis is now moved into the major groove side of the base pairs, the sugar conformation changes to C(3')-endo. The base pairs are tilted considerably to −20°, and there are 11 base pairs per pitch of 28.2 Å. Associated with these changes are changes in the grooves, the major groove becoming deeper and narrower and the minor groove wider and shallower (Table 20.1) [522, 663, 664].

Left-handed Z-DNA in alternating purine/pyrimidine sequences. If the synthetic polynucleotide poly(dG-dC) with alternating sequence is kept under low-salt conditions, it adopts a B-DNA type conformation. If the salt concentration is raised to >4 M NaCl or if alcohol is added, a transformation to the left-handed Z-DNA is induced. This is probably due to a change in the phosphate hydration (see Part IV, Chap. 24) [661, 662]. The G:C base-pair hydrogen bonding is still of the Watson-Crick type, but the conformation of individual nucleotides has changed: dG is in the unusual syn conformation with γ in the ap range and C(3')-endo sugar pucker, whereas dC is in the common anti with γ in sc range and C(2')-endo sugar (see Box 17.3). Biochemical studies have indicated that Z-DNA *may* occur in chromatin [665] and that it is not limited to G−C sequences but can also be obtained with A−T sequences flanked by G−C.

RNA occurs exclusively in the A-form. Constitutionally, RNA and DNA are very similar, with uracil substituting for thymine and ribose for deoxyribose. The conformational properties of these nucleic acids are very different, however, because RNA adopts only the A-form (and a closely related A'), and is unable to transform into the equivalent of the B-DNA due to sterical hindrance with the O(2')H hydroxyl group [666]. The A-type conformation is also observed in complementary DNA-RNA hybrids in which one chain of RNA in the double helix suffices to give both chains the same RNA-like double helical form [522].

Table 20.1. Helical parameters for right-handed A-, B-, C-DNA and for left-handed Z-DNA [522]

Structure type	Pitch (Å)	Base-pair tilt (°)	Number of nucleotides per pitch	Axial rise h and turn angle t per nucleotide		Groove width[b] (Å)		Groove depth[b] (Å)	
				h (Å)	t (°)	Major	Minor	Major	Minor
A	28.2	20.0	11	2.56	32.7	2.7	11.0	13.5	2.8
B	33.8	−5.9	10	3.38	36.0	11.7	5.7	8.5	7.5
C	31.0	−8.0	9.33	3.32	38.6	10.5	4.8	7.5	7.9
Z	45	−7	6×2[a]	3.7×2²	−30.0×2[a]	–	2.7	–	9.0

[a] Repeat unit is a dimer d(G−C).
[b] Groove width gives perpendicular van der Waals separation of helix strands drawn through phosphate groups; for definition of grooves, see caption Fig. 20.3.

For poly(dG-dC) with alternating guanine and cytosine sequence, a right-handed B-DNA form prevails at low salt conditions, whereas at higher salt concentrations a change occurs to a left-handed form called Z-DNA [661, 662]. In Z-DNA, the G:C base pairs are again of the Watson-Crick type and the two polynucleotide strands are antiparallel, but the nucleotide conformations are different (Box 20.1).

Major and minor grooves determine interactions between DNA or RNA and ligands. A consequence of having the glycosyl links and the sugar-phosphate chains attached at one side of the base pairs and not diametrically opposite, is the formation of two different indentations or grooves along the periphery of the double helices. They are called the "minor" and "major" grooves, according to the separation between the glycosyl bonds, the minor groove at the purine N(3), pyrimidine O(2) side of the base pair and the major groove at the purine N/O(6), pyrimidine N/O(4) side. The groove widths and depths vary dramatically with the conformation of the DNA (Table 20.1, Box 20.1). Depending on the shape of the double helix, there are different degrees of access for ligand binding to the base pairs [667]. This is of particular importance for the hydration of the nucleic acid double helices and will be further discussed in Part IV, Chapter 24.

Hydrogen bonding stabilizes DNA double helices across the helix axis but not in the direction of the axis[1]. In DNA, the deoxyribose-phosphate backbone can act as hydrogen-bond acceptor only through phosphate and sugar oxygen atoms. The bases contain hydrogen-bond donor groups, but these are engaged in base-pair interactions, which are horizontal with respect to the helix axis. There are no direct hydrogen bonds between adjacent base pairs and between bases and deoxyribose-phosphate backbone [see, however, the special case of poly(dA)·poly(dT), Sect. 20.5]. This implies that there is no axial stabilization of the DNA double helix by direct two-center intramolecular hydrogen bonding.

The RNA double helix is stabilized by hydrogen bonds in the axial direction. The situation is different in RNA where the ribose hydroxyl group O(2')H can act as hydrogen-bond donor and acceptor [667–672]. It gives rise to an axial stabilization through direct intramolecular hydrogen bonding to the O(4') atom of the adjacent nucleotide in the 3'-direction, as shown in the crystalline self-complementary, double helical oligoribonucleotide UUAUAUAUAUAUAA [666], and in tRNA [673]. As expected, the O(2')–H···O(4') hydrogen bonds in the oligonucleotide self-duplex are weak, with an average O(2')···O(4') distance of 3.3 Å, and they are formed in only half of the possible interactions. Computer-aided molecular modeling has shown that the O(2')–H hydroxyl would interfere sterically if the ribose were in other than C(3')-endo puckering mode, thereby stabilizing the A-form of RNA (see Box 20.1).

The propeller twist: bases in base pairs are rarely co-planar. When unsubstituted pyrimidine and purine bases crystallize in the form of homo or hetero base pairs (Part II, Chap. 16), these are generally planar due to symmetry requirements. If such symmetry constraints are absent, as in co-crystal of nucleosides or nucleotides or in crystals of oligonucleotides, the bases in the pairs are usually twisted

---

[1] This is in contrast to the α-helices in the proteins, where the hydrogen bonding stabilizes the structure *along* the helix axis.

about the long axis with respect to each other like the blades of a propeller. This twist is generally such that the rear base is twisted anticlockwise with respect to the front base, with a dihedral angle between the two bases in the range $-5°$ to $-20°$.

20.3 Systematic and Accidental Base-Pair Mismatches: "Wobbling" and Mutations

In nature, base pairs of the Watson-Crick type prevail [674] as described in the previous sections. However, there are some well-documented structures where other base pairs occur. In one type, illustrated in Fig. 20.4, the bases are in the canonical keto/amino forms but base pairs are in mismatch configurations called "wobble" base pairs [675]. They occur even systematically in interactions between messenger RNA (mRNA) and transfer RNA (tRNA), and appear to play a role in mutation processes during DNA replication. Another type of non-Watson-Crick base pairs is found if the bases occur in rare enol/imino forms. These are also believed to be involved in mutation processes, albeit by a mechanism other than "wobble" base pairs, vide infra.

Systematic wobble base pairing was predicted on the basis of the genetic code. According to the genetic code [674], each nucleotide triplet in DNA corresponds to one amino acid in protein. Since there are $4^3 = 64$ possible triplets but only 20 amino acids, there must be a redundancy in the code. This was first postulated by Crick [675] and later verified experimentally [676, 677].

The redundancy in the genetic code is settled on the tRNA anticodon side. For each codon on the mRNA, the first two nucleotides (counting from the

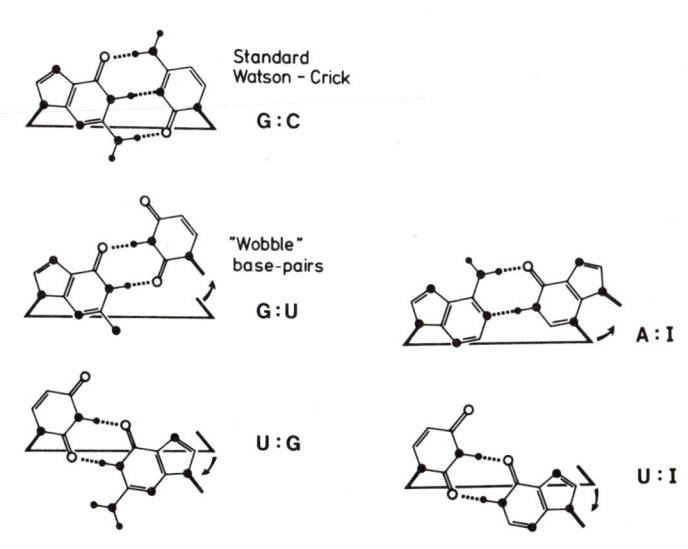

Fig. 20.4. Comparison of Watson-Crick (*top*) and wobble base pairs. Deviations from standard geometry are indicated by *arrows*. Atom designation: ○, oxygen; small filled circle, hydrogen; large filled circle, nitrogen; thick line, glycosidic Cl'−N bond

5'-position) are specific for an amino acid but the third nucleotide (in 3'-position) can vary. For instance, arginine is represented by the four mRNA codons CGG, CGA, CGC and CGU, reading in 5'→3' direction.

During translation, the anticodon (on tRNA) and the codon (on mRNA) are arranged in the form of a short antiparallel double helix of the Watson-Crick type so that the base in 3'-position of the codon forms a base pair with the base in 5'-position of the anticodon. Because the number of tRNA's (and the number of anticodons) is limited, the same tRNA anticodon has to base-pair with several of the possible mRNA codons which differ in their 3'-position. Nature solves this problem by allowing nonstandard (wobble) base pairs:

Base on the tRNA anticodon in 5'-position	Bases recognized on the mRNA codon	
	Standard Watson-Crick	Wobble
U	A	G
C	G	
G	C	U
I	C	U, A

(I, inosine, looks like G without NH_2)

The wobble base pairs are displayed in Fig. 20.4. In comparison with Watson-Crick base pairs, the positions of the glycosyl links differ, but their directions are more or less retained so that the short codon-anticodon double helix is smooth. Pyrimidine-pyrimidine base pairs U – U and U – C, which could also form (Part II, Chap. 16) are considered to be unlikely since their $C(1')\cdots C(1')$ distances are about 2 Å shorter than the 10.5 Å in a Watson-Crick base pair. The long A – I base pair with $C(1')\cdots C(1')$ distance of about 13.5 Å, as actually observed in a comparable crystalline base-pairing complex [678], is too long to be accommodated in a smooth double helix. It has been suggested therefore that inosine in *syn* form could mimic the Watson-Crick geometry [679].

The mismatch base pairs are incorporated into double helix in the A-, B- and Z-forms without disrupting the base-pair stacks. This was shown by a series of crystal structures of oligonucleotides with mismatch base pairs [680–688]. The mismatches are of the form A – C, G – T, G – A. Of these, G – T corresponds to the wobble G – U and G – A to the wobble I – A, because G and I are comparable if the amino group in position 2 of G is neglected. There are only minor distortions of the sugar-phosphate backbone at the sites where the mismatch base pairs are inserted. The more significant changes relative to normal Watson-Crick geometry are, in G – T, the shift of the guanine amino group into the minor groove, and of the thymine O(4) group into the major groove. Associated with these dispositions are changes in stacking patterns of mismatched base pairs and adjacent base pairs which, however, are still comparable to those observed in double helices with Watson-Crick geometry. A more pronounced difference is found in the *hydration* of the G – T and A – C mismatch base pairs. It involves a water

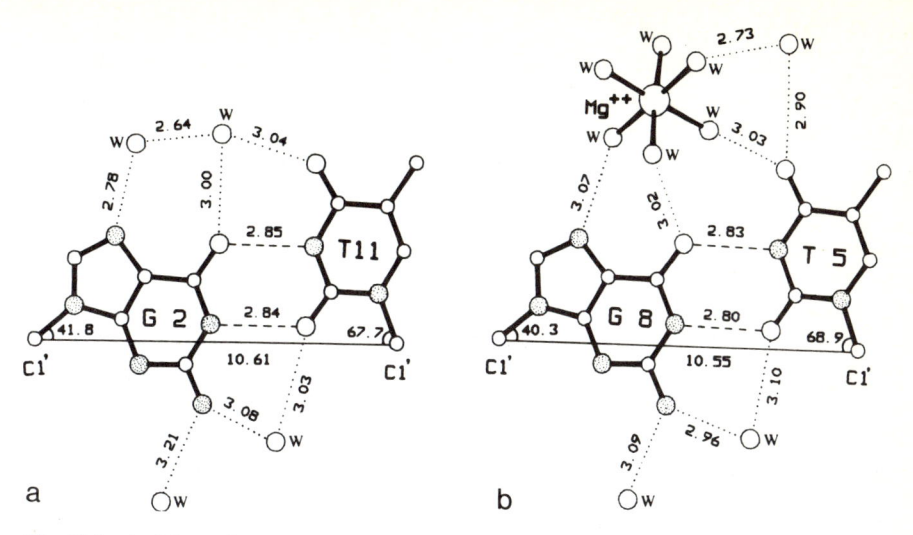

Systematic and Accidental Base-Pair Mismatches: "Wobbling" and Mutations

Fig. 20.5 a, b. Schematic representation of hydration of a G–T wobble base pair in the Z-DNA type crystal structure of d(CGCGTG). Hydrogen-bonding distances to water molecules (W) and C(1')···C(1') distances given in Å units. *Left*: hydration is only by water molecules; *right*: hydration involves a cluster [Mg(H$_2$O)$_6$]$^{2+}$ [683]

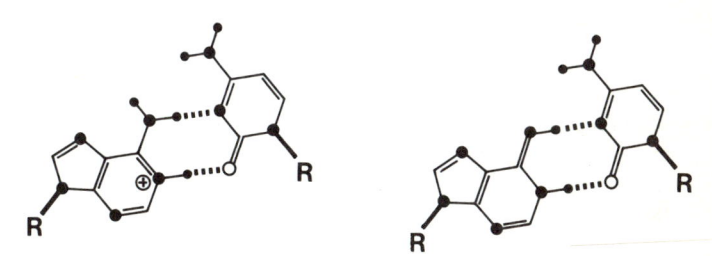

Fig. 20.6. Possible stabilization of the A:C base pair as seen in the crystal structure of d(CGCGAATTAGCG) by protonation of adenine (*left*) is more probable than the amino-imino tautomerism (*right*). Atom designation as in Fig. 20.4; R = sugar residue

molecule spanning systematically the bases in the major groove, thymineO(4) ···water···O(6)guanine and cytosineN(4)···water···N(6)adenine. In the G–T mismatched base pair, there is an additional water bridge in the minor groove, guanineN(2)···water···O(2)thymine (see Fig. 20.5).

Tautomeric enol and imino forms of bases occur only rarely, and can lead to mutations. It should be emphasized that in none of the above described mismatch base pairs is there any evidence for the existence of rare tautomeric forms. For the A–C pair, protonation at (adenine)N(1) appears more probable than the imino form (Fig. 20.6). However, conclusive evidence is still lacking because hydrogen atoms cannot be located at the attainable resolution of about 2 Å. Moreover, in none of the crystal structures of the nucleosides and nucleotides or of the bases themselves is there any evidence of the enol-imino tautomers (Part II, Chaps. 15, 16, 17).

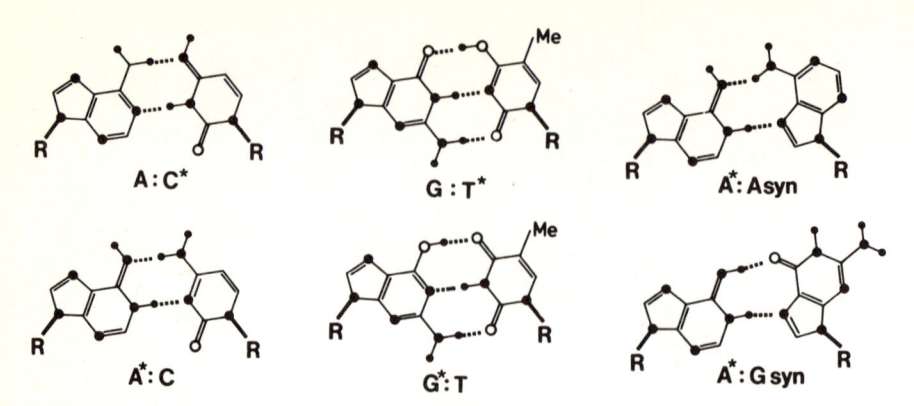

Fig. 20.7. Some mismatch base pairs with rare imino and enol tautomeric bases (marked *) likely to be involved in transition and transversion mechanism. For more possibilities, see [522, 678]; atomic designation as Fig. 20.4

There are two mutation processes where mispairing due to enol/imino tautomeric forms could be involved. In one, *transition,* a purine is replaced by another purine or a pyrimidine by another pyrimidine; in the other, *transversion,* a purine/pyrimidine or pyrimidine/purine exchange takes place. In these mutations, formation of base pairs of the type illustrated in Fig. 20.7 is postulated.

The rare enol/imino configurations of bases occur at concentrations of only 10^{-4} to 10^{-5} mol l^{-1} (Part II, Chap. 15). The *anti/syn* conversion of bases necessary for formation of some of the rare base pairs (see Fig. 20.7) decreases the probability by another factor of 10^{-1} to 10^{-2}. Therefore, the estimated mutation rates are of the order of 10^{-8} to 10^{-12} per incorporated nucleotide in the newly synthesized DNA strand, in overall agreement with experimental data [677, 679, 689–694].

20.4 Noncomplementary Base Pairs Have a Structural Role in tRNA

Just as main-chain NH···O=C hydrogen bonds are important for the stabilization of the α-helix and β-pleated sheet secondary structures of the proteins, the Watson-Crick hydrogen bonds between the bases, which are the "side-chains" of the nucleic acids, are fundamental to the stabilization of the double helix secondary structure. In the tertiary structure of tRNA and of the much larger ribosomal RNA's, both Watson-Crick and non-Watson-Crick base pairs and base triplets play a role. These are also found in the two-, three-, and four-stranded helices of synthetic polynucleotides (Sect. 20.5, see Part II, Chap. 16).

The secondary clover-leaf structure of tRNA is stabilized by Watson-Crick base pairs. The tRNA's are a large family of molecules consisting of 71 to 76 nucleotides, with about 10% of "rare" bases (Fig. 15.2). The known nucleotide sequences of over 200 tRNA's [695] can be arranged in a characteristic clover-leaf model with four double helical stems and three loop regions. In some of the positions, the same nucleotide occurs, called *invariant,* in others only the type is conserved, i.e.,

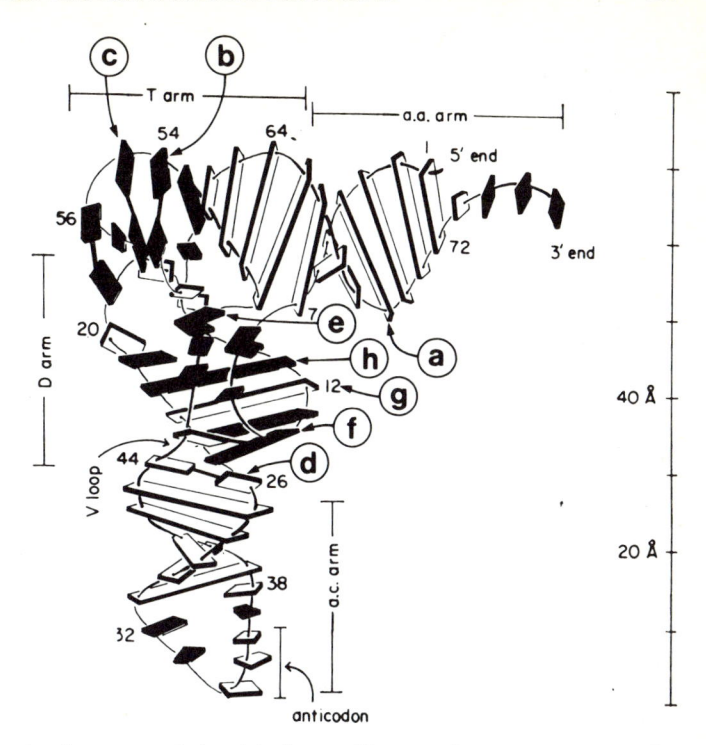

Fig. 20.8. Three-dimensional structure of phenylalanine specific tRNA from yeast. Watson-Crick type base pairs indicated by slabs, nonstandard base-base interactions that stabilize the tertiary structure are denoted *a* to *h*. Invariant and semi-invariant nucleotides are *shaded*, the four double helical regions are indicated by $a \cdot a \cdot$ (amino acid) *arm*, *T arm*, *D arm*, *a.c.* (anticodon *arm* [696]

purine or pyrimidine. This is called *semi-invariant*. These conserved nucleotide positions play an important role in folding all the tRNA's into a similar three-dimensional L-shape.

The three-dimensional L-shape of the tRNA's is stabilized by a number of non-Watson-Crick base pairs [673, 696, 697]. The folding of the tRNA's is such that stacking of base pairs is optimized by joining the anticodon arm and the D-arm into one longer helix and the T-arm and amino acid stem into another helix. Both helices are then arranged into the shape of an L, where the anticodon and the O(3')H terminus with the attached amino acid, are at opposite ends (Fig. 20.8). This configuration is similar for all tRNA's, and is stabilized by base-base interactions. These are formed mainly between the invariant or semi-invariant bases and are frequently not of Watson-Crick type, as illustrated in Fig. 20.9. Of particular interest are the base triplets, in which a Watson-Crick type double helix associates with a single-stranded RNA segment. There are also two base-base interactions stabilized by only one hydrogen bond (c and f in Fig. 20.9). Although this interaction is generally not considered sufficient to stabilize a base pair, the situation in tRNA is special because the base stacking is very strong, and even involves base intercalation, i.e., insertion of a base between two adjacent bases further away in

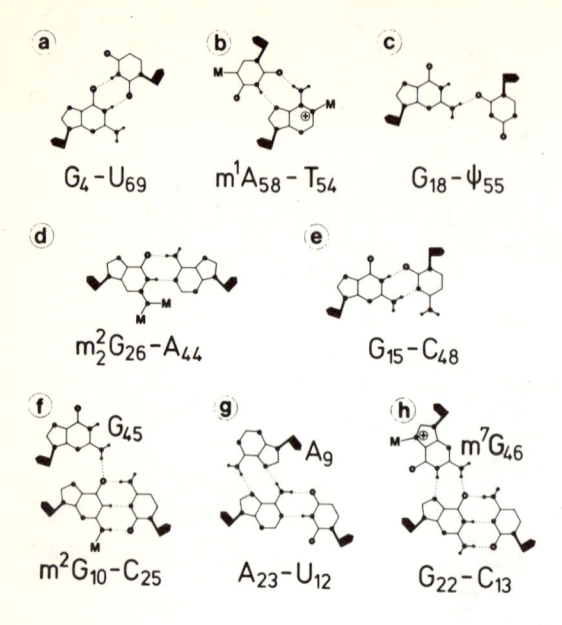

Fig. 20.9a–h. Base-base interactions **a** to **h** non-Watson-Crick-type indicated in Fig. 20.8. Riboses are marked by *flags*, atoms are indicated as • = nitrogen, ○ = oxygen, · = hydrogen; hydrogen bonds drawn as *dotted lines*. After [696] from [522]

the sequence. This stabilizes the single hydrogen-bond interactions occurring in base triplet f where A_9 is intercalated between G_{45} and m^7G_{46}, and in base pair c, where G_{57} is intercalated between bases G_{18} and G_{19}.

20.5 Homopolynucleotide Complexes Are Stabilized by a Variety of Base-Base Hydrogen Bonds – Three-Center (Bifurcated) Hydrogen Bonds in A-Tracts

Polynucleotides in the ribo- and deoxyribo series which contain only one type of base can form complexes with themselves and with other homopolynucleotides in which a variety of base-base hydrogen-bonding interactions discussed in Part II, Chapter 16 are observed.

Helices formed by homopolymer complexes can be parallel or antiparallel, depending on the geometry of the base pairs involved. In the hemiprotonated $C \cdot CH^+$ base pair formed by poly(C) at pH 4.5 [531] and in the $AH^+ \cdot AH^+$ base pair observed with poly(A) below pH 4 [532], the twofold rotation axis relating one base to the other is perpendicular to, instead of in the plane of the base-pairs, as shown in Fig. 20.10 (see also Part II, Chap. 16). Because the sugar-phosphate chains bonded with the glycosidic link to the bases are also symmetry-related by these twofold axes, they have to be parallel to each other and are not antiparallel, as in the Watson-Crick geometry.

A helix with parallel strands is also formed by poly(G), but in this structure four molecules are involved, [poly(G)]$_4$, with the base-quadruple illustrated in Fig. 20.10. The helix is nonintegral and repeats after two turns of 39.2 Å, comprising 11.5 nucleotides each. The tendency of guanine to self-associate is unique.

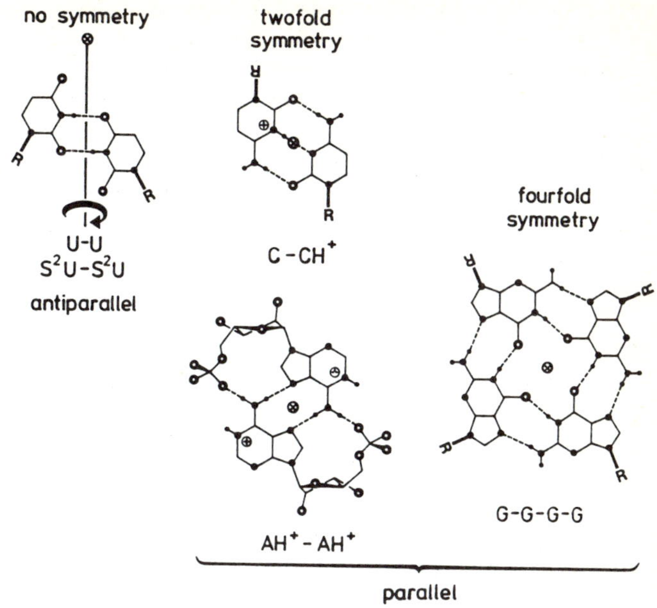

Fig. 20.10. Base pairs found in homopolymer nucleic acid self-complexes. The unsymmetrical uridine U–U and 2-thiouridine s^2U-s^2U base pairs lead to antiparallel orientation of polynucleotide chains similar to the RNA double helix. In the symmetrical AH^+-AH^+ and $C-CH^+$ base pairs, the dyad axis coincides with the helix axis and gives rise to parallel polynucleotide chains. For the same reason, the four polynucleotide chains in $[poly(G)]_4$ are related by a fourfold rotation axis and are also parallel; ⊗, location of helix axis [522]

Thus guanosine and several of its derivatives, including the 3'- and 5'-phosphates, form gels in which tetrameric hydrogen-bonded arrangements are stacked in a form reminiscent of the quadruple helix [534, 698].

In the homopolymers formed by uridine or 2-thiouridine, base pairs with twofold symmetry are feasible, as described in Part II, Chapter 16 [529, 699]. The situation is more complex, however, because the homopolymer folds back on itself, giving rise to a hairpin-like structure where the polynucleotide strands are necessarily antiparallel and the base pairs are of the form shown in Fig. 20.10. Not surprisingly, a comparable double helix is also observed for polyxanthylic acid poly(X) [700]. The purine base of xanthosine displays the same hydrogen-bonding functional groups as uracil and therefore, the same base pair can be formed, but the purine-purine pair requires a larger separation between the glycosidic links and consequently a wider helix radius.

Complexes formed between different homo-polynucleotides usually adopt the Watson-Crick base-pairing scheme as, for instance, poly(C)·poly(G), poly(C)·poly(I), poly(A)·poly(U), poly(dA)·poly(dT), etc. They are consequently right-handed double helices with antiparallel polynucleotide strands.

Poly(dA)·poly(dT) and its ribo analog do not transform into A-DNA, but disproportionate. As already discussed in Part II, Chapter 16.3 (see Fig. 16.16), these related homopolymer duplexes are special. If the salt concentration is raised, they do not change from the B-form double helix to the A-form, but they disproportionate into a single poly(dA) or poly(A) strand and a triple helix poly(U)·poly(A)·poly(U) or poly(dT)·poly(dA)·poly(dT). In the latter, a Watson-Crick double helix with antiparallel polynucleotide chains accommodates, in its major groove, an extra poly(dT) or poly(U) strand in Hoogsteen base-pairing mode, whose polynucleotide chain runs parallel to that of poly(dA) or poly(A) [701]. The reason for disproportionation appears to reside in the extra stabilization of the poly(dA)·poly(dT) duplex (or its ribo analog) by three-center hydrogen bonding and its very wide major groove, vide infra.

Three-center (bifurcated) hydrogen bonding between adjacent base pairs explains unusual properties of poly(dA)·poly(dT). The stability of the B-form of this duplex (and of its ribo analog) formed by two homopolymers might be a consequence of the extreme propeller twist (Box 20.1) reported for the central portions (the A-tracts) of the double helical dodecamer d(CGCAAAAAAGCG) [702] and d(CGCAAATTTGCG) [703]. In both crystal structures the DNA molecules show the same unusual three-center bonds linking adjacent base pairs.

All the base pairs are of the normal Watson-Crick form, but the A–T pairs display systematically larger propeller twists than normal. The average twist of $-14°$ observed for the G–C base pairs is as usually observed, whereas the average twist of $-20°$ for the A–T base pairs is much larger than found in other oligonucleotide structures which do not exhibit the AAATTT or AAAAAA base sequence. As a consequence of the large propeller twist, the N(6) amino group of adenine is positioned midway between the carbonyl O(4) atoms of the base-paired thymine and of the thymine in the adjacent base pair (see Fig. 20.11). This configuration is typical of three-center hydrogen bonds, with the major component of the Watson-Crick interaction and the minor component of the adjacent base pair [702]. The geometry around one of the N(6)–H hydrogen calculated from the C, N atomic coordinates is shown below:

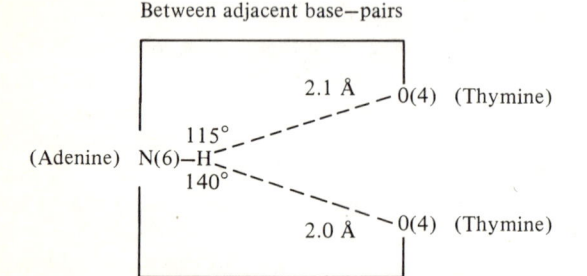

```
              Between adjacent base–pairs

                        2.1 Å      O(4)  (Thymine)
                  115°
(Adenine)  N(6)—H
                  140°
                        2.0 Å      O(4)  (Thymine)

              Watson-Crick base-pair
```

Based on the geometry of the central A-tract, a double helix was constructed. Compared with normal B-DNA, it displays a very narrow, deep minor groove and a wide, shallow major groove and has a pitch of 34 Å with 10.0 base pairs per turn. The three-center hydrogen bonding which links all the base pairs of the double helix in *axial* direction confers extra stabilization to this double helical structure, which explains several experimental findings not understood previously [702, 703].

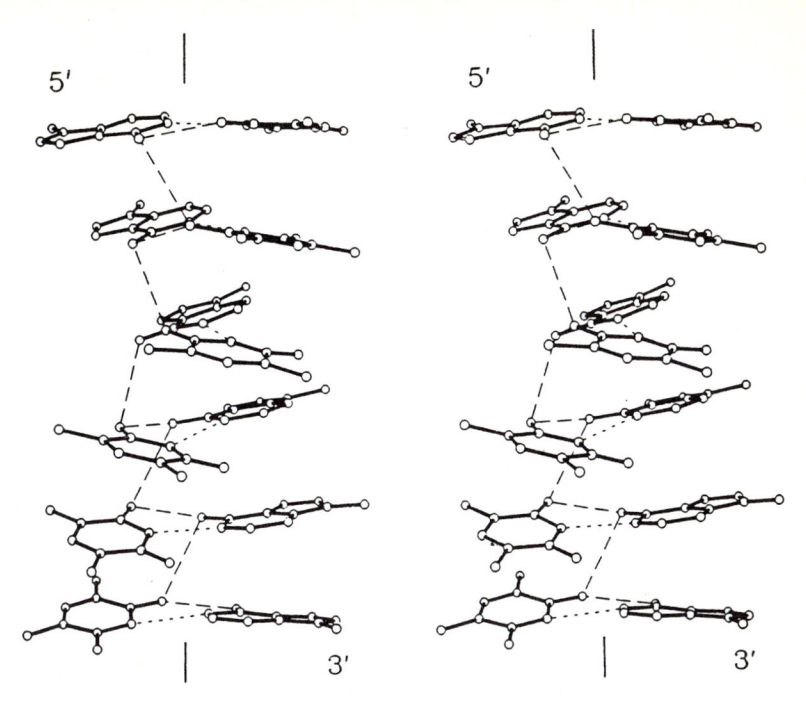

Fig. 20.11. Stereo drawing of d(CGCAAAAAAGCG). The amino N(6)–H groups of the central A-tract form three-center (bifurcated) hydrogen bonds with adenine N(6)–H acting as double donors due to the strong propeller twist of the A–T base pairs. Only bases are shown, sugar-phosphate backbone omitted for clarity [702]

20.6 Specific Protein-Nucleic Acid Recognition Involves Hydrogen Bonding

In molecular biology, the understanding of specific interactions and, notably, those involved in the recognition between nucleic acids and proteins, is of primary interest. Because the systems are complex, crystal structure analyses at good resolution are available for only a few protein-nucleotide, and protein-oligonucleotide complexes, for several DNA-binding proteins and for one tRNA-synthetase complex. Our detailed knowledge of the hydrogen bonding is therefore limited. There have been several model studies where complexes between nucleotides and amino acids or their constituents have been investigated. It is uncertain whether such complexes are really of biological relevance, and they are not considered in this text (for a review [522]).

Three levels of protein-nucleic acid recognition have been observed. Nature provides three examples of protein-nucleic acid interactions which we shall consider. The nucleic acid component can be (1) a single nucleotide, e.g., a coenzyme or a substrate, (2) a single-stranded DNA or RNA as in ribonucleases A and T_1, or (3) a double-stranded DNA or RNA as in the highly specific complexes with repressors in the tRNA-synthetase complex, or in the unspecific nuclease DNase I.

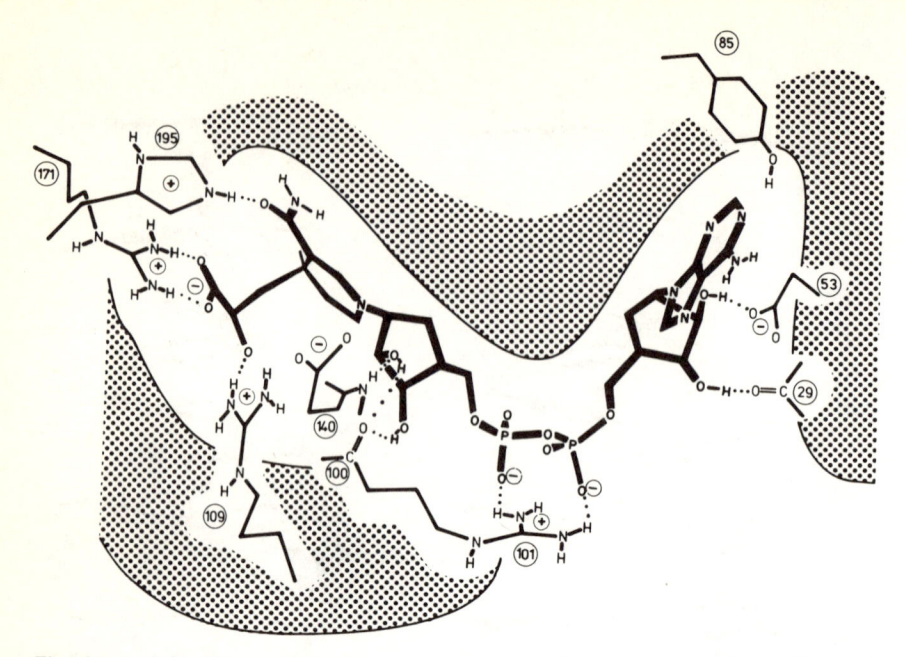

Fig. 20.12. Schematic representation of the binding of nicotin-amide-adenine-dinucleotide (NAD$^+$) to the active site in the ternary complex lactate dehydrogenase-NAD$^+$-pyruvate. Note salt bridges N$^+$–H$\cdots$O$^-$ between Arg101$\cdots$pyrophosphate and Arg171$\cdots$pyruvate [704]

Specific nucleotide recognition in the "ternary" complex between the enzyme lactate dehydrogenase and the modified coenzyme NAD$^+$-pyruvate (NAD$^+$ = nicotinamide-adenine-dinucleotide). NAD$^+$ consists of two nucleotides, adenosine-5'-phosphate and nicotinamide ribose-5'-phosphate, which are linked through a pyrophosphate bond.

The intermolecular contacts between lactate dehydrogenase and NAD$^+$ [704] are shown schematically in Fig. 20.12. There are three types of interactions: (1) in which the adenine heterocycle is held in a "hydrophobic pocket" by van der Waals contacts with a long hydrogen bond (not indicated in Fig. 20.12) to Tyr85, (2) by several salt-type hydrogen bonds between positively charged amino acid side chains (Arg101, Arg171) and negatively charged groups on NAD$^+$ (pyrophosphate and pyruvate carboxylate), (3) by a number of O–H$\cdots$O and N–H$\cdots$O hydrogen bonds between NAD$^+$ and protein main-chain peptide groups (29C=O, 100C=O, 140N–H) and side-chains, all of which are charged: Asp53, Arg109, His195.

In the complex with lactate dehydrogenase and with related enzymes, NAD$^+$ and, in general, the bound nucleotides do not adopt their most preferred conformations described in Box 17.2. Instead they occur in an "open" form with the torsion angle γ in the ap or $-$sc range. Consequently, the base, sugar, and phosphate moiety (attached at 5'-position) are fully exposed and more readily recognized by the respective protein. The energy required for this conformational change of the

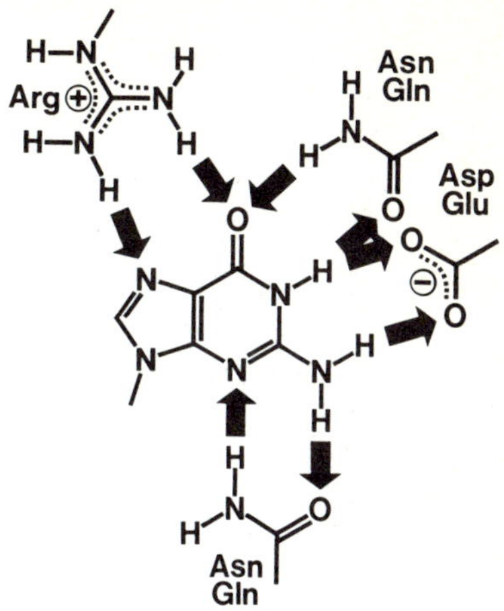

Fig. 20.13. Guanine can be recognized by the side-chains of arginine, aspartic and glutamic acids, asparagine, glutamine, in a variety of combinations. Only those interactions with two (cyclic) hydrogen bonds are drawn

nucleotide is presumably provided by the many hydrogen-bonding and Van der Waals contacts which anchor the nucleotide to the binding site of the protein.

Specific recognition of single-strand nucleic acid has been investigated in detail for the ribonucleases A and T₁ (RNase A and RNase T₁). While RNase A distinguishes pyrimidine nucleotides C and U from the purine nucleotides A and G, RNase T₁ specifically recognizes G.

In RNase A the recognition site is a trough (not a pocket) in which Thr45 is located [705]. Its peptide N–H group forms a hydrogen bond with O(2) of uracil *and* cytosine. The γ-hydroxyl group of this side-chain acts *either* as hydrogen-bond acceptor to recognize N(3)–H of uracil, *or* as donor for N(3) of cytosine. It is intriguing to see that both the main-chain and side-chain groups of the same amino acid are involved in the recognition and binding of the base, and that cyclic hydrogen bonds are formed which are more stable than monodentate interactions.

The situation for RNase T₁ is more complex than for RNase A [594]. There are several ways in which a protein can recognize guanine, using its amino acid side-chains to form cyclic hydrogen bonds, as shown in Fig. 20.13. In fact, only the interaction with glutamic acid glu 46 is used. In addition, the guanine base is specifically recognized by RNase T₁ by a combination of hydrogen-bonding contacts to main-chain N–H groups of the dipeptide Asn43-Asn44 and of the C=O group of Asn98, and by a stacking interaction which intercalates the guanine between the phenolic side-chains of Tyr42 and Tyr45 (Fig. 20.14).

Why are the bases of RNA not recognized by the side-chains of Asp, Glu, Asn, Gln, Arg, which are so suitable for hydrogen-bond formation? It seems that, on one hand, the end groups of these side-chains are too rigid and, on the other hand, their link to the main-chain skeleton through one, two, or three CH_2 groups is too flexible. Proper recognition of the correct base and discrimination of the incorrect bases therefore would require major conformational adjustments, which are too slow. It appears more favorable to mold the main-chain into a form suitable for

414   The Role of Hydrogen Bonding in the Structure and Function of the Nucleic Acids

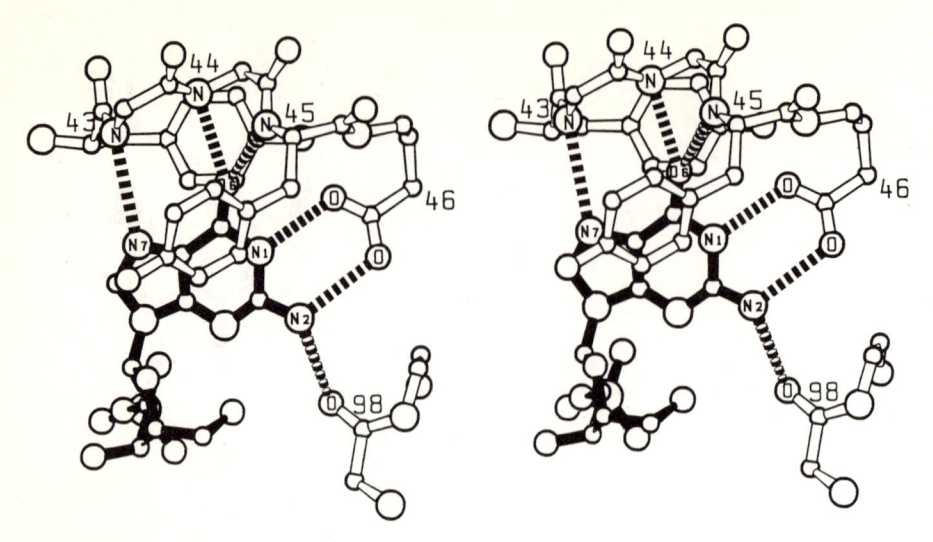

Fig. 20.14. Stereo drawing of a section of ribonuclease T$_1$. Guanine is specifically recognized by hydrogen bonding to main-chain and side-chain groups, and by intercalation of the guanine C(6)=O(6) carbonyl group between side-chains of Tyr42 and Tyr45. The hydrogen bonds are (guanine)N(1)–H and N(2)–H to the side chain carboxylate of Glu46, (guanine)N(2)–H to Asn98C=O, Asn44N–H and Tyr45N–H to guanine) O(6), Asn43N–H to guanine) N(7) [594]

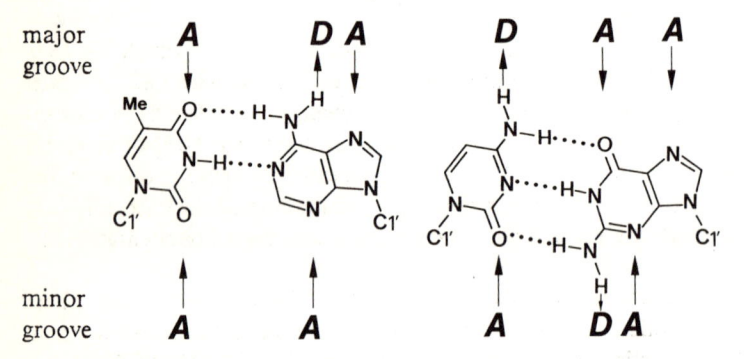

Fig. 20.15. Discrimination of base pairs A–T, T–A (*left*); G–C, C–G (*right*) by hydrogen-bonding interactions indicated by *D* (donor) and *A* (acceptor) can occur in the major (*wide*) groove, but not in the minor (*small*) groove, where only *G* can be discriminated from the other bases due to the donor function of its 2-amino group. After [706]

base recognition. This provides a well-defined framework for rapid checking of the correct base through complementary hydrogen-bonding patterns.

Specific recognition between proteins and nucleic acid double helices is most favorable in the major groove of the double helix. Unless the RNA and DNA double helices can be distinguished by their different backbone geometries (Box 20.1), the sugar-phosphate backbone can provide only an unspecific target for hydrogen bonding to proteins. For more specific recognition, the functional groups of the bases lining the major and minor grooves are better suited, since they reflect the nucleic acid sequence (Fig. 20.15). Model manipulations have shown that in the minor groove, hydrogen bonding can distinguish only guanine from the other bases because pyrimidine O(2) and purine N(3) are located in structurally isomorphous positions. In contrast, all possible base-pair combinations can be recognized in the major groove [706], which therefore provides the prime target for specific interactions with proteins.

Crystal structures of repressors suggest a common helix-turn-helix motif for DNA binding. Repressors are protein molecules which bind to and block specific nucleotide sequences in DNA (operators), thereby regulating gene expression. Some of the repressors act without co-factors, others need a co-repressor to bind to DNA, like the tryptophan (*trp*) repressor, or they do not bind to the DNA operator in presence of a co-repressor like tetracyclin (TET) repressor.

Several crystal structure analyses of individual repressors and of repressor-operator complexes disclosed a recurrent template on the protein side. It consists of an α-helix/turn/α-helix motif with an almost invariant glycine in the turn region [707–710]. Most of the repressors so far investigated display such a motif but the interactions with the operator DNA, which occur in the major groove, are individually different. Here we consider more closely the well-documented complexes between the N-terminal fragment of the repressor from phage 434 and its specific operator DNA [711], and the ternary complex formed between tryptophan repressor, tryptophan and DNA [712, 713].

Specific binding of 434 repressor distorts the operator DNA. Using limited proteolysis, the 434 repressor can be cleaved into two domains, one necessary for dimer formation in solution, and the other for specific DNA binding. The latter, comprising the 69 N-terminal amino acids, was crystallized with a 20 base pairs long DNA fragment. It had approximate 2-fold rotationally symmetric (palindromic) nucleotide sequence corresponding to the "consensus" sequence derived from comparison of the six naturally occurring operators to which 434 repressor binds:

```
       -3L  -2L  -1L  1L   2L   3L   4L   5L   6L   7L   7'R  6'R  5'R  4'R  3'R  2'R  1'R  -1'R -2'R -3'R
    5'  T  - A  - T  - A - C  - A  - A  - G  - A  - A  - A  - G  - T  - T  - T  - G  - T  - A  - C  - T      3'
    3'  T  - A  - T  - G - T  - T  - C  - T  - T  - T  - C  - A  - A  - A  - C  - A  - T  - G  - A  - A  5'
       -2'L -1'L 1'L  2'1  3'L  4'L  5'L  6'L  7'L  7R   6R   5R   4R   3R   2R   1R   -1R  -2R  -3R  -4R
```

In the crystal structure [711], a dimer formed by two N-terminal 434 repressor fragments is bound to the 20 base pairs DNA duplex so that the complex has overall 2-fold rotational symmetry. The polypeptide chain is folded into five α-helices H1 to H5, with helices H2 and H3 forming the helix/turn/helix motif (Fig. 20.16). Helices H3 and H3' of the 434 repressor dimer insert into two successive major grooves of the operator DNA whereas the N-termini of the flanking helices H2, H4 and H2', H4' contact the sugar-phosphate backbones.

The crystal structure of the 434 repressor fragment was also determined without DNA. There are only minor structural changes between free and DNA-bound repressor fragment which mainly involve rotation of three amino acid side chains interacting specifically with the DNA duplex. However, the DNA is bent like a "C" on an arc of 65 Å radius (Fig. 20.16). This compresses the minor groove

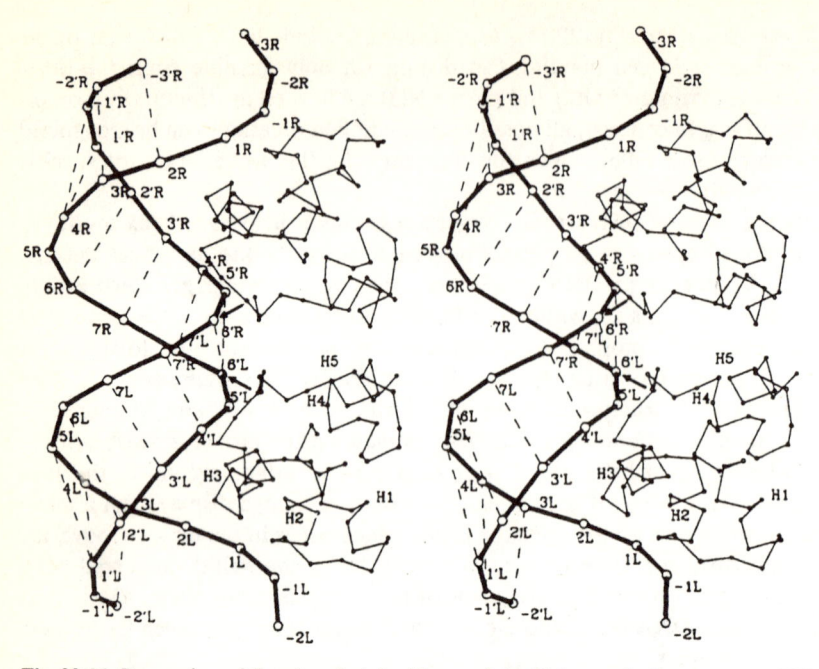

Fig. 20.16. Stereo view of the phosphate backbone of the 20 base-pairs long operator DNA fragment in complex with the α-carbon backbone of the dimeric 434 repressor fragment. The dashed lines connect phosphates in the minor groove, which is compressed in the center. H1 to H5 denote helices 1 to 5, arrows point at close protein-DNA contacts (from [711])

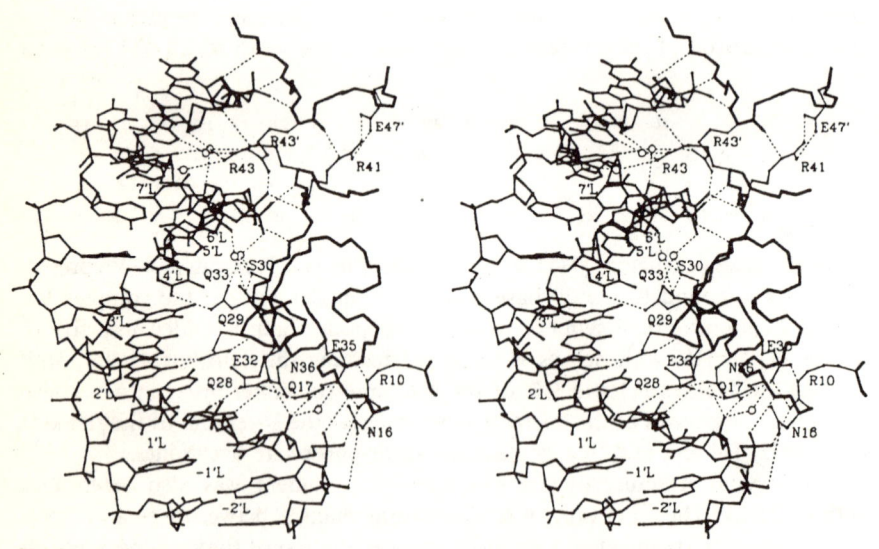

Fig. 20.17. Stereo view showing the specific recognition between 434 repressor fragment and operator DNA. Dotted lines are hydrogen bonds with X···A distances <3.5 Å, open circles are water molecules (from [711]).

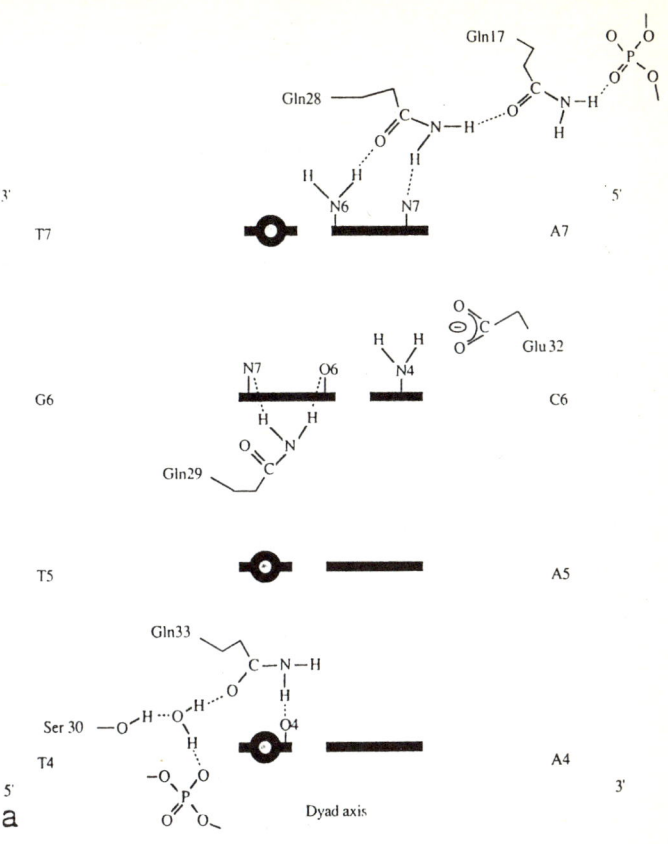

Fig. 20.18a, b. Schematic representation of base-specific interactions between one-half operator and (**a**) the 434 repressor fragment, (**b**) the *trp* repressor. The nucleotide numbering in (**a**) is different from Figs. 20.16, 20.17; numbers 7 to 4 should be replaced by 1'L to 4'L (*left*) and by 1L to 4L (*right*). In (**b**), there is only one direct Arg69-G9 interaction, all others are mediated by water molecules. Bases indicated by solid lines, methyl groups by circles. From [710]

by about 2.7 Å and as a consequence, the propeller twist increases so that three-center base-pair hydrogen bonds are formed, similar to those described in Fig. 20.11. These involve as donors the amino group of adenine AL7, [(A7L)N6-H···O4(T7'L), major component); O2(T7R, minor component)]; the N3-H group of thymine T6'L [(T6'L)N3-H···N1(A6L), major component); N1(A7L, minor component)]; the N2 amino group of G5L, [(G5L)N2-H···O2(C5'L), major component); 02(T4'L, minor component)].

This distortion of the DNA duplex facilitates formation of several hydrogen bonds and van der Waals contacts with the repressor protein (see Fig. 20.17). Direct recognition is through hydrogen bonds between 434 repressor fragment and the major groove base-pair atoms as indicated in Fig. 20.18a. There are additional direct hydrogen bonds to phosphate oxygens, and several tightly bonded water molecules serve as mediators.

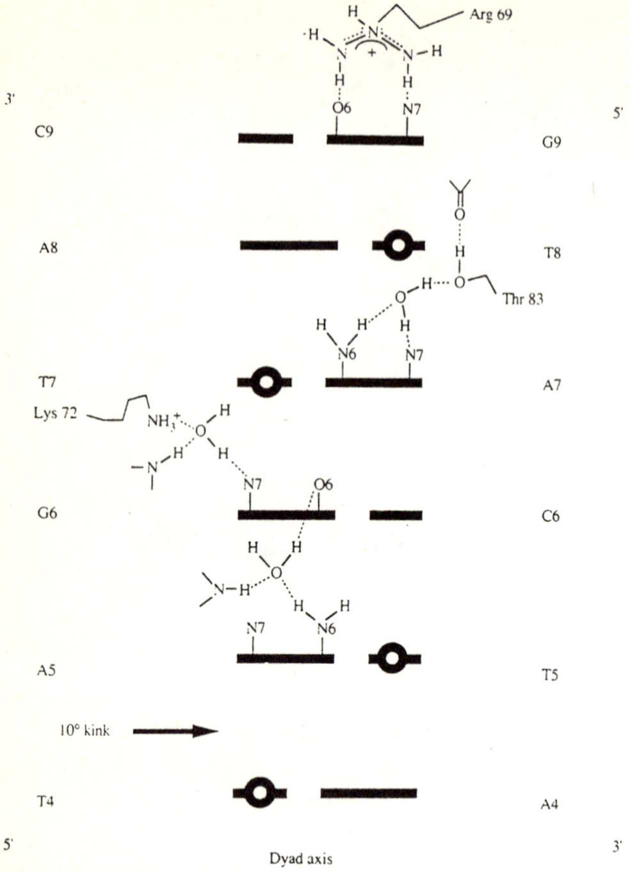

Fig. 20.18b

Conformational changes of *trp* (tryptophane) repressor and water-mediated contacts to DNA. *trp* Repressor is a 108 amino acids long polypeptide folded into 6 α-helices A to F. It was investigated by high resolution X-ray diffraction studies as *apo* repressor, as binary complex with its corepressor tryptophan, and as ternary complex bound to its specific operator DNA, a palindromic self-complementary 18-mer with overhanging T [712, 713]:

```
5'  T - G - T - A - C - T - A - G - T - T - A - A - C - T - A - G - T - A - C      3'
3'      C - A - T - G - A - T - C - A - A - T - T - G - A - T - C - A - T - G - T  5'
        9   8   7   6   5   4   3   2   1 - 1 - 2 - 3 - 4 - 5 - 6 - 7 - 8 - 9
```

In all these crystal structures, *trp* repressor occurs as a dimer with intertwined subunits related by a twofold rotation axis. When tryptophan binds to the *apo* repressor protein, a conformational change is induced through reorientation of several hydrogen bonds (Fig. 20.19). It rotates the helix/turn/helix motif such that

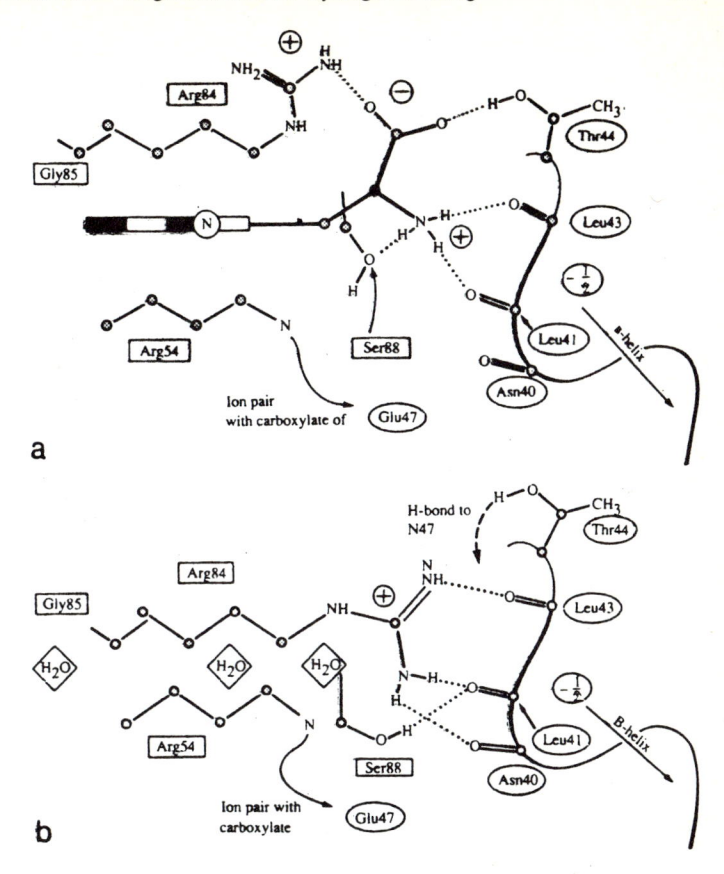

Fig. 20.19a, b. Schematic drawing describing changes in hydrogen-bonding pattern when tryptophan is (**a**) bound to *trp* repressor, and (**b**) removed and replaced by water molecules. From [712]

helices E and E' of the dimer can penetrate into adjacent major grooves of the operator DNA, associated with a 10^3 fold increase in binding affinity (Fig. 20.20).

Surprisingly, of the 14 direct hydrogen bonds between each *trp* repressor and the half-operator to which it binds, twelve are unspecific to non-ester oxygen atoms, and form salt bridges involving arginines. Only two are specific, between Arg69 and G9 at the "end" of the operator (Fig. 20.18b). As mutation experiments have disclosed, even these are not crucial for recognition.

In addition to these repressor-operator interactions, there are six water-mediated hydrogen bonds, three bridging to the major groove functional groups of A5, G6, A7 (Fig. 20.18b), and the other three to five phosphate oxygens. The water molecules can only be accommodated with G at position 6 and a purine at position 5, and a purine (adenine) at position 7. Recent studies have cast some doubts on the specificity of these interactions, however, and further experiments are necessary to show that this ternary complex is really of biological relevance [714].

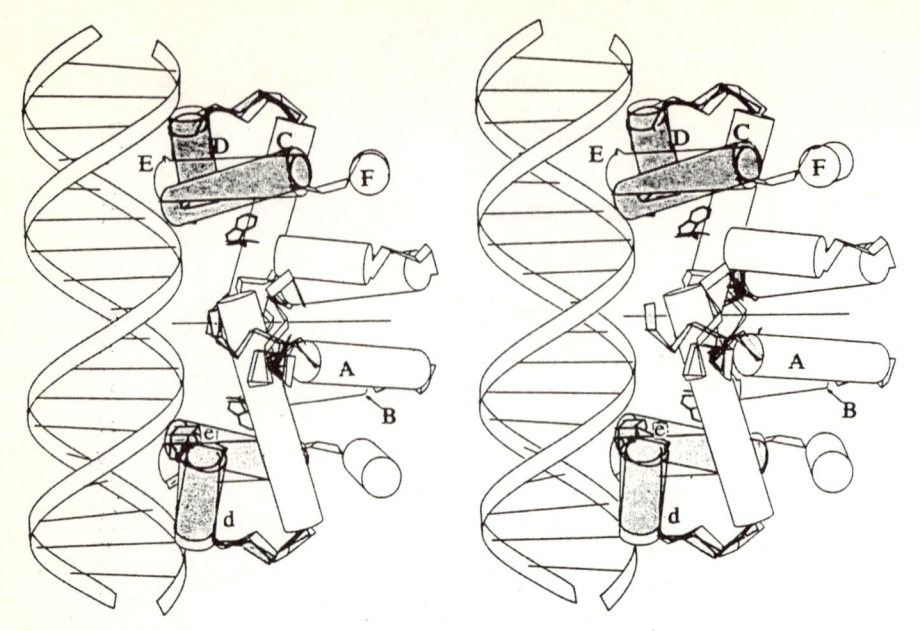

Fig. 20.20. Stereodiagram showing how the structure of the *trp* aporepressor (*shaded cylinders*) changes after binding to tryptophan and to operator DNA (*open cylinders*). From [712]

Opening of a base pair in tRNA^{Gln} accompanies complexation with its cognate glutaminyl-tRNA synthetase [715]. All the tRNAs have an L-shaped three-dimensional structure as shown in Fig. 20.8. At one end of the L is the anticodon triplet; it codes for the amino acid for which the respective tRNA is specific. At the other end of the L is the amino acid acceptor stem with the free O(3')H of the terminal invariant adenosine, A76 in tRNA^{Gln}. The amino acid is attached here through an ester linkage in a reaction which is catalyzed by enzymes called synthetases. In contrast to the uniform tRNA structure, synthetases vary in amino acid sequence, molecular weight and subunit composition.

The ternary complex between glutamine-specific E. coli tRNA^{Gln}, ATP and glutaminyl-tRNA-synthetase was crystallized and its structure determined at 2.5 Å resolution [715] (see Fig. 20.21). Compared with uncomplexed yeast tRNA^{Phe}, the sugar-phosphate backbone of tRNA^{Gln} in the complex shows several changes. They are most significant at the amino acid stem carrying the 3'-terminal adenosine to which glutamine is attached in the enzymatic reaction. The changes involve disruption of the terminal A72-U1 Watson-Crick base pair by insertion of the side chain of Leu136 (Fig. 20.22); the C74-C75-A76 terminus does not extend helically as in free tRNA (Fig. 20.8) but is folded back, with looped-out cytosine 74 interacting with a pocket in the protein; and the N2 amino group of G73 hydrogen bonds to the phosphate of A72. Two base pairs in the amino acid stem are directly recognized by hydrogen bonds in the minor groove side, with (G2)N2H···O=C Pro181 and (G3)N2H···O$_\varepsilon$ Asp235. This is supported by water-mediated hydrogen bonds between these two base pairs, the backbone pep-

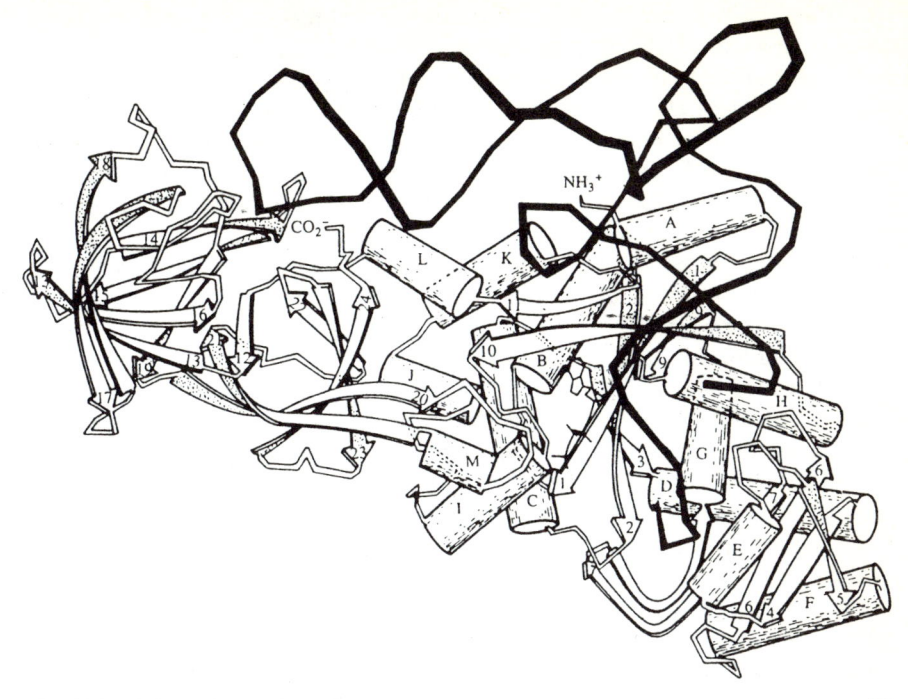

Fig. 20.21. Schematic drawing of the glutaminyl-tRNA-synthetase in complex with tRNA[Gln] (thick black line) and with ATP. From [715]

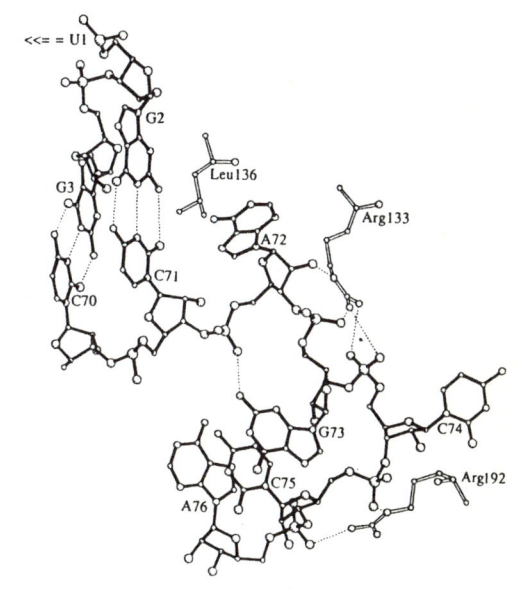

Fig. 20.22. Detail of Fig. 20.21, showing the amino acid stem of tRNA[Gln]. The side-chain Leu136 opens the U1-A72 base pair, and the structure of the C74-C75-A76 loop is stabilized by several hydrogen bonds. From [715]

tide of residue 183, and the Asp235 carboxylate. Another site for specific interactions is between the tRNA^{Gln} anticodon C34-U35-G36 and charged amino acid side chains of the synthetase. The hydrogen bonds are between two Arg and C34; Arg and Glu with U35; Arg with G36 [715].

Nonspecific protein-DNA interactions occur in the minor groove. Deoxyribonuclease I (DNase I) is an example. DNase I from bovine pancreas is a monomeric enzyme consisting of 257 amino acid residues. It acts like endonuclease on double-stranded DNA (much slower on single-stranded DNA), and hydrolyzes the P−O(3′) bond. It requires Ca^{2+} and Mg^{2+} or Ca^{2+} and Mn^{2+} for optimum activity. The cutting sites are neither base- nor sequence-specific, although the cleavage patterns and frequency of cutting sites suggest that DNase I recognizes certain variations in backbone geometry.

DNase I interacts with about one complete turn in the minor groove side of the DNA double helix [716, 717]. The contacts are predominantly between positively charged arginine and lysine side-chains of the enzyme and negatively charged phosphate oxygen atoms of the DNA, and they are also mediated by Ca^{2+}. In more detail, one strand of the DNA double helix interacts in two different locations about one turn away with DNase I, involving salt bridges between the side chains of Lys74, Arg38, and phosphate oxygens, and van der Waals-type interactions with Cys170, Thr174. The side-chain OH of Tyr73 forms a hydrogen bond with pyrimidine O(2) or purine N(3), which are in equivalent positions in the minor groove. Contacts with the other strand are shifted by about half a turn, i.e., five base pairs, and involve interactions between charged side-chains of Arg70, Arg108, Arg9 and phosphate oxygens. In addition, Ca^{2+} is in contact with a phosphate oxygen on DNA and, on the enzyme, with the side-chain of Gln36.

Part IV
Hydrogen Bonding by the Water Molecule

Chapter 21
Hydrogen-Bonding Patterns in Water, Ices, the Hydrate Inclusion Compounds, and the Hydrate Layer Structures

Water is an essential component of life processes. There are no plant or animal species that can exist in the complete absence of water[1]. This observation is apparent at both the macroscopic and atomic level. At the atomic level, proteins and nucleic acids lose their characteristic three-dimensional architecture and their biological function when water is removed. There is no question that the ubiquitous water molecule has an essential place in biological processes. From the hydrogen-bonding point of view, *the water molecule is unique* in having both double-donor and double-acceptor hydrogen-bond functionality, which is about one central atom, the oxygen. It is possible therefore for water molecules to exercise a wide range of orientational flexibility without substantial loss of hydrogen-bonding energy. It is this orientational freedom that makes it difficult to define the hydrogen-bond structure of an assembly of water molecules in the absence of precise information concerning the location of the hydrogen atoms and the dynamics of their motion.

The structure of water and the ices have been a subject of intense research since the last century and the literature on the subject is voluminous [31, 32, 423]. It is surprising, however, that the classical work on the high-pressure ices and the clathrate hydrates has not been carried to more precise conclusions with regard to the hydrogen atom positions now that the experimental tools are more readily available to do so.

In this chapter, we discuss the hydrogen bonding in structures where water is the sole or majority molecular species present. These are structures which are determined wholly or primarily by the hydrogen-bonding characteristics of the water molecules. They demonstrate the consequences of the dual hydrogen-bond donor-acceptor functionality, which, when combined with the cooperative and flip-flop dynamic properties of the hydroxyl groups, are essential for the formation of the hydration shells around the proteins and nucleic acids, and help to maintain their three-dimensional structures.

The dual hydrogen-bond functionality of water determines most of its physical properties. The structural flexibility that goes with this property is responsible, at least in part, for the difficulties associated with deriving a satisfactory structural

---

[1] An interesting example of anhydrobiosis is a trandgrade which appears to utilize the disaccharide trehalose for water regeneration [718].

and thermodynamic description of liquid water. For this reason, we will make some comments relating to the hydrogen bonding in liquid water, for which there is already a plentiful, if not conceptually very illuminating, literature.

In Chapter 22, we discuss the hydrogen bonding of the water molecules in the hydrates of some of the small biological molecules treated in Part II, Chapters 13 to 19, i.e., the carbohydrates, purines and pyrimidines, and nucleosides and nucleotides. These are crystal structures which are determined mainly by the packing forces and hydrogen bonding between the functional groups on the organic molecules. The water molecules play only a secondary role, occupying space between the molecules and adding to the hydrogen-bond energy of the crystal lattice, in competition with the molecular packing in the absence of water. It is notable that the disaccharides and nucleotides, which are awkward-shaped molecules from the packing point of view, form many more hydrates than do the spheroidal monosaccharides, which pack efficiently with optimum hydrogen bonding in the absence of water molecules. Similarly, the sugar alcohols, which pack most efficiently as straight or bent rods, form no hydrates, although they are generally very soluble in water.

A reason that hydrated crystals are so plentiful in the Cambridge Crystallographic Data Base is that water is the solvent of choice for crystallization, just as it is the solvent of choice for biological processes.

In the subsequent Chapters 23 and 24, we discuss what is known about the hydrogen bonding of water in proteins and nucleic acids and the dynamics of these processes. If biological molecules aggregate, or if a substrate enters the active site of an enzyme, water molecules have to move from the contact surface of the biological molecule in a coordinated and cooperative manner with the least expenditure of energy. Hydrogen bonding must play an important role in this substitution process, and the dynamic properties of both two- and three-center hydrogen bonds must be involved, as discussed in Chapter 25.

21.1 Liquid Water and the Ices

While there is general agreement that the structure of water is determined by hydrogen bonding, it has proved difficult to describe. The studies of liquid water by diffraction and spectroscopic and thermodynamic methods provide the basis for the broadly accepted views that the water molecules are hydrogen-bonded into a dynamic time-dependent nonperiodic assembly with a first nearest neighbor coordination of the water molecules which, on average, is fourfold and tetrahedral. The increase in density of water over that of ice is then achieved, as in the denser ice structures, by having closer second, third, and higher nearest neighbor distances that in ordinary ice [422].

To go beyond this description raises the conceptual question of what is meant by the term "structure". In solids, this refers to a particular topological arrangement, or configuration of atoms, which persists over the period of time necessary to observe it. In liquids, a particular topology exists for very short lifetimes, which can be observed only by methods which record the structures at even shorter times.

Liquid Water and the Ices

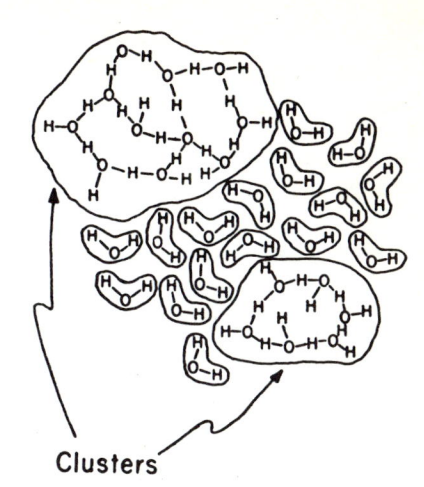

Fig. 21.1. The Frank model of water clusters in liquid water. In the clusters the water oxygens are three- and four-coordinated, while those outside the clusters are nonbonded monomers [102]

These are, with decreasing resolution on the time scale, the methods of infrared and Raman spectroscopy, dielectrical absorption and NMR spectroscopy, and inelastic neutron scattering. Thermodynamic properties depend on the long lifetime average structures, which are observed by X-ray and neutron scattering. For each of these methods, the descriptor "structure" has a different meaning.

Essentially two models have been developed for the "structure" of liquid water [719]. One is the uniform continuum models originally proposed by Bernal and Fowler [15][2], Lennard-Jones and Pople [720] and Pople [277]. In this model, all the oxygen atoms retain their four-coordination, but the hydrogen bonds can bend to such a degree that an "instantaneous" view from a central oxygen would perceive no order beyond the first nearest neighbor. These are four-coordinated random network models with very "soft" hydrogen bonds. They imply a hydrogen-bond-connected network with apex-linked polygons with four, five, six or more hydrogen bonds, as in the ices and high hydrates, but randomly arranged with no periodicity [721].

The alternative are "mixture" models. These invoke "clumps" or "clusters" of three- or four-coordinated water molecules with a mean lifetime of about a nanosecond [102, 722, 723]. These clusters are separated by boundaries in which the oxygen atoms accept only one, or less frequently, no hydrogen bonds, while other water molecules have "dangling" hydrogen bonds with no acceptor atoms, such as illustrated in Fig. 21.1. Such models imply that the hydrogen bonds do not bend so easily, and there is a significant energetic advantage associated with hydrogen-bond linearity. The evidence favoring this type of model comes mainly from thermodynamic and spectroscopic data [724]. For example, one interpretation of the Raman intensity measurements employs a model in which the hydrogen bonds "break" in a step-wise fashion. On a time scale of 10^{-12} s, both polymer

---

[2] The descriptor "hydrogen bond" is not used in this paper.

and monomeric water molecules would be observed, according to this model [725]. On the other hand, some spectroscopists favor the uniform continuum model [726].

While the thermodynamic evidence may favor the mixture models, the diffraction studies from the "static" crystalline state tend to support the continuum model. The water molecules in the ices and high hydrates are always four-coordinated.

In the crystal structures of the small molecule hydrates, discussed in Chapter 22, in which the water molecules are the minority species, the hydrogen bonds are seldom linear, and the O−Ĥ···O angles have a broad distribution between 90° and 180°. On the other hand, three-coordinated water molecules are more common than four-coordinated molecules in the majority of low hydrates, and small molecule studies indicate that more thought should have been given to including the three-center bond in water structure models. It has also been suggested that features of the Raman spectra can be better interpreted by considering this type of bond [306], and models for the hydrogen bonding in water, amorphous ice and alcohols involving three-center bonds have been proposed [726a, 726b].

More "crystallographic" variations of the water models are the vacant lattice, framework or cage models which include interstitial water molecules within a four-connected network. In one of these models, the void in the ice-like structure is occupied by a water molecule which at any particular instant may be nonbonded [727]. A similar model [728, 729] was proposed with an interstitial water molecule in the clathrate hydrate water structure. Both these models are now not seriously considered by the specialists in the field [730] because they are "too crystalline."

However, an instantaneous static view of the structure of an assemblage of water molecules in liquid water might reveal in particular localities examples of hydrogen bonding that correspond to all these different static models. When the concepts of three-center bonds, cooperativity of hydrogen bonding, cyclic structures and flip-flop dynamics are added to the picture, it is perhaps not surprising that attempts to describe the structure and thermodynamic properties of liquid water in terms of a single conceptual model have proved so elusive [731−733]. It is interesting to note that the concept of cooperativity, which is a very important factor in the hydrogen-bonded structure of polyhydroxy compounds such as the carbohydrates and cyclodextrins, is implicated in Frank's "flickering cluster" theory of liquid water [101, 330].

With the exception of the clathrate framework model, all these hypotheses appear to be qualitatively consistent with the available X-ray diffraction data on liquid water. It is argued by some investigators, however, that there are still significant inconsistencies between the most sophisticated statistical thermodynamic models for liquid water and the most sophisticated X-ray and neutron diffraction measurements [734−736]. The interpretation of these data from different experiments, using the concept of pair-correlation functions, shows discrepancies that are considered significant in terms of the instrumental precision, and the definitive answer seems not yet available [737].

The pair-correlation function curves from these experiments show peaks corresponding to the covalent O−H bond length of 1.00 Å, the hydrogen bond

H···O distance of 1.85 Å, and the nonbonded O···O distance of 2.85 Å, with approximately fourfold first-neighbor coordination. The H···O and O···O distances are ~0.1 Å longer than the ice values and ~0.1 Å shorter than the gas-phase water dimer values. This is qualitatively consistent with the concept that the liquid lies between the gas and the solid.

Amorphous ice has been studied in some detail by both X-ray and neutron diffraction [738–740]. The O···O pair-correlation functions are similar to those of liquid water, except that on condensing on very cold surfaces, i.e., 10 K, there is an extra sharp peak at 3.3 Å. This indicates some interpenetration of the tetrahedral disordered ice-like short-range structures. It appears that none of the many proposed atom-atom potential energy functions can simulate a structure for liquid water that predicts pair-correlation functions which are a satisfactory fit to the experimental data [741, 742]. Opinions seem to differ as to whether the discrepancy is in the theory or the experiments.

The concept of the "excluded region", which is useful for interpreting hydrogen bonding in macromolecular structures, applies also to liquid water. Attempts to interpret the hydrogen bonding in the hydrates of some large biological molecules [126, 127, 743] led to the concept of the excluded region, mentioned in Part IA, Chapter 2.10. This concept states that in an assemblage of water molecules, their orientation, which determines the position of the hydrogen atoms, is circumscribed by the repulsive interactions between the hydrogens and their first neighbor oxygens and other hydrogens. Their influence on the permissible hydrogen-bond geometries is shown by the plots of $O_W-H\cdots O$ angles versus $H\cdots O$ distances in Fig. 2.10.

These repulsive interactions are well defined with respect to structure because they arise from the repulsive component of the potential energy function, which has a large sensitivity of energy to distance. They will, of course, be included in all quantum mechanical calculations of water structure and all classical mechanical methods that make use of potential energy functions. This concept promises to be especially useful for limiting the conformational possibilities for hydrogen-bonded structures in the biological macromolecules, where the positions of the hydrogen atoms cannot be located experimentally, and the molecules are too complex for reliable conformational analysis computations (see Part IA, Chap. 4).

Polywater: a Nondiscovery [744]. The investigation of a reported form of water with a maximum density of 1.4 g cm^{-3}, which solidified below $-30\,°C$ to a solid which was not ice, lasted for 5 years, 1966 to 1971. Fueled by the popular press and a popular science fiction novel [745], it produced some of the most unscientific papers ever published in scientific journals by reputable investigators, except perhaps for phlogiston in the last century, when there was more excuse.

The polymorphic forms of ice illustrate the structural variety possible for the hydrogen-bonded polymers of four-coordinated water. Topologically, the ices belong to the same class of 4:2 framework structures as the silicates [746]. It is not surprising, therefore, that each of the ice polymorphs has its silica analog [747]. However, the hydrogen-bond $-O-H\cdots O-$ framework is weaker than the covalent $-\overset{|}{\underset{|}{Si}}-O-\overset{|}{\underset{|}{Si}}-$ framework. The structures based on the face- or edge-sharing of tetrahedra or octahedra, which involve close Si···Si nonbonding distances, cannot be sustained in the ices. The more open clathrate hydrate structures [748] do have their counterparts in the silicates, known as clathrasils [749].

Because of the analogy between $(H_2O)_n$ and $(O_2Si)_n$, the problems associated with the structural interpretation of liquid water have their "glassy" state analogies in the theories of glass structure. Even Pauling's clathrate hydrate model for liquid water has its counterpart in the "vitron" theory of glass [750]. Both theories are discredited by the experts in the field.

The atmospheric pressure ices, Ih and Ic, have regular structures with equal hydrogen-bond lengths and tetrahedral, or almost tetrahedral, O···Ô···O angles (isostructural with the Wurtzite and Zinc Blende, ZnS structures). In the high pressure ices, these regular lattices are distorted with respect to both hydrogen-bond lengths and angles, which are spread over a broad range of values, as shown in Fig. 21.2 [422]. With the very high pressure ices VI and VII, symmetrical struc-

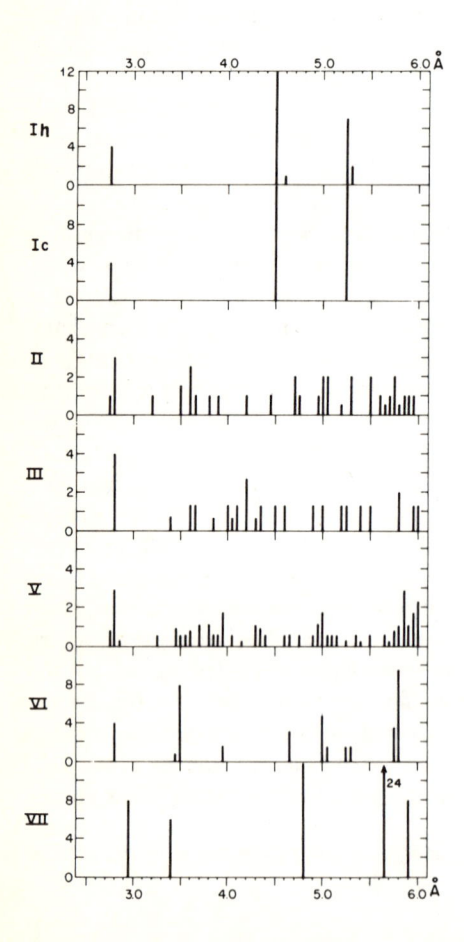

Fig. 21.2. Oxygen-oxygen interatomic distances in the ice polymorphs. Note the more symmetrical structures for Ih, Ic, VI and VII [422]

tures are again possible, due to the formation of interpenetrating lattices (self-clathrates) in which the nonbonded O···O separation is shorter than the hydrogen-bonded O−H···O separation. An interesting and noteworthy feature of the ice structures, in relation to the water structure in the hydration of the cyclodextrins, proteins and nucleic acids discussed in Part IV, is the occurrence of 4-, 5-, 6-, 7- and 8-membered cycles of hydrogen bonds [751].

In hexagonal ice, Ih, the six-membered rings of hydrogen bonds have the boat conformation, whereas in ice Ic, they have the chair conformation. In both structures, the protons are statistically disordered, giving rise to the residual entropy at 0 K. At low temperatures the protons are "frozen" into their disordered state. For ice Ih, a small thermal anomaly relating to the activation of thermal motion is reported at 100 K [752]. Similar "glass transitions" have been observed for ice Ic [753] and the other ices with disordered protons in the hydrogen bonds. It was first pointed out by Pauling that the number of possible configurations for the hydrogen atoms in ice Ih is approximately $(3/2)^n$, where n is the total number of water molecules in the crystal domain [115]. The **best** exact value is $(1.50685 \pm 0.00015)^n$ [116], resulting in the zero point entropy $S_0 = R\ln(1.50685) = 3.410$, where R is the universal gas constant. It is interesting to note that this model requires that ice is nonpolar. In the presence of point defects, a model in which the disorder of the hydrogen atoms takes place by tunneling across the O···O line is considered more probable than one involving the reorientation of the water molecules about the oxygen atom. In the absence of point defects in the crystals, this is an open question. There is an interesting hypothesis [754] that an antifreeze protein found in arctic fish which consists of a single α-helix inhibits ice formation by reducing the proton disorder in ice through a dipole-dipole interaction.

The hydrogen bonds are disordered in all the ices except II and VIII. As the lattices become less regular, the hydrogen-bond lengths tend to increase from 1.78 Å in ice I to 1.82 Å − 1.84 Å in ice VI. In the interpenetrating ice lattice, VIII, with a density of 1.5 g cm^{-3}, the hydrogen-bond length is 1.9 Å and the bonded O···O separation of 2.88 Å is longer than the nonbonded separation of 2.74 Å.

The determination of the precise crystal structure of ordinary ice, ice Ih, has been shown to be more complex than originally thought. The results of the first neutron diffraction single crystal study of D$_2$O Ih by Peterson and Levy [755] were puzzling relative to the molecular structure of water in two respects, i.e., a D−Ô−D angle of 109.5° versus 104.5°−104.8° in the gas phase, and an O−D covalent bond length of 1.015 Å versus values of 0.958−0.965 Å from gas-phase spectroscopy [756] and from the crystallographic data from the other forms of ice and the many hydrated structures that have been studied. Models involving thermal motion and disorder of the hydrogen atoms did not account for this discrepancy [757], which became regarded as a peculiarity of the ice structure. However, the most recent high-precision single-crystal neutron diffraction study of both H$_2$O and D$_2$O at 60 K, 123 K, and 223 K, using all the state-of-the-art corrections for absorption, extinction, and diffuse scattering, has shown that the determination of the precise structure of hexagonal ice is far from simple [424, 425, 758]. The result indicates that the only models which fit the experimental data require disorder of the oxygen atoms about their averaged crystallographic sites in the space group (P6$_3$/mcc). Unfortunately, there are three possible models, according to whether the oxygen atom is moved into a tetrahedral, octahedral, or trigonal space-averaged displacement pattern. With displacements of the oxygens from their mean crystallographic site of 0.06 Å, values for the O−D bond lengths and O−D̂−O angles were obtained which are more consistent with the gas-phase values and those observed in the other forms of ice.

It appears that the precise crystal structure of ordinary ice still eludes both the experimentalist and the theoretician who seek to explain the difference in bond length from that of the water dimer without the necessity for disorder [759].

21.2 The Hydrate Inclusion Compounds

"It is a pleasing thought for a crystallographer that when it snows on some distant planet, it might snow clathrate hydrates" [760, 761].

The hydrate inclusion compounds are crystalline material in which the water molecules are hydrogen-bonded to form regular, periodical three-dimensional four-connected arrangements of oxygen atoms which, like the ices, exist only in the solid state [432, 762]. Clearly they could occur as amorphous solids, analogous to amorphous ice, but hitherto this form has not been detected. All that is known about the structure of the hydrate inclusion compounds is based on X-ray crystal structure analyses [434] and NMR spectroscopy [762]. There have been only two published neutron crystal diffraction studies, those of ethylene oxide deuterohydrate [431] and of $3.5\,\text{Xe} \cdot 8\,\text{CCl}_4 \cdot 136\,\text{D}_2\text{O}$ [431 a].

At present, four main types of hydrate inclusion compounds are known: the gas hydrates, the alkylamine hydrates, the quaternary ammonium salt hydrates, and the polyhydrates of some strong acids. These compounds were all discovered separately in 1811 [763, 764], 1893 [765], 1940 [766], and most recently in 1987 [767]. The physical properties of the gas hydrates were studied extensively in the late 19th and early 20th century [768]. The structures, and raison d'être, of the gas hydrates were not discovered until 1951 [769−772] following Powell's discovery of the phenomenon of molecular clathration in 1947 [773]. The structural relationship between these hydrate inclusion compounds was revealed only by X-ray crystal structure analysis some 8 years later [434, 774]. The most recent additions to this class are some hydrates of the strong acids HAsF$_6$, HPF$_6$, HBF$_4$ and HClO$_4$ [767].

Large cages in ice-like structures are stabilized by hydrophobic guest molecules. As in the ices, the hydrogen-bonded water molecules in the clathrate hydrates are four-coordinated in a distorted tetrahedral arrangement. In this respect, these structures are correctly described as being "ice-like". They differ in having larger internal voids or cages. The void in hexagonal ice Ih and the smallest cage observed in the hydrate inclusion compounds are comparable in size, but in general, all the other cages in the inclusion hydrates are larger than the voids in the ices. The density of the $(H_2O)_n$ lattice in the clathrate hydrate structures ranges from 0.72 to 0.79 g cm^{-3}, compared with 0.93 g cm^{-3} in ice Ih [432].

Conditions for ice clathrate formation, hosts, and guests. With a few exceptions, the cages in the hydrate inclusion compounds are such that the van der Waals forces across their diameters of the order of 8 to 15 Å correspond to large collapsing interactions. These ice-like compounds, therefore, crystallize only out of water or aqueous solutions under conditions in which other molecules of suitable size occupy the otherwise unstable cages. With insoluble or only slightly soluble gases such as the hydrocarbons, clathrate hydrate formation is a *gas+liquid* reaction forming a solid. The solid crystallizes on the surface of the gas bubbles in water and large single crystals are not readily formed. With water-soluble compounds or liquids such as ethylene oxide, acetone, and tetrahydrofuran, the clathrate hydrate formation is a *liquid+liquid* reaction, and large single crystals form very readily.

As in the cyclodextrin inclusion compounds (Part III, Chap. 18), the framework of these complexes is called the *host* structure and the molecules which occupy the cavities are described as *guests*. Obviously, the conditions of temperature or pressure have to be such that the crystallization of the hydrate occurs, rather than the formation of ice. With some guest species, such as argon[3], this requires greater than atmospheric pressures. As in any well-run establishment, the guests are not such species as would interfere with the host structure. The guests in the clathrate hydrates are typically hydrophobic molecules with little or no hydrogen-bonding functionality. Biological molecules with many hydrogen-bonding functional groups, such as the amino acids, carbohydrates, purines and pyrimidines, or nucleosides and nucleotides, do not form clathrate hydrates.

Although the topology of the hydrogen-bonded water host structure is closely related in the gas hydrates, the alkylamine hydrates, and the alkylammonium salt hydrates, with the pentagonal dodecahedron shown in Fig. 21.3 playing a prominent role, the interactions between the host and guest species are different.

In the *gas hydrates,* which include some molecules which are liquids at room temperature, the host-guest interactions are solely van der Waals forces. These are true clathrates.

In the *alkylamine hydrates,* with one exception, the functional group of the amine is hydrogen-bonded to or included with the water host lattice. These compounds are not true clathrates; i.e., they are sometimes referred to as semi-clathrates [433].

In the *higher hydrates of the quaternary ammonium salts*, the cations are the guest species and the water and anions form the anionic host lattice. In the recently discovered strong acid hydrates such as $HClO_4 \cdot 6H_2O$ [767], the anions are the guests in a cationic water lattice.

The ability of water molecules to form either neutral host lattices or to combine with ions[4] to form ionic lattices implies that there is a much wider range of

---

[3] A clathrate hydrate of helium in ice II has recently been reported [775].
[4] Attempts to crystallize cation-water host lattices analogous to the per-alkyl ammonium salt hydrates, e.g., tetraalkyl borate hydrates, were unsuccessful [432].

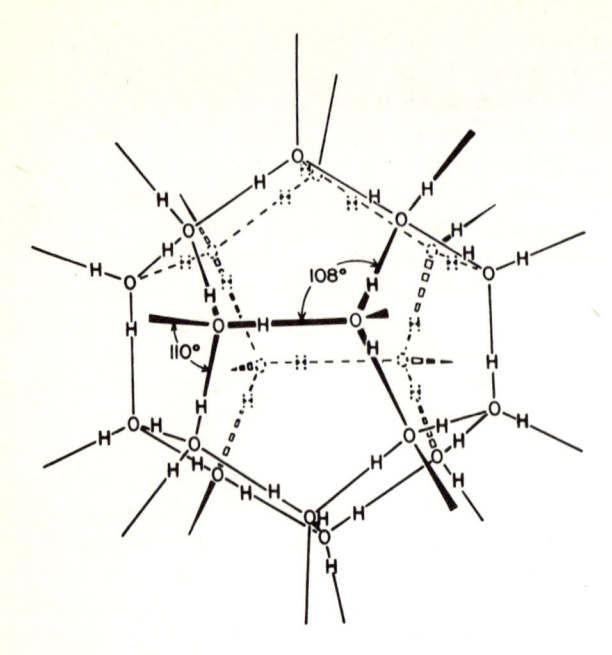

Fig. 21.3. The $(H_2O)_{20}$ pentagonal dodecahedron. This shows one of the ordered arrangements of the hydrogen atoms. In the gas hydrates, the hydrogen atoms are twofold disordered across the $O-(\frac{1}{2}H)\cdots(\frac{1}{2}H)-O$ edges. This unit of structure has 10 donor/10 acceptor hydrogen-bond functionality

hydrate inclusion compounds than has hitherto been recognized, particularly of the semi-clathrate or ionic type.[5]

The gas hydrates of type I and type II are formed by atoms or molecules ranging in size from argon[3] to carbon tetrachloride (Table 21.1). The smaller molecules form the type I structure and the larger molecules type II, with some intermediate-sized molecules which form both structures [762]. For example, CH_4 and C_2H_6 form type I, while the type II structures are formed by CH_3I to C_3H_8 and CBr_2F_2, C_2H_5Br, and cyclobutanone [762].

Until recently, it was assumed that argon, and the diatomic molecules such as oxygen and nitrogen, would resemble methane in forming the type I hydrate. However, it was predicted from thermodynamic considerations [776] that the smallest guest molecules, Ar, Kr, O_2, and N_2, would form type II hydrates. This was confirmed by powder neutron diffraction on D_2O hydrates [777, 778]. Molecules larger than butane do not form type I or type II clathrate hydrates. Neither do some small molecules of appropriate size but with more pronounced hydrogen-bonding functionality, such as NH_3, HCl. However, it is interesting to note that clathrate hydration suppresses the weak hydrogen-bonding donor functionality of

---

[5] Gas hydrates crystallized from saline solution, as in the hydrate desalination process, may contain anionic host lattices with the cationic counterions in the smaller voids.

Table 21.1. Molecules which form types I and II gas hydrates [762]

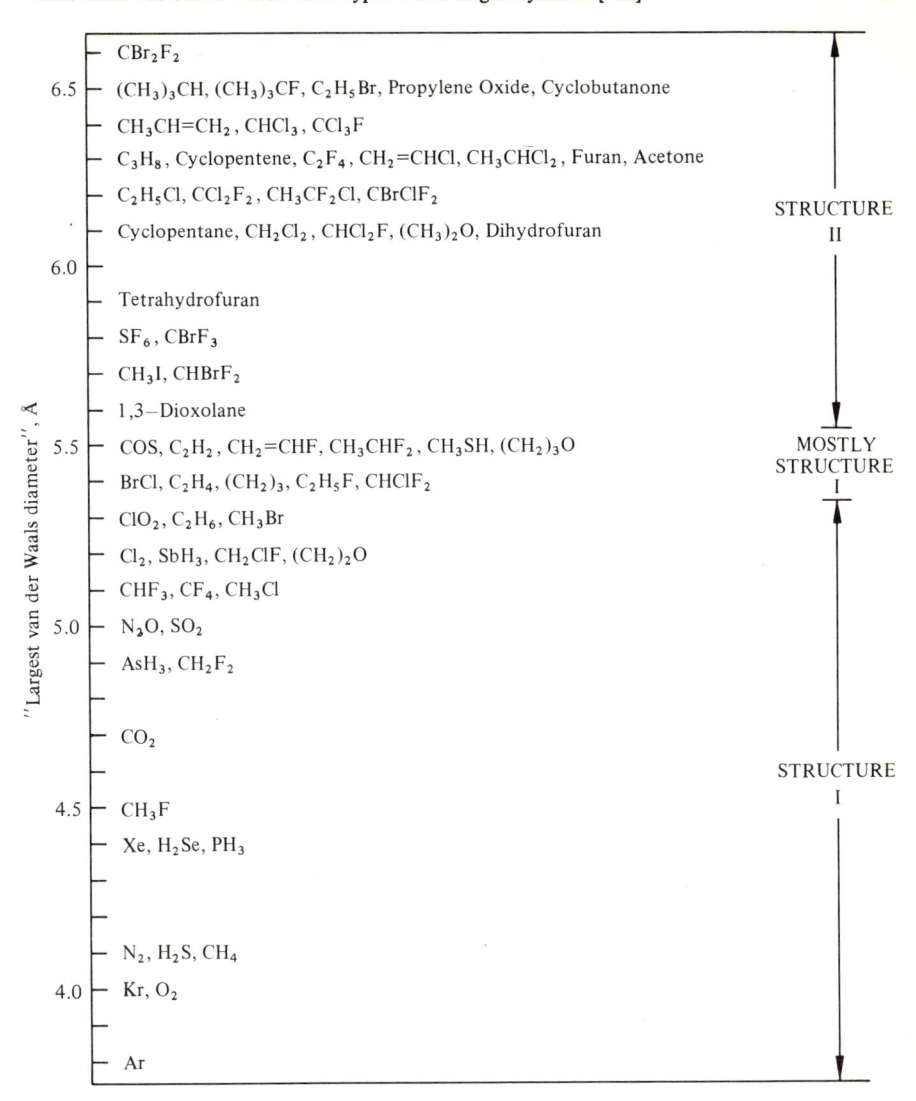

"Largest van der Waals diameter", Å	Molecules	Structure
6.5	CBr$_2$F$_2$	STRUCTURE II
	(CH$_3$)$_3$CH, (CH$_3$)$_3$CF, C$_2$H$_5$Br, Propylene Oxide, Cyclobutanone	
	CH$_3$CH=CH$_2$, CHCl$_3$, CCl$_3$F	
	C$_3$H$_8$, Cyclopentene, C$_2$F$_4$, CH$_2$=CHCl, CH$_3$CHCl$_2$, Furan, Acetone	
	C$_2$H$_5$Cl, CCl$_2$F$_2$, CH$_3$CF$_2$Cl, CBrClF$_2$	
	Cyclopentane, CH$_2$Cl$_2$, CHCl$_2$F, (CH$_3$)$_2$O, Dihydrofuran	
6.0	Tetrahydrofuran	
	SF$_6$, CBrF$_3$	
	CH$_3$I, CHBrF$_2$	
	1,3–Dioxolane	
5.5	COS, C$_2$H$_2$, CH$_2$=CHF, CH$_3$CHF$_2$, CH$_3$SH, (CH$_2$)$_3$O	MOSTLY STRUCTURE I
	BrCl, C$_2$H$_4$, (CH$_2$)$_3$, C$_2$H$_5$F, CHClF$_2$	
	ClO$_2$, C$_2$H$_6$, CH$_3$Br	
	Cl$_2$, SbH$_3$, CH$_2$ClF, (CH$_2$)$_2$O	
	CHF$_3$, CF$_4$, CH$_3$Cl	
5.0	N$_2$O, SO$_2$	
	AsH$_3$, CH$_2$F$_2$	
	CO$_2$	STRUCTURE I
4.5	CH$_3$F	
	Xe, H$_2$Se, PH$_3$	
	N$_2$, H$_2$S, CH$_4$	
4.0	Kr, O$_2$	
	Ar	

H$_2$S, or the weak hydrogen-bond acceptor properties of SO$_2$, CO$_2$, ethylene oxide, acetone, and tetrahydrofuran, all of which occupy voids as *nonhydrogen-bonded guests*.

The (H$_2$O)$_{20}$ pentagonal dodecahedron is a basic cage unit in many clathrate hydrates. The principal topological feature of the gas hydrates is the (H$_2$O)$_{20}$ pentagonal dodecahedron which is an Archimedean regular solid, with planar faces having equal edges and angles. When linked to other polyhedra, each oxygen is coordinated by four hydrogen bonds, two donor and two acceptor. In the ideal

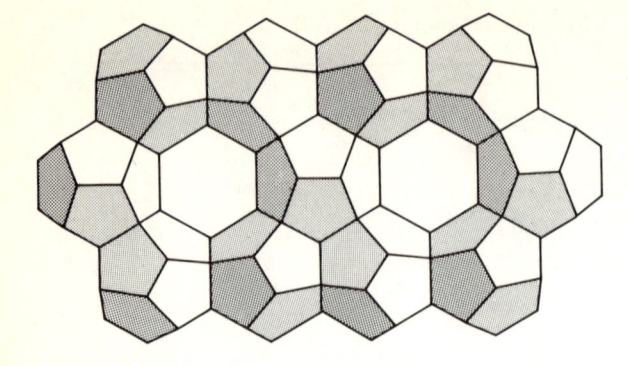

Fig. 21.4. Pattern of water structure formed by a face-sharing layer arrangement of pentagonal dodecahedra. The hexagonal "hole" cannot be filled by the pentagonal dodecahedron; it requires other polyhedra, as shown in Fig. 21.5

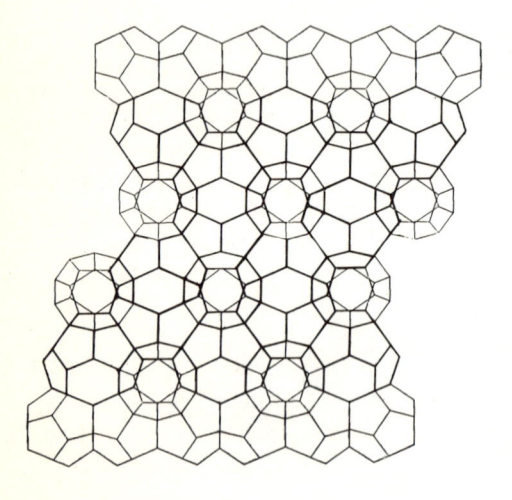

Fig. 21.5. A layer of hydrogen-bonded water molecules formed by a face-sharing arrangement of 5^{12}, $5^{12}6^2$ and $5^{12}6^3$ polyhedra

dodecahedron, the coordination is distorted tetrahedral with two of the O···Ô···O angles being 108° and two being 110°, as shown in Fig. 21.3. Of the 20 hydrogen bonds external to the pentagonal dodecahedron, half are donors and half acceptors. Unlike the other regular solids, the pentagonal dodecahedron is not a space-filling building block (see Figs. 21.4, 21.5).

When these polyhedra are linked by the two ways which can form four-connected nets, that is, by face-sharing or by connecting bonds, other polyhedra with 14, 15, and 16 faces are necessarily formed, of the type illustrated in Fig. 21.6. In a shorthand notation, pentagonal dodecahedra are described as 5^{12} (12 faces each with 5 vertices), the 14-hedron is $5^{12}6^2$ with two additional hexagonal faces, the 15-hedron is $5^{12}6^3$ with 12 pentagons and 3 hexagonal faces, and the 16-hedron, $5^{12}6^4$, consists of 12 pentagons and 4 hexagons. The polyhedra, other than the pentagonal dodecahedron, are sometimes referred to as *semi-regular solids*, because the faces cannot be strictly planar if they have equal edges and angles.

The Hydrate Inclusion Compounds

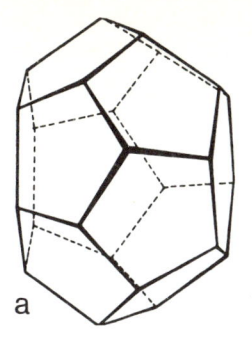

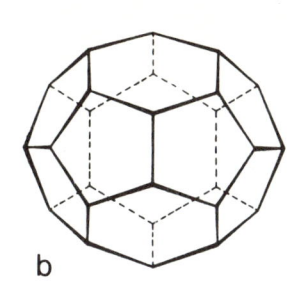

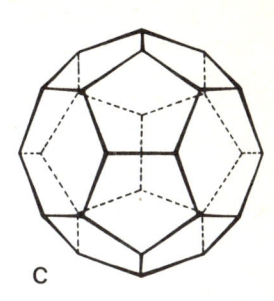

Fig. 21.6a–c. Other polyhedral voids observed in the clathrate hydrate structures. **a** tetrakaidecahedron, $5^{12}6^2$; **b** pentakaidecahedron, $5^{12}6^3$; **c** hexakaidecahedron, $5^{12}6^4$. Unlike the dodecahedron (Fig. 21.3) which is an regular solid, these are semi-regular solids. If the edges and angles in a face are equal, the faces cannot be exactly planar

In fact, in the actual crystal structures, the pentagonal dodecahedra are also not exactly regular, as found in the analysis of the ethylene oxide hydrate [431].

The gas hydrates occur in three types of structures, named type I, type II and type H. The type I has 5^{12} and $5^{12}6^2$ polyhedra associated by face-sharing in the ratio 1:3, whereas in type II, 5^{12} and $5^{12}6^4$ polyhedra are combined by face-sharing in the ratio 2:1. These structures have cubic symmetry. The recently discovered type H [779] has hexagonal symmetry and is isostructural with the hexagonal clathrasil-dodecasil 7–H [780]. The host lattice contains 5^{12}, $4^35^66^3$ and a larger $5^{12}6^8$ void which can accommodate larger organic molecules than the type II cubic structure.

There is a relationship between the size of the guest molecule and the stoichiometry of the type I and type II hydrate structures. In the type I hydrates, the smaller guest species can occupy both the 5^{12} hedra and the $5^{12}6^2$ hedra for a limiting stoichiometry of $8X \cdot 46H_2O$, while larger guest species which form type I structures can only enter the $5^{12}6^2$ hedra with a limiting stoichiometry of $6X \cdot 46H_2O$. Since it does not appear to be necessary to fill the 5^{12} hedra to form a stable crystal structure, a range of nonstoichiometry is possible between $6X \cdot 46H_2O$ and $8X \cdot 46H_2O$ for those molecules which are small enough to be accommodated within the 5^{12} hedra. This appears to be the case for methane, but not for Cl_2, which forms a stoichiometric $6Cl_2 \cdot 46H_2O$. Presumably the chlorine molecule is too large for the 5^{12} hedra. The larger guest molecules which form the type II hydrate cannot occupy the 5^{12} hedra, so the stoichiometry is necessarily $8X \cdot 136H_2O$, as in propane hydrate.

A "help gas" stabilizes double and mixed hydrates. There is some uncertainty whether these inclusion hydrates are in fact stable without a guest species in the small 5^{12} cages. All the clathrate hydrate structures are known to be stabilized and their melting points raised by including H_2S with the guest compounds, which for this reason is called the "helpgas" [782–786]. Since we know of no experiments where the gas hydrates were prepared in the absence of air or some other gas, this "help-gas" phenomenon may occur to a more or less degree in all laboratory preparations. When crystals of the liquid hydrates, such as those of ethylene oxide or tetrahydrofuran are melting, gas bubbles are generally observed. These are presumed to be air molecules occluded in the small 5^{12} voids, which may be necessary to stabilize these hydrates. A similar phenomenon was observed with crystals of the tetra butyl ammonium salt hydrates, where air could occupy the vacant pentagonal dodecahedra.

The ideal stoichiometric formulas for the hydrates prepared with the aid of H_2S are, for example, $6Cl_2 \cdot 2H_2S \cdot 46H_2O$, or $8C_3H_8 \cdot 16H_2S \cdot 136H_2O$ for types I and II hydrates, respectively. They are referred to as *double hydrates*. It is also possible to occupy the voids with mixtures of appropriate gases such as the mixture of methane, ethane, and propane, which occurs in natural gas, to yield *mixed hydrates*.

The details of the hydrogen bonding in clathrate hydrates are not well known. The original X-ray studies of the gas clathrate hydrate structures were based on powder diffraction data, since the gas hydrates are notoriously difficult to obtain as single crystals. Clathrate hydrates formed by compounds which are liquids above 0 °C, such as those of ethylene oxide and tetrahydrofuran, can be readily grown as large single crystals, and X-ray single crystal studies of both these hydrates have been carried out [787, 788].

A more precise neutron diffraction analysis was made for the ethylene oxide deuterohydrate [431]. In this structure, the expected disorder of the $O-(\frac{1}{2}D)\cdots(\frac{1}{2}D)-O$ bonds was verified, and geometrical details of the hydrogen bonds are given in Part IB, Chapter 7, Table 7.7. The covalent $O-(\frac{1}{2}D)$ bond lengths range from 0.972 to 0.998 Å and the $D-\hat{O}-D$ angles from 107.4° to 110.5°. The hydrogen-bond lengths are from 1.726 to 1.818 Å. The shortest bond is required by symmetry to be exactly linear and the longest bond has the greatest departure from linearity.

Hydrates with related clathrate-like structures are formed by tetramethyl and tetraethyl ammonium salts. The tetramethyl and tetraethyl ammonium salts form hydrates in which the cations occupy voids in hydrogen-bonded structures formed by the anions and the water molecules. Unlike the gas hydrates, the alkylamine hydrates, and the butyl and isoamyl ammonium salt hydrates discussed in the next section, these cage-like structures do not feature the pentagonal dodecahedron as a common structural component.

The structure of the low-temperature phase of $[(CH_3)_4N]^+[OH \cdot 5H_2O]^-$ [125, 435] is a clathrate hydrate with a water-anion host lattice which is based on a distorted close-packing of truncated octahedra. There is also a high-temperature phase which has a host lattice of undistorted close-packed $[4^46^4]$ octahedra. Unlike the structure of $[PF_6]^-[H_2F \cdot 5H_2O]^+$, with which it is isostructural [789], this structure is deficient in the number of protons necessary to form all the hydrogen bonds, leading to the concept of proton deficient two-center bonds discussed in Part IA, Chapter 2.9. The hydrates of $[(CH_3)_4N]^+[F \cdot 4H_2O]^-$ [436], shown in Fig. 21.7, $[(CH_3)_4N]_2^+[SO_4 \cdot 4H_2O]^{2-}$ [437], and $[(C_2H_5)_4N]^+ \cdot [Cl \cdot 4H_2O]^-$ [790], on the other hand, are not clathrates but are channel-type inclusion compounds formed from cross-linking nets and ribbons into framework structures, with anions integrated in the water structure. The lower hydrates of

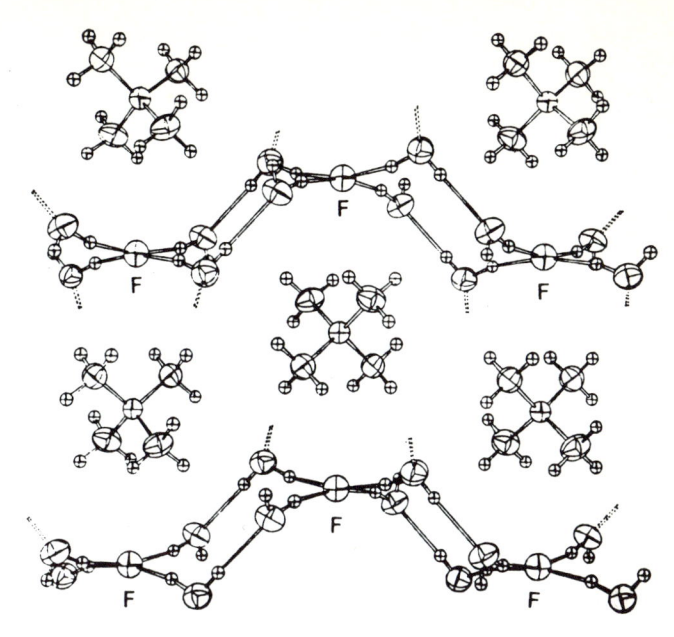

Fig. 21.7. Section of the framework anionic-water host lattice formed by linking almost planar $(H_2O)_4F^-$ groups in $(CH_3)_4N^+(F \cdot 4H_2O)^-$. The water molecules are three-coordinated with an almost planar configuration [436]

tetramethylammonium fluoride and hydroxide have also been studied by solid-state infra-red spectroscopy and shown to be anionic channel-framework hydrates [791, 792].

$[(C_2H_5)_4N]^+[Cl \cdot 4H_2O]^-$ has a "double channel" anion host structure formed by linking $(H_2O)_4Cl^-$ tetrahedra through their vertices [790]. The hydrogen bonding, shown in Fig. 21.8, includes a homodromic quadrilateral of four hydrogen-bonded water molecules. The structure of $4[(C_2H_5)_4N]^+[F \cdot 11H_2O]^-$ [438] is a channel clathrate with an anionic host structure formed by linear chains of edge-sharing $(H_2O)_4F^-$ tetrahedra and bridging water molecules shown in Fig. 21.9. $[(C_2H_5)_4N]^+[CH_3COO \cdot 4H_2O]^-$ [440], on the other hand, has the layer-like structure in which the acetate groups and water molecules form the anionic layers. In this structure, the water molecules donate two but accept only one bond, to form a pattern of pentagons and heptagons. A similar layer structure is found in pyridine trihydrate [793].

Here the layers contain only water molecules which form antidromic pentagons, quadrilaterals and homodromic hexagons (Fig. 21.10). Clathrate or semi-clathrate structures have been postulated for choline chloride hydrate, $(H_3C)_3N^+CH_2CH_2OH \cdot 2H_2O \cdot Cl^-$, on the basis of similarities in the solid-state infrared spectra [162], but this has not been confirmed by crystal structure analysis.

Interesting metallo-organic hydrates have been reported which contain ice-like columns of water molecules. The protons in the water structure are ordered and

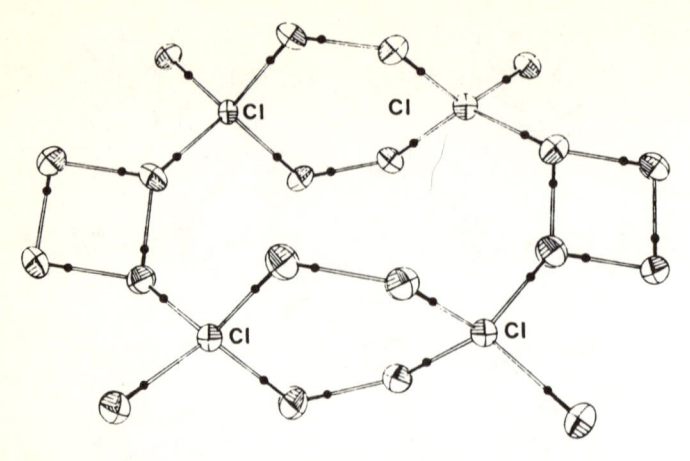

Fig. 21.8. A channel-type anionic-water host lattice formed from vertex-linked $(H_2O)_4Cl^-$ tetrahedra in the structure of $(C_2H_5)_4N^+ \cdot (H_2O)_4Cl^-$. The $(H_2O)_4$ homodromic quadrilaterals are almost planar. The water molecules are three-coordinated donating two and accepting one hydrogen bond. The protons appear to be ordered [790]

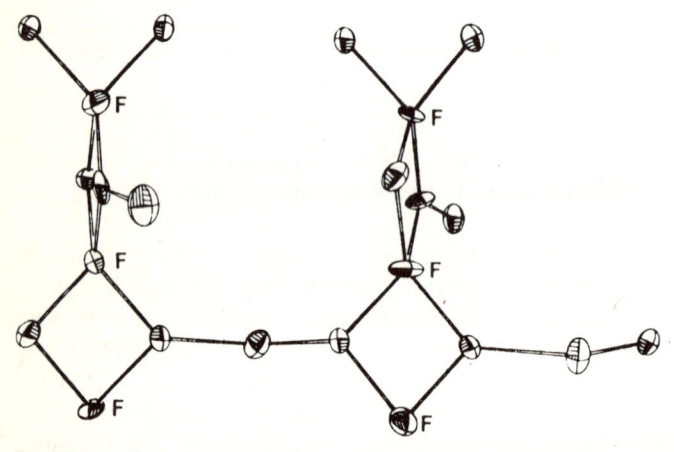

Fig. 21.9. A channel-type anionic-water host lattice formed by edge-sharing $(H_2O)_4F^-$ tetrahedra with bridging water molecules in the structure of $4(C_2H_5)_4N^+(F \cdot 11H_2O)^-$. In this structure, six waters are three-coordinated and five are only two-coordinated, i.e., are donors only [438]

the hydrogen bonding is homodromic. These hydrates are described as *exoclathrates* [794].

Larger hydrate cages are formed by bulkier quaternary alkyl ammonium salts. Until the more recent research described above, it was thought that, with one exception, crystalline alkyl ammonium clathrate hydrates were formed *exclusively* with the large tetra-n-butyl ammonium and tetra-isoamyl ammonium cations, with water contents of more than 50% by weight and melting points between 10° and 30°C. The increased viscosity of the solutions from which these salts are crystal-

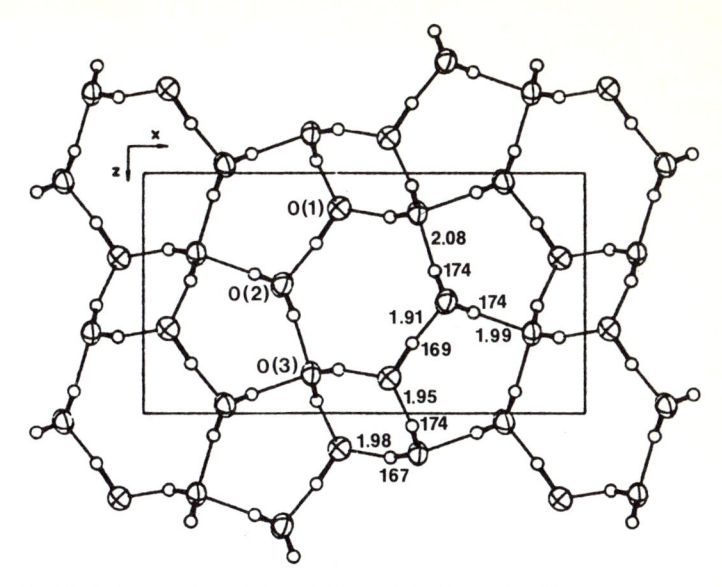

Fig. 21.10. The buckled hydrate layer in pyridine·3H$_2$O, with hydrogen-bonding geometry indicated. Note that the hydrogen-bonding pattern is homodromic in the hexagon and antidromic in the pentagon and quadrilateral. Pyridine molecules are stacked "above" and "below" this layer in nearly vertical orientation to the layer and bound to it by O(2)–H···N hydrogen bonds (not shown) [793]

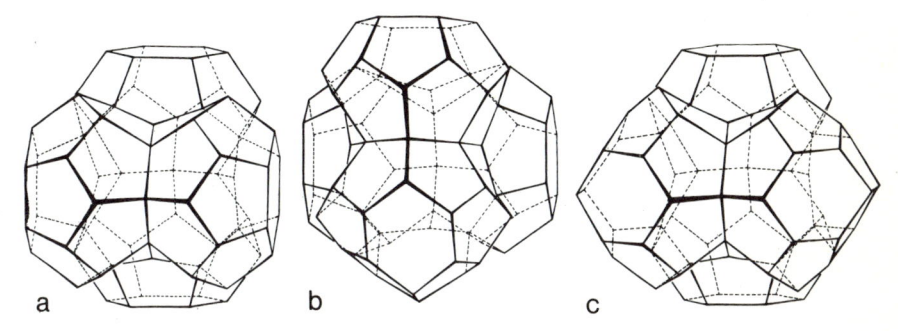

Fig. 21.11a–c. The polyhedral voids in the tetra-n-butyl and tetra-isoamyl ammonium salt hydrates. **a** 44-hedron, $5^{40}6^4$, formed from four 14-hedra; **b** 45-hedron, $5^{40}6^5$, formed from three 14-hedra and one 15-hedron; **c** 46-hedron, $5^{40}6^6$, formed from two 14-hedra and two 15-hedra

lized is observed with other bulky cations of comparable size, but no crystals were obtained. These hydrates are isostructural with the gas hydrates because larger cages can be formed by combining those occurring in the gas hydrates. As shown in Fig. 21.11a, four $5^{12}6^2$ polyhedra are combined by sharing hexagonal faces and omitting the vertex common to all four polyhedra. The result is a large cage with four compartments. These are filled with the tetrahedrally extended n-butyl or isoamyl chains, while the central nitrogens occupy the vacant vertex [795].

Table 21.2. Crystal data for some tetra-n-butyl and tetra-iso-amyl ammonium hydrates [774]

Salt	Melting point	Dimensions of unit cell a	b	c	Vol.	Meas. density g/cc at 5°C ±0.005	Hydration number (chemical)	Hydration number (crystal)	Number of formula units per unit cell
(C$_4$H$_9$)$_4$NF	24.9	23.78		12.53	7086	1.032	34.0	34.4	5
(C$_4$H$_9$)$_4$NCl	15.7	23.77		12.61	7125	1.026	32.1	33.8	5
(C$_4$H$_9$)$_4$NBr	12.5	23.65		12.50	6991	1.080	30.5	32.6	5
(C$_4$H$_9$)$_4$NCH$_3$CO$_2$	15.1	23.62		12.38	6907	1.040	30.7	31.3	5
[(C$_4$H$_9$)$_4$N]$_2$CrO$_4$	13.6	23.68		12.40	6953	1.059	68	65.1	2$\frac{1}{2}$
[(C$_4$H$_9$)$_4$N]$_2$WO$_4$	15.1	23.52		12.37	6843	1.131	60	62.7	2$\frac{1}{2}$
[(C$_4$H$_9$)$_4$N]$_2$C$_2$O$_4$	16.8	23.63		12.31	6874	1.043	67	64.0	2$\frac{1}{2}$
(C$_4$H$_9$)$_4$NHCO$_3$	17.8	23.77		12.46	7070	1.038	33	32.0	5
[(C$_4$H$_9$)$_4$N]$_2$HPO$_4$	17.2	23.55		12.34	6844	1.059		64.2	2$\frac{1}{2}$
		a ±0.02	b ±0.05	c ±0.02	V ±20Å				
(C$_5$H$_{11}$)$_4$NF	31.2	12.08	21.61	12.82	3347	1.021	38.0	40.0	2
(C$_5$H$_{11}$)$_4$NCl	29.8	11.98	21.48	12.83	3302	1.021	38.3	38.3	2
[(C$_5$H$_{11}$)$_4$N]$_2$CrO$_4$	21.6	12.18	21.53	12.67	3338	1.047	74.2	78.2	1
[(C$_5$H$_{11}$)$_4$N]$_2$WO$_4$	22.4	12.06	21.39	12.70	3276	1.109	80.7	75.4	1

These compounds have remarkably high melting points (see Table 21.2), considering their water content, e.g., $[(i-C_5H_{11})_4N]^+[F \cdot 38H_2O]^-$ has a melting point of 31 °C and contains 68 wt.% of water [795], which is comparable to the water content in protein crystals.

Similar large voids can be formed from the combinations of $5^{12}6^2$ and $5^{12}6^3$ polyhedra, illustrated in Fig. 21.11 b, c. To the present, there are no hydrates known which have voids formed by a combination of $5^{12}6^4$ polyhedra, which could accommodate cations with larger alkyl chains.

A very large cage is formed by $[n(C_4H_9)_3S]^+[F \cdot 23H_2O]^-$ [796]. In these monoclinic crystals[6], two cations are enclosed in a large void built up by 96 water molecules. The cage is bounded by 48 pentagons, 10 hexagons, and 4 quadrilaterals, i.e., $4^45^{48}6^{10}$, and has a volume of 1550 Å^3. The pair of cations is oriented with a center of symmetry midway between the S$^+$ atoms, which are separated by 3.49 Å, as illustrated in Fig. 21.12. This hydrate structure contains puckered layers of face-sharing pentagonal dodecahedra, in which the hydrogen bonds, as measured by the O$\cdots$O separations, are significantly shorter than between the layers. These layers of puckering pentagons are a common topological feature of hydrates of certain diamines and alkanediols as discussed in Chapter 21.3.

The alkylamine hydrate inclusion compounds form a large variety of structures. Phase diagram studies at the end of the last century [765] (see Table 21.3), and an extensive series of phase diagram studies carried out between 1970 and 1981 [798], showed that aliphatic amines form a large variety of hydrates. Relatively few of these have been studied by X-ray crystal structure analysis and none by neutron diffraction [433, 799]. These hydrates differ from the gas hydrates and the alkyl ammonium salt hydrates in that there appear to be no definite structural types into which several hydrates can be classified. Hitherto, a different crystal structure has been observed for every alkylamine hydrate studied. In this respect, they resemble low hydrated crystals.

Only the tert-butylamine hydrate is a true clathrate. In the crystal structure, of composition $16(CH_3)_3CNH_2 \cdot 156H_2O$ [800], the primary amine molecules occupy a 17-hedron which has square, pentagonal, hexagonal, and heptagonal faces. It is a $7^36^25^94^3$ polyhedron, shown in Fig. 21.13a. Since this is not a space-filling polyhedron, there are smaller 4^45^4 polyhedra in the crystal lattice which share quadrilaterals and pentagons with the larger polyhedra. This is the smallest cage observed in the clathrate hydrates, and is comparable in volume to the voids in the ice Ih structure. The polyhedral description is idealized, since only the quadrilaterals are exactly planar, whereas the other faces have displacements of atoms as large as ±0.17 Å from the mean planes. Since this is a true clathrate, it can be presumed that the hydrogen bonds are disordered, as in the gas hydrates, ethylene oxide hydrate, and the ices. There is no direct evidence on the hydrogen atom positions from this X-ray study. The O$\cdots$O separations range from 2.765 Å

---

[6] There is also a cubic modification, which appears to be isostructural with the type I gas hydrates [797]. Only one large crystal of the monoclinic form was grown, and it could not be repeated. As Henry Frank remarked at the time, such an example of cooperativity combined with periodicity is likely to have a small probability.

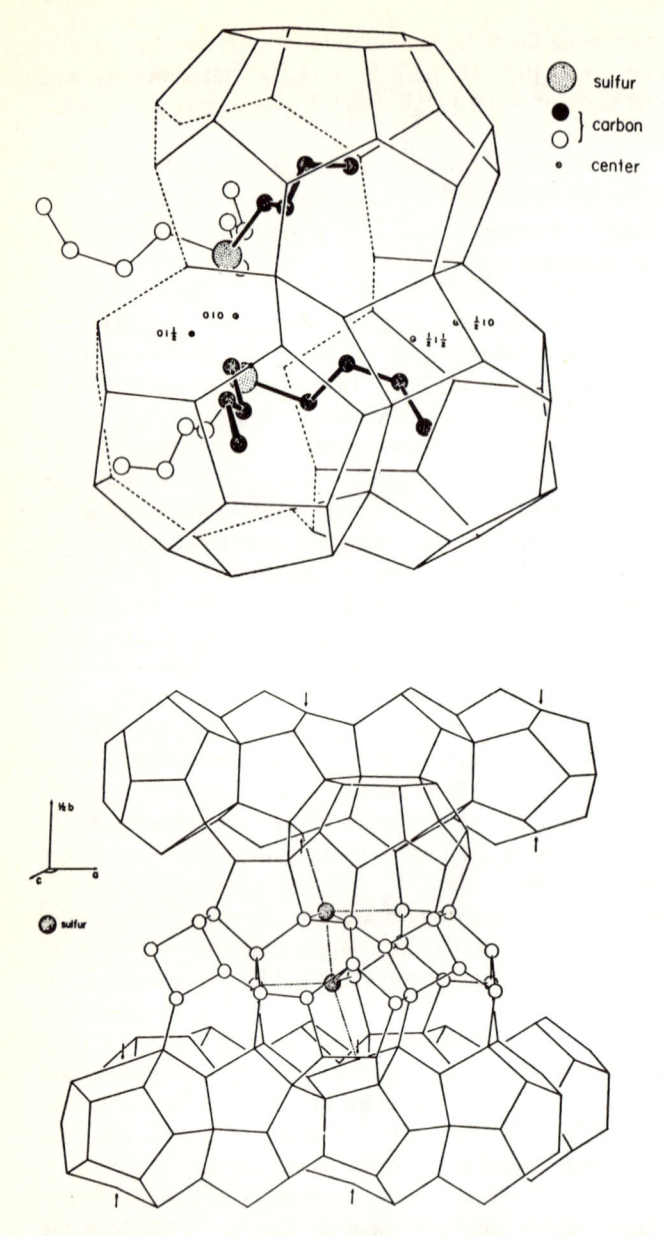

Fig. 21.12. The anionic-water host lattice in $[n(C_4H_9)_3S]^+[F \cdot 23H_2O]^-$, showing half the large $4^55^486^{10}$ polyhedral void that accommodates two anions. The other half is related by the center of symmetry at (0,1,0) sharing the dotted edges. The F^- anion was not distinguished from the oxygen in this X-ray analysis. It may be disordered over the water lattice. The top figure is the central part of the bottom figure [796]

The Hydrate Inclusion Compounds

Table 21.3. The alkylamine hydrates [432]

	Hydration no. reported from phase studies [765]						Hydration no. (mp, °C) from crystal structure studies
Methyl		3.0, 3.5		10			
Dimethyl	1			7			7.8 (−16.9)
Trimethyl	2			7	10, 11	20	10.25 (5.0)
Ethyl	0.5		5.5				7.6 (−7.5)
Diethyl	0.5	3.0	5.5	8	11	36	7.0, 8.67 (−6.0)
Triethyl	2			8		31	
Propyl	0.5	3.5		8			6.5, 10.0 (−13.5)
Isopropyl		3.5		8			8.0 (−4.0)
Dipropyl	0.5		5.5				
Isobutyl				7	14	36	
sec-Butyl	2				11	34	
t-Butyl							9.67 (−1.0)
Amyl			5.5			37	
Octyl		1.5	6.0				
Dodecyl	0.7	2.0		4			
Octadecyl	0.7	2.0					

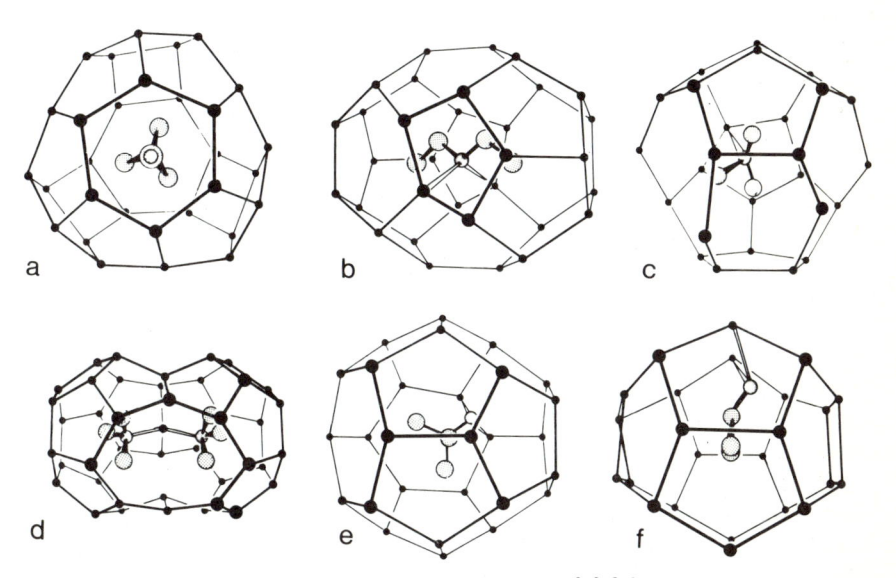

Fig. 21.13. a The guest $(CH_3)_3CNH_2$ molecule in a 17-hedra $7^36^25^94^3$ clathrate water cage. This is a true clathrate. The NH_2 and CH_3 groups are undergoing large oscillatory thermal motion relative to the water framework which is relatively rigid. The hydrogen atoms are not shown. **b–f** The semi-clathrate water polyhedral cages are shown as observed in the crystal structures of the amine hydrates: **b** 18-hedron, $5^{12}6^6$, in $12(CH_3CH_2)_2NH \cdot 104 H_2O$; **c** 15-hedron $5^{12}6^3$ and **d** 26-hedron $5^{24}6^2$ in $4(CH_3)_3N \cdot 41 H_2O$; **e** 16-hedron $5^{12}6^4$ and **f** 14-hedron $4^25^86^4$ in $10(CH_3)_2CHNH_2 \cdot 80 H_2O$. Hydrogen bonds involving guest nitrogen atoms are indicated by open bonds [433, 800]

to 2.847 Å with O···Ô···O angles between 87.9° and 133.2°. The tert-butylamine molecule is oriented with the C–N axis along the threefold symmetry axis of the polyhedron, as shown in Fig. 21.13a. The closest distances from the nitrogen atom to the oxygen atoms of the hydrate structure are 3.80 Å and 3.89 Å, which may correspond to weak hydrogen-bond interactions. If that is so, there may be, in fact, some partial ordering of the protons in the hydrogen bonds. To answer this question, a low-temperature neutron diffraction analysis is necessary.

The majority of alkylamine hydrates are "semi-clathrates". As Table 21.4 indicates, in all alkylamine hydrates except tert-butylamine, the amine molecule is hydrogen-bonded to the water framework, which, nevertheless, retains the cage-like structure. For this reason, these compounds are described as "semi-clathrates".

The secondary amino groups donate a hydrogen bond to a water molecule of the cage and accept one. The tertiary amine accepts two hydrogen bonds from the water structure in trimethylamine decahydrate [801]. It is curious that the secondary and tertiary amines are hydrogen-bonded, while the primary amine group in the tert-butylamine hydrate is not. This is, however, a special feature of this guest molecule and not a general property of primary amine hydrates, since the primary amine group is hydrogen-bonded in both the isopropylamine and n-propylamine hydrates.

Cages in amine hydrates display large structural variety. The topography of the cages which enclose the amine molecules is different for all the crystal structures hitherto analyzed (see Fig. 21.13b–f). The most complex of these amine hydrate cages is that of the low melting point (–35°C) modification of n-propylamine hydrate, with composition $16CH_3CH_2CH_2NH_2 \cdot 104H_2O$ [802]. The water framework consists of four large incomplete polyhedra which contain hydrogen-bonded amine molecules interspersed with small 11-hedra, $4^2 5^8 6^1$, which are vacant (they may contain occluded air molecules). The incomplete polyhedra form the largest polyhedral cage yet observed in the hydrate inclusion compounds, with a volume of approximately 2200 Å^3. The large polyhedron is difficult to describe in the terms of the face-sharing of smaller polyhedra as with the per-alkyl quaternary ammonium salt hydrates and the simple alkylamine hydrates.

This water structure illustrates the level of complexity of these periodic four-connected nets of water molecules that can occur when perturbed by molecular species that have both hydrophilic and hydrophobic character. If the condition of high-resolution periodicity throughout the water structure is relaxed, as in protein crystals, even greater complexity is possible.

The formation of inclusion compounds based on hydrogen-bonded anionic water frameworks with per-alkyl ammonium anions appears to be more widespread than originally thought [803–806]. Of particular interest are the bolaform salts [806] which have biquaternary alkyl ammonium cations, i.e., $[(n-C_4H_9)_3N-(CH_2)_x-N(n-C_4H_9)_3]^{2+}[2F \cdot mH_2O]^{2-}$, with x = 4, 6, 8, 10, and hydration numbers, m, ranging from 30 to 60. These compounds have melting points between 10° and 30°C, and are expected to have related structures, but no crystal structure analyses have been undertaken.

Do clathrate structures persist in their aqueous solutions. The increase in viscosity of the solutions prior to the formation of crystals suggests that this is so.

Table 21.4. Structural characteristics of the alkylamine hydrates [433]

Name, formula per unit cell	Melting point (°)	Amine-water relationship	Hydrogen-bonded n-hedra $(H_2O)_n$ [a]
t-Butylamine hydrate, $16(CH_3)_3CNH_2 \cdot 156H_2O$	−1	Nonbonded within 17-hedra	17-Hedra $(7^3 \cdot 6^2 \cdot 5^9 \cdot 4^3)$ 8-Hedra $(4^4 \cdot 5^4)$
Diethylamine hydrate, $12(CH_3CH_2)_2NH \cdot 104H_2O$	−6	Hydrogen-bonded in distorted 18-hedra and in an irregular cage	18-Hedra $(5^{12} \cdot 6^6)$ Irregular cage $(6^6 \cdot 5^8 \cdot 4^3)$
Trimethylamine hydrate, $4(CH_3)_3N \cdot 41H_2O$	+5	Hydrogen-bonded in very distorted 15- and 26-polyhedra	12-Hedra (5^{12}) [b] 15-Hedra $(5^{12} \cdot 6^3)$ 26-Hedra $(5^{24} \cdot 6^2)$ [c]
Isopropylamine hydrate, $10(CH_3)_2CHNH_2 \cdot 80H_2O$	−4	Hydrogen-bonded in very distorted 14- and 16-hedra	8-Hedra $(6^2 \cdot 4^6)$ 12-Hedra (5^{12}) [b] 14-Hedra $(4^2 \cdot 5^8 \cdot 6^4)$ 16-Hedra $(5^{12} \cdot 6^4)$
n-Propylamine $16CH_3CH_2CH_2NH_2 \cdot 104H_2O$ [d]	−35	Hydrogen-bonded in very distorted 14- and 16-hedra	11-Hedra $(4^2 \cdot 5^8 \cdot 6^1)$ [d] 14-Hedra $(5^{12} \cdot 6^2)$ 16-Hedra $(5^{12} \cdot 6^4)$

[a] m^n refers to n-sided faces, i.e., the 17-hedra have three heptagons, two hexagons, nine pentagons, and three squares. The notation used is that described in [746]. [b] The pentagonal dodecahedra, where occurring, are very regular. [c] These are derived from two 14-hedra $(5^{12} \cdot 6^2)$ sharing a common hexagonal face. [d] This polyhedron is empty.

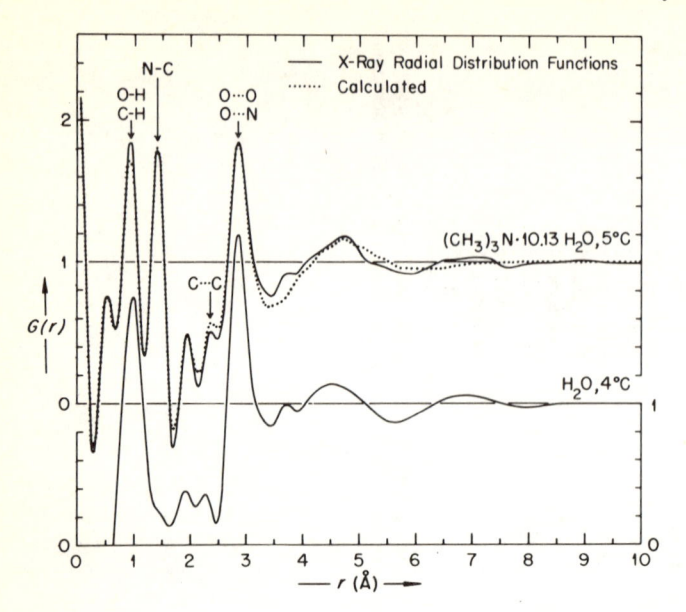

Fig. 21.14. Comparison of correlation functions from the X-ray diffraction pattern of liquid trimethylamine hydrate just above the melting point of 5 °C with that for water at 4 °C. (———) observed for the solution and for water; (· · ·) calculated from the clathrate hydrate crystal structure [807]

An X-ray liquid diffraction experiment on a solution corresponding to the stoichiometry of the trimethylamine decahydrate [807], $4(CH_3)_3N \cdot 40H_2O$, provided the radial distribution function shown in Fig. 21.14, which could be fitted equally well with models based on the hydrate crystal structure or on the ice I structure with interstitial amine molecules. This result clearly illustrates the insensibility of X-ray liquid diffraction data alone for distinguishing between competing models for aqueous solutions which have similar first- and second-order coordinations.

The existence of "clathrate-like" water structure adjacent to the hydrophobic surfaces of macromolecules is an attractive hypothesis. Models have been proposed which have received some support from thermodynamical arguments [808]. However, this concept has proved ineffective as a basis for the interpretation of the structure associated with the many electron density solvent peaks, separated by 2.8 Å to 3.0 Å, which are frequently observed on the Fourier X-ray maps close to the surface of a protein [809, 810]. Recently, however, some local clathrate-like water has been observed in special cases; in the high-resolution studies of the small plant protein, crambin [811], in a hydrated deoxydinucleoside-phosphate drug complex [812], in (Phe[4]Val[6]) antamanide hydrate [813] and in an oligodeoxynucleotide duplex [814].

The present evidence favors the view that clathrate-like water structure is observed in only a minority of biological structures. It is not clear, as yet, what particular features of these molecules induce this water structure. If it were simply the presence of hydrophobic side-chains, clathrate water structure would have been recognized more frequently in the many high and moderate resolution protein crystal structure analyses that have now been completed. It appears to be a combination of hydrophobic structure and suitably positioned hydrogen-bonding functional groups which contributes to the stabilization of clathrate-like water configurations at the surface of biological molecules (see Chap. 23, 24). Since such combinations are rarely realized, clathrates are seldom located in macromolecular crystal structures.

21.3 Hydrate Layer Structures

If organic molecules which form high hydrates are good hydrogen-bond donors or acceptors, and they are too large to be accommodated in the clathrate cages, the water molecules may form layers. The organic molecules are located between the layers and, most frequently, are hydrogen-bonded to the water layers on both sides. Hydrate layer structures are preferentially formed with molecules which have two hydrogen-bond functional groups well separated by a hydrophobic region. Piperazine [815], and pinacol [816] are such molecules, which form hexahydrates, as are tetraethylammonium acetate [440], 2,5-dimethyl-2,5-hexanediol [817] and 2,7-dimethyl-2,7-octanediol [818], which form tetrahydrates. High hydrates have also been reported for several alkyl diamines and for other alkanediols that are likely to have similar crystal structures with layer water structure [819].

The hydrogen-bonded water layers in the structures consist of buckled arrangements of edge-sharing polygons, in which there is a predominance of pentagons. The regular buckled layer consisting wholly of pentagons is shown in Fig. 21.15. (Edge-sharing pentagons cannot form a planar repetitive motif.) These structures are a variation of the alkylamine "semi-clathrates", since the functional groups of the organic molecules are hydrogen-bonded with the water structure. As in the clathrate and semi-clathrate quadrilaterals, hexagons and occasionally heptagons are also associated with the pentagons to form less regular patterns. In seeking connectivity in water structure in highly hydrated macromolecules, the order of occurrence is expected to be pentagons, hexagons, and squares, with heptagons and larger polygons in the minority.

Such an example is pyridine trihydrate [793] shown in Fig. 21.10, where there are 4-, 5-, and 6-membered hydrogen-bonded rings formed by water molecules. Of

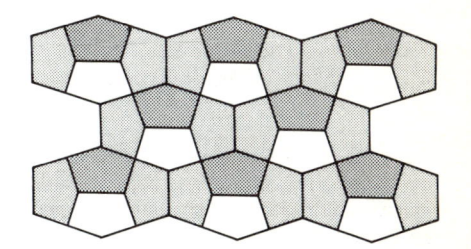

Fig. 21.15. Regular buckled pattern of pentagons formed by hydrogen-bonded water molecules in pinacol and piperazine hexahydrates and by water and hydroxyl groups in dimethyl hexane- and octanediol tetrahydrates

the three symmetry-independent water molecules, one O(3) is a double donor and double acceptor to other waters, another O(2) donates two and accepts one hydrogen bond, and the third O(1) accepts two bonds and donates two, one of which is to the N of pyridine.

In tetraethylammonium acetate tetrahydrate, $[(CH_3CH_2)_4N]^+ \cdot CH_3COO^- \cdot 4H_2O$ [440], the hydrogen-bonding scheme is somewhat more complex because the acetate ion is incorporated in the hydrogen-bonded water layer. There are six different cycles, of which only two pentagons are regular. All the other cycles are distorted because they contain the acetate carboxylate group. The water molecules donate two hydrogen bonds but they accept only one. The tetraethylammonium cations are accommodated between the acetate-water layers. They are not engaged in any obvious hydrogen-bonding contact and are only held by van der Waals and ionic forces.

The hydrate layer structures which display more regular buckled pentagonal nets belong to two main types. In one, the pentagonal nets consist exclusively of water molecules and are three-dimensionally connected by functional groups of the enclosed molecules. Thus, in piperazine hexahydrate [815], and in pinacol hexahydrate [816], the H_2N- and HO-groups respectively form hydrogen bonds to the

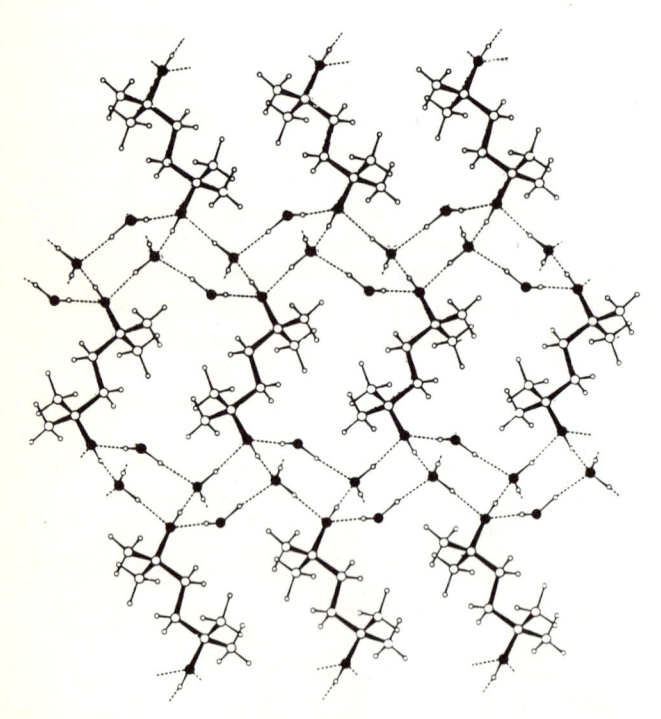

Fig. 21.16. Section of the crystal structure of 2,5-dimethyl-2,5-hexanediol·4H$_2$O. Oxygen atoms are drawn *black*, carbon and hydrogen atoms as *large* and *small circles*, respectively. Note the homodromic infinite chains running in opposite directions and linked in the form of antidromic pentagons [817]

buckled pentagonal layers above and below to produce a cage around the hydrophobic component of these molecules.

The other type of hydrate layer structure occurs when the hydrophobic part of the guest molecule is increased, as in the longer chain alkanediol hydrates such as 2,5-dimethyl-2,5-hexanediol tetrahydrate [817], and in 2,7-dimethyl-2,7-octanediol tetrahydrate [818]. The hydrogen-bonded layers of buckled pentagons are retained, but the gap between them cannot be bridged by hydrogen bonds. Thus the hydrogen-bond cage structure becomes a real hydrogen-bond layer structure, as illustrated in Fig. 21.16. In contrast to the first type of layer hydrates, where the pentagonal nets consist exclusively of water molecules, in the second type they are formed by water molecules *and* by hydroxyl groups of the guests.

Chapter 22
Hydrates of Small Biological Molecules: Carbohydrates, Amino Acids, Peptides, Purines, Pyrimidines, Nucleosides and Nucleotides

In low hydrates, the water molecules are not the major crystal structure-determining components. In salt hydrates, their role is generally to complete the coordination shell of the cations, as in the alkali and alkali-earth salts of organic cations and organic salt complexes which are almost invariably hydrated. In nonionic organic molecular structures, the water molecules fulfil a secondary role as a filler in a molecular packing that is determined primarily by the organic molecules. Molecules that are relatively simple in shape and pack very efficiently do not tend to form hydrates. The sugar alcohols are an example. Similarly, the pyranose monosaccharides, which are oblate molecules that pack efficiently, form many less hydrates than do the awkwardly shaped disaccharides, nucleosides or oligopeptides.

In contrast to the ices, the clathrate hydrates and layer structures described in the preceding chapter, where four-coordinated water is the majority species present in the structure, the water molecules in low hydrates may be three- or four-coordinated. While the water molecule rarely fails to donate two hydrogen bonds, it may accept one or two, but rarely none. This applies both to the organic hydrates discussed in this chapter and the inorganic salt hydrates where the water oxygen can be coordinated to one or two cations (cf. Table 7.8).

The relatively large number of crystal structure analyses of hydrates is, of course, a consequence of the use of water as the least expensive and most common solvent. The water molecules are small enough that they can stabilize crystal structures by filling voids between molecules for more efficient packing, while also contributing to the lattice energy by additional hydrogen bonds.

The relationship between solubility in water and hydrogen bonding in the crystal has never been systematically explored. In one of the few examples, it is reported that the "good solvent" properties of the methoxy alcohols, such as 2-methoxyethanol, can be associated qualitatively with strong hydrogen-bond donor-acceptor properties [820]. On the other hand, very strong hydrogen bonding in a crystal can reduce the solubility in water, as is the case for D-galactaric acid (mucic acid) [821].

A very informative review of water molecules in hydrated organic crystals was published in 1963 [822]. It reported the results of 62 crystal structure analyses, including the salts of organic cations. By necessity, it was based primarily on analysis of the nonhydrogen distances, $X \cdots O_W$ and $O_W \cdots A$. There was little informa-

Hydrates of Small Biological Molecules

Table 22.1. The distribution of water hydrogen-bond configurations observed in hydrates of the carbohydrates, amino acids and peptides, purines and pyrimidines, nucleosides and nucleotides (number of hydrates)

Configuration type (see Table 22.2)	Carbo- hydrates (22) No.	%	Amino acids and peptides (101) No.	%	Purines and pyrimidines (52) No.	%	Nucleosides and nucleotides (32) No.	%	Total % (207)
Three-coordinated pyramidal									
1A – 1/1D	1	3	25	22	14	20	19	37	24
1A – 1/2D	0	0	11	10	6	9	3	6	8
1A – 2/2D	0	0	0	0	4	6	1	2	2
1A – 3/CD	1	3	3	0	2	3	1	2	2
Three-coordinated planar									
1B – 1/1D	5	20	20	18	13	19	5	10	15
1B – 1/2D	2	8	11	10	3	4	4	8	7
1B – 2/2D	0	0	3	3	2	3	0	0	2
Four-coordinated									
2A – 1/1D	7	28	21	19	14	20	12	23	19
2A – 1/2D	4	15	13	12	4	6	3	6	9
2A – 2/2D	2	8	5	4	6	9	2	4	5
No acceptor									
OA – 1/1D	0	0	7	1	0	0	1	2	2
OA – 1/2D	0	0	4	1	1	1	0	0	1
Others (see Table 22.3)	4	15	20		0		0		4

tion available, or speculation, concerning the hydrogen atom positions, and no normalization of the O_W – H covalent bond lengths. Nevertheless, the comments on the individual structures were perceptive and the importance of the three-center (called bifurcated) hydrogen bond was predicted.

The carbohydrate crystal structures provide very limited data on hydrates, despite their ubiquitous solubility in water. Of the twenty simple pyranose and pyranoside crystal structures described in Part II, Chap. 13, only four are hydrates. Of these, three have four-coordinated water, and one, methyl-α-galactopyranoside monohydrate, has a three-coordinated water.

In the eight disaccharide hydrates, six water molecules are four-coordinated and three are three-coordinated. In the four trisaccharide hydrates, the ratio is reversed with eight three-coordinated and three four-coordinated. This suggests that the more complex the molecules, the greater trend toward three-coordination, but the sampling is too small for a definitive conclusion.

There is a much larger proportion of hydrated crystal structures in the other classes of small biological molecules. In these hydrates, the water molecules display a variety of configurations and three-coordination is favored over four-coordination by a factor of 1.7.

Table 22.2. Sixteen configurations observed for water molecules in hydrates of organic molecules (carbohydrates, amino acids, peptides, purines, pyrimidines, nucleosides and nucleotides). First number refers to acceptor bonds, $X-H \cdots O_W$. Second number refers to donor bonds $O_W H \cdots A$. CD indicates chelated configurations

$$\alpha + \alpha_1 + \alpha_2 \quad \begin{array}{l} \geqslant 350° \text{ planar} \\ < 350° \text{ pyramidal} \end{array}$$

Pyramidal

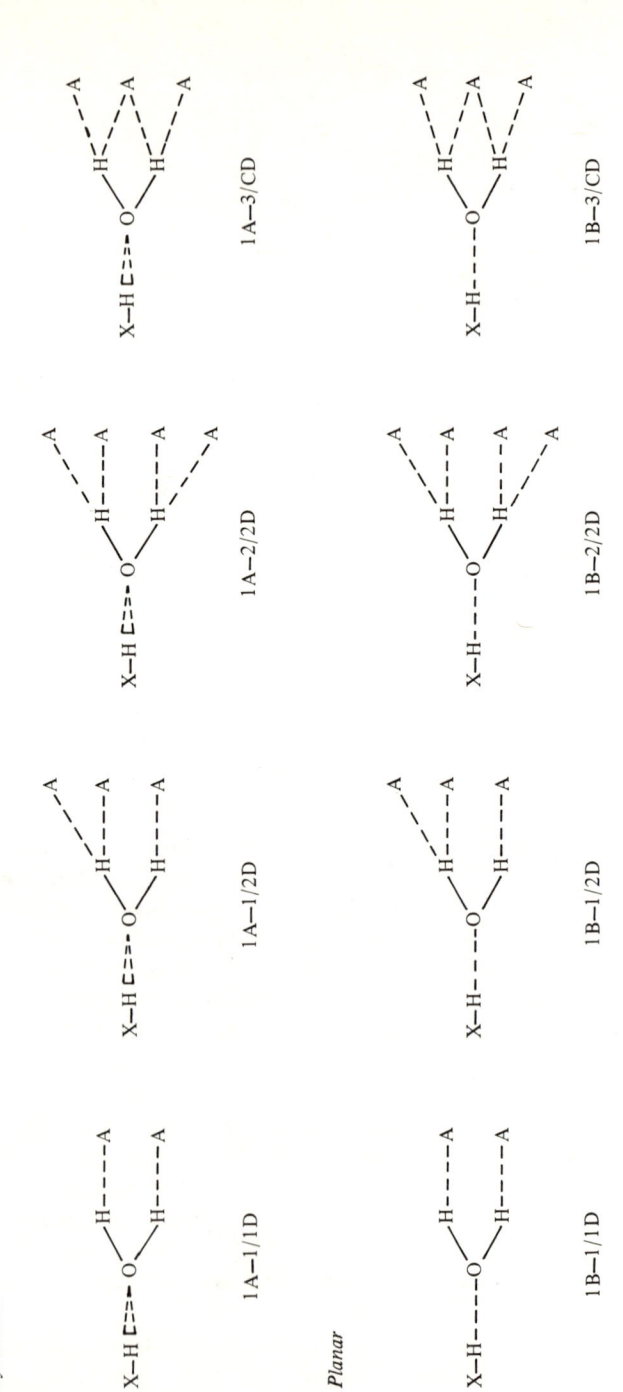

1A–1/1D   1A–1/2D   1A–2/2D   1A–3/CD

Planar

1B–1/1D   1B–1/2D   1B–2/2D   1B–3/CD

Hydrates of Small Biological Molecules   455

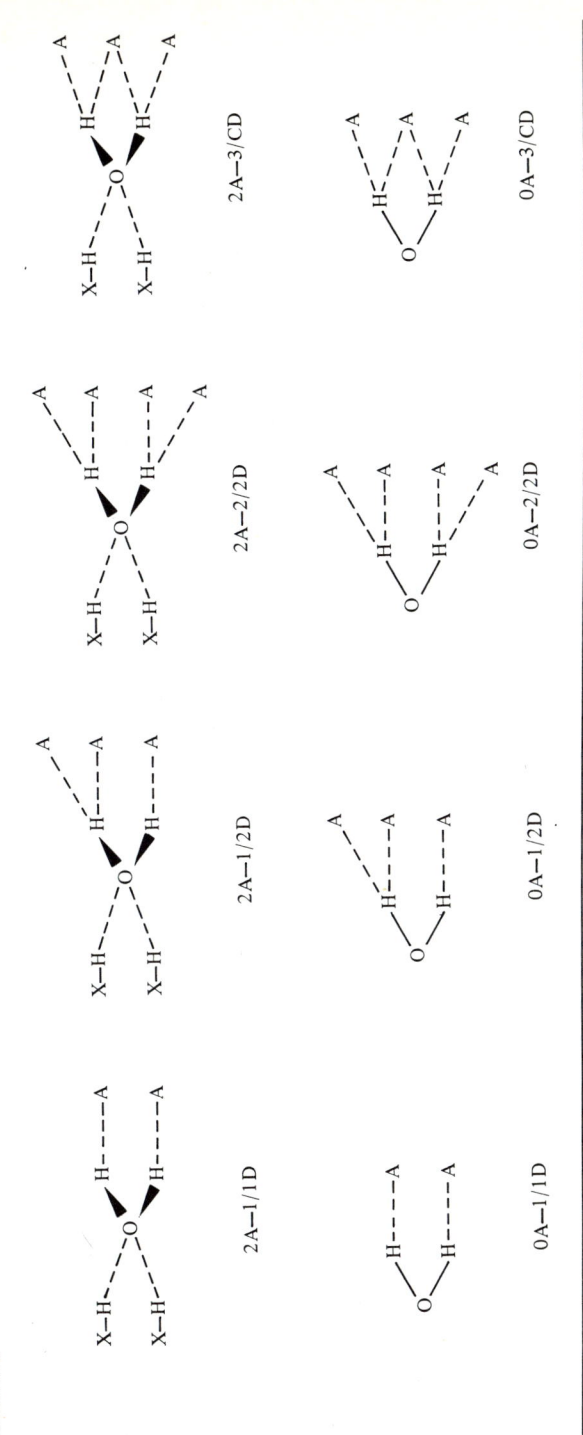

Tetrahedral

2A–1/1D   2A–1/2D   2A–2/2D   2A–3/CD

0A–1/1D   0A–1/2D   0A–2/2D   0A–3/CD

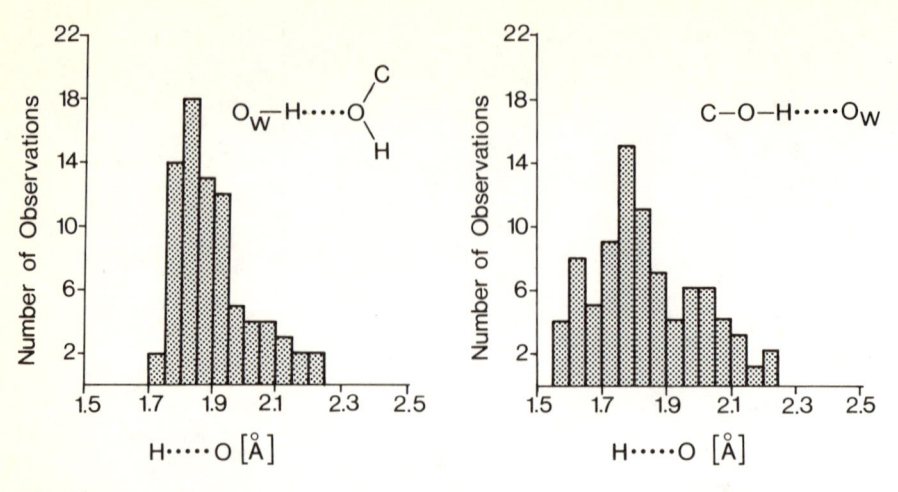

Fig. 22.1. H···O hydrogen-bond length distributions in carbohydrate, purine, pyrimidine, nucleoside, nucleotide and peptide hydrates. Note that OH···O$_W$ < O$_W$H···OH

An analysis of the water molecule configurations in hydrates of the amino acids, peptides, carbohydrates, purines, pyrimidines, nucleosides and nucleotides is presented in Table 22.1. The key to the 12 different observed configurations is given in Table 22.2. The water molecules, which are involved only in two-center bonds constitute 63% of the total number of bonds, the others are donors or acceptors of three-center bonds. Similar proportions are observed in all three different classes of molecules. The only possibly significant difference is a trend toward more three-coordinated pyramidal configurations with the nucleosides and nucleotides.

The bond length and bond angle distributions for the O$_W$H···O, and C–OH···O$_W$ bonds are shown in Fig. 22.1. The two-center bonds and the major components of the three-center bonds have a relatively narrow distribution with most probable H···O bond lengths of 1.75 Å and 1.87 Å respectively. The minor components of the three-center bonds extend from 1.9 to 2.25 Å.

The amino acid and peptide hydrates, where reasonably reliable data are available on the water hydrogen atom positions, show a greater variety of configurations around the water molecules. In addition to those shown in Table 22.2, there are a few examples of configurations in which the water oxygen accepts three hydrogen bonds, shown in Table 22.3. In these configurations, at least one of the H···O$_W$ distances exceeds 2.5 Å. All types of donor groups are represented, and there are a few examples of an NH$_2$ group forming a chelated donor; in these cases both H···O$_W$ distances exceed 2.3 Å. This greater variety in water environment is a consequence of the large proportion of the zwitterion groups, $\overset{+}{N}H_3$ and $\bar{C}OO$, which enhance the electrostatic character of the bonding. It gives rise to a higher proportion of three-center bonds, as noted in the amino acid crystal structures with or without the presence of water [74].

It is interesting to note that from the point of view of identifying water structure in macromolecules in the absence of information about the hydrogen atoms,

Table 22.3. Some water hydrogen bond configurations observed only in the amino acid and peptide hydrates. (Numbers in parentheses are the number of examples observed)

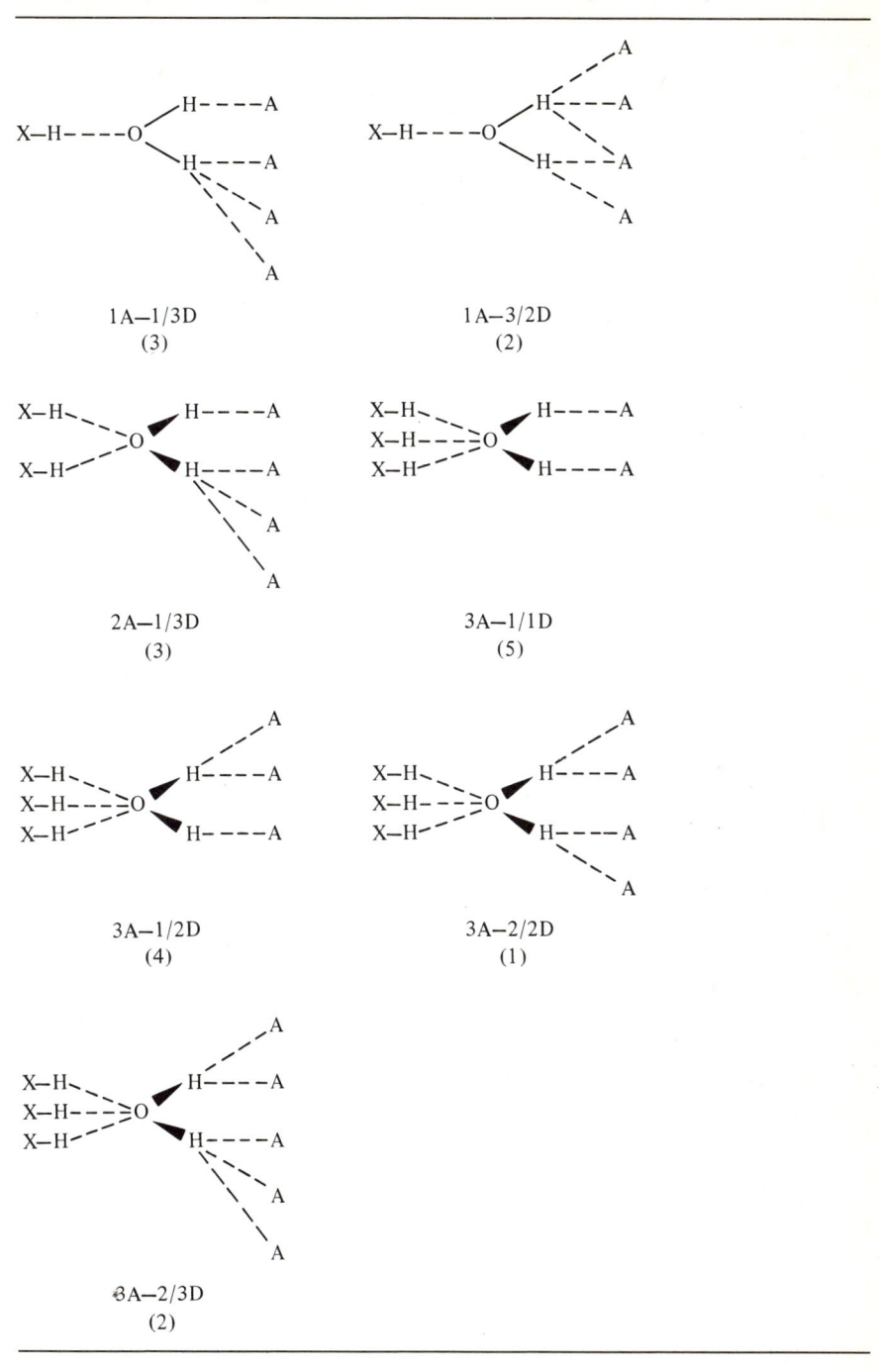

1A–1/3D
(3)

1A–3/2D
(2)

2A–1/3D
(3)

3A–1/1D
(5)

3A–1/2D
(4)

3A–2/2D
(1)

3A–2/3D
(2)

Table 22.4. Hydrates of larger biological molecules for which the water structure has been studied

Compound	Methods	Comments	Reference
Vitamin B_{12} monocarboxylic acid · 15 H_2O	X, N	All water oxygens and seven hydrogens located	[823]
[Phe^{4}Val^6] antamanide · 12 H_2O	X, X	All 12 water oxygens located; over 16 sites including pentagonal patterns, and bifurcated/three-center bonds	[813, 824]
Adenylyl-3',5'-uridine amino acridine complex · 15 H_2O	X	All water oxygens located Choice of two hydrogen-bonding schemes	[825]
Vitamin B_{12} co-enzyme · 14 – 18 H_2O	X, N	More than 140 water oxygens and 4 acetones located. Extensively disordered hydrogen-bonding schemes ambiguous	[826]
Deoxynucleotide phosphate d(CpG)-proflavine complex · 27 H_2O	X	All water oxygens located; Hydrogen-bonding scheme involving five pentagons identified	[812]

these configurations have different heavy atom signatures. The first neighbor coordination of nonhydrogen atoms around the water oxygen within a radius of 3.4 Å ranges from eight for 3A−2/3D to two for OA−1/1D. These signatures, together with the "excluded region" rule of Finney and Savage [127], would seem to form a useful methodology for identifying hydrogen-bond structure in heavily hydrated macromolecules.

Attempts have been made with limited success to interpret the hydrogen-bonding schemes in larger organic molecules. As the molecules become larger than those described previously, the hydration number increases, as does the complexity (Table 22.4). Few, if any, of the functional hydrogen atoms can be located experimentally by X-ray diffraction. Without this information, the hydrogen bonding involving water molecules is generally ambiguous. Even when combined X-ray and neutron diffraction is used, complete and unambiguous interpretation of the hydrogen bonding has not been achieved. Table 22.4 summarizes the results of some of these analyses.

Chapter 23
Hydration of Proteins

Hydration is so important for the structural, physical and biological properties of the proteins, that it has been studied by a variety of methods such as Raman, IR and NMR spectroscopy, calorimetry, gravimetry, molecular dynamics [621, 622, 810, 827–834].

The best and most direct methods to observe the hydration of a macromolecule are X-ray and neutron diffraction analyses carried out at high resolution to better than 1.8 Å. In the crystals of proteins, the macromolecules are heavily hydrated so that between 20% and 90% of the total volume are solvent. In fact, despite their well-defined crystal morphology, crystalline proteins more resemble concentrated protein solutions than the solid state.

As already pointed out in Part III, Chapter 19, in the crystal structures of proteins the amino acid main-chain and side-chain atoms at the periphery of the molecule display larger thermal motion (associated with larger B-factors [Å^2]) than those in the interior. Since the water of hydration molecules are bound primarily to these peripheral protein atoms, they display similar, and in general even more excessive thermal motion. As a consequence, the interpretation of the hydration water structure can be ambiguous, as outlined in Box 23.1.

Box 23.1. Water Molecules in Protein Crystallography

In the X-ray analysis of a protein crystal structure, "solvent molecules" appear as spheres of electron density in difference Fourier maps calculated at the end of a refinement. In a strict sense, the electron density map exhibits preferred *sites* of hydration which are occupied by freely interchanging solvent molecules. This electron density is well defined for the tightly bound solvent molecules and can be as spurious as "just above background" for ill-defined molecules which exhibit large temperature factors and/or only partly occupied atomic positions. Since these two parameters are correlated in least-squares refinement, this gives rise to methodological problems.

The evidence from X-ray analyses for solvent molecule sites can be ambiguous, because the calculation of the electron density distribution is much more sensitive to the structure factor phase angles than to the amplitudes which are measured. Hence the adage "you get out of a Fourier synthesis what you put into it". In protein crystallography, the initial structure model is in general based on phase angles derived from multiple isomorphous replacement, which are only reliable to medium resolution around 3 Å where individual solvent molecules are not yet "seen". In progressive cycles of refinement the structure model of the protein is improved, sidechains are located according to the known amino acid sequence, and data to higher resolution

Box 23.1 (continued)

limits are included stepwise, depending on the increasing quality of the model. If solvent peaks appear in hydrogen-bonding distance to the protein surface, they will be added to the model. Unless a correct or approximately correct solvent structure is included in the analysis, the structure factor phase angles will be biased to exclude it, resulting in the elimination or deemphasis of the solvent electron density in the Fourier maps.

The first serious attempt at interpreting the protein hydration by crystallographic methods was made 1978 on rubredoxin, a molecule consisting of only 54 amino acids [835]. The refinement by conventional (unconstrained) least squares methods included a total of 127 water sites and converged at an R-factor of 12.7% with 1.2 Å resolution data. The water oxygen atoms were added to the model only if their temperature factors were < 50 Å^2 and their electron density was > 0.3 e/Å^3, i.e., corresponding to the electron density expected for liquid water, 0.34 e/Å^3.

Since then, refinement methods were considerably improved with the introduction of constrained least-squares minimization, which significantly reduces the number of variables and increases the ratio data/parameters. Even with these methods, water structure more remote from the protein surface tends to be blurred or is featureless, rendering the interpretation of the less well-defined regions in solvent space more ambiguous or impossible.

Three other rubredoxins have been studied in which all, or almost all, of the water corresponds to discrete peaks and appears to be ordered. However, the complexity of the hydrogen bonding linking these solvent peaks is such that, hitherto, it has not been possible to deduce a rational bonding scheme analogous to that described in Fig. 18.5 for the cyclodextrin hydrates [835a].

The electron density distribution for solvent molecules can be improved if the contribution from bulk water to the X-ray scattering is included in the model. This affects the low-angle X-ray intensity data which are omitted in early stages of the least-squares refinement of protein crystal structures. If they are included in refinement and properly accounted for, the signal-to-noise ratio in the electron density maps is significantly improved and the interpretation of solvent sites is less ambiguous.

There are two ways for including the contribution from bulk water. One was first introduced in fiber diffraction analysis of polynucleotides. It subtracts the X-ray scattering contribution of bulk water from the individual atomic scattering factors used in the structure analysis [700, 836]. The other incorporates the continuous electron density of liquid water, 0.34 e/Å^3, in the electron density calculations. As a result the more localized solvent atoms are more clearly defined in difference electron density maps [837].

The assignment of solvent sites is sometimes ambiguous: Are they solvent or metal ions? In general, there is no clear distinction from a solvent peak as to type of solvent molecule if, for instance, water/alcohol mixtures were used as mother liquor. Moreover, the electron density for water oxygen atoms is initially indistinguishable from that attributed to cations such as Na$^+$, K$^+$, Ca^{2+}, Mg^{2+} because the differences in number of electrons can be compensated by variations in occupancy.

Therefore, criteria have to be followed which guide the protein crystallographer when he assigns electron density peaks to solvent molecules. A "water" position is accepted only if the contact distances to protein or other solvent atoms are of the magnitude expected for hydrogen bonding or van der Waals interactions. In many instances, cations can be clearly distinguished from solvent molecules (at least if they are well defined in the electron density) because their temperature factors are, in general, smaller than those of the solvent molecules, and their coordination distances are shorter than hydrogen bonds, with Na$^+\cdots$O and Ca$^{2+}\cdots$O around 2.3 Å instead of ~ 3 Å for O$_W\cdots$O.

23.1 Characterization of "Bound Water" at Protein Surfaces – the First Hydration Shell

If a protein is dissolved in water and the water is gradually removed, there remains a residual amount of water bound to the protein. The quantity of this "bound water" per gram of protein can be measured by the different methods described in Table 23.1a. It is immediately obvious that the results vary considerably, depending on the method used, with 0.25 to 0.52 g water g^{-1} lysozyme and 0.32 to 0.75 g water g^{-1} hemoglobin.

0.32 g Water g^{-1} lysozyme is a reasonable value for "bound water" and corresponds to a monolayer. The results become more consistent if only that fraction of the water is considered which does not freeze in the form of ice when aqueous films of the protein are cooled to liquid nitrogen temperature [810]. The value of 0.32 ± 0.02 g water g^{-1} lysozyme obtained with the three different methods listed in Table 23.1b is consistent and within the range described in Table 23.1a. This is an average of two water molecules per amino acid residue. To relate the observed molar ratio of bound water/protein surface on a molecular basis, the surface exposures of the various functional groups have to be assessed. This has been done for ribonuclease S [838] using the method of Lee and Richards [839] based on the known X-ray structure. A total of 245 possible theoretical water sites has been identified which are in the position to hydrogen-bond to ribonuclease S functional groups.

This corresponds to a hydration level of 0.31 g water g^{-1} protein, a value very close to that obtained experimentally for lysozyme (see Table 23.1). If we add another 130–150 water molecules to cover the lysozyme surface completely, a value of 0.46 g water g^{-1} protein is reached. In this range, IR and NMR experiments indicate a discontinuity suggestive of another kind of more mobile and less tightly bound water.

The hydration of a protein can be described in several steps [622, 810, 827–829]. If the molar ratio water/lysozyme is gradually increased, a number of discrete levels of hydration are observed [810, 828], as established by a variety of methods summarized in Fig. 23.1 and in Table 23.2:

a) The first major event is the ionization of charged or very hydrophilic groups which bind water in the sequence $COO^- > NH_3^+ > NH_2 > COOH$. Protons and water molecules are redistributed until this process is completed at around 0.1 g water g^{-1} lysozyme. The changes in heat capacity (Cp) at a ratio of 0.05 g water g^{-1} lysozyme may also reflect this redistribution, or some other structural or conformational change.

b) Changes in the IR amide I and II bands indicate that hydration of $C=O$ and NH groups occurs simultaneously, but the NH hydration is completed first. The reason might be that the NH group donates one hydrogen bond and binds only one water molecule, whereas the $C=O$ group can bind two water molecules, as described later (see also Part III, Chap. 19).

Table 23.1a. Fraction of "bound water" in hen egg white lysozyme and hemoglobin as measured by different experimental techniques [810]

Method	Viscosity	Diffusion	Sedimentation	NMR	DSC[a]	Dielectric	Isopiestic	Range
Fraction of g "bound water" g^{-1} protein								
Lysozyme	0.34	0.52	0.52	0.34	0.30	0.30	0.25	0.25 – 0.52
Hemoglobin	0.62	0.52	0.75	0.42	0.32		0.37	0.32 – 0.75

b. Fraction of nonfreezing water in films of hen egg white lysozyme determined by different experimental techniques [810]

Method	DSC[a]	IR	NMR	Average
Fraction of "non-freezing water" g^{-1} lysozyme	0.32 ± 0.02	0.31 ± 0.02	0.34 ± 0.02	0.32 ± 0.02

[a] DSC = Differential scanning calorimetry.

Fig. 23.1. Effect of hydration on lysozyme. *I* time-average properties and *II* dynamic properties. The curves are: *a* carboxylate absorbance (1580 cm^{-1}); *b* amide I shift (ca. 1660 cm^{-1}); *c* OD stretching frequency (ca. 2570 cm^{-1}); *d* apparent specific heat capacity; *e* diamagnetic susceptibility; *f* log rate of peptide hydrogen exchange; *g* □, enzymatic activity (log v_0), and ○, rotational relaxation time of an ESR probe (log τ^{-1}) [828]

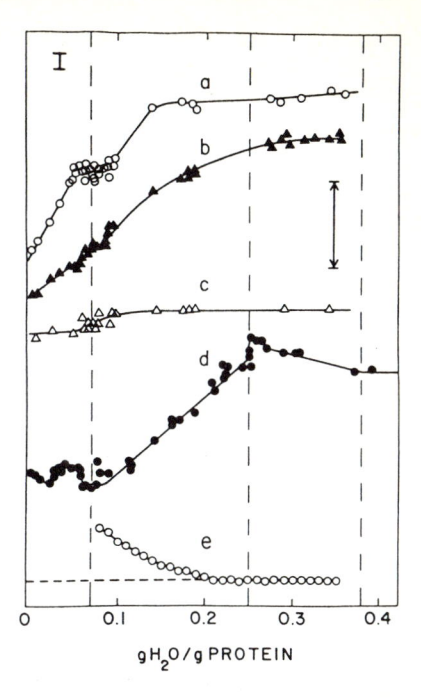

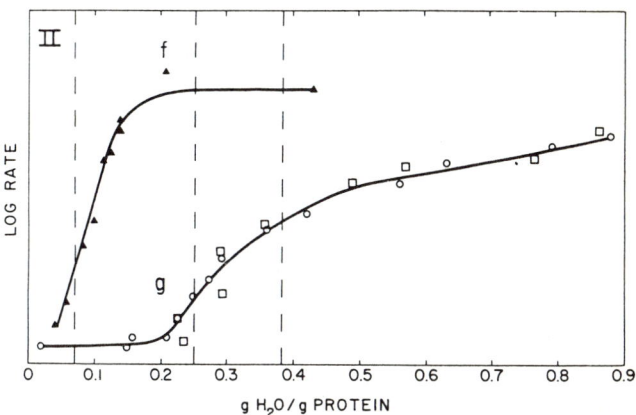

c) The onset of catalytic activity of lysozyme almost coincides with the total coverage of polar sites with water. This might give the enzyme the flexibility it needs for full activity.

d) At 0.38 g water g^{-1} lysozyme, the heat capacity, Cp, reaches the solution value, indicating that all polar and apolar groups of the protein surface are covered by water.

e) A steep increase in catalytic activity occurs above 0.9 g water g^{-1} lysozyme (not shown in Fig. 23.1) has been associated with increased mobility of the reaction product.

Table 23.2. Steps in the hydration of lysozyme. The hydration level is given in g water g^{-1} protein, and numbers in parentheses refer to number of water molecules per lysozyme molecule. Thermodynamic differences at full hydration are excess values per mole of water. The stability of the native conformation was determined by scanning calorimetry. ΔV was estimated from data on ovalbumin [827]

Hydration level	Thermodynamics	Structure	Motion
0–0.07 (0–60)	Large differences for transfer of water from bulk to hydration layer: average values, $\Delta \bar{G} = -1.5$, $\Delta \bar{H} = -17$ kcal mol^{-1}	Structure of protein at low hydration not detectably different from solution	
	Heat capacity between that of ice and liquid water: $\Delta \bar{C}p_1 = -0.2$ cal K^{-1} g^{-1}. Normalization of pK at 0.05 g g^{-1}. 'Knee' in adsorption isotherm. Native state very stable	Water bound principally to charged groups (ca. 1 water/charged atom site)	Water mobility ca. 100× less than for bulk water; increased motion with increased hydration. Bound ligand mobility τ constant from 0–0.2 g g^{-1}: $\tau = 4 \times 10^{-9}$ s. Enzymatic activity negligible.
		At 0.07 g g^{-1}: transition in surface water, from disordered to ordered and/or from dispersed to clustered state; seen in IR and EPR spectroscopic and thermodynamic properties; associated with completion of charged group hydration	
0.07–0.25 (60–220)	Differences for transfer of water small and decreasing with hydration: average values, $\Delta \bar{G} = -0.2$, $\Delta \bar{H} = -0.5$ kcal mol^{-1} $\Delta \bar{V} = 0$ at ≥ 0.15 g g^{-1}. Heat capacity greater for liquid water: $\Delta \bar{C}p = 0.4$ cal K^{-1} g^{-1} Plateau in adsorption isotherm. Stability of native state decreases strongly with increased hydration	Water bound principally to polar protein surface groups (ca. 1 water per polar site). Water clusters centered on charged and polar sites. Clusters fluctuate in size and/or arrangement, as in bulk liquid	Internal protein motion (H exchange) increases from 1/1000th at 0.04 g g^{-1} to full solution rate at 0.2 g g^{-1}

"Bound Water" at Protein Surfaces – the First Hydration Shell

0.25 – 0.38 (220 – 300)	Partial molar thermodynamic quantities close to bulk water values. Region of rapid rise in adsorption isotherm. Transition region in heat capacity	At 0.25 g g^{-1}: start of condensation of water onto weakly interacting, unfilled patches of surface; seen in dynamic and thermodynamic properties	Parallel increase in enzymatic activity and mobility of bound ligand. Fast water motion
0.38 – full hydration (300)	Differences between hydration and bulk water, averaged over all hydration water: $\Delta G = -0.5$ kcal mol^{-1}, $\Delta H = -3.0$ kcal mol^{-1} $\Delta Cp = 2.5$ cal K^{-1} mol^{-1}, $\Delta V = -1.7$ ml mol^{-1} Thermodynamics of unfolding close to dilute solution behavior	Monolayer of water covers surface. Interaction with charged and polar surface groups selects locally ordered arrangements of hydration water. Fluctuation between many instantaneous arrangements, as in liquid water. Selected arrangements combine with bulk solvent; Hydration water heterogeneous	Mobility close to bulk water for part and perhaps most of hydration water. Mobile bound ligand: $\tau = 7 \times 10^{-10}$ s. Enzymatic activity 1/10th solution value. Full internal motions of protein. Dynamic and thermodynamic coupling between hydration water and protein

23.2 Sites of Hydration in Proteins

Much of the following discussion is based on an original paper [840] and on a review [596]. In both articles, X–H···A hydrogen bonds are defined only if the X···A separation is 3.5 Å or less. Had such a cut-off value been used in the analysis of the hydrogen bonding of the small molecule crystal structures described in Part IB, Chapter 7, all of the NH···O=C or NH···O_W interactions would have been included, and only a few weaker ones would have been omitted where the –N(H)H group is the donor. Of the three-center bonds, however, about half would have been categorized as two-center because the minor, long X···A component was beyond 3.5 Å distance.

Oxygen atoms are more preferred ligands than nitrogens. In Table 23.3 are listed the number of protein ligands hydrogen-bonded to water molecules, as observed in 13 well-refined crystal structures (see Table 19.1 for details). Of the total of 2289 water molecules observed in these 13 protein structures, only 1685 are directly hydrogen-bonded to protein atoms, the remaining 604 water molecules are bonded to other waters and can be considered as belonging to the second hydration shell. The majority of the 1685 water molecules is in contact with only one protein ligand, 626 with oxygen, and 237 with nitrogen. This uneven distribution points to the observation made in Part III, Chapter 19 that carbonyl and carboxyl C=O groups are more accessible to hydrogen bonding than peptide or side-chain NH or NH_2 groups, and it suggests that *water has a greater innate tendency to act as donor than as acceptor* [54, 596, 621].

This is also reflected in the number of water molecules hydrogen-bonded to two protein ligands: 283 water molecules have two oxygen ligands, 235 have one oxygen and one nitrogen ligands, and only 28 have two nitrogen ligands. Likewise, among the water molecules bound to three ligands, the majority (132) are in contact with one nitrogen and two oxygens. These water molecules and those bound to four ligands are mostly internal and buried in the globular proteins. Therefore, they are not considered as hydration water.

Thermal B-factors of protein ligands and associated water molecules are comparable. In the structure refinements of human and tortoise egg white lysozyme (HL and TEWL respectively), a comparison was made of the average, apparent displacements $\langle \mu^2 \rangle^{1/2}$ of water molecules against the displacements of the hydrogen-bonded protein ligands [625] (Fig. 23.2). This plot clearly indicates that water molecules bound simultaneously to *two* ligands have a thermal B-factor comparable to those of the associated protein atoms, whereas those water molecules bound to only *one* ligand persistently have a higher displacement $\langle \mu^2 \rangle^{1/2}$, indicating less tight binding and therefore a more diffuse electron density. This is the reason that water molecules bound to only one ligand are more difficult to locate, especially those in the second or third layers.

Hydration of amino acid side-chain and polypeptide main-chain atoms. In Table 23.4, the hydration of the 15 well-refined proteins given in Table 19.1 is analyzed in terms of functional groups in main-chains and in side-chains. About 44% of the water molecules are bound to side-chain atoms, 42% to main-chain C=O and only 14% to main-chain N–H.

Sites of Hydration in Proteins

Table 23.3. Number of protein ligands hydrogen-bonded to water molecules [596]

Protein[a]	ACT	CPA	CYTC	DHFR	ECR	INS	LYS	MB	PLIP	PCY	RNASE	TLN	TRYP	Total
Total H$_2$O	272	279	54	264	94	325	140	388	99	44	78	173	154	2289
Number of water molecules in each category
Protein ligands
1O	91	65	12	74	31	80	31	111	22	13	20	45	31	626
1N	26	26	8	39	11	19	21	33	12	5	12	13	12	237
1O, 1N	27	30	10	24	7	23	16	30	14	7	8	24	15	235
2O	44	41	5	24	17	34	13	34	10	11	9	26	15	283
2N	—	4	2	3	5	4	3	5	1	—	—	1	—	28
1O, 2N	3	3	1	2	1	2	1	7	2	—	1	—	1	24
2O, 1N	14	19	8	10	8	10	5	14	7	5	3	16	13	132
3O	7	5	1	5	2	5	1	3	3	1	2	14	9	58
3N	—	—	—	—	—	—	—	—	—	—	—	—	—	0
1O, 3N	—	1	—	—	1	—	—	3	—	—	—	2	1	2
2O, 2N	2	1	—	—	1	2	—	2	—	1	1	3	1	13
3O, 1N	4	3	—	1	2	3	4	—	4	1	1	3	2	29
4O	—	—	—	—	—	—	1	—	1	—	—	4	—	8
4N	—	—	—	—	—	—	—	—	—	—	—	—	—	0
>4	—	2	1	—	1	—	2	1	—	—	—	2	1	10
H$_2$O bound to protein	218	200	48	182	87	182	98	243	76	44	56	150	101	1685

[a] For abbreviations of protein names, see Table 19.1. (ACT = Actinidin, CPA = carboxypeptidase, etc. in alphabetical order).

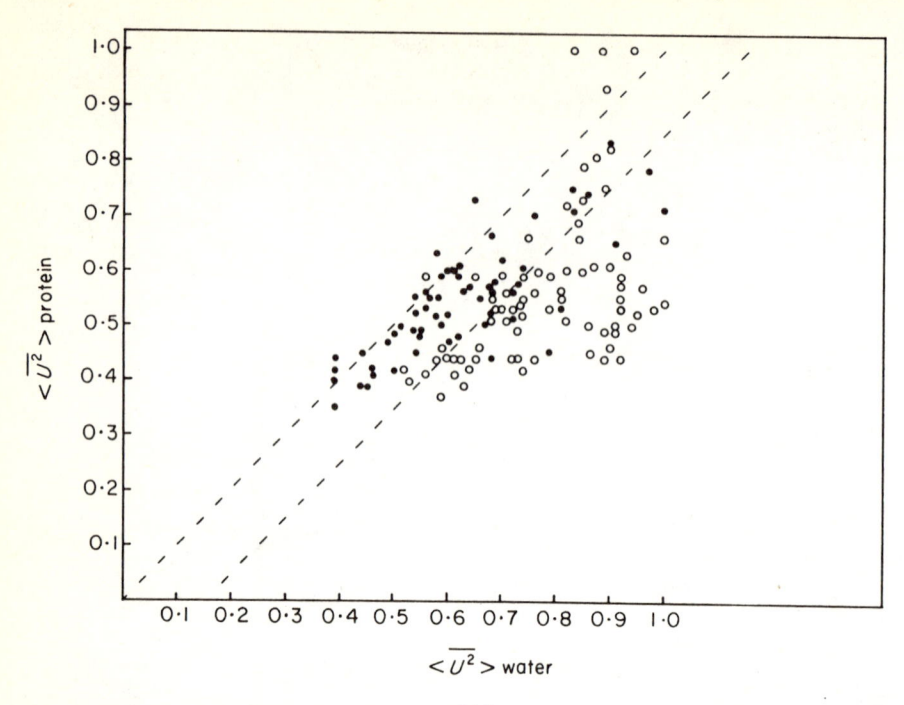

Fig. 23.2. Plot of the apparent displacement $\langle u^2 \rangle^{1/2}$ of the water molecules hydrogen-bonded to protein atoms against the displacement of the protein atoms to which they are bound. ○, Water molecules that make only one contact; ●, water molecules bound to two or more protein atoms. The *diagonal broken line* is the line of equivalence, the *other broken line* marks the waters that have displacements 0.15 Å^2 greater than the associated protein atoms. Data are based on human and turtle egg white lysozyme, the plot is taken from [625]

These data again reflect the differences in accessibility of main-chain C=O and NH groups. In a more quantitative study [625], the solvent accessibility [839] of the functional groups of human lysozyme was derived by rolling a spherical probe of 1.4 Å radius, to represent a water molecule, on the van der Waals surface of the protein and calculating the contact area. The plot obtained (Fig. 23.3) clearly indicates that main-chain NH groups are less accessible than C=O groups, and that the accessibility is especially low in regions of the secondary-structure elements such as the α-helix and the interior strands of β-sheets. The reverse β-turns are frequently located at the periphery of globular proteins and in contact with water molecules [596].

The hydration of amino acid side-chains is more difficult to evaluate because they display higher thermal motion relative to main-chain atoms. Therefore the thermal motion of the associated water molecules is also higher and they are frequently not observed in the electron density maps. For this reason, the apparent hydration of main-chain atoms is more extensive compared with that of side-chain atoms although, on average, the polar side-chains that cover the periphery of the

Sites of Hydration in Proteins   469

Table 23.4. Hydration of protein main-chain and side-chain functional groups. The sample consists of the 15 proteins in Table 19.1 where water molecules were located in the X-ray analysis ([596] in abbreviated form)

a) Total number of hydrogen bonds with water molecules

	Hydrogen bonds	
	Number of	Percentage of
Main-chain C=O	1157	42%
Main-chain N-H	391	14%
Side-chains (defined in **b**)	1231	44%
	2779	100%

b) Extent of hydration of protein side-chains

Side-chain	Number of side-chains in sample	Number of water molecules hydrogen-bonded	Number of water molecules		
			per side-chain	per polar atom	per hydrogen-bond site
Asp	119	225	1.89	0.94	0.47
Glu	92	191	2.08	1.04	0.52
Lys	135	107	0.79	0.79	0.26
Arg	72	88	1.22	0.41	0.24
Ser	176	117	0.66	0.66	0.22
Thr	140	114	0.81	0.81	0.27
Asn	133	119	0.89	0.44	0.22
Gln	94	109	1.16	0.58	0.29
Tyr	125	92	0.74	0.74	0.37
His	57	58	1.02	1.02	0.51
Trp	34	11	0.32	0.32	0.32
Total number or average	1177	1231	1.13[a]	0.74[a]	0.34[a]

[a] Average without Trp.

globular proteins contain more functional groups than the main-chain and therefore should be more hydrated.

The most important side-chains for hydration are Asp and Glu which bind, on average, 2 water molecules per carboxylate group (see Table 23.4b). Whereas Asp and Glu are preferentially located at the outside of the globular proteins and therefore in contact with solvent, the other amino acids are more buried in the interior. They hydrogen-bond not only with internal water molecules but also with protein main-chain and side-chain atoms. Consequently their functional groups are less accessible for water molecules, with only 0.34 water molecules bound per hydrogen-bond site on average.

If we compare the hydration of side-chains with the same functional groups but with different chain lengths, it is observed that, contrary to expectation, those with

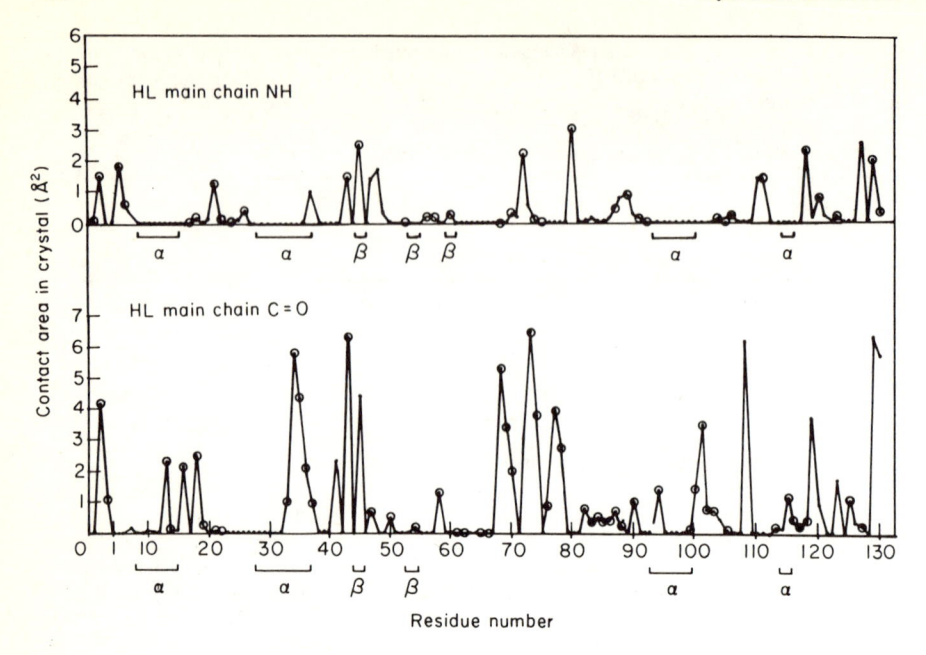

Fig. 23.3. Contact areas of main-chain N–H (*top*) and C=O (*bottom*) groups in human lysozyme. *Circles* indicate ordered water molecules hydrogen-bonded to the appropriate groups. α and β mark α-helical and β-pleated sheet regions [625]

the shorter chain length are less well hydrated than the "more flexible", longer sidechains. For the pairs Asp/Glu the number of water molecules per side-chain is 1.89/2.08 and for Asn/Gln, the ratio is 0.89/1.16. This feature can be explained by the back-folding of the side-chains and hydrogen bonding with adjacent mainchain C=O and N–H groups [596]. It occurs by far more frequently with the shorter side-chains Asp, Asn compared with Glu, Gln, which tend to hydrogenbond preferentially with water molecules. A ratio of 0.66/0.81 is obtained with Ser/Thr, where Ser is more flexible and less well defined than the bulkier Thr, as reflected in the number of water molecules per side-chain (see Table 23.4b).

23.3 Metrics of Water Hydrogen Bonding to Proteins

Water molecules are small probes. They can hydrogen-bond to protein functional groups with few steric limitations, in contrast to the hydrogen bonds between protein main-chain and side-chain atoms which have been the topic of Part III, Chapter 19. The hydrogen bonds involving water are therefore considered to reflect the intrinsic hydrogen-bonding properties of a protein.

Main-chain to water hydrogen bonds. If water molecules hydrogen-bond to the main-chain C=O groups, they associate preferentially in the direction of the oxygen lone-pair orbitals. This is clearly indicated in Fig. 23.4a and by the data given

Metrics of Water Hydrogen Bonding to Proteins

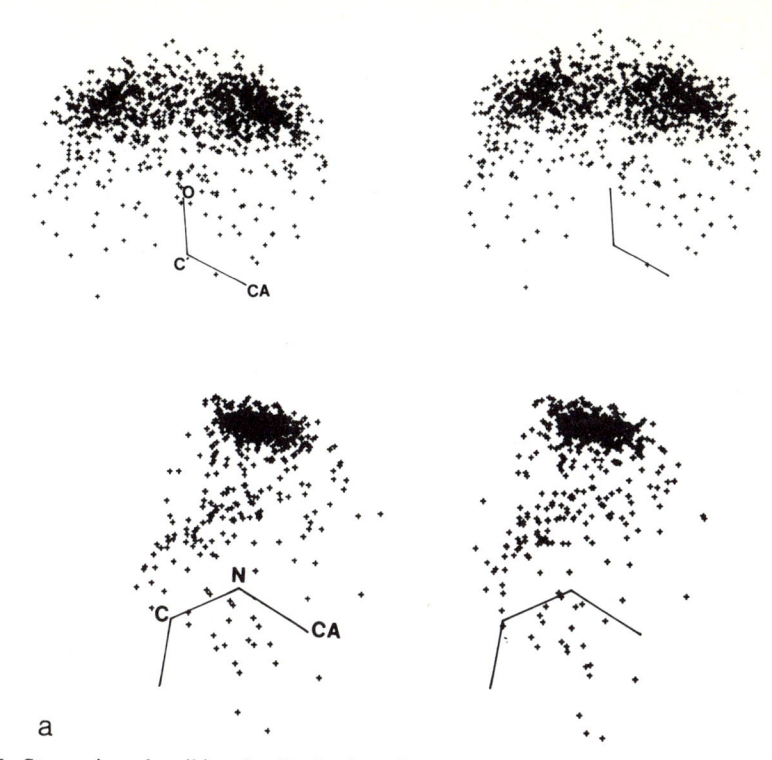

Fig. 23.4a, b. Stereo views describing the distribution of water molecules around protein atoms with an X···O_W distance of 3.5 Å, sometimes 5.0 Å. The 16 protein crystal structures used in this analysis were all refined to better than R = 26% at a resolution of ≤1.7 Å. **a** Main-chain C=O (*top*) and N–H (*bottom*) groups, **b** (next pages) side-chain atoms of Ser, Thr, Tyr (with views *on* and *along* the phenol ring), Asp, Glu, Asn, Gln, Arg, Lys, His [within 3.5 Å (*top*) and 5.0 Å (*bottom*)], Trp [within 3.5 Å (*top*) and 5.0 Å (*bottom*)]. Atom designations are *A, B, C, D, E, Z, H* for α, β, γ, δ, ε, ζ, η in Fig. 19.1 [840; see also ref. 840a]

in Table 23.5. The mean C=Ô···O_W angle between main-chain C=O groups and the hydrogen-bonded water molecules is 129(16)°. This value is consistent with mean values of C=Ô···H around 135° and C=Ô···O(N) in the range 132 to 140° for main-chain C=O to side-chain NH_2, $\overset{+}{N}H_3$ or OH [596] (see Table 19.6 for a more detailed description).

The linear C=O···O_W configuration with an angle close to 180° is not observed to an appreciable extent in proteins. This suggests that there might be sterical constraints imposed by the folding of the polypeptide chain, or that most C=O groups accept more than one hydrogen bond. In small molecule crystal structures the C=O···H angle is close to 180° only if the C=O group accepts only one hydrogen bond.

The main-chain N–Ĥ···O_W angles are approximately linear, 156±15° (Table 23.5b), with the 90% majority in the range 140°–180° (see also Fig. 23.4a). Angles below 140° occur mainly with three-center bonds where one N–H group donates to two acceptors.

472   Hydration of Proteins

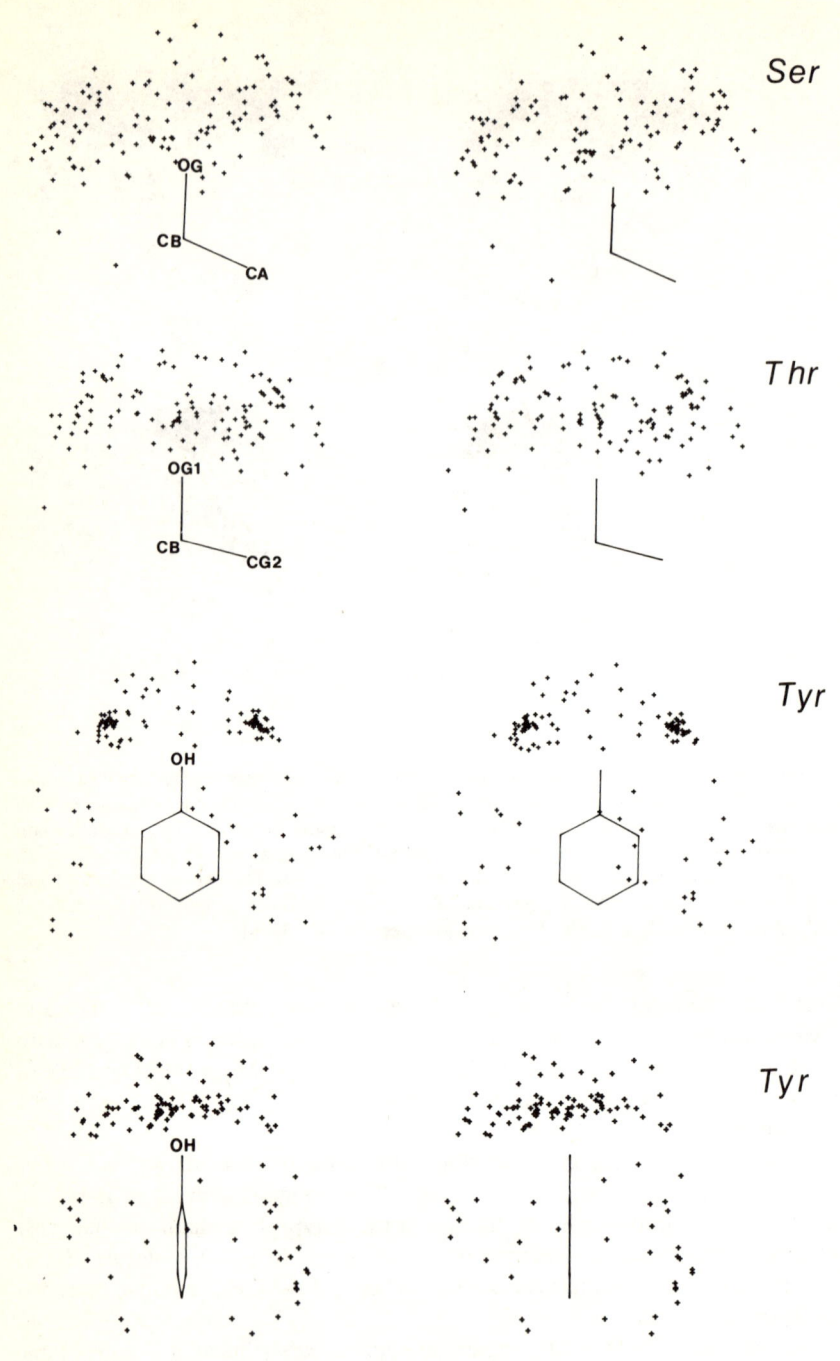

Ser

Thr

Tyr

Tyr

Fig. 23.4b I

Metrics of Water Hydrogen Bonding to Proteins

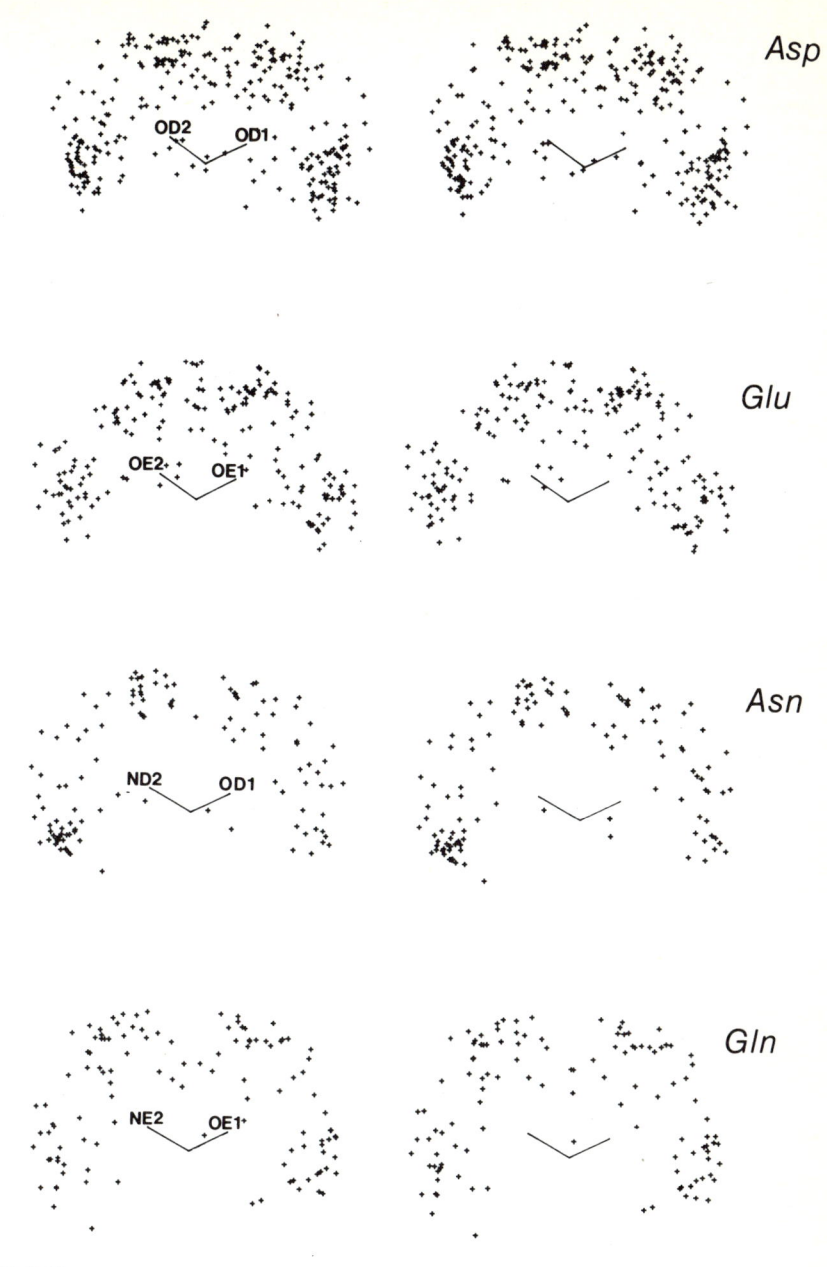

Fig. 23.4 b II

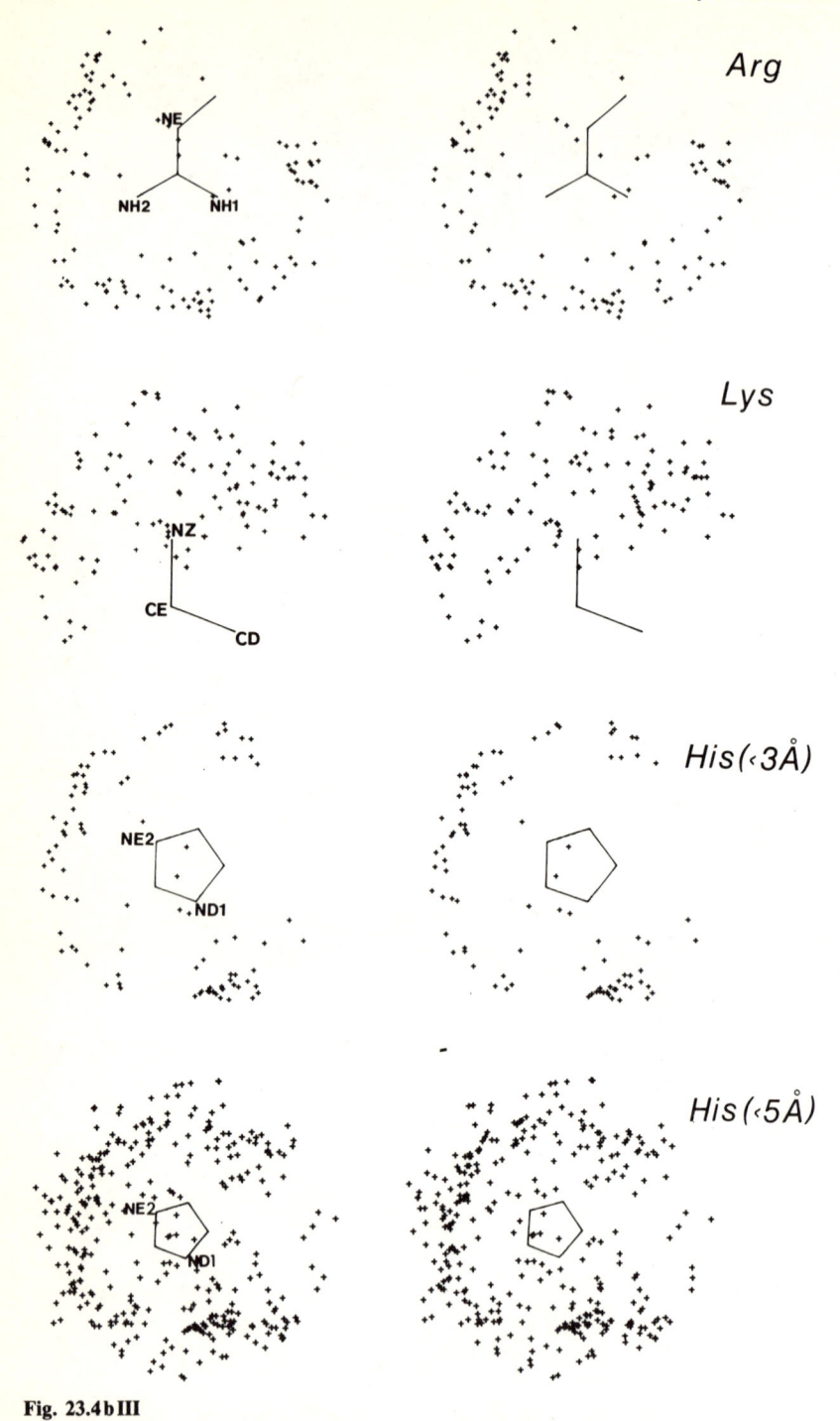

Fig. 23.4 b III

Metrics of Water Hydrogen Bonding to Proteins 475

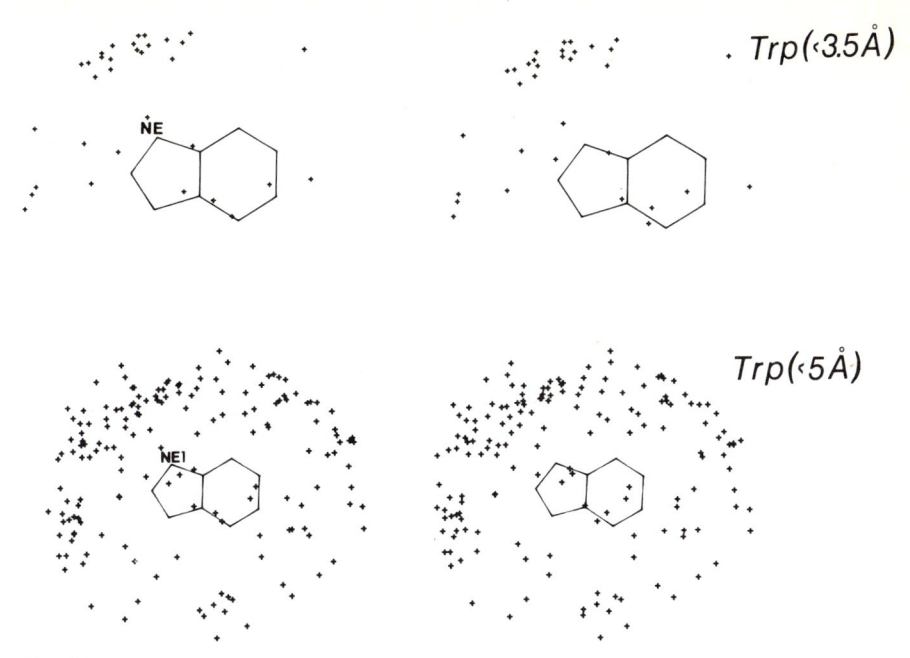

Trp(<3.5Å)

Trp(<5Å)

Fig. 23.4b IV

If distances between main-chain C=O and N–H groups and hydrogen-bonded water molecules are compared (Table 23.5 and Fig. 23.5a), it is apparent that the distribution is narrower about the N–H groups. The reason is that the C=O groups are more exposed and, in most cases, accessible to more than one or two hydrogen-bond donors. Where this results in overcrowding the C=O$\cdots$O$_W$ distances may be beyond the 3.5 Å cut-off limit used in the analysis [596]. In the major parts of both C=O$\cdots$O$_W$ and N–H$\cdots$O$_W$ distribution, the peaks in Fig. 23.5a are at 2.89 Å (2.94 Å in Table 23.5a) and 2.97 Å respectively, as expected from the data for the hydrogen bonds in small molecule crystal structures.

The geometry of side-chain – water hydrogen bonds is less well defined. This is because side-chains display wider thermal motion compared to main-chain atoms. Therefore, the information concerning the hydration of side-chain functional groups is less clear, as shown by the summary given in Table 23.5c and in Fig. 23.5b. This figure illustrates that for most of the side-chains, a considerable number of hydrogen bonds to water molecules are close to or beyond the cut-off limit, 3.5 Å for O$\cdots$O$_W$ and 2.5 Å for the H$\cdots$O$_W$ distance in N–H$\cdots$O$_W$, indicating that the distribution really exceeds these limits. The reason for this behavior must be seen in the less well-defined positions of side-chain atoms and of the water molecules associated with them. This also stresses again the point made in Part IA, Chap. 2.4 that the hydrogen bond is a "soft" long range force and that van der Waals cut-off limits should be avoided.

Nevertheless, some differences are observed from the distance and angle distribution histograms in Fig. 23.5b and from the stereo views in Fig. 23.4b. In the

Table 23.5. Metrical analysis of hydrogen bonds between water molecules and protein functional groups. The data are taken from the 15 protein crystal structures with water positions located, see Table 19.1, Part III, and [596]. Compare with Fig. 23.5 with somewhat different sampling

a) Main-chain C=O···water hydrogen bonds

	Number	Angle (°) $C=\hat{O}···O_W$	Distance (Å) $O···O_W$
Overall	1139	129 ± 16	2.94 ± 0.29
C=O bound to only 1 water	308	133 ± 14	2.91 ± 0.26
C=O bound to 2 waters	258	130 ± 16	2.94 ± 0.33

b) Main-chain N–H···water hydrogen bonds

	Number	N–$\hat{H}$···O_W (°)	H···O_W (Å)	N···O_W (Å)
Overall	389	156 ± 15	2.06 ± 0.20	2.97 ± 0.21

c) Side-chain···water hydrogen bonds

Side-chain	C–$\hat{O}$···O_W (°)	O···O_W (Å)	C–$\hat{N}$···O_W (°)	N–$\hat{H}$···O_W (°)	H···O_W (Å)	N···O_W (Å)
Ser	118 ± 16	2.94 ± 0.31				
Thr	122 ± 14	2.95 ± 0.32				
Asn, Gln	126 ± 16	2.92 ± 0.30		147 ± 19	2.10 ± 0.28	
Asp, Glu	124 ± 17	2.85 ± 0.32				
Lys			120 ± 20			2.93 ± 0.33
Arg				143 ± 23	2.13 ± 0.25	
His				153 ± 16	2.03 ± 0.28	
Tyr	122 ± 15	2.90 ± 0.33				
Trp				153 ± 13	2.11 ± 0.21	

side-chains with hydroxyl groups (Ser, Thr, Tyr), the C–$\hat{O}$···O_W angle peaks around 118°, and the water molecules are in circular distribution about the C–O vector in Ser, Thr (with some slight unsymmetry due to the bulky CγH$_3$ group). In Tyr, the distribution of water molecules about the OH group is bimodal (Fig. 23.4b) consistent with the partial π-character of the C–O bond which restricts the rotation of the C–OH hydroxyl group.

A clustering of water molecules is also indicated around the carboxylate groups of Asp and Glu (Fig. 23.4b), with preferred sites in direction of the electron lone-pair orbitals, as expected. Water molecules which bind simultaneously to both oxygen atoms of the carboxylate in a bifurcated configuration, $-C\begin{smallmatrix}\diagup\!\!\!\!O\\\diagdown\!\!\!\!O\end{smallmatrix}\cdots O_W$ are relatively rare, with ~11% of the total in Asp and ~15% in Glu [840]. A comparable distribution of water molecules is observed around the side-chain functional groups of Asn, Gln, with the O···O_W distance, 2.92 Å, shorter than the N···O_W distance of about 3.1 Å (as estimated from the H···O_W distance in Table 23.5c).

For the positively charged amino acid side-chains Lys and Arg, the circular distribution of water molecules around the vector of the C_ε–$\overset{+}{N}H_3$ group in Lys is

Metrics of Water Hydrogen Bonding to Proteins

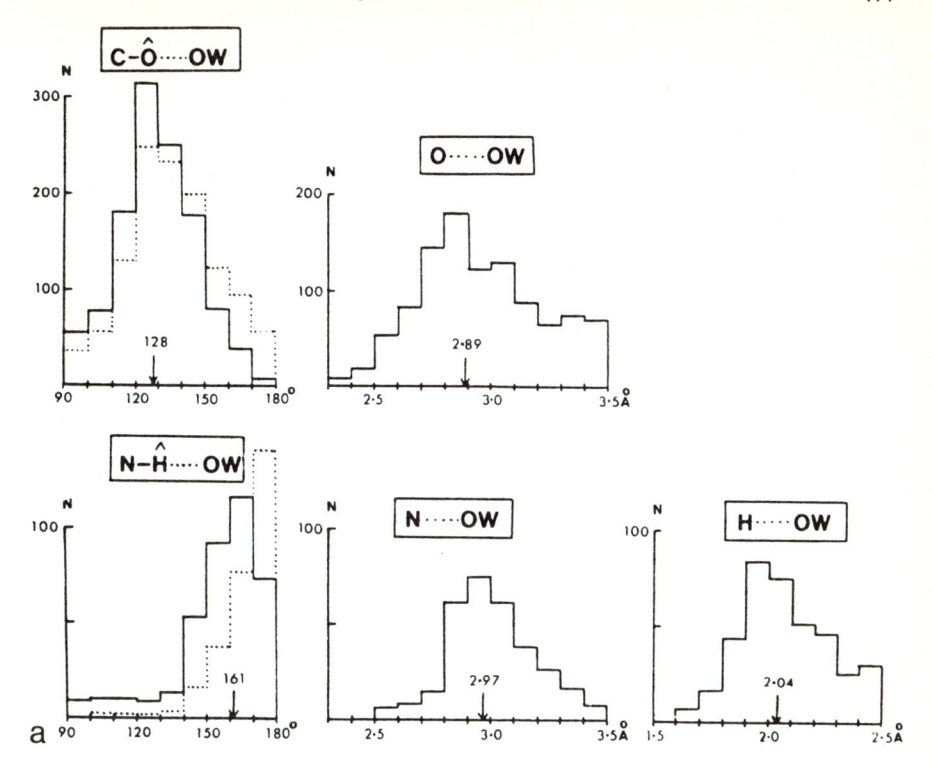

Fig. 23.5. a. Distribution of hydrogen-bond distances and angles in main-chain C=O and N–H contacts with water molecules. *Dotted lines* give population density distributions, i.e., the number of hydrogen bonds per unit solid angle, on an arbitrary scale. Data are for the 15 proteins listed in Table 19.1. **b** (next page) Distribution of hydrogen-bond distances and angles for protein side-chain to water interactions. The center of the main part of the distribution, although mostly ill-defined, is indicated. Numbers for Trp are too few and were omitted [596] (Compare also with Table 23.5 with different sampling but general agreement)

affected by the steric interference of the adjacent $C_\delta H_2$ group. This gives rise to a broad range of $C-\hat{N}\cdots O_W$ angles from 90° to 140°. As in the small molecule structures there is no indication of a trifurcated acceptor configuration

$$-\overset{\oplus}{\underset{H}{\overset{H}{N}}} \overset{\cdot}{\underset{\cdot}{\leftarrow}} H \overset{\cdot}{\underset{\cdot}{\cdot}} \cdot O_W,$$

with $C-\hat{N}\cdots O_W$ angle around 180°. In Arg, there is distinct clustering of water molecules around the $\overset{+}{N_\varepsilon}H$ atom and on either side of the NH_2 groups $N_{\eta 1}$ and $N_{\eta 2}$, close to the guanidyl plane. As in Asx and Glx, there is no preference for bifurcated water hydrogen bonding $-\underset{\overset{+}{NH}}{\overset{NH}{\underset{\cdot}{\leftarrow}}}\overset{\cdot}{\underset{\cdot}{\cdot}}O_W$ with only $\sim 7\%$ of the total water molecules between $N_{\eta 1}$ and $N_{\eta 2}$ and $\sim 3\%$ between $N_{\eta 2}$ and N_ε [840] (Fig. 23.5b).

In His, there is a clear indication for water molecules binding in the plane of the imidazol ring to $N_{\delta 1}$ and $N_{\varepsilon 1}$ (Fig. 23.4b). However, there are also water

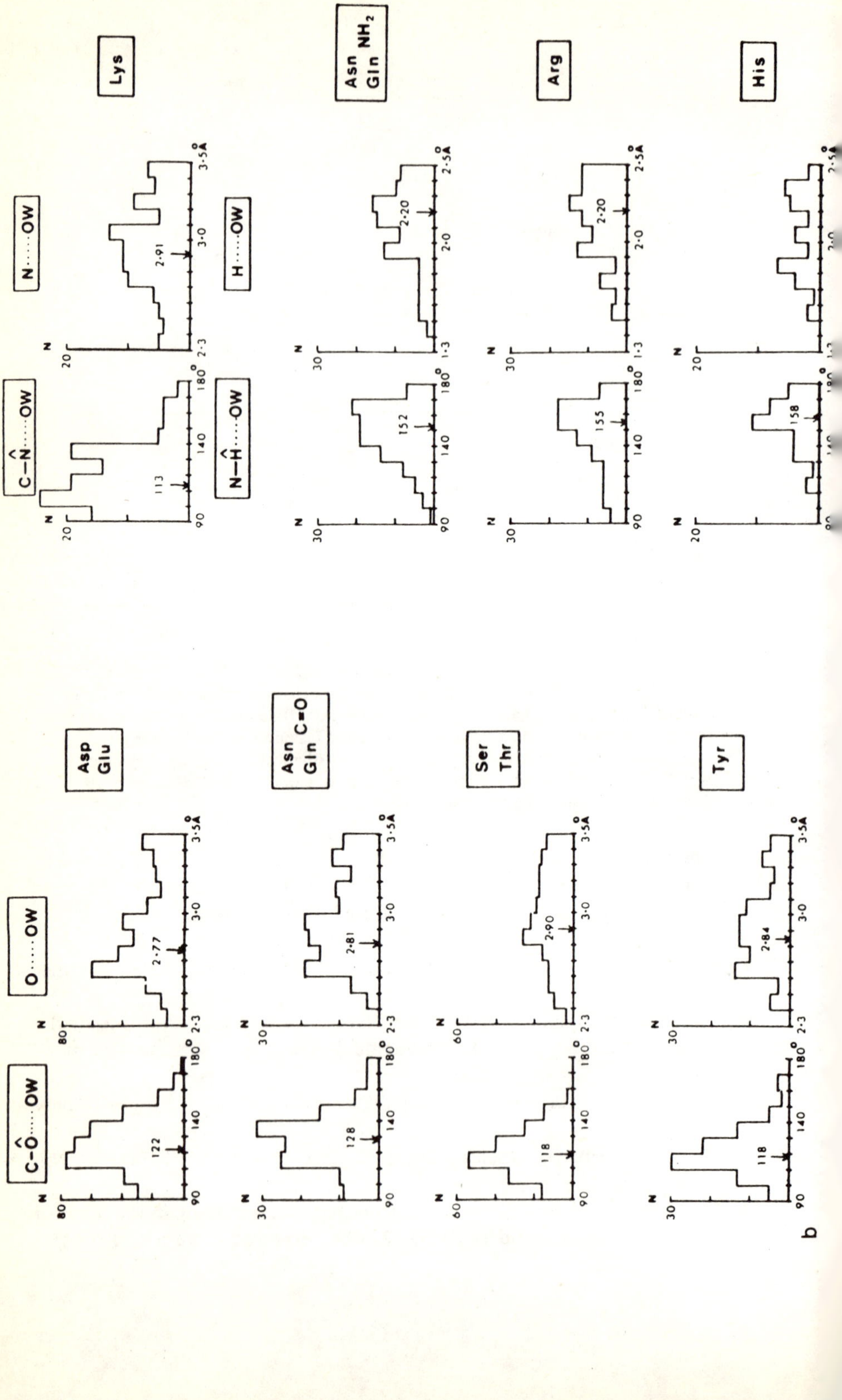
Lys
Asn Gln NH2
Arg
His
Asp Glu
Asn Gln C=O
Ser Thr
Tyr
C–N̂·····OW
N·····OW
N–Ĥ·····OW
H·····OW
C–Ô·····OW
O·····OW
113
2·91
152
2·20
155
2·20
158
122
2·77
128
2·81
118
2·90
118
2·84
b

molecules in contact with $C_{\delta 2}$ and $C_{\varepsilon 1}$. This could be due to ring rotation or to misinterpretation because N and CH_2 groups are not easily distinguished at ~1.8 Å resolution on the basis of the electron density. The $N_\varepsilon \cdots O_W$ distance distribution is bimodal, with peaks at 2.7 Å and 3.1 Å [840], see also Fig. 23.5b. In Trp, the $N \cdots HO_W$ distance is unimodal and flat, and water molecules are located close to the plane of the heterocycle.

The nonpolar side-chains of Ala, Pro, and Phe have only few water molecules within the 3.5 Å sphere, because the contact distance for CH_2 and OH_2 with van der Waals radii of 2.0 Å and 1.4 Å (for oxygen) respectively is ~3.4 Å and close to the 3.5 Å distance limit. Ala shows a distinct clustering between C_β and N–H, which indicates primary binding to N–H and secondary binding to the C_β methyl group. The 5 Å shell around Phe is, of course, more populated than the 3.5 Å shell, and shows no distinct water clustering for certain sites on the phenyl ring [840].

23.4 Ordered Water Molecules at Protein Surfaces – Clusters and Pentagons

The majority of water molecules identified from the X-ray electron density maps are individually bound to protein surfaces, with one or more hydrogen bonds formed with main-chain or side-chain functional groups. Less frequently, but still in significant numbers, two-water clusters are found hydrogen-bonded to the protein surface. Larger clusters are only rarely identified although they are of particular interest as hydration phenomena. In no cases are these clusters "iceberg"-like, a term first used by Frank and Evans to describe the hydration of argon:" ... where the water builds a microscopic iceberg around itself "[1].

The formation of ice clathrates occurs where ordered 'ice-like' water structures' are built around guest molecules *with no or only very weak* hydrogen-bonding character (see Chap. 21). On this premise, ordered water structures cannot form around protein surfaces where strong hydrogen-bonding side-chains predominate because *they* determine the organization of water molecules in the first hydration shell and therefore counteract the formation of clathrate-like structures. On the other hand, patches on protein surfaces of hydrophobic or only weak hydrogen-bonding character could, in principle, be covered by ordered water. It is a dilemma that such water structures will, in general, be loosely associated with the protein surface. They will therefore undergo large thermal motion which makes the electron density more or less uninterpretable. There are a few protein crystal structures,

---

[1] Indicates a footnote in [139] where the term "ice-like" or "iceberg" is more closely defined: "Throughout the rest of this paper we shall use these words, without quotation marks or apology, to represent a microscopic region, either of pure water or surrounding a solute molecule or ion, in which water molecules are tied together in some sort of quasi-solid structure. It is not implied that the structure is exactly like ice, nor is it necessarily the same in every case where the words are used".

480   Hydration of Proteins

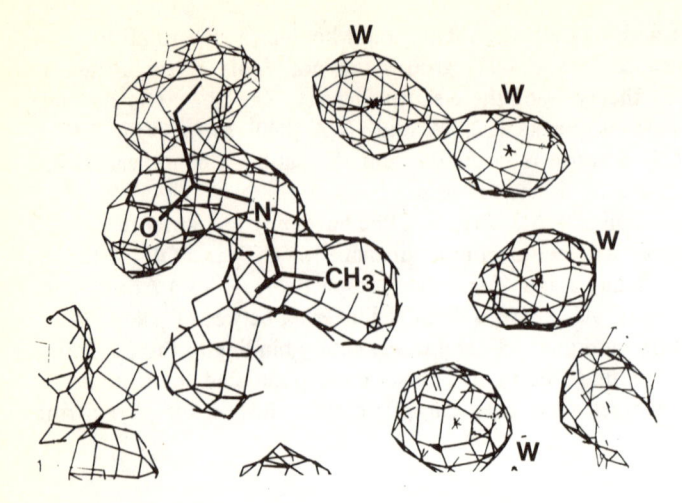

Fig. 23.6. Section of the electron density map of human lysozyme showing the 4 internal water molecules *W* located on a semi-circle around the methyl side-chain of Ala92 [625]

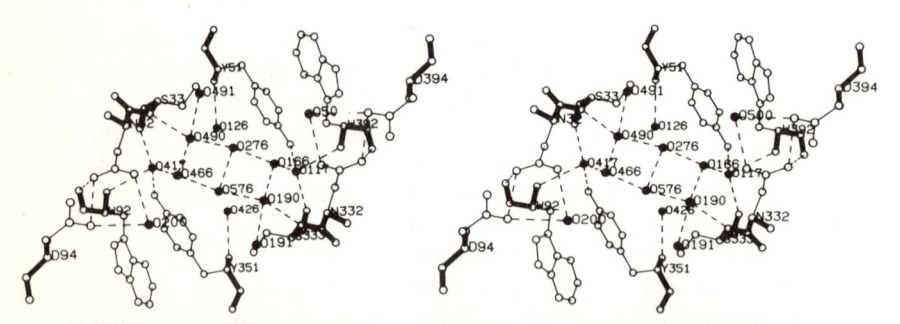

Fig. 23.7. Stereo view of two fused water quadrilaterals located at the interface between variable domains in the dimer of Bence Jones protein Rhe. The two quadrilaterals are related by a twofold axis located between water molecules 0276, 0576 [841]

however, where larger clusters of ordered water molecules have been observed internally or at the protein surface.

Hydration of an alanine side-chain. In human lysozyme, the methyl group of Ala92 is surrounded by four water molecules located on a semi-circle ([625], Fig. 23.6). These hydration waters are buried in the protein which might be the reason why they are so well ordered. We can also assume that such hydration schemes occur at the protein periphery but they are not seen in the X-ray analyses due to larger-thermal motion and/or disorder of the water molecules.

Two fused water quadrilaterals are found at the interface between variable domains in the dimer of Bence-Jones protein Rhe [841]. The quadrilaterals have one edge in common (Fig. 23.7) and are related by a twofold rotation axis perpendicular to this edge. The quadrilaterals are linked to the Bence-Jones protein by

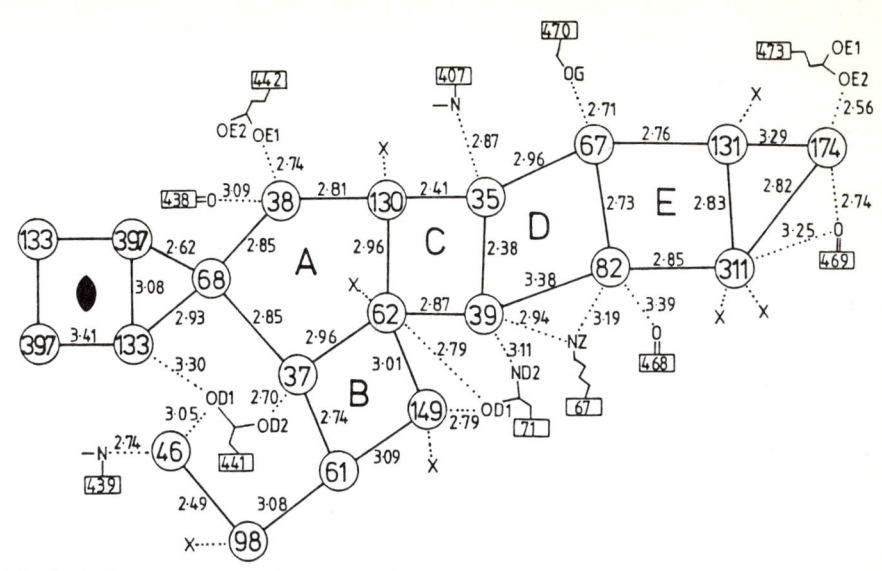

Fig. 23.8. Structure of the solvent network located in the cavity between the two subunits of glutathione reductase. Water molecules indicated by *circles*, amino acids by *boxed numbers* with functional main-chain or side-chain atoms indicated. Hydrogen bonds marked by *solid* or *dashed lines*, distances in Å units, the twofold symmetry axis is shown ♦. An X indicates a solvent molecule ≤ 3.5 Å away from one of the featured solvent molecules. Pentagonal (except W38) and quadrilateral arrangements are nearly planar, angles between the planes are AB = 115°, AC = 125°, CD = 120°, DE = 125° [842]

hydrogen bonding to the very tightly bound, symmetry-related water molecules 0117 and 0417, and to the peptide N–H's of the two symmetry-related Ser33 and Ser333. Because these water quadrilaterals are not buried within the protein but located at the dimer interface formed by two monomer units, they can be considered as hydration waters. Comparable quadrilateral water structures have also been found in cyclodextrin hydrates and in ice clathrates (see Part III, Chap. 18 and Part IV, Chap. 21).

A chain of two triangles, five quadrilaterals and one pentagon formed by water molecules is located in a cavity between two subunits in glutathione reductase [842] (see Fig. 23.8). One triangle is at one end of the chain, the pentagon is near the center and linked, through water molecule 68, to a quadrilateral with twofold symmetry. The water molecules constituting this network are all connected to protein main-chain or side-chain functional groups, or to the other water molecules, which serve as anchor points. There is no obvious structural similarity between this motif and the patterns observed in the clathrate or layer hydrates (Part IV, Chaps. 21, 22), in contrast to the pentagonal water structures found in insulin and crambin (vide infra).

Three fused water pentagons cap an apolar valine side-chain in insulin (Fig. 23.9a). They are anchored on the protein through oxygen atoms Thr8 O_γ and Glu4 $O_{\varepsilon 2}$ which are part of two pentagons, and further stabilized by the nearby

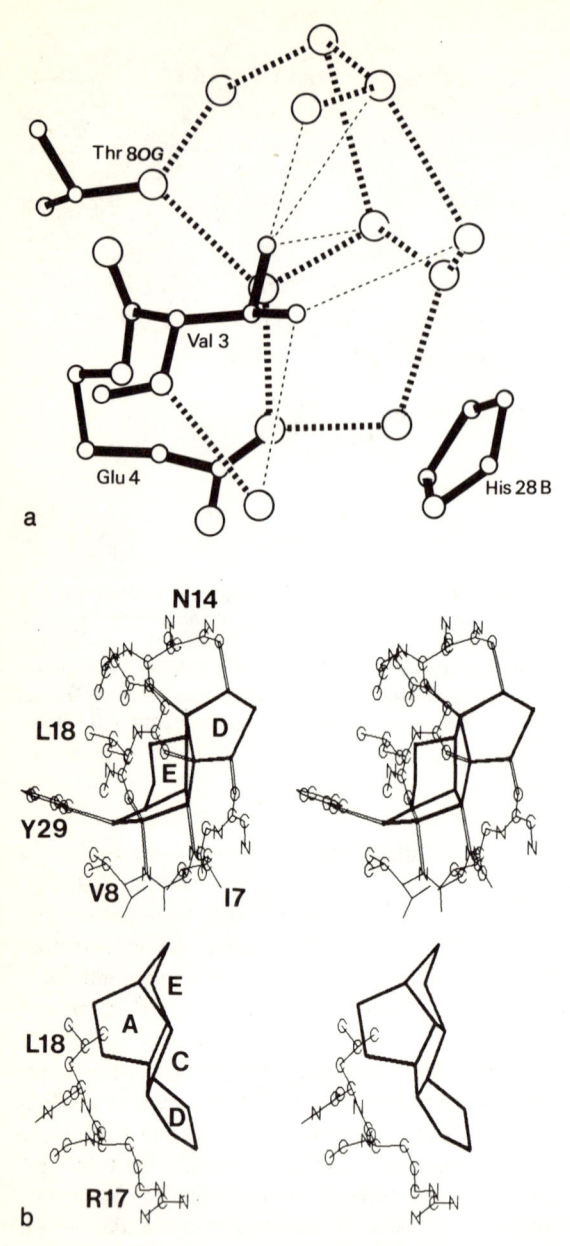

Fig. 23.9. a Three fused water/oxygen pentagons cover the apolar side-chain of valine 3 of the A-chain in insulin. Thin *broken lines* indicate short C−H···O contacts [843]. **b** Two stereo views of a cluster of five water pentagons observed at the hydrophobic surface region of crambin. Note that ring D is well anchored by hydrogen bonds to the protein as are three water molecules of ring A (pentagon B not shown) [811]

Fig. 23.10 a, b. Comparison of fused water pentagons **a** in crambin and **b** in the clathrates of ethylene oxide deutero hydrate and of n-propylamine hydrate. Pentagon arrangements which are comparable in both crystal structures are indicated by *thick lines* [844]

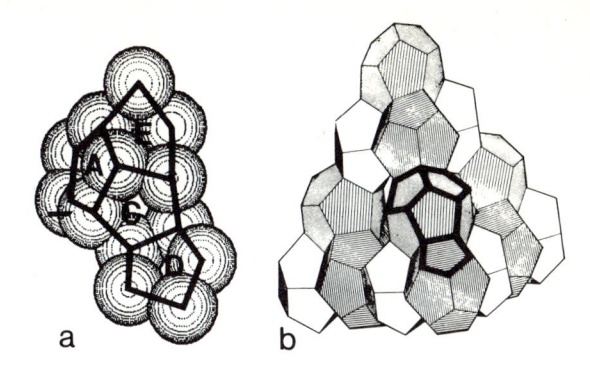

His28 side-chain. The third pentagon is formed exclusively by water molecules. The three pentagons have the near-ideal pentagonal dodecahedron geometry shown in Fig. 21.3 [843].

Five fused water pentagons forming a clathrate-like structure are the predominant feature in the hydration of crambin, a small hydrophobic protein of molecular weight 4720 [811, 844].

It was crystallized from 60% aqueous ethanol and contains, per asymmetric unit, about 85 water molecules (and 7% ethanol), of which 64 were located from the electron density map. Most of these water molecules link polar groups, but several of them are organized in the form of fused pentagons which cover a hydrophobic, intermolecular interface.

The cluster of water pentagons denoted A to E is illustrated in Fig. 23.9b. One of the pentagons, B, extends into the solvent (not shown in Fig. 23.9b), and the other four cover a hydrophobic patch on the surface of crambin. Three of these pentagons, A, C, E, share a common apex. They form a cap which covers the $C_{\delta 2}$ methyl group of Leu18, and pentagon A acts as a water monolayer between c-axis-related molecules. It is anchored by hydrogen bonds to the C=O groups of an α-helix on one side (top of Fig. 23.9b) and by terminal N-H groups of the same helix in symmetry-related position on the other side. Pentagon D is also attached by several hydrogen bonds to the surface of crambin whereas pentagons B, C, E are stabilized mainly by water-water hydrogen bonds.

Are these clusters of water pentagons unique? The crystals of crambin diffract to at least 0.88 Å resolution, far better than those of all the other known proteins, and the X-ray analysis of insulin is based on 1.5 Å data. It appears that these are two fortuitous examples where protein *and* water of hydration molecules are both well ordered. The water pentagons in crambin are arranged in a structural motif which also forms part of the clathrate cage in ethylene oxide deutero hydrate and in n-propylamine hydrate [431, 802] (see Fig. 23.10). It appears that such cage structures can occur in stable form if a basic unit, e.g., pentagons B, C in insulin and C, D in crambin, are anchored at suitable functional groups of the protein which then acts as a matrix. More water molecules can aggregate in additional pentagons which are favored if they can cover hydrophobic residues like the side-chain of Val3 in insulin or the methyl group of Leu18 in crambin. In this context

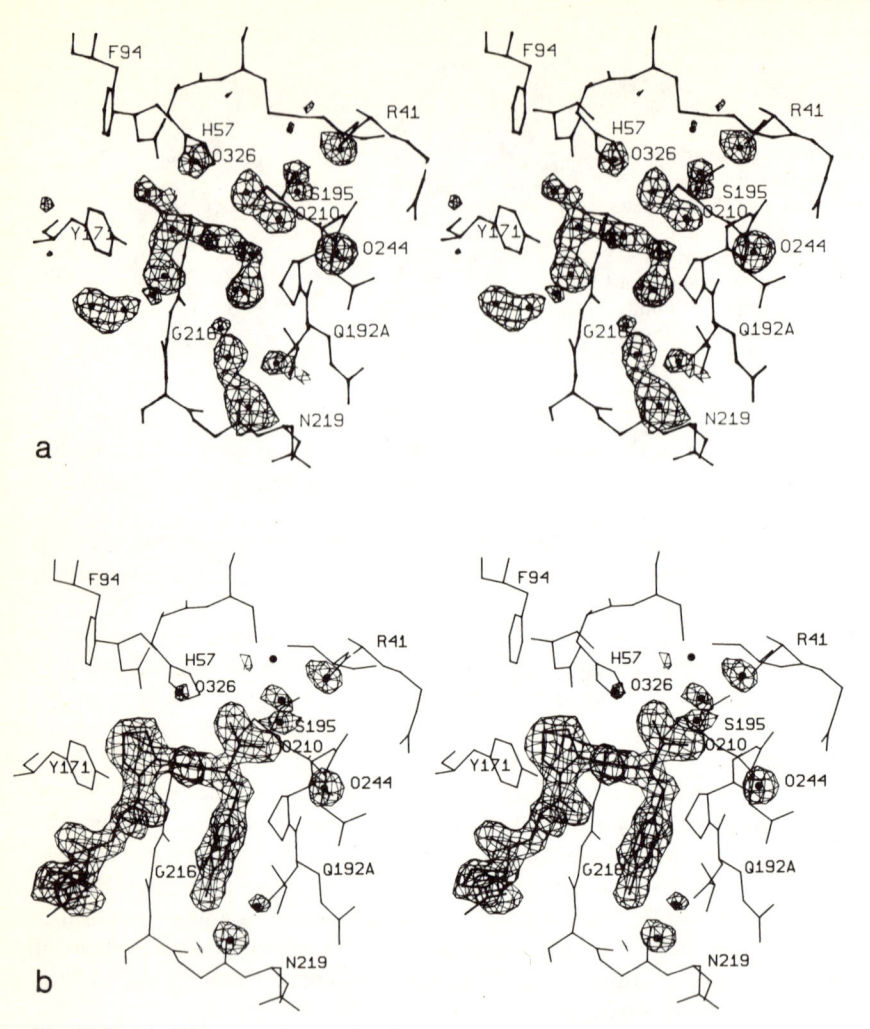

Fig. 23.11a, b. Difference electron density maps showing the hydration of the active site region of *S. griseus* protease A. **a** The native enzyme. **b** The enzyme in complex with the product tetrapeptide Ac-Pro-Ala-Pro-Tyr. Also shown are two internal water molecules O326, O244. Note similarity in electron densities due to bound solvent and to product [845]

the hydration of the methyl group of Ala92 in human lysozyme (Fig. 23.6) is further evidence for the tendency of water molecules to organize into regular motifs around hydrophobic cores.

It appears, then, that ordered structures formed by water of hydration molecules around hydrophobic surfaces may be more common. They are, however, rarely seen in X-ray crystal structures of proteins because the water molecules are only well ordered if the protein donates anchor points in suitable hydrogen-bonding positions. As we shall see in the next chapter, there are also water pentagons found

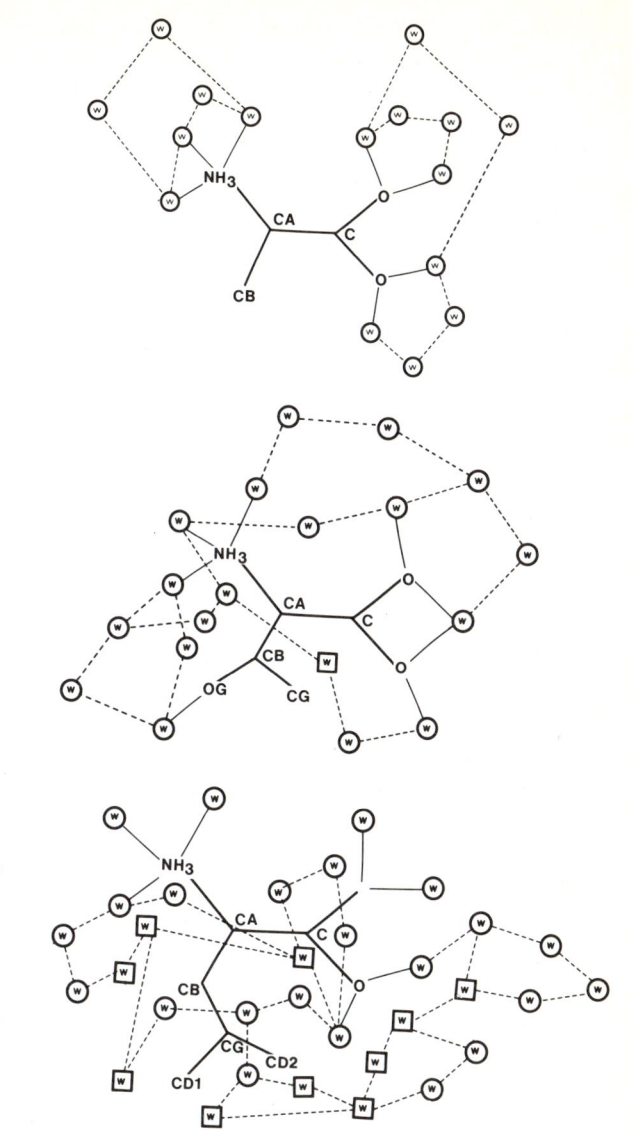

Fig. 23.12. Results of computer studies simulating the hydration of amino acids, (*top*) self-bridging loops of hydrogen-bonded water molecules around alanine; (*center*) polar bridging chains between polar solute atoms of threonine; (*bottom*) water networks associated with the apolar groups of leucine [847]

in crystal structures of oligonucleotides and we conclude that they are, in fact, one of the motifs in the hydration of biological macromolecules.

Hydration water molecules indicate substrate-binding sites. In crystal structures of native proteins, the active sites are usually hydrated if they are not in direct contact with symmetry-related protein molecules. Since the substrates or inhibitors are recognized by the protein and bound to its active site by hydrogen bonds and/or by insertion of hydrophobic residues into hydrophobic pockets, it is not surprising to find, in the native protein, water associated in positions which are the binding sites of substrates and inhibitors. The hydration of an active site consequently reflects the positions of atoms (especially functional groups) of a bound substrate and can be used as a guide for molecular modeling in drug or protein design.

A good example for such bound water molecules is *S. griseus* protease A, which in the native state has a cluster of water molecules associated with the active site [845] (see Fig. 23.11a). These water molecules are replaced by the product tetrapeptide Ac-Pro-Ala-Pro-Tyr (Fig. 23.11b). It is striking to see how the water cluster resembles the shape of the product in the hydrophilic and hydrophobic areas. Another example where a substrate mimics the water molecules, is found in proteinase K. A water molecule which is tightly bound by hydrogen bonds donated from Ser224NH and Asn161N$_\delta$H is located in the "oxyanion hole" of the active site. In this hole the C−O$^-$ oxygen of the transition state intermediate is located during the enzymatic reaction. The negative charge of the C−O$^-$ is stabilized by hydrogen bonds donated by Ser224NH and Asn161N$_\delta$H, the same residues, which also bind to the water molecule [846].

Water pentagons as structural motif in computer simulation studies. In recent years, a number of publications reported computer simulations of aqueous solutions of biological molecules [831−834]. In most studies the overall distribution of water molecules around a solute molecule was considered, and the organization of water molecules into distinct motifs was described [847]. In simulations of hydrated amino acids and a nucleic acid fragment, three motifs are frequently found in the water structure (Fig. 23.12):

1. Self-bridging loops (mostly pentagons) of hydrogen-bonded water molecules around each polar solute atom.
2. Polar bridging chains which extend from one polar solute atom to another.
3. Water networks around apolar solute atoms (methyl groups), or further from polar solute atoms (second and higher hydration layers) consistently included pentagonal motifs.

These studies, in combination with recent crystallographic data from cyclodextrins, proteins, and nucleic acids (see next chapter), clearly demonstrate that water of hydration around solute molecules is organized in certain recurrent patterns. The most favored motif is the pentagon, because the internal angle of 108° is close to the H−Ô−H angle in water, 105°. The four- and six-membered rings found in clathrates and cyclodextrin hydrates, are also expected to play a role in the hydration of biological macromolecules.

Chapter 24
Hydration of Nucleic Acids

The conformation of DNA, but not of RNA, is determined by water activity. If salts or organic polar solvents are added to an aqueous solution of double helical RNA or DNA, the water activity changes and water molecules are withdrawn from the outer hydration shell. This has almost no influence on the molecular structure of RNA which under all conditions maintains its A-type conformation. It transforms only to a comparable A'-RNA with 12 instead of 11 nucleotides per double helix turn if the salt concentration is raised beyond a certain level ([522, 751, 848–850], Fig. 24.1).

In contrast to RNA, DNA is polymorphic. Under low salt conditions or at high relative humidity, DNA adopts the B-form usually considered to be biologically active. With increasing addition of salt or of polar organic solvents (synonymous with reduced relative humidity or removal of available water of hydration), and with certain types of counterions, DNA and double-stranded synthetic polynucleotides transform from B-DNA to the A-, C-, D-, Z-forms (see Table 24.1 and Fig. 24.1. Only the A-, B- and Z-DNA structures, which have thus far been determined in detail by single crystal diffraction methods, are of structural interest and they are considered in the following sections.

24.1 Two Water Layers Around the DNA Double Helix

DNA is fully hydrated with about 20 water molecules per nucleotide [522, 853–862], as indicated in Fig. 24.2. The hydration depends strongly on the water activity a_w, which can be adjusted by addition of salt. At and above complete hydration, B-DNA prevails. It changes to some other conformation if the hydration is reduced, to an attainable minimum of 3.6 water molecules per nucleotide. If water is added in excess, the sample swells but remains in the B-DNA form [856, 857].

Primary and secondary hydration shells around DNA. The hydration of DNA is not homogeneous. It can be described in terms of two shells, as suggested by sedimentation equilibrium studies [857–859, 861], isopiestic measurements [860], gravimetric and infrared spectroscopic investigations [853–855, 862].

These experiments indicate that the structure of the primary hydration shell is not ice-like. It is impermeable to ions and consists of 11 to 12 water molecules per

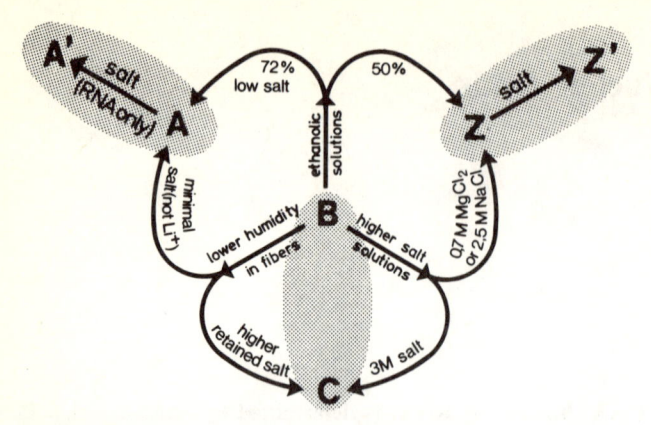

Fig. 24.1. Generalized scheme showing families of naturally occurring DNA's and of RNA. Transitions between members of one family (*within oval shading*) are influenced by salt concentrations. Transitions between families are induced by changing relative humidity and salt content in fibers or films, or changing ionic strength or solvent polarity in solution. Critical salt and ethanol concentrations give midpoints of transitions. Note that RNA is limited only to the A family and that the Z family with left-handed helical sense may be restricted only to certain alternating purine, pyrimidine sequences ([522] after [851])

nucleotide. These water molecules can be grouped in three classes with decreasing binding affinity for (1) phosphate, (2) phosphodiester plus sugar oxygen atoms, and (3) functional groups of bases (Fig. 24.3). It has been suggested that in fibers of DNA below 65% relative humidity, phosphate oxygens are hydrated with 5 to 6 water molecules, and phosphodiester and furanose O(4') oxygen atoms are also partly hydrated. Above 65% relative humidity, the base atoms become hydrated with the addition of 8 to 9 more water molecules. At about 80% relative humidity, the primary hydration of DNA is complete with approximately 20 water molecules per nucleotide (see also Fig. 24.2).

The primary hydration shell is different from bulk water. Of the 20 water molecules per nucleotide, only 11 to 12 are directly bound to DNA. They form a shell which is impermeable to cations [861] and does not freeze into an ice-like state [862]. It is these water molecules which are observed in crystal structure analyses and are hydrogen-bonded to DNA oxygen and nitrogen atoms.

The second hydration shell is permeable to cations. At subzero temperatures, it crystallizes in the form of ice I and therefore resembles bulk water [856]. Because Donnan-type equilibria could have an influence on the structure of this water layer around the DNA polyelectrolyte, it is believed that this second hydration layer is different from an outer layer of bulk water further away from the DNA.

Surface accessibility calculations were carried out on the DNA double helix and on extended single-stranded DNA. They suggest that in the double helical and single-stranded forms, phosphate oxygen atoms are about equally accessible to water. In contrast, bases become 80% buried if DNA folds into a double helix, and its polar character increases [863]. Of the total surface of a DNA double helix, phosphate oxygen atoms account for 45%, base atoms for 20% and the remaining

Table 24.1. Principal crystalline forms of DNA and RNA in fibers: dependence of form and helix symmetry on counterion and relative humidity (equivalent to salt concentration). Shortened from [852]

Polynucleotide	Counterion	Relative humidity (%)	Form	Helix symmetry
Native DNA	Na	75	A	11_1
	Na	92	B	10_1
	Li	57–66	C	9.33_1 (28_3)
	Li	44	C	9.33_1 (28_3)
	Li	66	B	10_1
Poly(dA)·poly(dT)	Na	70	β-B'	10_1
	Na	92	α-B'	10_1
Poly(dG)·poly(dC)	Na	75	A	11_1
	Na	92	B	10_1
Poly(dA–dT)·poly(dA–dT)	Na	75	D	8_1
	Na	up to 98	A	11_1
	Li	66	B	10_1
Poly(dA–dC)·poly(dG–dT)	Na	66	A	11_1
	Na	66–92	B	10_1
	Na	66	Z	6_5
Poly(dA–dG)·poly(dC–dT)	Na	66	C''	9_1
	Na	95	B	10_1
Poly(dG–dC)·poly(dG–dC)	Na	43	Z	6_5
	Na	up to 92	A	11_1
	Li	81	B	10_1
Poly(dA–dA–dT)·poly(dA–dT–dT)	Na	66	D	8_1
	Na	92	B	10_1
Poly(dA–dG–dT)·poly(dA–dC–dT)	Na	up to 98	A	11_1
	Li	98	B	10_1
	Li	66	C	9_1
Poly(dA–dI–dT)·poly(dA–dC–dT)	Na	66	D	8_1
	Na	81	C	9.33_1 (28_3)
	Na	92	B	10_1
Poly(dI–dC)·poly(dI–dC)	Na	66	B	10_1
	Na	75	D	8_1
Phage T2 DNA	Na	60	T	8_1
Native RNA (reovirus)	Na	up to 92	A	11_1
Poly(A)·poly(U)	Na	up to 92	A	11_1
Poly(I)·poly(C)	Na	up to 92	A'	12_1
Hybrid poly(rI)·poly(dC)[b]	Na	75	A'	12_1
Hybrid DNA-RNA	Na	33–92	A	11_1

35% is attributable to sugar atoms. If the total number of water molecules accessible to each nucleotide is calculated, values of 10.5 for A-DNA and 19.3 for B-DNA are obtained, in good agreement with the experimentally determined 20 water molecules for B-DNA previously mentioned.

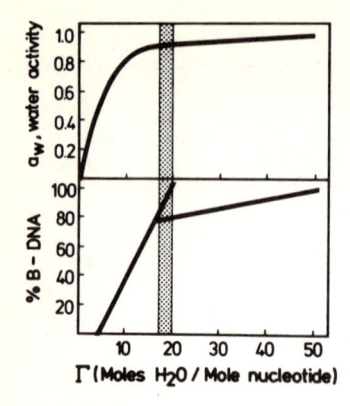

Fig. 24.2. Two plots describing the net hydration of DNA, Γ. *Top:* water activity a_w is changed by different Cs$^+$ salts. With about 20 water molecules per nucleotide, the hydration is complete for the Cs$^+$ salt of DNA (after [859]). *Bottom:* dependence of B-DNA content on net hydration Γ which was adjusted with different alkali chloride concentrations. Γ values for B-DNA = 0% and 100% indicate that minimum hydration of double helical DNA is 3.6 water molecules per nucleotide, and formation of B-DNA is complete with about 20 water molecules per nucleotide ([522] after [857, 859])

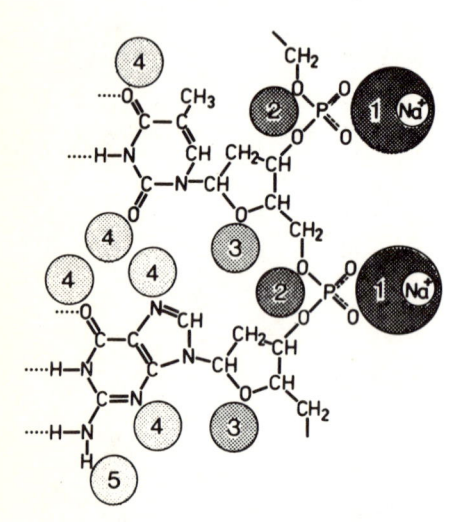

Fig. 24.3. Preferred hydration sites in B-DNA derived from physicochemical studies. Numbers *1* to *5* illustrate strength of binding, in decreasing order. Around phosphate groups, about five water molecules are found, as indicated by *larger circle* [853]

24.2 Crystallographically Determined Hydration Sites in A-, B-, Z-DNA. A Statistical Analysis

The data describing hydration in the nucleic acids are more limited than that of the proteins. In the RNA series, two dinucleoside phosphates [536, 538], one trinucleoside diphosphate [864] and one tetradecamer [666] have been studied by crystallographic methods; in the DNA series, many more oligonucleotides were investigated. The oligonucleotide sequences are such that the molecules are self-complementary and therefore can readily form double helices (Table 24.2). This implies that the duplexes exhibit inherent twofold symmetry which, however, only coincides with a crystallographic twofold axis in a few exceptional cases.

The diffraction qualities of the crystals with A-, B-, and Z-DNA differ, and therefore the results obtained are difficult to compare. Crystals of Z-DNA, in general, diffract to 1.2 Å resolution or better, whereas in A- and B-DNA, diffrac-

Table 24.2. Oligonucleotides which have been studied crystallographically at greater detail

Oligonucleotide sequence	Helix type	Resolution (Å) and R-factor (%)	Number of water molecules located	Some experimental details	Reference
d(GG^{Br}UA^{Br}UACC)	A	1.7; 14.8	68	Synchrotron radiation; crystallized from MPD[a]	[865]
d(GGGGCCCC)Mg	A	2.5; 14	106	Mg-, Na-salt, no spermine, no alcohol	[866]
d(CGCGAATTCGCG)	B	1.3; 17.3	114	Mg-salt, spermine, 60% MPD[a] data collection at 7 °C	[867–869]
d(CGCGCG)	Z	1.25; 18	100	Mg-salt, spermine, 10% MPD[a]	[869, 870]

[a] MPD = 2-methyl-2,4-pentanediol.

Table 24.3. Mean crystallographic temperature factors B (Å^2) for base, sugar and phosphate atoms in the DNA dodecamer d(CGCGAATTCGCG) investigated under different conditions [867]. The mean vibrational amplitudes $\bar{u}$ are obtained according to $B = 8\pi^2 \bar{u}^2$

Experimental conditions	Temperature factors B and mean vibrational amplitudes					
	Bases		Sugars		Phosphates	
	B (Å^2)	$\bar{u}$ (Å)	B (Å^2)	$\bar{u}$ (Å)	B (Å^2)	$\bar{u}$ (Å)
Dodecamer at 25 °C	28	0.60	42	0.73	51	0.80
Bromo derivative d(CGCGAATT^{Br}CGCG) at 7 °C	12	0.39	24	0.55	35	0.67
Dodecamer at 16 K	−3	—	11	0.37	17	0.46

Table 24.4. Total and average numbers of contacts made by the water molecules with other water molecules, with the nucleic acid, the phosphates, the sugars and the bases [871]

Form	To water	To nucleic acid	Phosphates	Sugars	Bases
A-form	176	110	57	10	43
	2.0	1.3	0.7	0.1	0.5
B-form	88	84	44	7	33
	1.4	1.3	0.7	0.1	0.5
Z-form	248	104	54	2	48
	3.0	1.3	0.7	0.0	0.6

tion does not extend beyond 1.7 Å, but can be improved on cooling to lower temperatures or if synchrotron radiation is employed (see Table 24.2). Since the refinement procedures for oligonucleotide crystal structures are similar to those used for proteins, the same problems arise with the location and definition of solvent molecules (or, better, solvent "sites"), from difference electron density maps (see Box 23.1). The (metal) counter ions could not yet be identified in A- and B-DNA crystal structures and were clearly indicated only in Z-DNA.

Temperature factors of atoms in DNA double helices increase from base to sugar to phosphate atoms (see Table 24.3). This holds for all three types of double helices. It implies that water molecules associated with functional groups of bases are better defined than those bound to sugar atoms, and water molecules around phosphate groups are the least well defined in the electron density maps.

This trend is somewhat counterbalanced because the negatively charged phosphate oxygen atoms have a greater affinity for hydration water than the uncharged atoms of base and sugar moieties (Table 24.4). In the three A-, B-, and Z-DNA forms, phosphate groups appear to be only slightly more hydrated than bases, and sugars are much less hydrated. This sequence of hydration affinity derived from crystallographic studies does not agree with the sequence illustrated in Fig. 24.3, which was obtained on the basis of physicochemical data.

The hydration numbers are comparable for A- and Z-DNA and smaller for B-DNA. They are given in Table 24.5 for minor and major groove base atoms, for backbone (sugar) oxygen atoms, and for the unesterified phosphate oxygens. In general, the hydration numbers are small for B-DNA compared to A- and Z-DNA. This trend reflects a greater stiffness and consequently better defined structure of the A- and Z-forms relative to B-DNA, which is known to be more flexible and dynamic [871].

Hydration numbers for phosphate and base atoms are comparable. This does not necessarily mean that they have the same affinity for hydration. Since, as shown in Table 24.3, the temperature factors of phosphate groups are more than twice those of the bases, they and the associated water molecules are less well defined and more difficult to locate from electron density maps. We have to assume that phosphate groups are, in fact, more hydrated than shown in Table 24.5.

Table 24.5. Average number of water contacts (within 3.4 Å) made by hydrophilic atoms in crystalline oligonucleotides in the three DNA forms[a] [871]

Atom type	Number		of	contacts
	A-form	B-form	Z-form	MC[b]
Minor groove	0.4	0.2	0.5	
O2 (Pyr)	0.9	0.4	2.2	2.5/4.1
N2 (Gua)	0.5	0.1	1.2	2.7
N3 (Pur)	1.0	0.4	0.0	2.1/1.4
Major groove	0.5	0.5	0.5	
N4 (Cyt)	1.0	0.7	1.0	2.3
O4 (Thy)	1.0	1.5	–	2.9
O6 (Gua)	1.5	0.4	1.7	3.8
N6 (Ade)	0.7	0.5	–	2.0
N7 (Pur)	1.1	0.4	2.0	1.2/2.3
Backbone	0.2	0.1	0.1	
O5' chain	0.2	0.3	0.1	0.7
O3' chain	0.1	0.0	0.3	0.7
O4'	0.6	0.3	0.2	–
Anionic oxygens	0.9	0.5	1.0	
O1P	1.9	0.6	2.2	2.8
O2P	1.4	0.9	2.3	2.6
O5' terminal	1.0	1.5	1.0	–
O3' terminal	2.5	0.5	1.0	–

[a] Ratios between the total number of water molecules within 3.4 Å of each hydrophilic atom and the number of occurrences of that atom in the crystal analysed.

[b] For purpose of comparisons, the statistical distribution of water molecules in the first hydration shell of nucleic acid constituents determined by Monte Carlo simulations [872] is shown (MC). For O2, first and second numbers refer to thymine and cytosine; for N3 and N7, they correspond to adenine and guanine. When the atom concerned is involved in Watson-Crick base pairing, the MC number should be divided by two.

Nevertheless, the data presented in Table 24.5 suggest that the negatively charged phosphate groups are well hydrated with Z-DNA ≈ A-DNA > B-DNA. Water molecules can be attached to phosphate oxygen atoms in monodentate mode, or they adopt a bidental binding where one bond is linked to one of the free phosphate oxygens and the other bridges to the ester oxygens O(3') or O(5'), or to one of the base atoms in the major groove. Those in the minor groove are too far away and the sugar atoms would interfere sterically. This latter bidentate binding is preferred especially in A-DNA, where phosphate O(1) is directed toward the major groove and better hydrated than O(2), which is exposed to bulk water. In Z-DNA, both phosphate oxygen atoms are equally well hydrated and in B-DNA, where both O(1) and O(2) point to bulk water, O(2) is slightly more hydrated than O(1).

Functional groups of bases are nearly fully hydrated. In general, water molecules are in hydrogen-bonding contact with at least one functional group; they are often involved in bidentate binding, viz. to N(7) and O(6) or N(6) of guanine or adenine, and to N(2) and N(3) of guanine. This is similar to that illustrated schematically in Fig. 20.5 for a G–T wobble base pair. There are also many instances, in all three types of DNA double helix, where one water molecule bridges the two bases in a Watson-Crick pair.

Because hydration water molecules are more tightly bound if they are located in cavities [see, for instance, the "internal" water molecules in proteins (Part III, Chap. 19)], water molecules in the deep major groove of A-DNA and in the narrow minor grooves of B- and Z-DNA are better defined and therefore easier to locate from the electron density maps. Presumably these grooves are, however, not more hydrated than the other, shallower grooves, where it is inherently more difficult to define the positions of water molecules.

In more detail, the hydration of base atoms is generally lower in B-DNA than it is in A- and Z-DNA (see Table 24.4). In A-DNA, major and minor groove functional atoms are well hydrated, with guanine N(2) exhibiting about half as many contacts as the other atoms, and guanine O(6) the maximum with an average of 1.5 water molecules. In Z-DNA, both grooves are also well hydrated except guanine N(3), which is shielded since the base is in the *syn* conformation. In B-DNA, thymine O(4) is most hydrated because the water molecule tends to be in bidentate hydrogen bonding with the methyl group ([871, 875]), similar to the hydration of alanine (Fig. 23.6). In the B-DNA minor groove, guanine N(2) is much less hydrated than pyrimidine O(2) and purine N(3) atoms, probably because this analysis is based on the dodecamer d(CGCGAATTCGCG) where guanine N(2) disrupts the "spine of hydration" located in the minor goove, vide infra.

Monte Carlo simulations partially predict the hydration as found in crystallographic studies. The simulation referred to in Table 24.5 (see also [359]) agrees with the crystallographic studies by predicting that the sugar oxygen atoms are least hydrated, with about $\frac{1}{3}$ to $\frac{1}{4}$ the amount observed for phosphate oxygens and base functional atoms. As in the proteins (Chap. 23), the base C=O groups are involved in more water contacts than N−H. The Monte Carlo simulation does not predict the low hydration in the DNA minor groove around guanine N(2) mentioned above.

24.3 Hydration Motifs in Double Helical Nucleic Acids

There are several recurrent, characteristic motifs in the hydration of A-, B-, Z-DNA. They can be broadly grouped according to [751]:

1. Sequence-independent motifs.
2. Sequence-dependent motifs.

24.3.1 Sequence-Independent Motifs

Motifs of this category are found in the major and minor grooves, at the sugars, and at the phosphates. The water molecules can bridge functional atoms in the *same* (intra) nucleotide such as purine N(7) with O(6)/N(6) groups or thymine O(4) with the C(5) methyl group, or they can occur between *different* (inter) nucleotides belonging to the same or to complementary polynucleotide strands denoted intra- and interstrand respectively in Table 24.6.

Intranucleotide water bridges between a free phosphate oxygen atom and a base atom are typical of A-DNA, where, as already mentioned, O(1)P points toward the major groove and is linked through one or preferentially two water molecules to purine N(7) or pyrimidine O(4). In B-DNA, a water molecule is frequently positioned between phosphate oxygen and thymine methyl group (vide infra) and in Z-DNA, the syn form of guanine favors a bridge N(2)···O$_W$···O(1)P.

The *inter*nucleotide water bridges that are observed most frequently can bridge nucleotides in one strand (intrastrand) or in different strands (interstrand). The most obvious trend in the water bridges is that A-DNA favors two-water bridges whereas one-water bridges dominate in Z-DNA. In B-DNA, both types of bridges are observed at about equal frequency.

Since the separation between adjacent base pairs ~3.4 Å is comparable to the 2.8 to 3.5 Å in water...water hydrogen bonding, the water molecules associated with one base pair can hydrogen bond to water molecules bound to adjacent base pairs. In this manner, extended filaments and nets composed of water molecules in the first hydration layer can be built up which cover minor and major grooves.

Hydration of sugar oxygen atoms is supported by phosphate or base hydrogen bonding. In crystal structures of monomeric nucleosides and nucleotides, the ring oxygen O(4') is less frequently involved in hydrogen bonding (see Part II, Chap. 17). This is also observed in A-, B-, and Z-DNA, where O(4') is located in the minor groove. It is close to N(3) of purine and to O(2) of pyrimidine bases in A- and B-DNA and can be hydrated if the water molecules are engaged in bidentate hydrogen bonding to these base atoms (Table 24.6).

In the minor groove of the B-DNA dodecamer d(CGCGAATTCGCG), such hydrogen bonding to O(4') is a characteristic feature of the "spine of hydration" discussed below.

The O(3') and O(5') oxygen atoms are stronger hydrogen-bond acceptors than O(4'). They can be engaged in monodentate hydrogen bonding to water molecules. Since the charged phosphate oxygen atoms are nearby, the water molecules tend to favor bidentate binding.

The extensive hydration of the phosphate backbone is sequence-independent and structure-dependent. In the three DNA forms, the sugar puckering modes differ in a characteristic manner. Since the phosphates are attached at the sugar O(3') and O(5') atoms, their respective positions are associated with the sugar puckering (see Fig. 24.4 and Table 24.7). In A-DNA with C(3')-endo pucker, adjacent phosphates along the DNA helix are closer together than in B-DNA with C(2')-endo pucker. In Z-DNA, where puckering modes alternate with pyrimidine C(2')-endo and purine C(3')-endo (with torsion angle γ in the trans range), there are two different phosphate···phosphate separations for purine-pyrimidine and for pyrimidine-purine steps which display both shorter and longer oxygen···oxygen separations compared with A- and B-DNA.

These interphosphate distances along the polynucleotide chain of the double helix are only marginally affected by base sequence and are determined solely by the conformation of the DNA-backbone, i.e., by the type of DNA structure. Since the different interphosphate separations are associated with different separations

Table 24.6. Frequently observed inter-residue water bridges [871]

Type of bridge	A-form	B-form	Z-form
Intrastrand			
One-water bridge in minor groove	N3...O$_w$...O4' O2...O$_w$...O4'	N3...O$_w$...O4' O2...O$_w$...O4'	N2...O$_w$...O1P
One-water bridge in major groove	None observed	O4...O$_w$...O4 N7...O$_w$...N7	N7...O$_w$...O4
Two-water bridge in minor groove	None observed	O2...O$_w$...O$_w$...O2	None observed
Two-water bridge in major groove	O6...O$_w$...O6 N7...O$_w$...N4 N4...O$_w$...N4 O6...O$_w$...O4 N7...O$_w$...O4 N7...O$_w$...O1P	O4...O$_w$...O$_w$...O4 N7...O$_w$...O$_w$...N4	None observed
Interstrand			
One-water bridge in minor groove	None observed	O2...O$_w$...O2 N3...O$_w$...O2	O2...O$_w$...O2
One-water bridge in major groove	None observed	None observed	None observed
Two-water bridge in minor groove	O2...O$_w$...O2 N3...O$_w$...O4'	N3...O$_w$...O4' N4...O$_w$...O$_w$...N4	None observed O6...O$_w$...O$_w$...N4
Two-water bridge in major groove	O4...O$_w$...O4 N6...O$_w$...N6 N6...O$_w$...O4	O6...O$_w$...O$_w$...O4 N7...O$_w$...O$_w$...O4 N6...O$_w$...O$_w$...O4	
Between phosphate anionic oxygens on the same strand			
	O1P...O$_w$...O1P O1P...O$_w$...O2P	None observed	O1P...O$_w$...O2P

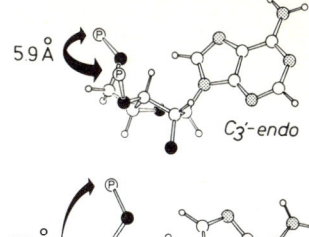

Fig. 24.4. Nucleotides in C(3')-endo and C(2')-endo conformation as observed in A- and B-type polynucleotide double helices. Phosphorus···phosphorus distances are indicated in Å units ([522] after [873])

Table 24.7. Average intrastrand separations between adjacent phosphate group in A-, B-, Z-DNA. Data based on [874]

DNA form	P···P	O(1)···O(1)	O(2)···O(2)	$^{3'}$O(2)···O(1)$^{5'}$	$^{3'}$O(1)···O(2)$^{5'}$
A	6.0(2)	5.8(4)	6.4(4)	7.5(2)	5.5(2)
B	6.6(1)	6.7(1)	7.2(2)	7.9(3)	6.7(1)
Z Step pCp	5.9(2)	6.8(3)	5.8(3)	4.7(4)	7.8(2)
Step pGp	6.6(6)	5.3(2)	8.8(3)	7.2(4)	7.1(4)

between oxygen atoms of adjacent phosphate groups, the hydration pattern around the phosphate groups is also affected.

In A-DNA, the average phosphate-phosphate distance, measured between phosphorous atoms, is small, 6.0(2) Å (Table 24.7), and distances between the free oxygen atoms O(1), O(2) vary between 5.5(2) Å and 7.5(2) Å. The shortest distance between adjacent oxygen atoms is $^{3'}$O(1)···O(2)$^{5'}$, where O(1) points into the major groove and O(2) is close to the minor groove. This distance is shorter than that between like atoms [O(1)···O(1), 5.8(4) Å and O(2)···O(2), 6.4(4) Å] because of the right-handed twist of the A-DNA double helix. We observe a systematic bidentate water hydrogen bonding between *adjacent* phosphate groups $^{3'}$O(1)···W···O(2)$^{5'}$ with directional specificity, i.e., O(1) is on the 3'-side of the polynucleotide chain and O(2) on the 5'-side (see Fig. 24.5a and Table 24.7).

In B-DNA, the phosphate···phosphate separation is larger, the shortest distance between oxygen atoms O(1)···O(1) and $^{3'}$O(1)···O(2)$^{5'}$ being 6.7(1) Å. This separation is too wide to anchor a water molecule in bidentate mode O(1)···O$_W$···(O(1) or O(2)) with good hydrogen-bonding geometry. Therefore, few bridges of this type are seen in the crystal structure of B-DNA, and most phosphate groups are hydrated individually (Fig. 24.5b).

In Z-DNA, the oxygen···oxygen distances differ widely. The short $^{3'}$O(2)···O(1)$^{5'}$ distance, 4.7(4) Å, is bridged by one water molecule O(2)···O$_W$···O(1), where O(1) is in 5'-direction and O(2) in 3'-direction, i.e., *in opposite sense* compared with the shortest $^{3'}$O(1)···O(2)$^{5'}$ distance in A-DNA because the helical twist is also opposite, i.e., left-handed (Fig. 24.5c).

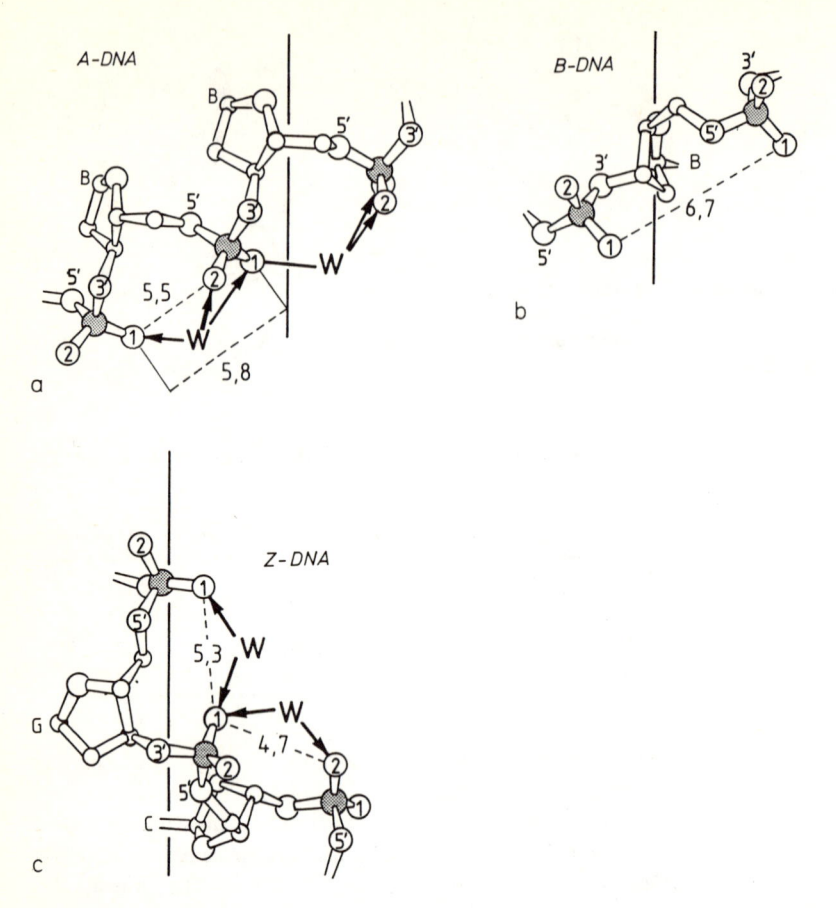

Fig. 24.5a–c. Schematic description of hydration of phosphate groups in A-, B-, Z-DNA. In B-DNA the distance between phosphate oxygen atoms, ~6.7 Å, is too wide to be bridged by a water molecule, in contrast to A- and Z-DNA where phosphate groups are bridged systematically [874]. Numbers are from Table 24.7

In B-DNA, hydration of A–T and G–C base pairs is different. Density gradient centrifugation studies [861] and theoretical considerations [875] suggest that an A–T base pair can bind one to two water molecules more than a G–C base pair. The reason is that the thymine methyl group is exposed and promotes water-water aggregation, as already described in Fig. 23.6 for the organization of water molecules around an alanine methyl group in the enzyme lysozyme. In the hydration of the A–T base pair, a water molecule is trapped between the thymine methyl group and the phosphate group attached at O(5') of the same thymine. For steric reasons, this requires the nucleotide to adopt the C(2')-endo sugar puckering mode characteristic for B-DNA because then base and phosphate groups are further apart than in A-DNA with C(3')-endo puckering, where this hydration of the thymine methyl group is not found.

24.3.2 Sequence-Dependent Motifs

At present, there are only a few cases known of sequence-specific hydration of double helical nucleic acids. With the availability of more crystal structure data, new information concerning sequence-specific hydration is expected. This kind of hydration can have an influence on the recognition of nucleic acids by proteins, and is therefore of great interest.

Water pentagons in the A-DNA major groove in sequences TATA. In the crystal structure analyses of the two isomorphous octanucleotides d(GGTATACC) and d(GG^{Br}UA^{Br}UACC), the same characteristic pattern of hydration is observed in the center of the major groove. It consists of four fused five-membered rings, as illustrated in Fig. 24.6 and in Fig. 24.9a (where only the two central rings are shown).

It appears that the sequence TATA induces the formation of these pentagons. The methyl groups (and the bromine atoms in the related sequence ^{Br}UA^{Br}UA) are hydrated by a ring of six water molecules and two phosphate O(1) oxygen atoms (compare with the hydration of the alanine methyl group in lysozyme, Fig. 23.6). All these water molecules and O(1) atoms are in hydrogen-bonding contact, and the water molecules are further connected with functional groups of the bases. In the center of the major groove are located two additional water molecules. Together with the rings of water molecules surrounding the methyl groups (and bromine atoms respectively), they form the edges of the two central fused five-membered rings.

It seems that this motif of hydration depends on the TATA sequence because the methyl groups are required for formation of the large rings consisting of water and O(1) atoms. In other sequences like TTAA or in G/C sequences, such pentagons are unlikely to occur, and they have not thus far been observed.

A "spine of hydration" in the minor groove of B-DNA has been found in the central AATT sequence of the dodecamer d(CGCGAATTCGCG) [867]. As illustrated in Fig. 24.7 and in Fig. 24.9b, water molecules span O(2) and N(3) atoms of bases in adjacent base pairs, i.e., O(2)···O$_W$···N(3), with additional close contacts to O(4') atoms. These water molecules form the first hydration layer. They are connected by water molecules in the second layer such that each of the water molecules in the first layer is tetrahedrally coordinated. Because the N(2) amino groups in G/C sequences would interfere sterically with this regular "spine of hydration", it is disrupted at both ends of the central AATT sequence in the dodecanucleotide.

In sequences other than A/T, the minor groove is wider. Adjacent nucleotides along the polynucleotide strand are bridged by intrachain water molecules in a systematic pattern (n)purine−N(3)···W···O(4')(n+1) or (n)pyrimidine−O(2)···W···O(4') (n+1) [see Fig. 24.7 (*bottom*)].

The minor groove of Z-DNA binds a "spine of hydration". The water molecules are primarily hydrogen-bonded to the O(2) atoms of the cytosine bases. These are cross-linked O(2)···W···O(2), and there are additional water molecules bridging these waters with phosphate oxygen atoms and with the guanine N(2) amino group (see Fig. 24.8 and 24.9c).

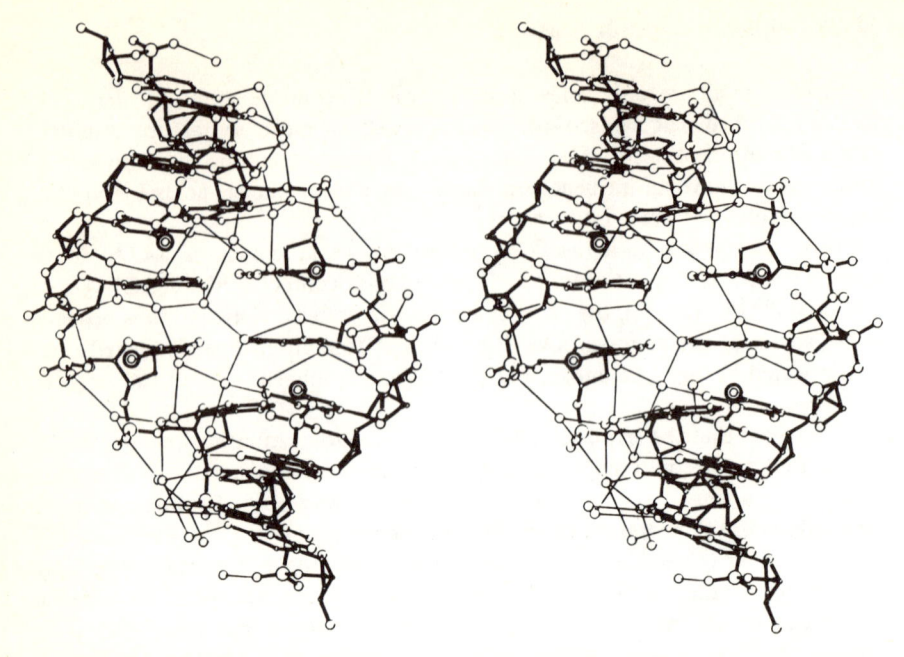

Fig. 24.6. Stereo diagram showing the four fused water pentagons in the major groove of the A-DNA-type octanucleotide d(GC^{Br}UA^{Br}UACC). Bromine atoms [or thymine methyl groups in the isomorphous d(GGTATACC)] are indicated by *concentric circles*. Note water molecules and phosphate oxygens surrounding the methyl groups (or bromine atoms) (see also Fig. 24.9a, [865])

24.4 DNA Hydration and Structural Transitions Are Correlated: Some Hypotheses

Because the addition of salt or organic solvents to aqueous solutions affects the water activity, we can expect a direct correlation between water activity and DNA conformation. Water activity, on the other hand, will influence the hydration of DNA, and therefore a connection between hydration and DNA structure can be expected. A few suggestions have been made concerning this correlation.

The **"spine of hydration"** in B-DNA is specific for A/T sequences in the minor groove and is disrupted by the N(2) amino group of guanine. In fact, the duplex poly(dA)·poly(dT) occurs only in the B-form. Upon addition of salt, it does not transform into A-DNA but disproportionates to give triple stranded poly(dA) ·2poly(dT) and single-stranded poly(dA) (see Part II, Chap. 16.3 and Part III, Chap. 20.5). Moreover, if we replace guanosine by inosine (which lacks the N(2) amino group) in polymers like poly(dI–dC)·poly(dI–dC) and poly(dI)·poly(dC), only the B-DNA form is observed and transformation into A-DNA does not take place. This is in agreement with the notion that the "spine of hydration" in the minor groove stabilizes the structure of B-DNA.

Because the "spine of hydration" in Z-DNA is so well defined and hydrogen-bonded to cytosine O(2), guanine N(2), and phosphate oxygen atoms, we can

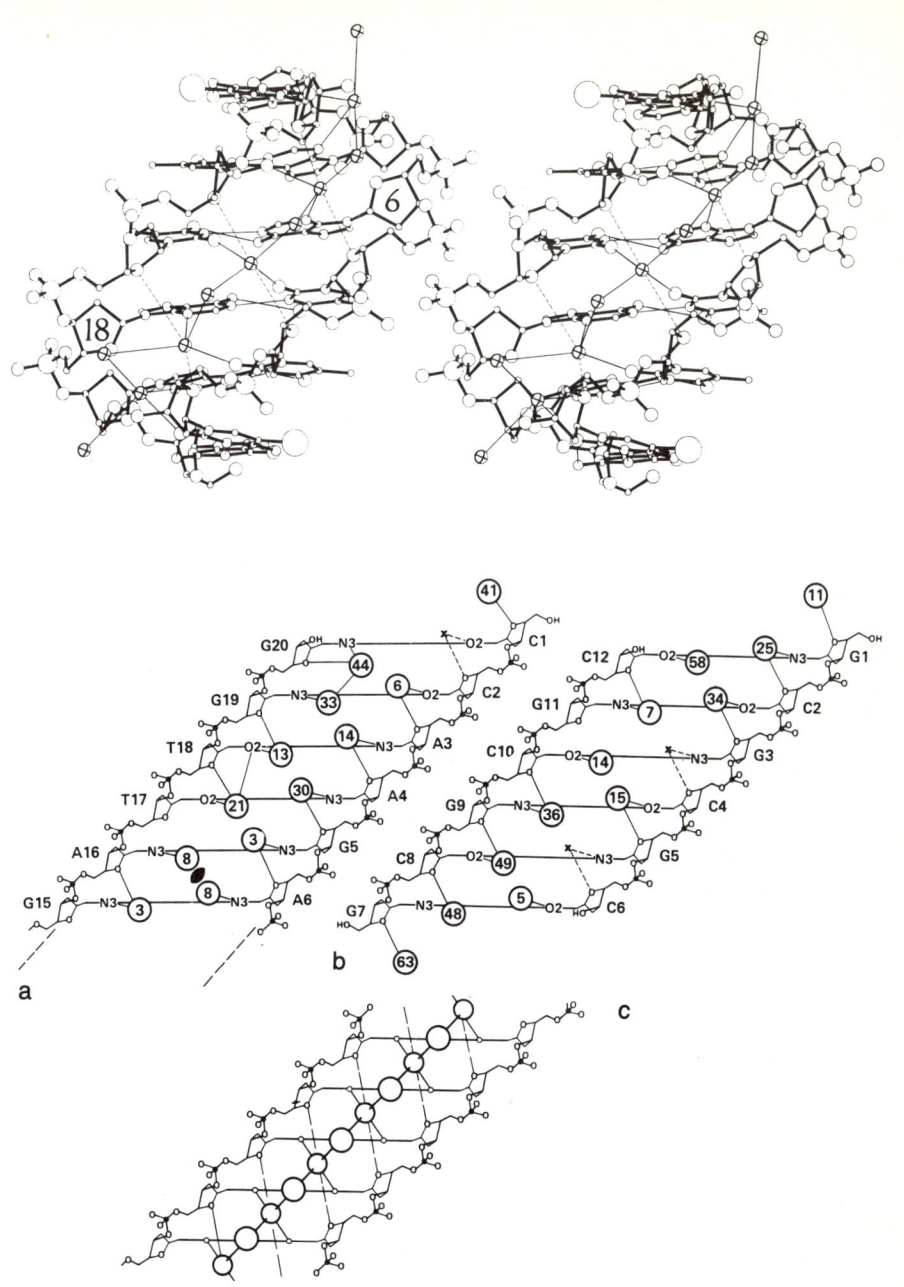

DNA Hydration and Structural Transitions Are Correlated: Some Hypotheses

Fig. 24.7a–c. (*Top*) The "spine of hydration" observed in the central part of the minor groove in the B-DNA dodecamer d(CGCGAATT^{Br}CGCG). Note close contacts of water molecules to sugar O(4') atoms indicated by *dashed lines*. Residues 6 and 18 are marked (see also Fig. 24.9b, [867]) (*Bottom*) Comparison of minor groove water structures in "mixed-sequence" B-DNA (**a**, CCAAGATTGG; **b**, phosphorothioate analog of d(GCGCGC) and **c** in an A/T sequence as in the central part of the stereo diagram obove [868]

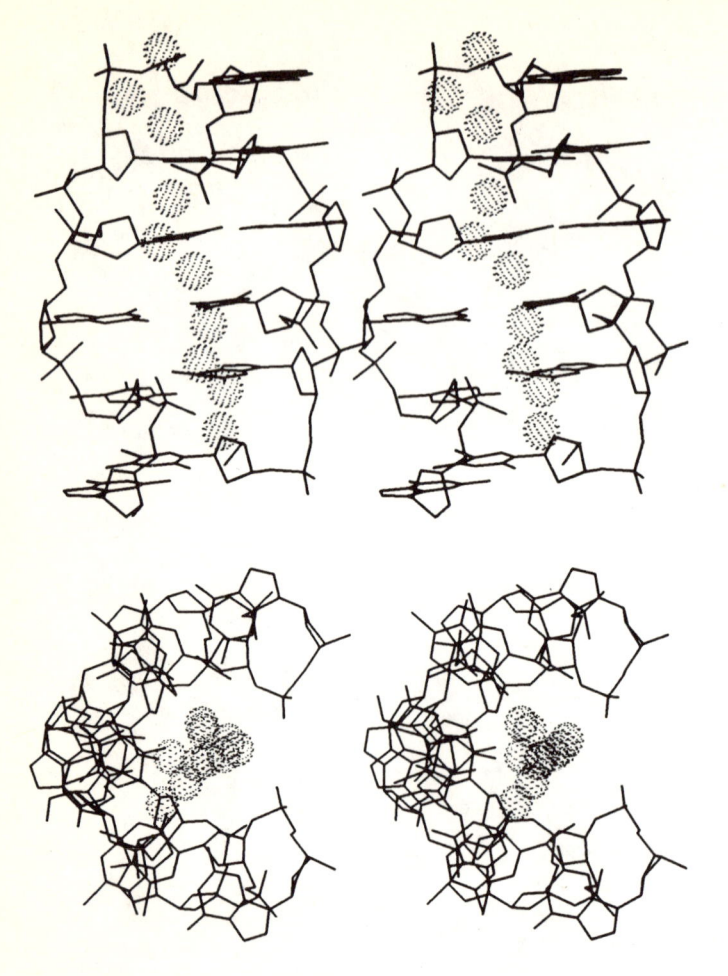

Fig. 24.8. Stereo diagram (side and top view) showing arrangement of spine of hydration in the minor groove of Z-DNA (see also Fig. 24.9c, [876])

assume that it also contributes significantly to the stabilization of this DNA conformation.

The water pentagons in the major groove of A-DNA do not appear to have a significant influence on the stabilization of this DNA conformation. Otherwise, polymers with alternating A–T sequence such as poly(dA–dT) should occur preferentially in the A-form. Because this polymer favors the B-form, the "spine of hydration" is probably a stronger stabilizer of structure than the water pentagons.

Phosphate hydration is more economical in A- and Z-DNA than in B-DNA, and therefore water activity determines DNA conformation. If sequence-specific features are excluded, the hydration of bases and sugars does not have any obvious influence on the DNA structure. This is different if we considere phosphate groups.

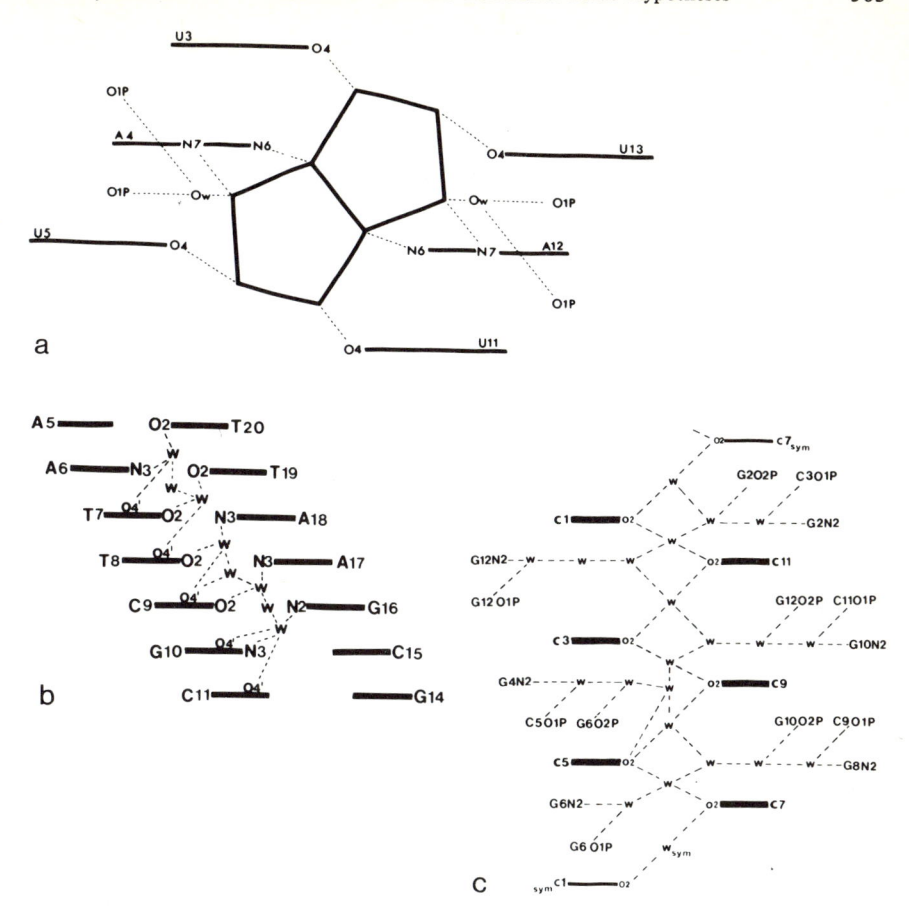

Fig. 24.9a–c. Schematic representation of Figs. 24.6, 24.7 (top), 24.8, showing the hydration in the major groove of **a** A-DNA, and in the minor grooves of **b** B-DNA, **c** Z-DNA. Water molecules indicated in **a** as O_W and corners of pentagons; in **b,c** as W, [871]

Because phosphate groups are far apart in B-DNA due to the C(2')-endo sugar pucker, their free oxygen atoms are hydrated *individually* (Table 24.7, Fig. 24.5). In A- and Z-DNA, intrastrand phosphate groups are so close together due to the C(3')-endo puckering modes that they can *share* water molecules in water bridges formed by one or two water molecules. This hydration is clearly more economical compared with B-DNA because fewer water molecules are necessary to achieve the same average hydration per phosphate oxygen atom [874].

Another feature associated with B-DNA, but not with A- and Z-DNA, is the extra water molecule located between free phosphate oxygen and thymine methyl groups. Apparently, this water molecule is loosely bound and disappears upon dehydration, allowing the sugar conformation to change from C(2')-endo to C(3')-endo.

Conformational transitions of DNA are cooperative. The phosphate and thymine methyl group hydration is intimately connected with the sugar puckering mode. At high water activity, C(2')-endo sugar puckering prevails, with methyl groups associated with a water molecule, and phosphate groups are hydrated individually. If salt or organic solvents are added to reduce the water available for hydration, we can envisage that hydration becomes more economical, with dehydration of the methyl groups and sharing of water molecules between adjacent phosphate groups. This, however, requires a change in sugar conformation from C(2')-endo to C(3')-endo which is cooperative in order to maintain a smooth double helical structure, as is indeed observed in a variety of experiments.

Chapter 25
The Role of Three-Center Hydrogen Bonds in the Dynamics of Hydration and of Structure Transition

We know from spectroscopic and crystallographic studies that proteins and nucleic acids are not "rigid" molecules [877–879]. They are subject to internal vibrations and structural fluctuations which range in frequencies from the ns range, if protein domain motion is concerned, to the ps-fs range if the dynamics of more local main-chain and side-chain atoms are considered. These movements were successfully simulated on the computer [273, 349, 353, 879]. In general, they are so subtle that the overall structures of the macromolecules are not disturbed, but in some cases they are vigorous enough that parts of the molecule cannot be seen in electron density maps derived from X-ray diffraction data. Irrespective whether the disorder in the crystal structure is of a dynamic or static nature, this indicates that the molecule is inherently flexible in solution.

As outlined in Chapters 23 and 24, water molecules are only loosely bound by hydrogen bonding at the peripheral atoms of the proteins and nucleic acids. Consequently, they are even more mobile than the atoms of the macromolecules to which they are coordinated. As we know from H_2O/D_2O exchange experiments, some water of hydration molecules are fully and easily replaced even in the crystalline state. The rate of dissociation must be diffusion-controlled and therefore at least in the ns time range.

If intermolecular associations are formed, hydration water molecules are removed from the interacting parts of the surface of a protein or a nucleic acid. This occurs, if the macromolecules form complexes with other macromolecules or with small substrates as, for instance, in enzymatic reactions (see Fig. 23.11). Because the contact between these interacting molecules is direct, water of hydration molecules have to be displaced from both partners. The question arises whether the water molecules located at the surface "jump" into the bulk water or whether they glide along the surface of the macromolecule and give way to the incoming, interacting molecules.

Three-center bonds facilitate movement of water of hydration molecules. Because the surface of globular, water-soluble proteins and of nucleic acids is covered by hydrophilic groups, the gliding movement of hydration water molecules along the surface is associated with breaking and formation of hydrogen bonds. This process will be greatly facilitated if it is cooperative, so that breaking and formation of hydrogen bonds occurs simultaneously and not sequentially.

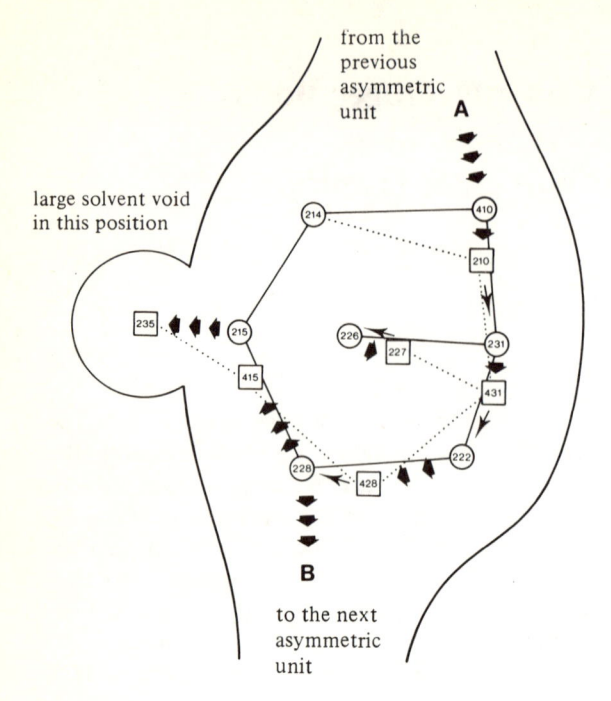

Fig. 25.1. Possible movements of the water molecules between the main water networks I and II in the channel region in the crystal structure of vitamin B_{12}. ○, Water network I; □, water network II, ♦, movements I→II; → movements II→I [880]

In macromolecular crystallography, this process cannot be observed because hydrogen atoms are not located and dynamic processes cannot be "seen". However, in the crystal structure of vitamin B_{12} studied by X-ray and neutron diffraction methods, an interpretation of the water movement could be derived [127, 880, 881].

In this crystal structure, the vitamin B_{12} molecules are well ordered (except for a side chain). Between the molecules there are wide channels which are filled with four acetone and with a total of >140 mostly disordered water molecules in the unit cell. Careful analysis of X-ray and neutron diffraction data suggested that the disorder can be described in terms of a movement of water molecules within the wide channels, as indicated schematically in Fig. 25.1. According to this interpretation, there exist two dynamically related water networks I (atoms in circles) and II (atoms in squares). Starting with network I, a water molecule can be either pushed into the asymmetric unit at position 410 (path A) or pulled out at position 228 (path B). If water molecule 228 is pulled out and moves into position 410 in the adjacent asymmetric unit, water 222 is pulled towards 428 and the void is filled by consecutive movements of water molecules 231, 226, 410 into positions 431, 227, 210. These movements are consistent with a transition from network I into network II. Conversely, if a water molecule enters the asymmetric unit through path A, then water molecules 210, 431, and 428 of network II move into positions

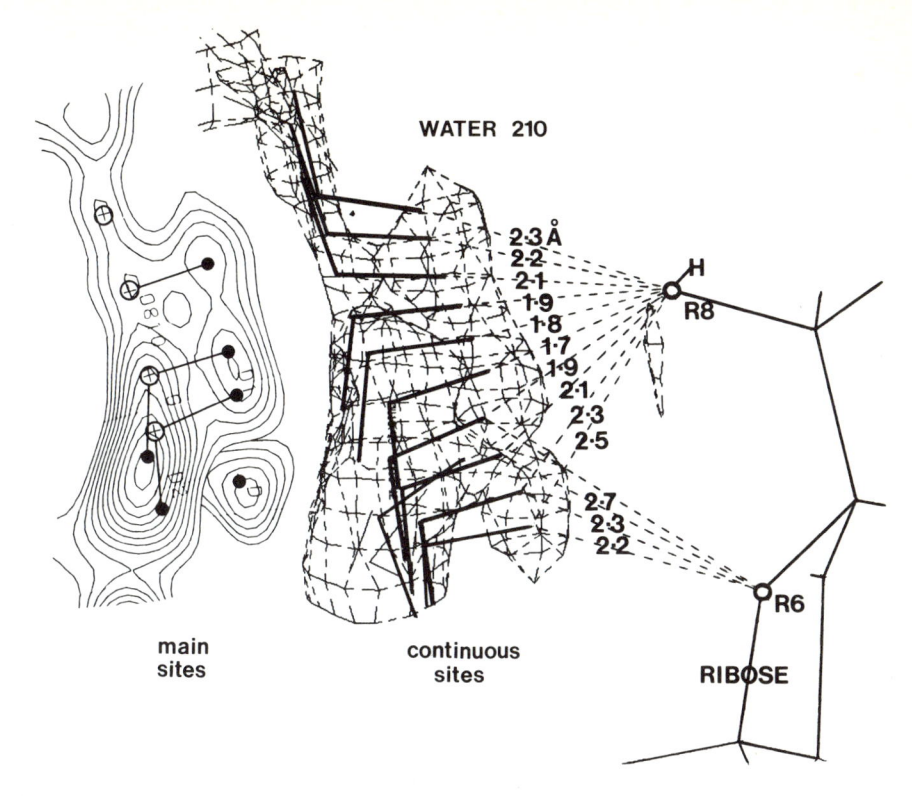

Fig. 25.2. *Left:* nuclear (Fermi) density of water molecule 210 (see Fig. 25.1); *right:* interpretation in terms of water molecules. Hydrogen-bonding distances H···O given in Å units, ribose drawn schematically, with oxygen atoms R6 and R8 as *circles* [880]

231, 222, 228 of network I so that the reverse transition is observed. All the movements, except into position 228, are of the order of 1 Å.

Water movement proceeds via three-center hydrogen bonds. As illustrated in Fig. 25.2 for water molecule 210, the main atomic sites observed in the electron and nuclear density maps can be interpreted in terms of a set of "continuous" sites. If water molecule 210 moves from "top" to "bottom", the two-center hydrogen bond $O_W-H\cdots O(R8)$ shortens from 2.3 Å to a minimum of 1.7 Å and then changes to a three-center bond $O_W-H\begin{smallmatrix}O(R8)\\ \\O(R6)\end{smallmatrix}$ with hydrogen-bond lengths between 2.1 and 2.7 Å. Further movement "down" would break the three-center bond and yield the two-center bond $O_W-H\cdots O(R6)$. Clearly, the three-center hydrogen bond is formed transiently when the water molecule passes along the surface of the vitamin B_{12} ribose.

We can envisage that this description of water movement along functional groups is not limited to the crystal structure of vitamin B_{12} but that it occurs generally in the hydration of biological macromolecules. If so, the concept of

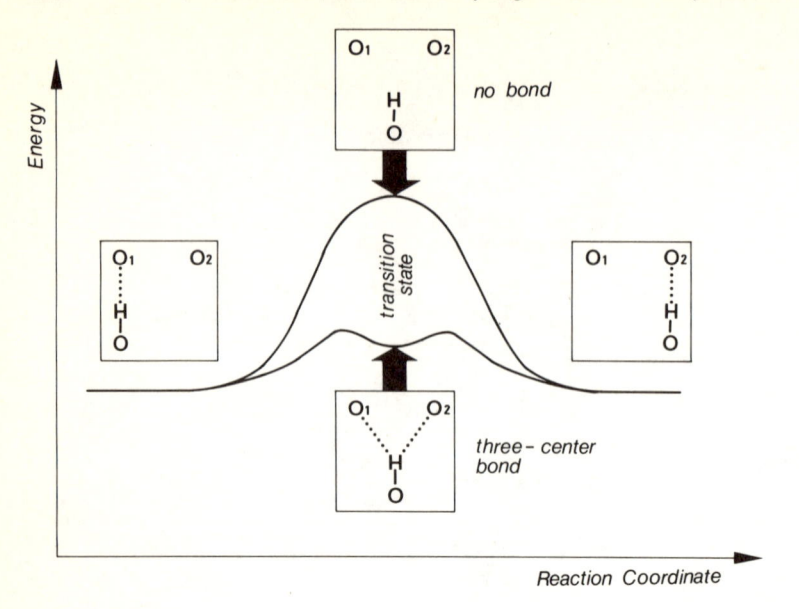

Fig. 25.3. Schematic description of energy distribution occurring during breaking and formation of O−H···O hydrogen bonds with (*lower line*) and without (*upper line*) a three-center transition intermediate

three-center hydrogen bonding is of major importance in the description of dynamical processes occurring during hydration. The same holds also for the structure of bulk water, where ribose atoms O(R6) and O(R8) in Fig. 25.2 are replaced by oxygen atoms of other water molecules. In terms of energetics, the transient formation of three-center hydrogen bonds should be favorable because the energy barrier that has to be overcome in the breaking of one and the formation of another two-center hydrogen bond is lowered if a three-center transition intermediate is involved, as shown in Fig. 25.3. On this basis, we anticipate that three-center bonds are equal in importance to the two-center hydrogen bonds in the dynamic processes which are associated with the hydration of macromolecules or with bulk water.

Three-center hydrogen bonds may be involved in α-helix folding and unfolding. As outlined in Fig. 19.9 and in Chapters 19 and 23, water molecules are frequently found hydrogen-bonded with the peptide carbonyl oxygen atoms of α-helices. This was analysed in more detail in a sample of 35 well-refined protein crystal structures which contained 315 α-helix-to-water interactions [882]. Of these, the majority, 262, are hydrogen-bonded "externally" (see Figs. 19.9 and 25.4a), 17 are acceptors of three-center hydrogen bonds (Fig. 25.4b), and 33 are incorporated in the α-helix (Fig. 25.4c). The internal water molecules are located preferentially at the amino termini, 26, and with lower frequency at the carboxy termini, 4, and within the α-helix, 3.

The observed α-helical segments with internal water molecules (of which only one is shown in Fig. 25.4c) in fact display the entire range of conformations from

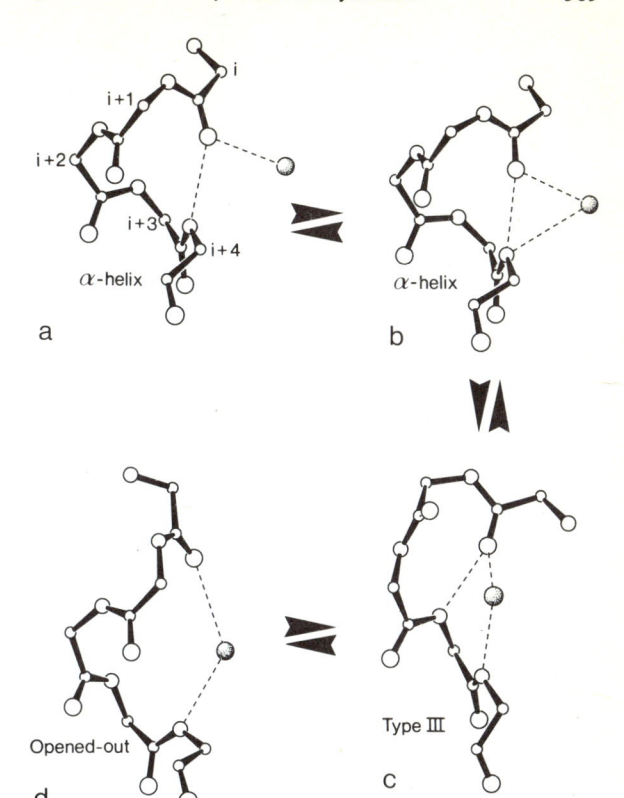

Fig. 25.4a–d. "Snapshots" of intermediates along the pathway of α-helix formation. (a) An external hydrogen-bonded water molecule in hemoglobin, (b) three-center hydrogen bonding in troponin c, (c) the α-helix is disrupted by an "internal" water molecule and the helix turn replaced by a type III turn in carboxypeptidase A, (d) the open turn with no hydrogen bond in azurin. Taken from [882]

the α-helix to the extended strand of the β-sheet and suggest a folding/unfolding pathway of the α-helix. We can assume for the unfolding that the α-helix with 5→1 hydrogen bonds is initially solvated at a peptide carbonyl oxygen (Fig. 25.4a), followed by formation of a three-center bond with peptide N–H as donor (Fig. 25.4b) which, as "transition intermediate", changes to a 4→1 hydrogen bond typical of the β-turns (see Fig. 19.8 and Table 19.5). As the type III β-turn (which is equivalent to a 3_{10} helix turn) is closest in conformation to the α-helix, it is expected to be formed initially and followed by type I or type II turn and by the "open" βturn which has no 4→1 hydrogen bond (Fig. 25.4d). The reverse of this pathway would lead to α-helix formation.

A comparable insertion of water molecules into α-helical structures was also observed in several peptides [883]. In a series of tripeptides, the fourth unit is substituted by a water molecule to yield, in the crystal lattice, "infinite α-helices" [884].

Water molecules and three-center hydrogen bonds as "lubricants" in structural transitions. If water participates and contributes to protein helix-to-coil transition as illustrated in Fig. 25.4, then one can also envisage comparable and more general schemes of tertiary- and secondary-structure unfolding and folding for proteins, nucleic acids and polysaccharides: water molecules are first bound to the surface as hydration water, they approach through three-center hydrogen bonds at positions with the proper (maybe transiently distorted) geometry, are finally inserted and locally pry open and deform the tertiary or secondary structure. This distortion then facilitates further insertion of water molecules which finally leads to the disruption of the three-dimensional structure of the molecule. These events, and all processes where different molecules interact, are associated with changes in hydration water structure which most probably are facilitated by the transient formation of three-center hydrogen bonds as shown in Fig. 25.3. The combination of water molecules and three-center hydrogen bonds may consequently be considered as "lubricants" in the dynamic processes of macromolecule folding, unfolding and interaction which are so important for life.

Dynamic hydrogen-bonding plays an important role in molecular recognition in an aqueous medium. In water, or in the aqueous component of biological systems, the hydrogen bonds form continuous networks which may be disordered in the solid state, as in a protein crystal, or dynamically changing in the liquid state. These networks are formed by the linkage of continuous chains of hydrogen bonds.

The displacement of the hydrogen electron from its proton gives rise to a dipole. In consequence, these chains form a sequence of monopoles or dipoles which will be highly sensitive to the electrostatic potential of solute or substrate molecules. These readily polarizable electronic chains provide a mechanism whereby information relating to the electrostatic potential of a molecule can be transmitted over long (atomic) distances to another molecule, or from a substrate to the active site of an enzyme. Such a mechanism for molecular recognition seems to be necessary to explain the lattice registration of protein molecules in highly hydrated protein crystals and the efficient approach of substrate molecules to particular active sites of the protein macromolecules.

References

1. Werner A (1902) Die Ammoniumsalze als einfachste Metallammoniake. Liebigs Ann 322:147–159
2. Werner A (1902) Über Haupt- und Nebenvalenzen und die Constitution der Ammoniumverbindungen. Liebigs Ann 322:261–297
3. Hantzsch A (1910) Über die Isomerie-Gleichgewichte des Acetessigesters und die sogenannte Isorrhopesis seiner Salze. Berichte 43:3049–3076
4. Moore TS, Winmill TF (1912) The state of amines in aqueous solution. J Chem Soc 101:1635–1676
5. Pfeiffer P (1913) Zur Theorie der Farblacke II. Liebigs Ann 398:137–197
6. Pfeiffer P (1914) Zur Kenntnis der sauren Salze der Carbonsäuren. Berichte 47:1580–1595
7. Huggins ML (1971) 50 Years of hydrogen bond theory. Angew Chem Int Ed Engl 10:147–152
8. Huggins ML (1922) Electronic structure of atoms. J Phys Chem 26:601–625
9. Latimer WM, Rodebush WH (1920) Polarity and ionization from the standpoint of the Lewis theory of valence. J Am Chem Soc 42:1419–1433
10. Herbine P, Dyke TR (1986) Rotational spectra and structure of the ammonia-water complex. J Chem Phys 83:3768–3774
11. Evans RC (1946) An introduction to crystal chemistry. Cambridge Univ Press, London, UK
12. Mark H, Weissenberg K (1923) Röntgenographische Bestimmung der Struktur des Harnstoffs und des Zinntetrajodids. Z Phys 16:1–22
13. Hendricks SB (1928) The crystal structure of urea and the molecular symmetry of thiourea. J Am Chem Soc 50:2455–2464
14. Wyckoff RWG, Corey RB (1934) Spectrometric measurements on hexamethylene tetramine and urea. Z Krist 89:462–468
15. Bernal JD, Fowler RH (1933) A theory of water and ionic solution, with particular reference to hydrogen and hydroxyl ions. J Chem Phys 1:515–548
16. Astbury WT, Street A (1931) Structure of hair, wool and related fibres, I. General. Trans R Soc A 230:75–101
17. Astbury WT, Woods HJ (1933) X-ray studies of the structure of hair, wool, and related fibres, II. Molecular structure and elastic properties of hair keratin. Trans R Soc A 232:333–394
18. Huggins ML (1936) Hydrogen bridges in organic compounds. J Org Chem 1:407–456
19. Huggins ML (1936) Hydrogen bridges in ice and liquid water. J Phys Chem 40:723–731
20. Bernal JD, Megaw HD (1935) Function of hydrogen in intermolecular forces. Proc R Soc (Lond) A 151:384–420
21. Huggins ML (1943) Structure of fibrous proteins. Chem Rev 32:195–218
22. Robertson JM (1935) Molecular map of resorcinol. Nature (Lond) 136:755–756
23. Llewellyn FJ, Cox EG, Goodwin TH (1937) Crystalline structure of the sugars. Part IV. Pentaerythritol. J Chem Soc 883–894
24. Pauling L (1939) The nature of the chemical bond. Cornell Univ Press, Ithaca, NY
25. Glasstone S (1937) The structure of some molecular complexes in the liquid phase. Trans Farad Soc 33:200–214

26. Pauling L, Corey RB, Branson HR (1951) The structure of proteins. Two hydrogen-bonded helical configurations of the polypeptide chain. Proc Nat Acad Sci USA 37:205–211
27. Watson JD, Crick FHC (1953) A structure for deoxyribose nucleic acid. Nature (Lond) 171:737–738
28. Pimentel GC, McClellan AL (1960) The hydrogen bond. Freeman, San Francisco
29. Uvarov EB, Chapman DR, Isaacs A (1971) The Penguin dictionary of science. Penguin, London, p 188
30. Watson JD (1965) The importance of the weak interactions. Molecular biology of the gene. Benjamin, New York, Chapter 4, pp 102–140
31. Dorsey NE (1940) Properties of ordinary water-substance. Am Chem Soc Monogr, Hafner, NY (facsimile of 1940 edit, Reinhold Publ. Corp. 673 pp). (This monograph claims to include every important article up to mid-1937 and the most important to mid-1938)
32. Franks F (ed) (1972–80) The chemistry and physics of water. A comprehensive treatise. Vols 1–7. Plenum Press, New York London
33. Vinogradov SN, Linnell RH (1971) Hydrogen bonding. Van Nostrand Reinhold, NY
34. Joesten MD, Schaad LJ (1974) Hydrogen bonding. Dekker, NY
35. Schuster P, Zundel G, Sandorfy C (eds) (1976) The hydrogen bond. Recent developments in theory and experiments. Vols I–III. North Holland, Amsterdam
36. Wang B-C (1985) Resolution of phase ambiguity in macromolecular crystallography. In: Wyckoff HW, Hirs CHW, Timasheff SN (eds) Diffraction methods for biological macromolecules, Part B, Methods in enzymology, vol 115. Academic Press, NY, pp 90–117
37. Karle J (1989) Direct methods in protein crystallography. Acta Cryst A 45:765–780
38. Pauling L (1966) The importance of being crystalline. Congress Lecture. Int Union Crystallogr, 1966
39. Allen FH, Bellard S, Brice MD, Cartwright BA, Doubleday A, Higgs H, Hummelink TWA, Hummelink-Peters BGMC, Kennard O, Motherwell WDS, Rodgers JR, Watson DG (1979) The Cambridge Crystallographic Data Center: Computer-based search, retrieval analysis and display of information. Acta Cryst B 35:2331–2339
40. Allen FH, Kennard O, Taylor R (1983) Systematic analysis of structural data as a research technique in organic chemistry. Accts Chem Res 16:146–153
41. Bernstein FC, Koetzle TF, Williams GJB, Meyer EF jr, Brice MD, Rodgers JR, Kennard O, Shimanouchi T, Tasumi M (1977) The Protein Data Bank: A computer-based archival file for macromolecular structure. J Molec Biol 112:535–542
42. Coulson CA (1952) Valence. Oxford Univ Press, UK
43. Haeberlen U (1976) Advances in magnetic resonance. Suppl Academic Press, NY
44. Tucker EE, Lippert E (1976) Structure and spectroscopy. In: Schuster P, Zundel G, Sandorfy C (eds) The hydrogen bond – Recent developments in theory and experiments, vol II, Chap 17. North Holland, Amsterdam
45. Blinc R (1976) Structure and spectroscopy. In: Schuster P, Zundel G, Sandorfy C (eds) The hydrogen bond – Recent developments in theory and experiments, vol. II, Chap. 18. North Holland, Amsterdam, pp 831–888
46. Bellamy LJ (1980) Associated XH frequencies. The hydrogen bond. Chap 8. In: The infra-red spectra of complex molecules, vol 2. Chapman and Hall, London, pp 240–284
47. Murthy ASN, Rao CNR (1968) Spectroscopic studies of the hydrogen bond. Appl Spectrosc Rev 2:69–191
48. Hamilton WC, Ibers JA (1968) Hydrogen bonding in solids. Benjamin, NY
49. Emsley J (1980) Very strong hydrogen bonding. Chem Soc Rev 9:91–124
50. Singh UC, Kollman PA (1985) A water dimer potential based on ab-initio calculations using Morokuma component analysis. J Chem Phys 83:4033–4040
51. Umeyana H, Morokuma K (1977) The origin of hydrogen bonding: An energy decomposition study. J Am Chem Soc 99:1316–1332
52. Pedersen B (1974) The geometry of hydrogen bonds from donor water molecules. Acta Cryst B 30:289–291
53. Kroon J, Kanters JA, Van-Duijneveldt-Van der Rijdt JGCM, Van-Duijneveldt FB, Vliegenthart JA (1975) O-H---O Hydrogen bonds in molecular crystals. Statistical and quantum-chemical analysis. J Molec Struct 24:109–129

54. Olovsson I, Jönsson PG (1976) X-ray and neutron diffraction studies of hydrogen bonded systems. Chap 8. The hydrogen bond – recent developments in theory and experiments. Schuster P (ed). North-Holland, Amsterdam, pp 393–456
55. Newton MD, Jeffrey GA, Takagi S (1979) Application of ab-initio molecular orbital calculations to the structural moieties of carbohydrates. 5. The geometry of the hydrogen bonds. J Am Chem Soc 101:1997–2002
56. Taylor R, Kennard O, Versichel W (1984) Geometry of the NH···O=C hydrogen bond. 2. Three-center ('bifurcated') and four-center ('trifurcated') bonds. J Am Chem Soc 106:244–248
57. Jeffrey GA, Takagi S (1978) Hydrogen-bond structure in carbohydrate crystals. Accts Chem Res 11:264–270
58. Ceccarelli C, Jeffrey GA, Taylor R (1981) A survey of O-H···O hydrogen bond geometries determined by neutron diffraction. J Molec Struct 70:255–271
59. Jeffrey GA, Mitra J (1983) The hydrogen bonding patterns in pyranose and pyranoside crystal structures. Acta Cryst B39:469–480
60. Jeffrey GA, Maluszynska H (1982) A survey of hydrogen-bond geometries in the crystal structures of amino acids. Int J Biol Macromol 4:173–185
61. Jeffrey GA, Maluszynska H (1986) A survey of the geometry of hydrogen bonding in barbiturates, purines and pyrimidines. J Molec Struct 147:127–142
62. Jeffrey GA, Maluszynska H, Mitra J (1985) Hydrogen bonding in nucleosides and nucleotides. Int J Biol Macromol 7:336–348
63. Albrecht G, Corey RB (1939) The crystal structure of glycine. J Am Chem Soc 61:1087–1103
64. Marsh RE (1958) Refinement of the crystal structure of glycine. Acta Cryst 11:654–660
65. Jönsson PG, Kvick A (1972) Precision neutron diffraction structure determination of protein and nucleic acid components. III. Crystal and molecular structure of the amino acid α-glycine. Acta Cryst B28:1827–1833
66. Hahn T, Buerger MJ (1957) The crystal structure of diglycine hydrochloride. Z Krist 108:419–453
67. Craven BM, Takei WJ (1964) The crystal structure of perdeuterated violuric acid monohydrate. The neutron diffraction analysis. Acta Cryst 17:415–420
68. Alagona G, Ghio C, Kollman PA (1983) Bifurcated vs. linear hydrogen bonds: dimethyl phosphate and formate anion interactions with water. J Am Chem Soc 105:5226–5230
69. Banerjee A, Dattagupta JK, Saenger W, Rabczenko A (1977) 1,3-Dimethyluracil: A crystal structure without hydrogen bonds. Acta Cryst B33:90–94
70. Parthasarathy R (1969) The crystal structure of glycylglycine hydrochloride. Acta Cryst B25:509–518
71. Iitaka Y (1960) The crystal structure of β-glycine. Acta Cryst 13:35–45
72. Iitaka Y (1958) The crystal structure of γ-glycine. Acta Cryst 11:225–226
73. Donohue J (1968) Selected topics in hydrogen bonding. In: Rich A, Davidson N (eds) Structural chemistry and molecular biology. Freeman, San Francisco, pp 443–465
74. Jeffrey GA, Mitra J (1984) Three-center (bifurcated) hydrogen bonding in the crystal structures of amino acids. J Am Chem Soc 106:5546–5553
75. Taylor R, Kennard O, Versichel W (1984) The geometry of the N-H---O=C hydrogen bond. 3. Hydrogen-bond distances and angles. Acta Cryst. B40:280–288
76. Eberhardt WH, Crawford BL, Jr, Lipscomb WN (1954) The valence structure of the boron hydrides. J Chem Phys 22:989–1101
77. Ceccarelli C, Jeffrey GA, McMullan RK (1980) A neutron diffraction refinement of the crystal structure of erythritol at 22.6 K. Acta Cryst B36:3079–3083
78. Chacko KK, Saenger W (1981) Topography of cyclodextrin inclusion complexes, 15. Crystal and molecular structure of the cyclohexaamylose 7.57 water complex. Form III. Four and six-membered circular hydrogen bonds. J Am Chem Soc 103:1708–1715
79. Jeffrey GA, Fasiska EY (1972) Conformation and intramolecular hydrogen-bonding in the crystal structure of potassium D-gluconate monohydrate. Carbohydr Res 21:187–199
80. Panagiotopoulos NC, Jeffrey GA, LaPlaca SJ, Hamilton WC (1974) The crystal structures of the A and B forms of potassium D-gluconate monohydrate by neutron diffraction. Acta Cryst B30:1421–1430

81. Wallwork SC (1962) Hydrogen-bond radii. Acta Cryst 15:758–759
82. Weiner SJ, Kollman PA, Case DA, Singh UC, Ghio C, Alagona G, Profeta S, Jr, Weiner P (1984) A new force field for molecular mechanical simulation of nucleic acids and proteins. J Am Chem Soc 106:765–784
83. Coppens P, Sabine TM (1969) Neutron diffraction study of hydrogen bonding and thermal motion in deuterated α and β oxalic acid dihydrate. Acta Cryst B25:2442–2451
84. Jeffrey GA (1989) Hydrogen bonding in crystal structures of nucleic acid components: purines, pyrimidines, nucleosides and nucleotides. In: Saenger W (ed), Landolt-Börnstein. Numerical Data and Functional Relationships in Science and Technology. New Series, Group VII, Vol. Ib. Springer, Berlin, pp 277–348
85. Bürgi HB, Dunitz JD (1988) Can statistical analysis of structural parameters from different crystal environments lead to quantitative energy relationships? Acta Cryst B44:445–448
86. Curtiss LA, Frurip DJ, Blander M (1979) Studies of molecular association in H_2O and D_2O vapors by measurement of thermal conductivity. J Chem Phys 71:2703–2711
87. Schuster P (1978) The fine structure of the hydrogen bond. In: Pullman B (ed) Perspectives in quantum chemistry and biochemistry, vol 2: Intermolecular interactions from diatomics to biopolymers. Chapt. 4. John Wiley & Sons, NY
88. Newton MD, Kestner NR (1983) The water dimer: theory versus experiment. Chem Phys Lett 94:198–201
89. Morokuma K (1971) Molecular orbital studies of hydrogen bonds III. C = O---H-O Hydrogen bond in H_2CO---H_2O and H_2CO---$2H_2O$. J Chem Phys 55:1236–1244
90. Jorgensen WL (1979) An intermolecular potential function for the methanol dimer from ab-initio calculations. J Chem Phys 71:5034–5038
91. Del Bene JE (1971) Theoretical study of open chain dimers and trimers containing CH_3OH and H_2O. J Chem Phys 55:4633–4636
92. Dreyfus M, Pullman A (1970) Nonempirical study of the hydrogen bond between peptide units. Theor Chim Acta 19:20–37
93. Spackman MA, Weber HP, Craven BM (1988) Energies of molecular interactions from Bragg diffraction data. J Am Chem Soc 110:774–782
93a. Radzika A, Wolfenden R (1988) Comparing the polarities of the amino acids: side-chain distribution coefficients between the vapor phase, cyclohexane, 1-octanol, and neutral aqueous solution. Biochemistry 27:1665–1670
93b. Fauchère J-L, Pliska V (1988) Hydrophobic parameters of amino-acid side chains from the partitioning of N-acetyl-aminoacid amides. Eur J Med Chem 18:369–375
94. Stout GH, Jensen LH (1968) X-ray structure determination. A practical guide. McMillan, London
95. Falk M, Knop O (1973) Water in stoichiometric hydrates. Chap 2. In: Franks F (ed) Water. A comprehensive treatise. Vol. 2. Plenum Press, NY, London, pp 55–113
96. Rahim Z, Barman BN (1978) The van der Waals criterion for hydrogen bonding. Acta Cryst A34:761–764
97. Bondi A (1964) Van der Waals volumes and radii. J Phys Chem 68:441–451
98. Allinger NL (1976) Calculation of molecular structure and energy for force-field methods. Adv Phys Org Chem 13:2–75
99. Price SL (1986) Model anisotropic intermolecular potentials for saturated hydrocarbons. Acta Cryst B42:388–401
99a. Nyburg SC, Faerman CH (1985) A revision of van der Waals atomic radii for molecular crystals: N, O, F, S, Cl, Se, Br and I bonded to carbon. Acta Cryst B41:274–279
100. Koetzle TF, Lehmann MS (1976) Neutron diffraction studies of hydrogen bonding in α-amino acids. Chapt 9. In: Schuster P, Zundel G, Sandorfy C (eds) The hydrogen bond, II. Structure and spectroscopy. North Holland, Amsterdam, pp 457–470
101. Frank HS (1958) Covalency in the hydrogen bond and the properties of water and ice. Proc R Soc A 247:481–492
102. Kavanau JL (1964) Water and solute-water interactions. Holden-Day, San Francisco
103. Del Bene J, Pople JA (1970) Theory of molecular interactions. I. Molecular orbital studies of water polymers using a minimal Slater type basis. J Chem Phys 52:4858–4866

104. Del Bene J, Pople JA (1973) Theory of molecular interactions. III. A comparison of studies of H_2O polymers using different molecular orbital basis sets. J Chem Phys 58:3605–3608
105. Hankins D, Moskowitz JW, Stillinger FH (1970) Water molecule interactions. J Chem Phys 53:4544–4554
106. Jeffrey GA, Gress ME, Takagi S (1977) Some experimental observations on H···O hydrogen bond lengths in carbohydrate crystal structures. J Am Chem Soc 99:609–611
107. Jeffrey GA, Lewis L (1978) Cooperative aspects of hydrogen bonding in carbohydrates. Carbohydr Res 60:179–182
108. Tse YC, Newton MD (1977) Theoretical observations on the structural consequences of cooperativity in H···O hydrogen bonding. J Am Chem Soc 99:611–613
108a. Gilli G, Bellucci F, Ferretti V, Bertolasi V (1989) Evidence for resonance-assisted hydrogen bonding from crystal-structure correlations on the enol form of the β-diketone fragment. J Am Chem Soc 111:1023–1028
109. Saenger W (1979) Circular hydrogen bonds. Nature (Lond) 279:343–344
110. Lesyng B, Saenger W (1981) Theoretical investigations on circular and chain-like hydrogen bonded structures found in two crystal forms of α-cyclodextrin hexahydrate. Models for hydration and water clusters. Biochim Biophys Acta 678:408–413
111. Koehler JEH, Lesyng B, Saenger W (1987) Cooperative effects in extended hydrogen bonded systems involving OH-groups. Ab initio studies of the cyclic S_4 water tetramer. J Comput Chem 8:1090–1098
112. Lundgren J-O, Olovsson I (1976) The hydrated proton in solids. Chap 10. In: Schuster P, Zundel G, Sandorfy C (eds) The hydrogen-bond, vol II. North Holland, Amsterdam, pp 471–526
113. Williams JM (1976) Spectroscopic studies of hydrated proton species, $H^+(H_2O)_n$ in crystalline compounds. Chap 14. In: Schuster P, Zundel G, Sandorfy C (eds) The hydrogen-bond, vol II. North Holland, Amsterdam, pp 655–682
114. Giaque WF, Stout JW (1936) The entropy of water and the third law of thermodynamics. The heat capacity of ice from 15 to 273 K. J Am Chem Soc 58:1144–1150
115. Pauling L (1935) The structure and entropy of ice and of other crystals with some randomness of atomic arrangement. J Am Chem Soc 57:2680–2684
116. Nagle JF (1966) Lattice statistics of hydrogen bonded crystals. I. The residual entropy of ice. J Math Phys 7:1484–1491
117. Lieb EH (1967) Residual entropy of square ice. Phys Rev 162:162–172
118. Saenger W, Betzel Ch, Hingerty BE, Brown GM (1982) Flip-flop hydrogen bonding in a partially disordered system. Nature (Lond) 296:581–583
119. Hunter L (1946) The hydrogen bond. Annu Rep Prog Chem 43:141–155, The Chemical Society, London, UK
120. Wolfenden RV (1969) Tautomeric equilibria in inosine and adenosine. J Mol Biol 40:307–310
121. Katritzky AR, Waring AJ (1962) Tautomeric azines. Part I. The tautomerism of 1-methyl-uracil and 5-bromo-1-methyl-uracil. J Chem Soc 1540–1544
122. Katritzky AR, Waring AJ (1963) Tautomeric azines. Part III. The structure of cytosine and its mono-cation. J Chem Soc 3046–3051
123. Pieber M, Kroon PA, Prestegard JH, Chan SI (1973) Erratum. Tautomerism of nucleic acid bases. J Am Chem Soc 95:3408
124. Mootz D, Seidel R (1990) Polyhedral clathrate hydrates of a strong base: phase relations of crystal structures in the system tetramethylammonium hydroxide-water. J Inclusion Phenom 8:139–157
125. McMullan RK, Mak TCW, Jeffrey GA (1966) Polyhedral clathrate hydrates XI. Structure of tetramethyl ammonium hydroxide pentahydrate. J Chem Phys 44:2338–2345
126. Savage H, Finney J (1986) Repulsive regularities of water structure in ices and crystalline hydrates. Nature (Lond) 322:717–720
127. Savage HFJ (1986) Water structure in crystalline solids: ices to proteins. Water Science Reviews 2:67–147
128. Némethy G (1967) Hydrophobe Wechselwirkungen. Angew Chem 79:260–271

129. Franks F (1975) The hydrophobic interaction. In: Franks F (ed) The chemistry and physics of water. A comprehensive treatise, vol 4. Plenum Press, New York, pp 1–94
130. Scheraga HA (1978) Interactions in aqueous solution. Acc Chem Res 12:7–14
131. Hildebrand JH (1979) Is there a 'hydrophobic effect'? Proc Nat Acad Sci USA 76:194
132. Tanford C (1980) The Hydrophobic effect. Formation of micelles and biological membranes. 2nd ed. Wiley, New York
133. Cantor CR, Schimmel PR (1980) Biophysical chemistry. Part I: Conformation of biological macromolecules. Chap 5. Freeman & Co, San Francisco, pp 279–288
134. Creighton TE (1984) Proteins. Structure and Molecular Principles, Chap 4. Freeman & Co, New York, pp 133–158
135. Jencks WP (1969) Catalysis in chemistry and enzymology. McGraw-Hill, New York
136. Chothia C (1974) Hydrophobic bonding and accessible surface area in proteins. Nature (Lond) 248:338–339
137. Richards FM (1977) Areas, volumes, packing and protein structure. Annu Rev Biophys Bioeng 6:151–176
138. Miller S, Janin J, Lesk AM, Chothia C (1987) Interior and surface of monomeric proteins. J Mol Biol 196:641–656
139. Frank HS, Evans MW (1945) Free volume and entropy in condensed systems. III. Entropy in binary liquid mixtures: partial model entropy in dilute solutions: structure and thermodynamics in aqueous electrolytes. J Chem Phys 13:507–532
140. Glew DN (1962) Aqueous solubility and the gas hydrates. The methane-water system. J Phys Chem 66:605–609
141. Hertz HG, Rädle C (1973) The orientation of water molecules in the hydration sphere of F^- and in the hydrophobic hydration sphere. Ber Bunsenges Phys Chem 77:521–531
142. Owicki JC, Scheraga HA (1977) Monte Carlo calculations in the isothermal-isobaric ensemble. 2 Dilute aqueous solutions of methane. J Am Chem Soc 99:7413–7418
143. Ravishanker G, Mezei M, Beveridge DL (1982) Monte Carlo simulation study of the hydrophobic effect. Potential of mean force for aqueous methane dimer at 25 and 50 °C. Farad Symp Chem Soc 17:79–91
144. Pangali C, Rao M, Berne BJ (1979) A Monte Carlo simulation of the hydrophobic interaction. J Chem Phys 71:2975–2981
145. Pangali C, Rao M, Berne BJ (1979) Hydrophobic hydration around a pair of apolar species in water. J Chem Phys 71:2982–2990
146. Jiang Xi-Kui (1988) Hydrophobic-lipophilic interactions. Aggregations and self-coiling of organic molecules. Accts Chem Res 21:362–367
147. Jeffrey GA (1986) Carbohydrate liquid crystals. Acc Chem Res 19:168–173
148. Chapman D, Williams RM, Ladbrooke BD (1967) Physical properties of phospholipids VI. Chem Phys Lipids 1:445–475
149. Abrahamsson S, Pascher I, Larsson K, Karlsson KA (1972) Molecular arrangements in glycosphingolipids. Chem Phys Lipids 8:152–179
150. Pascher I, Sundell S (1977) Molecular arrangements in sphingolipids. The crystal structure of cerebroside. Chem Phys Lipids 20:175–191
151. Chapman D (ed) (1968) Biological membranes. Physical fact and function. Academic Press, London New York
152. Abrahamsson S, Pascher I (eds) (1977) Structures of biological membranes. Plenum Press, New York
153. Davies M (1946) Physical aspects of the hydrogen bond. Ann Rep Prog Chem (Chem Soc Lond) 43:5–30
154. Hadži D (ed) (1957) The hydrogen bond. Pergamon Press, NY
155. Hadži D, Bratos S (1976) Vibrational spectroscopy of the hydrogen bond. Chapt 12. In: Schuster P, Zundel G, Sandorfy C (eds) The hydrogen bond. II. Structure and spectroscopy. North Holland, Amsterdam, pp 565–612
156. Badger RM (1934) A relation between internuclear distances and bond force constants. J Chem Phys 2:128–131
157. Falk M, Kemp O (1973) Water in stoichiometric hydrates. Chap 2. In: Franks F (ed) Water. A comprehensive treatise. Plenum Press, NY

158. Hussein MA, Millen DJ (1974) Hydrogen bonding in the gas phase. Part 1. Infra-red spectroscopic investigation of amine-alcohol systems. J Chem Soc Farad Trans II 70:685–692
159. Millen DJ, Mines GW (1974) Hydrogen bonding in the gas phase. Part II. Determination of thermodynamic parameters for amine-methanol systems from pressure, volume, temperature measurements. J Chem Soc Farad Trans II 70:693–699
160. Hussein MA, Millen DJ, Mines GM (1976) Hydrogen bonding in the gas phase. Part III. Infra-red spectroscopic investigation of complexes formed by phenol and 222 trifluoroethanol. J Chem Soc Farad Trans II 72:686–692
161. Hussein MA, Millen DJ (1976) Hydrogen bonding in the gas phase. Part 4. Infra-red spectroscopy investigation of the OH---O and C-H---N complexes; alcohol+ether and trichloromethane+amine systems. J Chem Soc Farad Trans II 72:693–699
162. Harmon KM, Günsel FA (1984) Hydrogen bonding. Part 17. IR and NMR study of the lower hydrates of choline chloride. J Molec Struct 118:267–275
162a. Van Duijneveldt-Van der Rijdt, JGCM, Duijneveldt FB, Kanters JA, Williams DR (1984) Calculations on vibrational properties of H-bonded OH groups, as a function of H-bond geometry. J Molec Struct (Theochem). 109:351–366
163. Orville-Thomas WJ, Legon AC (eds) (1988) Structure and properties of small molecules and hydrogen-bonded dimers. J Molec Struct 189:1–242
164. Rundle RE, Parasol M (1952) O-H stretching frequencies in very short and possibly symmetrical hydrogen bonds. J Chem Phys 20:1487–1488
165. Lord RD, Merryfield RE (1953) Strong hydrogen bonds in crystals. J Chem Phys 21:166–167
166. Lippincott ER, Schroeder R (1955) One-dimensional model of the hydrogen bond. J Chem Phys 23:1099–1106
167. Nakamoto K, Margoshes M, Rundle RE (1955) Stretching frequencies as a function of distances in hydrogen bonds. J Am Chem Soc 77:6480–6486
168. Pimentel GC, Sederholm CH (1956) Correlation of infra-red stretching frequencies and hydrogen bond distances in crystals. J Chem Phys 24:639–641
169. Bellamy LJ, Owen AJ (1969) A simple relationship between infra-red stretching frequencies and hydrogen bond distances in crystals. Spectrochim Acta 25A:329–333
170. Umemura J, Birnbaum GI, Bundle DR, Murphy WF, Bernstein HJ, Mantsch HH (1979) The correlation between O-H stretching frequencies and hydrogen bond distances in a crystalline sugar monohydrate. Can J Chem 57:2640–2645
171. Michell AJ (1988) Second derivative FT-IR spectra of celluloses I and II and related mono and oligosaccharides. Carbohydr Res 173:185–195
172. Martin AE (1959) Multiple differentiation as a means of band sharpening. Spectrochim Acta 14:97–103
173. Koenig JL (1981) Fourier transform infra-red spectroscopy of chemical systems. Acc Chem Res 14:171–178
174. Maddams WF, Mead WL (1982) The measurement of derivative IR spectra. I. Background studies. Spectrochim Acta 38A:437–444
175. Selp HM (1973) Electron diffraction theory and accuracy. In: Sutton LE (ed) Molecular structure by diffraction methods, vol 1. Chem Soc Lond, UK, Chap I, pp 7–57
176. Legon AC (1983) Pulsed-nozzle Fourier transform microwave spectroscopy of weakly bound dimers. Ann Rev Phys Chem 34:275–300
177. Dyke TR (1984) Microwave and radiofrequency spectra of hydrogen bonded complexes in the vapor phase. Top Curr Chem 120 (Fortschritte der Chemischen Forschung), Hydrogen-bonds, Schuster P (ed), Springer, Berlin Heidelberg New York Tokyo, pp 85–114
178. Weber A (1987) (ed) Structure and dynamics of weakly bound molecular complexes. NATO-ASI C212. Reidel Publ Co. Dordrecht, Holland
179. Legon AG, Millen DJ (1987) Directional character, strength, and nature of the hydrogen bond in gas phase dimers. Acc Chem Res 20:39–46
180. Luger P (1980) Modern X-ray analysis on single crystals. de Gruyter, Berlin
181. Glusker JP, Trueblood KN (1985) Crystal structure analysis. A primer. 2nd edn. Oxford Univ Press, London
182. Jones PG (1984) Crystal structure determination: A critical view. Chem Soc Rev 13:157–172

183. Koester L, Yelon WB (1983) Summary of low energy scattering lengths and cross sections. Neutron diffraction Newsletter, IUCr.
184. Becker PJ, Coppens P (1975) Extinction within the limit of validity of the Darwin transfer equations. III. Non-spherical crystals and anisotropy and extinction. Acta Cryst A31: 417–425
185. Thornley FR, Nelmes RJ (1974) Highly anisotropic extinction. Acta Cryst A30:748–757
186. Jeffrey GA (1984) The structures of some small molecules. Ab-inito molecular orbital calculations versus low temperature neutron diffraction crystal structure analyses. J Molec Struct (Theochem) 108:1–15
187. Jeffrey GA (1985) The structures of some small molecules. Ab-initio molecular orbital calculations versus low temperature neutron diffraction crystal structures, II, J Molec Struct 130:43–53
188. Wlodawer A, Savage H, Dodson G (1989) Structure of Insulin. Results of joint neutron and X-ray refinement, Acta Cryst B45:99–106
189. Allen FH (1986) A systematic pairwise comparison of geometric parameters obtained by X-ray and neutron diffraction. Acta Cryst B42:515–522
190. Cruickshank DWJ (1956) Errors in bond lengths due to rotational oscillations of molecules. Acta Cryst 9:757–758
191. Cruickshank DWJ (1961) Coordinate errors due to rotational oscillations of molecules. Acta Cryst 14:896–897
192. Busing WR, Levy HA (1964) The effect of thermal motion on the estimation of bond lengths from diffraction measurements. Acta Cryst 17:142–146
193. Willis BTM, Pryor AW (1975) Thermal vibrations in crystallography. Cambridge Univ Press, UK
194. Schomaker V, Trueblood KN (1968) On the rigid-body motion of molecules in crystals. Acta Cryst B24:63–76
195. Johnson CK (1970) An introduction to thermal motion analysis. In: Ahmed FR (ed) Crystallographic computing. Munksgaard, Copenhagen, pp 207–254
196. Coulson CA, Thomas MW (1971) The effect of molecular vibrations on apparent bond lengths. Acta Cryst B27:1354–1359
197. Thomas MW (1972) The effect of molecular vibrations on apparent bond lengths. III. Diatomic molecules. Acta Cryst B28:2206–2212
198. Cyvin SJ (1968) Molecular vibrations and mean square amplitudes. Elsevier, Amsterdam
199. Jeffrey GA, Ruble JR, Yates JH (1983) Neutron diffraction at 15 K and 120 K and ab-initio molecular orbital studies of the molecular structure of 1,2,4-triazole. Acta Cryst B39:388–394
200. Jeffrey GA, Ruble JR (1984, publ. 1985) Effect of thermal motion on carbon-hydrogen bond lengths. Trans Am Cryst Assoc 20:129–132
201. Craven BM, Swaminathan S (1984, publ. 1985) Neutron diffraction carbon-hydrogen bond lengths corrected for harmonic and anharmonic thermal vibration. Trans Am Cryst Assoc 20:133–135
202. Craven BM (1987) Studies of hydrogen atoms in organic molecules. Trans Am Cryst Assoc 23:71–81
203. Jeffrey GA, Ruble JR, McMullan RK, Pople JA (1987) The crystal structure of deuterated benzene. Proc R Soc A414:47–57
204. Jeffrey GA, Ruble JR, Yates JH (1984) π-Bond anisotropy in the molecular structure of thioacetamide. J Am Chem Soc 106:1571–1576
205. Bartell LS (1955) Effects of anharmonicity of vibration on the diffraction of electrons by free molecules. J Chem Phys 23:1219–1222
206. Kuchitsu K, Bartell LS (1961) Effects of anharmonicity of molecular vibrations on the diffraction of electrons. II. Interpretation of experimental structural parameters. J Chem Phys 35:1945–1949
207. Bartell LS (1963) Calculation of mean atomic positions in vibrating polyatomic molecules. J Chem Phys 38:1827–1833
208. Kuchitsu K, Morino Y (1965) Estimation of anharmonic potential constants. I. Linear XY_2 molecules. Bull Chem Soc Jpn 38:805–813

209. Eriksson A, Hermansson K (1983) Analysis of the thermal parameters of the water molecule in crystalline hydrates studied by neutron diffraction. Acta Cryst B39:703–711
210. Johnson CK (1970) Generalized treatments for thermal motion. Chap 9. In: Willis BTM (ed) Thermal neutron diffraction. Oxford Univ Press, London, pp 132–160
211. Stewart RF (1972) Valence structure from X-ray diffraction data. Physical properties. J Chem Phys 57:1664–1668
212. Cruickshank DWJ (1959) Fourier syntheses and structure factors, Sect 6. International tables for x-ray crystallography, vol II. Kynoch Press, UK, pp 318–340
213. Stewart RF (1973) Electron population analysis with generalized X-ray scattering factors: higher multipoles. J Chem Phys 58:1668–1676
214. Hirshfeld FL (1971) Difference densities by least-squares refinement: fumaramic acid. Acta Cryst B27:769–781
215. Hirshfeld FL (1976) Can X-ray data distinguish bonding effects from vibrational smearing? Acta Cryst A32:239–244
216. Hansen NK, Coppens P (1978) Testing aspherical atom refinements on small-molecule data sets. Acta Cryst A34:909–921
217. Stewart RF (1976) Electron population analysis with rigid pseudoatoms. Acta Cryst A32:565–574
218. Spackman MA, Stewart RF (1984) Electrostatic properties from accurate diffraction data. In: Hall SR, Ashida T (eds) Methods and applications in crystallographic computing. Clarendan Press, Oxford, pp 302–320
219. Kulda J (1984) A novel approach to dynamical neutron diffraction by a deformed crystal. Acta Cryst A40:120–126
220. Kulda J (1988) The RED extinction model I. An upgraded formalism. Acta Cryst A44:283–285
221. Stevens ED (1978) Low-temperature experimental electron density distribution of formamide. Acta Cryst B34:544–551
222. Craven BM, McMullan RK (1979) Charge density in parabanic acid from X-ray and neutron diffraction. Acta Cryst B35:934–945
223. He XM, Swaminathan S, Craven BM, McMullan RK (1988) Thermal vibrations and electrostatic properties of parabanic acid at 123 and 298 K. Acta Cryst B44:271–281
224. Epstein J, Ruble JR, Craven BM (1982) The charge density in imidazole by X-ray diffraction at 103 and 293 K. Acta Cryst B38:140–149
225. Krijn MPCM, Feil D (1988) Electron density distributions in hydrogen bonds: A local density functional study of α-oxalic acid dihydrate and comparison with experiment. J Chem Phys 89:4199–4208
226. Stewart RJ (1982) Mapping electrostatic potentials from diffraction data. God Jugosl. Cent. Kristologr. 17, 1–4, Zagreb
227. Kollman PA, McKelvey J, Johansson A, Rothenberg S (1975) Theoretical studies of hydrogen bonded dimers. Complexes involving HF, H_2O, NH_3, HCl, H_2S, PH_3, HCN, HNC, HCP, CH_2NH, H_2CS, H_2CO, CH_4, CF_3H, C_2H_2, C_2H_4, C_6H_6, F^-, and H_3O^+. J Am Chem Soc 97:955–965
228. Spackman MA (1986) A simple quantitative model of hydrogen bonding. J Chem Phys 85:6587–6601
229. Spackman MA (1987) A simple quantitative model of hydrogen bonding. Application to more complex systems. J Phys Chem 91:3179–3186
230. Windsor CG (1981) Pulsed neutron scattering. Taylor and Francis, London
231. David WIF (1987) The scope and possibilities of crystallography with pulsed neutrons. In: Carrondo MA, Jeffrey GA (eds) Chemical crystallography with pulsed neutrons and synchrotron radiation. NATO ASI, C221, Reidel, Dordrecht, pp 27–57
232. Rietveld HM (1969) A profile refinement method for nuclear and magnetic structures. J Appl Cryst 2:65–71
233. Young RA, Wiles DB (1982) Profile shape functions in Rietveld refinements. J Appl Cryst 15:430–438

234. Cheetham AK (1987) Pulsed neutron powder diffraction. In: Carrondo MA, Jeffrey GA (eds) Chemical crystallography with pulsed neutron and synchrotron X-rays. NATO ASI C221, Reidel, Dordrecht, pp 137–158
235. Kuhs WF, Finney JL, Vettier C, Bliss DV (1984) Structure and hydrogen ordering in ices VI, VII and VIII by neutron powder diffraction. J Chem Phys 81:3612–3623
236. Jörgensen JD, Beyerlein RA, Watanabe N, Worlton TG (1984) Structure of D_2O Ice VIII from in-situ powder neutron diffraction. J Chem Phys 81:3211–3214
237. Mehring M (1976) High resolution NMR spectroscopy in solids. Basic principles and progress. Springer, Berlin Heidelberg New York
238. Pines A, Gibby MG, Waugh JS (1973) Proton-enhanced NMR of dilute spins in solids. J Chem Phys 59:569–590
239. Jeffrey GA, Nanni R (1985) The crystal structure of anhydrous α,α-trehalose at $-150\,°C$. Carbohydr Res 137:21–30
240. Jeffrey GA, Wood RA, Pfeffer PE, Hicks KB (1983) Crystal structure and solid-state NMR analysis of lactulose. J Am Chem Soc 105:2128–2133
241. Wingert LM, Ruble JR, Jeffrey GA (1984) The crystal structure of ethyl-2,3-dideoxy-α-D-erythro-hex-2-enopyranose. Carbohydr Res 128:1–10
242. French AD, Roughhead WA, Miller DP (1986) The structures of cellulose: characterization of the solid state. Alalla RH (ed) ACS Symposium Series, No 340, pp 15–50
243. Berglund B, Vaughan RW (1980) Correlations between proton chemical shift tensors, deuterium quadrupole couplings and bond distances for hydrogen bonds in solids. J Chem Phys 73:2037–2043
244. Rohlfing CM, Alden LC, Ditchfield R (1983) Proton chemical shift tensors in hydrogen bonded dimers of RCOOH and ROH. J Chem Phys 79:4958–4966
245. Jeffrey GA, Yeon Y (1986) The correlation between hydrogen-bond lengths and proton chemical shifts. Acta Cryst B42:410–413
246. Naito A, McDowell CH (1984) Determination of the ^{14}N quadrupole coupling tensors and the ^{13}C chemical shielding tensors in a single crystal of L-aspargine monohydrate. J Chem Phys 81:4795–4803
247. Sastry DL, Takegoshi K, McDowell CH (1987) Determination of the ^{13}C chemical shift tensors in a single crystal of methyl α-D-glucopyranoside. Carbohydr Res 165:161–171
248. Roothaan CCJ (1951) New developments in molecular orbital theory. Rev Mod Phys 23:69–89
249. Roothaan CCJ (1960) Self-consistent field theory for open shells of electronic systems. Rev Mod Phys 32:179–185
250. Pople JA (1977) A priori geometry predictions. In: Schaefer III HF (ed) Modern theoretical chemistry, vol 4, Chap. 1. Plenum Press, NY, pp 1–27
251. Dunning TH, Jr, Hay PJ (1977) Gaussian basis sets for molecular calculations. In: Schaefer III HF (ed) Modern theoretical chemistry. Methods of electronic structure theory, vol 3, Chap 1. Plenum Press, NY, pp 1–30
252. Kollman PA (1977) Hydrogen-bonding and donor acceptor interactions. In: Schaefer III HF (ed) Modern theoretical chemistry, vol 4, Chapt 3. Plenum Press, NY, pp 109–152
253. Hehre WJ, Radom L, Schleyer PvR, Pople JA (1986) Ab-initio molecular orbital theory. John Wiley & Sons, NY
254. Pople JA, Segal GA (1966) Approximate self-consistent orbital theory. III. CNDO results for AB_2 and AB_3 systems. J Chem Phys 44:3289–3296
255. Pople JA, Beveridge DL, Dobosh PA (1967) Approximate self-consistent molecular orbital theory. V. Intermediate neglect of differential overlap. J Chem Phys 47:2026–2033
256. Dewar MJS, Thiel W (1977) Ground state of molecules. 38. The MNDO method. Approximations and parameters. J Am Chem Soc 99:4899–4907
257. Malrieu JP (1977) Semi-empirical methods of electronic structure calculation. In: Schaefer III HF(ed) Modern theoretical chemistry, vol 7, Chapt. 3. Plenum Press, NY, pp 69–104
258. Yadav JS, Barnickel G, Bradaczek H (1982) Quantum chemical studies on the conformational structure of bacterial peptidoglycan II. PCILO calculations on monosaccharides. J Theor Biol 95:151–166

259. Wrinn MC, Whitehead MA (1986) PCILO. Problems on predicting valid structure. J Molec Struct (Theochem) 137:197–205
260. Wiberg KB, Boyd RH (1972) Application of strain energy minimization to the dynamics of conformational change. J Am Chem Soc 94:8426–8430
261. Ermer O (1976) Calculation of molecular properties using force fields. Application to organic chemistry. In: Structure and bonding, vol 27. Springer, Berlin Heidelberg New York, pp 163–201
262. Burkert U, Allinger NL (1982) Molecular mechanics. Am Chem Soc Monogr 177, Am Chem Soc Washington, DC
263. Kroon-Batenburg LMJ, Kanter JA (1983) Influence of hydrogen bonds on molecular conformation. Molecular mechanics calculations on α–D-glucose. Acta Cryst B39:749–754
264. Murray W (ed) (1972) Numerical methods for unconstrained refinement. Academic Press, London
265. Melberg S, Rasmussen K (1980) Conformations of disaccharides by empirical force-field calculations. Part III. β-Gentiobiose. Carbohydr Res 78:215–224
266. Rohrer DC, Sarko A, Bluhm TL, Lee YN (1980) The structure of gentiobiose. Acta Cryst B36:650–654
267. Nanni RG, Ruble JR, Jeffrey GA, McMullan RK (1986) Neutron diffraction analysis at 15 K and ab-initio molecular orbital calculations for α-cyanoacetohydrazide. A crystal structure with a high energy conformer. J Mol Struct 147:369–380
268. White DNJ, Morrow C (1977) The global minimum energy conformation of cyclotetraglycyl. Tetrahedron Lett 3385–3388
269. Kuntz ID, Crippen GM, Kollman PA, Kimelman D (1976) Calculation of protein tertiary structure. J Mol Biol 106:983–994
270. White DNJ (1978) Molecular mechanics calculations. In: Sutton LE, Truter MR (eds) Molecular structures by diffraction methods. Vol. 6. The Chemical Society, London, pp 38–62
271. Pullman B, Pullman A (1974) Molecular orbital calculations on the conformation of amino acid residues of proteins. Adv Protein Chem 28:347–526
272. Brooks BR, Bruccoleri RE, Olafson BD, States DJ, Swaminathan S, Karplus M (1983) CHARMM. A program for macromolecular energy, minimization and dynamics calculations. J Comput Chem 4:187–217
273. Hermans J (1985) Molecular dynamics and protein structure. Polycrystal Book Service, Dayton, OH, USA
274. DeFrees DJ, Levi BA, Pollack SK, Hehre WJ, Binkley JS, Pople JA (1979) Effect of electron correlation on theoretical equilibrium geometries. J Am Chem Soc 101:4085–4089
275. Møller C, Plesset MS (1934) Note on an approximation treatment of many-electron systems. Phys Rev 46:618–622
276. van Duijneveldt-van der Rijdt JGCM, van Duijneveldt FB (1982) Gaussian basis sets which yield accurate Hartree-Fock electric moments and polarizabilities. J Mol Struct 89:185–201
277. Pople JA (1951) Molecular association in liquids II. A theory of the structure of water. Proc R Soc 205A:163–178
278. Morokuma K, Pedersen L (1968) Molecular-orbital studies of hydrogen bonds. An ab-initio calculation for dimeric H_2O. J Chem Phys 48:3275–3282
279. Neumann D, Moskowitz JW (1968) One electron properties of near Hartree-Fock wave functions. I. Water. J Chem Phys 49:2056–2070
280. Kollman PA, Allen LC (1969) Theory of the hydrogen bond: electronic structures of properties of the water dimer. J Chem Phys 51:3286–3293
281. Morokuma K, Winick JR (1970) Molecular orbital studies of hydrogen bonds: dimeric H_2O with a Slater minimal basis set. J Chem Phys 52:1301–1306
282. Kollman PA, Buckingham AD (1971) The structure of the water dimer. Molec Phys 21:567–570
283. Magnusson LB (1971) The structure of the water dimer. Molec Phys 21:571–575
284. Diercksen GHF (1971) SCF-MO-LCGO studies on hydrogen bonding. The water dimer. Theor Chim Acta 21:335–367
285. Kollman PA, Allen LC (1972) The theory of the hydrogen bond. Chem Rev 72:283–303

286. Shipman LL, Owichi JC, Scheraga HA (1974) Structure, energetics and dyamics of the water dimer. J Phys Chem 78:2055–2060
287. Popkie H, Kistenmacher H, Clementi E (1973) Study of the structure of molecular complexes IV. The Hartree-Fock potential for the water dimer and its application to the liquid state. J Chem Phys 59:1325–1336
288. Matsuoka O, Clementi E, Yoshimine M (1976) CI study of the water dimer potential surface. J Chem Phys 64:1351–1361
289. Williams DE, Craycroft DJ (1985) Estimation of dimer Coulombic intermolecular energy and site-charge polarization by the potential-derived method. J Phys Chem 89:1461–1467
290. Frisch MJ, Del Bene JE, Binkley JS, Schaefer III HF (1986) Extensive theoretical studies of the hydrogen-bonded complexes $(H_2O)_2$, $(H_2O)_2H^+$, $(HF)_2)$, $(HF)_2H^+$, F_2H^- and $(NH_3)_2$. J Chem Phys 84:2279–2289
291. Del Bene JE (1987) Basis set and correlation effects on computed hydrogen bond energies of the dimers $(AH_n)_2 : AH_2 = NH_3$, OH_2, FH. J Chem Phys 86:2110–2113
292. Dyke TR, Muenter JS (1974) Microwave spectrum and structure of hydrogen-bonded water dimer. J Chem Phys 60:2929–2930
293. Dyke TR, Mack KM, Muenter JS (1977) The structure of water dimer from molecular beam electric resonance spectroscopy. J Chem Phys 66:498–510
294. Odulota JA, Viswanathan R, Dyke TR (1979) Molecular beam electric deflection behavior and polarity of hydrogen-bonded complexes of ROH, RSH, RNH. J Am Chem Soc 101:4787–4792
295. Amos RD (1986) Structure, harmonic frequencies and infrared intensities of the dimers of H_2O and H_2S. Chem Phys 104:145–151
296. Spackman MA (1986) Atom-atom potentials via electron gas theory. J Chem Phys 85:6579–6586
297. Karpfen A (1984) Ab-initio studies of hydrogen bonded chain. IV. Structure and stability of formic acid chain. Chem Phys 88:415–423
298. Dreyfus M, Maigret B, Pullman A (1970) A non-empirical study of hydrogen bonding in the dimer of formamide. Theor Chim Acta 17:109–119
299. Jeffrey GA, Ruble JR, McMullan RK, De Frees DJ, Pople JA (1981) Neutron diffraction at 20 K and ab-initio molecular-orbital studies of the structure of monofluoroacetamide. Acta Cryst B37:1885–1890
300. Del Bene JE, Cohen I (1978) Molecular orbital theory of the hydrogen bond. 20. Pyrrole and imidazole as proton donors and proton acceptors. J Am Chem Soc 100:5285–5290
301. Sreerama N, Vishveshwara S (1985) An ab-initio study of (C-H---X)$^+$ hydrogen bonds including biological systems. J Molec Struct (Theochem) 133:139–146
302. Jasien PG, Stevens WJ (1986) Ab initio study of hydrogen bonding interactions of formamide with water and methanol. J Chem Phys 84:3271–3273
303. Newton MD (1983) Small water clusters as theoretical models for structural and kinetic properties of ice. J Phys Chem 87:4288–4292
304. Koehler JEH, Saenger W, Lesyng B (1987) Cooperative effects in extended hydrogen bonded systems involving O-H groups. Ab initio studies of the cyclic S_4 water tetramer. J Comput Chem 8:1090–1098
305. Dauben P, Hagler AT (1980) Crystal packing, hydrogen bonding, and the effect of crystal forces on molecular conformation. Acc Chem Res 13:105–112
306. Lesyng B, Jeffrey GA, Maluszynska H (1988) A model for the hydrogen bond length probability distributions in the crystal structures of small molecule components of the nucleic acids. Acta Cryst B44:193–198
307. McQuarrie DA (1973) Statistical thermodynamics. Harper & Row, New York, 343 pp
308. Newton MD (1983) Theoretical aspects of the OH---O hydrogen bond and its role in structural and kinetic phenomena. Acta Cryst B39:104–113
309. Giguère PA (1984) Bifurcated hydrogen-bonds in water. J Raman Spectros 15:354–359
310. Kollman PA, Allen LC (1970) An SCF partitioning scheme for the hydrogen bond. Theor Chim Acta 18:399–403
311. Kollman PA, Allen LC (1970) Hydrogen-bonded dimers and polymers involving hydrogen fluoride, water and ammonia. J Am Chem Soc 92:753–759

312. Yamabe S, Morokuma K (1975) Molecular orbital studies of hydrogen bonds. IX. Electron distribution analysis. J Am Chem Soc 97:4458–4465
313. Van Duijneveldt-Van der Rijdt JGCM, van Duijneveldt FB (1971) Perturbation calculation on the hydrogen bonds between some first-row atoms. J Am Chem Soc 93:5644–5653
314. Piela L, Delhalle J (1978) An efficient procedure to evaluate long-range Coulombic interactions within the framework of the LCAO-MO method for infinite polymers. Int J Quantum Chem 13:605–617
315. Westheimer FH (1965) Calculation of the magnitude of steric effects. Chap. 12. Steric effects in organic chemistry. Newman MS (ed). Wiley, NY
316. Rao VS, Sundararajan SR, Ramakrishnan PR, Ramachandran GN (1967) Conformational studies of amylose. In: Ramachandran GN (ed) Conformation of biopolymers. Academic Press, New York, pp 721–728
317. Bock K, Meldal M, Bundle DR, Iversen T, Garegg PJ, Norberg T, Lindberg AA, Svenson SB (1984) The conformation of *Salmonella* O-antigenic polysaccharide chains of serogroup A, B and D1 predicted by semi-empirical hard-sphere (HSEA) calculations. Carbohydr Res 130:23–34
318. French AD, French WA (1980) N-H mapping for polymers. Chap 14. In: French AD, Gardner KCH (eds) Fibre diffraction methods. ACS Symp Ser 141:239–250
319. Engler EM, Andose JD, Schleyer PvR (1973) Critical evaluation of molecular mechanics. J Am Chem Soc 95:8005–8025
320. Kollman PA (1977) A general analysis of noncovalent intermolecular interactions. J Am Chem Soc 99:4875–4894
321. Kollman PA (1978) A method of describing the charge distribution in simple molecules. J Am Chem Soc 100:2974–2984
322. Došen-Mićović L, Jeremić D, Allinger NL (1983) Treatment of electrostatic effects within the molecular mechanics method. J Am Chem Soc 105:1716–1722
323. Momany FA (1978) Determination of partial atomic charges from abinitio electrostatic potentials. Application to formamide, methanol and formic acid. J Phys Chem 82:592–601
324. Abraham RJ, Hudson B (1985) Charge calculations in molecular mechanics III. Amino acids and peptides. J Comput Chem 6:173–181
325. Singh UC, Kollman PA (1984) An approach to computing electrostatic charges for molecules. J Comput Chem 5:129–145
326. Kar T, Sannigrahi AB, Mukherjee DC (1987) Comparison of atomic charges, valencies and bond orders in some hydrogen-bonded complexes calculated from Mulliken and Löwdin SCF density matrices. J Molec Struct (Theochem) 153:93–101
327. Cox SR, Williams DE (1981) Representation of the molecular electrostatic potential by a net atomic charge model. J Comput Chem 2:304–323
328. Mulliken RS (1955) Electronic population analysis on LCAO-MO molecular wave functions, I. J Chem Phys 23:1833–1840
329. Williams DE, Jan J (1987) Point-charge models of molecules derived from least squares fitting of the electric potential. Adv Atomic & Mol Phys 23:87–129
330. Barnes P, Finney JL, Nicholas JD, Quinn JE (1979) Cooperative effects in simulated water. Nature (Lond) 282:459–464
331. Goodfellow JM, Finney JL, Barnes P (1982) Monte Carlo computer simulation of water-amino acid interactions. Proc R Soc Lond B214:213–228
332. Bader RFW (1985) Atoms in molecules. Acc Chem Res 18:9–15
333. Lippincott ER, Schroeder R (1955) One-dimensional model for the hydrogen bond. J Chem Phys 23:1099–1106
334. Schroeder R, Lippincott ER (1957) Potential function model of the hydrogen bond. II. J Phys Chem 61:921–928
335. McGuire RF, Momany FA, Scheraga HA (1972) Energy parameters in polypeptides. V. An empirical hydrogen bond potential function based on molecular orbital calculations. J Phys Chem 76:375–393
336. Momany FA, Carruthers LM, McGuire RF, Scheraga HA (1974) Interatomic potentials from crystal data. III. Determination of empirical potentials and application to the packing con-

figuration and lattice energies in crystals of hydrocarbons, carboxylic acids, amines and amides. J Phys Chem 78:1595–1620
337. Nemethy C, Scheraga HA (1977) Interatomic potentials from crystal data. V. Determination of empirical parameters of OH---O hydrogen bonds from packing considerations and lattice energies of polyhydric alcohols. J Phys Chem 81:928–931
338. Hagler AT, Huler E, Lifson S (1974) Energy functions for peptides and proteins. I. Derivation of consistent force field including the hydrogen bond from amide crystals. J Am Chem Soc 96:5319–5327
339. Lifson S, Hagler AT, Dauben P (1979) Consistent force field studies of intermolecular forces in hydrogen bonded crystals. I. Carboxylic acids, amides and the C-O---H hydrogen bonds. J Am Chem Soc 101:5111–5121
340. Derissen JL, Smit PH (1978) Intermolecular interactions in crystals of carboxylic acids. IV. Empirical interatomic potential functions. Acta Cryst A 34:842–853
341. Minicozzi WP, Bradley DF (1969) Determination of interaction energy surfaces. I. Formic and acetic acid dimers. J Comput Phys 4:118–137
342. Minicozzi WP, Stroot MT (1970) Determination of interaction energy functions. II. Crystalline formic acid. J Comput Phys 6:95–104
343. Melberg S, Rasmussen K, Scordamaglia R, Tosi C (1979) The nonbonded interactions in D-glucose and β-maltose: an ab-initio study of conformations produced by empirical force-field calculations. Carbohydr Res 76:23–37
344. Pertsin AJ, Kitaigorodsky AI (1987) The atom-atom potential method. Application to organic molecular solids. Springer, Berlin Heidelberg New York Tokyo
345. Taylor R (1981) An empirical potential for the O-H---O hydrogen bond. Part I. Comparison with ab-initio molecular orbital results. J Molec Struct 71:311–325
346. Taylor R (1982) An empirical potential for the O-H---O hydrogen bond. Part 2. Comparison with observed hydrogen bond geometries. J Molec Struct 72:125–136
346a. Kroon-Batenburg LMJ, Kanters JA (1983) Development of an empirical O-H---O hydrogen bond potential for MM2 force field calculations. J Molec Struct (Theochem) 105:417–418
347. Jeffrey GA, Taylor R (1980) The application of molecular mechanics to the structures of carbohydrates. J Comput Chem 1:99–109
348. Ciccotti G, Hoover WG (eds) (1986) Molecular-dynamics simulation of statistical-mechanical systems. Proceedings of the international school of physics "Enrico Fermi", course XCVII. North-Holland, Amsterdam
348a. Metropolis N, Rosenbluth AW, Rosenbluth MN, Teller AH, Teller E (1953) Equation of state calculations by fast computing machines. J Chem Phys 21:1087–1093
349. McCammon JA, Harvey SC (1987) Dynamics of proteins and nucleic acids. Cambridge Univ Press, Cambridge, UK
350. Morse MD, Rice SA (1982) Tests of effective pair potentials for water: predicted ice structures. J Chem Phys 76:650–660
350a. Boobbyer DNA, Goodford PJ, McWhinnie PM, Wade RC (1989) New hydrogen-bond potentials for use in determining energetically favorable binding sites on molecules of known structure. J Med Chem 32:1083–1089
351. Rees DA, Smith PJC (1975) Polysaccharide conformations. Part VIII. Test of energy functions by Monte Carlo calculations on monosaccharides. J Chem Soc Perkin II: 830–835
352. Brady JW (1986) Molecular dynamics simulations of α-glucose. J Am Chem Soc 108:8153–8160
353. Karplus M, McCammon JA (1983) Dynamics of proteins: elements and function. Annu Rev Biochem 52:263–300
354. Paine GH, Scheraga HA (1986) Prediction of the native conformation of a polypeptide by a statistical mechanical procedure II. Average backbone structure of enkephalin. Biopolymers 25:1547–1563
355. Swaminathan S, Beveridge DL (1977) A theoretical study of the structure of liquid water based on quasi-component distribution functions. J Am Chem Soc 99:8392–8398
356. Stillinger FH, Rahman A (1974) Improved simulation of liquid water by molecular dynamics. J Chem Phys 60:1545–1554

357. Berendsen HJC, Grigera JR, Straatsma TP (1987) The missing term in effective pair potentials. J Chem Phys 91:6269–6271
358. Mezei M, Beveridge DL, Berman HM, Goodfellow JM, Finney JL, Neidle S (1983) Monte Carlo studies on water in dCpG/proflavine crystal hydrate. J Biomol Struct Dynam 1:287–297
359. Subramanian S, Ravishanker G, Beveridge DL (1988) Theoretical considerations of the spine of hydration in the minor groove of d(CGCGAATTCGCG) d(CGCGAATTCGCG) Monte Carlo computer simulation. Proc Natl Acad Sci USA 85:1836–1840
360. Koehler JEH, Saenger W, van Gunsteren WF (1987) A molecular dynamics simulation of crystalline α-cyclodextrin hexahydrate. Eur Biophys J 15:197–210
361. Koehler JEH, Saenger W, van Gunsteren WF (1987) Molecular dynamics simulation of crystalline β-cyclodextrin dodecahydrate at 293 K and 120 K. Eur Biophys J 15:211–224
362. Wilcox GL, Quiocho FA, Levinthal C, Harvey SC, Maggiora GM, McCammon JA (1987) Symposium overview. Minnesota conference on supercomputing in biology: proteins, nucleic acids and water. J Computer-Aided Molecular Design 1:271–281
363. Lord RC, Merrifield RE (1953) Strong hydrogen bonds in crystals. J Chem Phys 21:166–167
364. Nakamoto K, Margoshes M, Rundle RE (1955) Stretching frequencies as a function of distances in hydrogen bonds. J Am Chem Soc 77:6480–6486
365. Joswig W, Fuess H, Ferraris G (1982) Neutron diffraction study of the hydrogen bond in trisodium hydrogen bissulphate and a survey of very short O-H---O bonds. Acta Cryst B 38:2798–2801
366. Almenningen A, Bastiansen O, Motzfeldt T (1969) Reinvestigation of the structure of monomer and dimer formic acid by gas diffraction technique. Acta Chem Scand 23:2848–2864
367. Derissen JL (1971) Reinvestigation of the molecular structure of acetic acid monomer and dimer by gas electron diffraction. J Molec Struct 7:67–80
368. Kwei GH, Curl RF (1960) Microwave spectrum of O^{18} formic acid and structure of formic acid. J Chem Phys 32:1592–1594
369. Jönsson PG, Hamilton WC (1972) Hydrogen-bond studies. LX. A single crystal neutron diffraction study of trichloroacetic acid dimer. J Chem Phys 56:4433–4439
370. Leviel J-L, Auvert G, Savariault JM (1981) Hydrogen bond studies. A neutron diffraction study of succinic acid at 30 and 77 K. Acta Cryst B 37:2185–2189
371. Jönsson PG (1971) Hydrogen-bond studies. XLIV. Neutron diffraction study of acetic acid. Acta Cryst B 27:893–898
372. Ellison RD, Johnson CK, Levy HA (1971) Glycolic acid: direct neutron diffraction determination of crystal structure and thermal motion analysis. Acta Cryst B 27:333–344
373. Sabine TM, Cox GW, Craven BM (1969) A neutron diffraction study of α-oxalic acid dihydrate. Acta Cryst B 25:2437–2441
374. Kitano M, Kuchitsu K (1974) Molecular structure of formamide as studied by gas electron diffraction. Bull Chem Soc Jpn 47:67–72
375. Ladell J, Post B (1954) The crystal structure of formamide. Acta Cryst 7:559–564
376. Ottersen T (1975) On the structure of the peptide linkage. The structures of formamide and acetamide, at −165 °C and an ab-initio study of formamide, acetamide and N-methylformamide. Acta Chem Scand A 29:939–944
377. Jeffrey GA, Ruble JR, McMullan RK, DeFrees DJ, Binkley JS, Pople JA (1980) Neutron diffraction at 23 K and ab-initio molecular orbital studies of the molecular structure of acetamide. Acta Cryst B 36:2292–2299
378. Jeffrey GA, Ruble JR, Pople JA (1982) Neutron diffraction at 9 K and abinitio molecular orbital studies of the molecular structure of glyoxime. Acta Cryst B 38:1975–1980
379. Jeffrey GA, Ruble JR, McMullan RK, DeFrees DJ, Pople JA (1982) Neutron diffraction at 15 K and ab-initio molecular orbital studies of the structure of N,N'-diformohydrazide. Acta Cryst B 38:1508–1513
380. Stevens ED, Rys J, Coppens P (1978) Quantitative comparison of theoretical calculations with the experimentally determined electron density distribution of formamide. J Am Chem Soc 100:2324–2328
381. Ramachandran GN, Ramakrishnan C (1968) Hydrogen bonding and conformation of polypeptides. In: Crewther WG (ed) Symp. Fibrous proteins. Butterworths, Sydney, Australia, pp 71–83

382. Ramakrishnan C, Prasad N (1971) Hydrogen bonds in amino acids and peptides. Int J Protein Res 3:209–231
383. Ramanadham M, Chidambaram R (1978) Amino acids: systematics of molecular structure, conformation and hydrogen bonding. In: Srinivasan R (ed) Adv. Crystallogr., Invited Rev Lect Natl Conf Crystallogr 1977, Oxford and IBH, New Delhi, pp 81–103
384. Mitra J, Ramakrishnan C (1977) Analysis of OH---O hydrogen bonds. Int J Pept Protein Res 9:27–48
385. Hopfinger AJ (1973) Conformational properties of macromolecules. Academic, New York, 348 pp
386. Dunitz JD, Waser J (1972) Geometric constraints in six- or eight-membered rings. J Am Chem Soc 94:5645–5662
387. Taylor R, Kennard O (1983) Comparison of X-ray and neutron diffraction results for the N-H---O=C hydrogen bond. Acta Cryst Sect B 39:133–138
388. McGaw BL, Ibers JA (1963) Nature of the hydrogen bond in sodium acid fluoride. J Chem Phys 39:2677–2684
389. De La Vega JR (1982) Role of symmetry in the tunnelling of the proton in double minimum potentials. Acc Chem Res 15:185–191
390. Küppers H, Kvick Å, Olovsson I (1981) Hydrogen bond studies: CXLII. Neutron diffraction study of the two very short hydrogen bonds in lithium hydrogen phthalate-methanol. Acta Cryst B 37:1203–1207
391. Takusagawa F, Koetzle TF (1979) Neutron diffraction study of quinolinic acid recrystallized from D_2O: evaluation of temperature and isotope effects in the structure. Acta Cryst B 35:2126–2135
392. Jones RDG (1976) The crystal structure of the enol tautomer of 1,3-diphenyl-1,3-propanedione (dibenzoylmethane) by neutron diffraction. Acta Cryst B 32:1807–1811
393. Jones RDG (1976) The crystal and molecular structure of the enol form of 1-phenyl-1,3-butanedione (benzoylacetone) by neutron diffraction. Acta Cryst B 32:2133–2136
394. Fuess H, Lindner HJ (1976) Intramolekulare Wasserstoffbrücke in 6-Hydroxyl-1-fulvencarbaldehyd. Chem Ber 108:3096–3104
395. Herbstein FH, Kapon M, Reisner GM, Lehmann MS, Kress RB, Wilson RB, Shiau W-I, Duesler EN, Paul IC, Curtin DY (1985) Polymorphism of naphthazarin and its relation to solid-state proton transfer. Neutron and X-ray diffraction studies on naphthazarin C. Proc R Soc Lond A 399:295–319
396. Hsu B, Schlemper EO (1980) X-N Deformation density studies of the hydrogen maleate ion and the imidazolium ion. Acta Cryst B 36:3017–3023
397. Ellison RD, Levy HA (1965) A centered hydrogen bond in potassium hydrogen chloromaleate: A neutron diffraction structure determination. Acta Cryst 19:260–268
398. Hermansson K, Tellgren R (1983) Neutron diffraction studies of potassium hydrogen diformate, $KH(HCOO)_2$, at 120 and 295 K. Acta Cryst C 39:1507–1510
399. Olovsson G, Olovsson I, Lehmann MS (1984) Neutron diffraction study of sodium hydrogen maleate trihydrate. $NaH[C_4H_2O_4] \cdot 3H_2O$, at 120 K. Acta Cryst C 40:1521–1526
400. Albertson J, Grenthe I (1973) A neutron diffraction study of potassium and rubidium hydrogen oxydiacetate. The dynamics of their hydrogen bonds. Acta Cryst B 29:2751–2760
401. McGregor DR, Speakman JC, Lehmann MS (1977) Crystal structures of some acid salts of monobasic acids. Part XIX. Potassium hydrogen dicrotonate. X-ray and neutron diffraction studies. J Chem Soc Perkin Trans 2:1740–1745
402. Hadži M, Leban I, Orel B, Iwata M, Williams JM (1979) Neutron diffraction and vibrational spectroscopic study of single crystals of $KH(D)(CHCl_2COO)_2$. J Cryst Mol Struct 9:117–134
403. Moore FH, Power LF (1971) The crystal structure of potassium hydrogen oxalate by neutron diffraction. Inorg Nucl Chem Lett 7:873–875
404. Delaplane RG, Tellgren T, Olovsson I (1984) Neutron diffraction study of sodium hydrogen oxalate monohydrate, $NaHC_2O_4 \cdot H_2O$, at 120 K. Acta Cryst C 40:1800–1803
405. Kostansek EC, Busing WR (1972) A single-crystal neutron-diffraction study of urea-phosphoric acid. Acta Cryst B 28:2454–2459

406. Thornley FR, Nelmes RJ, Rouse KD (1975) A neutron diffraction study of room-temperature monoclinic KD_2PO_4. Chem Phys Lett 34:175–177
407. Catti M, Ferraris G, Filhol A (1977) Hydrogen bonding in the crystalline state. $CaHPO_4$ (Monetite). P$\bar{1}$ or P1? A novel neutron diffraction study. Acta Cryst B33:1223–1229
408. Choudhary RNP, Nelmes RJ, Rouse KD (1981) A room-temperature neutron-diffraction study of NaH_2PO_4. Chem Phys Lett 78:102–105
409. Kennedy NSJ, Nelmes RJ (1980) Structural studies of RbH_2PO_4 in its paraelectric and ferroelectric phases. J Phys Chem 13:4841–4853
410. Ichikawa M (1972) The crystal structure and phase transition of ammonium hydrogen dichloracetate. I. The crystal structure of the paraelectric phase. Acta Cryst B28:755–760
411. Macdonald AL, Speakman JC, Hadži D (1972) Crystal structures of the acid salts of some monobasis acids. Part XIV. Neutron-diffraction studies of potassium hydrogen bis(trifluoroacetate) and potassium deuterium bis(trifluoroacetate): Crystals with short and symmetrical hydrogen bonds. J Chem Soc Perkin Trans II:825–832
412. Skinner JM, Stewart GMD, Speakman JC (1954) The crystal structure of the acid salts of some monobasic acids. Part III. Potassium hydrogen dibenzoate. J Chem Soc 180–184
413. Manojlović Lj, Speakman JC (1968) The crystal structure of, and hydrogen bond in, potassium hydrogen biphenylacetate: a redetermination. Acta Cryst B24:323–325
414. McAdam A, Currie M, Speakman JC (1971) The crystal structures of the acid salts of some dibasic acids. Part IV. Potassium hydrogen succinate: X-ray and neutron diffraction studies. J Chem Soc A:1994–1997
415. Sequeira A, Berkebile CA, Alan C, Hamilton WC (1968) Structure and dynamics in hydrogen bonding systems. I. A neutron diffraction study of potassium hydrogen diaspirinate (bisacetylsalicylate). J Molec Struct 1:283–294
416. Nilsson A, Liminga R, Olovsson I (1968) A neutron diffraction study of hydrazinium hydrogen oxalate, $N_2H_5HC_2O_4$. Acta Chem Scand 22:719–731
417. Manojlović Lj (1968) A reinvestigation of the crystal structure of potassium hydrogen di-p-hydroxybenzoate hydrate. Acta Cryst B24:326–330
418. Currie M, Speakman JC (1970) The crystal structures of the acid salts of some dibasic acids. Part III. Potassium hydrogen malonate: a neutron diffraction study. J Chem Soc A:1923–1926
419. Ichikawa M (1978) The O-H vs O$\cdots$O distance correlation, the geometric isotope effect in OHO bonds, and its application to symmetric bonds. Acta Cryst B34:2074–2080
420. Currie M, Speakman JC, Curry NA (1967) The crystal structures of the acid salts of some dibasic acids. Part I. A neutron-diffraction study of ammonium (and potassium) tetraoxolates. J Chem Soc A:1862–1869
421. Jönsson P-G (1972) Neutron and X-ray diffraction studies of the 1:1 addition compound of acetic acid with phosphoric acid. Acta Chem Scand 26:1599–1619
422. Kamb B (1968) Ice polymorphism and the structure of water. In: Rich A, Davidson N (eds) Structural chemistry and molecular biology. Freeman, San Francisco, pp 507–544
423. Hobbs PV (1975) Ice physics. Oxford Univ Press, Oxford, UK, 782 pp
424. Kuhs WF, Lehmann MS (1981) Bond-lengths, bond angles, and transition barrier in ice Ih by neutron scattering. Nature (Lond) 294:432–434
425. Kuhs WF, Lehmann MS (1983) The structure of ice Ih by neutron diffraction. J Phys Chem 87:4312–4313
426. Kamb B, Hamilton WC, LaPlaca SJ, Prakash A (1971) Ordered proton configuration in ice II from single-crystal neutron diffraction. J Chem Phys 55:1934–1945
427. Engelhardt H, Kamb B (1981) Structure of ice IV, a metastable high-pressure phase. J Chem Phys 75:5887–5899
428. Kamb B, Prakash A, Knobler C (1967) Structure of ice V. Acta Cryst 22:706–715
429. Jorgensen JD, Worlton TG (1985) Disordered structure of D_2O ice VII from in-situ powder neutron diffraction. J Chem Phys 83:329–333
430. LaPlaca SJ, Hamilton WC, Kamb B, Prakash A (1973) On a nearly proton-ordered structure for ice IX. J Chem Phys 58:567–580
431. Hollander F, Jeffrey GA (1977) Neutron diffraction study of the crystal structure of ethylene oxide deuterohydrate at 80 K. J Chem Phys 66:4699–4705

431a. McMullan RK, Kvick A (1990) Neutron diffraction study of the structure II clathrate hydrate: 3.5 Xe·8 CCl$_4$·136 D$_2$O at 13 and 100 K. Acta Crystallogr B46:390–399
432. Jeffrey GA, McMullan RK (1967) The clathrate hydrates. Prog Inorg Chem 8:43–108
433. Jeffrey GA (1969) Water structure in organic hydrates. Accts Chem Res 2:344–352
434. Jeffrey GA (1984) Hydrate inclusion compounds. Chap 5. In: Atwood JL, Davies JED, MacNicol DD (eds) Inclusion compounds, vol. 1. Academic Press, London, pp 135–190
435. McMullan RK, Mak TCW, Jeffrey GA (1966) Polyhedral clathrate hydrates. XI. Structure of tetramethylammonium hydroxide pentahydrate. J Chem Phys 44:2338–2345
436. McLean WJ, Jeffrey GA (1967) Crystal structure of tetramethylammonium fluoride tetrahydrate. J Chem Phys 47:414–417
437. McLean MJ, Jeffrey GA (1968) Crystal structure of tetramethylammonium sulfate tetrahydrate. J Chem Phys 49:4556–4564
438. Mak TCW (1985) Crystal structure of tetraethylammonium fluoride-water (4/11), 4(C$_2$H$_5$)$_4$N$^+$F$^-$·11H$_2$O, a clathrate hydrate containing linear chains of edge-sharing (H$_2$O)$_4$F$^-$ tetrahedra and bridging water molecules. J Inclusion Phenom 3:347–354
439. Loehlin JH, Kvick Å (1978) Tetraethylammonium chloride monohydrate. Acta Cryst B34:3488–3490
440. Mak TCW (1985) Tetraethylammonium acetate tetrahydrate, (C$_2$H$_5$)$_4$N$^+$CH$_3$COO$^-$·4H$_2$O, a layer structure with cations sandwiched between nets of puckered polygons formed by hydrogen bonded anions and water molecules. J Inclusion Phenom 4:273–280
441. Ferraris G, Franchini-Angela M (1972) Survey of the geometry and environment of water molecules in crystalline hydrates studied by neutron diffraction. Acta Cryst B28:3572–3583
442. Chiari G, Ferraris G (1982) The water molecule in crystalline hydrates studied by neutron diffraction. Acta Cryst B38:2331–2341
443. Newton MD (1986) Current views of hydrogen-bonding from theory and experiment. Structure, energetics and control of chemical behavior. Trans Am Cryst Assoc 22:1–17
444. Gallagher KJ (1959) The isotope effect in relation to bond lengths of hydrogen bonds in crystals. In: Hadži D (ed) Hydrogen bonding. Pergamon Press, New York, pp 45–54
445. Singh TR, Wood JL (1969) Isotope effect on the hydrogen-bond length. J Chem Phys 50:3572–3576
446. Thomas JO, Tellgren J, Olovsson I (1974) Hydrogen bond studies. LXXXIV. An X-ray diffraction study of the structures of KHCO$_3$ and KDCO$_3$ at 298, 219 and 95 K. Acta Cryst B30:1155–1166
447. Reuben J (1984) Isotopic multiplets in the carbon-13 NMR spectra of polyols with partially deuterated hydroxyls. Fingerprints of molecular structure and hydrogen bonding effects in the ^{13}C NMR spectra of monosaccharides with partially deuterated hydroxyls. J Am Chem Soc 106:6180–6186
448. Reuben J (1986) Intramolecular hydrogen bonding as reflected in the deuterium isotope effects on carbon-13 chemical shifts. Correlation with hydrogen bond energies. J Am Chem Soc 108:1735–1738
449. Christofides JC, Davies DB (1985) Simple ^{1}H NMR observation of cooperative hydrogen bonding in O–3′-sucrose derivatives. Magnetic Reson in Chem 23:582–584
450. Sass RL (1960) A neutron diffraction study of the crystal structure of sulfamic acid. Acta Cryst 13:320–324
451. Schlemper EO, Hamilton WC (1966) Neutron diffraction study of the structures of ferroelectric and paraelectric ammonium sulfate. J Chem Phys 44:4498–4509
452. Kuroda Y, Taira Z, Uno T, Osaki K (1975) Diacetamide (trans-cis form) C$_4$H$_7$NO$_2$. Cryst Struct Commun 4:325–328
453. Staab HA, Saupe T (1988) "Protonenschwamm"-Verbindungen und die Geometrie von Wasserstoffbrücken: Aromatische Stickstoffbasen mit ungewöhnlicher Basizität. Angew Chem 100:895–909
454. Betzel CH, Saenger W, Hingerty BE, Brown GM (1984) Circular and flip-flop hydrogen bonding in β-cyclodextrin undecahydrate: a neutron diffraction study. J Am Chem Soc 106:7545–7557

455. Zabel V, Saenger W, Mason SA (1986) Neutron diffraction study of the hydrogen bonding in β-cyclodextrin undecahydrate at 120 K: from dynamic flip-flops to static homodromic chains. J Am Chem Soc 108:3664–3673
456. Dippy JFJ (1939) Review of CH participation in H-bonds. Chem Rev 25:151–160
457. Green RD (1974) Hydrogen bonding by C-H groups. Wiley Interscience, NY
458. Allerhand A, Schleyer PvR (1963) A survey of C-H groups as proton donors in hydrogen bonding. J Am Chem Soc 85:1715–1723
459. Krimm S, Kurowa K, Rebare T (1967) Infra-red studies of C-H---O=C hydrogen bonding in polyglycine, II. In: Ramachandran GN (ed) Conformation of biopolymers. Academic Press, London, pp 439–447
460. Sutor DJ (1962) The C-H---O hydrogen bond in crystals. Nature (Lond) 195:68–69
461. Sutor DJ (1963) Evidence for the existence of C-H---O hydrogen bonds in crystals. J Chem Soc:1105–1110
462. Taylor R, Kennard O (1982) Crystallographic evidence for the existence of C-H---O, C-H---N and C-H---Cl hydrogen bonds. J Am Chem Soc 104:5063–5070
463. Amidon GL, Anik S, Rubin J (1975) An energy partitioning analysis of base-sugar intramolecular C-H---O hydrogen bonding in nucleosides and nucleotides. In: Sundaralingam M, Rao ST (eds) Structure and conformation of nucleic acids and protein-nucleic acid interactions. University Park Press, Baltimore, pp 729–744
464. Takusagawa F, Shimada A (1976) Isonicotinic acid. Acta Cryst B32:1925–1927
465. Krishnaswami S, Pattabhi V (1987) Nicotinoylglycine, a metabolic product. Acta Cryst C43:728–729
466. Takusagawa F, Koetzle TF, Srikrishnan T, Parthasarathy R (1979) C-H---O interactions and stacking of water molecules between pyrimidine bases in 5-nitro-l-(β-D-ribosyluronic acid)-uracil monohydrate, [1-(5-nitro-2,4-dioxopyrimidinyl)-β-ribofuranoic acid monohydrate]: a neutron diffraction study at 80 K. Acta Cryst B35:1388–1394
467. Seiler P, Weisman GR, Glendening ED, Weinhold F, Johnson VB, Dunitz JD (1987) Eine Csp³-gebundene Methylgruppe in ekliptischer Konformation; experimenteller und theoretischer Nachweis von C–H···O Wasserstoffbrücken. Angew Chem 99:1216–1218
468. Levitt M, Perutz MF (1988) Aromatic rings act as hydrogen bond acceptors. J Molec Biol 201:751–754
469. Burley SK, Petsko GA (1986) Amino-aromatic interactions in proteins. FEBS Lett 203:139–143
470. Burley SK, Petsko GA (1985) Aromatic-aromatic interaction: a mechanism of protein structure stabilization. Science 229:23–28; see also technical comment in Science 234, 1005 (1986)
471. Perutz MF, Fermy G, Abraham DJ, Poyart C, Bursaux E (1986) Hemoglobin as a receptor of drugs and peptides. X-ray studies of the stereochemistry of binding. J Am Chem Soc 108:1064–1078
472. Tüchsen E, Woodward C (1987) Assignment of asparagine-44 side-chain primary amide. ¹N NMR resonances and the peptide amide. N¹H resonance of glycine-37 in basic pancreatic trypsin inhibitor. Biochemistry 26:1918–1925
472a. Atwood JL, Hamada F, Robinson KD, Orr GW, Vincent RL (1991) X-ray diffraction evidence for aromatic π hydrogen bonding to water. Nature 349:683–684
473. Allerhand A, Schleyer PvR (1963) Halide anions as proton acceptors in hydrogen bonding. J Am Chem Soc 85:1233–1237
474. Murray-Rust P, Stallings WC, Monti CT, Preston RK, Glusker JP (1983) Intermolecular interactions of the C-F bond: the crystallographic environment of fluorinated carboxylic acids and related structures. J Am Chem Soc 105:3206–3214
475. Taylor R, Kennard O, Versichel W (1983) Geometry of the N-H---O=C hydrogen bond. 1. Lone-pair directionality. J Am Chem Soc 105:5761–5766
476. Murray-Rust P, Glusker JP (1984) Directional hydrogen-bonding to sp² and sp³ hybridized oxygen atoms and its relevance to ligand-macromolecular interactions. J Am Chem Soc 106:1018–1025
477. Vedani A, Dunitz JD (1985) Lone-pair directionality in hydrogen bond potential functions for molecular mechanics calculations. The inhibition of human carbonic anhydrase II by sulfonamides. J Am Chem Soc 107:7653–7658

478. Legon AC, Miller DJ (1987) Directional character, strength, and nature of the hydrogen bond in gas-phase dimers. Accts Chem Res 20:39–46
479. Jeffrey GA, Sundaralingam M (1974/86) Bibliography of crystal structures of carbohydrates, nucleosides and nucleotides. Adv Carbohydr Chem Biochem 30 (1974):445–466; 31 (1975):347–371; 32 (1976):353–384; 34 (1977):345–378; 37 (1980):373–436; 43 (1986):203–421
480. Lemieux RU, Bock K, Delbaere LTJ, Koto S, Rao VS (1980) The conformations of oligosaccharides related to the ABH and Lewis human blood group determinants. Can J Chem 58:631–653
481. Thogersen H, Lemieux RU, Boch K, Meyer B (1982) Further justification for the *exo*-anomeric effect. Conformational analysis based on nuclear magnetic resonance spectroscopy of oligosaccharides. Can J Chem 60:44–57
482. Jeffrey GA, French AD (1978) Mono-, oligo- and poly-saccharide crystal structures. Chap 8. In: Sutton LE, Truter MR (eds) Molecular structure by diffraction methods, vol. 6. The Chemical Society. London, UK, pp 183–223
483. French AD, Gardner KH (eds) (1980) Fibre diffraction methods. ACS Symp Ser 141 Am Chem Soc Washington, DC
484. Jeffrey GA, Kim HS (1970) Conformations of the alditols. Carbohydr Res 14: 207–216
485. Jeffrey GA, Robbins A (1978) 2,3-Dimethyl-2,3-butanediol (pinacol). Acta Cryst B34:3817–3820
486. Kauzman W, Clough FB, Tobais I (1961) The principle of pairwise interactions as a basis for an empirical theory of optical rotatory power. Tetrahedron 13:57–105
487. Angyal SJ (1968) Conformational analysis in carbohydrate chemistry. I. Conformational free energies. The conformation and $\alpha:\beta$ ratios of aldopyranoses in aqueous solution. Aust J Chem 21:2737–2746. Angew Chem Intern Edit 8:157–166 (1969)
488. Hockett RC, Hudson CS (1931) A novel modification of methyl-*d*-xyloside, a novel modification of lactose. J Am Chem Soc 53:4454–4456
489. Melberg S, Rasmussen K (1979) Conformations of disaccharides by empirical force field calculations, Part I. Carbohydr Res 69:27–30
490. Melberg S, Rasmussen K (1980) Conformations of disaccharides by empirical force field calculations, Part II. Carbohydr Res 71:25–34
491. Serianni AS, Chipman DM (1987) Furanose ring conformation: The application of ab-initio molecular orbital calculations to the structure and dynamics of erythrofuranose and threofuranose rings. J Am Chem Soc 109:5297–5303
492. Gritsan VN, Panov VP, Kachur VG (1983) Theoretical analysis of hydrogen bonds in carbohydrate crystals. Carbohydr Res 112:11–21
493. Boch K, Lemieux RU (1982) The conformational properties of sucrose in aqueous solution: intramolecular hydrogen bonding. Carbohydr Res 100:63–74
494. Eliel EL, Allinger NL, Angyal SJ, Morrison GA (1965) Conformational analysis. Wiley & Sons, New York
495. Lemieux RU, Pavia AA (1969) Substitutional and solvation effects on conformational equilibria. Effects on the interaction between opposing axial oxygen atoms. Can J Chem 47:4441–4446
496. Lemieux RU (1971) Effects of unshared pairs of electrons and their solvation on conformational equilibria. Pure Appl Chem 25:527–548
497. Perez S, Tanavel F, Vergelati C (1985) Experimental evidence of solvent-induced conformational changes in maltose. Nouv J Chim 9:561–564
498. Sundaralingam M (1968) Some aspects of stereochemistry and hydrogen bonding carbohydrates related to polysaccharide conformations. Biopolymers 6:189–213
499. Marchessault RH, Perez S (1979) Conformations of the hydroxymethyl groups in crystalline aldohexopyranoses. Biopolymers 18:2369–2374
500. IUPAC (1971) Tentative rules of the nomenclature of organic chemistry. Section E. Fundamental stereochemistry. Eur J Biochem 18:151–170
501. Hassel O, Ottar B (1947) The structure of molecules containing cyclohexane or pyranose rings. Acta Chem Scand 1:929–942

References

502. Perez S, St.-Pierre J, Marchessault RH (1978) Rotamers arising from restricted motions and electronic interactions of acetoxymethyl groups in carbohydrates. Can J Chem 56:2866–2871
503. McCain DC, Markley JL (1986) The solution conformation of sucrose: concentration and temperature dependence. Carbohydr Res 152:73–80
504. Gardner KH, Blackwell J (1974) Structure of native cellulose. Biopolymers 13:1975–2001
504a. Jeffrey GA, Huang D-B (1990) Hydrogen bonding in the crystal structure of raffinose pentahydrate. Carbohyd Res. In press
505. Asselineau C, Asselineau J (1978) Trehalose-containing glycolipids. Prog Chem Fats Other Lipids 16:59–99
506. Lederer E (1976) Cord factor and related trehalose esters. Chem Phys Lipids 16:91–106
507. Lederer E (1980) Synthetic immunostimulants derived from the bacterial cell wall. J Med Chem 23:819–825
508. Arnott S, Scott WE (1972) Accurate X-ray diffraction analysis of fibrous polysaccharides containing pyranose rings. Part I. The linked-atom approach. J Chem Soc Perkin II:324–335
509. Sarko A, Muggli R (1974) Packing analysis of carbohydrates and polysaccharides, III. *Valonia* cellulose and cellulose II. Macromolecules 7:486–494
510. Gardner KH, Blackwell J (1975) Refinement of the structure of β-chitin. Biopolymers 14:1581–1595
511. Claffey W, Blackwell J (1976) Electron diffraction of *Valonia* cellulose. A quantitative interpretation. Biopolymers 15:1903–1915
512. Stipanovic AJ, Sarko A (1976) Packing analysis of carbohydrates and polysaccharides, 6. Molecular and crystal structure of regenerated cellulose II. Macromolecules 9:851–857
513. Kolpak FJ, Blackwell J (1976) Determination of the structure of cellulose II. Macromolecules 9:273–278
514. Woodcock C, Sarko A (1980) Packing analysis of carbohydrates and polysaccharides, II. Molecular and crystal structure of native ramie cellulose. Macromolecules 13:1183–1187
515. Blackwell J, Gardner KH, Kolpak FJ, Minke R, Claffey WB (1980) Refinement of cellulose and chitin structures. Chapt 19. In: French AD, Gardner KH (eds) Fiber diffraction methods. Am Chem Soc Symp Ser 141, pp 315–334
516. Sarko A, Zugenmaier P (1980) Crystal structures of amylose and its derivatives. Chap. 28. In: French AD, Gardner KH (eds) Fiber diffraction methods. Am Chem Soc Symp Ser 141:459–482
517. Zugenmaier P (1974) Conformation and packing analysis of polysaccharides and derivatives, I. Mannan. Biopolymers 13:1127–1139
518. Elloway HF, Isaac DH, Atkins EDT (1980) Review of the structures of *Klebsiella* polysaccharides by X-ray diffraction. Chap. 27. In: French AD, Gardner KH (eds) Fiber diffraction methods. Am Chem Soc Symp Ser 141:429–458
519. Nieduszynski I, Marchessault RH (1972) The crystalline structure of poly-β-D-(1,4')-mannose: mannan I. Can J Chem 50:2130–2138
520. Davidson JN (1976) Biochemistry of the nucleic acids. 8th edn (revised by Adams RLP, Burdon RH, Campbell AM, Smellie RMS). Chapman and Hall, London
521. Suhadolnik RJ (1979) Nucleosides as biological probes. Wiley, New York
522. Saenger W (1984) Principles of nucleic acid structure. Springer, Berlin Heidelberg New York Tokyo
523. Agris PF (1980) The modified nucleosides of transfer RNA. Liss, New York
524. Szczepaniak K, Szczepaniak M (1987) Matrix isolation infrared studies of nucleic acid constituents. J Mol Struct 156:29–42
525. Sepiol J, Kazimierczuk Z, Shugar D (1976) Tautomerism of isoguanosine and solvent-induced keto-enol equilibrium. Z Naturforsch C 31:361–370
526. Stolarski R, Remin M, Shugar D (1977) Studies on prototropic tautomerism in neutral and monoanionic forms of pyrimidines by nuclear magnetic resonance spectroscopy. Z Naturforsch C 32:894–900
527. Crick FHC, Watson JD (1954) The complementary structure of deoxyribonucleic acid. Proc R Soc Lond Ser A 223:80–96
528. Zimmerman SB (1976) The polyuridylic acid complex with polyamines: An X-ray fiber observation. J Mol Biol 101:563–569

529. Mazumdar SK, Saenger W, Scheit KH (1974) Molecular structure of poly-2-thiouridylic acid, a double helix with non-equivalent polynucleotide chains. J Mol Biol 85:213–229
530. Kistenmacher TJ, Rossi M, Marzilli LG (1978) A model for the interrelationship between asymmetric interbase hydrogen bonding and base-base stacking in hemiprotonated polyribocytidylic acid: Crystal structure of 1-methylcytosine hemihydroiodide hemihydrate. Biopolymers 17:2581–2585
531. Borah B, Wood JL (1976) The cytidinium-cytidine complex: infrared and Raman spectroscopic studies. J Mol Struct 30:13–30
532. Rich A, Davies DR, Crick FHC, Watson JD (1961) The molecular structure of polyadenylic acid. J Mol Biol 3:71–86
533. Arnott S, Selsing E (1974) Structures of poly d(A)·poly d(T) and poly d(T)·poly d(A)·poly d(T). J Mol Biol 88:509–521
534. Zimmerman SB, Cohen GH, Davies DR (1975) X-ray fiber diffraction and model-building study of polyguanylic acid and polyinosinic acid. J Mol Biol 92:181–192
535. Pullman B, Pullman A (1971) Electronic aspects of purine tautomerism. Adv Heterocyc Chem 13:77–159
536. Seeman NC, Rosenberg JM, Suddath FL, Kim JJP, Rich A (1976) RNA double-helical fragment at atomic resolution. I. The crystal and molecular structure of sodium adenylyl-3′,5′-uridine. J Mol Biol 104:109–144
537. Frey MN, Koetzle TF, Lehmann MS, Hamilton WC (1973) Precision neutron diffraction structure determination of protein and nucleic acid components. XII. A study of hydrogen bonding in the purine-pyrimidine base pair 9-methyladenine and 1-methylthymine. J Chem Phys 59:915–924
538. Rosenberg JM, Seeman NC, Day RO, Rich A (1976) RNA double-helical fragment at atomic resolution. II. The crystal structure of sodium guanylyl-3′,5′-cytidine monohydrate. J Mol Biol 104:145–167
539. Blake RD, Massoulié J, Fresco JR (1967) Polynucleotides VIII. A spectral approach to the equilibria between polyriboadenylate and polyribouridylate and their complexes. J Mol Biol 30:291–308
540. IUPAC-IUB Joint Commission on Biochemical Nomenclature (1983). Abbreviations and symbols for the description of conformations of polynucleotide chains. Eur J Biochem 131:9–15
541. Altona C, Sundaralingam M (1972) Conformational analysis of the sugar ring in nucleosides and nucleotides. A new description using the concept of pseudorotation. J Am Chem Soc 94:8205–8212
542. Olson WK, Sussman JL (1982) How flexible is the furanose ring? 1. A comparison of experimental and theoretical studies. J Am Chem Soc 104:270–278; 2. An updated potential energy estimate. J Am Chem Soc 104:278–286
543. Saran A, Perahia D, Pullman B (1973) Molecular orbital calculations on the conformation of nucleic acids and their constituents. VII. Conformation of the sugar ring in β-nucleosides: the pseudorotational representation. Theor Chim Acta (Berlin) 30:31–44
544. Röder O, Lüdemann HD, von Goldammer E (1975) Determination of the activation energy for pseudorotation of the furanose ring in nucleosides by ^{13}C nuclear-magnetic resonance relaxation. Eur J Biochem 53:517–524
545. Yathindra N, Sundaralingam M (1973) Correlations between the backbone and side chain conformations in 5′-nucleotides. The concept of a "rigid" nucleotide conformation. Biopolymers 12:297–314
546. Suck D, Saenger W (1972) Molecular and crystal structure of 6-methyluridine. A pyrimidine nucleoside in the *syn* conformation. J Am Chem Soc 94:6520–6526
547. Birnbaum GI, Blonski WJP, Hruska FE (1983) Structure and conformation of the anticodon nucleoside 5-methoxyuridine in the solid state and in solution. Can J Chem 61:2299–2304
548. Buskov VI, Bushnev VN, Poltev VI (1980) Nuclear magnetic resonance study of C-H---O hydrogen bonds in nucleic acid base analogs. Mol Biol (USSR) 14:245–250 (in English)
549. Ts'o POP, Kondo NS, Schweizer MP, Hollis DP (1969) Studies of the conformation and interaction in dinucleoside mono- and diphosphates in proton magnetic resonance. Biochemistry 8:997–1029

550. Narayanan P, Berman H (1975) A crystallographic determination of a chemical structure: 6-Amino-10-(β-D-ribofuranosylamino)pyrimido-[5,4-d]pyrimidine, an example of an unusual D-ribose conformation. Carbohydr Res 44:169–180
551. Bender ML, Komiyama M (1978) Cyclodextrin chemistry. Springer, Berlin Heidelberg New York
552. Szejtli J (1982) Cyclodextrins and their inclusion complexes. Akademiai Kiado, Budapest
553. Huber O, Szejtli J (eds) (1988) Proceedings of the fourth international symposium on cyclodextrins. Kluwer-Academic Publ, Dordrecht
554. Uekama K, Otagiri M (1987) Cyclodextrins in drug carrier systems. CRC Crit Rev Therap Drug Carrier Systems 3:1–40
555. Cramer F (1954) Einschlußverbindungen. Springer, Berlin Heidelberg
556. Saenger W (1976) α-Cyclodextrin inclusion complexes: mechanism of adduct formation and intermolecular interactions. Jerusalem Symp Ser 8:265–308
557. Saenger W (1980) Cyclodextrins in research and industry. Angew Chem Int Ed Engl 19:344–362
558. Saenger W (1984) Structural aspects of cyclodextrins and their inclusion complexes. In: Atwood JL, Davies JED, MacNicol DD (eds) Inclusion compounds. Vol 2. Academic Press, London New York, pp 231–259
559. Stoddart JF (1971) Stereochemistry of carbohydrates. Wiley, New York
560. Brunck TK, Weinhold F (1979) Quantum-mechanical studies on the origin of barriers to internal rotation about single bonds. J Am Chem Soc 101:1700–1709
561. Tabushi I, Kiyosuke Y-I, Sugimoto T, Yamamura K (1978) Approach to the aspects of driving force of inclusion by α-cyclodextrin. J Am Chem Soc 100:916–919
562. Saenger W (1981) Cyclodextrins as catalysts. In: Eggerer H, Huber R (eds) Structural and functional aspects of enzyme catalysts. Springer, Berlin Heidelberg New York
563. Bergeron RJ, Pillor DM, Gibeily G, Roberts WP (1978) Thermodynamics of cycloamylose-substrate complexation. Bioorg Chem 7:263–271
564. Tabushi I (1982) Cyclodextrin catalysis as a model for enzyme action. Acc Chem Res 15:66–72
565. Saenger W (1985) Nature and size of included guest molecule determines architecture of crystalline cyclodextrin host matrix. Israel J Chem 25:43–50
566. Noltemeyer M, Saenger W (1980) Structural chemistry of linear α-cyclodextrin-polyiodide complexes. X-ray crystal structures of (α-cyclodextrin)$_2$·LiI$_3$·I$_2$·8H$_2$O and (α-cyclodextrin)$_2$·Cd$_{0.5}$·I$_5$·27H$_2$O. Models for the blue amylose-iodine complex. J Am Chem Soc 102:2710–2722
567. Harding MM, MacLennan JM, Paton RM (1978) Structure of the complex cycloheptaamylose-p-nitroacetanilide. Nature (Lond) 274:621–623
568. Lindner K, Saenger W (1980) Crystal structure of the γ-cyclodextrin-propanol inclusion complex: correlation of α-, β-, γ-cyclodextrin geometries. Biochem Biophys Res Commun 92:933–938
569. Kamitori S, Hirotsu K, Higuchi T (1987) Crystal and molecular structures of double macrocyclic inclusion complexes composed of cyclodextrins, crown ethers, and cations. J Am Chem Soc 109:2409–2414
570. Maclennan JM, Stezowski JJ (1980) The crystal structure of uncomplexed-hydrated cyclooctaamylose. Biochem Biophys Res Commun 92:926–932
571. Casu B, Reggiani M, Gallo GG, Vigevani A (1966) Hydrogen bonding and conformation of glucose and polyglucose in dimethylsulphoxide solution. Tetrahedron 22:3061–3082
572. Casu B, Reggiani M, Gallo GG, Vigevani A (1970) Conformation of acetylated cyclodextrins and amylose. Carbohydr Res 12:157–170
573. Rees DA (1970) Conformational analysis of polysaccharides. Part V. The characterization of linkage conformations (chain conformations) by optical rotation at a single wavelength. Evidence for a distortion of cyclohexaamylose in aqueous solution. Optical rotation and the amylose conformation. J Chem Soc B:877–884
574. Klar B, Hingerty B, Saenger W (1980) Topography of cyclodextrin inclusion complexes. XII. Hydrogen bonding in the crystal structure of α-cyclodextrin hexahydrate: the use of a multi-counter detector in neutron diffraction. Acta Cryst B36:1154–1165

575. Lindner K, Saenger W (1982) Topography of cyclodextrin inclusion complexes. XVI. Cyclic system of hydrogen bonds: structure of α-cyclodextrin hexahydrate, form (II): comparison with form (I). Acta Cryst B38:203–210
576. Koshland DE, Jr (1973) Protein shape and biological control. Sci Am 229 (Oct):52–64
577. Hingerty BE, Saenger W (1976) Topography of cyclodextrin inclusion complexes. 8. Crystal and molecular structure of the α-cyclodextrin-methanol-pentahydrate complex. Disorder in a hydrophobic cage. J Am Chem Soc 98:3357–3365
578. Saenger W, Noltemeyer M (1976) Röntgenstrukturanalyse des α-Cyclodextrin-Krypton-Pentahydrates. Zum Einschlußmechanismus des Modell-Enzyms. Chem Ber 109:503–517
579. Saenger W, Noltemeyer M, Manor PC, Hingerty B, Klar B (1976) "Induced-fit"-type complex formation of the model enzyme α-cyclodextrin. Bioorg Chem 5:187–195
579a. Steiner T, Saenger W, Kearly G, Lechner RE (1989) Dynamics of hydrogen bonding disorder in β-cyclodextrin undecahydrate. Physica B 156&157:336–338
580. Fujiwara T, Yamazaki M, Tomiza Y, Tokuoka R, Tomita K-I, Matsuo T, Suga H, Saenger W (1983) The crystal structure of a new form of β-cyclodextrin water inclusion compound and thermal properties of β-cyclodextrin inclusion complexes. Nippon Kagaku Kaishi 2:181–187
581. Hanabata H, Matsuo T, Suga H (1987) Calorimetric study of β-cyclodextrin undecahydrate. J Inclus Phenom 5:325–333
582. Pangborn W, Langs D, Pérez S (1985) Regular left-handed fragment of amylose: crystal and molecular structure of methyl-α-maltotrioside 4H$_2$O. Int J Biol Macromol 7:363–369
583. Hinrichs W, Büttner G, Steifa M, Betzel C, Zabel V, Pfannemüller B, Saenger W (1987) An amylose antiparallel double helix at atomic resolution. Science 238:205–208
583a. Hinrichs W, Saenger W (1990) Crystal and molecular structure of the hexasaccharide complex (p-nitrophenyl-maltohexaoside)$_2$·Ba(I$_3$)$_2$·27H$_2$O. J Am Chem Soc. 112:2789–2796
584. Wu H-CH, Sarko A (1978) The double-helical molecular structure of crystalline A-amylose. Carbohydr Res 61:27–40; The double-helical molecular structure of crystalline B-amylose. Carbohydr Res 61:7–25
585. Imberty A, Chanzy H, Pérez S, Buléon A, Tran V (1988) The double-helical nature of the crystalline part of A-starch. J Mol Biol 201:365–378
586. Alberts B, Bray D, Lewis L, Raff M, Roberts K, Watson JD (1983) Molecular biology of the cell. Garland Publ, Inc, New York, London
587. Dickerson RE, Geis I (1969) The structure and function of proteins. Benjamin, Menlo Park, USA
588. Schulz GE, Schirmer H (1979) Principles of protein structure. Springer, Berlin Heidelberg New York
589. Richardson JS (1981) The anatomy and taxonomy of protein structure. Adv Protein Chem 34:167–339
590. Pauling L, Corey RB (1951) Configurations of polypeptide chains with favored orientations around single bonds: two new pleated sheets. Proc Natl Acad Sci USA 37:729–740
591. Venkatachalam CM (1968) Stereochemical criteria for polypeptides and proteins. V. Conformation of a system of three linked peptide units. Biopolymers 6:1425–1436
592. Chou PY, Fasman GD (1974) Prediction of protein conformation. Biochemistry 13:222–245
593. Ramachandran GN, Sasisekharan V (1968) Conformation of polypeptides and proteins. Adv Prot Chem 23:283–437
593a. Takahashi K (1985) A revision and confirmation of the amino acid sequence of ribonuclease T1. J Biochem 98:815–817
594. Arni R, Heinemann U, Tokuoka R, Saenger W (1988) Three-dimensional structure of the ribonuclease T$_1$*2'GMP complex at 1.9Å resolution. J Biol Chem 263:15358–15368
595. Miller S, Janin J, Lesk AM, Chothia C (1987) Interior and surface of monomeric proteins. J Mol Biol 196:641–656
596. Baker EN, Hubbard RE (1984) Hydrogen bonding in globular proteins. Prog Biophys Molec Biol 44:97–179
597. Salemme FR, Weatherford DW (1981) Conformational and geometrical properties of β-sheets in proteins. I. Parallel β-sheets. J Mol Biol 146:101–117; II. Antiparallel and mixed β-sheets. J Mol Biol 146:119–141

References

598. Chou K-C, Pottle M, Némethy G, Ueda Y, Scheraga HA (1982) Structure of β-sheets. Origin of the right-handed twist and the increased stability of antiparallel over parallel sheets. J Mol Biol 162:89–112
599. Richardson JS, Getzoff ED, Richardson DC (1978) The β-bulge: a common small unit of nonrepetitive protein structure. Proc Natl Acad Sci USA 75:2574–2578
600. Hol WGJ, van Duijnen PT, Berendsen HJC (1978) The α-helix dipole and the properties of proteins. Nature (Lond) 273:443–446
601. Hol WGJ, Halie LM, Sander C (1981) Dipoles of α-helix and β-sheet: their roles in protein folding. Nature (Lond) 294:532–536
602. Cantor CR, Schimmel PR (1980) Biophysical Chemistry. Part I. Freeman, San Francisco
603. Matthews BW (1972) The γ-turn, evidence for a new folded conformation in proteins. Macromolecules 5:818–819
604. Baker EN, Dodson EN (1980) Crystallographic refinement of the structure of actinidin at 1.7 Å resolution by fast-Fourier least-squares methods. Acta Cryst A36:559–572
605. Rees DC, Lewis M, Lipscomb WN (1983) Refined crystal structure of carboxypeptidase A at 1.5 Å resolution. J Mol Biol 168:367–387
606. Takano T, Dickerson RE (1981) Conformation change of cytochrome c. I. Ferrocytochrome c structure refined at 1.5 Å resolution. J Mol Biol 153:79–94
607. Bolin JT, Filman DJ, Matthews DA, Hamlin RC, Kraut J (1982) Crystal structures of *Escherichia coli* and *Lactobacillus casei* dihydrofolate reductase refined at 1.7 Å resolution. J Biol Chem 257:13650–13662
608. Steigemann W, Weber E (1979) Structure of erythrocruorin in different ligand states refined at 1.4 Å resolution. J Mol Biol 127:309–338
609. Isaacs NW, Agarwal RC (1978) Experience with fast Fourier least squares in the refinement of the crystal structure of rhombohedral 2-Zn insulin at 1.5 Å resolution. Acta Cryst A34:782–791
610. Artymiuk PJ, Blake CCF (1981) Refinement of human lysozyme at 1.5 Å resolution. Analysis of nonbonded and hydrogen-bond interactions. J Mol Biol 152:737–762
611. Phillips SEV (1980) Structure and refinement of oxymyoglobin at 1.6 Å resolution. J Mol Biol 142:531–554
612. James MNG, Sielecki AR (1983) Structure and refinement of penicillopepsin at 1.8 Å resolution. J Mol Biol 163:299–361
613. Dijkstra BW, Kalk KH, Hol WGJ, Drenth J (1981) Structure of bovine pancreatic phospholipase A2 at 1.7 Å resolution. J Mol Biol 147:97–123
614. Guss JM, Freeman HC (1983) Structure of oxidized poplar plastocyanin at 1.6 Å resolution. J Mol Biol 169:521–563
615. Borkakoti N, Moss DS, Palmer RA (1982) Ribonuclease A: least-squares refinement of the structure at 1.45 Å resolution. Acta Cryst B38:2210–2217
616. Sielecki AR, Hendrickson WA, Broughton CG, Delbaere LTJ, Brayer GD, James MNG (1979) Protein structure refinement: *Streptomyces griseus* serine protease A at 1.8 Å resolution. J Mol Biol 134:781–804
617. Holmes MA, Matthews BW (1982) Structure of thermolysin refined at 1.6 Å resolution. J Mol Biol 160:623–639
618. Chambers JL, Stroud RM (1979) The accuracy of refined protein structures: comparison of two independently refined models of bovine trypsin. Acta Cryst B35:1861–1874
619. Blundell TL, Barlow D, Borkakoti N, Thornton J (1983) Solvent-induced distortions and the curvature of α-helices. Nature (Lond) 306:281–283
620. Gray TM, Matthews BM (1984) Intrahelical hydrogen bonding of serine, threonine and cysteine within α-helices and its relevance to membrane-bound proteins. J Mol Biol 175:75–81
621. Finney JL (1979) The organisation and function of water in protein crystals. In: Franks F (ed) The physics and chemistry of water. A comprehensive treatise, vol. 6. Plenum Press, New York, pp 47–122
622. Edsall JT, McKenzie HA (1983) Water and proteins, 2. The location and dynamics of water in protein systems and its relation to their stability and properties. Adv Biophys 16:53–183
623. Bode W, Schwager P (1975) The refined crystal structure of bovine β-trypsin at 1.8 Å resolution, II. Crystallographic refinement, calcium binding, benzamidine binding site and active site at pH 7.0. J Mol Biol 98:693–717

624. Sawyer L, Shotton DM, Campbell JW, Wendell PL, Muirhead H, Watson HC, Diamond R, Ladner RC (1978) The atomic structure of crystalline porcine pancreatic elastase at 2.5 Å resolution: Comparison with the structure of α-chymotrypsin. J Mol Biol 118:137–208
625. Blake CCF, Pulford WCA, Artymiuk PJ (1983) X-ray studies of water in crystals of lysozyme. J Mol Biol 167:693–723
626. Tainer JA, Getzoff ED, Beem KM, Richardson JS, Richardson DC (1982) Determination and analysis of the 2 Å structure of copper, zink superoxide dismutase. J Mol Biol 160:181–217
627. Chou PY, Fasman GD (1977) β-turns in proteins. J Mol Biol 115:135–175
628. Schoenborn BP (ed) (1984) Neutrons in biology. Plenum Press, New York
629. Raghavan NV, Schoenborn BP (1984) The structures of bound water and refinement of acid metmyoglobin. In: Schoenborn BP (ed) Neutrons in biology. Plenum Press, New York, pp 247–259
630. Phillips SEV (1984) Hydrogen bonding and exchange in oxymyoglobin. In: Schoenborn BP (ed) Neutrons in biology. Plenum Press, New York, pp 305–322
631. Hanson JC, Schoenborn BP (1981) Real space refinement of neutron diffraction data from sperm whale carbonmonoxymyoglobin. J Mol Biol 153:117–146
632. Wlodawer A, Sjölin L (1981) Hydrogen exchange in RNase A: Neutron diffraction study. Proc Natl Acad Sci USA 79:1418–1422
633. Wlodawer A, Miller M, Sjölin L (1983) Active site of RNase A: Neutron diffraction study of a complex with uridine vanadate, a transition-state analog. Proc Natl Acad Sci USA 80:3628–3631
634. Kossiakoff AA (1982) Protein dynamics investigated by the neutron diffraction-hydrogen exchange technique. Nature (Lond) 296:713–721
635. Kossiakoff AA (1985) The application of neutron crystallography to the study of dynamic and hydration properties of proteins. Annu Rev Biochem 54:1195–1227
636. Wlodawer A, Walter J, Huber R, Sjölin L (1984) Structure of bovine pancreatic trypsin inhibitor. Results of joint neutron and X-ray refinement of crystal form II. J Mol Biol 180:301–329
637. Mason SA, Bentley GA, McIntyre GJ (1984) Deuterium exchange in lysozyme at 1.4 Å resolution. In: Schoenborn BP (ed) Neutrons in biology. Plenum Press, New York, pp 323–334
638. Teeter MM, Kossiakoff AA (1984) The neutron structure of the hydrophobic plant protein crambin. In: Schoenborn BP (ed) Neutrons in biology. Plenum Press, New York, pp 335–384
639. Hvidt A, Nielsen SO (1966) Hydrogen exchange in proteins. Adv Protein Chem 21:287–386
640. Englander SW, Downer NW, Teitelbaum HA (1972) Hydrogen exchange. Annu Rev Biochem 41:903–924
641. Bentley GA, Delepierre M, Dobson CM, Mason SA, Poulsen FM, Wedin RE (1983) Exchange of individual hydrogens of a protein in a crystal and in solution. J Mol Biol 170:243–247
642. Englander SW, Calhoun DB, Englander JJ, Kallenbach NR, Liem H, Malin EL, Mandal C, Rogero JR (1980) Individual breathing reactions measured in hemoglobin by hydrogen exchange methods. Biophys J 32:577–589
643. Woodward CK, Hilton BD (1979) Hydrogen exchange kinetics and internal motions in proteins and nucleic acids. Annu Rev Biophys Bioeng 8:99–127
644. Richards FM (1979) Packing defects, cavities, volume fluctuation and access to the interior of proteins, including some general comments on surface area and protein structure. Carlsberg Res Commun 44:47–63
645. Bryan PN, Rollence ML, Pantoliano MW, Wood J, Finzel BC, Gilliland GL, Howard AJ, Poulos TL (1986) Proteases of enhanced stability: characterization of a thermostable variant of subtilisin. Proteins: Structure, Function and Genetics. 1:326–334
646. Alber T, Dao-pin S, Wilson K, Wozniak JA, Cook SP, Matthews BW (1987) Contributions of hydrogen bonds of Thr 157 to the thermodynamic stability of phage T4 lysozyme. Nature (Lond) 330:41–46
647. Fersht AR (1987) The hydrogen bond in molecular recognition. Trends Biol Sci 12:301–304
648. Fersht AR, Shi J-P, Knill-Jones J, Lowe DM, Wilkinson AJ, Blow DM, Brick P, Carter P, Wayne MMY, Winter G (1985) Hydrogen bonding and biological specificity analysed by protein engineering. Nature (Lond) 314:235–238

649. Bhat TN, Blow DM, Brick B, Nyborg J (1982) Tyrosyl-tRNA synthetase forms a mononucleotide-binding fold. J Mol Biol 158:699–709
650. Stryer L (1975) Biochemistry. Freeman, San Francisco
651. Lehninger AL (1975) Biochemistry, 2nd edn. Worth, New York
652. Kornberg A (1974) DNA synthesis. Freeman, San Francisco
653. Arnott S (1981) The secondary structures of polynucleotide chains as revealed by X-ray diffraction analysis of fibers. In: Neidle S (ed) Topics in nucleic acid structure. MacMillan, London, pp 65–82
654. Chandrasekaran R, Arnott S (1989) The structures of DNA and RNA in oriented fibers. In: Saenger W (ed) Nucleic acids, Landolt-Börnstein New Series Group VII. Biophysics, vol. 1 b. Springer, Berlin, pp 31–170
655. Sundaralingam M (1969) Stereochemistry of nucleic acids and their constituents, IV. Allowed and preferred conformations of nucleosides, nucleoside mono, di, tri, tetraphosphates, nucleic acids and polynucleotides. Biopolymers 7:821–830
656. Saenger W (1973) Structure and function of nucleosides and nucleotides. Angew Chem internat Edit. 12:591–601
657. Neidle S (1975) Nucleic acids and their constituents. In: Sutton LE, Truter MR (ed) Molecular structure by diffraction methods, vol 6. The chemical society, London, pp 224–239
658. Sundaralingam M, Westhof E (1979) The "Rigid" nucleotide concept in perspective. Int J Quant Chem 6:115–130
659. Sundaralingam M (1980) Nucleic acid principles and transfer RNA. In: Srinivasan R (ed) Biomolecular structure, conformation, function and evolution. Pergamon Press, New York, pp 259–283
660. Moras D (1989) Crystal structures of tRNAs. In: Saenger W (ed) Nucleic acids. Landolt-Börnstein New Series group VII. Biophysics, vol. 1 b. Springer, Berlin, pp 1–30
661. Wang AH-J, Quigley GJ, Kolpak FJ, van der Marel G, van Boom JH, Rich A (1980) Left-handed double helical DNA: Variations in the backbone conformation. Science 211:171–176
662. Dickerson RE, Drew HR, Conner BN, Wing RM, Fratini AV, Kopka ML (1982) The anatomy of A-, B- and Z-DNA. Science 216:475–485
663. Kennard O (1987) DNA structure. Current results from single crystal X-ray diffraction studies. In: Guschlbauer W, Saenger W (eds) DNA ligand interactions: from drugs to proteins. Plenum Press, New York, pp 1–22
664. Shakked Z, Kennard O (1985) The A-form of DNA. In: Jurnak F, McPherson A (eds) Biological macromolecules and assemblies, vol 2. Wiley, New York, pp 1–36
665. Nordheim A, Pardue ML, Lafer EM, Möller A, Stollar D, Rich A (1981) Antibodies to left-handed Z-DNA bind to interband regions of *Drosophila* polytene chromosomes. Nature (Lond) 294:417–422
666. Dock-Bregeon AC, Chevrier B, Podjarny A, Johnson J, deBear JS, Gough GR, Gilham PT, Moras D (1989) Crystallographic structure of an RNA helix: $[U(U-A)_6A]_2$. J Mol Biol 209:459–474
667. Guschlbauer W, Saenger W (eds) (1987) DNA ligand interactions: from drugs to proteins. Plenum Press, New York
668. Kölkenbeck K, Zundel G (1975) The significance of the 2'-OH group and the influence of cations on the secondary structure of the RNA backbone. Biophys Struct Mechanism 1:203–219
669. Young PR, Kallenbach NR (1978) Secondary structure in polyuridylic acid. Non-classical hydrogen bonding and the function of the ribose 2'-hydroxyl group. J Mol Biol 126:467–479
670. Zmudzka B, Shugar D (1969) Poly 2'-O-methylcytidylic acid and the role of the 2'-hydroxyl in polynucleotide structure. Biochem Biophys Res Commun 37:895–901
671. Bolton PH, Kearns DR (1979) Intramolecular water bridge between the 2'-OH and phosphate groups of RNA. Cyclic nucleotides as a model system. J Am Chem Soc 101:479–484
672. Ts'o POP, Barrett JC, Kan LD, Miller PS (1972) Proton magnetic resonance studies of nucleic acid conformation. Annu Rev NY Acad Sci 222:290–306
673. Jack A, Ladner JE, Klug A (1976) Crystallographic refinement of yeast phenylalanine transfer RNA at 2.5 Å resolution. J Mol Biol 108:619–649

674. Woese CR (1967) The genetic code. Harper and Row, New York
675. Crick FHC (1966) Codon-anticodon pairing: The wobble hypothesis. J Mol Biol 19:548–555
676. Söll D, RajBhandary UL (1967) Studies on polynucleotides. LXXVI. Specificity of tRNA for codon recognition as studied by amino acid incorporation. J Mol Biol 29:113–124
677. Lomant AJ, Fresco JR (1975) Structural and energetic consequences of noncomplementary base opposition in nucleic acid helices. Prog Nucl Acid Res Mol Biol 15:185–218
678. Sakore TD, Sobell HM (1969) Crystal and molecular structure of a hydrogen-bonded complex containing adenine and hypoxanthine derivatives: 9-ethyl-8-bromoadenine: 9-ethyl-8-bromohypoxanthine. J Mol Biol 43:77–87
679. Topal MD, Fresco JR (1976) Base-pairing and fidelity in codon-anticodon interaction. Nature (Lond) 263:289–293
680. Brown T, Kennard O, Kneale G, Rabinovich D (1985) High resolution structure of a DNA helix containing mismatched base pairs. Nature (Lond) 315:604–606
681. Kneale G, Brown T, Kennard O, Rabinovich D (1985) G·T base-pairs in a DNA helix: the crystal structure of d(GGGGTCCC). J Mol Biol 186:805–814
682. Brown T, Kneale G, Hunter WN, Kennard O (1986) Structural characterization of the bromouracilo-guanine base-pair mismatch in a Z-DNA fragment. Nucl Acids Res 14:1801–1809
683. Ho PS, Frederick CA, Quigley GJ, van der Marel G, van Boom JH, Wang AH-J, Rich A (1985) G·T wobble base-pairing in Z-DNA at 1.0 Å atomic resolution: the crystal structure of d(CGCGTG). EMBO J. 4:3617–3623
684. Kennard O (1985) Structural studies of DNA fragments. The G·T. base pair in A, B and Z-DNA, the G·A base pair in B-DNA. J Biomol Struct Dynam 3:205–225
685. Hunter WN, Brown T, Kennard O (1986) Structural features and hydration of d(CGCGAATTAGCG): a double helix containing two G·A. mispairs. J Biomol Struct Dynam 4:173–191
686. Hunter WN, Kennard O (1987) Structural features and hydration of a dodecamer duplex containing two C·A. mispairs. Nucl Acids Res 15:6589–6606
687. Kennard O (1988) Structural studies of base pair mismatches and their relevance to theories of mismatch formation and repair. In: Sarma MH, Sarma RH (eds) Structure and expression, vol 2. DNA and its drug complexes. Adenine Press, Schenectady, NY (USA), pp 1–25
688. Privé, GG, Heinemann U, Chandrasegaran S, Kan LS, Kopka ML, Dickerson RE (1988) A mismatch decamer as a model for general sequence B-DNA. In: Sarma MH, Sarma RH (eds) Structure and expression, vol 2. DNA and its drug complexes. Adenine Press, Schenectady, NY USA, pp 27–47
689. Chenon M-T, Pugmire RJ, Grant DM, Panzica RP, Townsend LB (1975) Carbon-13 magnetic resonance. XXVI. A quantitative determination of the tautomeric population of certain purines. J Am Chem Soc 97:4636–4642
690. Drake JW, Allen EF, Forsberg SA, Preparata R-M, Greening EO (1969) Spontaneous mutation. Nature (Lond) 221:1128–1132. Drake JW (1969) Appendix: comparative rates of spontaneous mutation. Nature (Lond) 221:1132
691. Topal MD, Fresco JR (1976) Complementary base-pairing and the origin of substitution mutations. Nature (Lond) 263:285–289
692. Brutlag D, Kornberg A (1972) Enzymatic synthesis of deoxyribonucleic acid. XXXVI. A proofeading function for the 3'-5' exonuclease activity in deoxyribonucleic acid polymerase. J Biol Chem 247:241–248
693. Fersht AR (1980) Enzymic editing mechanism in protein synthesis and DNA replication. Trends Biochem Sci 5:262–265
694. Poltev VI, Bruskov VI (1978) On molecular mechanism of nucleic acid synthesis. Fidelity aspects. 1. Contribution of base interactions. J Theor Biol 70:69–83
695. Sprinzl M, Gauss DH (1982) Compilation of tRNA sequences. Nucl Acids Res 10:r1–r55
696. Kim S-H (1981) Three-dimensional structure of transfer RNA and its functional implications. Adv Enzymol 46:279–315
697. Westhof E, Dumas P, Moras D (1985) Crystallographic refinement of yeast aspartic acid transfer RNA. J Mol Biol 184:119–145

698. Zimmerman SB (1976) X-ray study by fiber diffraction methods of a self-aggregate of guanosine-5'-phosphate with the same helical parameters as poly r(G). J Mol Biol 106:663–672
699. Arnott S, Chandrasekharan R, Leslie AGW, Puigjaner LC, Saenger W (1981) Structure of the poly(2-thiouridylic) acid duplex. J Mol Biol 149:507–520
700. Arnott S, Chandrasekharan R, Day AW, Puigjaner LC, Watts L (1981) Double helical structures for polyxanthylic acid. J Mol Biol 149:489–505
701. Arnott S, Bond PJ, Selsing E, Smith PJC (1976) Models of triple-stranded polynucleotides with optimised stereochemistry. Nucl Acids Res 3:2459–2470
702. Nelson HCM, Finch JT, Luisi BF, Klug A (1987) The structure of an oligo(dA)·oligo(dT) tract and its biological impliclations. Nature (Lond) 330:221–226
703. Coll M, Frederick CA, Wang AH-J, Rich A (1987) A bifurcated hydrogen-bonded conformation in the d(AT) base pairs of the DNA dodecamer d(CGCAAATTTGCG) and its complex with distamycin. Proc Natl Acad Sci USA 84:8385–8389
704. Chandrasekhar K, McPherson A, Jr, Adams MJ, Rossmann MG (1973) Conformation of coenzyme fragments when bound to lactate dehydrogenase. J Mol Biol 76:503–518
705. Wodak SY, Lin MY, Wyckoff HW (1977) The structure of cytidylyl(2',5')adenosine when bound to pancreatic ribonuclease S. J Mol Biol 116:855–875
706. Seeman NC, Rosenberg JM, Rich A (1976) Sequence-specific recognition of double helical nucleic acids by proteins. Proc Natl Acad SciUSA 73:804–808
707. Pabo CO, Sauer RT (1984) Protein-DNA recognition. Ann Rev Biochem 53:293–321
708. Travers AA (1989) DNA conformation and protein binding. Ann Rev Biochem 58:427–452
709. Brennan RG, Matthews BW (1989) The helix-turn-helix DNA binding motif. J Biol Chem 264:1903–1906
710. Steitz TA (1990) Structural studies of protein-nucleic acid interaction: the sources of sequence-specific binding. Quart Rev Biophys 23:205–280
711. Aggarwal AK, Rodgers DW, Drottar M, Ptashne M, Harrison SC (1988) Recognition of a DNA operator by the repressor of phage 434: a view at high resolution. Science 242:899–907
712. Zhang R-G, Joachimiak A, Lawson CL, Schevitz RW, Otwinowski Z, Sigler PB (1987) The crystal structure of *trp* repressor at 1.8 Å shows how binding tryptophan enhances DNA affinity. Nature 327:591–597
713. Otwinowski Z, Schevitz RW, Zhang R-G, Lawson CL, Joachimiak A, Marmorstein RQ, Luisi BF, Sigler PB (1988) Crystal structure of *trp* repressor/operator complex at atomic resolution. Nature 335:321–329
714. Staacke D, Walter B, Kisters-Woike B, v. Wilcken-Bergmann B, Müller-Hill B (1990) How Trp repressor binds to its operator. EMBO J 9:1963–1967
715. Rould MA, Perona JJ, Söll D, Steitz TA (1989) Structure of E. coli glutaminyl-tRNA synthetase complexed with tRNA[Gln] and ATP at 2.8 Å resolution: implications for tRNA discrimination. Science 246:1135–1142
716. Oefner C, Suck D (1986) Crystallographic refinement and structure of DNase I at 2 Å resolution. J Mol Biol 192:605–632
717. Suck D, Oefner C (1986) Structure of DNase I at 2.0 Å resolution suggests a mechanism for binding to and cutting DNA. Nature (Lond) 321:620–625
718. Crowe JH, Crowe LM, Chapman D (1984) Preservation of membranes in anhydrobiotic organisms. The role of trehalose. Science 223:701–703
719. Eisenberg D, Kauzman N (1969) Structure and properties of water. Oxford Univ Press, London, UK
720. Lennard-Jones J, Pople JA (1951) Molecular association in liquids, I. Molecular association due to lone-pairs. Proc R Soc A (Lond):155–162
721. Symons MCR (1981) Water structure and reactivity. Acc Chem Res 14:179–187
722. Frank HS, Wen WY (1957) Structural aspects of ion-solvent interactions in aqueous solutions. A suggested picture of water structure. Discuss Faraday Soc 24:133–140
723. Nemethy G, Scheraga HA (1962) A model for the thermodynamic properties of liquid water. J Chem Phys 36:3382–3400
724. Buis K, Choppin GR (1963) Near infrared studies of the structure of water. I. Pure water. J Chem Phys 39:2035–2041

725. Walrafen GE (1968) Structure of water. In: Covington AK, Jones P (eds) Hydrogen-bonded solvent systems. Taylor and Francis, London, pp 9–30
726. Stevenson DP (1968) Molecular species in liquid water. In: Rich A, Davidson N (eds) Structural chemistry and molecular biology. Freeman, San Francisco, pp 490–506
726a. Giguère PA (1987) The bifurcated hydrogen-bond model of water and amorphous ice. J Chem Phys 87:4835–4839
726b. Giguère PA, Pigeon-Gosselin M (1988) An electrostatic model for hydrogen bonds in alcohols. J Solution Chem 17:1007–1014
727. Narten AH, Levy HA (1969) Observed diffraction pattern and proposed models of liquid water. Science 165:447–454
728. Pauling L (1959) The structure of water. In: Hadži D (ed) Hydrogen bonding. Pergamon Press, London, pp 1–6
729. Frank HS, Quist AS (1961) Pauling's model and the thermodynamic properties of water. J Chem Phys 34:604–611
730. Stillinger FH (1980) Water revisited. Science 209:451–457
731. Stillinger FH, Rahman A (1972) Molecular dynamics study of temperature effects on water structure and kinetics. J Chem Phys 57:1281–1292
732. Lemberg HL, Stillinger FH (1975) Central force model for liquid water. J Chem Phys 62:1677–1690
733. Stillinger FH, David CW (1978) Polarization model for water and its ionic dissociation products. J Chem Phys 69:1473–1484
734. Narten AH, Levy HA (1971) Liquid water: molecular correlation functions from X-ray diffraction. J Chem Phys 55:2263–2269
735. Palinkas G, Kalman E, Kovaks P (1977) Liquid water II. Experimental atom-pair correlation functions for liquid deuterium oxide. Molec Phys 34:525–537
736. Thiessen WE, Narten AH (1982) Neutron diffraction study of light and heavy water mixtures at 25 °C. J Chem Phys 77:2656–2662
737. Dore JC (1985) Structural studies of water by neutron diffraction. In: Franks F (ed) Water Sci Rev 1:3–92, Cambridge Univ Press, UK
738. Boutron P, Alben R (1975) Structural model for amorphous solid water. J Chem Phys 62:4848–4853
739. Narten AH, Venkatesh CG, Rice SA (1976) Diffraction pattern and structure of amorphous solid water at 10 and 77 K. J Chem Phys 64:1106–1121
740. Chowdhury MR, Dore JC, Montague DC (1983) Neutron diffraction studies and CRN model of amorphous ice. J Phys Chem 87:4037–4039
741. Jorgensen WL, Chandrasekhar J, Madura JD, Impey RW, Klein ML (1983) Comparison of simple potential functions for simulating liquid water. J Chem Phys 79:926–935
742. Finney JL, Quinn JE, Baum JO (1985) The water dimer potential surface. Water Sci Rev 1:93–170
743. Savage HFJ (1986) Repulsive characteristics of hydrogen bonded water structures. Trans Am Cryst Assoc 22:19–29
744. Franks F (1981) Polywater. MIT Press, Cambridge, MA
745. Vonnegut K (1952) Cats' cradle. Gollencz, London, UK
746. Wells AF (1962) The third dimension in chemistry. Oxford Univ Press, London, UK
747. Kamb B (1973) Crystallography of ice. In: Whalley E, Jones SJ, Gold LW (eds) Physics and chemistry of ice. Royal Soc Canada, Ottawa, pp 28–41
748. Kamb B (1965) Clathrate crystalline form of silica. Science 148:232–234
749. Gies H, Liebau F, Gerke H (1982) "Dodecasils". A new series of polytype intercalation compounds of SiO_2. Angew Chem Int Ed 21:206–207
750. Hicks JFG (1967) Structure of silica glass. Science 155:459–461
751. Saenger W (1987) Structure and dynamics of water surrounding biomolecules. Ann Rev Biophys. Biophys Chem 16:93–114
752. Haida O, Matsuo T, Suga H, Seki S (1974) Calorimetric study of the glassy state, X. Enthalpy relaxation at the glass transition temperature of hexagonal ice. J Chem Thermodyn 6:815–825

753. Yamanuro O, Oguni M, Matsuo T, Suga H (1987) Heat capacity and glass transition of pure and doped ices. J Phys Chem Solid 48:935–942
754. Yang DSC, Sax M, Chakrabarty R, Hew CL (1988) Crystal structure of an antifreeze polypeptide and its mechanistic implications. Nature (Lond) 333:232–237
755. Peterson SW, Levy HA (1957) A single crystal neutron diffraction study of heavy ice. Acta Cryst 10:70–76
756. Harmony MD, Laurie VW, Kuczkowski RL, Schwendeman RH, Ramsay DA, Lovas FJ, Lafferty WJ, Maki AG (1979) Molecular structures of gas-phase polyatomic molecules determined by spectroscopic methods. J Phys Chem Ref Data 8:619–721
757. Chidambaram R (1961) A bent hydrogen bond model for the structure of ice I. Acta Cryst 14:467–468
758. Kuhs WF, Lehmann MS (1985) The structure of ice Ih. Water Science Rev 2:1–66
759. Yoon BJ, Morokuma K, Davidson ER (1985) Structure of ice Ih. Ab-initio two and three-body water-water potentials and geometry optimization. J Chem Phys 83:1223–1231
760. Miller SL (1961) The occurrence of gas hydrates in the solar system. Proc Natl Acad Sci USA 47:1798–1808
761. Klinger J, Berest D, Dollfus A, Smoluckowski R (eds) (1985) Ices in the solar system. NATO-ASI, Riedel, Dordrecht
762. Davidson DW (1973) Clathrate hydrates. In: Franks F (ed) The chemistry and physics of water. A comprehensive treatise, vol 2. Chap 3. Plenum Press, NY, pp 115–234
763. Faraday M (1823) On the hydrate of chlorine. J Sci Liter Arts 15:71–90
764. Davy H (1811) On some of the combinations of oxymuriatic acid gas and oxygen and on the chemical relations of the properties. Phil Trans R Soc (Lond) 101:30–55. Alembic Club Reprints No 9:98
765. Pickering SU (1893) The hydrate theory of solutions. Some compounds of the alkylamines and ammonia with water. Trans Chem Soc 63 (I):141–195
766. Fowler DL, Loebenstein WV, Pall DB, Kraus CA (1940) Some unusual hydrates of quaternary ammonium salts. J Am Chem Soc 62:1140–1144
767. Mootz D, Oellers E-J, Wiebcke M (1987) First examples of type I clathrate hydrates of strong acids: polyhydrates of hexafluorophosphoric, tetrafluoroboric, and perchloric acid. J Am Chem Soc 109:1200–1202
768. Schroeder W (1927) Die Geschichte der Gas Hydrate. Ahren's Sammlung Chem. Chem Tech Vorträge 29:1–98
769. Claussen WF (1951) Suggested structures of water in inert gas hydrates. J Chem Phys 19:259–260
770. Claussen WF (1951) A second water structure for inert gas hydrates. J Chem Phys 19:1425–1426
771. von Stackelberg M, Müller HR (1951) On the structure of the gas hydrates. J Chem Phys 19:1319–1320
772. Pauling L, Marsh RE (1952) The structure of chlorine hydrate. Proc Natl Acad Sci USA 38:112–118
773. Palin DE, Powell HM (1947) The structure of molecular compounds. Part III. Crystal structure of addition compounds of quinol with certain volatile compounds. J Chem Soc 208–221
774. McMullan RK, Jeffrey GA (1959) Hydrates of the tetra-n-butyl and tetra-i-amyl quaternary ammonium salts. J Chem Phys 31:1231–1234
775. Londomo D, Kuhs WF, Finney JL (1988) Enclathration of helium in ice II. Nature (Lond) 332:141–142
776. Holder GD, Manganiello DJ (1982) Hydrate dissociation pressure minima in multicomponent systems. Chem Eng Sci 37:9–16
777. Davidson DW, Handa VP, Ratcliffe CI, Tse JS, Powell BM (1984) The ability of small molecules to form clathrate hydrates of structure II. Nature (Lond) 311:142–143
778. Tse JS, Handa VP, Ratcliffe CI, Powell BM (1986) Structure of oxygen clathrate hydrate by neutron diffraction. J Incl Phenom 4:235–240
779. Ripmeester JA, Tse JS, Ratcliffe CI, Powell BM (1987) A new clathrate hydrate structure. Nature (Lond) 325:135–136

780. Gerke H, Gies H, Liebau F (1982) Dodecasil 1 H. Der einfachste Vertreter einer polytypen Reihe von SiO$_2$-Clathraten. Z Krist 159:52–54
781. Gerke H, Gies H (1984) Studies of clathrasils, IV. Z Krist 166:11–22
782. von Stackelberg M, Müller HR (1954) Feste Gashydrate II. Z Electrochem 58:25–39
783. von Stackelberg M, Meinhold W (1954) Feste Gashydrate III. Z Electrochem 58:40–45
784. von Stackelberg M, Frühbuss H (1954) Feste Gashydrate IV. Z Electrochem 58:99–104
785. von Stackelberg M (1954) Feste Gashydrate V. Z Electrochem 58:104–109
786. von Stackelberg M, Jahns W (1954) Feste Gashydrate VI. Z Electrochem 58:162–164
787. McMullan RK, Jeffrey GA (1965) Polyhedral clathrate hydrates. IX. Structure of ethylene oxide hydrate. J Chem Phys 42:2725–2732
788. Mak TCW, McMullan RK (1965) Polyhedral clathrate hydrates. X. Structure of the double hydrate of tetrahydrofuran and hydrogen sulfide. J Chem Phys 42:2732–2737
789. Bode H, Teufer G (1955) Die Kristallstruktur der Hexaflourophosphorsäure. Acta Cryst 8:611–614
790. Mak TCW, BruinsSlot HJ, Beurskens PT (1986) Tetraethylammonium chloride tetrahydrate, a double channel host lattice constructed from (H$_2$O)$_4$·Cl$^-$ tetrahedra linked between vertices. J Inclus Phenom 4:295–302
791. Harmon KM, Avci GF, Harmon J, Thiel AC (1987) Hydrogen-bonding, Part 23. Further studies on the stoichiometry, stability and structure of the lower hydrates of tetramethylammonium fluoride. J Molec Struct 160:57–66
792. Harmon KM, Avci GF, Gabriele JM, Jacks MJ (1987) Hydrogen-bonding, Part 22. Re-evaluation of stoichiometry and structure of lower hydrates of tetramethylammonium hydroxide. J Molec Struct 159:255–263
793. Mootz D, Wussow HG (1981) Crystal structures of pyridine and pyridine trihydrate. J Chem Phys 75:1517–1522
794. Nelson WO, Rettig SJ, Orvig C (1987) The exoclathrate Al(C$_7$H$_8$NO$_2$)$_3$·12H$_2$O. A facial geometry imposed by extensive hydrogen bonding with the ice I structure. J Am Chem Soc 109:4121–4123
795. Feil D, Jeffrey GA (1961) The polyhedral clathrate hydrates, II. Structure of the hydrate of tetra-iso-amyl ammonium fluoride. J Chem Phys 35:1863–1873
796. Beurskens PT, Jeffrey GA (1964) Polyhedral clathrate hydrates VII. Structure of the monoclinic form of the tri-n-butyl sulfonium fluoride hydrate. J Chem Phys 40:2800–2810
797. Jeffrey GA, McMullan RK (1962) Polyhedral clathrate hydrates, IV. The structure of the tri-n-butyl sulfonium fluoride hydrate. J Chem Phys 37:2231–2239
798. Favier R, Rosso J-C, Carbonnel L (1981) Etude des systèmes binaires eau-monoamines aliphatiques. Etablissement de onze diagrammes de phases, mise en évidence d'hydrates nouveaux. Bull Soc Chim 5–6 I:225–235
799. McMullan RK, Jordan TH, Jeffrey GA (1967) Polyhedral clathrate hydrates, XII. The crystallographic data on hydrates of ethylamine, dimethylamine, trimethylamine, n-propylamine (two forms), isopropylamine, diethylamine (two forms) and tert-butylamine. J Chem Phys 47:1218–1222
800. McMullan RK, Jeffrey GA, Jordan TH (1967) The polyhedral clathrate hydrates, Part XIV. The crystal structure of tert-butylamine hydrate. J Chem Phys 47:1229–1234
801. Panke D (1968) Polyhedral clathrate hydrates, XV. Structure of trimethylamine decahydrate. J Chem Phys 48:2990–2996
802. Brickenkamp CS, Panke D (1973) Polyhedral clathrate hydrates, XVII. Structure of the low melting hydrate of n-propylamine: a novel clathration framework. J Chem Phys 58:5284–5295
803. Dyadin Yu A, Udachin KA (1984) Clathrate formation in water-peralkylonium salts systems. J Inclus Phenom 2:61–72
804. Nakayama H, Watanabe K (1976) Hydrates of organic compounds. II. The effect of alkyl groups on the formation of quaternary ammonium fluoride hydrates. Bull Chem Soc Jpn 49:1254–1256
805. Nakayama H, Watanabe K (1978) Hydrates of organic compounds. III. The formation of clathrate-like hydrates of tetrabutylammonium dicarboxylates. Bull Chem Soc Jpn 51:2518–2522

806. Nakayama H (1979) Hydrates of organic compounds. IV. Clathrate hydrates of various bolaform salts. Bull Chem Soc Jpn 52:52–56
807. Folzer C, Hendricks RW, Narten AH (1971) Diffraction pattern and structure of liquid trimethylamine decahydrate at 5°C. J Chem Phys 54:799–805
808. Klotz IM (1962) Water. In: Kasha M, Pullman B (eds) Horizons in biochemistry. Academic Press, NY, pp 523–552
809. Finney JL (1977) The organization and function of water in protein crystals. Phil Trans R Soc Lond B 278:3–32
810. Finney JL, Goodfellow JM, Poole PL (1982) The structure and dynamics of water in globular proteins. In: Davies DB, Saenger W, Danyluk SS (eds) Structural molecular biology. Methods and applications. Plenum Press, NY, pp 387–426
811. Teeter MM (1984) Water structure of a hydrophobic protein at atomic resolution: pentagon rings of water molecules in crystals of crambin. Proc Natl Acad Sci USA 81:6014–6018
812. Neidle S, Berman HM, Shieh HS (1980) Highly structured water network in crystals of a deoxydinucleoside-drug complex. Nature (Lond) 288:129–133
813. Karle IL (1986) Water structure in [Phe^4 Ala^4] antamanide 12H$_2$O crystallized from dioxane. Int J Pept Protein Res 28:6–12
814. Shieh HS, Berman HM, Dabrow M, Neidle S (1980) The structure of a drug-deoxydinucleoside phosphate complex: generalized conformational behavior of intercalation complexes with RNA and DNA fragments. Nucleic Acid Res 8:85–97
815. Schwarzenbach D (1968) Structure of piperazine hexahydrate. J Chem Phys 48:4134–4140
816. Kim HS, Jeffrey GA (1970) Crystal structure of pinacol hexahydrate. J Chem Phys 53:3610–3615
817. Jeffrey GA, Shen MS (1972) Crystal structure of 2,5-dimethyl-2,5-hexanediol tetrahydrate: A water-hydrocarbon layer structure. J Chem Phys 57:56–61
818. Jeffrey GA, Mastropaolo D (1978) The crystal structure of 2,7-dimethyl-2,7-octanediol tetrahydrate. Acta Cryst B34:552–556
819. Hatt HH (1956) The crystalline hydrates of alcohols and glycols. Rev Pure Appl Chem 6:153–190
820. Moye CJ (1972) Non-aqueous solvents for carbohydrates. Adv Carbohydr Chem Biochem 27:85–125
821. Jeffrey GA, Wood RA (1982) The crystal structure of galactaric acid (mucic acid) at −147°C: an unusually dense hydrogen-bonded structure. Carbohydr Res 108:205–211
822. Clark JR (1963) Water molecules in hydrated organic crystals. Rev Pure Appl Chem 13:50–90
823. Moore FH, O'Connor BH, Willis BTM, Hodgkin DC (1984) X-ray and neutron diffraction studies of the crystal and molecular structure of the predominant monocarboxylic acid obtained by mild acid hydrolysis of cyanocobalamin. Part III. Neutron diffraction studies of wet crystals. Proc Ind Acad Soc (Chem Sci) 93:235–260
824. Karle IL, Duesler E (1977) Arrangement of water molecules in cavities and channels of the lattice of [Phe^{4}Val^6] antamanide dodecahydrate. Proc Natl Acad Sci USA 74:2602–2606
825. Seeman NC, Day RO, Rich A (1975) Nucleic acid-mutagen interactions: crystal structure of adenylyl-3',5'-uridine plus 9-aminoacridine. Nature (Lond) 253:324–326
826. Savage HFJ, Lindley PF, Finney JL, Timmins PA (1987) High resolution neutron and X-ray refinements of vitamin B$_{12}$ coenzyme, C$_{72}$H$_{100}$CoN$_{18}$O$_{17}$P·17H$_2$O. Acta Cryst B43:280–295
827. Kuntz ID, Kauzmann W (1974) Hydration of proteins and polypeptides. Adv Protein Chem 28:239–345
828. Rupley JA, Gratton E, Careri G (1983) Water and globular proteins. Trends Biol Sci 8:18–22
829. Edsall JT, McKenzie HA (1983) Water and proteins. II. The location and dynamics of water in protein systems and its relation to their stability and properties. Adv Biophysics 16:53–183
830. Picullel L, Halle B (1986) Water spin relaxation in colloidal systems. Part 2. ^{17}O and ^{2}H relaxation in protein solutions. J Chem Soc Faraday Trans I 82:401–414
831. Berndt M, Kwiatkowski JS (1986) Hydration of biomolecules. In: Náray-Szabó G (ed) Theoretical chemistry of biological systems. Studies in physical and theoretical chemistry, vol. 41. Elsevier, Amsterdam, pp 349–422

832. Beveridge DL, Mezei M, Mehrotra PK, Marchese FT, Thirumalai V, Ravishanker (1981) Liquid state computer simulations of biomolecular solvation problems. Ann NY Acad Sci 367:108–131
833. Mezei M, Mehrotra PK, Beveridge DL (1984) Monte Carlo computer simulation study of the aqueous hydration of the glycine zwitterion at 25 °C. J Biomol Struct Dynam 2:1–27
834. Finney JL, Goodfellow JM, Howell PL, Vovelle F (1985) Computer simulation of aqueous biomolecular systems. J Biomol Struct Dynam 3:599–622
835. Watenpaugh KD, Margulis TN, Sieker LC, Jensen LH (1978) Water structure in a protein crystal: rubredoxin at 1.2 Å resolution. J Mol Biol 122:175–190
835a. Jensen LH, Adman ET, Sieker LC, Stenkamp RE (1989) Water structure at high resolution. Abst HA4, Am Cryst Assoc Meeting, Seattle, July 1989
836. Langridge R, Marvin DA, Seeds WE, Wilson HR, Hooper CW, Wilkins MHF, Hamilton LD (1960) The molecular configuration of deoxyribonucleic acid. J Mol Biol 2:38–64
837. Bolin JT, Filman DJ, Matthews DA, Hamlin RC, Kraut J (1982) Crystal structures of E. coli and L. casei dihydrofolate reductase refined at 1.7 Å resolution. J Biol Chem 257:13650–13662
838. Finney JL (1978) Volume occupation, environment, and accessibility in proteins. Environment and molecular area of RNase-S. J Mol Biol 119:415–441
839. Lee B, Richards FM (1971) The interpretation of protein structures: estimation of static accessibility. J Mol Biol 55:379–400
840. Thanki N, Thornton JM, Goodfellow JM (1988) Distribution of water around amino acid residues in proteins. J Mol Biol 202:637–657
840a. Ippolito JA, Alexander RS, Christianson DW (1990) Hydrogen bond stereochemistry in protein structure and function. J Mol Biol 215:457–471
841. Furey W, Jr, Wang BC, Yoo CS, Sax M (1983) Structure of a novel Bence-Jones protein (Rhe) fragment at 1.6 Å resolution. J Mol Biol 167:661–692
842. Karplus PA, Schulz GE (1987) Refined structure of glutathione reductase at 1.54 Å resolution. J Mol Biol 195:701–729
843. Baker T, Dodson E, Dodson G, Hodgkin D, Hubbard R (1985) The water structure in 2 Zn insulin crystals. In: Moras D, Drenth J, Strandberg B, Suck D, Wilson K (eds) Crystallography in molecular biology. Plenum Press, New York, pp 179–192
844. Teeter MM, Whitlow MD (1987) Hydrogen bonding in the high resolution structure of the protein crambin. Trans Am Cryst Assoc 22:75–88
845. James MNG, Sielecki AR, Brayer GD, Delbaere LTJ, Bauer C-A (1980) Structures of product and inhibitor complexes of *Streptomyces griseus* protease A at 1.8 Å resolution. J Mol Biol 144:43–88
846. Betzel C, Pal GP, Saenger W (1988) Three-dimensional structure of proteinase K at 0.15 nm resolution. Eur J Biochem 178:155–171
847. Goodfellow JM, Howell PL, Elliott R (1986) Water at biomolecule interfaces. In: Moras D, Drenth J, Strandberg B, Suck D, Wilson K (eds) Crystallography in molecular biology. Plenum, New York, pp 167–177
848. Texter J (1978) Nucleic acid-water interactions. Prog Biophys Mol Biol 33:83–97
849. Berman HM (1986) Hydration of nucleic acid crystals. Ann NY Acad Sci 482:166–178
850. Westhof E (1988) Water: an integral part of nucleic acid structure. Ann Rev Biophys Biophys Chem 17:125–144
851. Drew H, Takano T, Tanaka S, Itakura K, Dickerson RE (1980) High-salt d(CpGpCpG): a left-handed Z′ DNA double helix. Nature (Lond) 286:567–573
852. Leslie AGW, Arnott S, Chandrasekaran R, Ratliff RL (1980) Polymorphism of DNA double helices. J Mol Biol 143:49–72
853. Falk M, Hartmann KA Jr, Lord RC, (1962) Hydration of deoxyribonucleic acid. I. A gravimetric study. J Am Chem Soc 84:3843–3846
854. Falk M, Hartmann KA Jr, Lord RC (1963) Hydration of deoxyribonucleic acid. II. An infrared study. J Am Chem Soc 85:387–391
855. Falk M, Hartmann KA Jr, Lord RC, (1963) Hydration of deoxyribonucleic acid. III. A spectroscopic study of the effect of hydration on the structure of deoxyribonucleic acid. J Am Chem Soc 85:391–394

856. Wolf B, Hanlon S (1975) Structural transitions of deoxyribonucleic acid in aqueous electrolyte solutions. I. Reference spectra of conformational limits. Biochemistry 14:1648–1660
857. Wolf B, Hanlon S (1975) Structural transitions of deoxyribonucleic acid in aqueous solutions. II. The role of hydration. Biochemistry 14:1661–1670
858. Cohen G, Eisenberg H (1968) Deoxyribonucleate solutions: Sedimentation in a density gradient, partial specific volumes, density and refractive index measurements, and preferential interactions. Biopolymers 6:1077–1100
859. Hearst JE, Vinograd J (1961) The net hydration of deoxyribonucleic acid. Proc Natl Acad Sci USA 47:825–830
860. Hearst JE (1965) Determination of the dominant factors which influence the net hydration of the native sodium deoxyribonucleate. Biopolymers 3:57–68
861. Tunis M-JB, Hearst JE (1968) On the hydration of DNA. I. Preferential hydration and stability of DNA in concentrated trifluoracetate solution. Biopolymers 6:1325–1344; II. Base composition dependence of net hydration of DNA. Biopolymers 6:1345–1353
862. Falk M, Poole A-G, Goymour CG (1970) Infrared study of the state of water in the hydration shell of DNA. Can J Chem 48:1536–1542
863. Alden CJ, Kim S-H (1979) Solvent-accessible surfaces of nucleic acids. J Mol Biol 132:411–434
864. Suck D, Manor PC, Saenger W (1976) The structure of a trinucleoside diphosphate: adenylyl-(3′,5′)-adenylyl-(3′,5′)-adenosine hexahydrate. Acta Cryst B32:1727–1737
865. Kennard O, Cruse WBT, Nachman J, Prangé T, Shakked Z, Rabinovich D (1986) Ordered water structure in an A-DNA octamer at 1.7 Å resolution. J Biomol Struct Dynam 5:623–648
866. McCall M, Brown T, Kennard O (1985) The crystal structure of d(GGGGCCCC). A model for poly(dG)·poly(dC). J Mol Biol 183:385–396
867. Kopka ML, Fratini AV, Drew HR, Dickerson RE (1983) Ordered water structure around a B-DNA dodecamer. A quantitative study. J Mol Biol 163:129–146
868. Privé GG, Heinemann U, Chandrasegaran S, Kan L-S, Kopka ML, Dickerson RE (1987) Helix geometry, hydration and G·A mismatch in a B-DNA decamer. Science 238:498–504
869. Westhof E, Prangé T, Chevrier B, Moras D (1985) Solvent distribution in crystals of B- and Z-oligomers. Biochimie 67:811–817
870. Gessner RV, Quigley GJ, Wang AH-J, van der Marel GA, van Boom JH, Rich A (1985) Structural basis for stabilization of Z-DNA by cobalt hexaammine and magnesium cations. Biochemistry 24:237–240
871. Westhof E (1987) Hydration of oligonucleotides in crystals. Int J Biol Macromol 9:186–192
872. Beveridge DL, Maye PV, Jayaram B, Ravishanker G, Mezei M (1984) Aqueous hydration of nucleic acid constituents: Monte Carlo computer simulation studies. J Biomol Struct Dynam 2:261–270
873. Sundaralingam M (1974) Principles governing nucleic acid and polynucleotide conformations. In: Sundaralingam M, Rao ST (eds) Structure and conformation of nucleic acids and protein-nucleic acid interactions. Univ Park Press, Baltimore, pp 487–524
874. Saenger W, Hunter W, Kennard O (1986) Conformation of DNA is determined by economics in the hydration of phosphate groups. Nature (Lond) 324:385–388
875. Goldblum A, Perahia D, Pullman A (1978) Hydration scheme of the complementary basepairs of DNA. FEBS Lett 91:213–215
876. Chevrier B, Dock AC, Hartmann B, Leng M, Moras D, Thuong MT, Westhof E (1986) Solvation of the left-handed hexamer d(5BrC-G-5BrC-G-5BrC-G) in crystals grown at two temperatures. J Mol Biol 188:707–719
877. Frauenfelder H, Petsko GA, Tsernoglou D (1979) Temperature-dependent X-ray diffraction as a probe of protein structural dynamics. Nature (Lond) 280:558–563
878. Sternberg MIE, Grace DEP, Phillips DC (1979) Dynamic information from protein crystallography. J Mol Biol 130:231–253
879. Clementi E, Corongiu G, Sarma RH, Sarma MH (eds) (1984) Structure and motion: Membranes, nucleic acids and proteins. Adenine, Guilderland, NY
880. Vovelle F, Goodfellow JM, Savage HFJ, Barnes P, Finney JL (1985) Solvent structure in vitamin B_{12} coenzyme crystals. Eur Biophys J 11:225–237

881. Savage H (1986) Water structure in vitamin B_{12} coenzyme crystals. I. Analysis of the neutron and X-ray solvent densities. J Biophys Soc 50:947–980
882. Sundaralingam M, Sekharudu YC (1989) Water-inserted α-helical segments implicate reverse turns as folding intermediates. Science 244:1333–1337
883. Karle IL, Flippen-Anderson JL, Uma K, Balaram P (1989) Solvated helical backbones: X-ray diffraction study of Boc-Ala-Leu-Aib-Ala-Leu-Aib-OMe·H_2O. Biopolymers 28:773–781
884. Parthasarathy R, Chatuvredi S, Go K (1990) Design of crystalline helices of short oligopeptides as a possible model for nucleation of α-helix: role of water molecules in stabilized helices. Proc Natl Acad Sci (USA) 87:871–875

Refcodes

ABINOR01. β-D,L-Arabinopyranose (neutron study) ($C_5H_{10}O_5$). Takagi S, Nordenson S, Jeffrey GA (1979) Acta Crystallogr, Sect B 35:991

ABINOS01. β-L-Arabinose (neutron study) ($C_5H_{10}O_5$). Takagi S, Jeffrey GA (1977) Acta Crystallogr, Sect B 33:3033

ACGLUA11. N-Acetyl-α-glucosamine ($C_8H_{15}NO_6$). Mo F, Jensen LH (1975) Acta Crystallogr, Sect B 31:2867

ACYGLY. N-Acetylglycine ($C_4H_7NO_3$). Donohue J, Marsh RE (1962) Acta Crystallogr 15:941

ACYGLY02. N-Acetylglycine (neutron study) ($C_4H_7NO_3$). Peterson SW, Levy HA, Schomaker V (1957) Acta Crystallogr 10:844

ACYGL11. N-Acetylglycine (neutron study) ($C_4H_7NO_3$). Mackay MF (1975) Cryst Struct Commun 4:225

ACYTID. α-Cytidine ($C_9H_{13}N_3O_5$). Post ML, Birnbaum GI, Huber CP, Shugar D (1977) Biochim Biophys Acta 479:133

ADBURM. Adenosine-5-bromouridine monohydrate $C_{10}H_{13}N_5O_4$ $C_9H_{11}BrN_2O_6$, H_2O. Haschemeyer AEV, Sobell HM (1965) Acta Crystallogr 18:525

ADENBH. Adeninium hydrobromide hemihydrate [$C_5H_6N_5^+$, Br^-, $0.5(H_2O)$]. Langer V, Huml K (1978) Acta Crystallogr, Sect B 34:1881

ADENCH02. Adenine hydrochloride hemihydrate [$2(C_5H_6N_5^+)$, $2(Cl^-)$, H_2O]. Kistenmacher TJ, Shigematsu T (1974) Acta Crystallogr, Sect B 30:166

ADENIC. N-(3-(Aden-9-yl)propyl)-3-carbamoylpyridinium bromide hydrobromide dihydrate [$C_{14}H_{17}N_7O^{++}$, $2(Br^-)$, $2(H_2O)$]. Johnson PL, Frank JK, Paul IC (1973) J Am Chem Soc 95:5377

ADENOH10. Bis(adeninium) dinitrate monohydrate [$2(C_5H_6N_5^+)$, $2(NO_3^-)$, H_2O]. Hingerty BE, Einstein JR, Wei CH (1981) Acta Crystallogr, Sect B 37:140

ADENOS10. Adenosine (absolute configuration) ($C_{10}H_{13}N_5O_4$). Lai TF, Marsh RE (1972) Acta Crystallogr, Sect B 28:1982

ADENPH. Adenium phosphate $C_5H_6N_5^+$, $H_2O_4P^-$. Langer V, Huml K, Zachova J (1979) Acta Crystallogr, Sect B 35:1148

ADENSH. 8,5′-Cycloadenosine monohydrate ($C_{10}H_{11}N_5O_4$, H_2O). Haromy TP, Raleigh J, Sundaralingam M (1980) Biochemistry 19:1718

ADESON10. Adenine-N-1-oxide sulfuric acid ($C_5H_7N_5O^{++}$, O_4S^{--}). Prusiner P, Sundaralingam M (1972) Acta Crystallogr, Sect B 28:2142

ADESUL. Adeninium hemisulfate monohydrate [$C_5H_6N_5^+$, $0.5(O_4S^{--})$, H_2O]. Langer V, Huml K, Lessinger L (1978) Acta Crystallogr, Sect B 34:2229

ADGALA01. α-D-Galactose ($C_6H_{12}O_6$). Ohanessian J, Gillier-Pandraud H (1976) Acta Crystallogr, Sect B 32:2810

ADLFUC. α-D,L-Fucopyranose ($C_6H_{12}O_5$). Longchambon F, Gillier-Pandraud H, Becker P (1977) Acta Crystallogr, Sect B 33:2094

ADLMAN. α-D,L-Mannose ($C_6H_{12}O_6$). Planinsek F, Rosenstein RD (1967) ACA (Summer) 70

ADOSHC. Adenosine hydrochloride ($C_{10}H_{14}N_5O_4^+$, Cl^-). Shikata K, Ueki T, Mitsui T (1973) Acta Crystallogr, Sect B 29:31

ADPOSD. Adenosine-3'-phosphate dihydrate [$C_{10}H_{14}N_5O_7P$, 2(H_2O)]. Sundaralingam M (1966) Acta Crystallogr 21:495

ADPROP. 3-(Adenin-9-yl)-propionamide ($C_8H_{10}N_6O$). Takimoto M, Takenaka A, Sasada Y (1981) Bull Chem Soc Jpn 54:1635

ADPRTR. 3-(Adenin-9-yl)propiontryptamide ($C_{18}H_{19}N_7O$). Ohki M, Takenaka A, Shimanouchi H, Sasada Y (1977) Bull Chem Soc Jpn 50:2573

ADTALO01. 1A,2A,3E,4A,5E-α-D-Talose ($C_6H_{12}O_6$). Hansen LK, Hordvik A (1977) Acta Chem Scand, Ser A 31:187

ADTALO10. α-D-Talopyranose ($C_6H_{12}O_6$). Ohanessian J, Avenel D, Kanters JA, Smits D (1977) Acta Crystallogr, Sect B 33:1063

AFCYDP. 2,2'-Anhydro-1-β-D-arabinofuranosyl-cytosine-3',5'-diphosphate monohydrate ($C_9H_{13}N_3O_{10}P_2$, H_2O). Yamagata Y, Suzuki Y, Fujii S, Fujiwara T, Tomita K (1979) Acta Crystallogr, Sect B 35:1136

AFUTHU. 2,2'-Anhydro-1-β-D-arabinofuranosyl-2-thio-uracil ($C_9H_{10}N_2O_4S$). Yamagata Y, Yoshimura J, Fujii S, Fujiwara T, Tomita K-I, Ueda T (1980) Acta Crystallogr, Sect B 36:343

AGALAM01. N-Acetyl-α-D-galactosamine ($C_8H_{15}NO_6$). Gilardi RD, Flippen JL (1974) Acta Crystallogr, Sect B 30:2931

AGALAM10. N-Acetyl-α-D-galactosamine ($C_8H_{15}NO_6$). Neuman A, Gillier-Pandraud H, Longchambon F, Rabinovich D (1975) Acta Crystallogr, Sect B 31:474

AGLYSL01. Ammonium glycinium sulfate (neutron study) ($C_2H_6NO_2^+$, H_4N^+, O_4S^{--}). Vilminot S, Philippot E, Lehmann M (1976) Acta Crystallogr, Sect B 32:1817

AGMTHY. (1-Methyöthymine)-silver(I) ($C_6H_7O_2^-$, Ag^+. Guay F, Beauchamp AL (1979) J Am Chem Soc 101:6260

AHGALP. 1,6-Anhydro-β-D-galactopyranose ($C_6H_{10}O_5$). Ceccarelli C, Ruble JR, Jeffrey GA (1980) Acta Crystallogr, Sect B 36:861

AHGLPY01. 1,6-Anhydro-β-D-glucopyranose levoglucosan ($C_6H_{10}O_5$). Lindberg KB (1974) Acta Chem Scan, Ser A 28:1181

AHGLPY10. 1,6-Anhydro-β-D-glucopyranose levoglucosan ($C_6H_{10}O_5$). Park YJ, Kim HS, Jeffrey GA (1971) Acta Crystallogr, Sect B 27:220

AHLPRO. Allo-4-hydroxy-L-proline dihydrate [$C_5H_9NO_3$, 2(H_2O)]. Shamala N, Row TNG, Venkatesan K (1976) Acta Crystallogr, Sect B 32:3267

AHXGLP. 3-Amino-1,6-anhydro-3-deoxy-β-D-glucopyranose (neutron study) ($C_6H_{11}NO_4$). Noordik JH, Jeffrey GA (1977) Acta Crystallogr, Sect B 33:403

ALAHCL. L-Alanine hydrochloride ($C_3H_8NO_2^+$, Cl^-). Blasio B di, Pavone V, Pedone C (1977) Cryst Struct Commun 6:745

ALFUCO. α-L-Fucose ($C_6H_{12}O_5$. Longchambon F, Ohannesian J, Avenel D, Neuman A (1975) Acta Crystallogr, Sect B 31:2623

ALITOL01. Allitol ($C_6H_{14}O_6$). Azarnia N, Jeffrey GA, Shen MS (1972) Acta Crystallogr, Sect B 28:1007

AMAFAP. 8,2'-Anhydro-8-mercapto-9-β-D-arabinofuranosyl-adenine 5'-monophosphate trihydrate [$C_{10}H_{12}N_5PS$, 3(H_2O)]. Tanaka K, Fujii S, Fujiwara T, Tomita K (1979) Acta Crystallogr, Sect B 35:929

AMANOF. 1,6-Anhydro-β-D-mannofuranose ($C_6H_{10}O_5$). Lechat J, Jeffrey GA (1972) Acta Crystallogr, Sect B 28:3410

AMCYTS. N-4-Aminocytosine ($C_4H_6N_4O$). Takayanagi H, Ogura H, Hayatsu H (1980) Chem Pharm Bull 28:2614

AMDOAD. α-D-2'-Amino-2'-deoxyadenosine monohydrate ($C_{10}H_{14}N_6O_3$, H_2O). Rohrer DC, Sundaralingam M (1970) J Am Chem Soc 92:4956

AMOADA. 3'-Amino-3'-deoxyadenosine ($C_{10}H_{14}N_6O_3$). Sheldrick WS, Morr M (1980) Acta Crystallogr, Sect B 36:2328

AMOROT. Ammonium orotate monohydrate, ammonium uracil-6-carboxylate monohydrate ($C_5H_3N_2O_4^-$, H_4N^+, H_2O). Solbakk J (1971) Acta Chem Scand 25:3006

AMPYRM01. 2-Aminopyrimidine (at 107 K) ($C_4H_5N_3$). Furberg S, Grogaard J, Smedsrud (1979) Acta Chem Scand Ser B 33:715

AMPYRN. 2-Aminopyrimidine hydrochloride hemihydrate (at 110 deg.K) [$C_4H_6N_3^+$, Cl^-, 0.5(H_2O)]. Furberg S, Grogaard J (1980) Acta Chem Scand, Ser A 34:695

AMURID. 5-Aminouridine ($C_9H_{13}N_3O_6$). Egert E, Lindner HJ, Hillen W, Gassen HG (1978) Acta Crystallogr, Sect B 34:2204

ANALPR. 1,6-Anhydro-β-D-allopyranose allosan ($C_6H_{10}O_5$). Norrestam R, Bock K, Pedersen C (1981) Acta Crystallogr, Sect B 37:1265

APAPAD10. Adenylyl-(3',5')-adenylyl-(3',5')-adenosine hexahydrate (form I) [$C_{30}H_{37}N_{15}O_{16}P_2$, 6($H_2O$)]. Suck D, Manor PC, Saenger W (1976) Acta Crystallogr, Sect B 32:1727

APRTAM. 3-(9-Adenyl)-propionyl-tyramine dihydrate [$C_{16}H_{18}N_6O_2$, 2(H_2O)]. Ohki M, Takenaka A, Shimanouchi H, Sasada Y (1977) Acta Crystallogr, Sect B 33:2956

APSURD. α-Pseudouridine monohydrate ($C_9H_{12}N_2O_6$, H_2O). Rohrer DC, Sundaralingam M (1970) J Am Chem Soc 92:4950

ARABOL. D,L-Arabinitol ($C_5H_{12}O_5$). Hunter FD, Rosenstein RD (1968) Acta Crystallogr, Sect B 24:1652

ARACYP. Arabinosyl-cytidine-2',5'-cyclic phosphate ($C_9H_{12}N_3O_7P$). Kung W, Marsh RE, Kainosho M (1977) J Am Chem Soc 99:5471

ARADEN10. 9-β-D-Arabinofuranosyl-adenine ($C_{10}H_{13}N_5O_4$). Bunick G, Voet D (1974) Acta Crystallogr, Sect B 30:1651

ARBCYT10. 1-β-D-Arabinofuranosyl-cytosine ($C_9H_{13}N_3O_5$). Tougard P, Lefebvre-Soubeyran O (1974) Acta Crystallogr, Sect B 30:86

ARBIMC10. 5-Amino-1-β-D-ribofuranosyl-imidazole-4-carboxamide ($C_9H_{14}N_4O_5$). Adamiak DA, Saenger W (1979) Acta Crystallogr, Sect B 35:924

ARFCYT10. 1-(β-D-Arabinofuranosyl)cytosine hydrochloride (absolute configuration) ($C_9H_{14}N_3O_5^+$, Cl^-). Sherfinski JS, Marsh RE (1973) Acta Crystallogr, Sect B 29:192

ARFHCY. 6,2'-Anhydro-1-β-D-arabinofuranosyl-6-hydroxy-cytosine ($C_9H_{11}N_3O_5$). Yamagata Y, Fujii S, Kanai T, Ogawa K, Tomita K (1979) Acta Crystallogr, Sect B 35:378

ARFMAD. 9β-D-Arabinofuranosyl-8-morpholino-adenine dihydrate [$C_{14}H_{20}N_6O_5$, 2(H_2O)]. Swaminathan V, Sundaralingam M, Chattopadhyaya JB, Reese CB (1980) Acta Crystallogr, Sect B 36:828

ARFUAD01. Arabinofuranosyl-adenine hydrochloride ($C_{10}H_{14}N_5O_4^+$, Cl^-). Hata T, Sato S, Kaneko M, Shimizu B, Tamura C (1974) Bull Chem Soc Jpn 47:2758

ARGHCL10. L-Arginine hydrochloride monohydrate ($C_6H_{15}N_4O_2^+$, Cl^-, H_2O). Dow J, Jensen LH, Mazumdar SK, Srinivasan R, Ramachandran GN (1970) Acta Crystallogr, Sect B 26:1662

ARGIND11. L-Arginine dihydrate (neutron study) [$C_6H_{14}N_4O_2$, 2(H_2O)]. Lehmann MS, Verbist JJ, Hamilton WC, Koetzle TF. J Chem Soc, Perkin 2:133

ASPARM02. L-Asparagine monohydrate (neutron study) ($C_4H_8N_2O_3$, H_2O). Verbist JJ, Lehmann MS, Koetzle TF, Hamilton WC (1972) Acta Crystallogr, Sect B 28:3006

ASPARM03. L-Asparagine monohydrate (neutron study) ($C_4H_8N_2O_3$, H_2O). Ramanadham M, Sikka SK, Chidambaram R (1972) Acta Crystallogr, Sect B 28:3000

ASPARM05. L-Asparagine monohydrate (at 100 deg.K) ($C_4H_8N_2O_3$, H_2O). Wang JL, Berkovitch-Yellin Z, Leiserowitz L (1985) Acta Cryst, B (Str Sci) 41:341

ASPART10. D,L-Aspartic acid hydrochloride D,L-1-amino-1,2-dicarboxyethane hydrochloride ($C_4H_8NO_4^+$, Cl^-). Dawson B (1977) Acta Crystallogr, Sect B 33:882

ATPPZN10. Bis[(adenosine-5'-triphoshato)-(2,2-bipyridine)-zinc(II)] tetrahydrate [$C_{40}H_{44}N_{14}O_{26}P_6Zn_2$, 4($H_2O$)]. Orioli P, Cini R, Donati D, Mangani S (1981) J Am Chem Soc 103:4446

AZTYMD. 6-Azathymidine ($C_9H_{13}N_3O_5$). Banerjee A, Saenger W (1978) Acta Crystallogr, Sect B 34:1294

AZURID10. 6-Azauridine ($C_8H_{11}N_3O_6$). Schwalbe CH, Saenger W (1973) J Mol Biol 75:129

BABBOV. 5-Amino-2-thiocytosine dihydrochloride dihydrate [$C_4H_8N_4S^{++}$, 2(Cl^-), 2(H_2O)]. Sagstuen E, Nordenson S (1981) Acta Crystallogr, Sect B 37:1777

BABRUR. 2-Amino-5-bromo-1-hydroxy-6-phenyl-4-pyrimidinone hemihydrate [$C_{10}H_8BrN_3O_2$, 0.5(H_2O)]. Wierenga W, Skulnick HI, Dow RL, Chidester CG. Heterocycles 16:563

BABXAD. N-6-Benzoyladenine ($C_{12}H_9N_5O$). Raghunathan S, Pattabhi V (1981) Acta Crystallogr, Sect B, 37:1670

BAGXEM. 9-Methylhypoxanthinium chloride monohydrate ($C_6H_7N_4O^+$, Cl^-, H_2O). Belanger-Gariepy F, Beauchamp AL (1981) Cryst Struct Commun 10:1165

BAGZEO. 1-O-α-D-Glucopyranosyl-D-mannitol dihydrate [$C_{12}H_{24}O_{11}$, 2(H_2O)]. Lindner HJ, Lichtenthaler FW (1981) Carbohydr Res 93:135

BAVCAC. 6-O-(α-D-Glucopyranosyl)-D-glucitol isomaltitol ($C_{12}H_{24}O_{11}$). Lichtenthaler FW, Lindner HJ (1981) Justus Liebigs Ann Chem :2372

BCYTGA. 5-Bromocytosine N-tosyl-L-glutamic acid ($C_4H_4BrN_3O$, $C_{12}H_{15}NO_6S$). Ohki M, Takenaka A, Shimanouchi H, Sasada Y (1976) Bull Chem Soc Jpn 49:3493

BDDIGX. β-D-Digitoxose ($C_6H_{12}O_4$). Kanters JA, Batenburg LMJ, Gaykema WPJ, Roelofsen G (1978) Acta Crystallogr, Sect B 34:3049

BDGHEP. β-D-Glucoheptose D-glycero-β-D-gulo-heptopyranose ($C_7H_{14}O_7$). Kanters JA, Kock AJHM, Roelofsen G (1978) Acta Crystallogr, Sect B 34:3285

BDGLOS01. β-D-Galactose ($C_6H_{12}O_6$). Longchambon F, Ohannesian J, Avenel D, Neuman A (1975) Acta Crystallogr, Sect B 31:2623

BDGLOS10. β-D-Galactose ($C_6H_{12}O_6$). Sheldrick B (1976) Acta Crystallogr, Sect B 32:1016

BDGPGL. 4-O-β-D-Glucopyranosyl-D-glucitol ($C_{12}H_{24}O_{11}$). Gaykema WPJ, Kanters JA (1979) Acta Crystallogr, Sect B 35:1156

BECGUL. 2-Deoxy-β-D-arabino-hexapyranose (at −150 deg.C) ($C_6H_{12}O_5$). Maluszynska H, Ruble JR, Jeffrey GA (1981) Carbohydr Res 97:199

BEPRAP. Inosine cyclic-(3'-5')-monophosphate monohydrate ($C_{10}H_{11}N_4O_7P$, H_2O). Sundaralingam M, Haromy TP, Prusiner P (1982) Acta Crystallogr, Sect B 38:1536

BEURID10. Uridine ($C_9H_{12}N_2O_6$). Green EA, Rosenstein RD, Shiono R, Abraham DJ, Trus BL, Marsh RE (1975) Acta Crystallogr, Sect B 31:102

BIDRUB10. Adeninium dinitrate [$C_5H_7N_5^{++}$, 2(NO_3^-)]. Hardgrove GL Jr, Einstein JR, Hingerty BE, Wei CH (1983) Acta Cryst, C (Cr Str Comm) 39:88

BIFYOE. 9-Methyladenine 2-thiohydantoin $C_6H_7N_5$, $C_3H_4N_2OS$. Cassady RE, Hawkinson SW (1982) Acta Crystallogr, Sect B 38:2206

BIRMEU. 5-Fluoro-cytosine monohydrate ($C_4H_4FN_3O$, H_2O). Louis T, Low JN, Tollin P (1982) Cryst Struct Commun 11:1059

BIYRIK. 1-(2-Carboxyethyl)-uracil $C_7H_8N_2O_4$. Fujita S, Takenaka A, Sasada Y (1982) Acta Crystallogr, Sect B 38:2936

BIZHIB. 4-O-α-D-Glucopyranosyl-D-glucitol maltitol ($C_{12}H_{24}O_{11}$). Ohno S, Hirao M, Kido M (1982) Carbohydr Res 108:163

BLACTO. β-Lactose ($C_{12}H_{22}O_{11}$) (1974) Hirotsu K, Shimada A (1974) Bull Chem Soc Jpn 47:1872

BOBKUY10. 4-O-β-D-Galactopyranosyl-β-D-fructofuranose (at −150 deg.C) lactulose ($C_{12}H_{22}O_{11}$). Jeffrey GA, Wood RA, Pfeffer PE, Hicks KB (1983) J Am Chem Soc 105:2128

BOHJIR. 3-(7-Adeninyl)-propionamide monohydrate ($C_8H_{10}N_6O$, H_2O). Takimoto M, Takenaka A, Sasada Y (1983) Acta Cryst C (Cr Str Comm) 39:73

BRCPDG. 5-Bromocytosine-phthaloyl-DL-glutamic acid hemihydrate [$C_4H_4BrN_3O$, $C_{13}H_{11}NO_6$, 0.5(H_2O)]. Ohki M, Takenaka A, Shimanouchi H, Sasada Y (1977) Bull Chem Soc Jpn 50:90

BRCYTS. 5-Bromo-cytosine ($C_4H_4BrN_3O$). Kato M, Takenaka A, Sasada Y (1979) Bull Chem Soc Jpn 52:49

BRINOS10. 8-Bromoinosine ($C_{10}H_{11}BrN_4O_5$). Sternglanz H, Thomas JM, Bugg CE (1977) Acta Crystallogr, Sect B 33:2097

BURBAD. 1-Methyl-5-bromouracil-9-ethyl-8-bromo-adenine complex ($C_5H_5BrN_2O_2$, $C_7H_8BrN_5$). Tavale SS, Sakore TD, Sobell HM (1969) J Mol Biol 43:375

BUREAP. 1-Methyl-5-bromouracil-9-ethyl-2-aminopurine complex ($C_5H_5BrN_2O_2$, $C_7H_9N_5$). Mazza F, Sobell HM, Kartha G (1969) J Mol Biol 43:407

CADVUY. D,L-Glutamic acid monohydrate ($C_5H_9NO_4$, H_2O). Ciunik Z, Glowiak T (1983) Acta Cryst, C (Cr Str Comm) 39:1271

CADZAI. 1-β-D-5-Methylarabinosyl-cytosine monohydrate ($C_{10}H_{15}N_3O_5$, H_2O). Birnbaum GI, Gentry GA (1983) J Am Chem Soc 105:5398

CAFFCD. 1,3,7-Trimethyl-2,6-purine-dione hydrochloride dihydrate, caffeine hydrochloride dihydrate [$C_8H_{11}N_4O_2^+$, Cl^-, 2(H_2O)]. Mercer A, Trotter J (1978) Acta Crystallogr, Sect B 34:450

CDURID. 2'-Chloro-2'-deoxyuridine ($C_9H_{11}ClN_2O_5$). Suck D, Saenger W, Hobbs J (1972) Biochem Biophys Acta 259:157

CELLOB02. β-Cellobiose ($C_{12}H_{22}O_{11}$). Chu SSC, Jeffrey GA (1968) Acta Crystallogr, Sect B 24:830

CITSIH10. O-α-D-Galactopyranosyl-(1-4)-D-galactopyranose galabiose ($C_{12}H_{22}O_{11}$). Svensson G, Albertsson J, Svensson C, Magnusson G, Dahmen J (1986) Carbohydr Res 146:29

CLDOUR. 5-Chloro-2'-deoxyuridine ($C_9H_{11}ClN_2O_5$). Young DW, Morris EM (1973) Acta Crystallogr, Sect B 29:1259

CLPURB. 6-Chloro-purine-D-riboside (absolute configuration) ($C_{10}H_{11}ClN_4O_4$). Sternglanz H, Bugg CE (1975) Acta Crystallogr, Sect B 31:2888

CLURID10. 5-Chlorouridine ($C_9H_{11}ClN_2O_6$). Hawkinson SW, Coulter CL (1971) Acta Crystallogr, Sect B 27:34

COFMEP10. O-β-D-Mannopyranosyl-(1-4)-O-β-D-mannopyranosyl-(1-4)-O-α-D-mannopyranose trihydrate, mannotriose trihydrate [$C_{18}H_{32}O_{16}$, $3(H_2O)$]. Mackie W, Sheldrick B, Akrigg D, Perez S (1986) Int J Biol Macromol 8:43

COFOMY10. Coformycin sesquihydrate (absolute configuration by internal comparison), 3-(β-D-Ribofuranosyl)-6,7,8-trihydroimidazo(4,5-D)(1,3)diazepin-8(R)-ol sesquihydrate [$C_{11}H_{16}N_4O_5$, $1.5(H_2O)$]. Nakamura H, Koyama G, Umezawa H, Iitaka Y (1976) Acta Crystallogr, Sect B 32:1206

COKBIN. β-D-Allose ($C_6H_{12}O_6$). Kroon-Batenburg LMJ, van der Sluis P, Kanters JA (1984) Acta Cryst, C (Cr Str Comm) 40:1863

CTSGLM. Cytosine N,N-phthaloyl-D,L-glutamic acid dihydrate [$C_{13}H_{10}NO_6^-$, $C_4H_6N_3O^+$, $2(H_2O)$] (1980) Takenaka A, Ohki M, Sasada Y (1980) Bull Chem Soc Jpn 53:2724

CXMURD. 5-Carboxymethyl-uridine ($C_{11}H_{14}N_2O_8$). Berman HM, Marcu D, Narayanan P, Fissekis JD, Lipnick RL (1978) Nucleic Acids Res 5:893

CYACET. Cytosine-5-acetic acid ($C_6H_7N_3O_3$). Marsh RE, Bierstedt R, Eichhorn EL (1962) Acta Crystallogr, 15:310

CYCYPH10. Cytidine 2',3'-cyclic-phosphate ($C_9H_{12}N_3O_7P$). Reddy BS, Saenger W (1978) Acta Crystallogr, Sect B 34:1520

CYSCLM10. L(+)-Cysteine hydrochloride monohydrate ($C_3H_8NO_2S^+$, Cl^-, H_2O). Ayyar RR (1968) Z Kristallogr 126:227

CYSMEC. L-Cystine dimethyl ester dihydrochloride monohydrate [$C_8H_{18}N_2O_4S_2^{++}$, $2(Cl^-),H_2O$]. Vijayalakshmi BK, Srinivasan R (1975) Acta Crystallogr, Sect B 31:993

CYSTAC01. L-Cysteic acid monohydrate (neutron study) ($C_3H_7NO_5S$, H_2O). Ramanadham M, Sikka SK, Chidambaram R (1973) Acta Crystallogr, Sect B 29:1167

CYSTCL01. L-Cystine dihydrochloride (neutron study) [$C_6H_{14}N_2O_4S_2^{++}$, $2(Cl^-)$]. Gupta SC, Sequeira A, Chidambaram R (1974) Acta Crystallogr, Sect B 30:562

CYSTCL02. L-Cystine dihydrochloride (neutron study) [$C_6H_{14}N_2O_4S_2^{++}$, $2(Cl^-)$]. Jones DD, Bernal I, Frey MN, Koetzle TF (1974) Acta Crystallogr, Sect B 30:1220

CYSTEA. L-Cysteic acid ($C_3H_7NO_5S$). Konishi H, Ashida T, Kakudo M (1968) Bull Chem Soc Jpn 41:2305

CYSURC10. 5-S-Cysteinyluracil monohydrate ($C_7H_9N_3O_4S$, H_2O). Williams GJB, Varghese AJ, Berman HM (1977) J Am Chem Soc 99:3150

CYTIAC. Cytidine 3'-phosphate (orthorhombic form) cytidylic acid B ($C_9H_{14}N_3O_8P$). Sundaralingam M, Jensen LH (1965) J Mol Biol 13:914

CYTIAC01. Cytidine-3'-phosphate (monoclinic form) cytidylic acid B ($C_9H_{14}N_3O_8P$). Bugg CE, Marsh RE (1967) J Mol Biol 25:67

CYTIDI10. Cytidine ($C_9H_{13}N_3O_5$). Furberg S, Petersen CS, Romming C (1965) Acta Crystallogr 18:313

CYTIDN. Cytidinium nitrate ($C_9H_{14}N_3O_5^+$, NO_3^-). Guy JJ, Nassimbeni LR, Sheldrick GM, Taylor R (1976) Acta Crystallogr, Sect B 32:2909

CYTOSC. Cytosine hydrochloride ($C_4H_6N_3O^+$, Cl^-). Mandel NS (1977) Acta Crystallogr, Sect B 33:1079

CYTOSH. Trideutero-cytosine deuterium oxide solvate (neutron study at 82 K) ($C_4H_2D_3N_3O$, D_2O). Weber HP, Craven BM, McMullan RK (1980) Acta Crystallogr, Sect B 36:645

CYTOSH01. Trideutero-cytosine deuterium oxide solvate (at 82 deg.K, X-ray and neutron studies) ($C_4H_2D_3N_3O$, D_2O). Craven BM, Benci P, Epstein J, Fox RO, McMullan RK, Ruble JR, Stewart RF, Weber H-P (1979) Am Cryst Assoc, Ser 2, 7:42

CYTOSM11. Cytosine monohydrate ($C_4H_5N_3O$, H_2O). McClure RJ, Craven BM (1973) Acta Crystallogr, Sect B 29:1234

CYTSIN01. Cytosine ($C_4H_5N_3O$). McClure RJ, Craven BM (1973) Acta Crystallogr, Sect B 29:1234

CXMURD. 5-Carboxymethyl-uridine ($C_{11}H_{14}N_2O_8$). Berman HM, Marcu D, Narayanan P, Fissekis JD (1978) Nucleic Acids Res 5:893

DACHIY. 2-Deoxy-β-D-lyxo-hexose 2-deoxy-β-D-galactose ($C_6H_{12}O_5$). Puliti R, Mattia CA, Barone G (1984) Carbohydr Res 135:47

DAZCYT10. 3-Deazacytidine 1-(β-D-ribofuranosyl)-4-amino-2-pyridone ($C_{10}H_{14}N_2O_5$). Hutcheon WLB, James MNG (1977) Acta Crystallogr, Sect B 33:2224

DCHIST. D,L-Histidine hydrochloride dihydrate [$C_6H_{10}N_3O_2^+$, Cl^-, 2(H_2O)]. Bennett I, Davidson AGH, Harding MM, Morelle I (1970) Acta Crystallogr, Sect B 26:1722

DEKYEX. α-D-Glucopyranosyl-α-D-glucopyranoside (at -150 deg.C) α-Trehalose ($C_{12}H_{22}O_{11}$). Jeffrey GA, Nanni R (1985) Carbohydr Res 137:21

DGLYHC. Diglycine hydrochloride ($C_2H_5NO_2$, $C_2H_6NO_2^+$, Cl^-). Hahn T (1960) Z. Kristallogr 113:26

DHTHYD10. Dihydrothymidine ($C_{10}H_{16}N_2O_5$). Konnert J, Karle IL, Karle J (1970) Acta Crystallogr, Sect B 26:770

DHTURC. 5,6-Dihydro-2-thiouracil ($C_4H_6N_2OS$). Kojic-Prodic B, Ruzic-Toros Z, Coffou E (1976) Acta Crystallogr, Sect B 32:1099

DHURAC10. Dihydrouracil ($C_4H_6N_2O_2$). Rohrer DC, Sundaralingam (1970) Acta Crystallogr, Sect B 26:546

DIHTUJ. O-β-D-Mannopyranosyl-(1-4)-α-D-mannopyranose ($C_{12}H_{22}O_{11}$). Sheldrick B, Mackie W, Akrigg D (1984) Carbohydr Res 132:1

DLALNI. D,L-Alanine ($C_3H_7NO_2$). Donohue J (1950) J Am Chem Soc 72:949

DLASPA10. D,L-Aspartic acid ($C_4H_7NO_4$). Rao ST (1973) Acta Crystallogr, Sect B 29:1718

DLEUHC. Di-L-Leucine hydrochloride ($C_6H_{13}NO_2$, $C_6H_{14}NO_2^+$, Cl^-). Golic L, Hamilton WC (1972) Acta Crystallogr, Sect B 28:1265

DLGLAC. D,L-Glutamic acid hydrochloride ($C_5H_{10}NO_4^+$, Cl^-). Dawson B (1953) Acta Crystallogr 6:81

DLHIST. D,L-Histidine ($C_6H_9N_3O_2$). Edington P, Harding MM (1974) Acta Crystallogr, Sect B 30:204

DLILEU. D,L-Isoleucine ($C_6H_{13}NO_2$). Benedetti E, Pedone C, Sirigu A (1973) Acta Crystallogr, Sect B 29:730

DLLEUC. D,L-Leucine ($C_6H_{13}NO_2$). Di Blasio B, Pedone C, Sirigu A (1975) Acta Crystallogr, Sect B 31:601

DLLYSC10. D,L-Lysine hydrochloride ($C_6H_{15}N_2O_2^+$, Cl^-). Bhaduri D, Saha NN (1979) J Cryst Mol Struct 9:311

DLMETA02. D,L-Methionine (α form, at 333 deg.K) ($C_5H_{11}NO_2S$) (1980) Taniguchi T, Takaki Y, Sakurai K (1980) Bull Chem Soc Jpn 53:803

DLMETA03. D,L-Methionine (β form, at 293 deg.K) ($C_5H_{11}NO_2S$). Taniguchi T, Takaki Y, Sakurai K (1980) Bull Chem Soc Jpn 53:803

DLPROL. D,L-Proline hydrochloride ($C_5H_{10}NO_2^+$, Cl^-). Mitsui Y, Tsuboi M, Iitaka Y (1969) Acta Crystallogr, Sect B 25:2182

DLPROM. D,L-Proline monohydrate ($C_5H_9NO_2$, H_2O). Fox RO J, Rosenstein RD (1976) ACA (Summer) 50

DLSERN11. D,L-Serine (neutron study) ($C_3H_7NO_3$). Frey MN, Lehmann MS, Koetzle TF, Hamilton WC (1973) Acta Crystallogr, Sect B 29:876

DLTYRS. D,L-Tyrosine ($C_9H_{11}NO_3$). Mostad A, Romming C (1973) Acta Chem Scand 27:401

DLVALC. D,L-Valine hydrochloride ($C_5H_{12}NO_2^+$, Cl^-). Di Blasio B, Napolitano G, Pedone C (1977) Acta Crystallogr, Sect B 33:542

DMADEN10. N(6),N(9)-Dimethyl-adenine ($C_7H_9N_5$). Sternglanz H, Bugg CE (1978) J Cryst Mol Struct 8:263

DMANTL. D-Mannitol (β form) ($C_6H_{14}O_6$). Berman HM, Jeffrey GA, Rosenstein RD (1968) Acta Crystallogr, Sect B 24:442

DMURAC. 1,3-Dimethyluracil ($C_6H_8N_2O_2$). Banerjee A, Dattagupta JK, Saenger W, Rabczenko A (1977) Acta Crystallogr, Sect B 33:90

DMURID. 5-Dimethylamino-uridine ($C_{11}H_{17}N_3O_6$). Egert E, Lindner HJ, Hillen W, Gassen HG (1979) Acta Crystallogr, Sect B 35:920

DMANTL01. D-Mannitol (K form) ($C_6H_{14}O_6$). Kim HS, Jeffrey GA, Rosenstein RD (1968) Acta Crystallogr, Sect B 24:1449

DOCYPO. Deoxycytidine 5'-phosphate monohydrate (diffractometer data) ($C_9H_{14}N_3O_7P$, H_2O). Viswamitra MA, Swaminatha Reddy B, Lin GH-Y, Sundarlingam M (1971) J Am Chem Soc 93:4565

DOCYTC. 2'-Deoxycytidine hydrochloride (absolute configuration) ($C_9H_{14}N_3O_4^+$, Cl^-). Subramanian E, Hunt DJ (1970) Acta Crystallogr, Sect B 26:303

DOXADM. Deoxyadenosine monohydrate ($C_{10}H_{13}N_5O_3$, H_2O). Watson DG, Sutor DJ, Tollin P (1965) Acta Crystallogr 19:111

DTURAC. 2,4-Dithiouracil ($C_4H_4N_2S_2$). Shefter E, Mautner HG (1967) J Am Chem Soc 89:1249

DUDGUE. Sodium α-L-guluronate dihydrate (at 87 deg.K) [$C_6H_9O_7^-$, Na^+, $2(H_2O)$]. Mo F, Brobak TJ, Siddiqui IR (1985) Carbohydr Res 145:13

DXCYTD. 2'-Deoxycytidine ($C_9H_{13}N_3O_4$). Young DW, Wilson HR (1975) Acta Crystallogr, Sect B 31:961

DXSORF10. 6-Deoxy-α-L-sorbofuranose ($C_6H_{12}O_5$). Rao ST, Swaminathan P, Sundaralingam M (1981) Carbohydr Res 89:151

EADIND. 9-Ethyyladenine indole complex ($C_7H_9N_5$, C_8H_7N). Kaneda T, Tanaka J (1976) Bull Chem Soc Jpn 49:1799

EADPBA. 9-Ethyladenine-parabanic acid ($C_7H_9N_5$, $C_3N_2O_3$). Shieh H-S, Voet D (1976) Acta Crystallogr, Sect B 32:2361

EBAEBH. 9-Ethyl-8-bromoadenine-9-ethyl-8-bromo-hypoxanthine ($C_7H_8BrN_5$, $C_7H_7BrN_4O$). Sakore TD, Sobell HM (1969) J Mol Biol 43:77

EBURCL10. 1-Ethyl-5-bromouracil (form I) ($C_6H_7BrN_2O_2$). Mizuno H, Fujiwara T, Tomita K (1972) 45:905

EBURCL11. 1-Ethyl-5-bromouracil (form II) ($C_6H_7BrN_2O_2$). Tsukihara T, Ashida T, Kakudo M (1972) Bull Chem Soc Jpn 45:909

EGMCYT10. 9-Ethylguanine-1-methylcytosine complex ($C_7H_9N_5O$, $C_5H_7N_3O$). O'Brien EJ (1967) Acta Crystallogr 23:92

EGUMBC. N-Ethylguanine-N-methyl-5-bromocytosine ($C_7H_9N_5O$, $C_5H_6BrN_3O$). Sobell HM, Tomita K, Rich A (1963) Acta Crystallogr 16:A79

EMURAC. 5-Ethyl-6-methyluracil ($C_7H_{10}N_2O_2$). Reeke GN jr, Marsh RE (1966) Acta Crystallogr 20:703

ETABFU. 9-Ethyladenine-1-methyl-5-fluorouracil ($C_7H_9N_5$, $C_5H_5FN_2O_2$). Tomita K, Katz L, Rich A (1967) J Mol Biol 30:545

ETGUAN. 9-Ethylguanine ($C_7H_9N_5O$). Destro R, Kistenmacher TJ, Marsh RE (1974) Acta Crystallogr, Sect B 30:79

FORMYB01. Formycin B 8-aza-9-deaza-inosine ($C_{10}H_{12}N_4O_5$). Koyama G, Nakamura H, Umezawa H, Iitaka Y (1976) Acta Crystallogr, Sect B 32:813

FRUCTO02. β-D-Fructopyranose (neutron study) ($C_6H_{12}O_6$). Takagi S, Jeffrey GA (1977) Acta Crystallogr, Sect B 33:3510

FURACL. 5-Fluorouracil ($C_4H_3FN_2O_2$). Fallon III L (1973) Acta Crystallogr, Sect B 29:2549

FUREAP. 1-Methyl-5-fluorouracil-9-ethyl-2-aminopurine complex ($C_5H_5FN_2O_2$, $C_7H_9N_5$). Mazza F, Sobell HM, Kartha G (1969) J Mol Biol 43:407

FURMCY. 5-Fluorouracil-1-methylcytosine complex ($C_4H_3FN_2O_2$, $C_5H_7N_3O$). Kim S-H, Rich A (1969) J Mol Biol 42:87

GALACT. Galactitol dulcitol ($C_6H_{14}O_6$). Berman HM, Rosenstein RD (1968) Acta Crystallogr, Sect B 24:435

GAPRHM10. 4-O-β-Galactopyranosyl-L-rhamnitol ($C_{12}H_{24}O_{10}$). Takagi S, Jeffrey GA (1977) Acta Crystallogr, Sect B 33:2377

GENTBS01. Gentiobiose β-D-Glucopyranosyl-(1-6)-β-D-glucopyranose ($C_{12}H_{22}O_{11}$). Rohrer DC, Sarko A, Bluhm TL, Lee YN (1980) Acta Crystallogr, Sect B 36:650

GLCICH01. Glycylglycine hydrochloride monohydrate (neutron diffraction) ($C_4H_9N_2O_3^+$, Cl^-, H_2O). Koetzle TF, Hamilton WC, Parthasarathy R (1972) Acta Crystallogr, Sect B 28: 2083

GLCTSM. α-D-Galactosamine-1-phosphate monohydrate ($C_6H_{14}NO_8P$, H_2O). Sundaralingam M, Fries DC (1980) Acta Crystallogr, Sect B 36:2342

GLPMAC10. Trans-O-β-D-glucopyranosyl methyl acetoacetate ($C_{11}H_{18}O_8$). Ruble J, Jeffrey GA (1974) Carbohydr Res 38:61

GLTLYR10. Glycyl-L-tyrosine dihydrate [$C_{11}H_{14}N_2O_4$, $2(H_2O)$]. Cotrait M, Bideau J-P (1974) Acta Crystallogr, Sect B 30:1024

GLUCIT01. D-Glucitol (form A, neutron study) ($C_6H_{14}O_6$). Park YJ, Jeffrey GA, Hamilton WC (1971) Acta Crystallogr, Sect B 27:2393

GLUCMH. α-D-Glucose monohydrate ($C_6H_{12}O_6$, H_2O). Killean RCG, Ferrier WG, Young DW (1962) Acta Crystallogr 15:911

GLUCMH11. α-D-Glucose monohydrate (absolute configuration) ($C_6H_{12}O_6$, H_2O). Hough E, Neidle S, Rogers D, Troughton PGH (1973) Acta Crystallogr, Sect B, 29:365

GLUCSA. α-D-Glucose (neutron study) ($C_6H_{12}O_6$). Brown GM, Levy HA (1965) Science 147:1038

GLUCSA01. α-D-Glucose (neutron study) ($C_6H_{12}O_6$). Brown GM, Levy HA (1979) Acta Crystallogr, Sect B 35:656

GLUCSE01. β-D-Glucose ($C_6H_{12}O_6$). Chu SSC, Jeffrey GA (1968) Acta Crystallogr, Sect B 24:830

GLUTAM01. L-Glutamine (neutron study) ($C_5H_{10}N_2O_3$). Koetzle TF, Frey MN, Lehmann MS, Hamilton WC (1973) Acta Crystallogr, Sect B 29:2571

GLUTAN. L-Glutamine hydrochloride ($C_5H_{11}N_2O_3^+$, Cl^-). Shamala N, Venkatesan K (1972) Cryst Struct Commun 1:227

GLYCIN. Glycine (β form) ($C_2H_5NO_2$). Iitaka Y (1960) Acta Crystallogr 13:35

GLYCIN03. Glycine (α form, neutron study) ($C_2H_5NO_2$). Jönsson P-G, Kvick A (1972) Acta Crystallogr, Sect B 28:1827

GLYCIN05. Glycine (α form, neutron study) ($C_2H_5NO_2$). Power LF, Turner KE, Moore FH (1976) Acta Crystallogr, Sect B 32:11

GLYCIN15. Glycine (γ form, neutron study) ($C_2H_5NO_2$). Kvick A, Canning WM, Koetzle TF, Williams GJB (1980) Acta Crystallogr, Sect B, 36:115

GLYCIN16. Glycine (γ form, neutron study, at 83 deg.K) ($C_2H_5NO_2$). Kvick A, Canning WM, Koetzle TF, Williams GJB (1980) Acta Crystallogr, Sect B 36:115

GLYCIN17. Glycine (α form, at 120 deg.K, deformation electron density study) ($C_2H_5NO_2$). Legros J-P, Kvick A (1980) Acta Crytallogr, Sect B 36:3052

GLYCIN18. γ-Glycine (absolute configuration) ($C_2H_5NO_2$). Shimon LJW, Lahav M, Leiserowitz L (1986) Nouv J Chim 10:723

GLYGLP. Glycylglycine phosphate monohydrate ($C_4H_9N_2O_3^+$, $H_2O_4P^-$, H_2O). Freeman GR, Hearn RA, Bugg CE (1972) Acta Crystallogr, Sect B, 28:2906

GLYGLY01. Glycylglycine (β form) ($C_4H_8N_2O_3$). Hughes EW, Moore WJ (1949) J Am Chem Soc 71:2618

GLYGLY04. Glycyclglycine (α form, at 82 K, neutron study) ($C_4H_8N_2O_3$). Kvick A, Al-Karaghouli AR, Koetzle TF (1977) Acta Crystallogr, Sect B 33:3796

GLYHCL. Glycine hydrochloride (neutron study) ($C_2H_6NO_2^+$, Cl^-). Al-Karaghouli AR, Cole FE, Lehmann MS, Miskell CF, Verbist JJ, Koetzle TF (1975) J Chem Phys 63:1360

GUANBM. Guanine hydrobromide monohydrate ($C_5H_6N_5O^+$, Br^-, H_2O). Wei CH (1977) Cryst Struct Commun 6:525

GUANCD. Guanine hydrochloride dihydrate [$C_5H_6N_5O^+$, Cl^-, $2(H_2O)$]. Iball J, Wilson HR (1965) Proc R Soc London, Ser A, 288:418

GUANMH10. Guanine monohydrate (C_5H_5O, H_2O). Thewalt U, Bugg CE, Marsh RE (1971) Acta Crystallogr, Sect B 27:2358

GUANPH01. Guanosine-5'-monophosphate trihydrate [$C_{10}H_{14}N_5O_8P$, $3(H_2O)$]. Emerson J, Sundaralingam M (1980) Acta Crystallogr, Sect B 36:1510

GUANSH10. Guanosine dihydrate [C$_{10}$H$_{13}$N$_5$O$_5$, 2(H$_2$O)]. Thewalt U, Bugg CE, Marsh RE (1970) Acta Crystallogr, Sect B, 26:1089

GUNPIC10. Guanine picrate monohydrate (C$_5$H$_6$N$_5$O$^+$, C$_6$H$_2$N$_3$O$_7^-$, H$_2$O). Bugg CE, Thewalt U (1975) Acta Crystallogr, Sect B 31:121

HBXTCT. D,L-2-(α-Hydroxybenzyl)-oxythiaminechloride hydrochloride trihydrate [C$_{19}$H$_{23}$N$_3$O$_3$S^{++}, 2(Cl$^-$), 3(H$_2$O)]. Shin W, Pletcher J, Sax M (1979) J Am Chem Soc 101:4365

HBZTAM10. D,L-2-(α-Hydroxybenzyl)thiamine chloride hydrochloride trihydrate [C$_{19}$H$_{24}$N$_4$O$_2$S^{++}, 2(Cl$^-$), 3(H$_2$O)]. Pletcher J, Sax M, Blank G, Wood M (1977) J Am Chem Soc 99:1396

HDTURD10. 5,6-Dihydro-2,4-dithiouridine (C$_9$N$_{14}$N$_2$O$_4$S$_2$). Kojic-Prodic B, Kvick A, Ruzic-Toros Z (1976) Acta Crystallogr, Sect B 32:1090

HIPADS. 8-(α-Hydroxyisopropyl)-adenosine dihydrate [C$_{13}$H$_{19}$N$_5$O$_5$, 2(H$_2$O)]. Birnbaum GI, Shugar D (1978) Biochim Biophys Acta 517:500

HIPPAC02. Hippuric acid (neutron study) n-benzoyl-glycine (C$_9$H$_9$NO$_3$). Currie M, MacDonald AL (1974) +J Chem Soc, Perkin 2, 784

HISTCM12. L-Histidine hydrochloride monohydrate (neutron study) (C$_6$H$_{10}$N$_3$O$_2^+$, Cl$^-$, H$_2$O). Fuess H, Hohlwein D, Mason SA (1977) Acta Crystallogr, Sect B 33:654

HISTDC10. Histidine dihydrochloride [C$_6$H$_{11}$N$_3$O$_2^{++}$, 2(Cl$^-$)]. Kistenmacher TJ, Sorrell T (1974) J Cryst Mol Struct 4:419

HOPROL12. 4-Hydroxy-L-proline (neutron study) (C$_5$H$_9$NO$_3$). Koetzle TF, Lehmann MS, Hamilton WC (1973) Acta Crystallogr, Sect B 29:231

HXURID. 5-Hydroxyuridine (C$_9$H$_{12}$N$_2$O$_7$). Thewalt U, Bugg CE (1973) Acta Crystallogr, Sect B 29:1393

ICYTIN. Isocytosine (C$_4$H$_5$N$_3$O). Sharma BD, McConnell JF (1965) Acta Crystallogr, 19:797

IDITOL. D-Iditol (C$_6$H$_{14}$O$_6$). Azarnia N, Jeffrey GA, Shen MS (1972) Acta Crystallogr, Sect B 28:1007

IMATUL. Isomaltulose monohydrate (C$_{12}$H$_{22}$O$_{11}$, H$_2$O). Dreissig W, Luger P (1973) Acta Crystallogr, Sect B 29:514

INOSIN10. Inosine (monoclinic form) (C$_{10}$H$_{12}$N$_4$O$_5$), Munns ARI, Tollin P (1970) Acta Crystallogr, Sect B 26:1101

INOSIN11. Inosine (orthorhombic form) (C$_{10}$H$_{12}$N$_4$O$_5$). Subramanian E (1979) Cryst Struct Commun 8:777

INOSND01. Inosine dihydrate [C$_{10}$H$_{12}$N$_4$O$_5$, 2(H$_2$O)]. Munns ARI, Tollin P, Wilson HR, Young DW (1970) Acta Crystallogr, Sect B 26:1114

INOSND10. Inosine dihydrate [C$_{10}$H$_{12}$N$_4$O$_5$, 2(H$_2$O)]. Thewalt U, Bugg CE, Marsh RE (1970) Acta Crystallogr, Sect B, 26:1089

KDGLUM01. Potassium D-Gluconate monohydrate (form A, neutron study) (C$_6$H$_{11}$O$_7^-$, K$^+$, H$_2$O). Panagiotopoulos NC, Jeffrey GA, La Placa SJ, Hamilton WC (1974) Acta Crystallogr, Sect B 30, 1421

KDGLUM02. Potassium D-gluconate monohydrate (form B, neutron study) (C$_6$H$_{11}$O$_7^-$, K$^+$, H$_2$O). Panagiotopoulos, NC, Jeffrey GA, La Placa SJ, Hamilton WC (1974) Acta Crystallogr, Sect B 30:1421

KESTOS. 1-Kestose (C$_{18}$H$_{32}$O$_{16}$). Jeffrey GA, Park YJ (1972) Acta Crystallogr, Sect B 28:257

KGULAM. 2-Keto-L-gulonic acid monohydrate (C$_6$H$_{10}$O$_7$, H$_2$O). Hvoslef J, Bergen B (1975) Acta Crystallogr, Sect B 31:697

KTHYMT. Potassium thyminate trihydrate [C$_5$H$_5$N$_2$O$_2^-$, K$^+$, 3(H$_2$O)]. Lock CJL, Pilon P, Lippert B (1979) Acta Crystallogr, Sect B 35:2533

LACTOS10. 4-O-β-D-Galactopyranosyl-α-D-glucopyranose monohydrate, α-lactose monohydrate (C$_{12}$H$_{22}$O$_{11}$, H$_2$O). Fries DC, Rao ST, Sundaralingam M (1971) Acta Crystallogr 27:994

LALNIN12. L-Alanine (neutron study) (C$_3$H$_7$NO$_2$). Lehmann MS, Koetzle TF, Hamilton WC (1972) J Am Chem Soc 94:2657

LAMBIO. Laminarabinose hydrate O-β-D-glucopyranosyl-(1-3)-β-D-glucopyranose hydrate [$C_{12}H_{22}O_{11}$, 0.19(H_2O)]. Takeda H, Yasuoka N, Kasai N (1977) Carbohydr Res 53:137

LARGIN. L-Arginine hydrochloride ($C_6H_{15}N_4O_2^+$, Cl^-). Mazumdar SK, Venkatesan K, Mez H-C, Donohue J (1969) Z Kristallogr 130:328

LARGPH01. L-Arginine phosphate monohydrate ($C_6H_{15}N_4O_2^+$, $H_2O_4P^-$, H_2O). Saenger W, Wagner KG (1972) Acta Crystallogr, Sect B 28:2237

LASPRT. L-Aspartic acid ($C_4H_7NO_4$). Derissen JL, Endeman HJ, Peerdeman AF (1968) Acta Crystallogr, Sect B 24:1349

LCYSTI10. L-Cystine (hexagonal form) ($C_6H_{12}N_2O_4S_2$). Oughton BM, Harrison PM (1959) Acta Crystallogr 12:396

LCYSTI11. L-Cystine (tetragonal form) ($C_6H_{12}N_2O_4S_2$). Chaney MO, Steinrauf LK (1974) Acta Crystallogr, Sect B 30:711

LCYSTN. L-Cysteine (monoclinic form) ($C_3H_7NO_2S$). Harding MM, Long HA (1968) Acta Crystallogr, Sect B 24:1096

LCYSTN12. L-Cysteine (orthorhombic form, neutron study) ($C_3H_7NO_2S$). Kerr KA, Ashmore JP, Koetzle TF (1975) Acta Crystallogr, Sect B 31:2022

LCYSTN21. L-Cysteine (orthorhombic form) ($C_3H_7NO_2S$). Kerr KA, Ashmore JP (1973) Acta Crystallogr, Sect B 29:2124

LEUCIN. L-Leucine ($C_6H_{13}NO_2$). Harding MM, Howieson RM (1976) Acta Crystallogr, Sect B 32:633

LEUCIN01. L-Leucine ($C_6H_{13}NO_2$). Coll M, Solans X, Font-Altaba M, Subirana JA (1986) Acta Crystallogr, C (Cr Str Comm) 42:599

LGLUAC03. L-Glutamic acid (α form, neutron study) ($C_5H_9NO_4$). Lehmann MS, Nunes AC (1980) Acta Crystallogr, Sect B 36:1621

LGLUAC11. L-Glutamic acid (β form, neutron study) ($C_5H_9NO_4$). Lehmann MS, Koetzle TF, Hamilton WC (1972) J Cryst Mol Struct 2:225

LGLUCA. Calcium di-L-glutamate tetrahydrate [2($C_5H_8NO_4^-$, Ca^{++}, 4(H_2O)]. Einspahr H, Bugg CE (1979) Acta Crystallogr, Sect B 35:316

LGLUTA. L-Glutamic acid hydrochloride (neutron study) ($C_5H_{10}NO_4^+$, Cl^-). Sequeira A, Rajagopal H, Chidambaram R (1972) Acta Crystallogr, Sect B 28:2514

LHISTD02. L-(+)-Histidine (monoclinic form, data set of Harding and Hoy) ($C_6H_9N_3O_2$). Madden JJ, McGandy EL, Seeman NC, Harding MM, Hoy A (1972) Acta Crystallogr, Sect B 28:2382

LHISTD13. L-Histidine (orthorhombic form, neutron study) ($C_6H_9N_3O_2$). Lehmann MS, Koetzle TF, Hamilton WC (1972) Int J Pept Protein Res 4:229

LILEUC10. L-Isoleucine hydrochloride monohydrate (absolute configuration) ($C_6H_{14}NO_2^+$, Cl^-, H_2O). Varughese KI, Srinivasan R (1976) Pramana, 6:189

LISLEU. L-Isoleucine ($C_6H_{13}NO_2$). Torii K, Iitaka Y (1971) Acta Crystallogr, Sect B 27:2237

LMETON10. L-Methionine ($C_5H_{11}NO_2S$). Torii K, Iitaka Y (1973) Acta Crystallogr, Sect B 29:2799

LSERIN01. L-(−)-Serine ($C_3H_7NO_3$). Kistenmacher TJ, Rand GA, Marsh RE (1974) Acta Crystallogr, Sect B 30:2573

LSERMH10. L-Serine monohydrate (neutron study) ($C_3H_7NO_3$, H_2O). Frey MN, Lehmann MS, Koetzle TF, Hamilton WC (1973) Acta Crystallogr, Sect B 29:876

LTHREO02. L-Threonine (neutron study) ($C_4H_9NO_3$). Ramanadham M, Sikka SK, Chidambaram R (1973) Pramana 1:247

LTYRHC10. L-Tyrosine hydrochloride (neutron study) ($C_9H_{12}NO_3^+$, Cl^-). Frey MN, Koetzle TF, Lehmann MS, Hamilton WC (1973) J Chem Phys 58:2547

LTYROS11. L-Tyrosine (neutron study) ($C_9H_{11}NO_3$). Frey MN, Koetzle TF, Lehmann MS, Hamilton WC (1973) J Chem Phys 58:2547

LVALIN. L-Valine ($C_5H_{11}NO_2$). Torii K, Iitaka Y (1970) Acta Crystallogr, Sect B 26:1317

LYSASP. L-Lysine L-aspartate ($C_6H_{15}N_2O_2^+$, $C_4H_6NO_4^-$). Bhat TN, Vijayan M (1976) Acta Crystallogr, Sect B 32:891

LYSCLH. L-Lysine monohydrochloride dihydrate [$C_6H_{15}N_2O_2^+$, Cl^-, 2(H_2O)]. Wright DA, Marsh RE (1962) Acta Crystallogr 15:54

LYSCLH02. L-Lysine monohydrochloride dihydrate (neutron study) [$C_6H_{15}N_2O_2^+$, Cl^-, 2(H_2O)]. Bugayong RR, Sequeira A, Chidambaram R (1972) Acta Crystallogr, Sect B, 28:3214

LYSCLH11. L-Lysine monohydrochloride dihydrate (neutron study) [$C_6H_{15}N_2O_2^+$, Cl^-, $2(H_2O)$]. Koetzle TF, Lehmann MS, Verbist JJ, Hamilton WC (1972) Acta Crystallogr, Sect B 28:3207

LYXOSE01. β-L-Lyxopyranose (neutron study) ($C_5H_{10}O_5$). Nordenson S, Takagi S, Jeffrey GA (1978) Acta Crystallogr, Sect B 34:3809

MACPTL. 7-(Methyl-2-acetamido-2,3,4-trideoxy-α-D-erythro-hex-2-enopyranosid-4-yl)-theophylline monohydrate ($C_{16}H_{21}N_5O_6$, H_2O). Ruzic-Toros Z, Kojic-Prodic B, Coffou E (1981) Acta Crystallogr, Sect B 37:877

MADEND. 1,9-Dimethyl-adeninium chloride ($C_7H_{10}N_5^+$, Cl^-). Chiang CC, Epps LA, Marzilli LG, Kistenmacher TJ (1979) Acta Crystallogr, Sect B 35:2237

MALARA10. Methyl α-L-Arabinopyranoside ($C_6H_{12}O_5$). Takagi S, Jeffrey GA (1978) Acta Crystallogr, Sect B 34:1591

MALTOS10. β-Maltose monohydrate ($C_{12}H_{22}O_{11}$, H_2O). Quigley GJ, Sarko A, Marchessault RH (1970) J Am Chem Soc 92:5834

MALTOS11. β-Maltose monohydrate (neutron study) ($C_{12}H_{22}O_{11}$, H_2O). Gress ME, Jeffrey GA (1977) Acta Crystallogr, Sect B 33:2490

MALTOT. α-Maltose ($C_{12}H_{22}O_{11}$). Takusagawa F, Jacobson RA (1978) Acta Crystallogr, Sect B 34:213

MALTPY. Methyl α-D-altropyranoside ($C_7H_{14}O_6$). Gatehouse BM, Poppleton BJ (1971) Acta Crystallogr, Sect B 27:871

MALTPY01. Methyl α-D-altropyranoside (neutron study) ($C_7H_{14}O_6$). Poppleton BJ, Jeffrey GA, Williams GJB (1975) Acta Crystallogr, Sect B 31:2400

MANGAL. Methyl 3,6-anhydro-α-D-galactoside ($C_7H_{12}O_5$). Campbell JW, Harding MM (1972) Chem Soc. Perkin 2, 1721

MARAFC. 3'-O-Methyl-1-β-D-arabinofuranosylcytosine ($C_{10}H_{15}N_3O_5$). Birnbaum GI, Darzynkiewicz E, Shugar D (1975) J Am Chem Soc 97:5904

MBDGAL. Methyl β-D-galactopyranoside ($C_7H_{14}O_6$). Sheldrick B (1977) Acta Crystallogr, Sect B 33:3003

MBDGAL02. Methyl β-D-galactopyranoside (neutron study) ($C_7H_{14}O_6$). Takagi S, Jeffrey GA (1979) Acta Crystallogr, Sect B 35:902

MBDGPH10. Methyl β-D-glucopyranoside hemihydrate [$C_7H_{14}O_6$, $0.5(H_2O)$]. Jeffrey GA, Takagi S (1977) Acta Crystallogr, Sect B 33:738

MBDRIP. Methyl 4,6-O-benzylidene-2-deoxy-3-O-[(methylthio)-thiocarbonyl)-α-D- ribopyranoside ($C_{16}H_{20}O_5S_2$). Luger P, Elvers B, Paulsen H (1979) Chem Ber 112:3855

MBLARA10. Methyl β-L-arabinopyranoside ($C_6H_{12}O_5$). Takagi S, Jeffrey GA (1978) Acta Crystallogr, Sect B 34:1591

MBUMAD10. 1-Methyl-5-bromouracil-9-methyladenine ($C_5H_5BrN_2O_2$, $C_6H_7N_5$). Kondrashev YuD (1966) Zh Strukt Khim 7:399

MBURAC. 1-Methyl-5-bromouracil ($C_5H_5BrN_2O_2$). Mizuno H, Morita K, Fujiwara T, Tomita K-I (1972) Chem Lett :965

MCELOB. Methyl β-cellobioside methanol solvate (at −193 deg.C) ($C_{13}H_{24}O_{11}$, CH_4O). Ham JT, Williams DG (1970) Acta Crystallogr, Sect B 26:1373

MCYTIM10. Bis(1-methylcytosine) hydroiodide monohydrate ($C_5H_7N_3O$, $C_5H_8N_3O^+$, I^-, H_2O). Kistenmacher TJ, Rossi M, Caradonna JP, Marzilli LG (1979) Adv Mol Relax Int Proc 15:119

MDHURC10. 6-Methyl-5,6-dihydrouracil ($C_5H_8N_2O_2$). Kou WWH, Parthasarathy R (1977) Acta Crystallogr, Sect B 33:934

MDRIBP02. Methyl β-D-ribopyranoside (neutron study) ($C_6H_{12}O_5$). James VJ, Stevens JD, Moore FH (1978) Acta Crystallogr, Sect B 34:188

MDTRPY20. Methyl 1,5-dithio-α-D-ribopyranoside hydrate [$C_6H_{12}O_3S_2$, $0.25(H_2O)$]. Girling RL, Jeffrey GA (1974) Acta Crystallogr, Sect B 30:327

MEADEN02. 9-Methyl-adenine (at 126 deg.K, neutron study)($C_6H_7N_5$). McMullan RK, Benci P, Craven BM (1980) Acta Crystallogr, Sect B 36:1424

MECTSI. Bis(1-methylcytosine)bis(1-methylcytosinium) hexafluorosilicate dihydrate [$2(C_5H_8N_3O^+)$, $2(C_5H_7N_3O)$, F_6Si^{--}, $2(H_2O)$]). Kistenmacher TJ, Rossi M, Chiang CC, Caradonna JP, Marzilli LG (1980) Adv Mol Relax Int Proc 17:113

MECYTO10. 1-Methylcytosine hydrochloride ($C_5H_8N_3O^+$, Cl^-). Trus BL, Marsh RE (1972) Acta Crystallogr, Sect B 28:1834

MECYTS01. 5-Methylcytosine hemihydrate [$C_5H_7N_3O$, 0.5(H_2O)]. Grainger CT, Bailey D (1981) Acta Crystallogr, Sect B 37:1561

MEDOUR. 6-Methyl-2'-deoxyuridine ($C_{10}H_{14}N_2O_5$). Birnbaum GI, Hruska FE, Niemczura WP (1980) J Am Chem Soc 102:5586

MELEZT. Melezitose monohydrate (form II) ($C_{18}H_{32}O_{16}$, H_2O). Hirotsu K, Shimada A (1973) Chem Lett 83

MELEZT02. O-α-D-glucopyranosyl-(1-3)-β-D-fructofuranosyl-α-D-glucopyranoside monohydrate (form II) melezitose monohydrate ($C_{18}H_{32}O_{16}$, H_2O). Becquart J, Neuman A, Gillier-Pandraud H (1982) Carbohydr Res 111:9

MELIBM01. α,β-Melibiose monohydrate 6-O-α-D-galactopyranosyl-D-glucosemonohydrate ($C_{12}H_{22}O_{11}$, H_2O). Hirotsu K, Higuchi T (1976) Bull Chem Soc Jpn 49:1240

MELIBM02. α,β-Melibiose monohydrate 6-O-α-D-galactopyranosyl-D-glucose monohydrate ($C_{12}H_{22}O_{11}$, H_2O). Kanters JA, Roelofsen G, Doesburg HM, Koops T (1976) Acta Crystallogr, Sect B 32:2830

MELIBM03. α-Melibiose monohydrate 6-O-α-D-galactopyranosyl-D-glucose monohydrate ($C_{12}H_{22}O_{11}$, H_2O). Neuman A, Gillier-Pandraud H (1976) CR Acad Sci Ser C 283:667

MELIBM10. 6-O-Galactopyranosyl-(α,β)-glucopyranose monohydrate α,β-melibiose monohydrate ($C_{12}H_{22}O_{11}$, H_2O). Gress ME, Jeffrey GA, Rohrer DC (1978) Acta Crystallogr, Sect B 34:508

MEMANP. Methyl α-D-Mannopyranoside ($C_7H_{14}O_6$). Gatehouse BM, Poppleton BJ (1970) Acta Crystallogr, Sect B 26:1761

MEMANP11. Methyl-D-mannopyranoside (neutron study) ($C_7H_{14}O_6$). Jeffrey GA, McMullan RK, Takagi S (1977) Acta Crystallogr, Sect B 33:728

MERPUM. 6-Mercaptopurine monohydrate (immunosuppressant drug) ($C_5H_4N_4S$, H_2O). Sletten, E., Sletten J, Jensen LH (1969) Acta Crystallogr, Sect B 25:1330

METCYT01. 1-Methylcytosine ($C_5H_7N_3O$). Rossi M, Kistenmacher TJ (1977) Acta Crystallogr, Sect B 33:3962

METHCL. L-Methionine hydrochloride ($C_5H_{12}NO_2S^+$, Cl^-). Di Blasio B, Pavone V, Pedone C (1977) Cryst Struct Commun 6:845

METHYM01. 1-Methylthymine (neutron study) ($C_6H_8N_2O_2$). Kvick A, Koetzle TF, Thomas R (1974) J Chem Phys 61:2711

METRBP10. Methyl 5-thio-α-D-ribopyranoside (absolute configuration) ($C_6H_{12}O_4S$). Girling RL, Jeffrey GA (1973) Acta Crystallogr, Sect B 29:1102

MEURID. 5-Methyluridine hemihydrate [$C_{10}H_{14}N_2O_6$, 0.5(H_2O)]. Hunt DJ, Subramanian E (1969) Acta Crystallogr, Sect B 25:2144

MEYRID. 6-Methyluridine ($C_{10}H_{14}N_2O_6$). Suck D, Saenger W (1972) J Am Chem Soc 94:6520

MFRMYC. 2-Methyl-Formycin ($C_{11}H_{15}N_5O_4$). Abola JE, Sims MJ, Abraham DJ, Lewis AF, Townsend LB (1974) J Med Chem 17:62

MFXHUR. 1-Methyl-5-fluoro-6-methoxy-5,6-dihydrouracil ($C_6H_9FN_2O_3$). James MNG, Matsushima (1976) Acta Crystallogr, Sect B 32:957

MGALPY. Methyl α-D-galactopyranoside monohydrate ($C_7H_{14}O_6$, H_2O). Gatehouse BM, Poppleton BJ (1971) Acta Crystallogr, Sect B 27:654

MGALPY01. Methyl α-D-galactopyranoside monohydrate (neutron study) ($C_7H_{14}O_6$, H_2O). Takagi S, Jeffrey GA (1979) Acta Crystallogr, Sect B 35:902

MGLUCP. Methyl α-D-glucopyranoside ($C_7H_{14}O_6$). Berman HW, Kim SH (1968) Acta Crystallogr, Sect B 24:897

MGLUCP11. Methyl α-D-glucopyranoside (neutron study) ($C_7H_{14}O_6$). Jeffrey GA, McMullan RK, Takagi S (1977) Acta Crystallogr, Sect B 33:728

MHCYTC. 1-Methyl-N-4-hydroxy-cytosine hydrochloride ($C_5H_8N_3O_2^+$, Cl^-). Birnbaum GI, Kulikowski T, Schugar D (1979) Can J Biochem 57:308

MHURAC. 1-Methyl-5,6-dihydrouracil ($C_5H_8N_2O_2$). Gallezot P, Nofre C (1968) Bull Soc Chim Fr :4057

MMALTS. Methyl β-maltoside monohydrate ($C_{13}H_{24}O_{11}$, H_2O). Chu SSC, Jeffrey GA (1967) Acta Crystallogr 23:1038

MMCPUR. 6-Methylmercaptopurine trihydrate [$C_6H_6N_4S$, $3(H_2O)$]. Cook WJ, Bugg CE (1975) J Pharm Sci 64:221

MOROTD10. Methyl orotate trans-syn-dimer dihydrate [$C_{12}H_{12}N_4O_8$, $2(H_2O)$]. Birnbaum GI (1972) Acta Crystallogr, Sect B 28:1248

MRFBZI10. 5,6-Dimethyl-1-(α-D-ribofuranosyl)-benzimidazole ($C_{14}H_{18}N_2O_4$). Ruzic-Toros Z (1979) Acta Crystallogr, Sect B 35:1277

MTHMAD. 9-Methyladenine-1-methylthymine complex ($C_6H_7N_5$, $C_6H_8N_2O_2$). Hoogsteen K (1963) Acta Crystallogr, 16:907

MTHYMD. 1-Methylthymine trans-anti dimer ($C_{12}H_{16}N_4O_4$). Einstein JR, Hosszu JL, Longworth JW, Rahn RO, Wei CH (1967) Chem Commun :1063

MTURAC. 1-Methyl-4-thiouracil ($C_5H_6N_2OS$). Hawkinson SW (1975) Acta Crystallogr, Sect B 31:2153

MTRIBP10. Methyl 1-thio-α-D-ribopyranoside (absolute configuration) ($C_6H_{12}O_4S$). Girling RL, Jeffrey GA (1973) Acta Crystallogr, Sect B 29:1006

MURCAC. 6-Methyl-uracil-5-acetic acid ($C_7H_8N_2O_4$). Destro R, Marsh RE (1972) Acta Crystallogr, Sect B 28:2971

MXFGPY. Methyl 4-deoxy-4-fluoro-α-D-glucopyranoside ($C_7H_{13}FO_5$). Choong W, Stephenson NC, Stevens JD (1975) Cryst Struct Commun 4:491

MXLPYR. Methyl α-D-xylopyranoside ($C_6H_{12}O_5$). Takagi S, Jeffrey GA (1978) Acta Crystallogr, Sect B 34:3104

MXURID01. 5-Methoxyuridine ($C_{10}H_{14}N_2O_7$). Birnbaum GI, Blonski WJP, Hruska FE (1983) Can J Chem 61:2299

NACMAN 10. N-Acetyl-β-D-mannosamine ($C_8H_{15}NO_6$). Neuman A, Gillier-Pandraud H, Longchambon F (1975) Acta Crystallogr, Sect B 31:2628

NIMURC10. 5-Nitro-6-methyluracil ($C_5H_5N_3O_4$). Parthasarathy R, Srikrishnan T (1977) Acta Crystallogr, Sect B 33:1749

NRURAM11. 5-Nitro-1-(β-D-ribosyluronic acid)-uracil monohydrate (neutron study, at 80 deg.K) 1-(5-nitro-2,4-dioxopyrimidinyl)-β-D-ribofuranoic acid monhydrate ($C_9H_9N_3O_9$, H_2O). Takusagawa F, Koetzle TF, Srikrishnan T, Parthasarathy R (1979) Acta Crystallogr, Sect B 35:1388

NRURAM20. 5-Nitro-1-(β-D-ribosyluronic acid)-uracil monohydrate ($C_9H_9N_3O_9$, H_2O). Srikrishnan T, Parthasarathy R (1978) Acta Crystallogr, Sect B 34:1363

OMCYTD20. 2'-O-Methylcytidine ($C_{10}H_{15}N_3O_5$). Hingerty B, Bond PJ, Langridge R, Rottman F (1977) Acta Crystallogr, Sect 33:1349

OXOFMB. Oxoformycin B ($C_{10}H_{12}N_4O_6$). Koyama G, Nakamura H, Umezawa H, Iitaka Y (1976) Acta Crystallogr, Sect B 32:813

PHALNC01. L-Phenylalanine hydrochloride (neutron study) ($C_9H_{12}NO_2^+$, Cl^-). Al-Karaghouli AR, Koetzle TF (1975) Acta Crystallogr, Sect B 31:2461

PLANTE10. Planteose dihydrate α-D-galactopyranosil-(1,6)-β-D-fructofuranosil-(2,1)-α-D-glucopyranoside dihydrate $C_{18}H_{32}O_{16}$, $2(H_2O)$. Rohrer DC (1972) Acta Crystallogr, Sect B 28:425

PMTADN10. N-6-(C_δ-2-Isopentenyl)-2-methylthioadenine ($C_{11}H_{15}N_5S$). McMullan RK, Sundaralingam M (1971) J Am Chem Soc 93:7050

PROLIN. L-Proline ($C_5H_9NO_2$). Kayushina RL, Vainshtein BK (1965) Kristallografiya 10:833

PROLNH. DL-Proline hemihydrochloride ($C_5H_{10}NO_2^+$, $C_5H_9NO_2$, Cl^-). Swaminathan S, Chacko KK (1981) Cryst Struct Commun 10:469

PRSARH. L-Prolyl-sarcosine monohydrate ($C_8H_{14}N_2O_3$, H_2O). Kojima T, Kido T, Itoh H, Yamane T, Ashida T (1980) Acta Crystallogr, Sect B 36:326

PSCYTD. Pseudo-isocytidine hydrochloride ($C_9H_{14}N_3O_5^+$, Cl^-). Birnbaum GI, Watanabe KA, Fox JJ (1980) Can J Chem 58:1633

PUCGLR10. N-6-(N-Glycylcarbonyl)adenosine ($C_{13}H_{16}N_6O_7$). Parthasarathy R, Ohrt JM, Chheda GB (1977) Biochemistry 16:4999

PYMDSD. Bis-pyrimidyl-2,2'-disulfide dihydrate [$C_8H_6N_4S_2$, $2(H_2O)$]. Furberg S, Solbakk J (1973) Acta Chem Scand 27:2536

QQQBTP01. DL-Tryptophan (monoclinic form) ($C_{11}H_{12}N_2O_2$). Bakke O, Mostad A (1980) Acta Chem Scand, Ser B 34:559

RAFINO. Raffinose pentahydrate [$C_{18}H_{32}O_{16}$, $(5(H_2O)$]. Berman HM (1970) Acta Crystallogr, Sect B 26:290

RBFORM. Rubidium 5-fluoro-orotate monohydrate ($C_5H_2FN_2O_4^-$, RB^+, H_2O). Macintyre WM, Zirakzadeh M (1964) Acta Crystallogr 17:1305

RBFROX. 3-Amino-6-(β-D-ribofuranosyl)-6H-1,2,6-thiadiazine-1,1-dioxide ($C_8H_{13}N_3O_6S$). Su TL, Bennua B, Vorbruggen H, Lindner HJ (1981) Chem Ber 114:1269

RFURPD. 3-Deaza-4-deoxyuridine 1-β-D-ribofuranosyl-2-pyridone ($C_{10}H_{13}NO_5$). Egert E, Lindner HJ, Hillen W, Gassen HG (1977) Acta Crystallogr, Sect B 33:3704

RHAMAH12. α-L-Rhamnose monohydrate (neutron study) ($C_6H_{12}O_5$, H_2O). Takagi S, Jeffrey GA (1978) Acta Crystallogr, Sect B 34:2552

RIBTOL. Ribitol ($C_5H_{12}O_5$). Kim HS, Jeffrey GA, Rosenstein RD (1969) Acta Crystallogr, Sect B 25:2223

RPPYPY20. 6-Amino-10-(β-D-ribofuranosylamino)pyrimido(5,4-D)pyrimidine ($C_{11}H_{14}N_6O_4$). Narayanan P, Berman HM (1975) Carbohydr Res 44:169

SOPROS. O-β-D-Glucosyl-1(1-2)-α-D-glucose monohydrate sophorose monohydrate ($C_{12}H_{22}O_{11}$, H_2O). Ohanessian J, Longchambon F, Arene F (1978) Acta Crystallogr, Sect B 34:3666

SORBOL01. α-L-Sorbopyranose (neutron study) ($C_6H_{12}O_6$). Nordenson S, Takagi S, Jeffrey GA (1979) Acta Crystallogr, Sect B 35:1005

STACHY10. O-α-D-Galactopyranosyl-(1-6)-O-α-D-galactopyranosyl-(1-6)-O-α-D-glucopyranosyl-(1-2)-α-D-fructofuranoside pentahydrate stachyose pentahydrate [$C_{24}H_{42}O_2$, $5(H_2O)$]. Gilardi R, Flippen-Anderson JL (1987) Acta Cryst C (Cr Str Comm) 43:806

SUCROS04 Sucrose (neutron study, refinement) ($C_{12}H_{22}O_{11}$). Brown GM, Levy HA (1973) Acta Crystallogr, Sect B 29:790

SURMAD01. 1-Methyl-4-thiouracil-9-methyladenine (further discussion) ($C_5H_6N_2OS$, $C_6H_7N_5$). Saenger W, Suck D (1973) Eur J Biochem 32:473

SURMAD10. 1-Methyl-4-thiouracil-9-methyladenine ($C_5H_6N_2OS$, $C_6H_7N_5$). Saenger W, Suck D (1971) J Mol Biol 60:87

TAGTOS. α-D-Tagatose ($C_6H_{12}O_6$). Takagi S, Rosenstein RD (1969) Carbohydr Res 11:156

TCYPIC. 2-Thiocytosine picrate ($C_4H_6N_3S^+$, $C_6H_2N_3O_7^-$. Delucas LJ, Hearn RA, Bugg CE (1977) Acta Crystallogr, Sect B 33:2611

TCYTDH. 2-Thiocytidine dihydrate [$C_9H_{13}N_3O_4S$, $2(2H_2O)$]. Lin GH-Y, Sundaralingam M, Arora SK (1971) J Am Chem Soc 93:1235

TGLYSU11. Triglycine sulfate (ferroelectric form, neutron study, structure A) [$C_2H_5NO_2$, $2(C_2H_6NO_2^+)$, O_4S^{--}. Kay MI, Kleinberg R (1973) Ferroelectrics 5:45

TGUANS10. 6-Thioguanosine monohydrate ($C_{10}H_{13}N_5O_4S$, H_2O). Thewalt U, Bugg CE (1972) J Am Chem Soc 94:8892

THCYTO10. Thiocytosine ($C_4H_5N_3S$). Furberg S, Jensen LH (1970) Acta Crystallogr, Sect B 26:1260

THGUAN10. 6-Thioguanine ($C_5H_5N_5S$). Bugg CE, Thewalt U (1970) J Am Chem Soc 92:7441

THPROL. Thiamine picrolonate dihydrate thiamine 3-methyl-4-nitro-1-(p-nitro-phenyl)-2-pyrazolin-5-onate dihydrate [$C_{12}H_{17}N_4OS^+$, $C_{10}H_7N_4O_5^-$, $2(H_2O)$]. Shin W, Pletcher J, Blank G, Sax M (1977) J Am Chem Soc 99:3491

THYBNZ10. Thymine-p-benzoquinone complex ($C_5H_6N_2O_2$, $C_6H_4O_2$). Sakurai T, Okunuki M (1971) Acta Crystallogr, Sect B 27:1445

THYMIN. Thymine ($C_5H_6N_2O_2$). Ozeki K, Sakabe N, Tanaka J (1969) Acta Crystallogr, Sect B 25:1038

THYDIN. Thymidine ($C_{10}H_{14}N_2O_5$). Young DW, Tollin P, Wilson HR (1969) Acta Crystallogr, Sect B 25:1423

THYMMH. Thymine monohydrate ($C_5H_6N_2O_2$, H_2O). Gerdil R (1961) Acta Crystallogr 14:333

TOADEN. 8-Thioxo-adenosine monohydrate ($C_{10}H_{13}N_5O_4S$, H_2O). Mizuno H, Kitamura K, Miyao A, Yamagata Y, Wakahara A, Tomita K, Ikehara M (1980) Acta Crystallogr, Sect B 36:902

TPATAA. Tryptaminium 1-thyminyl-acetate ($C_{10}H_{13}N_2^+$, $C_7H_7N_2O_4^-$). Ishida T, Inoue M, Tomita K (1979) Acta Crystallogr, Sect B 35:1642

TREHAL01. α,α-Trehalose dihydrate (data of Rohrer and Berking) [$C_{12}H_{22}O_{11}$, $2(H_2O)$]. Brown GM, Rohrer DC, Berking B, Beevers CA, Gould RO, Simpson R (1972) Acta Crystallogr, Sect B 28:3145

TRYPTC. L-Tryptophan hydrochloride ($C_{11}H_{13}N_2O_2^+$, Cl^-). Takigawa T, Ashida T, Sasada Y, Kakudo M (1966) Bull Chem Soc Jpn 39:2369

TRYPTC01. L-Tryptophan hydrochloride (neutron study) ($C_{11}H_{13}N_2O_2^+$, Cl^-). Andrews LC, Farkas R, Frey MN, Lehmann MS, Koetzle TF (1974) ACA (Spring) 61

TURANS01. O-α-D-Glucopyranosyl-(1-3)-β-D-fructopyranose turanose ($C_{12}H_{22}O_{11}$). Kanters JA, Gaykema WPJ, Roelofsen G (1978) Acta Crystallogr, Sect B 34:1978

TURIDN10. 2-Thiouridine ($C_9H_{12}N_2O_5S$). Hawkinson SW (1977) Acta Crystallogr, Sect B 33:80

TYMCXA. Thymidine 5'-carboxylic acid ($C_{10}H_{12}N_2O_6$). Suck D, Saenger W, Rohde W (1974) Biochim Biophys Acta 361:1

URACIL. Uracil ($C_4H_4N_2O_2$). Stewart RF, Jensen LH (1967) Acta Crystallogr 23:1102

URARAF10. 1-β-D-Arabinofuranosyl-uracil ($C_9H_{12}N_2O_6$). Tollin P, Wilson HR, Young DW (1973) Acta Crystallogr, Sect B 29:1641

URIDMP10. Uridine-3'-monophosphate monohydrate ($C_9H_{13}N_2O_9P$, H_2O). Srikrishnan T, Fridey SM, Parthasarathy R (1979) J Am Chem Soc 101:3739

VALEHC10. L-Valine hydrochloride ($C_5H_{12}NO_2^+$, Cl^-). Ando O, Ashida T, Sasada Y, Kakudo M (1967) Acta Crystallogr 23:172

VALEHC11. L-Valine hydrochloride (neutron study) ($C_5H_{12}NO_2^+$, Cl^-). Koetzle TF, Golic L, Lehmann MS, Verbist JJ, Hamilton WC (1974) J Chem Phys 60:4690

VALHCL10. L-Valine hydrochloride monohydrate ($C_5H_{12}NO_2^+$, Cl^-, H_2O). Rao ST (1969) Z Kristallogr 128:339

VALIDL. DL-Valine ($C_5H_{11}NO_2$). Mallikarjunan M, Rao ST (1969) Acta Crystallogr, Sect B 25:296

XANTOS. Xanthosine dihydrate [$C_{10}H_{12}N_4O_6$, $2(H_2O)$]. Koyama G, Nakamura H, Umezawa H, Iitaka Y (1976) Acta Crystallogr, Sect B 32:969

XANTOS01. Xanthosine dihydrate [$C_{10}H_{12}N_4O_6$, $2(H_2O)$]. Herceg M, Fischer J, Weiss R (1976) IZV Jug Cent Krist, Ser A 11:130

XFMANP. 2-Deoxy-2-fluoro-β-D-mannopyranose ($C_6H_{11}FO_5$). Choong W, Craig DC, Stephenson NC, Stevens JD (1975) Cryst Struct Commun 4:111

XFURCC10. 1-D-Xylofuranosyl-cytosine hydrochloride ($C_9H_{14}N_3O_5^+$, Cl^-). Post ML, Huber CP, Birnbaum GI, Shugar D (1981) Can J Chem 59:238

XYFCYT10. 1-α-D-Xylofuranosyl-cytosine ($C_9H_{13}N_3O_5$). Post ML, Huber CP, Birnbaum GI, Shugar D (1981) Can J Chem 59:238

XYLOBM01. Methyl β-D-xylopyranoside (neutron study) ($C_6H_{12}O_5$). Takagi S, Jeffrey GA (1977) Acta Crystallogr, Sect B 33:3033

XYLOSE01. α-L-Xylopyranose (neutron study I) ($C_5H_{10}O_5$). Takagi S, Jeffrey GA (1979) Acta Crystallogr, Sect B 35:1482

XYLOSE02. α-L-Xylopyranose (neutron study II) ($C_5H_{10}O_5$). Takagi S, Jeffrey GA (1979) Acta Crystallogr, Sect B 35:1482

XYLTOL. Xylitol (orthorhombic form) ($C_5H_{12}O_5$). Kim HS, Jeffrey GA (1969) Acta Crystallogr, Sect B 25:2607

Subject Index

ab-initio MO calculations 71–77
acceptor geometry 164
accessible surface 46
adenine
–, ApU 263
–, base triplet A·2U 267, 268
–, derivatives 244, 245, 278
–, hetero base pairs 260, 264, 266
– homo base pairs 249, 252, 256
–, 9-methyladenine·1-methylthymidine 261
adenosine 297
–, derivatives 298, 299
–, 3'-phosphate 300
active site, hydration 484
Albrecht 20
alditols 172–178, 186
alkanediol hydrates 449, 450
alkylamine hydrates 443–447
Allinger 30
alpha helix, see proteins α-helix
amino acids 115–117, 129, 141–145, 147, 162, 220–231, 353
– –, bent hydrogen bonds 224
– –, cyclic configuration 231
– –, functional groups 223
– –, hydrochlorides 221–222
– –, three-center bonds 225, 230
– –, zwitterion bridges 224
amino group conjugated 234
amino-imino tautomerism 42, 128, 235–236, 406–407
amyloses 200, 347
anharmonic motion 62
anhydrobiosis 425
anionic water host lattice 433, 439, 444
anomeric effect 123–125
– mixtures 179–180, 208
anticodon 398, 404–406
antidromic 38, 168, 320–323, 330–333, 439–441
antifreeze protein 431

aromatic ring acceptors 159
Astbury 3, 4, 5, 351
atom-pair interactions 87
atomic charges 88
– radii 26

Badger's rule 50
Baker 360–361, 364
barbiturates 233–234, 253
base-pairs
–, hemi-protonated 251
–, hetero 247–248, 259–268
–, homo 247–259, 410
–, *Hoogsteen* 259, 260–261, 267–269
–, mismatch 404–407
–, in nucleic acids 232–234, 410–412
–, reverse *Hoogsteen* 261
–, reverse *Watson-Crick* 260–264
–, *Watson-Crick* 7, 8, 37, 132, 247–248, 259–269, 396, 398–399, 404–409
–, wobble 265–266, 404–406
base quadruplets 266–267
– triplets 266–267
bases, modified 234
basis sets 74–76
– –, superposition 74, 76
Bence-Jones protein, hydration 480
Bernal 3, 4, 6, 427
beta turn, see proteins β-turn
bifurcated, see three-center
– dimer 83
Bjerrum defect 83
Bondi atomic radii 30
bound water in proteins 461–465
Branson 8
Brookhaven Protein Data Base 14

cage structures 313–315
Cambridge Crystallographic Data Base 12, 13, 57, 111, 160, 170, 319, 426

Subject Index

carbohydrates 167–219
–, alditols (sugar alcohols) 172–178, 186
–, anomeric effect of 123–125
–, – mixtures 179–180, 208
–, cellobiose/oside 151, 198–199
–, cellulose 216–219
–, cocrystallization 179, 180
–, configurational heterogeneity 178
–, cooperativity 172, 174, 177, 191
–, disaccharides 169–210
–, enantiomeric conformers 175
–, enantiomers (D, L), racemates 170
–, galabiose 185
–, galactopyranosylrhamnitol 209–210
–, gentiobiose 198, 205, 207
–, glucopyranosylglucitol 209–210
–, homodromic cycle 175
–, interresidue hydrogen bonds 198
–, intramolecular hydrogen bonds 176, 181–184, 198
–, isomaltulose 202–203
–, kestose 185, 210
–, lactose 198, 202–203
–, lactulose 180
–, laminarabiose 198, 208–209
–, maltohexaoside polyiodide 343–344
–, maltose/oside 135, 151, 195, 198–201
–, mannobiose 198
–, mannotriose 198, 211–212
–, melibiose 205, 207
–, melizitose 211, 213
–, orientational disorder 185
–, patterns in disaccharides 196–197
–, – in monosaccharides 188
–, planteose 185, 211, 214
–, primary OH group 184–185
–, pyranoses/osides 187–195
–, raffinose 214, 215
–, ribopyranosides 181–183
–, rules for hydrogen bonding 187
–, sophorose 207–208
–, stachyose 214
–, sucrose 195, 198, 201
–, three-center bonds 187
–, trehalose 204, 206
–, turanose 198, 204
–, vicinal hydroxyls 183–184
carboxylic acids 95, 96, 115–116
catalysis 312
cationic water host lattice 433
cellulose 216–219
central dogma in molecular biology 398
chain-stopper 125
channel structures 313–315, 319, 438–440
charge, atomic 88
– interactions 86
– transfer 87
charge-density studies 16, 28, 63–67
choline chloride hydrates 51, 439
clover-leaf structures 407–409
CNDO 84
co-crystallization 175, 179, 180
coherent neutron scattering 60
complementarity 397
computer technology 11, 13, 34, 73, 93
configurational heterogeneity 178
conformational enantiomorphism 175
– mutarotation 175
– polymorphism 175–177
conjugated amino group 234
contact areas 470
cooperativity 9, 33–37, 79–81, 88, 94, 121–125, 172–176, 186, 235, 320–322, 340–344, 352
Corey 8, 20, 351
crambin hydration 482–483
Crick 7, 8
crystal field distortions 80
– packing 313
– structure analysis 11
cut-off criteria 29–33
– – in proteins 360, 362, 475
cyclic dimers 83
cyclodextrins 11, 24, 85, 153–154, 309–346
–, cage and channel structures 313–315
–, chemical structure 310
–, hydrates 309–330
–, hydrogen bonding 315ff
–, –, cooperativity 330–344
–, –, flip-flop disorder 333–344
–, –, intramolecular 317, 332–333
–, –, theoretical studies 330–332
–, inclusion complexes 309, 312
–, induced fit 312, 332–335
–, intramolecular hydrogen bonds 317
–, physical parameters 312
cytidine 290
–, derivatives 292–295
–, phosphates 280, 296, 297
cytosine 240
–, derivatives 241–244, 256, 257
–, hemi-protonated base pair 251
–, hetero base pairs 266
–, homo base pairs 249, 251, 255, 256
–, isocytosine 257
–, 1-methylcytosine·5-fluorouracil 267

Debye temperature 83
deformation density 5–6, 66, 98
deuteration 60, 69, 386–390
diaxial interactions 315
dihedral angles 185

Subject Index

diketones 37
dipole interactions 86
– moments 64, 253, 344
direct methods 11
disaccharides, see carbohydrates
disorder 40–42, 53, 114, 185, 333–344
dispersion forces 87
DNA 7, 8, 396–422, 480, 487–488
– A, B, Z 399–402
–, crystal forms 489
–, helical parameters 402
–, hydration 487–509
–, polymorphism 401
–, protein interactions 415–422
DNase I 422
Donohue 22
double channel host lattice 439
dynamics of hydration 505–510

*E*dgeworth expansion 63
electric field gradient 64
– – –, correlation 74
– – potential 64
electron density deformation 65–66, 98
– – distribution 16, 90
– diffraction 52
electronegativity 15
electrostatic potentials 63, 66, 89
– properties 19, 64
empirical force fields 71, 85–93
energy partitioning 84
Evans 478
excluded region 43–44, 429
exo-clathrates 440
extinction 55, 56

fibre diffraction, see X-ray
flickering clusters 428
flip-flop disorder 40–42, 333–344
formamide 65, 96–98
four-center bonds 23, 42, 137, 145–146, 202
Fowler 3, 427
Frank 427–428, 478
Franklin 8

gamma turns 378
gas electron diffraction 50
– hydrates 432–435
gas-phase studies 51, 95
Gaussian M.O. method 74, 75
genetic code 396, 403–404
– engineering 388–393
Glasstone 7
global searching 72–73
Gram-Charlier expansion 63
Gruneisen constant 82

guanine 245–258
–, hetero base pairs 260–265
–, homo base pairs 250, 252, 257, 258
–, 6-thio 258
–, wobble 265
guanosine 300
–, GpC 263
– phosphates 279, 301
–, 6-thio 301
guest molecules 314, 435
Gulliver 9

*H*antzsch 3
Hartree-Fock 19, 71
Hassel-Ottar effect 315
H/D exchange 383–388
– isotope effect 134–135
– ratio 55, 59
heat capacity, hydration 463–465
helical parameters, DNA 402
helices – 7, 9, 352–355, 359
–, folding dynamics 508–509
help-gas 437
heterodromic 38
hexakaidecahedron 437
homodromic 38–39, 79, 80, 168, 175, 183, 205, 320–324, 330–333, 340, 439–441
homopolymer nucleic acids 409–412
Hoogsteen 259–260
Hubbard 360–361
Hudson 179
Huggins 3, 4, 6, 8, 16, 24, 111, 351
hydrates
– of acids 111, 113
– of acid salts 111
– of alkylamines 443–448
– of alkyl ammonium salts 440–442
– of (small) biological molecules 452–458
– of carbohydrates 192–193, 453–458
–, channel 439–440
–, clathrate 432–451
–, configuration 454–458
–, double 437
– of gases 432–435
–, inclusion 432–449
–, layered 449–451
–, mixed hydrates 437
hydration, see also water
– of DNA 487, 504
–, dynamics of 505–510
–, ice-like 116, 192, 432, 478
– of lysozyme 461–465
– of nucleic acids 487–504
– of nucleosides and nucleotides 432–457
– of oligonucleotides 491

hydration (cont.)
- of proteins 459–478
- –, active site 484
- –, clusters and pentagons 479–486
- –, computer simulation 486
- –, metrics 470–479
- of purines and pyrimidines 432–457
- of RNA 488–504
- shells in proteins 461–465
- spine of 499–501
hydrogen bond
- –, acceptor 15, 16, 106, 126, 159, 161–163, 223, 269
- –, – geometry 164–165
- –, bent 224
- –, bifurcated 20–21, 83
- –, C–H– – –O 17, 156–160, 254
- –, chains 192–195
- –, chelated 142
- – to C=O in protein 476
- –, complexes 77
- –, compression of 81–82
- –, configurations 20
- –, cooperativity, see cooperativity
- –, cut-off criteria 29–33, 360, 362, 475
- –, definitions 15
- –, dimers 77–79, 99
- –, distribution of 140, 505–510
- –, donor 15, 16, 126, 223, 269
- –, effect on molecular structure 94
- –, energies 27–28, 66–73, 351–393
- –, experimental studies 47–70
- –, F– – –H– – –F 17–18
- –, flip-flop, see flip-flop disorder
- –, four center 23, 42, 137, 145–146, 202
- –, functional groups 16, 106, 126, 169, 223, 270–272
- –, geometry of 57, 71, 103–105
- –, group pair porperties 25
- – to halides 161–163
- –, non-base pairing 237
- –, phenomena 49
- –, P–OH– – –O 18, 271
- –, potentials 89, 91
- –, proton deficient 42
- –, resonance-assisted 37
- –, short 112, 114
- –, statistics 26, 104–105
- –, symmetrical 112, 114
- –, tandem 20, 24, 40
- –, theoretical calculations 71–93
- –, three-center 20, 23, 42, 83, 110, 136–143, 162, 187, 238–239, 382–383, 411–412, 505–508
- –, two-center 20
hydrogen bond angles 109

– – – in four-center bonds 145
– – – in intramolecular bonds 112, 148–154
– – –in three-center bonds 137–143, 162
– – –in two-center bonds 96, 162–163, 224
hydrogen-bond energy 15–20, 27, 28, 51, 63, 66
hydrogen-bond lengths 26–28, 51, 108
– – in amino acids 31–32, 115, 117–119, 141, 146
– – in amino acid hydrochlorides 162
– – in carbohydrates 124–125, 139, 149, 150, 169–214
– – in carboxylic acids 95, 96, 116
– –, C–H– – –O 158, 159
– –, chelated 142–143
– – in β-cyclodextrin 153–164
– – in dimers 78–79, 96, 99
– – in disaccharides 150, 152
– –, effect of cations 123
– – in formamide 96
– –, four-centered 145–146
– – in high hydrates 456
– – in hydrogen carboxylic acids 114
– – – phosphates 113
– – in ices 118–120
– –, N–H– – –N 132
– –, N–H– – –O 25, 32, 128–132, 380
– –, N$^+$–H– – –Cl$^-$ 162–163
– –, normalization 107–110
– – in nucleosides and nucleotides 123, 127, 140, 147, 148, 163, 270–353
– –, O–D– – –O 26, 153–154
– –, O–H– – –N 133
– –, O–H– – –O 95, 108–109, 112–127, 149–152, 456
– –, O$^+$–H– – –O 17–18
– –, O–H– – –O$^-$ 17–18
– –, O$_W$H– – –O 123, 456
– –, P–OH– – –O 18, 272
– – in proteins 374–380, 476–479
– – in purines and pyrimidines 123, 129–130, 163
– –, sequences of 133–134
– – in three-center bonds 136–144, 382–383
hydrogen-bond patterns
– – in amino acids 220–231, 457
– –, antidromic 38, 118, 320–323, 330–333
– – in base pairs 247–269
– – in carbohydrates 58–59, 181–192
– –, catenary 238
– –, cyclic 7, 24, 33, 38–39, 238, 278–284, 330

Subject Index

– – in cyclodextrins 313–345
– – in disacharidese 195–210
– –, flip-flop 40–42, 333–344
– –, heterodromic 38
– –, homodromic 38–39, 80, 168, 175, 183, 205, 320–324, 330–333, 340
– – in hydrates 453–457
– – in monosaccharides 187–195
– – in nucleic acids 396–422, 494–496
– – in nucleosides and nucleotides 277–308
– – in proteins 363–371
– – in purines and pyrimidines 232–246
– –, sequence dependent 499
– –, three-center 21, 23, 137, 139, 140–143, 382–383, 411–412, 505–508
– – in tri- and tetrasaccharides 210–214
– –, very strong 18
hydrogen bond strenght, see hydrogen-bond energy
hydrophobic effect 44–48
– guests 432
– surface 448
hydrophobicity scale 45–47, 353
hypoxanthine
–, hetero base pairs 264–266

ice 4, 27, 40, 69, 116, 118, 121, 429–432
–, amorphous 429
–, clathrates 432–449
–, polymorphism 429–430
–, zero point energy 431
iceberg model 45, 49
icelike structure 116, 192, 432, 478
inclusion complexes 312, 432–449
INDO 84
induced fit 312–314, 332–337
infra-red spectroscopy 50–51, 204, 461–463
inosine 302
–, derivatives 277, 303, 304
–, 3′, 5′-cyclic phosphate 305
insulin hydration 481–482
internal waters 374–376
interpenetrating lattices 120
intramolecular hydrogen bonds 147–150, 181–183, 198, 239, 272, 302, 317–318, 322, 332, 347–348
isotope effect 134

keto-enol tautomerism 42, 235–236

Latimer 3
Lennard-Jones 87, 90, 427, 479
Levy 431
Lewis 3

liquid water 93, 426–429, 448
lysozyme 389–392, 462–465, 480

magic-angle spinning 69
major groove 403, 415
McClellan 17, 136, 144
Megaw 4
membrane structure 47
microwave spectroscopy 51
MINDO 14
minor groove 403
mismatch base pairs 404–407
MMI, MMI-CARB 91
MNDO 84
molecular dynamics 86, 93, 96
– mechanics 71, 85–93
Møller-Plesset 74
Monomoto 19
Monte-Carlo calculations 86, 92–93, 494
Morse function 89, 90
Mulliken population 88
mutagenesis 388–391
mutations 388, 404, 407

Nebenvalenz 3, 4
neutron
–, absorption coefficients 55
–, comparison with X-ray 54
–, diffraction 11, 20, 52–63, 383
–, extinction corrections 54, 56
–, incoherent scattering 55
–, powder diffraction 67–69
–, refinement data 56
–, riding motion 62
–, Rietveld method 68
–, scattering-cross sections 55, 60
–, spallation source 67–69
–, thermal motion correction 60–63
Newtonian Laws 86
NH – – – O=C hydrogen bond 97
NMR
–, C^{13} CP-MAS spectroscopy 70, 180, 204
–, chemical shift tensor 69
–, solid-state 69–70, 94, 135
non-additivity 36, 79, 80
non-bonding interactions 86
nuclear atomic coordinates 60
nucleic acid 12, 394–422
– –, base pairs 7, 37, 232–234, 247–268, 404–407, 410–412
– –, bases 233–234
– – in biological processes 394–396
– –, components 232, 270–276
– –, data base 13
– –, double helices 397–403

nucleic acid (cont.)
– –, major, minor groove   402, 414
– –, mismatches   404
– –, mutations   403–406
– –, nomenclature   269–275, 395, 400
– –, noncomplementary base pairs   406
– –, polymorphism   400
– –, propeller twist   402
– –, protein-nucleic acid interactions   411
– –, tautomerism   405–406
– –, Watson-Crick, see Watson-Crick
nucleosides, nucleotides   126–133, 140, 146, 163, 270–353
–, –, conformation   270–275
–, –, cyclic patterns   277–280
–, –, nomenclature   270–274, 275

Offset correction   76
oligoamyloses   311
oligosaccharides   344
order-disorder transitions   115
orientational disorder   185
oxyanions   111

pair-correlation function   428
Parthasarathy   22
Pasteur   175
Pauling   6, 8, 15, 16, 17, 26, 30, 31, 32, 351, 430
PCILO   84, 85, 331
pentagonal dodecahedron   434–437
pentakaidecahedron   437
peptide link   12, 37, 97, 355
peri-interaction   176, 315
Petersen   431
phosphate distances   497
Pimentel   17, 136, 144
pinacol hexahydrate   175, 449, 450
piperazine hexahydrate   449, 450
pleated sheet   7, 9, 352–360, 375–377
polarization   36, 80, 87
polyguanylic acid   257
polynucleotides   267–268
polypeptide chains   4, 355–359
polysaccharides   195, 214–219
polywater   429
Pople   427
potential energy function   19, 25, 91
propeller twist   403
protein   10, 12, 351–395
–, data base   13
–, flexibility   385
–, H/D exchange   383, 388
–, α-helix   352–358, 385
–, hydrogen bond patterns   362–371
–, internal water   372–374

–, metrical analyses   374–379
–, neutron diffraction   383
–, nomenclature   352–353
–, nucleic acid complexes   415–422
–, β-pleated sheet   7, 9, 362–360, 375–377
–, primary structure   351
–, quaternary structure   351
–, secondary structure   351, 355, 365
–, site-directed mutagenesis   388–393
–, structural organization   352
–, tertiary structure   355
–, thermal motion   466
–, – stability   388–389
–, three-center bonds   382–383
–, α-turn   378
–, β-turns   352, 359, 368–369, 377
proton
–, chemical shift   70, 94
–, deficiency   22, 42–43, 141
–, transfer   115
–, tunnelling   41, 111
prototropic tautomerism   237
purines, pyrimidines   128–133, 140, 145, 148, 163–165, 232–268
pyridine trihydrate   449

Ramachandran plot   85, 354
Raman spectroscopy   49–50
Raoult law   49
recognition   412–422
reduplication   397
refcodes   57, 111
replication   398
repressor-operator complexes   415–420
residual entropy   40
resonance-assisted bonding   37
ribonuclease T1   354–356, 413–414
riding motion   62
Rietveld profile refinement   68–69
rigid-body motion   61
RNA crystalline forms   396–415
Rodebush   3

S. griseus protease A hydration   486
salt bridges   128, 371
secondary structure   355, 358, 362, 365
self-association   49
semi-clathrates   446
semi-empirical calculations   71, 84–85
sequential bonding   38
β-sheet, see pleated sheet
short hydrogen bonds   111
solvation   26, 28, 49, 459–460
spallation neutron source   67
spine of hydration   499–501
starch   347

Subject Index

sucrose 144, 202
sugar alcohols 172–179
surface accessibility 487–488
Swift 9
synchroton X-rays 11
syndiaxial orientation 149, 176, 183

tandem bonds 20, 24, 34, 40
tautomerism 235–238, 406
tetraalkylammonium salt hydrates 42–43, 51, 438–443
tetrakaidecahedron 437
theoretical calculations 71–93
thermal ellipsoids 61
– –, motion 60–62, 466–468, 492
– –, stability 388–389
three-center bonds 20–23, 42, 83, 110, 136–143, 162, 187, 238–239, 382–383, 411–412, 505–508
thymidine 288
–, 6-aza 289
–, 5′-carboxylic acid 289
thymine
–, hetero base pairs see uracil
–, homo base pairs, see uracil
–, 1-methyl 255
–, 1-methylthymine·9-methyladenine 261
time of flight (TOF) method 67, 68
torsion angles 86, 185–186, 275
transcription 398
tRNA 396, 403, 406–408
–, synthetase 420–421
trp repressor 415, 418–420
truncated octahedron 43
turns 352, 359, 368–369, 377

uracil
–, base triplet A·2U 267, 268
–, dimethyl 254
–, 5-fluorouracil·1-methylcytosine 267
–, hetero base pairs 260–266
–, homo base pairs 248, 249, 254

–, wobble 265
uridine 277, 281
–, derivatives 276–287
–, 3′-phosphate 280, 288

van der Waals
–, forces 7
–, radii 16, 29, 30–33, 156
vitamin B12 hydration 506–508
vitron theory 430

water, see also hydration
–, bound 461–463
–, bridges 494–497
–, clathrate 428, 448, 481
–, clusters 427, 479
–, configurations 122, 454–458
–, dimer 19, 27, 52, 72, 77
– in DNA-*trp*repressor complex 418
–, internal in proteins 372–374
–, liquid 93, 426–429, 448
–, pentagons 479, 481–486
–, polymers 79, 80–81
–, properties of 10
–, role of 10
Watson-Crick 7, 8, 37, 132, 247–248, 259–268, 394–399, 404–409
Werner 8
wobble base-pairs 265–266, 403–440

X-ray
–, comparison with neutron 54
–, correlation functions 448
–, difference maps 52
–, diffraction 52–63
–, fiber diffraction 92, 214–219, 347, 489
–, refinement 53

zero-point energy 431
– entropy 431
zwitterion 220, 224, 296, 306
– bridges 223